Wave and Tidal Energy

Wave and Tidal Energy

Edited by

Deborah Greaves

Professor of Ocean Engineering, School of Engineering,
University of Plymouth, UK

and

Gregorio Iglesias

Professor of Coastal Engineering, School of Engineering,
University of Plymouth, UK

This edition first published 2018
© 2018 John Wiley & Sons Ltd

The right of Deborah Greaves and Gregorio Iglesias to be identified as the authors of the editorial material in this work has been asserted in accordance with law.

Registered Offices
John Wiley & Sons, Inc., 111 River Street, Hoboken, NJ 07030, USA
John Wiley & Sons Ltd, The Atrium, Southern Gate, Chichester, West Sussex, PO19 8SQ, UK

Editorial Office
The Atrium, Southern Gate, Chichester, West Sussex, PO19 8SQ, UK

For details of our global editorial offices, customer services, and more information about Wiley products visit us at www.wiley.com.

Wiley also publishes its books in a variety of electronic formats and by print-on-demand. Some content that appears in standard print versions of this book may not be available in other formats.

Library of Congress Cataloging-in-Publication Data

Names: Greaves, Deborah, editor. | Iglesias, Gregorio, 1969– editor.
Title: Wave and tidal energy / edited by Deborah Greaves (University of Plymouth), Gregorio Iglesias (University of Plymouth).
Description: Hoboken, NJ : Wiley, 2018. | Includes bibliographical references and index. |
Identifiers: LCCN 2017042798 (print) | LCCN 2017051143 (ebook) | ISBN 9781119014454 (pdf) | ISBN 9781119014478 (epub) | ISBN 9781119014447 (cloth)
Subjects: LCSH: Ocean wave power. | Tidal power. | Water-power. | Renewable energy sources.
Classification: LCC TC147 (ebook) | LCC TC147 .W37 2018 (print) | DDC 621.31/2134–dc23
LC record available at https://lccn.loc.gov/2017042798

Cover design by Wiley
Cover images: (Waves) © kateafter/Gettyimages;(Rays) barbol88/Gettyimages

Set in 10/12pt Warnock by SPi Global, Pondicherry, India

Printed in Great Britain by TJ International Ltd, Padstow, Corwall

10 9 8 7 6 5 4 3 2 1

Contents

List of Contributors

Javier Abanades Tercero
Offshore Renewable Energy Consultant,
TYPSA, Spain
Associate Researcher, School of
Engineering, University of Plymouth,
UK

Francisco Acuña
Chief Executive Officer, InTrust Global
Investments LLC, Washington, D.C.
USA

Sharay Astariz
Associate Researcher, University of
Santiago de Compostela, Spain

Bradley Buckham
Department of Mechanical Engineering,
University of Victoria, BC, Canada

Chenyu
Researcher, Ocean University of China,
Qingdao, China

Curran Crawford
Department of Mechanical Engineering,
University of Victoria, BC, Canada

Erica Cruz
Senior Researcher, WavEC – Offshore
Renewables, Lisboa,
Portugal

Boel Ekergard
Seabased Industry AB, Sweden

Deborah Greaves
Professor of Ocean Engineering,
School of Engineering, University of
Plymouth, UK

John Griffiths
Associate, EMEC Ltd, Chair of UK
National Committee on Wave & Tidal
Standards

Martyn Hann
Lecturer in Coastal Engineering,
School of Engineering, University of
Plymouth, UK

Andrew J. Hillis
Senior Lecturer in Mechanical
Engineering, Centre for Power
Transmission and Motion Control,
University of Bath, UK

Brian Holmes
MaREI Centre, Environmental
Research Institute,
University College Cork, Ringaskiddy,
Ireland

Gregorio Iglesias
Professor of Coastal Engineering,
School of Engineering, University of
Plymouth, UK

Lars Johanning
Professor of Ocean Technology, University
of Exeter, Penryn Campus, UK

Kwangsoo Lee
Principal Research Scientist,
Institute of Ocean Science & Technology,
Korea

Mats Leijon
Professor, Uppsala University, Sweden

Inês Machado
Senior Researcher, WavEC – Offshore
Renewables, Lisboa, Portugal

Vanesa Magar
Centro de Investigación Científica y
Educación Superior de Ensenada
(CICESE), México

Dorleta Marina
BIMEP,
Lemoiz, Spain

Allan Mason-Jones
School of Engineering, Cardiff University,
Wales, UK

Daphne M. O'Doherty
School of Engineering, Cardiff University,
Wales, UK

Tim O'Doherty
School of Engineering, Cardiff University,
Wales, UK

Anne Marie O'Hagan
Senior Research Fellow, MaREI Centre,
Environmental Research Institute,
University College Cork, Ringaskiddy,
Ireland

Carlos Perez-Collazo
PRIMaRE Research Fellow, School of
Engineering, University of Plymouth, UK

Andrew R. Plummer
Professor of Machine Systems,
Centre for Power Transmission and
Motion Control, University of Bath, UK

Hongda Shi
Professor, Ocean University of China,
Qingdao, China

Sungwon Shin
Research Professor, Kangwon National
University, Korea

Teresa Simas
Senior Researcher, WavEC – Offshore
Renewables, Lisboa, Portugal

Phillip R. Thies
Senior Lecturer Renewable Energy,
University of Exeter, Penryn Campus,
UK

Yago Torre-Enciso
BIMEP, Lemoiz, Spain

Angela Vazquez
Associate Researcher, University of
Santiago de Compostela, Spain

Paul Vigars
Director of Teobi Engineering
Associates Ltd, UK (formerly Research
& Technology Manager at Alstom Ocean
Energy, UK)

Thomas Vyzikas
Associate Researcher,
School of Engineering, University of
Plymouth, UK

Sam D. Weller
Senior Research Fellow, University of
Exeter, Penryn Campus, UK

Foreword

Since the 1990s the importance of developing renewable energies has been well recognised worldwide. At the time of writing, onshore wind, solar and hydropower are mature and making relevant contributions to the energy mix. However, the untapped potential of these land-based forms of renewable energy is not unlimited; therefore, new renewable energies, including wave, tidal and offshore wind, must be developed if carbon-based energy production is to be further reduced, in the spirit of the recent Treaty of Paris and previous agreements on climate change.

Offshore wind is technologically more mature than wave and tidal energy, arguably thanks to its similarities with its onshore counterpart. Indeed, as offshore wind moves into deeper waters, those facets that are not shared with onshore wind turbines, such as floating systems or hybrid (wave–wind or tidal–wind) systems warrant the greatest research effort at present.

Wave and tidal energy, the focus of this book, are technologically more challenging, not least because of the aggressive marine environment. Because of this, and the fact that their development began more recently, they are further away from full market commercialisation. Their trajectory has been similar to that of any nascent technology, with initial successes and failures.

Arguably the harsh marine environment has hindered the technological development of both wave and tidal energy, not least in relation to wind energy, the main elements of which were developed for a less aggressive environment. This also made possible the application of wind energy at different scales, from the domestic to the industrial, and its stepwise progression towards the large wind turbines that we see today. Nevertheless, the faster development of wind energy that we have witnessed does not detract in the least from the potential of wave and tidal energy. Given the intensive research efforts and the level of international interest in the field, there can be little doubt that the vast, so far untapped, wave and tidal resource in the ocean will be exploited within the next decades.

This new book aims to provide a reference text for students and practitioners in the wave and tidal energy industry. It presents a holistic view of the sector, the state of the art and the perspectives for future development. The main tools of physical and numerical modelling are explained, together with the technical aspects of device design and development, the environmental effects and the consent and legal processes. These are then illustrated with a series of case studies and a review of regional project developments.

Wave and tidal energy is a fascinating field with many exciting research challenges. Driven by the passion of the researchers and practitioners involved, the momentum in the sector is poised to transform wave and tidal energy from its present research and development status into a fully fledged renewable contributing substantially to the energy mix.

Acknowledgements

This book would not have been possible without the collaboration and wholehearted support of the many authors of the individual chapters. Their expertise in the many facets of wave and tidal energy has been central to the project, and we are greatly indebted to them.

We are grateful to Wiley for suggesting the idea of the book and their consistent support throughout the processes of writing and editing the chapters and producing the book in its final form.

We also acknowledge the contributions and support for this project from the members of the COAST Engineering Research Group and the COAST Laboratory at the University of Plymouth. The Laboratory has been essential to the development of marine renewable energy research at Plymouth. The experience in physical and numerical modelling gained through the activity of the Laboratory and Research Group, working together with industry and academic partners on numerous European and national research and development projects, has resulted in the scientific environment that has ultimately crystallised in this book.

1

Introduction

Deborah Greaves[a] and Gregorio Iglesias[b]

[a] *Professor of Ocean Engineering, School of Engineering, University of Plymouth, UK*
[b] *Professor of Coastal Engineering, School of Engineering, University of Plymouth, UK*

1.1 Background

More than 83% of the energy conversion in the world is today based on fossil fuels; meanwhile scientists all over the world are debating the topic of peak oil [1] and the secondary effects of the emissions from the fossil fuels [2, 3]. Fossil fuels are a finite resource; burning them generates significant carbon dioxide emissions that are changing the world's climate. The impact of climate change is thought to be changing habitats at a rate faster than many species can adapt, and the level of pollution in many of the world's cities is today causing concern. As a future worldwide shortage of useful energy supply can have devastating consequences on the political stability and economy of the world, there is a growing consensus that the world needs to switch to a more sustainable energy system. The focus and requirement for clean and cheap renewable energy conversion techniques has therefore increased.

The Paris Summit of 2015 [4] has driven further impetus for finding alternative sources of energy, and a deal was agreed to attempt to limit the rise in global temperatures to less than 2 °C. The Paris agreement is the first to commit all countries to cut carbon emissions, and is partly legally binding and partly voluntary. The measures in the agreement include [5]: to peak greenhouse gas emissions as soon as possible and achieve a balance between sources and sinks of greenhouse gases in the second half of this century; to keep global temperature increase 'well below' 2 °C (3.6 °F) and to pursue efforts to limit it to 1.5 °C; to review progress every 5 years; and $100 billion a year in climate finance for developing countries by 2020, with a commitment to further finance in the future. There is clear acknowledgement of climate change and also a clearly stated will to address the anthropogenic causes of climate change and to reduce emissions and seek alternative sustainable and environmentally benign sources of energy. How this new agreement will be implemented within individual countries will be influenced by local factors.

Renewable sources of energy are essential alternatives to fossil fuels and to nuclear energy, which also has a finite resource as well as long-term safety concerns. Renewable energy sources include solar, wind, geothermal and marine renewable energy (MRE).

Wave and Tidal Energy, First Edition. Edited by Deborah Greaves and Gregorio Iglesias.
© 2018 John Wiley & Sons Ltd. Published 2018 by John Wiley & Sons Ltd.

Their use reduces greenhouse gas emissions, diversifies energy supply and reduces dependence on unreliable and volatile fossil fuel markets. The world is moving on renewables, and they have become the cornerstone of any low-carbon economy today, not just in the future. The USA is targeting a 32% cut in power sector emissions by 2030, India plans 100 GW of solar by 2022, and China is investing heavily in wind and renewable energy: the transition to a low-carbon energy system is well under way.

Within this drive for renewable energy, MRE is poised to play a major role [6], in particular in certain countries where these resources are vast. Renewable energy from the sea is generated by the sun, wind and tides, and may be exploited through various technologies such as wave energy, tidal stream, tidal range, offshore wind energy and ocean thermal energy currents (OTEC). MRE, also often termed 'ocean energy', has a major part to play in closing the world's energy gap and lowering carbon emissions. Key global challenges that remain for MRE relate to technology, grid infrastructure, cost and investment, environmental impact, and marine governance. Of these technologies, offshore wind is mature and many commercial projects exist in shallow waters, although new offshore wind technology is needed to develop sites further offshore in deeper water. Technologically, the development of offshore wind in shallower water is a natural extension of onshore wind, and typical difficulties for onshore wind in gaining social acceptability and approval are often less problematic if turbines are located offshore. Also, the wind resource offshore is greater due to lack of obstructions to the wind flow. Offshore wind turbines are typically similar to those used onshore and consist of three blades rotating about a hub, and in shallower water the wind turbine structures are typically on piled foundations or fixed jackets. However, as development of wind farms moves further offshore and into deeper water, other solutions need to be sought involving floating structures and the costs increase significantly. Although offshore wind technology is rapidly being implemented, there remain many fascinating engineering problems to overcome. These include: offshore foundations and floating support structures; alternative turbine designs based on three-dimensional computational fluid dynamics; use of advanced materials for blades; ship manoeuvring for safe maintenance; and shared offshore platform applications (such as energy production, storage, and marine aquaculture).

Tidal power is approaching commercial maturity, and recent investments and commercial developments have been made. Tidal range projects exist, but there are concerns about the extent of the environmental impact they bring, and tidal lagoon technology is emerging as an attractive alternative. Tidal steam technologies have seen great advances in recent years. On the other hand, wave energy encompasses emerging technologies that are currently not economically competitive, but still attract engineering interest thanks to the significant resource in high power density sea waves and its potential exploitation [7].

Within Europe, ocean energy is considered to have the potential to be an important component of Europe's renewable energy mix, as part of its longer-term energy strategy. According to the recent studies [8,9], the potential resource of wave and tidal energy is 337 GW of installed capacity by 2050[8] globally, with 36 GW quoted as the practically extractable wave and tidal resource by 2035 in the UK, representing a marine energy industry worth up to £6.1 billion per annum. Today 45% of wave energy companies and 50% of tidal energy companies from the EU [9,10] have been tested in EU test centres [11,12], and the global market is estimated to be worth up to €53 billion annually by 2050 [13].

The need to address climate change and concerns over security of supply has driven European policy-makers to develop and implement a European energy policy. In 2009, the European Commission set ambitious targets for all member states through a directive on the promotion of the use of energy from renewable sources (2009/28/EC). This requires the EU to reach a 20% share of energy from renewable sources by 2020. The directive required member states to submit national renewable energy action plans (NREAPs), that establish pathways for the development of renewable energy sources, to the Commission by June 2010. From their NREAPs, it is clear that many member states predict a significant proportion of their renewable energy mix to come from wave and tidal energy by 2020. This commitment should act as a strong driver at national level to progress the sector.

MRE can significantly contribute to a low-carbon future. Ambitious development targets have been established in the EU, including an installed capacity of 188 GW and 460 GW for ocean (wave and tidal) and offshore wind energy, respectively, by 2050 [10]. To comprehend how challenging these targets are it is sufficient to consider the corresponding targets for 2020: 3.6 GW and 40 GW for ocean and offshore wind energy, respectively. It is clear that for the 2050 targets to be met, a major breakthrough must happen – and there are huge benefits to be reaped if these targets are met, such as the reduction of our carbon footprint.

1.2 History of Wave and Tidal Energy

Although MRE and ocean energy can be interpreted to include all energy conversion technologies located in the ocean environment, including offshore wind, OTEC as well as wave and tidal, in this book we focus on wave and tidal energy. Tidal energy converts the energy obtained from tides into useful forms of power, mainly electricity. Tides are more predictable than wind energy and solar power. Among the sources of renewable energy, tidal power has traditionally suffered from relatively high cost and limited availability of sites with sufficiently high tidal ranges or flow velocities, thus constricting its total availability. However, significant learning has been gained through relatively long-term deployments of tidal turbines [14], and together with developments in tidal lagoon technology [15], and first array scale deployments [16], it is expected that the total availability of tidal power is significant, and that economic and environmental costs may be brought down to competitive levels.

Historically, tide mills [17] have been used both in Europe and on the Atlantic coast of North America for milling grain, and in the nineteenth century the use of hydropower to create electricity was introduced in the USA and Europe [18]. Tidal range projects include the world's first large-scale tidal power plant, the La Rance Tidal Power Station in France, which became operational in 1966 [19]. It was the largest tidal power station in terms of power output, before Sihwa Lake Tidal Power Station in South Korea (described in Chapter 12) surpassed it. Many innovative tidal stream energy devices have been proposed. An example is Salter's cross-flow turbine [20], which has blades arranged vertically, supported at each end on what are rather like enormous bicycle wheels. Although tidal power assessment seems easy, the very presence of tidal turbines alters the flow field, and in turn this affects power availability.

Tidal energy technology is dominated by in-sea/estuarine tidal stream devices; however, a significant number of developers have also been developing smaller in-river devices.

There is certainly potential for tidal energy to consolidate technologies and progress from small-scale to larger developments within the full-scale prototype field. The last few years to 2016 have seen the total number of globally active developers fall, perhaps as the technology naturally converges. Leading developers are actively testing at EMEC [21] and moving strongly towards commercial readiness and preparing for transition to large-scale commercial generation in the UK Crown Estate lease areas, north-west France and Canada's Bay of Fundy. Alongside the progress to full-scale device deployment technology activity, there has been clear progress on site development, with the consent and finance secured for a 6 MW tidal array off the north of Scotland by MeyGen and the subsequent news of Atlantis Resources Ltd. having purchased the project. This is the first example of real value being attributed to a site and associated development consent [22].

The Severn Estuary holds the second highest tidal range in the world, and within this Swansea Bay benefits from an average tidal range during spring tides of 8.5 m. Plans to construct a tidal lagoon [15] to harness this natural resource would be the world's first, man-made, energy-generating lagoon, with an expected 320 MW installed capacity and 14 hours of reliable generation every day. In a bid to overcome potential socio-economic and environmental concerns, the development also offers community and tourism opportunities in sports, recreation, education, arts and culture, conservation, restocking and biodiversity programmes as well as the added benefit of coastal flood protection.

Wave energy converter technology is a thriving area in which new inventions keep appearing. Here, engineers must find ways to maximise power output, improve efficiency, cut environmental impact, enhance material robustness and durability, reduce costs, and ensure survivability. Theoretical predictions of the power generated by wave energy converters require validation through laboratory-scale physical model studies and field tests. The latest simulation methods involve wave to wire modelling of arrays of wave energy converters, which integrates wave hydrodynamics, body responses, power take-off (PTO), real-time control, and electricity production.

There are more than one thousand patents for devices for capturing and transforming wave energy into useful energy. The first wave energy converter was patented in France in 1799, and oscillating water column navigation buoys have been commercialised in Japan since 1965 [6]. The oil crisis in 1973 raised interest in wave energy in Europe, but interest dwindled in 1980s and it was not until the 1990s that interest increased again.

Wave energy has the largest potential in Europe and worldwide, and can be captured in a number of ways through the use of different converters, such as point absorbers, attenuators, overtopping, oscillating wave surge convertors, and oscillating water columns. The technology has not yet reached the stage of commercial scale development [23], but progress continues to be made, as evidenced by the growing number of test sites and pilot zones being established across Europe [11]. Many different types of wave energy converters have been designed, but only a small proportion of these so far have reached the full-scale prototype stage. Wave energy has many advantages over other forms of renewable energy, being much more predictable than, for instance, wind, giving more scope for short-term planning of grid usage.

In the past, the wave energy industry faced some failures that delayed its development, for example the device in Toftestallen wrecked during a heavy storm [24] or the external wall of the Mutriku device that was damaged by a storm [25]. Attempting to set a framework for assessing the progress of potential developers on their way to commercial applications, Weber [26] introduced the technology readiness level and technology performance level matrix, so that fewer failures occur in the future.

1.3 Unknowns and Challenges Remaining for Wave and Tidal Energy

Access to ocean energy systems is expensive and hazardous. Present and future challenges include remote monitoring, control systems, robotics for operational support, and real-time weather forecasting for predictive maintenance to ensure devices can survive in extreme sea states as they arise. Wave and tidal energy has huge potential, but demanding global challenges have to be met before the seascape will give up its precious energy resources. As in the Industrial Revolution, a new generation of engineers is required with the ingenuity, wisdom, and boldness to meet these interdisciplinary challenges. The unknowns and challenges still remaining in wave and tidal energy can be considered to fall within ten different technical research themes as identified by PRIMaRE [27]: materials and manufacture; fluid dynamics and hydrodynamics; survivability and reliability; environmental resources; devices and arrays; power conversion and control; infrastructure and grid connection; marine operations and maritime safety; socio-economic implications; and marine planning and governance.

1.3.1 Materials and Manufacture

The development of new materials and manufacturing processes is a key element in reducing costs and ensuring the survivability of MRE devices. Any technology submerged or in contact with the sea is likely to be affected by biofouling. The interaction of the devices or their components with marine growth is crucial as it affects the device performance and design conditions, and therefore the development of new materials to avoid or minimise biofouling is key. Use of steel or metallic alloys is common practice in the MRE industry. Correct understanding of the corrosion processes, of the use of new coatings and manufacture techniques, and of how to adapt the operation and maintenance inspections to maximise the lifetime and operability of MRE devices will help reduce their total cost. Application of novel materials and construction techniques that will reduce costs, improve reliability and extend the lifetime of devices is an active research area, necessary to move the sector forward – for example, novel materials such as reinforced concrete and composites, novel construction techniques, disposable materials are being investigated.

1.3.2 Fluid Dynamics and Hydrodynamics

As technology devices that harness energy from fluids in motion and are affected by the extreme forces produced by these motions, a proper understanding of the fluid dynamics and hydrodynamics of MRE devices is crucial to their development. In particular, turbulence and its effects on single and multiple devices is important in understanding how devices will interact and perform in arrays. In the real sea environment MRE devices commonly face the effects of combined waves, tidal currents and wind. The combined action of these forces on MRE devices makes characterisation of their response at laboratory scale and with numerical models of special relevance to obtain a better understanding of how they perform under such circumstances. One of the particularities of MRE is that the devices need to face extreme loads and survive storms. Thus, the development of novel evaluation techniques to model these extreme loads appropriately at laboratory scale and by numerical models is required.

When deployed in real sea conditions, MRE devices are subjected to irregular waves and variable tidal currents. A feature of these variable resources is that the differences between maximum and mean values are particularly high, especially for wave energy. The standard engineering techniques to model the behaviour and response of MRE devices consider linear models in order to simplify the problems and obtain faster solutions. However, the reality is often far from the linear model and nonlinear effects must be considered to achieve a proper understanding of the performance of the devices in real conditions. Thus, the development of nonlinear models and tools to assess these effects is of special relevance. Advanced numerical models able to simulate accurately the response of full-scale devices require long computational times and resources. The development of validated tools and resources that optimise simulation times is necessary for the development of MRE.

1.3.3 Survivability and Reliability

The survivability and reliability of MRE devices in the marine environment need to be proven for the industry to become commercial. Ensuring the survivability of devices under the high loads occurring during extreme events is essential to reduce the risk of failure and increase their range of operability. The dynamic nature of many MRE devices means that traditional oil and gas or seakeeping mooring concepts are usually not valid, due to either their high cost or the different loading conditions. A competitive cost of energy which allows MRE to become viable in comparison with other renewable energies is fundamental for the development of the sector. This means that a compromise between reliability and cost of energy throughout the lifetime of the device should be found. The weakest of its components defines the entire reliability of a MRE device. This, together with the harshness of the marine environment, the frequent exposure to extreme loads, and near-constant exposure to varying cyclic loads, makes the design of all components crucial. Research is needed to assess each individual component and adapt it to the MRE industry needs, redesigning components where necessary and making use of available technology where possible, for example from the oil and gas sector. Furthermore, MRE devices are subject to potential impacts, for example, the impact of a marine mammal striking the rotor of a tidal turbine, or collision between wave energy converters due to a mooring failure, and these impacts could severely damage the integrity of the device.

1.3.4 Environmental Resources

Resource assessment for wave and tidal energy is described in Chapter 2, and a thorough understanding of the environmental resources is imperative to harnessing them in an economic and efficient manner. Even though wave, tidal currents and offshore winds are well understood at medium and large scales, there are still multiple physical processes related to them that require further study, especially when energy extraction is involved: turbulence, atmospheric thermal gradient, oceanic fronts, salinity, spatial (and spectral) variability of waves, sediment transport, etc. Wave, tidal and wind macro-scale models are well known and have been validated in recent years [28,29]; however, to transfer the input from these large-scale models to more reduced-scale regional models still requires further research [30]. Monitoring MRE resources at the sea is expensive and sometimes risky or impossible due to the weather conditions. The development of novel remote sensing techniques which allow monitoring of MRE resources further from the site or to cover larger areas will help to reduce uncertainties about the resource and its costs, such as floating LIDARs, high-frequency radars [31], and satellite monitoring. Forecasting MRE resources with high resolution is important not only to be able to predict the energy production, but also to inform control strategies that require being able to predict and respond precisely to the incoming resource.

1.3.5 Devices and Arrays

The level of technological development among MREs is very variable, depending on the type of energy being harnessed; thus on the one hand floating offshore wind and tidal turbines have successfully achieved full-scale demonstrations and are moving towards arrays, while on the other hand wave energy converters are less developed in their progression towards array deployments. Accessibility of MRE devices at sea is not always possible due to the weather conditions, and therefore developing remote monitoring and control techniques and operation procedures becomes an important requirement to increase operation times and reduce operation & maintenance (O&M) costs. Mooring systems for MRE devices are a crucial subsystem that influences both the global performance of the device and its survivability. However, for arrays of MRE devices, and in particular interconnected arrays, these become even more important because the layout of the array will be highly dependent on the mooring dynamics and interactions. Few MRE devices have been tested in arrays so far, which means that significant unknowns remain around how the energy extraction of some devices will affect the performance of the other devices in the array. This is an important research topic as it will influence significantly the layout of future arrays. One aspect of arrays is that they may lead to sharing of some infrastructure, such as the mooring lines and PTO, with the consequent cost reduction. Future research is needed in order to assess whether sharing of these subsystems is appropriate or not [32].

1.3.6 Power Conversion and Control

Ultimately MRE devices are meant to produce electricity at an affordable cost. This means that power conversion and control subsystems are fundamental parts of an MRE device. A PTO is a key component of every MRE device, as it is through this that the power is ultimately harnessed. Increasing the performance of PTOs, making them more reliable and extending their lifetime are required steps to reduce MRE costs. Due to the irregular character of MRE resources, and in particular of wave energy, control strategies are required in order to maximise the power output. At present the relatively high cost of MRE, together with the high variability in power production, means that additional research is needed to tackle these issues. Novel power electronics control systems able to transform the variable energy production obtained from the MRE resources into conditions required for the electric grid may be part of the solution and are required for successful development of the devices.

1.3.7 Infrastructure and Grid Connection

For offshore energy projects, the infrastructure and grid connection costs represent one of the largest components of the total project costs. The development of novel concepts of subsea electric connectors which will allow reduced times for installation, decommissioning and standard plug and unplug operations during O&M is required. The development of specific protocols and tools for O&M procedures at MRE test centres is of special interest in order to extrapolate them to other future projects. Furthermore, these test centres can also be used to develop and test novel O&M techniques. The combination of more than one MRE technology at the same site is a novel approach that takes advantage of the multiple synergies arising from deployment of more than one technology at the same location. Understanding the implications of this practice is crucial in order to further develop these kinds of projects. Integrating the energy harnessed at MRE parks into the energy grid is not a simple task due to the existence of multiple challenges, such as the lack of grid infrastructure inland near the parks, the variable character of the energy production, and the lack of coupling between the production of and the demand for energy.

1.3.8 Marine Operations and Maritime Safety

Operations and safety at sea are not simple due to the variable conditions and harsh working environment, which means that operations to access an offshore device or to deploy a wind turbine may be high-risk challenges. Further research in order to simplify and reduce the risk in these tasks is crucial to further develop the sector. The consideration of the decommissioning phase of a MRE device at its development phase is very relevant in order to save time and resources. Further research is needed in order to include these aspects during the development phase or the design of novel materials or devices. To access an offshore device is not a simple task, especially when the sea conditions differ. This means that at some sites device access times are reduced to below 60% of the year, resulting in an increase in O&M costs and in performance reduction. Design of novel access techniques or unmanned O&M protocols is one of the main challenges for the MRE sector to accomplish in order to reduce operation costs. The deployment of more than one MRE technology at the same site has the associated risk of possible interference between technologies (e.g., reducing the energy production, collision or impact between devices). In order to make this promising solution a reality, these challenges must be further addressed in the coming years. An accurate forecast of the weather windows to access offshore sites to perform O&M and deployment activities is highly important to reduce costs and deployment times. Future research is needed in order to produce new short- and long-term models to forecast weather windows.

1.3.9 Socio-Economic Implications

As MRE develops beyond the demonstration scale, improved knowledge of the potential socio-economic implications of larger arrays is required. In particular, a better understanding is needed of the social, economic and environmental costs and benefits and how they are distributed spatially and temporally between local, regional and national levels. The findings of this empirical research can then be applied to support marine planning, licensing, and governance and stakeholder engagement.

The foundation for understanding socio-economic issues is to determine how stakeholders and the wider public perceive the development and impacts of MRE, and which socio-demographic factors shape their attitudes. At present, the public has very limited awareness of MRE, and it is therefore important to understand how different messages and media types are perceived as sources of credible information. There is a growing policy trend to seek to understand environmental impacts in terms of their societal implications, which is achieved through the framework of ecosystem services. Ecosystem service approaches also facilitate monetary valuation of natural capital impacts, and hence support enhanced cost–benefit analysis of MRE developments and natural capital accounting. For these reasons, ecosystem service approaches need to be embedded in full life-cycle analysis of MRE.

Commercial-scale MRE developments have the potential to bring economic benefits through, for example, job creation and supply chain development, although other sectors (fisheries, recreation) may be negatively affected. At present, the scale of these effects and their implications for regional and national economies are poorly understood. Also key to planning and policy decisions is understanding how the benefits of MRE development, such as clean, renewable energy and job creation, are traded off against natural capital and ecosystem service impacts at the level of individual energy consumers, local communities, and regional and national economies.

Traditionally the marine and maritime sectors have been characterised by the absence of coordinated planning and governance, as opposed to what happens onshore, where coordinated regulations and planning have been in place for many years (e.g., land use planning, mining exploitation concessions and natural protected areas). However, in recent years some EU initiatives, such the Maritime Policy, Maritime Spatial Planning, Integrated Coastal Protection Management, Marine Strategy Framework and Water Framework Directives (via consideration of impacts of MRE on the targets of good environmental/ecological status), are driving the need for more coordinated planning of the marine and coastal space. Integration of the uses of different resources is key to ensuring sustainable development.

The engagement of stakeholders and especially local communities is an important aspect of the successful development of an MRE project. Many examples show that disregarding these actors at early stages of project development has been damaging for the projects. Developing novel tools to facilitate this engagement and to increase the communication between developers and stakeholders is of special relevance for the success of future MRE projects. An appropriate assessment and planning of the complete life cycle of an MRE device is important to increase the sustainability of the technology. The development of new generation concepts and new tools to tackle this at full scale is necessary.

1.3.10 Marine Planning and Governance, Environmental Impact

Appropriate planning of the spatial distribution of maritime resources and uses is vital for sustainable development of the seas and its coexistence with traditional and new users. Significant effort has been made by European and national administrations to start this large endeavour; however, future research is needed to propose novel approaches to planning and managing use of the sea's resources. Multiple marine resources are usually available at the same site or geographical location, and synergies between the different users of these resources exist. Sustainable development of the exploitation requires multi-use of these marine resources, for which further research and development to prove its viability is needed. Research and demonstration of the options for combinations of wave and offshore wind energy with other uses, such as aquaculture or maritime leisure, is required for successful implementation of multi-use of space.

Understanding the effects on the physical coastal processes of the deployment of MRE infrastructure, such as mooring anchors and solid foundations, is of special relevance to understanding how these technologies affect the environment. Energy extraction from harnessing waves or tidal currents affects the wave and flow field and sediment transport processes occurring in them. This may have a significant effect on coastal erosion phenomena for large and cumulative scale effects of MRE deployments. The interaction between solid foundations of MRE devices and currents and waves produces erosion of the seabed in the surroundings of the structure. Known as 'scour', this phenomenon may alter the baseline sediment transport processes, and further research is needed in order to understand the significance of this effect. There is a lack of international standards or common industry practice concerning the required burying depth for electrical marine cables, and the effects of the electromagnetic fields over the ecosystem are not well understood [33].

The deployment of MRE devices inevitably involves contact with the seabed and so has a clear effect on the benthic flora and fauna. The presence of new structures altering the environment could potentially act as barriers or generate displacements or migrations in benthic communities. In order to understand these possible effects further research should be conducted during future deployments of MRE devices. The energy extraction caused by harnessing marine resources could potentially affect the trophic chain, and the presence of MRE devices could affect the production of nutrients, displace species, etc. Deploying MRE devices creates new habitat with a potential risk of introducing non-native invasive species that can significantly alter the ecosystem. New research should be conducted to understand this risk and develop contingency measures to avoid it.

In addition to the research needs for the benthic communities, there are others that affect local communities such as fishes, birds and marine mammals. Development of novel active and passive acoustic monitoring techniques and instruments is an important area of research needed to acquire a better understanding of how MRE devices affect local communities. As MRE device deployment continues, the study of such interactions using novel techniques and the development of new instrumentation will be of particular relevance (e.g., new hydrophones or lateral sonars able to record the interaction of local communities with MRE devices). Modelling the behaviour of marine mammals, as individuals in the upper levels of food chains and of societal importance, is also needed to understand possible effects of MRE devices and so that accurate behaviour models for marine mammals may be developed.

One of the possible positive impacts of deploying MRE devices is the creation of artificial reefs; however, this needs to be monitored in order to fully understand the effects. In order to do this, new research including monitoring of new deployments of MRE devices and the use of computer models is needed. Large individuals such as marine mammals may be at risk of collision or entanglement with the MRE devices or some of their subsystems (e.g., impact of a marine mammal with a tidal turbine or entanglement with mooring lines). The development of models incorporating behavioural characteristics of marine mammals is needed to assess and simulate their response to MRE developments.

In summary, MREs are among the renewables with the greatest potential; however, to succeed in the quest to develop the MRE energy sector, some significant challenges need to be tackled in the coming years. Wave and tidal energy provide an unlimited 'clean' resource from which to generate electrical power and thus make a significant contribution to the renewables mix. Realisation of this potential, however, will require continuing advances in marine engineering technology in order to achieve economic viability and ensure minimal environmental impact.

1.4 Synopsis

The aim of this book is to establish an authoritative, up-to-date reference text that will assist advanced study and research in wave and tidal current energy to maintain progress in the field. In Chapter 2, the marine resource is described, including wave and tidal energy resource assessment, a discussion of the alternative approaches, and acknowledgement that there is still some lack of consensus on resource assessment, while making informed recommendations on methodology, *in situ* measurement and data analysis techniques.

In the next two chapters, the fundamentals of wave and tidal technology are given. In Chapter 3 wave energy converter (WEC) technology is described, and the fundamental theory, the history of development and the categorisation of WECs are discussed. Tidal energy technologies described in Chapter 4 include horizontal and vertical axis turbines; the chapter also covers both tidal stream and tidal lagoon technologies.

Device design is tackled in Chapter 5, in which the development and application of standards is described together with structural design, moorings and reliability. The power systems developed for wave and tidal energy are explored in Chapter 6, and theory is given for power take-off and control.

Key components of the design and development of any MRE system are physical and numerical modelling. In Chapter 7, scale model testing of single devices and arrays in laboratory facilities and larger-scale device testing at sea are described. Testing protocols are also discussed, as well as the different scales appropriate for different stages of design. Chapter 8 gives a thorough overview of numerical modelling for wave and tidal energy. It includes a discussion of the role of numerical modelling in the design process, a brief discussion of the numerical models available, their governing equations and their use within MRE. These include resource models, device design models, farm design models (energy yield and environmental impact), power train models (and control).

In order for a wave or tidal project to be realised, the potential environmental impacts need to be considered and mitigated if appropriate. Chapter 9 describes environmental impact assessment for both physical and biological systems. Each aspect includes a detailed discussion on the near-field and far-field effects of arrays, the short- and long-term impacts, as well as impact assessment and monitoring techniques.

Consenting and legal aspects of wave and tidal energy development are considered in Chapter 10, with a discussion of consenting processes and marine planning procedures, national perspectives and of experience gained from MRE consenting processes.

In Chapter 11, the economics of wave and tidal energy is covered, including the levelised cost of energy approach and the externalities – an important aspect which, if applied to the different energy sources, can contribute to the implementation of optimal electricity production and prices. In Chapter 12, the different stages of project development are illustrated through four representative case studies: two in tidal energy and two in wave energy. With a structure articulated around these case studies, Chapter 12 brings together device design, array planning, policy and economics, drawing material from other chapters, and including detailed discussions of technical aspects (e.g., access and deployment strategies, installation, and reliability and maintenance) and non-technical aspects (e.g., planning, consenting, economic assessment and externalities, financing).

Chapter 13 gives an up-to-date review of developments worldwide, covering regional activities in Europe, North America, Latin and Central America, Asia-Pacific and China. The book concludes with an epilogue on the future of wave and tidal energy.

References

1 Kharecha, P. and Hansen, J. (2008) Implications of 'peak oil' for atmospheric CO_2 and climate. *Global Biogeochemical Cycles* 22(3), GB3012.

2 Nel, W.P. and Cooper, C.J. (2009) Implications of fossil fuel constraints on economic growth and global warming. *Energy Policy* 37(1), 166–180.

3 Pacala, S. and Socolow, R. (2004) Stabilization wedges: Solving the climate problem for the next 50 years with current technologies. *Science* 305(5686), 968–972.

4 Willis, R. (2015) Paris 2015: getting a global agreement on climate change. http://www. green-alliance.org.uk/resources/Paris

5 http://www.theguardian.com/environment/2015/dec/12/paris-climate-deal-key-points

6 Clément, A., McCullen, P., Falcão, A., Fiorentino, A., Gardner, F., Hammarlund, K., Lemonis, G., Lewis, T., Nielsen, K., Petroncini, S., Pontes, M.-T., Schild, P., Sjöstrom, B.-O.; Sørensen, H.C. and Thorpe, T. (2002) Wave energy in Europe: Current status and perspectives. *Renewable and Sustainable Energy Reviews* 6(5), 405–431.

7 Enferad, E. and Nazarpour, D. (2013) Implementing double fed induction generator for converting ocean wave power to electrical. *4th Power Electronics, Drive Systems and Technologies Conference (PEDSTC)*. Piscataway, NJ: IEEE.

8 IEA-Ocean Energy Systems, Annual Report 2013. https://www.ocean-energy-systems. org/documents/82577_oes_annual_report_2013.pdf

9 Magagna, D. and Uihlein, A. (2015) Ocean energy development in Europe: Current status and future perspectives. *International Journal of Marine Energy* 11, 84–104.

10 JRC (2015) *2014 Ocean Energy Status Report*. European Commission. https://setis. ec.europa.eu/sites/default/files/reports/2014-JRC-Ocean-Energy-Status-Report.pdf

11 Osta Mora-Figueria, V., Huerta Olivares, C., Holmes, B., O'Hagan, A.M., Torre-Enciso, Y. and Leeney, R.H. (2011) Catalogue of Wave Energy Test Centres, Plymouth. https://www. plymouth.ac.uk/research/coast-engineering-research-group/sowfia-project.

12 O'Callaghan, J., O'Hagan, A.M., Holmes, B., Muñoz Arjona, E., Huertas Olivares, C., Magagna, D., Bailey, I., Greaves, D., Embling, C., Witt, M., Godley, B., Simas, T., Torre Enciso, Y. and Marina, D. SOWFIA Work Package 2 Final Report. https://www.plymouth.ac.uk/research/coast-engineering-research-group/sowfia-project.

13 House of Commons Energy and Climate Change Committee (2012) *The Future of Marine Renewables in the UK. Government Response to the Committee's Eleventh Report of Session 2010–12.* London: The Stationery Office. http://www.publications.parliament.uk/pa/cm201213/cmselect/cmenergy/93/93.pdf

14 Douglas, C.A., Harrison, G.P. and Chick, J.P. (2008) Life cycle assessment of the SeaGen marine current turbine. *Proceedings of the Institution of Mechanical Engineers, Part M: Journal of Engineering for the Maritime Environment* 222(1), 1–12.

15 Baker, C. and Leach, P. (2006) *Tidal lagoon Power Generation Scheme in Swansea Bay: A Report on Behalf of the Department of Trade and Industry and the Welsh Development Agency.* London: DTI.

16 Lewis, M., Neill, S.P., Robins, P.E. and Hashemi, M.R. (2015) Resource assessment for future generations of tidal-stream energy arrays. *Energy* 83, 403–415.

17 Minchinton, W.E. (1979) Early tide mills: Some problems. *Technology and Culture* 20(4), 777–786.

18 Dorf, R. (1981) *The Energy Factbook.* New York: McGraw-Hill.

19 Andre, H. (1978) Ten years of experience at the 'La Rance' tidal power plant. *Ocean Management* 4(2), 165–178.

20 Salter, S.H. (2012) Are nearly all tidal stream turbine designs wrong? Paper presented to 4th International Conference on Ocean Engineering.

21 www.emec.org.uk, accessed 20/01/2016

22 Marine Energy Global Technology Review 2015. http://www.marine-energy-matters.com/globaltechnologyreview2015, accessed 06/09/2017

23 Strategic Initiative for Ocean Energy (SI OCEAN) https://ec.europa.eu/energy/intelligent/projects/en/projects/si-ocean, accessed 19/01/2016

24 Charlier, R.H., Chaineux, M-C.P., Finkl, C.W. and Thys, A.C. (2010) Belgica's Antarctic toponymic legacy.' *Journal of Coastal Research* 26(6), 1168–1171.

25 Torre-Enciso, Y., Marqués, J. and López de Aguileta, L.I. (2010) Mutriku. Lessons learnt. In *Proceedings of the 3rd International Conference on Ocean Energy.*

26 Weber, Jochem. "WEC Technology Readiness and Performance Matrix–finding the best research technology development trajectory." *International Conference on Ocean Energy.* 2012.

27 www.Primare.org, accessed 19/01/2016.

28 van Nieuwkoop, J.C., Smith, H.C., Smith, G.H. and Johanning, L. (2013). Wave resource assessment along the Cornish coast (UK) from a 23-year hindcast dataset validated against buoy measurements. *Renewable Energy* 58, 1–14.

29 Blunden, L.S. and Bahaj, A.S. (2007). Tidal energy resource assessment for tidal stream generators. *Proceedings of the Institution of Mechanical Engineers, Part A: Journal of Power and Energy,* 221(2), 137–146.

30 Roc, T., Conley, D.C. and Greaves, D. (2013) Methodology for tidal turbine representation in ocean circulation model. *Renewable Energy* 51, 448–464.

31 Lopez, G., Conley, D., and Greaves, D. (2015) Calibration, validation and analysis of an empirical algorithm for the retrieval of wave spectra from HF radar sea-echo. *Journal of Atmospheric and Oceanic Technology* 33(2), 245–261.

32 WETFEET H2020 Project (2014) http://www.wetfeet.eu/, accessed 19/01/2016.

33 Greaves, D., Conley, D., Magagna, D., Aires, E., Chambel Leitão, J., Witt, M., Embling, C.B., Godley, B.J., Bicknell, A., Saulnier,J-B., Simas, T., O'Hagan, A.M., O'Callaghan, J., Holmes, B., Sundberg, J., Torre-Enciso,Y. and Marina, D., (2016) Environmental Impact Assessment: Gathering experience at wave energy test centres in Europe. *International Journal of Marine Energy* 14, 68–79.

2

The Marine Resource

Gregorio Iglesias

Professor of Coastal Engineering, School of Engineering, University of Plymouth, UK

2.1   Introduction

This chapter is concerned with the characterisation of the energy resource from wind waves and tides – a fundamental step towards its exploitation. The term 'characterisation' goes beyond a mere assessment of the resource, and implies an in-depth analysis of its characteristics [1]. For instance, in the case of the wave resource, it is not sufficient to state that the mean wave power at a given site in an average year is $50 \, \text{kW} \, \text{m}^{-1}$; this overall figure, for all its interest, must be complemented with the analysis of the characteristic parameters of the sea states providing the energy: their significant wave heights, energy periods, mean wave directions, etc. This information is essential in determining the actual power output that can be obtained from a particular Wave Energy Converter (WEC), as well as I selecting the WECs that are best suited to the characteristics of the wave resource at the site in question [2]. The wave and tidal resources are covered in Sections 2.2 and 2.3, respectively.

2.2   The Wave Resource

Wind waves, infragravity waves, tsunamis, tides, etc. – there exist many types of waves in the ocean, with different driving mechanisms and properties. However, the term *wave energy* does not refer – at least in its common usage – to all of them, but to a particular type of waves: wind waves, those generated by the wind blowing over the sea [3]. Winds are ultimately the result of solar energy acting on the atmosphere, and therefore wave energy may be regarded as a concentrated form of solar energy – and the oceans, as a gigantic energy conversion system transforming one type of mechanical energy (wind energy) into another, more concentrated type (wave energy).

Considering that the oceans cover 71% of the Earth's surface and that wind waves can propagate over long distances with little energy loss, it is not surprising that a substantial amount of wave power reaches almost continuously any stretch of coastline open to the ocean. This quasi-ubiquity of wave power inshore constitutes one of its principal

Wave and Tidal Energy, First Edition. Edited by Deborah Greaves and Gregorio Iglesias.
© 2018 John Wiley & Sons Ltd. Published 2018 by John Wiley & Sons Ltd.

advantages, and explains the enormous, virtually unlimited global resource – which has been estimated at 32,000 TWh per year [4].

Indeed, wave energy has great potential for development, and its present status is reminiscent of wind energy in the 1980s. The deployment of arrays of WECs, also known as *wave farms*, to harness this vast, untapped resource is hindered by a number of factors, not least the lack of maturity of the technology and the concomitant high costs, which curtail the competitiveness of wave energy *vis-à-vis* other renewables. For wave energy to truly take off, development on four fronts is necessary. First, robust and efficient WECs must be developed (e.g. [5–8]). Second, the resource must be thoroughly characterised so that the areas of interest can be selected and the energy output of prospective wave farms determined [1, 4, 9–13]). Third, the economics of these farms must be assessed, which requires – in addition to the characterisation of the resource – a thorough understanding of the lifetime costs of the WECs themselves and the other components of a wave farm, including their installation, maintenance and decommissioning, as well as the balancing costs of accommodating a fluctuating power source into the network [14–17]. The cost of wave energy may be reduced by realising synergies with other renewables, most notably wind energy [18–24]. Finally, the impacts of wave energy on marine and, in particular, coastal environments must be assessed at different time scales, from the short to the long term; in fact, some of these impacts may turn out to be positive – in the form of a reduction in the amount of wave energy reaching vulnerable coastlines, and research is under way on the application of wave farms to coastal erosion management in synergy with their primary purpose, carbon-free energy generation [25–29].

The motivation for characterising the wave resource is clearly practical, with a view to its exploitation by means of wave farms; these must be situated near the coastline – in some cases, *on* the coastline[1] – for reasons to do with economic viability. Indeed, a number of relevant costs spiral with increasing distance to the coastline and water depth, including those of the submarine cable, mooring system, and maintenance. However, there are also advantages to locating the farm some distance from the shoreline, such as avoiding the harsh environment of the surf zone (a practical necessity, for breaking waves would compromise the survival of the WECs), reducing the visual impact from the coast and, generally speaking, tapping into a greater resource. All these factors must be pondered locally, for each project, and a compromise struck. In principle it may be assumed that wave farms will not be deployed in water depths above 100–150 m, and will consist primarily of floating WECs – bottom-fixed alternatives being too expensive; therefore, it is the *nearshore* wave resource that we are primarily interested in.

Generally speaking, the nearshore wave resource is less uniform than its offshore counterpart, the reason being that waves do not interact with the seabed in deep water. Ocean waves propagate over deep water until, eventually, they approach a coast. When the water depth decreases to half the wavelength, i.e. when the relative water depth, h/L, is ½ (with h the water depth and L the wavelength), the wave begins to interact with the seabed and, by definition, leaves deep water to enter transitional water. To visualise this it is useful to consider the concept of *wave base* – a reference horizontal plane at a depth

---

1 Onshore WECs are indeed a possibility, and may be of interest in particular cases, for instance mounted on (new) coastal structures, as in Mutriku (Spain), but it is doubtful that they will ever be deployed in numbers large enough to provide substantial power, not least because of the impact that this would have on coastal landscapes.

of a half wavelength, $L/2$, beneath which the effects of the wave on the water column (orbital motions of the water particles, pressure fluctuations, etc.) are negligible.[2] In other words, for all practical purposes the wave extends vertically from the surface to the wave base. In deep water, the wave base is above the seabed, implying that there can be no interaction between the wave and the seabed. At some point in the wave propagation towards the coast, the water depth becomes equal to, then smaller than, $L/2$; from that point onwards the wave is constrained vertically by the seabed, and this will affect its properties. That point marks the end of deep water and the beginning of transitional water – an altogether different regime in which the *phase celerity* (the speed at which the wave propagates) is determined no longer by the wave period alone but also by the water depth.[3] We shall see below that this all-important control of the phase celerity by the water depth and the wave period is governed by the dispersion relationship. In essence, as the properties of the medium (water depth) through which the wave propagates change, so does the propagation velocity. This change in the phase celerity causes a change in the direction of propagation – a process known as *refraction*, which, incidentally, affects not only waves in the ocean but also other types of waves, and even light. In the case of ocean waves, the change in the direction of propagation leads to changes in wave height, power, etc. Furthermore, as the wave travels over transitional water the varying water depths lead to changes in the *group celerity* (the speed at which wave *energy* propagates), which in turn also lead to changes in wave height, power, etc. – a process known as *shoaling*. In sum, in transitional water varying water depths affect the fundamental wave parameters (with the exception of the wave period, nonlinearities excluded). As a result of refraction and shoaling, and other processes such as friction with the seabed, the relative uniformity of the wave field in deep water is gradually transformed into ever greater spatial variability – in particular, where the bathymetry (the seabed contours) is highly irregular.

We have seen why the nearshore wave resource presents significant spatial variability. Whereas certain areas – the *nearshore hotspots* [30] – have a substantial resource, nearby areas have a much lower resource. This spatial variability is a crucial element that must be accounted for in the characterisation of the resource and, particularly, in the selection of prospective wave farm sites [31]. Similarly, the temporal variability must be considered, for it will determine the performance and energy output of the wave farm, as well as the survivability of its WECs [2, 31–34]. For these reasons, a thorough characterisation of the spatial and temporal variability of the nearshore wave resource in the area of interest is a prerequisite to assessing the economic viability of a wave farm project [14, 16].

A brief note on the structure and contents of this section. Ideally, the reader should be familiar with wave theory before delving into the characterisation of the wave resource; a brief summary of the theory is presented in Sections 2.2.1 (linear wave theory) and

---

2 This value of $L/2$ is admittedly somewhat arbitrary, and related to the interpretation, or quantification, of "negligible"; strictly speaking, the effects of the wave on the water column (orbital motions of water particles, pressure fluctuations, etc.) decrease exponentially with water depth and therefore do not become zero at any finite depth, but tend to zero at infinity.

3 Later on, as the wave continues to propagate towards the coast, a second threshold of relative water depth, $h/L = 1/20$, will eventually be reached, marking the end of transitional water and the beginning of shallow water – yet a different regime in which waves cease to be dispersive, that is, the phase celerity is controlled exclusively by the water depth.

2.2.2. (random waves). The characterisation of the wave resource proper is covered in Sections 2.2.3 (offshore resource) and 2.2.4 (nearshore resource).

2.2.1 Fundamentals of Linear Wave Theory

In linear wave theory, also known as Airy wave theory, the displacement of the free surface (the sea surface) is given by

$$\eta = \frac{H}{2}\cos\left(\frac{2\pi}{L}x - \frac{2\pi}{T}t\right),$$ (2.1)

where η is the vertical excursion of the free surface, H is the wave height, L the wavelength, and T the wave period. If the amplitude, wavenumber and angular frequency are denoted by a, k and ω, respectively, then

$$a = \frac{H}{2},$$ (2.2)

$$k = \frac{2\pi}{L},$$ (2.3)

$$\omega = \frac{2\pi}{T},$$ (2.4)

and the sinusoidal displacement of the free surface, equation (2.1), may be rewritten as

$$\eta = a\cos\left(kx - \omega t\right).$$ (2.5)

The wavenumber, angular frequency and water depth (h) are related through the dispersion relationship,

$$\omega = kg\tanh\left(kh\right),$$ (2.6)

with g the gravitational acceleration. Alternatively, the dispersion relationship may be expressed in terms of the wavelength, period and water depth:

$$L = \frac{gT^2}{2\pi}\tanh\left(\frac{2\pi h}{L}\right).$$ (2.7)

Of importance here is the fact that, for a given wave period at a certain point (water depth), there exists only one wavelength that satisfies equation (2.7). In other words, at a certain point the wave period determines the wavelength, or vice versa.

For a given wave period, the greater the water depth, the longer the wave. As waves approach the coast, the wave period may be assumed to remain constant;[4] therefore, as the water depth reduces, so does the wavelength, as determined by equation (2.7). Eventually, the water depth becomes too small for the wave to be viable, at which point it breaks.

---

4 *Stricto sensu* this holds only if a steady state can be assumed – which is often the case.

Phase celerity (from *celeritas*, Latin for speed or velocity) is the speed at which the wave form (i.e., the wave crest, trough, etc., *but not the wave energy*) travels,

$$c = \frac{L}{T}. \tag{2.8}$$

Combining equations (2.7) and (2.8) yields

$$c = \frac{gT}{2\pi} \tanh\left(\frac{2\pi h}{L}\right). \tag{2.9}$$

For a given water depth, the greater the wave period, the greater the wave celerity, i.e. the faster the wave. This property, *dispersion*, is of great importance in wave theory.[5]

Wave energy density is the amount of energy per unit surface area averaged over the wave period,

$$E = \frac{1}{8}\rho g H^2, \tag{2.10}$$

where ρ is seawater density (as a first approximation, $\rho \sim 1025\,\mathrm{kg\,m^{-3}}$). In the SI wave energy density is expressed in joules per square metre ($\mathrm{Jm^{-2}}$). The term *density* alludes to the fact that the energy is measured *per unit surface area* (per square metre in the SI) in the reference plane (i.e., the quiescent sea surface). It can be proven [35] that the wave energy density is composed of kinetic and potential energy, with equal shares: 50% of kinetic and 50% of potential energy. Importantly, the wave energy density varies with the square of the wave height, so a 2 m wave has four times as much energy density as a 1 m wave.

Group celerity is the speed at which wave energy propagates,

$$c_g = nc, \tag{2.11}$$

with

$$n = \frac{1}{2}\left(1 + \frac{2kh}{\sinh(2kh)}\right). \tag{2.12}$$

Wave power, also known as wave energy flux, is the amount of energy that passes per unit time through a vertical section of unit width perpendicular to the direction of propagation (i.e., parallel to the wave fronts); it is given by

$$P = Ec_g. \tag{2.13}$$

In the SI wave power has units of watts or kilowatts per metre *of wave front* ($\mathrm{Wm^{-1}}$ or $\mathrm{kWm^{-1}}$).

---

5 In particular in the evolution from a *wind sea* to a *swell*. A *wind sea* occurs when waves are being generated by the wind. In a wind sea, waves of different periods and directions coexist (the wave spectrum is broad-banded). By contrast, in a *swell*, waves have similar periods and directions (the wave spectrum is narrow-banded). (These concepts can be quantified through spectral theory as explained below.) As waves travel away from the area of active wave generation by wind, they do so at different celerities, according to their periods. For this reason, and others of a nonlinear nature, a wind sea transforms gradually into a swell as waves propagate away from their generation area.

So far we have covered linear wave theory. As mentioned above, this theory applies only to certain wave conditions. Other wave theories (Stokes, solitary wave, cnoidal, etc.) are beyond the scope of this chapter, and the interested reader may consult, for example, [36,37].

The wave power over a certain width, b, of wave front may be obtained from

$$P_b = Ec_g b. \tag{2.14}$$

This quantity is useful in establishing the amount of wave power that is available to a WEC, and is the basis of the concept of *capture width*.

2.2.2 Random Waves

In the linear, or Airy, wave theory described in the previous subsection, a wave consists of a sinusoidal oscillation of the free surface. It suffices to observe the surface of the ocean for a couple of minutes or to look at the records of a wave buoy or other wave measuring device to realise that, in fact, the sea surface oscillates following not a sinusoidal curve (a regular wave) but a much more complex pattern (random or irregular waves).

The analysis of random waves is based on the concept of sea state, which may be analysed in the time or the frequency domain. In the time domain we consider a time series of surface elevations, $\eta(t)$, corresponding to a point in space (e.g. the observations from a wave buoy), with a typical duration of between 15 min and 30 min. For the statistical analysis, the hypothesis is made that the surface elevation is a stationary process; for this reason, the time series cannot be too long. Nor can it be too short, or the statistics derived from it (e.g., the significant wave height) would not represent the variability in the sea state properly.

Mathematically the sea state may be described as a Fourier series [38], i.e. a sum of an infinite (in practice, a large) number of harmonic components,

$$\eta(x,t) = \sum_{i=1}^{\infty} a_i \cos(k_i x - \omega_i t), \tag{2.15}$$

with a_i, k_i and ω_i the amplitude, wavenumber and angular frequency, respectively, of the ith harmonic component (or harmonic for short). Each harmonic is, in its own right, a linear regular wave of the type described in the previous subsection. This description of the sea state by means of a Fourier series is based on the principle of linear superposition, a mathematical tool of great potency that forms the basis of the random phase–amplitude model, itself the basis of our wave analysis [39].

The amount of energy available can be quantified in terms of the average wave power or the total energy (total resource) over a certain period of time, e.g. a (typical) year or a (typical) winter. The calculation of the total resource over a certain period based on a time series of wave data is straightforward. Typically, the time series stems from a wave buoy or numerical model and has a three-hourly frequency. In some cases the data consist of a discretised form of the spectral variance or energy density curve, i.e. the values of the spectral curve at a series of frequencies, which give the distribution of the energy in the wave field across frequencies and, if the buoy is directional rather than scalar, also across directions; in other cases, only the characteristic values of the wave parameters are given, e.g. the significant wave height and the energy period.

If spectral density values are given, calculating the characteristic wave parameters is straightforward. Assuming that wave heights are Rayleigh distributed, the significant wave height can be computed from

$$H_s = 4.004\sqrt{m_0} \approx 4\sqrt{m_0}, \tag{2.16}$$

where m_0 is the zeroth-order moment of the variance spectrum,

$$m_0 = \int_0^\infty S(f)\,df, \tag{2.17}$$

with $S(f)$ the spectral variance density. The significant wave height thus obtained can be denoted by H_s or H_{m_0}, the latter notation emphasising the fact that it was calculated from the zeroth-order spectral moment rather than from a time series.

Although the significant wave height is the parameter used most commonly to give the vertical scale of the waves, in certain energy-related applications it may be preferable to use the root-mean-square wave height, H_{rms}, which has a direct interpretation as the wave height of a sinusoidal (Airy) wave with the same energy density as the sea state. Assuming wave heights are Rayleigh distributed [39], the relationship between the significant and root-mean-square wave heights is

$$H_s = \sqrt{2}H_{rms}. \tag{2.18}$$

The energy period, T_e or T_{m-10}, also has a direct interpretation as the period of the sinusoidal (Airy) wave with the same wave energy flux or power as the sea state in question; it may be obtained from

$$T_e = \frac{m_{-1}}{m_0}. \tag{2.19}$$

Alternatively, the energy period, T_e, may be obtained from the peak period, T_p, assuming a certain theoretical spectral shape. For instance, if a JONSWAP spectrum is assumed, then $T_e = T_p/1.11$. Having defined the significant wave height and the energy period, the wave power, or wave energy flux, of a deep-water sea state may be expressed as

$$P = \frac{\rho g^2}{64\pi} H_s^2 T_e. \tag{2.20}$$

P is the wave power per unit width of wave front, with SI units of watts or kilowatts per metre (Wm^{-1} or kWm^{-1}). The accuracy of this expression depends on wave heights being Rayleigh distributed. It is therefore valid for most applications in deep water. Using SI units and assuming a typical value of $\rho = 1025\,kg\,m^{-3}$ for seawater density, the above expression may be approximated as

$$P \approx 0.4906 H_s^2 T_e \approx \frac{1}{2} H_s^2 T_e. \tag{2.21}$$

In this expression, if H_s and T_e are input in m and s, respectively, P is obtained in kWm^{-1}. The above expressions allow the wave power in a sea state to be calculated based on two of its characteristic values, the significant wave height and energy period.

Once wave power is known, the *wave energy* or *wave resource* over a certain period of time, from $t = t_1$ to $t = t_2$, can be obtained by integration with respect to time:

$$E = \int_{t_1}^{t_2} P(t)\,dt, \tag{2.22}$$

where $P(t)$ is wave power as a function of time. As in the case of P, E refers to a unit length of wave front, i.e. it is the wave energy or wave resource *per unit length of wave front* between $t = t_1$ and $t = t_2$, and may be expressed in joules per metre (J m^{-1}) or kilowatt-hours per metre (kWh m^{-1}). The annual resource, i.e. the total energy available from the waves over an entire year, can be calculated by setting t_1 and t_2 to correspond with the beginning and end, respectively, of the year in question.

The mean wave power over a certain period of time, between $t = t_1$ and $t = t_2$, is given by

$$P_{\text{mean}} = \frac{1}{t_2 - t_1} \int_{t_1}^{t_2} P(t)\,dt. \tag{2.23}$$

Usually the analysis is performed on the basis of a time series of characteristic wave parameters (e.g. H_s, T_e) or spectral density curves from which the wave parameters can be obtained. This time series consists of discrete datapoints with a certain frequency, e.g. every 3 hours, and therefore the integrals in the preceding equations must be replaced by discrete summations; for instance, in the case of the mean power, equation (2.23) is replaced by

$$P_{\text{mean}} = \frac{1}{N} \sum_{i=1}^{N} P_i, \tag{2.24}$$

where P_i is the wave power at time

$$t_i = (i-1)\Delta t, \quad \text{for} \quad i = 1, 2, 3, \dots, N, \tag{2.25}$$

and Δt is the time interval between consecutive data points. If the database has a three-hourly frequency, for instance, then $\Delta t = 10{,}800$ s. By using equation (2.24) the assumption is made that the power value P_i calculated for each data point (each sea state) applies to that time interval, 3 hours in the example, even though the sea state itself is considerably shorter for the reasons outlined above.

2.2.3 Offshore Wave Resource

As mentioned, the quantification provided by overall parameters such as the total annual resource and the mean wave power does not convey sufficient information for the purposes of harnessing the wave resource by means of WECs. It is necessary to *characterise* the resource, i.e. to describe it in terms of the characteristics of the sea states that provide the energy. The most usual characteristic values are the significant wave height (H_s), energy period (T_e) and mean direction (θ_m). If the WECs to be installed are not sensitive to the direction of the waves (e.g. heaving point absorbers) then the situation is simpler, and the main parameters to be considered are the significant wave height and energy period.[6] This information can be obtained based on offshore wave buoy data or, absent these, hindcast data, i.e. time series of wave parameters covering a number of years in the past, generated by a numerical wave model through reanalysis of meteorological data and global, or at least large-scale, atmospheric models [1].

To illustrate the procedure for characterising the wave resource, we present a case study off the Death Coast (*Costa da Morte*, in the vernacular) in Galicia, NW Spain, so called after the large number of wrecks caused by its most energetic wave climate (Figure 2.1). Additional details on the resource in this area may be found in previous work [13,33]. The analysis here is based on data from the directional wave buoy located at 43.49°N and 9.21°W, in 388 m of water depth. Operated by Spain's State Ports, this deep-water wave buoy transmits data at an hourly frequency. The period analysed extends from 15/05/1998 to 26/01/2016. For each data point, the following parameters are available, *inter alia*: significant wave height, energy period, and mean wave direction.

It is convenient to analyse the bivariate distribution of energy period and significant wave height (T_e, H_s) through a *resource characterisation matrix* (the numerical information in Figure 2.2). Each greyed square in the matrix corresponds to a bivariate interval of T_e and H_s, e.g. $T_e = (9.5, 10.5)$ s and $H_s = (1.75, 2.25)$ m. These bivariate intervals, with lengths of 1.0 s (ΔT_e) and 0.50 m (ΔH_s) in the example, will be referred to henceforth as *energy bins* [13]. The number in each greyed square in the matrix gives the occurrence of the corresponding energy bin or, to be more precise, of the sea states whose values of T_e and H_s fall within the corresponding bivariate interval in the period considered, which may be a particular year, season (e.g. winter 2015), month (e.g. January 2016), or a typical year (defined e.g. as the average of a number of years) or season. In Figure 2.2, the period considered is a typical year in the series from 15/05/1998 to 26/01/2016, obtained by averaging and trimming the data series as appropriate. The occurrence can be expressed in units of time, as in Figure 2.2 (hours), or as a percentage of the total time in the period considered.

The resource characterisation matrix may be regarded as a scatter plot onto which the grid defining the energy bins; is superimposed. Assuming that each data point represents, as in the case study, one hour of real time, the number in each bin in Figure 2.2 is the number of data points in that bin in the scatter plot. Thus, the figure in each energy bin is the occurrence in number of hours in an average year of sea states with values of T_e and H_s that fall into the respective interval.

The curves in Figure 2.2 are wave power isolines, i.e. lines of constant wave power, from 2 kW m^{-1} to 1000 kW m^{-1}, computed based on the deep-water approximation, equation (2.20). The highest values of wave power occur in the upper right-hand corner of the matrix, corresponding to high values of the energy period and, most importantly, the significant wave height. The dependence of wave power on energy period and significant wave height is linear and quadratic, respectively (equation (2.20)).

The resource characterisation matrix may be complemented to form a resource characterisation diagram with information on the contribution of the energy bins to the total resource, which is calculated by adding the energy (wave power times duration) provided by every sea state in the energy bin. This is represented by the grey scale in Figure 2.2. The darker hues, for instance, indicate energy bins providing between 10 and

---

6 Even in the case of heaving buoys some degree of sensitivity to wave directions will always exist due to the layout of the wave farm and the shadow effect of each WEC on the other WECs in its lee.

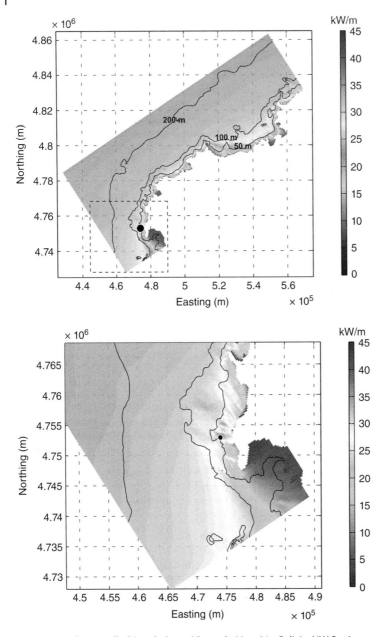

Figure 2.1 The so-called Death Coast (*Costa da Morte*) in Galicia, NW Spain: general map (above) and detailed map of the area around the nearshore hotspot (below), with average wave power values. The nearshore hotspot considered in Section 2.4 is depicted by a black circle.

12 MWh m^{-1} in a typical year. Conversely, the light grey energy bins contribute less than 2 MWh m^{-1}. The advantage of the grey scale is that it makes it easy to visualise immediately which ranges of period and height provide the bulk of the resource in a typical year – and it is on this basis that the WECs to be installed should be selected or designed. This is the *resource characterisation diagram* in Figure 2.2, which consists of

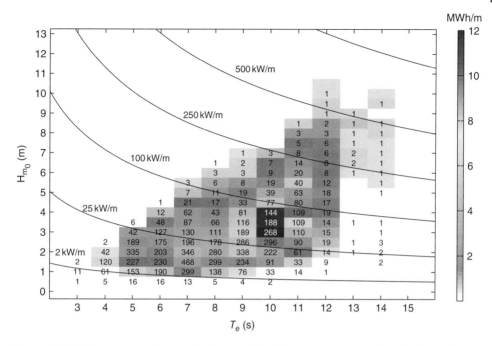

Figure 2.2 Offshore resource characterisation matrix including wave power isolines for the Death Coast (*Costa da Morte*), Galicia, NW Spain. The numbers in the greyed squares represent the occurrence of the respective *energy bins*, expressed in hours in a typical year. The greyscale indicates the contribution of each *energy bin* to the annual resource.

the resource characterisation matrix (the occurrence of the different energy bins), the grey scale (the contribution of each energy bin to the total resource of the period considered) and the wave power isolines.

The contribution of the different energy bins is governed by two factors: the occurrence and wave power of the sea states whose values of T_e and H_s fall within the corresponding interval. The energy bins in the upper right-hand corner of the matrix have high power but low occurrence, therefore contribute little to the overall resource. This does not mean, however, that their most energetic sea states are irrelevant – they need to be taken into account not least with respect to the survivability of the WECs. At the other extreme, the energy bins in the lower left-hand corner of the graph, occurrence values are far higher but wave power is low; therefore, the contribution of these sea states to the total resource is also rather limited. The energy bins that contribute most to the resource (the darkest hues in the central area of the matrix) represent a compromise between occurrence and wave power. In the case study, the bulk of the resource is provided by sea states with significant wave heights between 2.25 m and 5.25 m, and energy periods between 8.5 s and 11.5 s. These high periods are indicative of the oceanic provenance of the waves – to be expected in an area exposed to the long Atlantic fetch. Finally, the energy bins grey with some contribution to the total resource form a triangle with its apex tilted towards the right-hand side (towards the larger periods), which reflects the well-known correlation between wave heights and periods.

For the purposes of calculating the power output from a particular WEC, it is recommended that the resource characterisation matrix for the deployment site be obtained

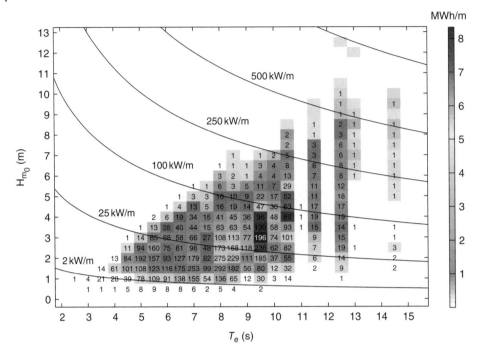

Figure 2.3 Offshore resource characterisation matrix for the Death Coast (NW Spain), with improved resolution: 0.5 s (ΔT_e) × 0.5 m (ΔH_s).

with (at least) the same resolution as the power matrix of the WEC. To expedite the calculation procedure, the offshore resource characterisation matrix should also have the same resolution. In other words, the lengths of the energy bins (ΔT_e, ΔH_s) should match those in the power matrix of the WEC, as should their boundaries. This is of course merely a matter of selecting the energy bins appropriately. For instance, if the resolution of the resource characterisation matrix in Figure 2.2, 1.0 s (ΔT_e) × 0.50 m (ΔH_s), were not enough for a particular WEC, it would be straightforward to produce a refined matrix, e.g. with a resolution of 0.5 s (ΔT_e) × 0.5 m (ΔH_s), as in Figure 2.3.

2.2.4 Nearshore Wave Resource

Having established the offshore wave resource, the next step in characterising the inshore wave resource is the propagation from offshore to the nearshore area of interest. In deep water, by definition, waves do not interact with the seabed. As waves travel towards the coast, water depth decreases; eventually waves reach a water depth at which they begin to 'feel' the seabed. As indicated in Section 2.1, in linear wave theory this water depth is conventionally assumed to be equal to half the wavelength – the depth of the so-called *wave base*. When the wave base 'touches' the seabed, the wave passes from deep water to intermediate or transitional water; in this new regime, it is constrained vertically by the seabed as it continues to propagate towards the shore. This interaction of the wave with the seabed gives rise to a number of physical phenomena, notably refraction and shoaling [35], which affect the wave properties, including the length, celerity, height and power. Importantly, the wave period remains unchanged.

As a result of these processes, the relative spatial uniformity of the resource offshore is transformed into spatial variability nearshore, especially where the bathymetry is highly irregular. In this case, the wave resource is concentrated in particular areas known as n [30], which alternate with areas of comparatively little wave energy. Often drastic changes in the resource occur over distances of the order of hundreds of metres, hence the importance of a fine characterisation of the spatial variability of the nearshore resource. In addition to wave interaction with the seabed, other processes may contribute to the differences between the nearshore and offshore resources, including wave diffraction at headlands or coastal structures, nonlinear wave-wave interactions, energy input from the wind (wave generation) and energy dissipation through whitecapping.

For the purposes of this chapter we assume that the term 'nearshore' signifies an area close to the shore but not so close that waves break because of limited water depth. In other words, we explicitly exclude the surf zone, in which the breaking process dissipates a substantial amount of energy – more or less gradually depending on the type of breaker; for this reason, the resource in the surf zone is typically smaller than elsewhere. In any case, the surf zone, with its highly complex hydrodynamics and high levels of turbulence, would appear to be too harsh an environment to deploy WECs.

Leaving aside bays sheltered by headlands or areas in the lee of structures, there are essentially two reasons why the nearshore wave resource differs from its offshore counterpart: wave interaction with the seabed (refraction, shoaling, etc.) and energy transfer from the wind as waves propagate towards the coast. Assessing the inshore wave resource requires, therefore, that the offshore resource be 'propagated'. This is typically undertaken by means of spectral coastal wave models such as SWAN (Simulating WAves Nearshore), based on the wave action equation [40]. (Wave action density has the advantage over wave energy density that it is conserved in the presence of a current field [41]).

The spectral wave model is implemented onto a computational grid covering the inshore area of interest and extending offshore well into deep water. The geometry of the grid should be adapted to the coastline shape. Cartesian or curvilinear grids may be used; although Cartesian grids are far more common, curvilinear grids have been used advantageously where the coastline is curved [13]. The area of interest should be distant from the grid boundaries – not only from the offshore (ocean) boundary, but also from the lateral boundaries – to prevent numerical disturbances arising at the boundary from affecting the model results in that area.

For the sake of computational efficiency, the resolution of the grid is often not uniform – the grid spacing is smaller in the area of interest, inshore, where wavelengths are usually smaller, than in the deeper sections of the grid offshore. Alternatively, or complementarily, nested grids may be used to reduce the computational cost of the model.

In the case study, the computational grid used for wave propagation off the Death Coast (Figure 2.4) is a Cartesian grid with its x-axis parallel to the general coastline orientation and $\Delta x = 249.9$ m. The y-axis has a variable resolution, from $\Delta y = 888.9$ m at the offshore boundary to $\Delta y = 284.8$ m inshore. The number of nodes along the x and y directions is $m = 577$ and $n = 193$, respectively, with the total number of nodes equal to 111,361.

It is important for the bathymetry to have sufficient resolution, especially in areas where the seabed contours are highly irregular. Bathymetric data are generally available from hydrographic (nautical) charts, and in some cases *ad hoc* surveys of the area of interest may be undertaken. Once this information has been gathered, the water depths at the nodes of the computational grid can be obtained by interpolation.

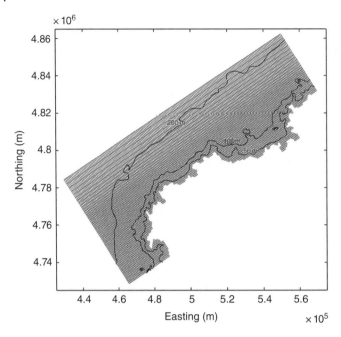

Figure 2.4 Computational grid for the Death Coast case study. For the sake of clarity, only one in three nodes is represented.

Having implemented the numerical model, a number of *wave cases* are then propagated. The number of wave cases should be sufficient to cover the complexity (temporal variability) of the offshore wave resource; in other words, the wave cases should encompass the combinations of wave height, period, direction, etc. that are relevant in the area, so that a substantial proportion of the total wave resource and of the total time in the period considered is covered. In each case the offshore boundary conditions are prescribed based on a certain set of wave parameters (significant wave height, peak period, mean direction, directional spreading, etc.) and spectral shape (e.g. JONSWAP). The straightforward approach is to consider the offshore trivariate distribution of wave height, energy period and mean wave direction, which may be visualised as a 3D resource characterisation matrix similar to that in Figure 2.3, but with an added axis or dimension for the mean wave direction. This is used to select for the propagation one *wave case* for each energy bin, starting with the energy bin with the greatest contribution to the resource and continuing in order of decreasing contribution, until a predetermined percentage of the total resource is covered.

For instance, in the case study, with energy bins of 0.5 s (ΔT_e) × 0.5 m (ΔH_s) × 22.5° ($\Delta\theta$), a total of 787 wave cases were selected for propagation, representing 95% of the total resource and 88.7% of the total time in the period considered. It may be seen that the offshore resource characterisation matrix that may be constructed based on these wave cases (Figure 2.5) is not dissimilar to the (total) offshore resource matrix (Figure 2.3), which can serve as a basic quality assurance that the number of cases propagated cover large enough proportions of the total resource and of the total time. It would obviously be possible to propagate even more cases and thus cover an even greater percentage of the total resource; however, the important point here is that a compromise between computational cost and accuracy must be struck.

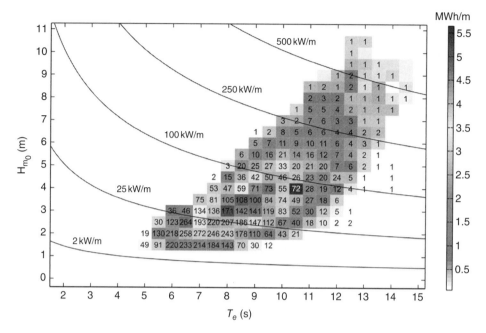

Figure 2.5 Offshore resource characterisation matrix limited to the wave cases selected for propagation, i.e. 95% of the total resource.

The parameters of each wave case for the propagation are then determined so as to represent the sea states in the corresponding energy bin. For the energy period and mean direction, the mid-interval values are usually selected. The significant wave height, however, is another matter: given that the relationship between significant wave height and wave power is quadratic (equation (2.20)), it is recommended that the value selected to represent the interval (H_{s1}, H_{s2}) be $H_{s1} + a(H_{s2} - H_{s1})$, with $a = 2^{-1/2} = 0.7071$.

A numerical model, SWAN [40] in the case study, is now used to propagate the wave cases thus defined. The resource characterisation matrix corresponding to any node of the computational grid is then constructed by adding the contribution of the different cases weighted by their occurrence in the period of time considered (in the case study, a typical year). If the point of interest nearshore does not coincide with a grid node, the results may be interpolated from the values at the surrounding nodes. The calculations are best done by means of a database combined with a decision-aid tool for site selection [33, 42]. In a previous characterisation of the wave resource off the Death Coast [13], the nearshore hotspot in Figure 2.1 was identified. The resource characterisation matrix for this hotspot is presented in Figure 2.6. Here yields a characterisation of nearshore terms.

We have seen that the resource characterisation matrix is a convenient means of representing the bivariate distribution of (T_e, H_s), the main characteristic parameters of the wave resource, whether offshore or nearshore (Figures 2.3 and 2.6, respectively). However, there are cases in which the bivariate distribution (T_e, H_s) is not sufficient, e.g. in designing a wave farm, and must be complemented with the trivariate distribution (T_e, H_s, θ_m), where θ_m is the mean wave direction. The corresponding resource characterisation matrix is now three-dimensional, and therefore not so easy to interpret visually; in fact, 2D representations taking the parameters in pairs may be preferable. For

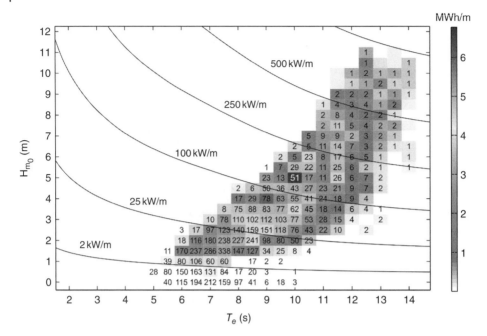

Figure 2.6 Resource characterisation matrix at the nearshore hotspot in Figure 2.1.

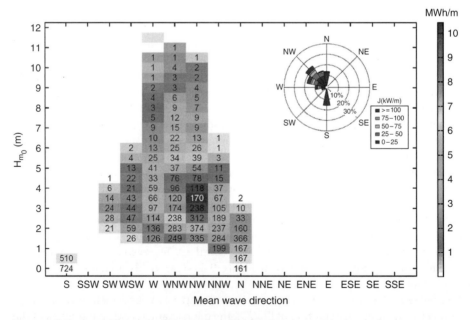

Figure 2.7 Resource characterisation matrix at the nearshore hotspot in terms of mean wave direction and significant wave height (θ_m, H_s). A wave power rose is also included.

illustration, the resource characterisation matrices for the nearshore hotspot in terms of (θ_m, H_s) and (θ_m, T_e) are presented in Figures 2.7 and 2.8, respectively, complemented with a wave power rose. It is apparent that the lion's share of the resource is provided by waves from the IV quadrant, primarily from the northwest.

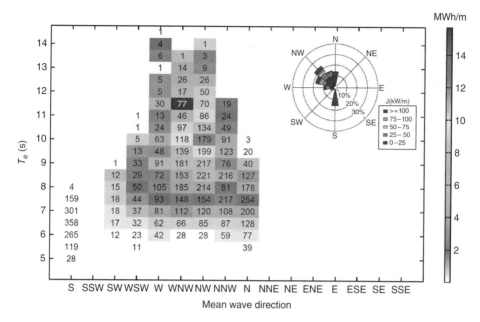

Figure 2.8 Resource characterisation matrix at the nearshore hotspot in terms of mean wave direction and energy period (θ_m, T_e). A wave power rose is also included.

2.3 The Tidal Stream Resource

2.3.1 Fundamentals of the Tide

The tide is the periodic oscillation of the sea level caused by the gravitational attraction of the Moon and Sun acting on the water particles in the hydrosphere. The force driving the tide is controlled by the complex motion of the Earth–Moon–Sun system. In spite of the much larger mass of the Sun, solar attraction plays a lesser role, approximately 46% of lunar attraction, due to the greater distance from the Earth to the Sun than to the Moon. Tides are different across the globe, with many coastal regions experiencing only very weak tides (e.g. enclosed seas such as the Mediterranean) while others are subjected to large tidal oscillations. This is due to the fact that the oscillation driven by this astronomical forcing propagates as a long wave in an ocean or sea with particular characteristics (water depths, boundaries) and under the effects of the Coriolis force. Indeed, this long wave is refracted over the continental shelf, reflected at land boundaries, amplified by the tapering banks of estuaries or by resonance, dampened by friction with the seabed, etc. In sum, the tide is a complex phenomenon driven by astronomical forces but controlled locally by the bathymetry and coastline geometry of the estuary, gulf, or sea concerned, and of the ocean basin to which it is connected.

Tides have been known since Antiquity, but their explanation has eluded scientists for many centuries, including eminent names such as Galileo. It was not until Newton and his theory of gravity that the primary cause of the tide was understood. However, the differences in the tide between regions could not be explained until much later, when Laplace introduced the concept of the tide as a wave propagating over a sea with certain water depths and boundaries (coastlines). A complete description of the theory of tides is beyond the scope of this book, and the interested reader may consult [43].

As a periodic oscillation, the tide is described as a long wave that propagates over oceans, seas and estuaries, with the usual properties of waves, e.g. wave length and amplitude. When the crest or trough of the wave reaches a particular point, this point is at high or low water, respectively. The period of the tide, or time interval between two consecutive high waters, is approx. 12 h 25 min in the case of semidiurnal tides (usual in the Atlantic Ocean and other regions), and 24 h 50 min in the case of diurnal tides (in some areas of the Pacific Ocean, Gulf of Mexico, etc.) The terms semidiurnal and diurnal allude to the duration of one tide, approximately a half day or one day, respectively.

The difference in elevation between consecutive high and low water is called the tidal range, which varies in space and time, i.e. at any given point two tides will have different tidal ranges. The tidal range in the open ocean is never really large; it can, however, become large near shore, particularly in semi-enclosed seas and estuaries, due to resonance and convergence of the land boundaries [43]. A conspicuous case is the Bay of Fundy, Canada, where the tidal range reaches 16.3 m. Coastlines are classified as microtidal, mesotidal or macrotidal when the tidal range is below 0.5 m, between 0.5 m and 4.0 m, and above 4.0 m, respectively.

Semidiurnal tides vary with a fortnightly period (14.8 days, or half the lunar month). At new or full moon, when the Moon, Sun and Earth are aligned (in syzygy), the solar tide reinforces the lunar tide, which results in particularly large tidal ranges (spring tides) shortly afterwards. The reason for the lag between the syzygy (the full or new moon) and the maximum tidal range is the inertia of the water mass. Half a lunar month, i.e. 14.8 days, later, at the first or last quarters, the Moon and Sun are orientated at a 90° angle relative to the Earth (in quadrature) and the opposite occurs: the solar effects reduce the lunar tide, which leads to particularly small tidal ranges (neap tides).

Another relevant aspect of the astronomical forcing of the tide is the varying distance between the Moon and the Earth, leading to increased and decreased tidal ranges at the lunar perigee and apogee – when the Moon is closest and farthest from the Earth in its orbit, respectively. The period of this oscillation is 27.55 (solar) days. The largest tides occur, therefore, when spring tides coincide with the lunar perigee.

The periodic oscillation of the sea level at a given point may be described mathematically by means of a Fourier series

$$\zeta(t) = \sum_i a_i \cos(\omega_i t + \phi_i), \tag{2.26}$$

where t is time, $\zeta(t)$ is the tidal level, and a_i, ω_i and ϕ_i are the amplitude, angular frequency and phase of the ith harmonic component, respectively. The summation is performed over a large number of harmonic components or *tidal constituents* – the larger the number, the greater the accuracy. Seven or eight may suffice in most cases, although in certain areas more may be required, for instance in estuaries with tidal asymmetry.[7]

The angular frequency of a given tidal constituent is the same at any point on the planet: it is determined by the relative motions of the Earth–Moon–Sun system. However, the amplitudes and phases of the constituents vary from point to point, as does the tidal range. The main tidal constituents at the mouth of the Severn Estuary are shown in Table 2.1.

The semidiurnal or diurnal character of the tide depends on the relative amplification of certain tidal constituents, which is measured by the tidal form factor, also known as the F-factor [45],

$$F = \frac{K_1 + O_1}{M_2 + S_2}, \tag{2.27}$$

where K_1, O_1, M_2 and S_2 represent the amplitudes of the corresponding tidal constituents (Table 2.1). It may be seen that the constituents in the numerator and denominator are, respectively, the principal diurnal and semidiurnal constituents. It follows that high and low values of F will correspond to diurnal and semidiurnal tides. To be more precise, for $F < 0.25$ or $F > 3$, the tidal regime is semidiurnal or diurnal, respectively. For the intermediate values an intermediate regime exists, predominantly semidiurnal or diurnal for $0.25 < F < 1.5$ or $1.5 < F < 3$, respectively.

So far we have discussed the astronomical tide, but this is not the only cause of variations in sea level. The meteorological tide is the change in sea level due to winds and changes in atmospheric pressure. When the wind blows onshore, shear stress on the sea surface causes the surface level to rise near shore; we could say that water 'piles up' against the coast; conversely, when the wind blows offshore, nearshore surface tidal levels drop. As regards atmospheric pressure, the effects of its variations on surface levels are easy to understand if we consider that the sea surface is, in reality, the interface between two fluids, air and seawater. When atmospheric pressure rises above its standard value at sea

Table 2.1 Tidal constituents at the mouth of the Severn Estuary [44].

Constituent	Description	Amplitude (cm)	Phase (°)
M_2	Principal lunar semidiurnal	235.24	156.87
S_2	Principal solar semidiurnal	84.17	201.21
N_2	Larger lunar elliptic semidiurnal	44.79	138.48
K_2	Lunisolar semidiurnal	24.45	195.80
K_1	Lunar diurnal	6.77	127.34
O_1	Lunar diurnal	6.70	351.17
P_1	Solar diurnal	2.23	121.81
Q_1	Larger lunar elliptic diurnal	1.95	305.66
M_4	Shallow water overtides of principal lunar constituent	3.69	290.99

---

7 In which case it is important to include the M_4 (overtide) harmonic.

level (1013 HPa) – in an anticyclone or high pressure – surface levels drop; conversely, when the pressure decreases below 1013 HPa – in a depression or low pressure – surface levels rise. With 1 HPa of pressure variation causing 1 cm of tidal level change, the meteorological tide is generally well below ±0.5 m; it can, however, reach much larger values under storm conditions (storm surge). In a hurricane the storm surge can reach several metres, as it did during the infamous Hurricane Katrina (2005) [46].

For a given point, a harmonic analysis can be carried out based on a time series of tidal levels (e.g. from a tide gauge) to determine the amplitudes and phases of the tidal harmonics. A tide gauge records the oscillations of sea level over time, which are caused by the astronomical tide and other factors, notably the meteorological tide, infragravity waves and wind waves. For this reason, the data series should be pre-treated prior to the harmonic analysis so that the influence of the non-astronomical factors is removed. Removing short waves is straightforward by means of a low-pass filter; other techniques (e.g. wavelets) may also be applied [47–49]. Involved the effects of atmospheric pressure changes may be extracted easily if a time series of atmospheric pressure at or near the tide gauge is available, but removing the effects of winds before the harmonic analysis may be more complicated [50–52].

Once the effects of the main non-astronomical factors on the time series of tidal level have been extracted, the harmonic analysis may be performed. Essentially, it is a means of translating information from the time domain into the frequency domain. The fast Fourier transform (FFT) [47,53] is the most popular tools. Based on the time series of tidal level (ζ_i, for $i = 1, 2, 3, \ldots$), the FFT yields the amplitudes and phases of the harmonics or tidal constituents.

For the harmonic analysis to yield accurate values, the time series must be sufficiently long. In theory, 18.6 years (the nodal period of the Moon) would be the minimum time to cover the full complexity of the motions of the Earth–Moon–Sun system. In practice, however, such a long a time series for a particular location may not be available, and 8–10 years may be considered sufficient for many applications. Of course, the difficulty is that at many sites of interest there is no tide gauge, so that no time series of observed tidal levels is available. In this case, the alternative is to use a global database such as TOPEX/Poseidon, which gives the amplitudes and phases of the tidal components at any point, based on a global ocean model and satellite altimetry [54–57]. Having obtained the amplitudes and phases of the main constituents, the truncated Fourier series enables tidal energy developers or coastal engineers to predict tidal level oscillations at a point of interest with considerable accuracy. This predictability is arguably one of the great advantages of tidal energy relative to other renewable energy sources.

2.3.2 Tidal Barrage or Lagoon vs. and Tidal Stream

There are essentially two methods for harnessing the energy of the tide. The first consists of building a tidal barrage or lagoon in an estuary or other semi-enclosed body of water with a significant tidal range. With the barrage sluices closed, a difference in elevation is established between the two sides of the barrage or lagoon as the tidal level rises or falls on the sea (outer) side. Eventually, the sluice gates are open, and the flow through them drives turbine-generator groups. The second method consists of deploying turbines in the free flow of water caused by the tide (the tidal stream) to harness its kinetic energy.

The potential energy of the water mass impounded by a tidal barrage or lagoon has been used since antiquity. Modern-day applications to generate electricity include the tidal power stations at La Rance (France) and Sihwa Lake (South Korea), which have operated since 1966 and 2011, respectively, with rated power of 240 MW and 254 MW. For all their interest, tidal barrages have a number of disadvantages which may hinder prospective developments: the number of potential sites is very limited, their environmental impact is potentially considerable and a large capital investment is required before the first kilowatt-hour is produced. These disadvantages are mitigated in the case of tidal lagoons, and they are currently under consideration for certain areas (S Wales, UK). In any case, many more sites are available for tidal stream farms, their environmental impact is lower, and the financial aspects are less challenging. Consequently, the focus of this chapter is on the tidal stream resource.

2.3.3 The Tidal Stream Resource

The tidal stream resource may be defined as the kinetic energy in tidal currents, i.e. horizontal motions of water caused (primarily) by the tide; other agents may also play a role, e.g. riverine flow, winds, salinity or temperature gradients.

The tidal stream power through a cross-section Ω normal to the flow is the flux of kinetic energy through that section,

$$P_\Omega = \frac{1}{2}\alpha\rho\bar{v}^3 A_\Omega, \tag{2.28}$$

where $\bar{v}$ is the magnitude of the flow velocity (the speed of the current) averaged over Ω, A_Ω is the surface area of section Ω, ρ is the density of water, and α is the energy coefficient [45],

$$\alpha = \frac{1}{\bar{v}^3 A_\Omega} \int_\Omega v^3 dA, \tag{2.29}$$

with v the magnitude of the flow velocity at a generic point of the section Ω and dA the differential area. P_k has units of watts in the SI.

The tidal stream power given by equation (2.28) is the amount of kinetic energy that crosses the section considered (Ω) per unit time. This is the power *available* in the tidal stream flow; however, not all of this tidal stream power can be transformed into mechanical power (and later on into electrical power) by the turbine-generator group, due to Betz's law and the mechanical and electrical losses in the system.

The energy coefficient, α, takes into account the variations in flow speed over the section considered, Ω. If the speed is uniform within the section, then $\alpha = 1$; this case is, however, only of academic interest. In the general case, i.e. when the magnitude of flow velocity flow speed does vary within the section, α is higher than unity, for the average of the velocity cubed is always higher than the average velocity cubed. The actual value of α will depend on the flow characteristics, the section and point in time considered, with typical values slightly above unity; for instance, if the section is a transect of a wide channel, a value of $\alpha = 1.03$ is recommended [58]. In the absence of better data, a value $\alpha \approx 1$ is often assumed as an approximation, which amounts to neglecting the variation of v within the section considered, and leads to the following conservative estimate of tidal stream power:

$$P_\Omega \approx \frac{1}{2}\rho v^3 A_\Omega. \tag{2.30}$$

Importantly, area A_Ω must be normal to the flow.

On these grounds, *tidal stream power density* may be defined as the tidal stream power passing through a unit surface area normal to the flow,

$$p = \frac{1}{2}\rho v^3,$$ (2.31)

where p has SI units of watts per square metre. The tidal stream power available to a turbine may be obtained by multiplying the tidal stream power density by the rotor (swept) area.

If the variation of the flow speed, v, with time is known (typically through numerical modelling), then the power density may be integrated with respect to time to obtain the *tidal stream energy density*:

$$e = \int_{t_1}^{t_2} p(t)\,dt,$$ (2.32)

where the interval of integration considered, (t_1, t_2), may be e.g. a spring-neap tidal cycle or a typical one year. The tidal stream energy density may be expressed in joules per square metre or kilowatt-hours per square metre.

If the interval of integration is one year, the *annual tidal stream energy density* is obtained; to reduce the computational cost for of the numerical model, an approximation may be obtained by performing the integration over a tidal (spring-neap) cycle, and extrapolating the result to one year.

In certain cases it may be of interest to consider the *tidal stream power per unit width*, i.e. the flux of kinetic energy through a vertical section normal to the flow of unit width, extending from the seabed to the surface [59]; this is given by

$$P_u = \frac{1}{2}\int_{-h}^{0} \rho(z)\left[v(z)\right]^3 dz,$$ (2.33)

where $v(z)$ and $\rho(z)$ are, respectively, the flow velocity magnitude and water density as a function of z, the vertical coordinate; $z = 0$ at the reference plane (the quiescent water level) and $z = -h$ at the seabed, i.e. the z-axis is positive upwards. P_u has SI units of watts per metre.

The variation of density with depth may be caused by differences in salinity and temperature (especially in stratified estuaries) or by the weight of the water column above the level considered; in nearshore waters, where tidal stream sites are located, water depths are not large, of the order of $O(10^1\,\mathrm{m})$, and the latter effect is negligible. In any case, the variations in $\rho(z)$ nearshore are generally below 3%, and can be disregarded for our purposes here. Under the assumption $\rho(z) \approx \mathrm{const.} = \rho$, equation (2.33) simplifies to

$$P_u = \frac{1}{2}\rho\int_{-h}^{0} \left[v(z)\right]^3 dz.$$ (2.34)

An energy coefficient could be easily defined for the vertical section of unit width along the lines of α in the preceding section [59]. It is also possible to neglect the variation of $v(z)$ altogether, i.e. to replace $v(z)$ by the depth-averaged flow velocity magnitude,

$$\overline{v_u} = \frac{1}{h}\int_{-h}^{0} v(z)\,dz, \tag{2.35}$$

which yields

$$P_u \approx \frac{1}{2}\rho h \left(\overline{v_u}\right)^3. \tag{2.36}$$

Similar to the previous case of an arbitrary section Ω, neglecting the variability of the flow velocity magnitude in the vertical section of unit width, i.e. replacing equation (2.34) by equation (2.36) in the calculation of the tidal stream power per unit width, leads to a conservative estimate.

2.3.4 Selection of Potential Tidal Stream Sites

Multiple aspects must be considered in the selection of a site for installing a tidal stream farm, i.e. an array of tidal stream turbines or, more generally, tidal energy converters, including the tidal stream resource, the environmental impact on the area, other uses of the marine space, and the connection to the electricity network. The tidal stream resource available in the area is fundamental among these aspects. In the previous section we have seen that tidal stream power is a function of the velocity of the current cubed.

For tidal currents to be significant, the tidal range must be relevant. For instance, in the Mediterranean Sea, where most coastlines have tidal ranges below 0.5 m, the tidal stream resource is negligible. However, although a significant tidal range is a *conditio sine qua non*, it is not sufficient for strong tidal currents to occur. Indeed, on coasts open to the ocean tidal currents are typically negligible, even if the tidal range is not. Tidal flows only become relevant when a substantial tidal range occurs in a semi-enclosed body of water, such as a sea, bay, estuary, ria or fjord.[8] Examples are the Severn Estuary and the Bristol Channel (UK) [17,44,60–62], the North Sea, in particular in areas such as the Pentland Firth (UK) [63,64] or Orkney (UK) [65], the Bay of Fundy (Canada) [66] or the estuaries (rias) of Ribadeo [67], Ortigueira [68,69], Muros [45] and Arousa [70] in NW Spain. Within these semi-enclosed bodies of water tidal currents tend to be particularly strong at certain transects where the coastline geometry constrains the flow.

At many of these constrictions, water depths are limited, and may preclude in some cases tidal stream energy exploitation. From equation (2.36) it is clear that tidal stream power per unit width depends on the magnitude of the flow cubed and the water depth. For this reason, it makes sense to consider tidal flow and water depth conjointly in assessing the suitability of an area for tidal stream exploitation. This is the rationale behind the Tidal Stream Exploitability (TSE) index [59], which combines information on the tidal flow intensity and water depths with certain technical constraints relating to the exploitation of tidal stream power by means of tidal stream turbines. In certain areas it may be important to consider wave action in assessing the tidal stream energy resource [71]. The role of the type of turbine in the actual power than can be produced

---

8 Rias and fjords are two particular types of estuary.

at a given site is also important, and significant differences in the power output may exist between floating and fixed (bottom-mounted) turbines [72].

In the equations in the previous subsection the magnitude of the flow velocity is cubed; therefore, a small difference in velocity gives rise to a great difference in power [68]. For instance, an increase of a mere 20% in the speed of the tidal current implies 72.8% more tidal stream power. For this reason, in characterising the tidal stream resource of an area it is crucial to determine the velocity field and its fluctuations with great accuracy. Without numerical modelling, this would require measuring tidal currents at a large number of stations in the area of interest over long periods of time, which would be an expensive, time-consuming procedure. Fortunately, it is possible to reduce the number of stations and the duration of the observations thanks to numerical modelling.

To characterise the tidal stream resource in an area of interest, a numerical model of the hydrodynamics of the area is implemented, calibrated and validated based on field data. Once validated, the model results can be trusted to represent faithfully the actual flow field and its variability. The application of numerical modelling to the characterisation of the tidal resource is the topic of the following subsections.

2.3.5 Implementation of the Numerical Model

The numerical models used to characterise the tidal resource are based on the Navier–Stokes equations:

$$\frac{\partial u}{\partial x} + \frac{\partial v}{\partial y} + \frac{\partial w}{\partial z} = Q, \tag{2.37}$$

$$\left.\begin{array}{l} \dfrac{Du}{Dt} = fv - g\dfrac{\partial \zeta}{\partial x} - \dfrac{g}{\rho_0}\displaystyle\int_{z'=z}^{z'=\varsigma}\dfrac{\partial \rho}{\partial x}dz' + \upsilon_h\left(\dfrac{\partial^2 u}{\partial x^2} + \dfrac{\partial^2 u}{\partial y^2}\right) + \upsilon_v\left(\dfrac{\partial^2 u}{\partial z^2}\right) \\[4mm] \dfrac{Dv}{Dt} = -fu - g\dfrac{\partial \zeta}{\partial y} - \dfrac{g}{\rho_0}\displaystyle\int_{z'=z}^{z'=\varsigma}\dfrac{\partial \rho}{\partial y}dz' + \upsilon_h\left(\dfrac{\partial^2 v}{\partial x^2} + \dfrac{\partial^2 v}{\partial y^2}\right) + \upsilon_v\left(\dfrac{\partial^2 v}{\partial z^2}\right) \end{array}\right\}, \tag{2.38}$$

$$\frac{\partial p}{\partial z} = -\rho g, \tag{2.39}$$

$$\frac{Dc}{Dt} = D_h\left(\frac{\partial^2 c}{\partial x^2} + \frac{\partial^2 c}{\partial y^2}\right) + D_v\frac{\partial^2 c}{\partial z^2} - \lambda_d c + R_s, \tag{2.40}$$

where the coordinates x, y and z are directed along the east, north and vertical directions, respectively, with u, v and w the components of the flow velocity along the respective directions; ζ is the elevation of the free surface with respect to the reference level; ρ and ρ_0 are the density and reference density of seawater, respectively; Q is the intensity of mass sources per unit area; f is the Coriolis parameter; g is the gravitational acceleration; υ_h and υ_v are the horizontal and vertical eddy viscosity coefficients, respectively; c may stand for salinity or temperature; D_h and D_v are the horizontal and vertical eddy diffusivity coefficients, respectively; λ_d represents the first-order decay process; and R_s is the source term.

Equation (2.37) is the continuity equation, which expresses the conservation of fluid mass. Equations (2.38) express the momentum balance (Newton's second law) in the horizontal directions (x, y). Equation (2.39) expresses the momentum balance in the vertical direction under the Boussinesq approximation. Finally, equation (2.40) is the transport equation, which is solved for both salinity and temperature.

The Boussinesq approximation assumes that even though the density does vary along the vertical, the primary influence of this variation as regards fluid stability is on the body force rather than on the acceleration term, so that the latter may be neglected; in other words, the variation of density may be ignored in the momentum equations except in respect of the gravitational term [73].

Equations (2.37)–(2.40) are in three dimensions; in some cases, where the three-dimensional features of the flow are negligible (e.g. in well-mixed estuaries), these equations may be replaced by a set of equations in the two horizontal dimensions (plus time) by integrating vertically the velocity and other relevant variables from the seabed to the surface [45, 74].

With the exception of certain highly simplified cases, the solution of the Navier–Stokes equations can only be accomplished through a numerical model. To this end, both the equations and the computational domain must be discretised. The details of the discretisation of the governing equations are discussed further in Chapter 8. A number of 'off-the-shelf' software packages exist to solve the above set of equations, e.g. Delft3D or MIKE21, which can operate in either two-dimensional horizontal (2DH) or 3D mode.

As regards the discretisation of the computational domain, the grid is generally Cartesian in the horizontal plane (coordinates x, y), although curvilinear grids can be used if they are better suited to the geometry of the area to be modelled. For the vertical coordinate, the standard Cartesian approach of a z coordinate would lead to different numbers of vertical layers across the domain to accommodate the changing water depth, which might compromise numerical efficiency. For this reason, the so-called σ-coordinate approach is generally preferred [75, 76].

A Dirichlet condition is usually prescribed along the outer (ocean) boundary, in which the variation with time of sea surface elevation is prescribed, alongside salinity and temperature. The elevation itself is calculated based on the tidal harmonics or tidal constituents, which are typically obtained from a global database, itself obtained from large-scale (global) numerical modelling and satellite altimetry (e.g. TOPEX/Poseidon [54, 55, 57]). Alternatively, the Dirichlet boundary condition can be combined with Neumann boundary conditions, in which the rate of change of elevation rather than the elevation itself is prescribed, to form a mixed open boundary condition. This approach has been shown to reduce the generation of numerical disturbances at the boundary [77, 78].

Importantly, the number of tidal harmonics must be sufficient to capture the complexity of the tide in the area. At least seven harmonics must generally be taken into account. In the case of estuaries with significant tidal asymmetry, the relevant harmonics must be included, notably the M4 (overtide) [65, 79]. Salinity and temperature at the ocean boundaries may be prescribed based on measurements or large-scale modelling. At the seabed, bed shear stress is typically accounted for by means of a quadratic stress law, and where relevant, the wind stress at the surface can also be prescribed [76, 80].

For the inner (land–water) boundaries, a null flow, free slip boundary condition is usually adopted [81]. At a river mouth, the flow into the computational domain is taken into account by specifying the time variation of the riverine inflow and its temperature at the boundary.

In the inner sections of estuaries, intertidal areas are common, which flood and dry with the tide. This means that the computational domain and its boundaries change with the tide, and the numerical scheme must be capable of handling this without compromising its stability. Typically grid points are removed when the tide falls and added again when it rises.

Having prescribed the boundary conditions, the initial condition is usually the so-called *cold start*, in which the flow velocity and surface elevation are assumed to be null throughout the computational domain. This is obviously unrealistic, and is only assumed for the sake of simplicity. As a result, in the initial period of the simulation the model variables do not represent the real hydrodynamics in the domain, and care must be taken to exclude the results from this *spin-up* period from the analysis. As a conservative assumption, the spin-up period can be assumed to extend 1 month, after which the model results are assumed to be correct, subject to calibration and validation.

Importantly, the model must be calibrated and validated based on field measurements. Calibration consists of running the model, comparing the results with the observations (of tidal level, flow velocity, salinity, temperature), tweaking certain model parameters (the calibration parameters), then running the model and comparing its results again, and repeating this cycle until the differences between the model results and the observations, which can be quantified by means of e.g. the correlation coefficient or the root-mean-square error (RMSE), are negligible. At this point the model has been validated, and its results can be assumed to represent the actual hydrodynamics and salinity and temperature fields in the computational domain. Typical calibration parameters are the eddy viscosity or the seabed roughness [82, 83].

As regards the hydrodynamics, the field data for calibration and validation should ideally comprise both surface elevation (tidal level) and flow velocity data [69]. These data can be collected by means of acoustic Doppler current profilers (ADCPs) deployed on the seabed, attached to some structure (e.g. a pier or bridge pile) and even mounted on a boat, in which case the boat motion must be accounted for. ADCPs yield profiles of flow velocity, which can be vertical or horizontal depending on the orientation of the device, and pressure at the location of the instrument, which can be converted into tidal level. ADCP data often require de-noising, which can be carried out by means of wavelets (e.g. stationary wavelet transforms [82]), among other methods; an in-depth revision of data analysis methods is provided by [47]. If there are ports with tidal gauges within the domain, these are naturally a useful means to obtain tidal level data. Certain (simplified) current velocity data can also be obtained from hydrographic charts (e.g. *tidal diamonds*). Wherever possible, the calibration and validation of the model should be based on data not only of tidal level but also of flow velocity, ideally from observations at two or more points in the model domain. An alternative approach is to carry out the calibration and validation by comparing the tidal harmonics obtained from a harmonic analysis of tidal gauge data with those calculated based on model results [82]. As in the previous approach, two or more points should be used. As regards the validation of salinity and temperature, vertical profiles can be collected from a boat by means of a CTD sensor – which measures conductivity, temperature and pressure.

2.3.6   Case study I: Bristol Channel and Severn Estuary

In this subsection, the procedure is illustrated through a case study in the Bristol Channel and Severn Estuary, an area of great potential for tidal stream [17, 44, 60–62] (and tidal barrage [84]) development, in which a model based on the 2DH equations is implemented. A summary is presented here; more details can be found in [17]. The study area extends from the mouth of the River Severn to the Celtic Sea, with its ocean boundary between Trevose Head and St. Govan's Head (Figure 2.9).

The computational grid had a resolution of 500×500 m. Water depths were interpolated from the GEBCO data set. As regards the boundary conditions, the tidal forcing was prescribed along the ocean boundary based on the main nine constituents, which were obtained for eight nodes of the boundary by means of the TPXO 7.2 global database [85]. Values at intermediate nodes were interpolated. Salinity and temperature

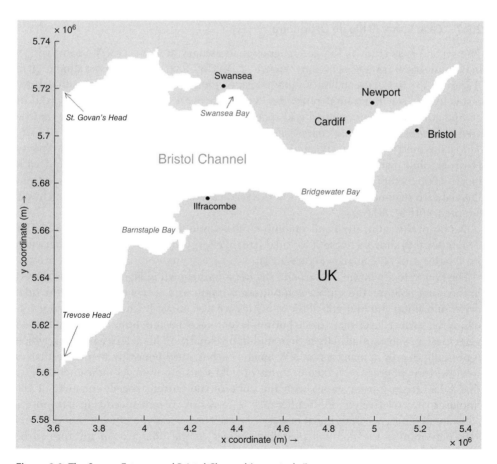

Figure 2.9 The Severn Estuary and Bristol Channel (case study I).

values at the boundary were obtained from the British Oceanographic Data Centre (BODC). At the land boundaries, null flow, free slip conditions were prescribed.

The period of interest considered for the simulation was a complete tidal cycle extending from 14/03/2011 to 28/03/2011. As explained in the preceding subsection, the model simulation covered not only the period of interest but also a spin-up period of 31 days. The initial conditions were null surface elevation and velocity throughout the grid (cold start). The calibration parameter was the horizontal eddy viscosity; on the basis of previous work [17], a value of $30 \, m^2 \, s^{-1}$ was selected.

For validation, the model results were compared with tidal levels from four tidal gauges operated by the UK Tide Gauge Network (BODC) and tidal stream data from five tidal diamonds from Admiralty Chart No. 1165. The comparison of tidal levels (Figure 2.10) shows excellent agreement (with correlation coefficient values $R > 0.97$). As regards velocities, the goodness of fit between model results and field data is quantified in Table 2.2, again with excellent results [17]. Once validated, the model was used to calculate the annual energy density (Figure 2.11). In certain hotspots values are well above 60 MWh m^{-2}, reaching up to 90 MWh m^{-2} [62].

2.3.7 Case Study II: Ria de Ortigueira

The second case study is Ria de Ortigueira, an estuary in Galicia (NW Spain) with a maximum tidal range of 4.5 m (Figure 2.12) [69, 82]. The computational grid (Figure 2.13) is Cartesian, its x- and y-axis orientated west–east and south–north, respectively. The horizontal resolution varies through the domain for the sake of computational efficiency, with a grid spacing of 50×50 m ($\Delta x \times \Delta y$) within the ria (the area of interest), becoming coarser along the y-axis outside the ria, up to 50×150 m. The bathymetry was interpolated from nautical charts 408 and 4083 of Spain's Hydrographic Institute. In the intertidal areas the charts were complemented with local 1:5000 cartography – more precisely, with maps 00718d, 00188d and 00281d from the Galician regional government. On the vertical axis the σ-coordinate approach was adopted, with 12 σ-layers.

Details of the calibration and validation of the model may be found in [69, 82]. The TSE index [59] shows a hotspot for tidal stream energy in the narrow passage between Pt. Postiña and Pt. Cabalar (Figure 2.14).

The flow velocity in the area around the hotspot is shown in Figure 2.15 at two levels in the water column, the surface and bottom σ-layers, and at two moments of the tidal cycle, mid-flood and mid-ebb. Peak velocities are well above 1.5 m s^{-1}, with greater values in the surface layer than in the bottom layer – as expected. Importantly, mid-flood velocities are substantially larger than mid-ebb velocities. This tidal asymmetry, which is present not only in many rias in NW Spain [79] but, more generally, in many estuaries worldwide, is of great relevance in terms of tidal stream power. As indicated, the fact that tidal stream power varies with the cube of the current speed, equation (2.31), implies that a relatively modest difference in terms of current speed translates into a substantial difference in terms of power.

In determining the nominal (installed) power of the tidal stream turbines to be deployed at a tidal farm, the developers should not aim to fully exploit the power peaks in the flow, for this would lead to rated power being attained only a small percentage of

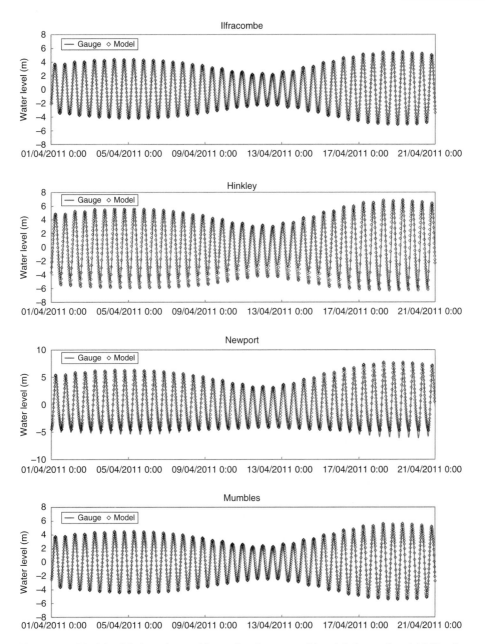

Figure 2.10 Model validation: observed (gauge) and computed (model) time series of tidal levels at four stations in the Bristol Channel.

the time, and consequently to low capacity factors and higher costs [68]. Indeed, the selection of the turbines and generators must take into account not only technical but also economic and functional constraints [17, 62]. The variability of the flow may be considered in selecting the turbines by means of a parametric approach [70].

Table 2.2 Correlation coefficient (*R*) and RMSE values of predicted and measured currents.

Location		*R*	RMSE[a]
Site F	Spring tide	0.9054	0.1406
(51° 33.3′ N,	Neap tide	0.8908	0.1568
3° 56.9′ W)	Direction	0.9956	4.5546
Site G	Spring tide	0.8770	0.2241
(51° 27′ N,	Neap tide	0.8853	0.1543
3° 55.6′ W)	Direction	0.9948	9.9818
Site H	Spring tide	0.8960	0.1646
(51° 32.5′ N,	Neap tide	0.9170	0.1239
3° 53.5′ W)	Direction	0.8966	13.4300
Site K	Spring tide	0.8581	0.2833
(51° 20.1′ N,	Neap tide	0.9062	0.1475
3° 50.3′ W)	Direction	0.8911	15.231
Site L	Spring tide	0.8689	0.3436
(51° 16′ N,	Neap tide	0.8702	0.2163
3° 47.4′ W)	Direction	0.9946	9.4107

a) RMSE values in metres per second for the spring and neap tidal velocities, and in degrees for the direction.

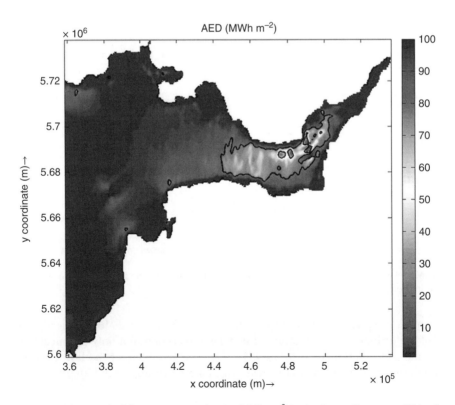

Figure 2.11 Annual tidal stream energy density (MWh m^{-2}) in the Severn Estuary and Bristol Channel (UK).

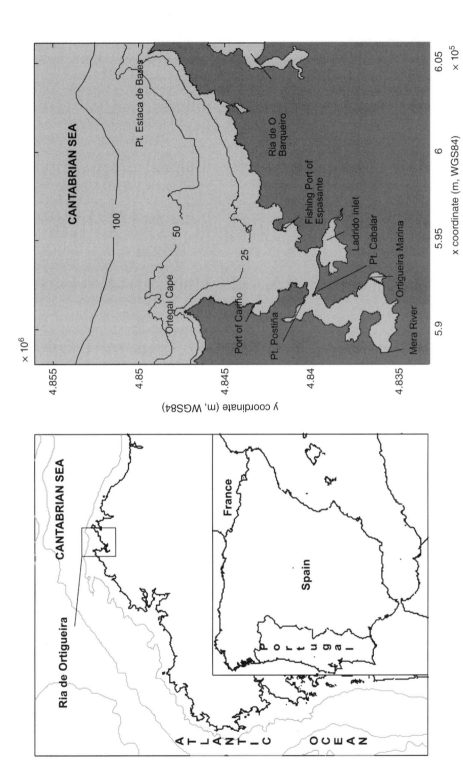

Figure 2.12 Ria de Ortigueira, Galicia (NW Spain).

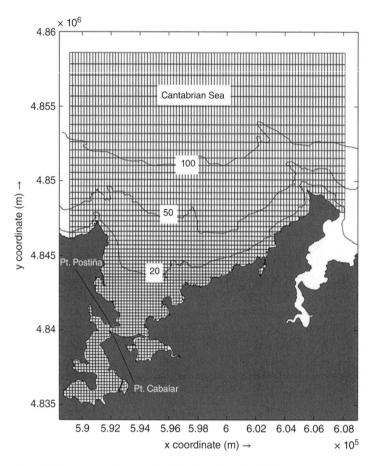

Figure 2.13 Computational grid of Ria de Ortigueira (NW Spain).

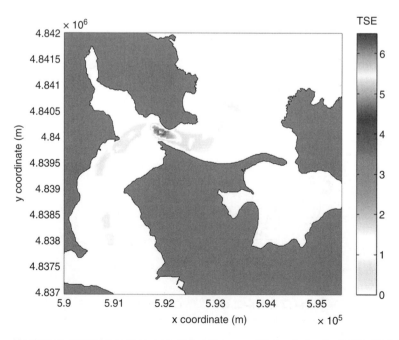

Figure 2.14 TSE index in the inner estuary. The area with greatest potential for tidal stream exploitation between Pt. Postiña and Pt. Cabalar (Figure 2.13) is conspicuous in the graph.

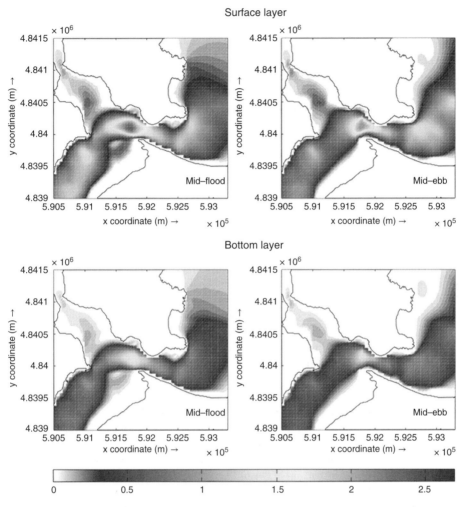

Figure 2.15 Flow velocity at mid-flood and mid-ebb in the surface and bottom layers of the inner estuary, Ria de Ortigueira (NW Spain).

Acknowledgements

The author is indebted to N. Arean, M. Sanchez and Dr R. Carballo for their help with the figures and computation, and to Spain's State Ports and the British Oceanographic Data Centre for kindly providing data.

References

1 Iglesias G, López M, Carballo R, Castro A, Fraguela JA, Frigaard P. Wave energy potential in Galicia (NW Spain). *Renewable Energy* 2009;34:2323–33.
2 Carballo R, Sánchez M, Ramos V, Fraguela J, Iglesias G. The intra-annual variability in the performance of wave energy converters: A comparative study in N Galicia (Spain). *Energy* 2015;82:138–46.

3   Kinsman B. *Wind Waves: Their Generation and Propagation on the Ocean Surface.* Dover; 2002.

4   Mørk G, Barstow S, Kabuth A, Pontes MT. Assessing the global wave energy potential. *ASME 2010 29th International Conference on Ocean, Offshore and Arctic Engineering* (pp. 447–54). American Society of Mechanical Engineers; 2010.

5   Fernandez H, Iglesias G, Carballo R, Castro A, Fraguela J, Taveira-Pinto F, *et al.* The new wave energy converter WaveCat: Concept and laboratory tests. *Marine Structures* 2012;29:58–70.

6   Iglesias G, Fernández H, Carballo R, Castro A, Taveira-Pinto F. The WaveCat© – Development of a new wave energy converter. *World Renewable Energy Congress 2011: Marine and Ocean Technology* 2011;9:2151–8.

7   Vicinanza D, Margheritini L, Kofoed JP, Buccino M. The SSG wave energy converter: Performance, status and recent developments. *Energies* 2012;5:193–226.

8   Margheritini L, Vicinanza D, Frigaard P. SSG wave energy converter: Design, reliability and hydraulic performance of an innovative overtopping device. *Renewable Energy* 2009;34:1371-80.

9   Rusu E, Guedes Soares C. Numerical modelling to estimate the spatial distribution of the wave energy in the Portuguese nearshore. *Renewable Energy* 2009;34:1501–16.

10 Vicinanza D, Cappietti L, Ferrante V, Contestabile P. Estimation of the wave energy in the Italian offshore. *Journal of Coastal Research* 2011; SI 61:613–617.

11 Vicinanza D, Contestabile P, Ferrante V. Wave energy potential in the north-west of Sardinia (Italy). *Renewable Energy* 2013;50:506–21.

12 Pontes MT. Assessing the European wave energy resource. *Journal of Offshore Mechanics and Arctic Engineering* 1998;120:226–31.

13 Iglesias G, Carballo R. Wave energy potential along the Death Coast (Spain). *Energy* 2009;34:1963–75.

14 Astariz S, Iglesias G. The economics of wave energy: A review. *Renewable and Sustainable Energy Reviews* 2015;45:397–408.

15 Astariz S, Iglesias G. Wave energy vs. other energy sources: A reassessment of the economics. *International Journal of Green Energy* 2016;13:747–55.

16 Astariz S, Vazquez A, Iglesias G. Evaluation and comparison of the levelized cost of tidal, wave, and offshore wind energy. *Journal of Renewable and Sustainable Energy* 2015;7:053112.

17 Vazquez A, Iglesias G. LCOE (levelised cost of energy) mapping: A new geospatial tool for tidal stream energy. *Energy* 2015;91:192–201.

18 Astariz S, Iglesias G. Enhancing wave energy competitiveness through co-located wind and wave energy farms. A review on the shadow effect. *Energies* 2015;8:7344–66.

19 Astariz S, Iglesias G. Accessibility for operation and maintenance tasks in co-located wind and wave energy farms with non-uniformly distributed arrays. *Energy Conversion and Management* 2015;106:1219–29.

20 Astariz S, Iglesias G. Output power smoothing and reduced downtime period by combined wind and wave energy farms. *Energy* 2016;97:69–81.

21 Astariz S, Perez-Collazo C, Abanades J, Iglesias G. Hybrid wave and offshore wind farms: A comparative case study of co-located layouts. *International Journal of Marine Energy* 2016;15:2–16.

22 Astariz S, Perez-Collazo C, Abanades J, Iglesias G. Co-located wind-wave farm synergies (operation & maintenance): A case study. *Energy Conversion and Management* 2015;91:63–75.

23 Astariz S, Perez-Collazo C, Abanades J, Iglesias G. Co-located wave-wind farms: Economic assessment as a function of layout. *Renewable Energy* 2015;83:837–49.

24 Perez Collazo C, Astariz S, Abanades J, Greaves D, Iglesias G. Co-located wave and offshore wind farms: A preliminary case study of an hybrid array. *International Conference in Coastal Engineering (ICCE)* 2014.

25 Abanades J, Greaves D, Iglesias G. Coastal defence through wave farms. *Coastal Engineering* 2014;91:299–307.

26 Abanades J, Greaves D, Iglesias G. Wave farm impact on the beach profile: A case study. *Coastal Engineering* 2014;86:36–44.

27 Abanades J, Greaves D, Iglesias G. Coastal defence using wave farms: The role of farm-to-coast distance. *Renewable Energy* 2015;75:572–82.

28 Abanades J, Greaves D, Iglesias G. Wave farm impact on beach modal state. *Marine Geology* 2015;361:126–35.

29 Rusu E, Onea F. Study on the influence of the distance to shore for a wave energy farm operating in the central part of the Portuguese nearshore. *Energy Conversion and Management* 2016;114:209–23.

30 Iglesias G, Carballo R. Wave energy and nearshore hot spots: The case of the SE Bay of Biscay. *Renewable Energy* 2010;35:2490–500.

31 Carballo R, Iglesias G. A methodology to determine the power performance of wave energy converters at a particular coastal location. *Energy Conversion and Management* 2012;61:8–18.

32 Carballo R, Sánchez M, Ramos V, Castro A, Taveira F, Iglesias G. Wave resource characterisation for energy production computations of WECs. *Renewable Energies Offshore* 2015:71.

33 Carballo R, Sánchez M, Ramos V, Fraguela J, Iglesias G. Intra-annual wave resource characterization for energy exploitation: A new decision-aid tool. *Energy Conversion and Management* 2015;93:1–8.

34 Ringwood J, Brandle G. A new world map for wave power with a focus on variability. *Proceedings of the 11th European Wave and Tidal Energy Conference* 2015.

35 Dean RG, Dalrymple RA. *Water Wave Mechanics for Engineers and Scientists*. World Scientific; 1991.

36 USACE. *Coastal Engineering Manual*. Washington, DC: U.S. Army Corps of Engineers; 2002.

37 Mei CC, Stiassnie M, Yue DKP, Mei CC. *Theory and Applications of Ocean Surface Waves*. World Scientific; 2005.

38 Kreyszig E. *Advanced engineering mathematics*. New York: Wiley; 1972.

39 Holthuijsen LH. *Waves in Oceanic and Coastal Waters*. Cambridge University Press; 2007.

40 Booij N, Ris RC, Holthuijsen LH. A third-generation wave model for coastal regions 1. Model description and validation. *Journal of Geophysical Research C: Oceans* 1999;104:7649–66.

41 Svendsen IA. *Introduction to Nearshore Hydrodynamics*. World Scientific; 2006.

42 Carballo R, Sanchez M, Ramos V, Taveira-Pinto F, Iglesias G. A high resolution geospatial database for wave energy exploitation. *Energy* 2014;68:572–83.

43 Pugh DT. *Tides, Surges, and Mean Sea-Level: A Handbook for Engineers and Scientists*. John Wiley & Sons; 1996.

44 Vazquez A, Iglesias G. Device interactions in reducing the cost of tidal stream energy. *Energy Conversion and Management* 2015;97:428–38.

45 Carballo R, Iglesias G, Castro A. Numerical model evaluation of tidal stream energy resources in the Ría de Muros (NW Spain). *Renewable Energy* 2009;34:1517–24.

46 Fritz HM, Blount C, Sokoloski R, Singleton J, Fuggle A, McAdoo BG, *et al*. Hurricane Katrina storm surge distribution and field observations on the Mississippi Barrier Islands. *Estuarine, Coastal and Shelf Science.* 2007;74:12–20.

47 Thomson RE, Emery WJ. *Data Analysis Methods in Physical Oceanography*. Newnes; 2014.

48 López M, López I, Iglesias G. Hindcasting Long Waves in a Port: An ANN Approach. *Coastal Engineering Journal* 2012:1550019.

49 Pawlowicz R, Beardsley B, Lentz S. Classical tidal harmonic analysis including error estimates in MATLAB using T_TIDE. *Computers & Geosciences* 2002;28:929–37.

50 Lopez M, Iglesias G. Artificial intelligence for estimating infragravity energy in a harbour. *Ocean Engineering* 2013;57:56–63.

51 López M, Iglesias G. Long wave effects on a vessel at berth. *Applied Ocean Research* 2014;47:63–72.

52 Lopez M, Iglesias G, Kobayashi N. Long period oscillations and tidal level in the Port of Ferrol. *Applied Ocean Research* 2012;38:126–34.

53 Press WH. *Numerical recipes: The art of scientific computing* (3rd edn). Cambridge University Press; 2007.

54 Matsumoto K, Takanezawa T, Ooe M. Ocean tide models developed by assimilating TOPEX/POSEIDON altimeter data into hydrodynamical model: A global model and a regional model around Japan. *Journal of Oceanography* 2000;56:567–81.

55 Egbert GD, Bennett AF, Foreman MGG. Topex/Poseidon tides estimated using a global inverse model. *Journal of Geophysical Research* 1994;99:24821–52.

56 Kalra R, Deo MC. Derivation of coastal wind and wave parameters from offshore measurements of TOPEX satellite using ANN. *Coastal Engineering* 2007;54:187–96.

57 Le Provost C, Bennett AF, Cartwright DE. Ocean tides for and from Topex/Poseidon. *Science* 1995;267:639–47.

58 Te Chow V. *Open Channel Hydraulics*. McGraw-Hill; 1959.

59 Iglesias G, Sanchez M, Carballo R, Fernandez H. The TSE index – A new tool for selecting tidal stream sites in depth-limited regions. *Renewable Energy* 2012;48:350–7.

60 Ahmadian R, Falconer RA. Assessment of array shape of tidal stream turbines on hydro-environmental impacts and power output. *Renewable Energy* 2012;44:318–27.

61 Hashemi M, Abedini M, Neill S, Malekzadeh P. Tidal and surge modelling using differential quadrature: A case study in the Bristol Channel. *Coastal Engineering* 2008;55:811–9.

62 Vazquez A, Iglesias G. A holistic method for selecting tidal stream energy hotspots under technical, economic and functional constraints. *Energy Conversion and Management* 2016;117:420–30.

63 Draper S, Adcock TA, Borthwick AG, Houlsby GT. Estimate of the tidal stream power resource of the Pentland Firth. *Renewable Energy* 2014;63:650–7.

64 Baston S, Harris R. Modelling the hydrodynamic characteristics of tidal flow in the Pentland Firth. *9th European Wave and Tidal Energy Conference*, Southampton, UK, 2011.

65 Neill SP, Hashemi MR, Lewis MJ. The role of tidal asymmetry in characterizing the tidal energy resource of Orkney. *Renewable Energy* 2014;68:337–50.

66 Karsten RH, McMillan J, Lickley M, Haynes R. Assessment of tidal current energy in the Minas Passage, Bay of Fundy. *Proceedings of the Institution of Mechanical Engineers, Part A: Journal of Power and Energy* 2008;222:493–507.

67 Vazquez A, Iglesias G. Public perceptions and externalities in tidal stream energy: A valuation for policy making. *Ocean & Coastal Management.* 2015;105:15–24.

68 Sanchez M, Iglesias G, Carballo R, Fraguela JA. Power peaks vs. installed capacity in tidal stream energy. *IET Renewable Power Generation* 2013;7:246–253.

69 Sanchez M, Carballo R, Ramos V, Iglesias G. Floating vs. bottom-fixed turbines for tidal stream energy: A comparative impact assessment. Energy 2014;72:691–701.

70 Ramos V, Iglesias G. Performance assessment of Tidal stream turbines: A parametric approach. *Energy Conversion and Management* 2013;69:49–57.

71 Lewis M, Neill S, Hashemi M, Reza M. Realistic wave conditions and their influence on quantifying the tidal stream energy resource. *Applied Energy* 2014;136:495–508.

72 Sánchez M, Carballo R, Ramos V, Iglesias G. Energy production from tidal currents in an estuary: A comparative study of floating and bottom-fixed turbines. *Energy* 2014;77:802–11.

73 Graebel W. *Advanced fluid mechanics.* Academic Press; 2007.

74 Iglesias G, Carballo R, Castro A. Baroclinic modelling and analysis of tide- and wind-induced circulation in the Ría de Muros (NW Spain). *Journal of Marine Systems* 2008;74:475–84.

75 Carballo R, Iglesias G, Castro A. Residual circulation in the Ria de Muros (NW Spain): A 3D numerical model study. *Journal of Marine Systems* 2009;75:116–30.

76 Iglesias G, Carballo R. Seasonality of the circulation in the Ría de Muros (NW Spain). *Journal of Marine Systems* 2009;78:94–108.

77 Iglesias G, Carballo R. Effects of high winds on the circulation of the using a mixed open boundary condition: The Ría de Muros, Spain. *Environmental Modelling & Software* 2010;25:455–66.

78 Iglesias G, Carballo R. Corrigendum to 'Effects of high winds on the circulation of the using a mixed open boundary condition: the Ría de Muros, Spain' [*Environ. Model. Software* 25 (2009) 455–466]. *Environmental Modelling & Software* 2010;25:889–.

79 Iglesias G, Carballo R. Can the seasonality of a small river affect a large tide-dominated estuary? The case of Ría de Viveiro, Spain. *Journal of Coastal Research* 2011;27:1170–82.

80 Ramos V, Carballo R, Sanchez M, Veigas M, Iglesias G. Tidal stream energy impacts on estuarine circulation. *Energy Conversion and Management* 2014;80:137–49.

81 Ramos V, Carballo R, Alvarez M, Sanchez M, Iglesias G. Assessment of the impacts of tidal stream energy through high-resolution numerical modeling. *Energy* 2013;61:541–54.

82 Sánchez M, Carballo R, Ramos V, Iglesias G. Tidal stream energy impact on the transient and residual flow in an estuary: A 3D analysis. *Applied Energy* 2014;116:167–77.

83 Carballo R, Iglesias G, Castro A. Residual circulation in the Ría de Muros (NW Spain): A 3D numerical model study. *Journal of Marine Systems* 2009;75:116–30.

84 Ahmadian R, Falconer R, Lin B. Hydro-environmental modelling of proposed Severn barrage, UK. *Proceedings of the Institution of Civil Engineers – Energy* 2010;163:107–17.

85 Dushaw BD, Egbert GD, Worcester PF, Cornuelle BD, Howe BM, Metzger K. A TOPEX/POSEIDON global tidal model (TPXO. 2) and barotropic tidal currents determined from long-range acoustic transmissions. *Progress in Oceanography.* 1997;40:337–67.

3

Wave Energy Technology

Deborah Greaves

Professor of Ocean Engineering, School of Engineering, University of Plymouth, UK

3.1 Introduction

The sheer power in the sea is immense, as is evident from the effect of waves on the motion of the largest ships and tankers and as can be seen in the aftermath of any storm by its impact on the coastline. Over centuries, mankind has been fascinated with harnessing power from the sea; in 1889, Thomas Edison, inventor of the light bulb, spent hours on deck studying the waves during a journey across the Atlantic, afterwards stating: 'It made me positively savage to think of all that power going to waste' [1]. The first patent for wave energy utilisation was established in 1799 by the French father and son, Messrs Girard, and since then thousands of patents for wave energy utilisation have been filed [2]. Modern research into harnessing energy from waves was stimulated by the emerging oil crisis of the 1970s [3], but later lost its impetus as the crisis passed [4].

During the 1970s when the oil crisis hit, an ambitious research and development (R&D) programme on wave energy was set up by the British government. The target for the programme was to develop a cost-effective 2000 MW wave power plant and several projects were funded, some of which are illustrated in Figure 3.1, including Salter's Duck [3] in Figure 3.1 (a), the Cockerell Raft [5] in Figure 3.1 (b), the Bristol Cylinder [6] in Figure 3.1 (c) and the NEL oscillating water column (OWC) [7] in Figure 3.1 (d). During the course of the programme a large number of devices were considered but unfortunately found to be uneconomic [8], although with hindsight the objective of that programme was perhaps over ambitious. Aiming for such a high-capacity power plant resulted in massive devices being designed, with corresponding high capital and generating costs. Following the conclusion of the UK programme in the early 1980s, the Department of Trade and Industry continued to support R&D on smaller devices that were potentially more economic, allowing wave energy to be developed in a more systematic way. A number of different devices emerged, including the PS Frog, the Circular SEA Clam, the Solo Duck and the Shoreline OWC [9]. As a consequence, a 75 kW OWC device was installed and tested on the island of Islay in Scotland [10], and provided a large amount of useful information during its service life, which has benefited the design of future shoreline OWCs. However, an

Wave and Tidal Energy, First Edition. Edited by Deborah Greaves and Gregorio Iglesias.
© 2018 John Wiley & Sons Ltd. Published 2018 by John Wiley & Sons Ltd.

(a)　　　　　　　　　　　　　　　　(b)

(c)　　　　　　　　　　　　　　　　(d)

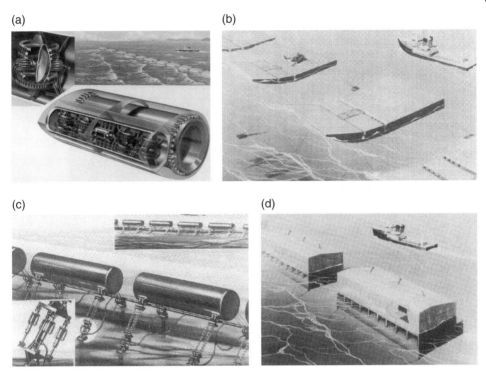

Figure 3.1 Early UK wave energy converter concepts. (a) Salter's duck [3] (b) Cockerell raft [5] with kind permission of dover publications (c) Bristol cylinder [6] (d) NEL OWC [7] with kind permission of dover publications.

independent review [9] concluded that, despite improvements in the technology, electricity from wave energy was still too expensive.

Due to a change in government policy, in 1982 the British Wave Energy Programme came practically to a halt, without any full-sized prototype having been constructed and tested [11]. At the same time the less ambitious Norwegian programme resulted in the construction, in 1985, of two shoreline prototypes shown in Figure 3.2, the TAPCHAN [12] in Figure 3.2 (a) and the Kværner OWC [13] in Figure 3.2 (b) with vertical axis air turbine, deployed on the coast, at Toftestallen near Bergen.

Wave energy development is determined by the vagaries of government support, but the situation in Europe changed dramatically in 1991 [14], when the European Commission included Wave Energy in their research and development programme on Renewable Energies. Since then, many projects have been funded by the European Commission, involving a large number of European teams. With global attention now being drawn to climate change and the rising level of CO2, the focus on generating electricity from renewable sources is once again an important and active area of research and is being explored world-wide.

It is estimated that the potential worldwide wave power resource is 2 TW [11], and that wave energy could make a real contribution to future electricity generation globally. The exploitable resource is difficult to quantify as it depends on the technology deployed and many other factors, but has been estimated to be approximately 5% of the total [15].

(a)   (b)

Figure 3.2 Early Norwegian wave energy converter concepts. (a) TAPCHAN [12] (b) Kværner OWC [13].

In particular for countries with extensive coastline, such as the UK, it is expected that wave and tidal energy together could meet up to 20% of the UK demand [16]. Ocean waves are generated by the wind blowing over the ocean surface, so wave energy may be seen as a concentrated form of wind energy [11, 17]. Relative to other renewable energies, wave energy presents a number of advantages: high resource predictability, high power density, relatively high utilisation factor, and low environmental and visual impact. In comparison with tidal stream energy, there are more potential sites for wave farms globally as strong tidal currents occur only in a relatively small number of areas [18]. For these reasons, wave energy is widely regarded as one of the renewable energy sources with the greatest potential for development over the next few years [19]. Intensive work is being devoted to the assessment of the wave energy resource and to developments in wave energy technology in many regions worldwide [20–26]. A discussion of the regional developments in MRE is given in Chapter 12.

The challenge for wave energy is how it can be harnessed and converted to useful power (e.g. electricity or desalination) in a cost-effective manner. Over the centuries numerous patents and designs for wave energy conversion have been proposed; and whereas in the wind sector, designs have converged to the three-bladed horizontal axis turbine, in wave energy many very different approaches are still being explored. Wave energy absorption is a complex hydrodynamic process that is often counter-intuitive, and many early failures in wave energy lacked a proper understanding of these processes. The hydrodynamics of wave energy converters (WECs) is a sub-area of surface-wave hydrodynamics and naval architecture in which the theory of marine hydrodynamics have been developed over centuries for the design of ships and offshore oil and gas structures. However, wave energy extraction and how it may be maximised brings in new challenges, and WECs have some clear differences from these traditional offshore structures for which the theory has been developed and applied. For example, WECs tend to be far more dynamic than oil and gas structures. Usually the dynamics of the WEC is what drives the energy conversion system and so systems may be designed to maximise response in waves by designing the WEC to be close to resonance in typical wave conditions. For offshore oil and gas structures and ships, however, the typical design strategy is to ensure that natural frequencies of the systems are far from the wave frequency in order to prevent and restrict unwanted motions, which may limit drilling and offloading operations in offshore structures and cause discomfort in ships. WEC response is often further refined during operation in order to match encountered wave conditions, and this means that tuning and control should be included in the hydrodynamics of WEC design. Furthermore, whereas a ship will be able to take action to avoid storms, a wave farm is designed to operate in an energetic environment and to survive extreme conditions.

In order to be able to predict whether a WEC will be able to deliver the desired performance in a given sea state, it is first necessary to understand the dynamics of the sea state and then how this can be made use of in a machine designed to convert the wave motions to mechanical motion driving a generator to generate electricity. The main factors that need to be considered are: assessment of the wave energy resource (discussed in Chapter 2); how the wave energy resource will be extracted (operation mode of the WEC); survivability in extreme conditions; power take-off (PTO, discussed in Chapter 6); maintenance and serviceability. The success and commercial potential of a WEC is determined by its levelised cost of electricity (LCOE) and this will be influenced by each of these factors. The LCOE is a measure of the cost of a power source over its lifetime and is used as a standard measure to compare different methods of electricity generation. It is an economic assessment of the average total cost to build and operate a power-generating asset over its lifetime divided by the total power output of the asset over that lifetime. The LCOE can also be regarded as the cost at which electricity must be generated in order to break even over the lifetime of the project.

Design for survivability may either involve employing a survivability strategy to avoid extreme loads, or increasing the durability of the device to withstand extremes. Each of these may reduce the efficiency of the device and will add to the design cost, although would be expected to reduce the cost of maintenance and serviceability. PTO design is complex and is described in more detail in Chapter 6. The oscillatory motion of ocean waves is slow but highly energetic; typically the forces are large and the velocity low, and this is contrary to traditional turbine generators, where the forces are low and velocity high. A PTO using auxiliary fluid is commonly used in wave energy, the fluid being air, oil or water, and the efficiency of the overall electricity generation is highly influenced by the choice of physical components. Furthermore, ocean waves have a wide variation in amplitude and period, which is likely to reduce the mean efficiency considerably, regardless of which PTO system is used.

The different aspects will be considered as the WEC goes through its design and development stages, and at each stage physical experiments are recommended to validate the numerical predictions. The process is discussed in Chapter 5 on device design. There have been a number of recommendations for development pathways and testing protocols, and some are described in the Equimar project [27]. These describe how the device may proceed in its development through technology readiness levels (TRLs) and the appropriate scale of testing and type of modelling analysis to be considered at each stage. The TRLs were originally established by NASA, as described by Moorhouse [28]. Weber [29] discussed the application of TRL to wave energy and proposed an additional indicator: the technology performance level.

In this chapter the general theory of hydrodynamics as applied to wave energy conversion is discussed. First the fundamental concepts are explained in Section 3.2, then the hydrodynamics of wave energy conversion theory is introduced in Section 3.3. This is followed in Section 3.5 by the classification of wave energy converters and in Sections 3.6–3.8 the three main categories of WEC are discussed; the operating principle and theory for each of the categories is described and examples of WECs in each category are given.

3.2 Fundamentals

3.2.1 Simple Wave Theory

The operating principle of wave energy devices relies on interaction with the wave conditions, and so in order to design a wave energy device one must first understand the sea state and then the loads transferred to the device from the sea state in order to assess the response of the device in waves. A discussion of the sea state and its analysis as part of the wave resource characterisation is given in Chapter 2. As seen there, the sea state may be considered to be a combination of individual waves with individual wave height, period, phase and direction, which using linear theory may be superposed together. The first step in understanding the interaction of the wave energy device with such a complex system is to understand simple wave theory for a single wave.

Simple wave theory is also known as linear wave theory and Airy wave theory [30], and describes two-dimensional periodic waves in the x–z plane of time period T, propagating in water of constant depth, h. It relies on many fundamental assumptions: that surface tension can be neglected, that the fluid is incompressible and inviscid, and the wave height, H, is much smaller than the wavelength, λ. The inviscid fluid assumption implies that, away from the boundary layers, vorticity is not created by viscous stresses and therefore the flow is irrotational. Thus we can describe the fluid motion using a velocity potential, ϕ, of the form

$$u = -\frac{\partial \phi}{\partial x} = \phi_x, \quad w = -\frac{\partial \phi}{\partial z} = \phi_z, \tag{3.1}$$

where u and w are the velocity components. Mass conservation, expressed as the continuity equation for $\mathbf{u}$, leads to Laplace's equation for ϕ:

$$\nabla \cdot \mathbf{u} = u_x + w_z = \nabla^2 \phi = 0, \tag{3.2}$$

where u_x and w_z are velocity gradients with respect to x and z, respectively. With surface elevation given by η, and gradients with respect to t and x given by η_t and η_x, the bottom, kinematic surface, dynamic surface and periodic boundary conditions can also be expressed in terms of ϕ, leading to the equations

$$\phi_z = 0 \quad \text{at } z = -h, \tag{3.3}$$

$$\eta_t + \phi_x \eta_x - \phi_z = 0 \quad \text{at } z = \eta, \tag{3.4}$$

$$g\eta + \frac{1}{2}\left(\phi_x^2 + \phi_y^2\right) + \phi_t = 0 \quad \text{at } z = \eta, \tag{3.5}$$

$$\phi_x(0,z,t) = \phi_x(\lambda,z,t) \quad (-h < z < 0), \tag{3.6}$$

These equations are nonlinear and cannot be solved analytically, but by linearising the boundary conditions and assuming that the wave height is small compared with the wavelength, the problem may be further simplified. In this way, the kinematic and dynamic surface boundary conditions are applied at the still water level (SWL) rather than at the surface itself. The linearised equations can be solved analytically yielding the solutions

$$\phi(x,z,t) = -\frac{Hg}{2\omega} \frac{\cosh\left[k(h+z)\right]}{\cosh\left[kh\right]} \sin(kx - \omega t + \theta), \tag{3.7}$$

$$\eta = \frac{H}{2}\cos(kx - \omega t + \theta), \tag{3.8}$$

where k is the wavenumber, ω is the wave frequency and θ is the phase. Based on the definition and solution for ϕ, given by equations (3.1) and (3.7), the velocity flow (u, w) at a set point (x, z) is derived from the velocity potential:

$$u(x,z,t) = \frac{H}{2}\omega \frac{\cosh\left[k(h+z)\right]}{\sinh\left[kh\right]} \cos(kx - \omega t + \theta), \tag{3.9}$$

$$w(x,z,t) = \frac{H}{2}\omega \frac{\sinh\left[k(h+z)\right]}{\sinh\left[kh\right]} \sin(kx - \omega t + \theta). \tag{3.10}$$

According to Sorensen [31], the velocity consists of three components: the deep water particle speed, $\pi H/T$; a hyperbolic function causing an exponential decrease with depth; and a periodic component dependent on the phase of the wave. Note that $h + z$ is the distance of the free surface measured up from the bottom, and the horizontal and vertical velocity components are 90° out of phase. The pressure beneath the regular surface wave is given by

$$
\begin{aligned}
p(x,z,t) &= -\rho \frac{\partial \phi}{\partial t} - \rho g z \\
&= \rho g \frac{H}{2} \frac{\cosh[k(h+z)]}{\cosh[kh]} \cos(kx - \omega t + \theta) - \rho g z \\
&= \rho g \zeta \frac{\cosh[k(h+z)]}{\cosh[kh]} - \rho g z.
\end{aligned}
\tag{3.11}
$$

The wave slope is given by

$$\alpha = \frac{\partial \zeta}{\partial x} = \frac{kH}{2}\sin(kx - \omega t + \theta),$$

(3.12)

and by integrating the velocity components in equations (3.9) and (3.10) with respect to time, the horizontal, x_P, and vertical, z_P, displacement of a particle is obtained:

$$x_P(x,z,t) = -\frac{H}{2}\frac{\cosh[k(h+z)]}{\sinh[kh]}\sin(kx - \omega t + \theta),$$

(3.13)

$$z_P(x,z,t) = \frac{H}{2}\frac{\sinh[k(h+z)]}{\sinh[kh]}\cos(kx - \omega t + \theta).$$

(3.14)

These equations indicate that the particles move in elliptical orbits, and that the radius of the orbit decreases exponentially with depth. In deep water, $\cosh(k(z+h)) \approx \sinh(k(z+h))$, the particles move in circular orbits, whereas in shallow water the elliptical orbits become more and more stretched in the x-direction as the water becomes shallower with respect to the wavelength. This has significant implications for the design of wave energy devices for shallow and deep-water locations.

The horizontal component of particle acceleration a_x may be written

$$a_x(x,z,t) = u\frac{\partial u}{\partial x} + w\frac{\partial u}{\partial z} + \frac{\partial u}{\partial t},$$

(3.15)

where the first two terms on the right-hand side are the convective acceleration and the third term is the local acceleration. The magnitude of the convective acceleration for a small-amplitude wave is of the order of the wave steepness (H/λ) squared, while the magnitude of the local acceleration is of the order of the wave steepness. Since the wave steepness is much smaller than unity, we can usually neglect the higher-order convective acceleration term in determining the particle acceleration. This yields

$$a_x(x,z,t) = \frac{H}{2}\omega^2\frac{\cosh[k(h+z)]}{\sinh[kh]}\sin(kx - \omega t + \theta)$$

(3.16)

and

$$a_z(x,z,t) = -\frac{H}{2}\omega^2\frac{\sinh[k(h+z)]}{\sinh[kh]}\cos(kx - \omega t + \theta)$$

(3.17)

for the horizontal and vertical components of acceleration. The terms in brackets are the same for both the particle velocity and acceleration components. The cosine/sine terms indicate that the particle velocity components are 90° out of phase with the acceleration components. This is easily seen by considering a particle following a circular orbit. The velocity is tangent to the circle and the acceleration is directed towards the centre of the circle or normal to the velocity.

For the wave to be free, i.e. only propagating under the influence of gravity, the pressure at the free surface must be zero. For this to hold, it can be shown that the wave must satisfy [30]

$$\omega^2 = gk \tanh(kh), \tag{3.18}$$

which is known as the dispersion relationship. The dispersion relation can be rearranged to give the wavelength in terms of wave period

$$\lambda = \frac{gT^2}{2\pi} \tanh\left(\frac{2\pi}{\lambda}h\right). \tag{3.19}$$

In deep water the dispersion relationship is approximately equivalent to

$$\lambda = \frac{gT^2}{2\pi}, \tag{3.20}$$

and in shallow water to

$$\lambda = T\sqrt{gh}. \tag{3.21}$$

The conditions for water depth approximations are: for shallow water, $0 < h/\lambda < 1/20$; intermediate, $1/20 < h/\lambda < 1/2$; deep water, $1/2 < h/\lambda < \infty$. The speed of the water surface, known as the phase speed or celerity of the wave, c, can be calculated by substituting $c = \lambda/T = \omega/k$ into the dispersion relationship, yielding for arbitrary depth,

$$c = \frac{g}{k}\tanh(kh), \tag{3.22}$$

for the deep water approximation,

$$c = \sqrt{\frac{g}{k}},$$

and for shallow water,

$$c = gh. \tag{3.23}$$

Thus for shallow water waves, the phase speed does not change with either the wavenumber or the frequency and waves of this type are known as non-dispersive. However, for arbitrary depth, if there are multiple waves with different frequencies travelling in the same direction, then each wave is moving at a different speed through the water. As they propagate the wave components are summed together so a group of waves is formed which has a maximum wave height when the wave components are in phase and a minimum when they are out of phase. The envelope of this wave group moves at a different velocity to the individual waves, and this velocity is known as the group velocity. It can be shown that the group velocity is given by

$$c_g = \frac{\partial \omega}{\partial k} = \frac{c}{2}\left(1 + \frac{2kh}{\sinh 2kh}\right). \tag{3.24}$$

3.2.2 Wave Energy

The total mechanical energy in a surface gravity wave is the sum of the kinetic and potential energies. The kinetic energy, E_k, for a unit width of wave crest and for one wavelength is equal to

$$E_k = \int_0^\lambda \int_{-h}^0 \frac{1}{2} \rho \, dx \, dz \left(u^2 + w^2 \right), \tag{3.25}$$

where the upper limit of the vertical integral is taken as zero in accordance with the assumptions of the small-amplitude wave theory. Inserting the velocity terms (equations (3.9) and (3.10)), integrating, and performing the required algebraic manipulation yields the kinetic energy

$$E_k = \frac{1}{16} \rho g H^2 \lambda. \tag{3.26}$$

If we subtract the potential energy of a mass of still water (with respect to the bottom) from the potential energy of the wave form, we will have the potential energy due solely to the wave form. This gives the potential energy per unit wave crest width and for one wavelength E_p as

$$E_p = \int_0^\lambda \rho g (h+\eta) \left(\frac{h+\eta}{2} \right) dx - \rho g \lambda h \left(\frac{h}{2} \right). \tag{3.27}$$

Substituting for the surface elevation as a function of x and performing the integration and simplifying yields

$$E_p = \frac{1}{16} \rho g H^2 \lambda. \tag{3.28}$$

Thus, the average total energy per unit wave crest width and for one wavelength is the sum of the kinetic and potential energy, total energy (E) = potential energy (E_p) + kinetic energy (E_k):

$$E = E_p + E_k = \frac{1}{8} \rho g H^2 \lambda, \tag{3.29}$$

where none of E, E_p, E_k depend on the wave period. The average energy per unit surface area is given by

$$\bar{E} = \frac{E}{\lambda} = \frac{1}{8} \rho g H^2, \tag{3.30}$$

and is usually known as the energy density or specific energy of a wave. Equations (3.29) and (3.30) apply for deep to shallow water within the limits of the small-amplitude wave theory.

3.2.3 Wave Power

The wave energy is propagated forward at the group velocity, and the incident wave power P_{inc} is the wave energy per unit time transmitted in the direction of wave propagation. Wave power can be written as the product of the force acting on a vertical plane normal to the direction of wave propagation times the particle flow velocity across this plane. The wave-induced force is provided by the dynamic pressure (total pressure minus hydrostatic pressure) and the flow velocity is the horizontal component of the particle velocity. Thus

$$P_{inc} = \int_{0}^{T}\int_{-h}^{0} (p + \rho gz)u\,dz\,dt, \tag{3.31}$$

where the term in parentheses is the dynamic pressure. Inserting the dynamic pressure from equation (3.11) and the horizontal component of velocity from equation (3.9) and integrating leads to

$$P_{inc} = \frac{\rho g H^2 \lambda}{16T}\left(1 + \frac{2kh}{\sinh 2kh}\right), \tag{3.32}$$

or

$$P_{inc} = \frac{E}{2T}\left(1 + \frac{2kh}{\sinh 2kh}\right). \tag{3.33}$$

The incident power is often expressed per unit length of wave crest,

$$P_{inc} = \frac{E\lambda}{2T}\left(1 + \frac{2kh}{\sinh 2kh}\right) = \frac{EC}{2}\left(1 + \frac{2kh}{\sinh 2kh}\right) = EC_g = \frac{1}{8}\rho g H^2 C_g, \tag{3.34}$$

for deep water, and written in terms of the wave amplitude, A, this gives

$$P_{inc} = \frac{1}{2}\rho g A^2 C_g. \tag{3.35}$$

The incident wave power, P_{inc}, equals the group velocity multiplied by the propagating incident wave's time-average energy per unit of horizontal sea surface. Half of this energy is potential energy related to water being lifted against gravity from wave troughs to wave crests, while the remaining half is kinetic energy associated with the water's oscillating velocity. For regular waves in deep water, the deep water approximation gives

$$P_{inc} \approx H^2 T. \tag{3.36}$$

For irregular waves in deep water of wave energy period T_J and significant wave height, H_s,

$$P_{inc} \approx \frac{H_s^2 T_J}{2}. \tag{3.37}$$

3.2.4 Capture Width

There are a number of different approaches to allow comparison between various devices so as to assess their fulfilment of the requirements of a wave energy converter. The principal quantity used to evaluate a device's performance is its capture width. Capture width, C_w, is defined as the ratio of total average power extracted, P_{out}, to average power per unit crest width of the incident wave train P_{inc}, where the two powers are averaged over a wave period:

$$C_w = \frac{P_{out}}{P_{inc}}. \tag{3.38}$$

Average power per unit crest width over a wave period is defined by equation (3.36) for a regular wave and equation (3.37) for an irregular wave. When using irregular waves the power extracted is averaged over the energy period. Capture width has units of metres, and the capture width ratio, C_{wr}, is a non-dimensional version of capture width, which is calculated by dividing capture width by the width of the device (i.e. the dimension of the device perpendicular to the propagation direction of the incident waves):

$$C_{wr} = \frac{P_{out}}{P_{inc}D}, \tag{3.39}$$

where D is the characteristic dimension of the device. Using the power matrix of sea conditions, the capture width matrix is built from the capture width ratios for each wave period and height.

The useful power extracted, P_{out}, will depend on the PTO and operation of the WEC. However, another important measure is the power absorbed by the device, and the absorption width, a_w, is the ratio of the absorbed power to the incident power, also equivalent to the energy removed from the environment as a proportion of the incident wave:

$$a_w = \frac{P_{abs}}{P_{inc}} = \frac{E_{abs}}{E_{inc}}. \tag{3.40}$$

It is found that there is a limit to the maximum energy which may be absorbed by a heaving axisymmetric body; this is the wave energy transported by the incident wave front of width equal to

$$a_{w\,max} = \frac{\lambda}{2\pi}, \tag{3.41}$$

and this is explained further in Section 3.3.

3.2.5 Wave Loading

Wave energy conversion involves interaction of a wave energy converter with water waves. To understand how the device responds to waves it is necessary to solve the equation of motion of the structure, and to do this we first need to understand the forces on the structure. The computation of water wave forces on an offshore structure is difficult since it involves interaction of the waves with the structure. In the case of a wave energy device the motion response of the structure is also important and adds further to the complexity. Much of the theory for wave structure interaction has developed in the offshore engineering sector and much of the theory for motion responses of floating structures is derived in naval architecture [32]. There are a large variety of offshore structures, including piled jacket structures consisting of small-diameter tubular members and large-volume structures, such as tension legged platforms and semisubmersibles. Different formulations for calculating the wave forces on offshore structures are available, depending on the type and size of the structural members. Wave forces on structures may be conveniently classified under three headings (inertia, drag and diffraction) and calculated in three different ways (Morison equation, Froude–Krylov theory and diffraction theory [32]). The relative importance of these in a particular case depends on the type and size of the structure and the nature of the wave conditions. The characteristic dimension, D, of the structure, the wave height H (peak to trough), and the wavelength λ serve to define the flow regime.

The Morison equation,

$$F = \rho C_M V a + \frac{1}{2} \rho C_D A u |u|,$$

where V is the volume of the structure and A is the cross-sectional area perpendicular to the flow direction, assumes the force to be composed of inertia and drag forces linearly added together. The inertia coefficient (C_M) and drag coefficient (C_D) are determined by experiment or numerical modelling. The Morison equation applies when the drag force is significant. The term involving C_D dominates the term involving C_M, and viscous forces generated by boundary layer separation on the structures are a significant component of the total fluid loading. This is usually the case when a structure is small compared to the wavelength, $D/H < 0.1$, and thus wave loads on the very small-diameter components of offshore and coastal structures are usually drag-dominated. In the inertia regime where the structure is still relatively small, $0.5 < D/H < 1.0$, but the term involving C_M in Morison's equation dominates the term involving C_D, the Froude–Krylov theory may be applied. This takes the incident wave pressure (without altering the wave field) and applies this to the area of the surface of the structure to compute the force. In the diffraction regime, the size of the structure is comparable to the wavelength, $D/\lambda > 0.2$, and the wave field is altered in the vicinity of the structure. In this case, the diffraction of the waves from the surface of the structure should be taken into account and diffraction theory is used to calculate diffraction forces due to scattering of the incident wave by the structure. In the diffraction regime, inertia forces dominate and viscous forces are often ignored. This approach leads to the application of diffraction analysis in which the inviscid Laplace equation is assumed to apply. Solutions based on linear theory can be derived and numerical approaches are common for deriving the hydrodynamic coefficients (added mass and radiation damping) describing the forces and motion responses.

In order to assess which fluid regime applies for a particular case and thus which analysis is appropriate, a dimensional analysis may be performed [33]. The dimensionless force due to waves on a structure may be expressed as a function of four non-dimensional quantities: the dimensionless time, t/T, the Keulegan–Carpenter parameter, $u_0 T/D$, the Reynolds number, u_0D/ν, and the diffraction parameter, $\pi D/\lambda$, where t is the time, T is the wave period, λ is the wavelength, u_0 is the maximum horizontal water particle velocity, ρ is the density of water and ν is the kinematic viscosity.

In wave energy, the variety of WEC designs and the range of device and component sizes mean that designs will often incorporate both diffraction and viscous regimes for analysis. Estimation of WEC operation performance and loading is carried out by numerical analysis and laboratory experiments. These two aspects are discussed in detail in Chapters 7 and 8. Different approaches will be appropriate for different types of devices, different stages of design and different aspects of the design. For example, a frequency domain diffraction analysis approach may well be taken to assess device performance in operating conditions. However, in order to assess performance, control and efficiency with the PTO included, a time domain approach is necessary and this will commonly be derived from the frequency domain solution. For more nonlinear wave conditions, extreme loads and survivability conditions a full computational fluid dynamics (CFD) approach may be appropriate. Each of these methods will provide different levels of approximation with appropriate assumptions, and will entail different levels of computational expense.

3.3 Hydrodynamics of Wave Energy Conversion

In this section the existing theory for WECs operating in a single mode of motion in the case of a rigid body motion moving against a fixed reference is first summarised and it is shown that power output is maximised by impedance matching. This is achieved by simultaneously balancing the inertia and restoring forces so that the system is in resonance, while matching the radiation damping of the WEC with the externally imposed damping.

3.3.1 The Equation of Motion

Understanding the motion response of bodies in waves is fundamental to understanding wave energy conversion and can be considered in its simplest form as the mechanics of an oscillating body that is driven by wave motion. The simple spring–mass system can be used to describe the oscillatory response of a floating body in waves. The basic equation of motion for the simple spring–mass system is given by

$$m\frac{\mathrm{d}^2 x}{\mathrm{d}t^2} + b\frac{\mathrm{d}x}{\mathrm{d}t} + Kx = F(t), \tag{3.42}$$

where m is the mass, b is the system damping, K is the spring stiffness and $F(t)$ is the external force. The natural angular frequency of the system is given by

$$\omega_N = \sqrt{\frac{K}{m} - \frac{b^2}{4m^2}}, \tag{3.43}$$

and the amplitude decay function is

$$\exp\left(-\frac{b}{2m}t\right). \tag{3.44}$$

Figure 3.3 Six degrees of freedom motion of a floating body.

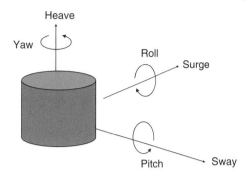

For this system, the solution for free motion response leads to

$$x(t) = \exp\left(-\frac{b}{2m}t\right)\left[a_1 \cos\omega_N t + \frac{1}{\omega_N}a_2 \sin\omega_N t\right]. \tag{3.45}$$

When in forced harmonic motion, $F(t) = F_0 \cos(\omega t)$, the solution can be found from the sum of the free and forced response, $x(t) = x(t)^{\text{free}} + x(t)^{\text{forced}}$. The free motion oscillates with the natural frequency of the system, ω_N, and then decays as a result of the damping in the system, whereas the forced motion oscillates with ω and is sustained by the applied force. After some time, the system will oscillate at the frequency of excitation force, not at the natural frequency, and if the excitation force is sustained it will reach steady state. If the excitation frequency matches the natural frequency of the system, then it will resonate with magnified amplitude oscillations.

The spring–mass system provides the cornerstone for understanding the responses of an object to applied forces. Only one degree of freedom is considered for the spring–mass system, i.e. the mass could exhibit only one translatory motion; however, a floating object in water, such as a WEC, has six degrees of freedom (surge, sway, heave, roll, pitch, yaw) as shown in Figure 3.3, and these motions are all coupled to different degrees. For example, the pitch and heave are closely coupled, and an alteration in the heave of an object is likely to change its trim angle and thus will cause the body to pitch. The motion of a freely floating object is rather involved, but we can obtain some answers in a simple way by making some assumptions. If the magnitude of roll, pitch or heave may be considered 'small' and the wave height may also be considered 'small', then this mathematically allows us to use linear theory in which we can apply the superposition of regular waves. Furthermore, if we assume that the six degrees of motion of a WEC are decoupled, then we can consider any one motion individually, assuming all other motions to be trivial.

If we consider a one-dimensional system in which a WEC is free to move in heave, z, only,

$$m\ddot{z}(t) = F_h(t) + F_m(t), \tag{3.46}$$

where F_h is the force on the wetted surface of the WEC, and F_m is the external force acting on the device, provided by the PTO for example [34]. F_h is a function of frequency and is composed of the wave excitation force, F_e, due to the action of the

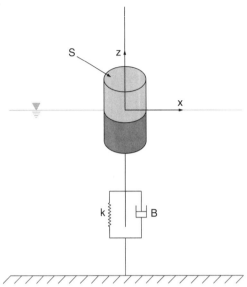

Figure 3.4 Schematic of heaving buoy.

incident wave on the fixed body; the wave radiation force, F_r, which results from the radiated wave generated due to the motion of the body; and the hydrostatic force, F_{hs}, resulting from the position of the WEC in the water. F_e includes the so-called Froude–Krylov force, F_i, due to the incident wave (unaffected by the body), and the diffraction force, F_d, caused by the diffracted or scattered wave (which occurs whether the body is fixed or floating). The hydrostatic force is given by

$$F_{hs}(t) = -\rho g S z(t),$$ (3.47)

where S is the waterplane area as shown in Figure 3.4. The basic equation of motion can then be written as

$$m\ddot{z}(t) = F_e(t) + F_r(t) - \rho g S z(t) + F_m(t).$$ (3.48)

The hydrostatic force acts as a restoring force with spring stiffness in the z-direction and is analogous to K in equation (3.42) [14]. The radiation force, which is the wave force acting on the device due to the radiated waves generated by its own motion, can be written as

$$F_r(t) = -m_a\ddot{z}(t) - B_{rad}\dot{z}(t).$$ (3.49)

The motion of the body results in motion of the water surrounding the body, and m_a is the added mass coefficient, which is used to describe the additional force needed to move the water surrounding the body. Similarly B_{rad} is the wave radiation damping coefficient (analogous to b in equation (3.42)), which is used to describe the force needed to overcome the wave radiation impedance:

$$(m + m_a)\ddot{z}(t) + B_{rad}\dot{z}(t) + \rho g S z(t) = F_i(t) + F_d(t) + F_m(t).$$ (3.50)

In general, an oscillating body in waves has six degrees of freedom – heave, surge, sway, roll, pitch and yaw – which can be represented by the vector **X**. If we include all six degrees of freedom, the equation of motion is then

$$\left(m+m_a\left(\omega\right)\right)\ddot{\mathbf{X}}(t)+B_{\mathrm{rad}}\left(\omega\right)\dot{\mathbf{X}}(t)+K\mathbf{X}(t)=F_i\left(t\right)+F_d\left(t\right)+F_m\left(t\right), \tag{3.51}$$

where m_a is the added mass, B_{rad} is the radiation damping and K is the hydrostatic stiffness for the floating body in waves of frequency ω. When the PTO is included, the equation of motion becomes

$$\left(m+m_a\left(\omega\right)+m_{\mathrm{PTO}}\right)\ddot{\mathbf{X}}(t)+\left(B_{\mathrm{rad}}\left(\omega\right)+B_{\mathrm{PTO}}\right)\dot{\mathbf{X}}(t)+\left(K+K_{\mathrm{PTO}}\right)\mathbf{X}(t)=F_i\left(t\right)+F_d\left(t\right). \tag{3.52}$$

Since the system is linear and the input incident wave is represented by a harmonic function of time, the body displacement and forces on the body are also harmonic functions of time and we can write

$$x(t)=Xe^{i\omega t},\quad F_i(t)=F_{0i}e^{i\omega t},\quad F_d(t)=F_{0d}e^{i\omega t} \tag{3.53}$$

where X is the complex amplitude of the body displacement **X** and F_{0i} and F_{0d} are the complex amplitudes of incident and diffraction wave forces, respectively. Although these amplitudes are complex, the real part is taken when the complex expression is equated to a physical quantity.

The governing equation (3.52) for fluid motion may be written as

$$-\omega^2(m+m_a\left(\omega\right)+m_{PTO})Xe^{i\omega t}=\left(F_{0i}+F_{0d}-(K+K_{PTO})X-i\omega\left(B_{rad}\left(\omega\right)+B_{PTO}\right)\right)e^{i\omega t} \tag{3.54}$$

or more simply as

$$-\omega^2\left(m+m_a\left(\omega\right)+m_{\mathrm{PTO}}\right)X+i\omega\left(B_{\mathrm{rad}}\left(\omega\right)+B_{\mathrm{PTO}}\right)+\left(K+K_{\mathrm{PTO}}\right)X=F_{0i}+F_{0d}. \tag{3.55}$$

In this way the time dependence has been removed and the equation may be solved in the frequency domain, and the solution will represent steady-state conditions once the initial transients have died down. Solution of the fully coupled equation of motion in six degrees of freedom involves a complex matrix solution, with 36 added mass and 36 radiation damping coefficient terms. To find the fluid motion and pressure field it is convenient to use the velocity potential, and to solve for Laplace's equation in the fluid domain with appropriate boundary conditions. Various numerical and semi-analytical methods of solution are available, typically using Green's function with the boundary element method (BEM) [35]. However, in order to gain some insight to the hydrodynamics of wave energy conversion, the solution for a simple one-dimensional system is considered here [36].

For the case of a vertical wavemaker shown in Figure 3.5, moving in the horizontal direction, $x(t)$ only, with wavemaker mass, m, stiffness, K, and external PTO radiation damping B_{PTO}, the equation of motion can be written as

$$\left(m+m_a\left(\omega\right)\right)\ddot{x}+\left(B_{\mathrm{rad}}\left(\omega\right)+B_{\mathrm{PTO}}\right)\dot{x}+\left(K\right)x=F_e\left(t\right), \tag{3.56}$$

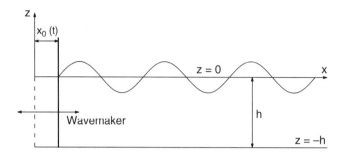

Figure 3.5 Schematic of wavemaker.

where $m_a(\omega)$ is the added mass, $B_{rad}(\omega)$ is the radiation damping and $f_d(t)$ is the wave excitation force on the plate when held immobile. If the wavemaker were to act as a perfect wave absorber and absorb all of the energy of the incident wave, then the plate must move so as to generate a wave that exactly opposes in amplitude and phase the wave reflected by the fixed plate. A key question that arises is whether this can be achieved by adjusting the terms of mass, stiffness and damping in the equations of motion. For the vertical plate considered here, the added mass and damping coefficients may be obtained through linear theory and are given by

$$m_a = \rho \sum_{n=1}^{\infty} \frac{A_n}{k_n^2} \sin k_n h, \tag{3.57}$$

$$B_{rad} = \rho \omega \frac{A_0}{k_0^2} \tanh k_0 h, \tag{3.58}$$

where the coefficients are given by

$$A_0 = \frac{2 \sinh 2k_0 h}{2k_0 h + \sinh 2k_0 h} \tag{3.59}$$

and

$$A_n = \frac{4 \sin k_n h}{2k_n h + \sin 2k_n h}. \tag{3.60}$$

The velocity potential of the incident wave can be written as (see equation (3.7))

$$\Phi_I(x,z,t) = \frac{A_I g}{\omega} \frac{\cosh k_0 (z+h)}{\cosh k_0 h} \sin(k_0 x + \omega t) = \Re\left(\frac{-iA_I g}{\omega} \frac{\cosh k_0 (z+h)}{\cosh k_0 h} e^{-ik_0 x - i\omega t} \right), \tag{3.61}$$

where A_I is the incident wave amplitude. If the plate is held fixed then the velocity potential in front of the plate is composed purely of incident and diffracted wave parts,

$$\phi_I + \phi_D = \frac{-iA_I g}{\omega} \frac{\cosh k_0 (z+h)}{\cosh k_0 h} \cos k_0 x, \tag{3.62}$$

and the wave excitation force is

$$F_e(t) = \Re\left(F_e e^{-i\omega t}\right) = \Re\left(2\rho A_I g \frac{\tanh k_0 h}{k_0} e^{-i\omega t}\right). \tag{3.63}$$

The motion of the plate is given by

$$x(t) = \frac{F_e}{K - (m + m_a)\omega^2 - i\omega(B_{\text{rad}} + B_{\text{PTO}})} \tag{3.64}$$

or

$$x(t) = \Re\left\{xe^{-i\omega t}\right\} = X_0(\cos\omega t + \theta'). \tag{3.65}$$

The energy converted in one period is given by

$$E_{\text{PTO}} = \int_{t_0}^{t_0+T} B_{\text{PTO}} \dot{X}^2 dt = B_{\text{PTO}}\omega^2 X_0^2 \frac{T}{2} = B_{\text{PTO}}\omega^2 \frac{T}{2} xx^*, \tag{3.66}$$

where x^* is the complex conjugate of x, and the equivalent average power recovered is

$$P_{\text{PTO}} = \frac{1}{2} B_{\text{PTO}}\omega^2 X_0^2 = \frac{1}{2} \frac{B_{\text{PTO}}\omega^2 F_e F_e^*}{\left[K - (m + m_a)\omega^2\right]^2 + \omega^2 \left[B_{\text{rad}} + B_{\text{PTO}}\right]^2}. \tag{3.67}$$

In order to maximise the power recovered, we can consider the parameters K, M and B_{PTO}, and reduce the denominator towards zero. If the first square bracket is set to zero, this is equivalent to choosing K and M to be in resonance, thus

$$-(m + m_a(\omega))\omega^2 + K = 0, \tag{3.68}$$

and

$$\omega_N = \sqrt{\frac{K}{(m + m_a(\omega))}}, \tag{3.69}$$

and the power expression (3.65) becomes

$$P_{\text{PTO}} = \frac{1}{2} \frac{B_{\text{PTO}} F_e F_e^*}{\left[B_{\text{rad}} + B_{\text{PTO}}\right]^2}. \tag{3.70}$$

Maximum power, P_{max}, is then achieved when the PTO damping, B_{PTO}, is equal to the wave radiation damping, B_{rad}, and is given by

$$P_{\text{max}} = \frac{1}{8} \frac{F_e F_e^*}{B_{\text{rad}}}. \tag{3.71}$$

Replacing with the terms above gives

$$P_{max} = \frac{1}{2}\rho g^2 A_I^2 \frac{\tanh k_0 h}{\omega A_0} = \frac{1}{2}\rho g^2 A_I^2 \frac{\tanh k_0 h}{\omega} \frac{2k_0 h + \sinh 2k_0 h}{2\sinh sk_0 h} = \frac{1}{2}\rho g c_g A_I^2, \quad (3.72)$$

which is equivalent to the available incident power, P_{inc}, in equation (3.35). So we find that in theory and in two dimensions we can recover the entire power of the incident waves, which means that by moving the plate in a certain way, it will generate waves that algebraically oppose the wave that would be reflected from the plate if held fixed.

In two dimensions (e.g. a device of constant cross-section spanning a wave tank under normally incident waves) the quantities F_e and B_{rad} are connected by the formula [37]

$$\frac{F_e F_e^*}{B_{rad}} = 8P_{inc} \frac{\left|A_i^-\right|^2}{\left|A_i^+\right|^2 + \left|A_i^-\right|^2}, \quad (3.73)$$

where A^+ and A^- are the complex wave amplitudes of the waves generated towards $x = +\infty$ and $x = -\infty$, respectively, by the forced motion of the device in the absence of incident waves. Using equations (3.71) and (3.73), in two dimensions, the efficiency of absorption, $\eta = P/P_{inc}$, and the maximum efficiency is

$$\eta_{max} = \frac{P_{max}}{P_{inc}} = \frac{\left|A_i^-\right|^2}{\left|A_i^+\right|^2 + \left|A_i^-\right|^2}, \quad (3.74)$$

where P_{inc} is the mean power incident per crest length. This result was obtained independently by Mei [38], Evans [39] and Newman [37] and the formula helps to explain the high efficiency of Salter's Duck [3] whose shape is such that $|A^+| << |A^-|$.

3.3.2 Power Absorption Limits

In two dimensions a small heaving or surging body will radiate waves of equal size that propagate fore and aft of the body, symmetrically for a heaving body, and anti-symmetrically for a surging body. Thus, under interaction with an incident wave the body can move so as to radiate a wave downstream that cancels the incident wave completely behind the body, but will also simultaneously create a wave upstream that is the same size as the incident wave. Its net absorption is zero as it has effectively reflected the entire wave upstream of the body. The maximum absorption occurs when the body moves to create waves propagating upstream and downstream that are each half the amplitude of the incoming wave. The wave it radiates downstream cancels half the amplitude of the incident wave behind the body, releasing three-quarters of its power. But the wave it radiates upstream provides no cancellation and is simply wasted; since it has a quarter of the power of the incident wave the net absorption is half. So, an absorbing body free to move in a single degree of freedom has a maximum efficiency of 50%. A body able to move in two degrees of freedom (e.g. heaving and surging, or heaving and pitching) can combine both the symmetric and anti-symmetric modes to radiate a wave in the same direction as the incident wave. It can then cancel the incident wave with no reflection, thus achieving 100% absorption [40].

In three dimensions the theoretical limits are associated with the shape of the converter's far-field radiation pattern and how it interacts with the incident wave, with a greater focused radiated wave field giving a larger theoretical limit. Consequently, in three dimensions, the heaving and surging bodies no longer have the same efficiency. The theoretical limits are attained when power is extracted under the optimal control of the PTO system. The optimal control is that which both (a) induces resonance in the motion of the WEC by providing a reactive power which matches and cancels out its external reactance (i.e. stiffness and inertia terms), and (b) matches the external wave radiation damping of the converter. This is termed 'complex conjugate control'. For a three-dimensional device, we define the capture width, C_w, as the ratio of the absorbed power to the mean incident power per unit crest length, P_{inc},

$$C_w = \frac{P}{P_{inc}}, \tag{3.75}$$

which represents the equivalent length of incident wave from which all power is taken. For an axisymmetric device oscillating in heave, Newman [37] showed that

$$2\pi F_e F_e^* = 8 B_{rad} \lambda P_{inc}, \tag{3.76}$$

where λ is the incident wavelength; thus the maximum capture width is

$$C_{w\,max} = \frac{P_{max}}{P_{inc}} = \frac{1}{8} \frac{FF^*}{B_{rad}} \frac{8 B_{rad} \lambda}{2\pi FF^*} = \frac{\lambda}{2\pi}. \tag{3.77}$$

This result was first proved independently by Budal and Falnes [41], Evans [39] and Newman [37]. Remarkably, it demonstrates that the maximum theoretical capture width is independent of the physical dimensions of the device, provided the incident wave amplitude is below a certain value.

The maximum capture widths theoretically possible for some other modes have been derived and are used as part of a method for categorisation of WECs in Section 3.4. These are summarised in Table 3.1 and are proposed and discussed in more detail by Equimar [40]. Point absorber limits [42] are applicable to machines which are small compared to the wavelength or to axisymmetric machines of any size. Line absorber results [43, 44] apply to machines with length comparable to the wavelength and small width and have been shown to increase with line length. It should be noted that the

Table 3.1 Theoretical limits to capture width [42–44].

		Heave	Surge	Pitch	Heave and pitch	Surge and pitch
Point absorber	Shoreline	λ/π	λ/π	λ/π		
	Offshore	$\lambda/2\pi$	λ/π	λ/π	$3\lambda/2\pi$	$3\lambda/2\pi$
Line absorber	Offshore	$\lambda/2\ (L=\lambda)$				
		$3\lambda/4\ (L=2\lambda)$				

theoretical limits are derived under the assumptions of linear theory in infinitesimal waves. In practice there may be severe limitations to achieving these values. In long waves the converter may be too small to generate the required radiated pattern, and for wave periods away from the natural periods of WEC the reactive power requirement for optimal control may be too large to implement without a detrimental amplification of losses in the power take of system.

The capture width limits due to radiation mode have been derived with no regard for the size of the body. To radiate waves of given amplitude the amplitude of motion of the body is inversely related to the body's size. At one extreme, for a very small body, the amplitude of motion must be very large – with the obvious practical limit for a heaving system that the vertical stroke must be less than its own vertical dimension. At the other extreme a large body will only require small amplitude motion to generate large waves – with the added benefit that the larger it is the lower its resonant frequency will be. For a hemispherical body, this approaches the wave frequency as the diameter approaches half the wavelength. However, for economic reasons and to survive extreme waves, the device cannot be too large. Consequently, practical machines are of an intermediate size – which will require motion amplitude greater than the wave amplitude for good absorption. This in turn requires resonance, but since the body is now too small to have a natural resonance similar to the wave frequency, it must be achieved with other means, such as through reactive control. Kurniawan *et al.* [45] found that a lower natural frequency for a heaving device may also be achieved through use of deformable flexible fabric structures.

Optimal complex conjugate control requires reactive power that can be several times greater than the absorbed real power, and power train losses may then cancel the net benefit. (Mechanical latching is a substitute for reactive control but at an engineering cost and has not been demonstrated on any but the smallest devices.) Without resonance, whether natural or induced by control, limited body motion will much reduce power capture. Evans [41] showed that for a heaving sphere, if the amplitude of motion does not exceed the wave amplitude, the capture width does not exceed 70% of the diameter of the sphere – much less than the radiation capture width limits. Similar amplitude limits apply to all WECs.

As seen here, the capture width is commonly used to describe the performance of a WEC. The annual absorbed or converted power or energy may then be obtained by combining this with the wave power occurrence in a particular location. However, Babarit *et al.* [46] investigated alternative performance measures which are designed to include cost indicators of the device. The annual absorbed energy represents potential income from a WEC project and may be estimated once the external dimensions, working principle, machinery function and local wave resource are known. However, to assess the overall viability of the project, its LCOE is required, which is the average economic cost per kilowatt-hour over the device life span, and this will dictate whether a project goes ahead. LCOE requires full cost estimation, including the costs of design, fabrication, installation, operation, maintenance, and eventually decommissioning. However, these costs are complex to determine and there are significant uncertainties in the information they rely on, and so Barbarit *et al.* [46] performed a comparative study of different WEC types using a set of performance measures that provide a good indication of cost. They considered: annual absorbed energy; annual absorbed energy per characteristic mass; annual absorbed energy per characteristic surface area; and annual absorbed energy per unit of characteristic PTO force. For sea-bottom-referencing devices the characteristic mass is taken to include the mass of foundations, and for self-referencing floating systems the displaced mass is increased slightly to account for the mooring system. For the characteristic wetted surface area, foundations are included in the same way, but no additional surface area has been attributed to mooring systems. The PTO system is seen as a significant source of cost and the higher the PTO forces, the more expensive the PTO system will be. Therefore, the amount of absorbed energy per unit of characteristic PTO force is expected to be a relevant measure for the cost of the PTO system.

3.4   Classification of Wave Energy Converters

There are several reviews of WEC concepts [7,47–49] and a review of technologies and working principles is given by Falcão [14]. Wave energy devices have traditionally been classified into three categories [11, 50]. *Point absorbers*, such as WaveBob [11], are small with respect to the incoming wavelength and are usually symmetrical about the vertical axis. *Terminators*, such as Salter's Duck [3], have one dimension much greater than the other and greater than the incoming wavelength. This larger dimension lies perpendicular to the direction of the incoming wave. *Attenuators*, such as Pelamis [51], also have one longer dimension but this lies parallel to the propagation direction of incoming waves.

However, with more than a thousand patented devices these descriptions have become increasingly non-inclusive. For example, overtopping devices such as the Wave Dragon [52] do not fall into any of the categories and it is unclear where shore-mounted devices fit into this scheme. Device classification has therefore more recently been attempted according to energy capture principles and/or positioning with respect to the shore [14, 49, 53]. Polinder and Scuotto [53] identify four classes of device:

- *Oscillating water columns*: partially submerged structures, open below the water surface and with air trapped above the water surface. Incoming waves make the water surface within the device oscillate, pushing the air through a turbine. Examples include Limpet [54] and Pico [55].
- *Hinged contour devices*: devices with two or more separate parts that move relative to one other as a wave passes under them. Energy is extracted from the reaction between the individual components. Examples include Pelamis [51], Salter's Duck [3] and the McCabe Wave Pump [56].
- *Buoyant moored devices*: Buoyant devices where energy is extracted from the motion induced as waves pass underneath. Examples include Aqua BuOY [57] and WaveBob [58].
- *Overtopping devices*: Reservoirs which waves fill with water. The water then drains out via a turbine. Examples include Wave Dragon [52] and Mighty Whale [59].

Nevertheless, some devices do not fit into any of these categories. A proposed fifth category would be for devices incorporating flexible materials which somehow change shape due to a wave. This would incorporate not only the Anaconda [60] but also devices such as the Lancaster Flexible Bag [61] and the SEA Clam [62].

Another approach is to classify devices according to location, and in this case, there are three distinct zones [11,53]: offshore, nearshore and shoreline. Devices such as the

WaveBob [58], Powerbuoy [63] and Pelamis [51] fall into the offshore category. This region has the most energetic wave climate of the three. However, the associated distance from the shore results in the highest set-up and associated maintenance costs, as well as potentially significant losses during onshore transmission of generated electricity. The large costs associated with offshore devices mean that large arrays will probably be required to make the offshore zone economically viable. The nearshore category incorporates devices in 10–20 m of water and up to a kilometre from the shore. Examples include the Oyster [64], Waveroller [65] and Seabased [66]. Compared to offshore devices, installation and maintenance costs are reduced. Offshore devices can only realistically be moored to the seabed where in the nearshore zone bottom-fixed structures can be used which exploit full wave motion [53].

The final category, shoreline devices, includes devices mounted onshore (e.g. Pico [55] and Limpet [54]) or attached to coastal structures (e.g. Mutriku [67] and Boccotti [68]). These devices clearly offer the cheapest installation and maintenance costs. The wave climate is the least energetic, however, and 'sweet spots' where energy is focused by the surrounding coastline and bathymetry generally have to be found, thereby limiting the available locations. Their proximity to shore may also lead to greater regulatory constraints and issues with visual and noise pollution. In some cases, dual use of wave energy devices for coastal protection in addition to electricity production is under investigation. Boccotti [68] proposed an OWC integrated into a sea wall, and at Mutriku [67] OWC are built into the new harbour wall. A classification summary given by Falcão [14] is shown in Figure 3.6.

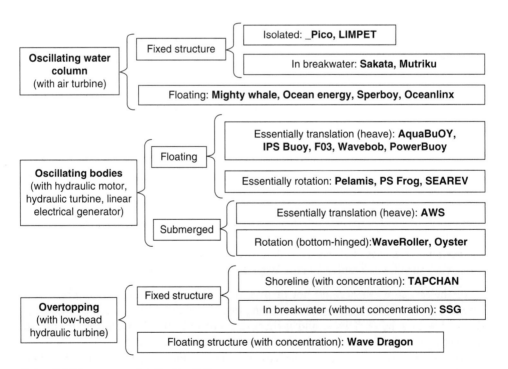

Figure 3.6 Wave energy classification [14].

An alternative approach to classification of WECs in which distinct classes are precisely defined and able to accommodate new instances is proposed by Equimar [40]. The categorisation is based on clear physical principles and identifies both the WEC's means of referencing for PTO, and its modes of motion. When the classification is performed in this way, with the subcategories as detailed below, the different categories relate to known theoretical absorption limits. While many WECs are accommodated within the scheme there are still some device concepts that cannot be captured. Terminating devices that have a fixed station with no moving parts are not generally referenced to another body. They rely on focusing the wave energy and manipulation of the waves in order to focus energy onto or into the PTO subsystem. OWCs also do not fit the taxonomy perfectly and so are described as a separate element in the top layer of the device description.

3.4.1 Classification with Referencing Configuration

In order to extract energy from a wave an appropriate damping force must be applied by a component of the WEC in contact with and providing resistance to the wave. This is called the primary conversion component of the WEC, which must in turn be referenced, via a PTO mechanism, to the seabed, the shoreline, or to another component of the WEC which will not simply move along with the primary conversion component. It is possible for two primary components to react against each other so long as the wave forces acting on them are out of phase. Floats and flaps are the most common types of primary conversion components. In the case of OWC devices the primary conversion component is the pressurised air chamber which is referenced to atmospheric air pressure via a turbine for PTO.

In the WEC categorisation based on referencing configurations proposed by Equimar [40], a further distinction is made between buoyancy and inertia components of a WEC, as the buoyancy force is proportional to the relative wave elevation, whereas the inertia force is proportional to its acceleration. For a float which is reasonably small compared to the wavelength the strong buoyancy force greatly dominates the weaker inertia force, so it is classed as a buoyancy component. For a submerged volume or flap the inertia force will dominate, so is classed as an inertia component. An OWC is predominately buoyancy-driven because the system is pressurised by the water level rise, but some floating OWCs and very large WEC components may have both buoyancy and inertia characteristics, so these can be categorised as mixed.

Seabed/shoreline referencing: shoreline OWCs, WECs with a float or a flap referencing the seabed. Engineering difficulties arise in designing a seabed or shoreline attachment which is sufficiently rigid to provide a reference for power extraction in small and moderate seas but which can economically accommodate the large loads and excursions in severe seas.

Buoyancy–inertia self-referencing: The vertical acceleration in a wave is in anti-phase with the elevation; consequently many machines have been designed with a buoyancy component reacting against an inertia component. This has the advantage that the whole system is self-referencing and can therefore move with a large storm wave, requiring only a compliant (and relatively low-cost) mooring system to keep the machine roughly on station. The disadvantage of using an inertia component for reaction is that it must be several times larger than the buoyancy component in order to provide a balance between the strong buoyancy force and the weaker inertia force.

Buoyancy–buoyancy self-referencing: This configuration also has the advantage of compliance to give good survivability in a large storm waves when coupled with a compliant mooring. However, for the buoyancy components to react against each other, they must be located in different parts of the wave. This implies the machine must be roughly the same length (in the direction of wave propagation) as the wave, which is one reason why such machines have attracted limited attention over the years. Nevertheless, in terms of the overall volume to power-absorption ratio, buoyancy–buoyancy self-referencing is considerably more favourable than buoyancy–inertia self-referencing.

The detailed classification given by Equimar [40] includes three layers: the overarching device description, the specification of PTO and the specification of reaction and control subsystems. This system identifies the radiation mode of the device and its reaction technique, it is able to describe nearly all WECs, and divides them into categories with practical, useful distinctions. However, for simplicity, the devices are presented here under the overarching mode of operation descriptors: OWCs, oscillating bodies, overtopping devices and other.

3.5 Oscillating Water Columns

Of the very many WEC patents registered worldwide, OWC technology appears to be one of the most successful, having reached the stage of full-scale prototypes [69, 70]. This category of devices consists of an air chamber which is exposed to wave action. Under the action of waves, the water column oscillates and the air in the chamber is forced to pass through an air turbine, usually a bi-directional turbine placed in a duct, to convert the energy transmitted to the air into useful power. A schematic diagram of a simple OWC is shown in Figure 3.7.

OWC designs exist for both offshore (floating) and shoreline (fixed) deployment, and recently OWCs have been integrated within breakwater and sea wall structures [68, 71]. Offshore OWCs are floating devices located in deep water where the wave power is relatively high, and were first developed by Masuda and commercialised in Japan in 1965 [72, 73]. The OWC was the first type of wave energy device to be developed and to reach the large-scale prototype stage, with several bottom-standing OWCs being built in 1985–90, such as the OWC at Sanze, Japan, in 1985 [72] and at Trivandrum, India, in 1990 [74]. The first theoretical model of a floating OWC was established by McCormick [75] and recently a 1 : 4 scale buoy converter was deployed in Galway Bay, Ireland [76].

Most OWC prototypes built so far use a self-rectifying air turbine, the most popular of which is the Wells turbine [77]. In 1991, a small (75 kW) OWC (Figure 3.8a) was constructed on Islay, Scotland [54], and another on the island of Pico [55], Azores, Portugal, in 1999 (Figure 3.8b). In 2008, 16 OWCs were integrated into the new breakwater constructed at Mutriku in northern Spain [67] (Figure 3.8c). An alternative to the conventional OWC is the U-OWC device [68], which claims to be more efficient and able to resonate in greater frequency bandwidth, making it more suitable for realistic sea conditions.

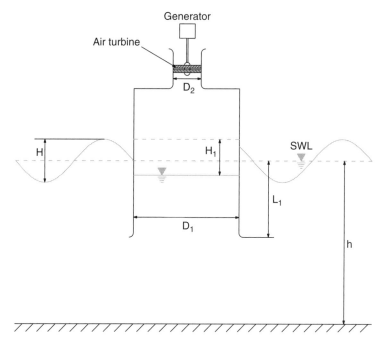

Figure 3.7 Schematic of oscillating water column.

(a)

(b)

(c)

Figure 3.8 Shoreline OWCs. (a) Pico, Azores, Portugal [55] (b) Limpet, Islay, Scotland [54] (c) Mutriku Spain [67.]

(a)

(b)   (c)

Figure 3.9 Floating OWCs. (a) Oceanlinx (Australia) [78] (www.oceanlinx.com) (b) Mighty whale, Gokasho Bay (Japan) [59] (c) The OE12 quarter-scale device on test at sea in Galway Bay (Ireland). Courtesy of Ocean Energy Ltd [76].

Offshore OWCs are usually free to float and to respond in waves in heave and pitch. These motions will combine with the water column motion to determine the performance of the device. Examples of floating devices are shown in Figure 3.9 and include the Oceanlinx [78] in Figure 3.9a, which is a floating platform with eight air chambers [79], the Ocean Energy Buoy [76] in Figure 3.9c, which is a backward bent duct buoy (BBDB) tested at 1 : 4 scale in Galway Bay in 2008, the Mighty Whale [59,80] in Figure 3.9b, comprising three OWCs with frontward facing openings deployed in Japan in 1998, and the Sparbuoy developed by Falcão *et al.* [81].

In the case of shoreline devices and those built into coastal structures, the device is fixed in position and activation of the water column drives the air flow across the turbine. These devices are located along the coast in shallow water. A large part of the available wave energy, around 70%, is lost through bottom friction as the depth decreases [82] and shoreline device performance is also affected by tidal range. However, the onshore or nearshore OWCs have some advantages compared to offshore: mooring lines and wet power-transmission cables are not required; they operate in a safer sea environment, which increases their survivability; they can be easily accessed and maintained; and they can have a dual purpose – electricity generation and coastal protection [83]. The latter is a considerable advantage for OWCs embedded within breakwaters by setting a cost-sharing basis for a project and thus increasing the chances for funding and profit.

OWC devices have been examined extensively with physical, analytical and numerical models. Some milestone experimental studies have been undertaken [84–86] and theoretical models for the hydrodynamic efficiency of an OWC using linear potential flow theory have been derived [87]. The analytic description of a U-OWC under the assumption of linear wave theory was suggested by Boccotti *et al.* [88] and has since been further developed [89] with a more accurate description of the wave field and the dynamics of the OWC. Potential flow models have been applied to OWC simulation because of their relatively low computational cost [90], for example, by Josset and Clément [91] and Ning *et al.* [92], who used a high-order BEM with mixed Eulerian–Lagrangian techniques. Potential flow models cannot simulate viscous flows, however, and viscous flow separation is an important component of the flow at the OWC chamber inlet. More advanced and generalised numerical modelling approaches may be used, which solve the Navier–Stokes equations with nonlinear free surface models. CFD codes have been used by many researchers [93–95] and an integrated approached that combines potential flow solvers and CFD models is used by Bouhrim and El Marjani [96].

3.5.1 Operating Principle: Shoreline Device

The oscillatory pressure at the chamber inlet due to the incident waves causes the column of water inside the chamber to oscillate and act as a piston driving air in and out across the turbine. In order for the amplitude of oscillation of the water column to be maximised, the water column should be driven at its natural frequency and so the OWC should be designed so that its natural frequency is close to the incident wave frequency. If the chamber is narrow relative to λ, the incident wavelength and L_1, the draft of the chamber, then the column of water of length L_1 inside the chamber will behave like a rigid body and will resonate with a natural frequency, f_c (in hertz), given by

$$f_c = \frac{1}{2\pi}\sqrt{\frac{g}{L_1 + L_1'}},$$ (3.78)

where L' is an effective length due to the added mass excited by the water column [97]. Further theoretical description of the OWC, including derivation of the maximum capture width and relation to the oscillating body theory, are given by Evans and Porter [34].

The flux (Q_w) across the internal free surface due to the incident waves is transferred to flux across the turbine and can be written as

$$Q_w = Q_s + (i\omega D - B)p,$$ (3.79)

where p is the dynamic air pressure (relative to atmospheric pressure) on the internal free surface, Q_s is the volume flux across the internal free surface when p is taken to be zero, corresponding to an internal free surface open to the atmosphere, and the second term on the right-hand side is the volume flux across the free surface due to the induced pressure, p, acting upon it. Here D and B are coefficients dependent only on ω, respectively termed the radiation susceptance and radiation conductance by Falnes [98]. They play a role in OWC devices similar to the added mass and radiation damping terms for the rigid body theory, but there is a fundamental difference in their dependence on ω. Whereas for a rigid body WEC the resonance condition is generally satisfied by a single value of frequency, so that the power absorbed peaks at this frequency and falls off on either side, Evans and Porter [34] show that the radiation susceptance has a number of zeros which can affect the performance of the device by increasing the number of peaks in power output.

The water column acts as a piston and causes oscillatory motion in the air column. At resonance the water column usually oscillates with a double amplitude, H_1, which is greater than the wave height, H. The vertical displacement of the water column is given by

$$z_1 = \frac{H_1}{2}\cos(\omega t). \tag{3.80}$$

Here, H_1 is the average vertical displacement of the water column and ω is the wave frequency. This motion of the water column acts as a fluid piston on the air above it, applying a force, F_c, of

$$F_c = pA_1 = p\frac{\pi D_1^2}{4}, \tag{3.81}$$

where p is the air pressure inside the chamber and A_1 is the cross-sectional area of the chamber. The equation assumes a circular cross-section of diameter D_1.

The energy transferred to the air is

$$E = F_c z_1 = pA_1 z_1, \tag{3.82}$$

and the power is

$$P = F_c \frac{dz_1}{dt} = F_c v_1 = pA_1 v_1. \tag{3.83}$$

The velocity and acceleration of the water column are given by

$$v_1 = \frac{dz_1}{dt} = -\frac{\omega H_1}{2}\sin(\omega t), \tag{3.84}$$

and

$$a_1 = \frac{d^2 z_1}{dt^2} = -\frac{\omega^2 H_1}{2}\cos(\omega t) = -\omega^2 z_1. \tag{3.85}$$

The air above the water column will have the same vertical motion as the water column and this forces air through the air turbine in the duct. Assuming the air to be incompressible, the air velocity in the duct of cross-sectional area A_2 is determined from the continuity equation

$$v_2 = v_1 \frac{A_1}{A_2}, \tag{3.86}$$

where A_1 is the cross-sectional area of the water column.

3.5.2 Example Calculation: Shoreline OWC

A vertical cylinder OWC (Figure 3.8) has internal diameter of $D_1 = 2.75$ m and duct diameter of $D_2 = 0.65$ m. The water column has length $L_1 = 6$ m, and the water column effective length is assumed to be 20% of the column length. If the double amplitude of motion is $H_1 = 1.53$ m at resonance, calculate the natural frequency of the internal water motion and the velocity of the air flow through the duct.

The natural frequency of the water column is

$$f_c = \frac{1}{2\pi}\sqrt{\frac{g}{L_1 + L_1'}} = \frac{1}{2\pi}\sqrt{\frac{9.81}{6 + 0.2 \times 6}} = 0.1858 \text{ Hz}, \qquad (3.87)$$

Thus, the natural period is

$$T_c = \frac{1}{f_c} = \frac{1}{0.1858} = 5.3828 \text{s}. \qquad (3.88)$$

At resonance, the wave frequency equals the water column frequency,

$$\omega = \omega_c = \frac{2\pi}{T_c} = \frac{2 \times 3.1415}{5.3828} = 1.1673 \text{ rads}^{-1}, \qquad (3.89)$$

The water column velocity amplitude is

$$v_1 = \frac{\omega H_1}{2} = \frac{1.1673 \times 1.53}{2} = 0.8930 \text{ ms}^{-1}, \qquad (3.90)$$

And so the air velocity through the duct is

$$v_2 = v_1 \frac{A_1}{A_2} = v_1 \frac{D_1^2}{D_2^2} = 0.8930 \times \frac{2.75^2}{0.65^2} = 15.9835 \text{ ms}^{-1}. \qquad (3.91)$$

3.5.3 Operating Principle: Floating OWC Device

An offshore floating OWC will typically have a wide section at the SWL to create the buoyancy needed for the OWC to float. A definition sketch is given in Figure 3.10.

If the OWC is freely floating, the water column natural period, ω_c, can be designed to match the heave natural period, ω_z, to enhance the resonance amplitude,

$$\omega_c = \omega_z. \qquad (3.92)$$

The natural frequency of the device in heave is given by

$$f_z = \frac{1}{2\pi}\sqrt{\frac{\rho g A_{wp}}{m + m_w}}, \qquad (3.93)$$

where for an axisymmetric float with outer diameter, D_0, and inner diameter, D_1, the waterplane area is

$$A_{wp} = \frac{\pi}{4}\left(D_0^2 - D_1^2\right). \qquad (3.94)$$

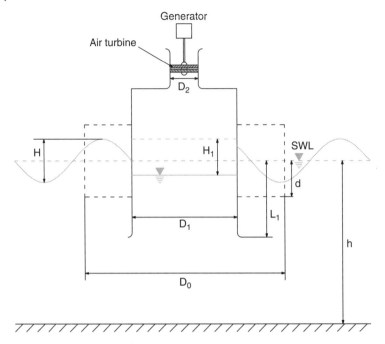

Figure 3.10 Schematic of floating OWC.

3.5.4 Example Calculation: Floating OWC

For the OWC geometry used in Section 3.5.2 and Figure 3.11, and assuming the mass, m, and added mass, m_w, are the same and equal to 5 tonnes, calculate the outer diameter and draught of the buoyancy sleeve needed in order to achieve heave resonance at the same frequency as the water column resonance.

The mass of the OWC is approximated as

$$m = \rho d A_{wp}, \tag{3.95}$$

where d is the draught of the sleeve, and D_0 is the sleeve outer diameter. Substituting equation (3.94), we have

$$m = \rho d \frac{\pi}{4}\left(D_0^2 - D_1^2\right). \tag{3.96}$$

Now

$$f_z = f_c = 0.1858\text{Hz} = \frac{1}{2\pi}\sqrt{\frac{\rho g \frac{\pi}{4}\left(D_0^2 - D_1^2\right)}{2\rho d \frac{\pi}{4}\left(D_0^2 - D_1^2\right)}} = \frac{1}{2\pi}\sqrt{\frac{g}{2d}}, \tag{3.97}$$

$$d = \frac{g}{2\left(2\pi \times 0.1858\right)^2} = \frac{9.81}{2\left(2\pi \times 0.1858\right)^2} = \frac{9.81}{2.7257} = 3.5990 \text{ m}, \tag{3.98}$$

Therefore the outer diameter of the float,

$$D_0 = \sqrt{\frac{4m}{\rho d \pi} + D_1^2} = \sqrt{\frac{4 \times 5000,000}{1025 \times 3.5990 \pi} + 2.75^2} = 41.6329 \text{ m.} \qquad (3.99)$$

3.6 Overtopping Systems

An overtopping device captures seawater of incident waves in a reservoir above the sea level, then releases the water back to sea through low-head turbines. Overtopping devices may be fixed to the sea bed or floating depending on the location. They may also be designed with concentration, in which case the waves are directed into a narrowing channel to increase wave heights and thus overtopping volume. An early example of a shoreline Overtopping device with concentration is the TAPCHAN [12] (Tapered Channel Wave Power Device, Figure 3.2), which was built at Toftestallen in Norway in 1985 and incorporates a 350 kW vertical-axis water turbine in a natural channel [12]. An example of a shoreline overtopping device without concentration is the Seawave Slot-Cone Generator (SSG) described by Vicinanza and Frigaard [99], and an artist's illustration of the device is given in Figure 3.11.

Offshore floating overtopping concepts have also been designed with concentration of the incoming waves, including the Wave Dragon [52] and the WaveCat [100] (Figure 3.12). The Wave Dragon uses a pair of large curved reflectors to gather waves into the central receiving part, where they flow up a ramp and over the top into a raised reservoir, from which the water is allowed to return to the sea via a number of low-head turbines. The Wave Dragon was developed and tested under the Danish Wave Energy Programme. Unlike other WEC technologies, the performance of overtopping devices is not dependent on resonance with the waves and therefore they may be constructed very large. Central issues for floating overtopping WEC's are to control and stabilise the floating structure to optimise power output.

Figure 3.11 Seawave Slot-Cone Generator (SSG) [99].

(a) (b)

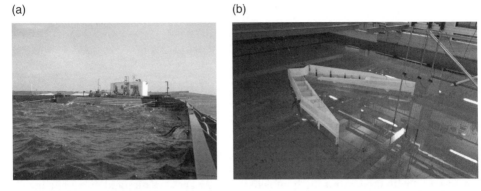

Figure 3.12 Floating overtopping devices. (a) Wave Dragon [52] (www.wavedragon.net) (b) Schematic of the WaveCat concept [100].

WaveCat is an offshore floating WEC whose principle of operation is oblique wave overtopping. It consists of two hulls, like a catamaran (hence its name). Unlike a catamaran, however, the hulls are not parallel but converging, forming a wedge in the plan view; they are joined at the stern by a hinge, which allows the angle between them to be varied depending on the sea state. The freeboard and draught can also be varied according to the sea state. Moreover, the freeboard varies along with the length of each hull, decreasing towards the stern so that overtopping continues as the wave propagates between the hulls (with a constant freeboard, the decrease in the height of the wave crest due to the initial overtopping could interrupt the overtopping before the wave reaches the stern reservoirs).

The traditional linear systems approach used by most other WECs is usually not considered appropriate for use with overtopping devices. Such an approach would consider the water oscillating up and down the ramp as the excited body, and overtopping the highly nonlinear PTO system. However, due to the nonlinearities it is generally considered to be too computationally demanding to model usefully and so an empirical approach is taken. Depending on the current wave state (H_s, T_p) and the crest freeboard R_c (height of the ramp crest above mean water level, MWL) of the device, water will overtop into the reservoir with overtopping flow rate q. The power absorbed by the reservoir is a product of this overtopping flow, the crest freeboard and gravity. For the WaveCat, the extracted power, P_a, is obtained from

$$P_a = \rho g \left(q_a F_a + q_f F_f \right), \tag{3.100}$$

where q_a and q_f are respectively the average overtopping rates of the aft and fore reservoirs; and F_a and F_f are the freeboards at the centre of the aft and fore reservoirs. The useful hydraulic power converted by the turbines, P_{ext}, is then the sum over n turbines of the product of turbine flow, Q_t, the head across the turbine, H_t, water density and gravity:

$$P_{ext} = \rho g \sum_{i=1}^{n} Q_{ti} H_{ti}. \tag{3.101}$$

Within the field of coastal engineering there is a considerable body of work concerning overtopping rates on rubble-mound breakwaters, sea walls and dykes. Van der Meer and Janssen [101] provide the basis of the theory on the average expected overtopping rate and De Gerloni *et al.* [102] investigate the time distribution of the flow. However, this work was focused on structures designed to minimise the rate of overtopping, whereas the overtopping WECs are designed to maximise the rate of overtopping. It is found by Kofoed [52] and Fernandez *et al.* [100] that the level of overtopping depends on the wave period.

3.7   Oscillating Bodies

WECs involving oscillating bodies involve the motion response of single or multiple oscillators in waves. The action of the waves is used to drive the oscillator, and its motion, relative to the seabed or to another oscillator of different characteristics, is used to drive a generator and convert the motion into electrical power. Oscillating bodies can be categorised according to their complexity. In the simplest case, oscillators are attached to and react against the seabed, such as heaving buoy devices connected to the seabed through a PTO. Devices within this category include the Archimedes Wave Swing (AWS, Figure 3.13a), which is a submerged body oscillating in heave with a linear electric generator, developed in The Netherlands, and was tested at sea at Aguçadoura, Portugal, in 2004 [103]. Seabased [66] (Figure 3.13b) was originally developed by Uppsala University, Sweden; it is a seabed-referencing WEC with a linear electric generator located between the float and foundation. The BOLT [104], a Norwegian device designed by Fred Olsen, is taut moored to the seabed with a winch and generator PTO system (Figure 3.13c).

Two-body WECs are self-referencing and have oscillating parts that move relative to one another. For example, the PowerBuoy [63] (Figure 3.14a) has a high-pressure-oil hydraulic circuit for the PTO which lies between the float and reaction plate. It is developed by OPT in the USA and a 40 kW unit was deployed in Santoño (Spain) in 2008. The WaveBob [58] (Figure 3.14b), developed in Ireland, has a high-pressure-oil hydraulic circuit PTO connecting two bodies of different mass, and a 1 : 4 scale model was tested in Galway Bay (Ireland) in 2008. The AquaBuOY [57] (Figure 3.14c) has a high-head hydraulic turbine PTO between surface buoy and submerged masses and was tested in Oregon, USA, in 2007.

Examples of floating multibody, heaving devices include FO3 [105] from Norway, shown in Figure 3.15a during 1 : 3 scale model testing in Norway, and the Wave Star [106] (Figure 3.15b) developed in Denmark and tested in the North Sea off Hanstholm in 2009.

Floating oscillating body pitching devices include Salter's Duck [3], UK, 1979 (Figure 3.1a), in which an asymmetric float oscillates at the free surface and the motion is resisted by an internal gyroscope PTO. The McCabe Wave Pump [56] (Figure 3.16) has three floating pontoons; the centre one is attached to a submerged damper plate and is connected via hydraulic pump PTO to the oscillating arms. It was deployed from 1996 for nine continuous years in the Shannon Estuary, off the coast of Ireland.

In a further category of oscillating wave energy converters, the floater reacts against an internal body. Examples of this category are shown in Figure 3.17 and include the Searev [107], developed in France, and the Penguin WEC [108], developed by Wello in Finland.

(a)

(b)

(c)

Figure 3.13 Seabed reacting point absorbers. (a) Archimedes wave swing (AWS) [103] (www.
teamwork.nl) (b) Seabased [66] (http://www.seabased.com) (c) BOLT, Norway (Fred Olsen) [104].

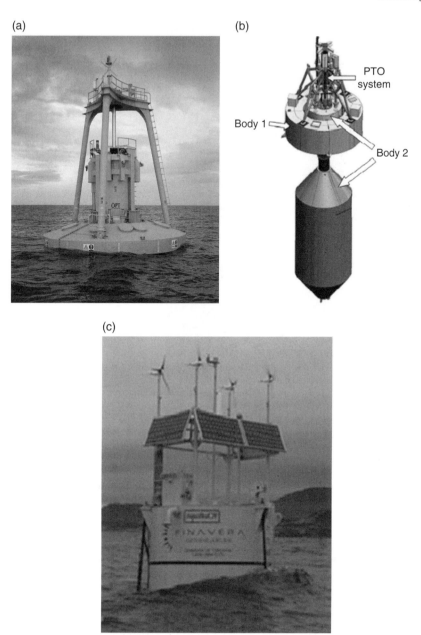

Figure 3.14 Two-body floating WECs. (a) PowerBuoy [63], USA with kind permission of ocean power technologies, Inc. (b) WaveBob [58] Ireland (en.wikipedia.org/wiki/Wavebob) (c) AquaBuOY [57] (http://nnmrec.oregonstate.edu/finavera-point-absorber).

(a)

(b)

Figure 3.15 Floating multibody heaving WECs. (a) O3 [105] Norway (www.seewec.org) (b) Wavestar [106], Denmark (http://wavestarenergy.com).

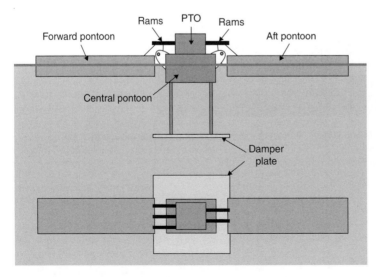

Figure 3.16 McCabe Wave Pump [56] (http://oceanenergysys.com).

(a)

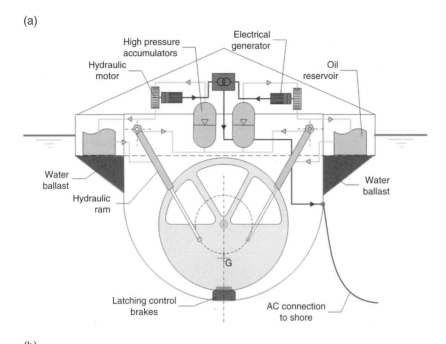

(b)

Figure 3.17 Oscillating WECs with internal reacting body. (a) Searev (Courtesy of ECN) [107] (b) Penguin [108], Wello (www.wello.eu).

Figure 3.18 Pelamis [51] (courtesy of Pelamis Wave Power).

The Pelamis [51] (Figure 3.18) is a floating, multibody, pitching and yawing device; it has a hydraulic PTO system and was deployed in a three-unit farm in Portugal in 2008 and at the European Marine Energy Centre, EMEC, in 2010–2014. The M4 [109], shown in Figure 3.19 during laboratory testing at Plymouth University's COAST laboratory [110], is a three-body heave, pitch and surge device.

Figure 3.19 M4 [109].

Flap-type devices can be categorised as submerged body, bottom-hinged, pitching devices. Nearshore devices in this category include the Oyster [64], developed in the UK (Figure 3.20a), which has a high-pressure seawater PTO and was tested at EMEC, and Waveroller [65], Finland (Figure 3.20b), which has a high-pressure oil PTO and was tested at sea at Peniche in Portugal in 2007.

3.7.1 Operating Principle: Oscillating Device

The hydrodynamics theory discussed in Section 3.3 can be applied to analyse the forces and motion response of oscillating bodies in waves [112]. Analytical solutions are available and have been developed through potential flow and linear theory for simple shapes and simplified cases, for example the vertical plate with one degree of freedom described in Section 3.3. However, in general, the shape of WECs may be complex,

(b)

(a)

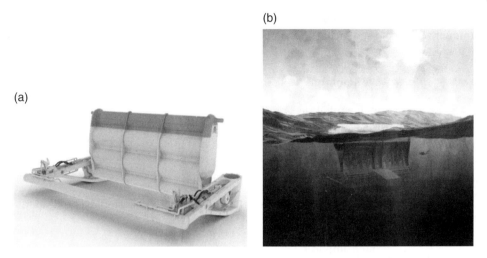

Figure 3.20 Flap-type WECs. (a) Oyster [111], UK (b) Waveroller [65], Finland (aw-energy.com/).

multiple degrees of freedom may be excited, and the device may comprise multiple bodies, and so analytical solutions are not generally available. For this reason, numerical methods are used to determine the motion responses in waves (see Chapter 8), such as BEMs solving for potential fluid flow in the frequency domain for linear conditions and CFD for highly nonlinear situations. Alternatively, motion responses can be determined through model testing as described in Chapter 7.

The motion responses of the WEC may be described in terms of transfer functions called response amplitude operators (RAO) that relate the motion of the WEC to the exciting wave amplitude without considering the relative phases of the signals (discussed further in Chapter 7). RAOs are widely used to study the relationship between different input and output or response parameters of a linear system. Note that this can be defined to analyse the relationship between different parameters (i.e., wave amplitude, horizontal or vertical velocities at a fixed depth, pressure, etc.) and several degrees of freedom [113]. Figure 3.4 defines the six degrees of freedom for a floating body and the following equation defines the RAO as a non-dimensional operator:

$$RAO_i(\omega) = \frac{|\Xi_i|}{a/l^\gamma} \tag{3.102}$$

where (Ξ_i) is the complex amplitude of body displacement in oscillation mode i (i.e., surge, sway, heave, roll, pitch and yaw); l is the characteristic dimension of the WEC, typically its width or diameter; and the exponent γ is 0 for the translation modes and 1 for the rotation modes. Usually a transfer function or the RAO spectrum is composed from individual values of the RAO obtained from a regular waves test series. Assuming linear conditions and that the motions are not coupled, this spectral response can then be multiplied directly by any regular or irregular wave power spectrum to obtain the response of the WEC under a wide set of possible conditions. For a floating structure they will need to be calculated for all six motions and for each separately moving body. RAOs are functions of the wave frequency and are uniquely defined, for a given WEC geometry, as follows:

$$RAO_{\text{roll}} = \frac{\text{roll amplitude}}{\text{wave amplitude}} = Y_{\phi\varsigma}(\omega), \tag{3.103}$$

$$RAO_{\text{pitch}} = \frac{\text{pitch amplitude}}{\text{wave amplitude}} = Y_{\theta\varsigma}(\omega), \tag{3.104}$$

$$RAO_{\text{heave}} = \frac{\text{heave amplitude}}{\text{wave amplitude}} = Y_{z\varsigma}(\omega), \tag{3.105}$$

Note that RAOs for roll and pitch, defined this way, are not dimensionless. Instead, the RAOs for angular motion may be defined by

$$RAO_{\text{pitch}} = \frac{\text{pitch amplitude}}{\text{wave amplitude}/l} = Y_{\theta\varsigma}(\omega), \tag{3.106}$$

where l is the characteristic length of the device. In order to determine the motion response spectrum, it may be calculated from the sea spectrum using

$$S_{\phi}(\omega) = \left[Y_{\phi\varsigma}(\omega)\right]^2 S_{\varsigma}(\omega) \tag{3.107}$$

for roll,

$$S_{\theta}(\omega) = \left[Y_{\theta\varsigma}(\omega)\right]^2 S_{\varsigma}(\omega) \tag{3.108}$$

for pitch, and

$$S_z(\omega) = \left[Y_{z\varsigma}(\omega)\right]^2 S_{\varsigma}(\omega) \tag{3.109}$$

for heave.

Once the motion spectra are obtained, then it is possible to perform stochastic analysis to consider the probability of certain conditions and extreme motions. For any spectrum, the variance, m_0, can be calculated. As an example, for the heave spectrum given by $S_z(\omega)$, the variance is

$$m_0 = \int_0^{\infty} S_z(\omega)d\omega. \tag{3.110}$$

Observations show that both wave height and heave excursion in a realistic sea state can be approximated by the Rayleigh distribution, and in this case the mean ($1.25\sqrt{m_0}$) and significant ($2\sqrt{m_0}$) amplitude motions can be determined from the variance. Furthermore, the probability of amplitude a exceeding a certain value, say a_0, is given by

$$P(a \geq a_0) = \exp\left(-\frac{a_0^2}{2m_0}\right). \tag{3.111}$$

It is important to know where resonance of the floating system will occur. For the axisymmetric heaving float, the equation for undamped heaving in still water is given by

$$\frac{\Delta_w}{g}\frac{d^2z}{dt^2} = -(\rho g)A_{wp}z, \tag{3.112}$$

where Δ_w is the displacement weight, A_{wp} is the waterplane area, and ρ is the density of the water. The hydrostatic stiffness is thus $\rho g A_{wp}$ and the natural frequency of undamped heaving is

$$f_z = \frac{1}{2\pi}\sqrt{\frac{\rho g A_{wp}}{m+m_w}}. \tag{3.113}$$

The heave force may be written as

$$F_z = F_0 \cos(\omega t), \tag{3.114}$$

and the heave displacement of the buoy is given by

$$z = \frac{(F_0/\rho g A_{wp})\cos(\omega t + \gamma - \sigma_z)}{\sqrt{(1-\omega^2/\omega_z^2)^2 + (2\Delta_z\omega/\omega_z)^2}}. \tag{3.115}$$

This may be written as

$$z = Z_0\cos(\omega t + \gamma - \sigma_z), \tag{3.116}$$

in which F_0 is the force amplitude and Z_0 is the motion amplitude, γ and σ_z are phase angles and Δ_z is the dimensionless system damping factor. The velocity and acceleration of the heaving body are

$$v_z = \frac{dz}{dt} = -\omega Z_0 \sin(\omega t + \gamma - \sigma_z) \tag{3.117}$$

and

$$a_z = \frac{d^2z}{dt^2} = -\omega^2 Z_0 \sin(\omega t + \gamma - \sigma_z) = -\omega^2 z. \tag{3.118}$$

The kinetic energy of the heaving float is

$$E_{kz} = \frac{1}{2}(m+m_w)v_z^2 = \frac{1}{2}(m+m_w)\omega^2 Z_0^2 \sin^2(\omega t + \gamma - \sigma_z), \tag{3.119}$$

and its potential energy is

$$E_{Pz} = \frac{1}{2}\rho g A_{wp}z^2 = \frac{1}{2}\rho g A_{wp}Z_0^2 \cos^2(\omega t + \gamma - \sigma_z), \tag{3.120}$$

thus the total energy for the heaving float is

$$E_z = E_{kz} + E_{Pz} = \frac{1}{2}\left[(m+m_w)\omega^2 \sin^2(\omega t + \gamma - \sigma_z) + \rho g A_{wp} \cos^2(\omega t + \gamma - \sigma_z)\right]Z_0^2,$$

(3.121)

The mechanical power available from the body in pure heave is the wave induced force, F_z multiplied by heave velocity, v_z. For a heaving circular float, which is symmetric about the xz and yz planes, such that $\gamma = 0$, and at resonance, such that $\omega/\omega_z = 1$ and $\sigma_z = 90°$, the wave induced force can be shown from strip theory [114] to be equal to

$$F_z = F_0 \cos(\omega t) = \frac{\rho g H \pi R^2}{4}\left(1 - \frac{\pi^2 R^2}{2\lambda^2}\right)\left(e^{-2\pi d/\lambda} + 1\right)\cos(\omega t),$$

(3.122)

where R is the radius of the waterplane and d is the draught of the float. The power of the heaving symmetric body in resonance with the wave is thus

$$P_z = F_z \frac{dz}{dt} = F_z v_z,$$

(3.123)

and the average power over one wave period is then

$$\hat{P}_z = \frac{1}{T}\int_0^T P_z \, dt = \frac{F_0 \omega Z_0}{2}.$$

(3.124)

3.7.2 Example Calculation: Oscillating Device

Calculate the natural period of a cylindrical heaving float that has diameter, $D = 12.5$ m and floats at draught, $d = 2.45$ m in seawater.

Assuming the added mass for a semi-submerged cylinder,

$$m_w = 0.167 \rho D^3 = 0.167 \times 1025 \times 12.5^3 = 334.326 \times 10^3 \text{ kg},$$

the natural period is given by

$$T_z = 2\pi\sqrt{\frac{m+m_w}{\rho g A_{wp}}}.$$

The weight of the float is equal to its buoyancy, given by

$$\rho\Delta = \rho A_{wp} d = \rho \frac{\pi D^2}{4} d.$$

We have

$$A_{wp} = \frac{\pi D^2}{4} = \frac{3.14 \times 12.5^2}{4} = 122.718 \text{ m}^2,$$

so

$$\rho\Delta = \rho A_{wp}d = 1025 \times 122.718 \times 2.45 = 308.176 \times 10^3 \, \text{kg}$$

Therefore, the natural period is given by

$$T_z = 2\pi\sqrt{\frac{334.326 \times 10^3 + 308.176 \times 10^3}{1025 \times 9.81 \times 122.718}} = 4.534 \, \text{s.}$$

3.8 Other Technologies

There are a number of technologies that do not fit easily into the categories above. The distensible anaconda [60] is an attenuator device that uses the bulge wave principle, in which the wave action causes a bulge to travel the length of the device, and pump water through a PTO. The AWS-III [62] is a multi-cell array of flexible membrane absorbers which covert wave power to pneumatic power through compression of air within each cell. The Lancaster Floating Bag [61] was investigated first in the 1970s, and has a ship-like hull lying roughly head-on to the prevailing seas, with rows of flexible bags down each side. The bags were partially filled with air, which was forced through a set of non-return valves and across a turbine by the action of the waves on the bags, although the structure was later found to be too costly [115]. The use of flexible fabrics in WEC design and deformable bodies is explored as a means of altering the hydrodynamic response [45] and by incorporating dielectric generators as a direct power conversion [116].

3.9 The Wave Energy Array

It is expected that wave farms of the future will comprise arrays of large numbers of WECs [46, 117]. The array layout and mooring configuration need to be designed with the performance of the array in mind, but there are also other factors that will influence the layout of the array. For example, array planning will aim to maximise performance and minimise environmental impact, to benefit from cost savings possible due to infrastructure sharing, and will be constrained by electrical connections, construction, installation and maintenance access.

In modelling the effect of interactions between components of the array [118, 119], the gain factor, q, is defined as

$$q = \frac{\text{power absorbed by the array of } n \text{ devices}}{\text{power absorbed by } n \text{ isolated devices}}. \tag{3.125}$$

If there were no hydrodynamic interactions between the devices within the array then q would be equal to 1. Values of q in excess of 1 indicate that the hydrodynamic interactions increase the power absorption capability of the array, and values less than 1 indicate that the interactions decrease the overall power absorbed.

In an array, constructive or destructive interference between the devices is due to diffraction and radiation from each device. Optimising the spacing is fundamental in order to maximise the power absorbed by the array which depends on the spacing between the devices [120, 121]. For example, in regular waves, the optimal spacing is often a multiple of the half wavelength. However, in irregular waves, Myers *et al.* [121] note that it is more difficult to achieve optimal configurations for a broad frequency range because some spacing achieves positive interference for some frequencies but generates negative interference for others. It is generally recommended to avoid interference during the first deployments by keeping devices sufficiently sparse. The effect of device spacing on wave scattering and interaction between devices has been investigated by Babarit [123] for 10 m diameter devices and it was concluded that: if the device spacing is less than 100 m, then interactions have to be considered; if it is 100–500 m, then interactions depend on the configuration, especially when the waves are not aligned with the array; and if it is greater than 500 m, then interactions are negligible.

For some terminator devices, such as the Wave Dragon, the wave shadow effect should also be considered. Beels *et al.* [124] found a large decrease in wave power behind the wave energy device, especially in long-crested waves. Even if devices are non-directional, such as axisymmetric point absorbers, an array of devices may be directional. Interactions within the array can create disparities in power absorption, such that some devices work at their full capacity whereas others work at reduced capacity [125]. In order to reduce the directional effect in a row of devices, the row should be perpendicular to the main wave direction [121]. In order to stabilise array power outputs across wave direction, configurations may be designed to minimise directional effects.

Wave energy arrays can benefit from shared infrastructure such as moorings, offshore substation, electrical cable and connection to shore. Shared moorings may lead to cost savings, but the coupling between the array components through the shared mooring and its effect on performance need to be carefully assessed [126, 127]. If the devices are sparse, a long interconnecting power cable will need to be used [127]. Distances between devices should be chosen to avoid extreme mechanical loads on power cables or excessive power loss in the cable.

One of the most important criteria for array installations is the need for access and the ability to manoeuvre between the devices for installation and maintenance. This also needs consideration for anchor deployment, inspection and maintenance and for safe removal and transportation/towing of devices. Service vessels require space for manoeuvring, and access to devices for maintenance purposes can be best facilitated if all the devices of the array can be approached from outside the array.

The overall aim of arrays is to lower wave energy costs through economies of scale for building, installation and maintenance, and the sharing of infrastructure within the array [128]. This would reduce the costs of shore transmission and optimise the use of the area, which cannot be used for other economic activities. First arrays for demonstration of technology are likely to be at small scale and have low interaction between components. Once the technology is proven, arrays of increased density are likely to be developed by reducing the distance between array components, and these larger, more closely spaced arrays will be more complex and include more interference effects [125].

Two array generations are foreseen, based on their complexity [125]. First-generation arrays are considered to be a single row of devices perpendicular to the incoming wave with the possibility of a second row with devices staggered relative to the first row. First-generation arrays allow access to all devices from outside the array, minimise device interaction and allow the deployment of a large array (but this would require wide rows). Distances between the devices in a row and between the rows have to be chosen to limit device interaction and depend on device type and wave climate. These early first-generation arrays will inform later designs of second-generation arrays. Second-generation arrays are considered to consist of more than two rows of devices with a large overall number of devices in the array. The issues with this type of array are interaction effects which may lead to reduced performance and reduced access to some devices within the array.

In the Equimar project [125], arrays are categorised in terms of their size and purpose. Demonstrator arrays are the first array type, composed of up to 10 devices. Such arrays can be used to demonstrate the availability of the renewable power source and grid connection and may be used to test installation method and maintenance operations. Small arrays are considered to comprise between 10 and 50 devices and are expected to be the next step in upscaling from demonstration arrays to small commercial arrays. In this step, the cost reduction through economies of scale and evolution of components is expected to be proven. Medium arrays will comprise between 50 and 200 devices, and at this level the cost of electricity is expected to have seen considerable reduction so as to be comparable with existing electricity generation technologies. At this stage, the design will be stabilised for metocean conditions and for economic installation and lifetime operation and maintenance.

Future deployments in large arrays are expected to be greater than 200 devices. These will be competitive with other electricity generation technologies and will be next-generation designs, located in more challenging and energetic conditions. The early-stage demonstration and small arrays are likely to use existing deployment vessels and be limited potentially to small deployment windows; whereas the medium and larger arrays are likely either to require custom-built or heavily modified deployment vessels or to be designed to suit existing vessels. In either case, the design of future wave farms will move from a focus on individual device performance towards design of the entire system of multiple devices and their interaction with the wave field, with their mooring/anchoring infrastructure, with the electrical connection and grid infrastructure, with the environment and with their through-life installation, operation and maintenance requirements.

References

1 Straume, (2010) Straumekraft AS: Durable and profitable wave power. Paper presented to 3rd International Conference on Ocean Energy, Bilbao, 6 October.

2 Ross D (1995) *Power from the Waves*. Oxford: Oxford University Press.

3 Salter SH (1974) Wave power. *Nature*, 249(5459), 720–724.

4 Miller C (2004) A brief history of wave and tidal energy experiments in San Francisco and Santa Cruz. http://www.outsidelands.org/wave-tidal.php (accessed 15/11/15)

5 Platts WJ (1979) The development of the wave contouring raft. In *International Conference on Future Energy Concepts* (Vol. 1, pp. 160–163). London: Institute of Electrical Engineers.

6 Clare R, Evans DV, Shaw TL (1982). Harnessing sea wave energy by a submerged cylinder device. *ICE Proceedings*, 73(3), 565–585.

7 Moody GW (1979) The NEL oscillating water column: recent developments. Paper presented to First Symposium on Wave Energy Utilization.

8 Davies PG (1985) Wave Energy: the Department of Energy's R&D Programme 1974-1983, ETSU Report Number R-26, March.

9 Thorpe TW (1992). A Review of Wave Energy, Vols. 1 and 2. ETSU Report Number R-72, December.

10 Whittaker TJT *et al.* (1995). Implications of operational experiences of the Islay OWC for the design of Wells Turbines. In *Proceedings of the Second European Wave Power Conference*, Lisbon, Portugal, 8–10 November.

11 Cruz J (2008) *Ocean Wave Energy*. Heidelberg: Springer.

12 Tjugen KJ (1993) TAPCHAN ocean energy project. In *Proceedings of the European Wave Energy Symposium*, Edinburgh, UK, July 1993 (pp. 265–276).

13 Malmo O, Reitan A (1986) Development of the Kværner multiresonant OWC. In *Hydrodynamics of Ocean Wave-Energy Utilization* (pp. 57–67). Springer: Berlin.

14 Falcão AFO (2010) Wave energy utilization: A review of the technologies. *Renewable and Sustainable Energy Reviews*, 14, 899–918.

15 Gunn K, Stock-Williams C (2012) Quantifying the global wave power resource, *Renewable Energy*, 44, 296 – 304.

16 Huertas Olivares C, Domínguez Quiroga JA, Holmes B, O'Hagan AM, Torre-Enciso Y, Leeney R, Conley D, Greaves D (2011) SOWFIA Deliverable D.2.1, Catalogue of Wave Energy Test Centres and Review of National Targets.

17 Falnes J. (2007) A review of wave-energy extraction. *Marine Structures*, 20, 185–201.

18 Carballo R, Iglesias G, Castro A. (2009) Numerical model evaluation of tidal stream energy resources in the Ría de Muros (NW Spain). *Renewable Energy*, 34, 1517–1524.

19 European Science Foundation (2010) Marine board vision document 2.

20 Rusu E, Pilar P, Guedes Soares C (2008) Evaluation of the wave conditions in Madeira Archipelago with spectral models. *Ocean Engineering*, 35, 1357–1371.

21 Iglesias G, López M, Carballo R, Castro A, Fraguela JA, Frigaard P (2009) Wave energy potential in Galicia (NW Spain). *Renewable Energy*, 34, 2323–2333.

22 Folley M, Whittaker TJT (2009) Analysis of the nearshore wave energy resource. *Renewable Energy* 34, 1709–1715.

23 Hughes MG, Heap AD (2010) National-scale wave energy resource assessment for Australia. *Renewable Energy*, 35, 1783–1791.

24 Wilson JH, Beyene A (2007) A California wave energy resource evaluation. *Journal of Coastal Research*, 23, 679–690.

25 Cornett AM (2008) A global wave energy resource assessment. In *International Offshore and Polar Engineering Conference*, Vancouver (pp. 318–26).

26 Henfridsson U, Neimane V, Strand K, Kapper R, Bernhoff H, Danielsson O, *et al.* (2007) Wave energy potential in the Baltic Sea and the Danish part of the North Sea, with reflections on the Skagerrak. *Renewable Energy*, 32, 2069–2084.

27 Ingram DM (2011) *Protocols for the Equitable Assessment of Marine Energy Converters*. Lulu.com.

28 Moorhouse DJ (2002) Detailed definitions and guidance for application of technology readiness levels. *Journal of Aircraft*, 39(1), 190–192.

29 Weber J (2012) WEC Technology Readiness and Performance Matrix: finding the best research technology development trajectory. In *International Conference on Ocean Energy*, Dublin.

30 Holthuijsen LH (2007) *Waves in Oceanic and Coastal Waters*. Cambridge University Press.

31 Sorensen RM (2006). *Basic Coastal Engineering* (3rd edn). Springer.

32 Faltinsen O (1993) *Sea Loads on Ships and Offshore Structures* (Vol. 1). Cambridge University Press.

33 Chakrabarti, SK (1987) *Hydrodynamics of Offshore Structures*. Southampton: Computational Mechanics.

34 Evans DV, Porter R (2012) Wave energy extraction by coupled resonant absorbers. *Philosophical Transactions of the Royal Society A*, 370, 315–344

35 Teng B, Taylor RE. (1995) New higher-order boundary element methods for wave diffraction/radiation. *Applied Ocean Research*, 17(2), 71–77.

36 Molin B 1999. Numerical and Physical Wavetanks – Making Them Fit: 22. Georg-Weinblum-Gedächtnisvorlesung. Arbeitsbereiche Schiffbau, Techn. Univ..

37 Newman JN (1976) The interaction of stationary vessels with regular waves. In *11th Proc. Symp. Naval Hydrodyn., London* (pp. 491–501).

38 Mei CC (1976) Power extraction from water waves. *Journal of Ship Research*, 20, 63–66.

39 Evans DV (1976) A theory for wave-power absorption by oscillating bodies. *Journal of Fluid Mechanics*, 77, 1–25.

40 Equimar (2010) Deliverable D5.2 Device classification template. http://www.wiki.ed.ac.uk/display/EquiMarwiki/EquiMar (last accessed 15/11/2015)

41 Budal K, Falnes J (1975) A resonant point absorber of ocean-wave power. *Nature*, 256, 478–479; corrigendum 257, 626.

42 Evans DV (1988) The maximum efficiency of wave-energy devices near coast lines. *Applied Ocean Research*, 10(3).

43 Farley FJM (1982) Wave energy conversion by flexible resonant rafts. *Applied Ocean Research*, 4(1).

44 Rainey RCT (2001) The Pelamis wave energy converter: it may be jolly good in practice, but will it work in theory? Paper presented to 16th International Workshop on Water Waves and Floating Bodies, Japan.

45 Kurniawan A, Greaves D, Chaplin J (2014) Wave energy devices with compressible volumes. *Proceedings of the Royal Society A*, 470. 20140559.

46 Babarit A, Hals J, Muliawan MJ, Kurniawan A, Moan T, Krokstad J (2012) Numerical benchmarking study of a selection of wave energy converters, *Renewable Energy*, 41, 44e63.

47 Duckers L (2004) Wave energy. In RG Boyle (ed.), *Renewable energy* (2nd edn, Ch. 8). Oxford: Oxford University Press.

48 Callaghan J, Boud R (2006) *Future Marine Energy: Results of the Marine Energy Challenge: Cost Competitiveness and Growth of Wave and Tidal Stream Energy*, Technical Report. The Carbon Trust, January.

49 Clément A, McCullen P, Falcão A, Fiorentino A, Gardner F, Hammarlund K, Lemonis G, Lewis T, Nielsen K, Petroncini S, Pontes M-T, Schild P, Sjöstrom B, Sorensen HC, Thorpe T (2002) Wave energy in Europe: Current status and perspectives. *Renewable and Sustainable Energy Reviews*, 6(5), 405–431.

50 Evans DV (1981) Maximum wave-power absorption under motion constraints. *Applied Ocean Research*, 3(4).

51 Yemm R (1999) The history and status of the Pelamis Wave Energy Converter. In *Wave Power–Moving towards Commercial Viability*, IMECHE Seminar, London.

52 Kofoed JP (2002) Wave overtopping of marine structures: utilization of wave energy. PhD thesis, Aalborg University, Denmark.

53 Polinder H, Scuotto M (2005) Wave energy converters and their impact on power systems. *International Conference on Future Power Systems.*

54 Heath T, Whittaker TJT, Boake CB (2000) The design, construction and operation of the LIMPET wave energy converter (Islay, Scotland). In *Proceedings of the Fourth European Wave Energy Conference*, Aalborg, Denmark.

55 Falcão, ADO (2000). The shoreline OWC wave power plant at the Azores. In *Proceedings of the Fourth European Wave Energy Conference*, Aalborg, Denmark (pp. 4–6).

56 Kraemmer DRB, Ohl COG, McCormick ME (2000) Comparison of experimental and theoretical results of the motions of a McCabe wave pump. In *Proceedings of the Fourth European Wave Energy Conference*, Aalborg, Denmark.

57 Weinstein A, Fredrikson G, Parks MJ, Nielsen K (2004) AquaBuOY – the offshore wave energy converter numerical modeling and optimization. In *OCEANS'04. MTTS/IEEE TECHNO-OCEAN'04* (Vol. 4, pp. 1854–1859). IEEE.

58 Weber J, Mouwen F, Parish A, Robertson D (2009) Wavebob—research & development network and tools in the context of systems engineering. In *Proceedings of the Eighth European Wave and Tidal Energy Conference*, Uppsala, Sweden.

59 Washio Y, Osawa H, Nagata Y, Fujii F, Furuyama H, Fujita T (2000) The offshore floating type wave power device 'Mighty Whale': open sea tests. In *The Tenth International Offshore and Polar Engineering Conference*. International Society of Offshore and Polar Engineers.

60 Chaplin JR, Farley FJM, Prentice ME, Rainey RCT, Rimmer SJ, Roach AT (2007) Development of the ANACONDA all-rubber WEC. *Proc. 7th EWTEC.*

61 French MJ (1979) The search for low cost energy and the flexible bag device." *Proc. 1st Symposium on Wave Energy Utilization*, Gothenburg, Sweden. 1979.

62 Duckers LJ (2012) Towards a prototype floating circular clam wave energy converter. *Renewable Energy, Technology and the Environment*, p. 2541.

63 Chozas JF, Soerensen HC (2009) State of the art of wave energy in Spain. In *2009 IEEE Electrical Power & Energy Conference (EPEC)*. IEEE.

64 Whittaker T, Collier D, Folley M, Osterried M, Henry A, Crowley M (2007) The development of Oyster—a shallow water surging wave energy converter. In *Proceedings of the 7th European Wave and Tidal Energy Conference*, Porto, Portugal (pp. 11–14).

65 Mäki T, Vuorinen M, Mucha T (2014) WaveRoller – one of the leading technologies for wave energy conversion. *5th International Conference on Ocean Energy*, 4–6 November, Halifax, NS, Canada.

66 Boström C, Svensson O, Rahm M, Lejerskog E, Savin A, Strömstedt E, Engström J, Gravråkmo H, Haikonen K, Waters R, Björklöf D (2009) Design proposal of electrical system for linear generator wave power plants. In *Industrial Electronics, 2009. IECON'09. 35th Annual Conference of IEEE* (pp. 4393–4398). IEEE.

67 Torre-Enciso Y, Ortubia I, López de Aguileta LI, Marqués J (2009). Mutriku Wave Power Plant: from the thinking out to the reality. In *Proceedings of the 8th European Wave and Tidal Energy Conference* (pp. 319–329).

68 Boccotti P (2003) On a new wave energy absorber. *Ocean Engineering*, 30(9), 1191–1200.

69 Boake CB, Whittaker TJT, Folley M, Ellen H (2002). Overview and initial operational experience of the LIMPET Wave Energy Plant. *Proceedings of the 12th International Offshore and Polar Engineering Conference*, 3, 586–594.

70 Pecher A, Le Crom I, Kofoed J (2011) Performance assessment of the Pico OWC power plant following the EquiMar methodology. In *Proceedings of International Offshore and Polar Engineering Conference* (Vol. 8, pp. 548–556).

71 Torre-Enciso Y, Marqués J, López de Aguileta LI (2010) Mutriku. Lessons learnt. In *3rd International Conference on Ocean Energy*, Bilbao, Spain.

72 Falcão AFO (2013). Modelling of wave energy conversion. Technical report. https://fenix.tecnico.ulisboa.pt/downloadFile/3779580606646/ Chapter%25201%25282014%2529.pdf

73 Masuda Y (1990) The progress of wave power generator in Japan. In *The First ISOPE Pacific/Asia Offshore Mechanics Symposium*. International Society of Offshore and Polar Engineers.

74 Graw KU (1996). Wave energy breakwaters: a device comparison. In *Proc Conference in Ocean Eng. IIT*, Madras, India.

75 McCormick ME (1974). Analysis of a wave energy conversion buoy. *Journal of Hydronautics*, 8(3), 77–82.

76 Rourke FO, Boyle F, Reynolds A (2009) Renewable energy resources and technologies applicable to Ireland. *Renewable and Sustainable Energy Reviews*, 13(8), 1975–1984.

77 Raghunathan S (1995) The Wells air turbine for wave energy conversion. *Progress in Aerospace Sciences*, 31(4), 335–386.

78 www.oceanlinx.com/.

79 Gareev A (2011) Analysis of variable pitch air turbines for oscillating water column (OWC) wave energy converters. PhD thesis, School of Mechanical, Materials and Mechatronic Engineering, University of Wollongong.

80 Osawa H, Washio Y, Ogata T, Tsuritani Y, Nagata Y (2002) The offshore floating type wave power device "Mighty Whale" Open Sea Tests Performance of The Prototype–. In *The Twelfth International Offshore and Polar Engineering Conference*. International Society of Offshore and Polar Engineers.

81 Falcão AFO, Henriques JCC, Cândido JJ (2012) Dynamics and optimization of the OWC spar buoy wave energy converter. *Renewable Energy*, 48, 369–381.

82 Stagonas D, Müller GU, Maravelakis N, Magagna D, Warbrick D (2010). Composite seawalls for wave energy conversion: 2D experimental results. In *3rd International Conference on Ocean Energy*, Bilbao, Spain.

83 Hammons TJ (2008) Energy potential of the oceans in Europe and North America: Tidal, wave, currents, OTEC and offshore wind. *International Journal of Power and Energy Systems*, 28(4).

84 López I, Castro A, Iglesias G (2015) Hydrodynamic performance of an oscillating water column wave energy converter by means of particle imaging velocimetry. *Energy*, 83, 89–103.

85 Morris-Thomas MT, Irvin RJ, Thiagarajan KP (2007) An investigation into the hydrodynamic efficiency of an oscillating water column. *Journal of Offshore Mechanics and Arctic Engineering*, 129, 273–278.

86 Ram K, Faizal M, Rafiuddin Ahmed M, Lee Y-H (2010) Experimental studies on the flow characteristics in an oscillating water column device. *Journal of Mechanical Science and Technology*, 24(10), 2043–2050.

87 Evans DV, Porter R (1995) Hydrodynamic characteristics of an oscillating water column device. *Applied Ocean Research*, 17(3), 155–164. http://doi.org/10.1016/0141-1187(95)00008-9

88 Boccotti P, Filianoti P, Fiamma V, Arena F (2007) Caisson breakwaters embodying an OWC with a small opening. Part II: A small-scale field experiment. *Ocean Engineering*, 34(5–6), 820–841.

89 Malara G, Arena F (2013) Analytical modelling of an U-oscillating water column and performance in random waves. *Renewable Energy*, 60, 116–126.

90 Coe RG, Neary VS (2014) Review of methods for modeling wave energy converter survival. In *The 2nd Marine Energy Technology Symposium*. Seattle, WA.

91 Josset C, Clément AH (2007) A time-domain numerical simulator for oscillating water column wave power plants. *Renewable Energy*, 32(8), 1379–1402.

92 Ning D-Z, Shi J, Zou Q-P, Teng B (2015) Investigation of hydrodynamic performance of an OWC (oscillating water column) wave energy device using a fully nonlinear HOBEM (higher-order boundary element method). *Energy*, 83, 177–188.

93 Liu PL-F, Losada IJ (2002) Wave propagation modeling in coastal engineering. *Journal of Hydraulic Research*, 40(3), 229–240.

94 Teixeira PRF, Davyt DP, Didier E, Ramalhais R (2013) Numerical simulation of an oscillating water column device using a code based on Navier–Stokes equations. *Energy*, 61, 513–530.

95 Zhang Y, Zou Q-P, Greaves D (2012) Air–water two-phase flow modelling of hydrodynamic performance of an oscillating water column device. *Renewable Energy*, 41, 159–170.

96 Bouhrim H, El Marjani A (2014) On numerical modeling in OWC Systems for wave energy conversion. *IEEE*

97 McCormick ME (2007) *Ocean Wave Energy Conversion*. Dover Publications.

98 Falnes J (2002) *Ocean Waves and Oscillating Systems*. Cambridge University Press.

99 Vicinanza D, Frigaard P (2008) Wave pressure acting on a seawave slot-cone generator. *Coastal Engineering*, 55(6), 553–568.

100 Fernandez H, Iglesias G, Carballo R, Castro A, Fraguela JA, Taveira-Pinto F, Sanchez M (2012) The new wave energy converter WaveCat: Concept and laboratory tests. *Marine Structures*, 29, 58–70.

101 van der Meer JW, Janssen JPFM (1994) Wave run-up and wave overtopping at dikes and revetments. Delft Hydraulics.

102 De Gerloni M, Cris E, Franco L 1992. 20. The safety of breakwaters against wave overtopping. In *Coastal Structures and Breakwaters: Proceedings of the Conference Organized by the Institution of Civil Engineers, and Held in London on 6–8 November 1991* (p. 335). London: Thomas Telford.

103 Valério D, Beirão P, Sá da Costa J (2007) Optimisation of wave energy extraction with the Archimedes Wave Swing. *Ocean Engineering*, 34, 2330–2344.

104 Bjerke IK, Sjolte J, Hjetland E, Tjensvoll G (2011) Experiences from field testing with the BOLT Wave Energy Converter. In *Proceedings of the European Wave and Tidal Energy Conference (EWTEC)*, Southampton, UK, September.

105 Taghipour R, Moan T (2008) Efficient frequency-domain analysis of dynamic response for the multi-body wave energy converter in multi-directional wave. In *The Eighteenth International Offshore and Polar Engineering Conference*. International Society of Offshore and Polar Engineers.

106 Hansen RH, Kramer MM, Vidal E (2013) Discrete displacement hydraulic power take-off system for the wavestar wave energy converter. *Energies*, 6(8), 4001–4044.

107 Clément A, Babarit A, Gilloteaux JC, Josset C, Duclos G (2005) The SEAREV wave energy converter. In *Proceedings of the 6th Wave and Tidal Energy Conference,* Glasgow.

108 Boren BC, Batten B, Paasch RK (2014) Active control of a vertical axis pendulum wave energy converter. In *American Control Conference (ACC), 2014* (pp. 1033–1038). IEEE.

109 Stansby P, Moreno EC, Stallard T (2015) Capture width of the three-float multi-mode multi-resonance broadband wave energy line absorber M4 from laboratory studies with irregular waves of different spectral shape and directional spread. *Journal of Ocean Engineering and Marine Energy,* 1(3), 287–298.

110 Collins K, Greaves D, Iglesias G, Stripling S, Toffoli A (2013) The new COAST laboratory at Plymouth University: A world-class facility for marine energy. ICE Coasts, Marine Structures and Breakwaters 2013: From Sea to Shore - Meeting the Challenges of the Sea, Edinburgh, 18–20 September.

111 Schmitt P, Windt C, Nicholson J, Elsässer B (2016) Development and validation of a procedure for numerical vibration analysis of an oscillating wave surge converter. *European Journal of Mechanics B,* 58, 9–19.

112 Falnes J, Kurniawan A (2015) Fundamental formulae for wave-energy conversion. *Royal Society Open Science,* 2(3), 140305.

113 Lopes MFP (2011) Experimental development of offshore wave energy converters. PhD thesis, Instituto Superior Tecnico.

114 Newman JN (1978) *The Theory of Ship Motions.* New York: Academic Press.

115 French M, Bracewell R (xxxx) The systematic design of economic wave energy converters. *The Sixth International Offshore and Polar Engineering Conference.* International Society of Offshore and Polar Engineers.

116 Vertechy R, Fontana M, Papini GR, Bergamasco M (2013) Oscillating-water-column wave-energy-converter based on dielectric elastomer generator. In *SPIE Smart Structures and Materials + Nondestructive Evaluation and Health Monitoring* (pp. 86870I-86870I). International Society for Optics and Photonics.

117 Borgarino B, Babarit A, Ferrant P (2012) Impact of wave interactions effects on energy absorption in large arrays of wave energy converters. *Ocean Engineering,* 41, 79–88.

118 Folley M, Babarit A, O'Boyle L, Child, B, Forehand, D, Silverthorne, K, Spinneken, J, Stratigaki, V, Troch, P (2012) A review of numerical modeling of wave energy converter arrays. In *Proceedings of the 31st International Conference on Offshore Mechanics & Arctic Engineering,* Rio de Janeiro, Brazil, 10–15 June.

119 Li Y, Yu Y-H (2012) A synthesis of numerical methods for modeling wave energy converter-point absorbers. *Renewable and Sustainable Energy Reviews,* 16, 4352–4364.

120 Falnes J (1984) Wave-power absorption by an array of attenuators oscillating with unconstrained amplitudes; *Applied Ocean Research,* 6(1), 16–22

121 Fitzgerald J (2009) Position mooring of wave energy converters. PhD thesis, Chalmers University of Technology,

122 Myers L, Bahaj AS, Retzler C, Ricci P, Dhedin J-F (2010) Inter-device spacing issues within wave and tidal energy converter arrays. In *Proceedings of International Conference on Ocean Energy (ICOE),* 6–8 October, Bilbao, Spain

123 Babarit A (2013) On the park effect in arrays of oscillating wave energy converters. *Renewable Energy,* 58, 68–78.

124 Beels C, Troch P, De Visch K, Kofoed J-P, De Backer G (2010) Application of the time-dependent mild-slope equations for the simulation of wake effects in the lee of a farm of Wave Dragon wave energy converters. *Renewable Energy*, 35(8), 1644–1661.

125 Equimar (2010) Deliverable D5.5 Pre-deployment and operational actions associated with marine energy arrays. http://www.wiki.ed.ac.uk/display/EquiMarwiki/EquiMar, (last accessed 15/11/2015).

126 Gao Z, Moan T (2009) Mooring system analysis of multiple wave energy converters in a farm configuration. In *Proceedings of the 8th European Wave and Tidal Energy Conference (EWTEC)*, 7–10 September 2009, Uppsala, Sweden

127 Vicente PC, Falcão AFO, Gato LMC, Justino PAP (2009) Dynamics of arrays of floating point-absorber wave energy converters with inter-body and bottom slack-mooring connections. *Applied Ocean Research*, 31(4), 267–281.

128 WETFEET Project Deliverables D6.1,6.2, http://www.wetfeet.eu/documentation/

4

Tidal Energy Technology

Tim O'Doherty, Daphne M. O'Doherty and Allan Mason-Jones

School of Engineering, Cardiff University, Wales, UK

4.1 General Introduction

With a growing demand for energy from both the industrial and domestic sectors there is an ever increasing demand for cleaner energy, in the bid to reduce carbon emissions from the current global levels of an estimated 31.6 Gt of carbon dioxide in 2011 [1]. To mitigate the potential environmental impacts associated with global pollution many countries are now introducing policies to reduce their emissions; such policies must be credible in their long-term goal if they are to have a sustained impact. The mainstay of these policies is to increase the share of renewable energy. For example, the European Union is aiming to generate 21% of its electricity from renewable sources by 2020 [2].

There are a number of alternative energy solutions which can help towards lowering CO_2 emissions; these include nuclear, wind, solar, hydro and wave and tidal. Possibly the most resourceful and yet least developed of these methods is tidal energy. In addition, tidal energy has the major advantage of being highly predictable, so the security of supply is high.

In practice, there are two forms of tidal energy generation, tidal range and tidal stream, and these are described in detail later in this chapter.

4.2 Location of Operation

While clearly tidal energy devices are water-based, there is the possibility for tidal devices of any specification to be deployed at various depths and distances from the shore. Tidal range devices are often integrated into onshore structures such as harbour walls and breakwaters or nearshore lagoons which are easily accessible and also allow for simple power transmission, so reducing the costs such as those involved in offshore maintenance.

The location of tidal stream devices is normally classified as shallow waters or deep waters, with shallow waters being less than 40 m deep. These classifications reflect the different methodology that is needed to deploy a tidal turbine, rather than the design of

Wave and Tidal Energy, First Edition. Edited by Deborah Greaves and Gregorio Iglesias.
© 2018 John Wiley & Sons Ltd. Published 2018 by John Wiley & Sons Ltd.

the turbine itself which can be the same except for the support and mooring structures. It is envisaged, and evidenced by the current deployment sites of various demonstration projects, that shallow-water sites will be the easiest in the short term and that deep-water sites will start to be utilised once the technology has been proven and the financial viability improved.

In addition, the water surface has a direct effect on the depth–velocity profile from factors such as surface wave height and wake to surface interaction. It is generally accepted that the lower portion of the velocity profile should be avoided due to increasing shear rates or possibly bathymetry-generated turbulence. In a detailed study on the energy yield potential from the Alderney Race in the Channel Islands, Bahaj and Myers [3], suggested that the lowest point of the swept area of a tidal turbine should not fall inside a depth band of 25% of the overall depth if high levels of cyclic power generation and blade loading from the high shear rates were to be avoided. However, as the technology is expanded into regions where restrictions from local shipping apply, these depths may be considerably increased. The tidal phase difference between different sites around the coastline would play a critical role in balancing out the national energy supply. However, due to the existence of major ports such as Newport and Bristol, there are a large number of shipping lanes which could potentially pass through potential energy extraction sites, requiring the turbine to be positioned closer to the seabed.

4.3 Environmental Impacts

Many tidal energy technologies are still in development stages, and as a result of this there have been few studies into the environmental effects of such technologies. The studies which have been carried out are mostly either predicted or unverified [4]. SeaGen, however, is one of the few companies to have conducted a substantial environmental report on its device [5].

The quality of the water surrounding tidal energy generation devices could change significantly, for example due to changes in turbidity as a result of sediment transportation, and deposition caused by the altered tidal flows and current streams will occur around these structures. The effect this would have on marine life is uncertain. Thriving species could be affected and vacate the area, whereas the opposite effect could also be possible where new species are attracted.

The effect on shipping is a great concern with all types of marine energy conversion techniques; the most obvious is with the tidal barrage system which could potentially prevent any vessels from entering the bay or estuary which it encloses. This problem can be overcome through the installation of shipping locks; however, these are a slow alternative and impose a restriction on the size of vessels which may pass through. Although the barrage obstructs shipping it creates new opportunities for the crossing of estuaries and bays as roads, railways and walkways can be constructed along the top of the structure, as well as providing the opportunity for hybrid schemes, that is, introducing wind and/or solar systems. Nearshore and offshore devices may also pose navigational problems for such vessels, especially devices which are proposed for deployment in arrays consisting of up to several hundred devices and consist of low-lying buoys and semi-submerged structures. Possible advantages of having large arrays of floating devices include the ability to reduce coastal erosion, though one of the downsides might be to reduce commercial fishing in those areas. Pollution from the machinery used within the marine energy devices should also be considered such as the release of hydraulic fluids, lubricants and toxic antifouling coatings into the surrounding water. Electromagnetic fields (EMFs) from the generating equipment and transmission lines could possibly have an effect on the migratory and feeding habits of some mammals and fish such as sharks, eels and rays which use the Earth's geomagnetic field for navigational purposes. The effects of EMFs on marine life are a highly controversial topic and no evidence has yet been found to support either argument [6].

The interaction between marine life and energy conversion devices is also an area which is concerning. There is a possibility of devices, mainly turbines, striking creatures such as fishes, mammals or even diving birds. This risk is thought to be greater with ducted turbines which have higher flow and blade velocities, especially those used in tidal range methods. There are also concerns about the cavitation effects from rotors which cause sudden pressure changes that could harm nearby fishes. There are devices which have accounted for the risk to marine life in their designs, such as the Open Centre Turbine which creates a passage for marine life that may be pulled through the turbine [7]. Tidal range methods to help reduce fish fatalities include the installation of 'fish ladders' which are essentially a series of flooded steps which allow migratory fish to traverse the barrage; however, the effectiveness of this alternative is questionable [8].

The installation of devices can have an effect on the surrounding environment. The deployment of mooring systems, such as anchors and foundations, and also the laying of transmission cables and pipes can cause damage to the seabed, though this is thought to be only temporary. These features could eventually become thriving habitats for marine creatures, as observed in the offshore wind industry [9]. The installation and general operation of devices has the potential to produce noise pollution which could affect surrounding areas both above and below the water.

4.4 Tides

The physics of astronomical tides has long been understood, and full details can be found in the book by Pugh [10]. However, a summary of the key points in relation to tidal power generation are given below.

The characteristics of a tide at any location are defined by its range and period, that is, the height between successive high and low tides and time between one high/low tide and the next. Each tidal cycle typically takes an average of just over 12 hours, so that there are two tides for each lunar day. This leads to such tides being commonly described as semidiurnal tides. The range of these semidiurnal tides fluctuates over a 14-day period, with the maximum range occurring at a spring tide which occurs a few days after either a new or a full moon. The minimum range occurs at a neap tide, which occurs shortly after the first or last quarter. Given the lunar cycle of 28 days, the range of the spring tides and neap tides also fluctuates such that the spring tide is at a maximum if it coincides with when the Moon is closest to the Earth and the neap tide is at a minimum when the Moon is furthest from the Earth.

The tidal changes are the results of multiple influences that act over different periods, with the two main constituents arising from the Moon and Sun. This gives rise to a principal lunar constituent, known as M_2 and a principal solar constituent known as S_2, where the subscript 2 identifies the semidiurnal nature of the tide. For diurnal tides, the lunar and solar constituents are referred to as O_1 and K_1, respectively.

This periodic rise and fall of the water levels gives rise to tidal currents, and it is the exploitation of these currents that leads to tidal power generation. The natural bathymetry of coastal areas means that these tidal currents can be accelerated, for example where the flow is driven through narrow channels or around headlands. These currents normally flow in two directions, landwards when they are termed flood currents, and seawards when they are termed ebb currents. Therefore tidal velocities normally range from zero at slack water, when the current is changing from flood to ebb or vice versa, to a maximum between successive slack waters.

Tidal currents can further be classified in terms of the hydrodynamic mechanism that gives rise to them. These are:

- Tidal streaming (TS). This is when the current is forced through a constriction and is accelerated.
- Hydraulic current (HC). This occurs when the pressure gradient created between two adjoining bodies of water, which are either out of phase or have a different tidal range, drives the flow.
- Resonant system (RES). This occurs when an incoming tidal wave[1] constructively interferes with a reflected tidal wave.

4.5 Tidal Range Generation

Tidal range can be defined as the difference in vertical height of the water level between its highest and lowest points, and tidal range generation extracts energy by harnessing the potential energy which is available through the difference in water height, defined as

$$E_P = \frac{1}{2}\rho Agh^2 \qquad (4.1)$$

where ρ is the density (kg m^{-3}), A is the area of the basin (m^2), g is acceleration due to gravity (m s^{-2}) and h is the head (m).

There are two types of civil tidal range systems: barrages and lagoons. These differ in that a barrage creates an impoundment across an estuary, while a lagoon is a stand-alone impoundment and so can be sited completely offshore. Tidal barrages are not new and have been used most notably in France (the 240 MW La Rance Barrage which was commissioned in 1966) and South Korea (the 254 MW Sihwa Lake Tidal Power Station which was completed in 2011) [11, 12].

Due to the elliptical shape of the Earth as well as the bathymetry and topographical features, tidal ranges vary across the globe, which results in only a few areas which are suitable for this method of power generation. The larger tidal ranges occur closer to the shoreline in shallow waters and the areas with the highest tidal ranges. The largest tidal ranges recorded are around the Bay of Fundy in Canada, which has a tidal range of approximately 12 m. Other areas with large tidal ranges include the Bristol Channel, Normandy, Kamchatka, the Magellan Strait and the Cook Inlet, Alaska [13, 14].

---

1 A tidal wave is one that results from the daily tides; as such it is caused by the imbalanced gravitational influences of the Moon and Sun and most pronounced in narrow bays.

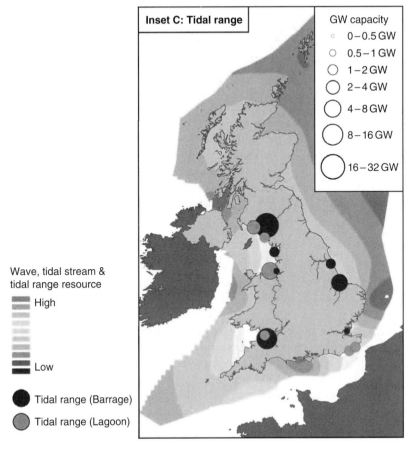

Figure 4.1 Tidal range resources around the UK [16].

Within the UK it has been estimated that there is the potential for 45 GW to be generated from tidal barrage schemes and a further 14 GW from lagoons. These are located at key sites (Figure 4.1), with the Bristol Channel offering the greatest potential. These sites are chosen because they all have a mean tidal range greater than 4 m, which is commonly taken as the benchmark for economic viability [15].

4.5.1 Tidal Barrages

Tidal barrage generation is one of the most mature forms of ocean energy, dating back to the 1960s. Barrages are similar in structure to dams, except that flow through a barrage is controlled via sluices and turbines and they are generally constructed across estuaries. There are two main methods of generation used, these are ebb generation and flood generation, though other generation systems have been utilised.

Ebb generation. This works on the principle of generating electricity as the tide ebbs. Water is allowed to flow into the basin through sluice gates as the tide rises during the flood tide; once the tide has reached its maximum height the sluice gates are closed, forming a basin of water. During the ebb tide the sea level falls, producing a head of water and thus potential energy. The potential energy stored in the basin is captured by releasing the water through turbines. The Annapolis Royal Generating Station is one example of a barrage which operates using this method of generation; it opened in 1984 and spans the River Annapolis. The plant operates through a single turbine rated at 20 MW [17].

Flood generation. This method captures the energy on the flood tide. By closing the sluice gates as the tide rises, a head is produced as the water level in the basin is lower than that of the sea level. Energy is captured by allowing the water to flood through turbines into the basin once the sea level is at an optimal height. This method of generation tends to be less efficient than ebb generation due to the lower head levels achievable. The largest tidal power barrage in the world, Sihwa Tidal Power Plant, uses flood generation to power ten 25.4 MW bulb turbines. The barrage was constructed in 1994 for fresh water and reclamation purposes, although due to pollution of the water, installation of the power plant was later carried out which was completed in 2011 [18].

Dual generation. Utilising both flow and ebb generation techniques, dual generation is a very attractive form of tidal generation due to the longer generating periods available. The system works by initially generating on the flood tide until there is an insufficient head; the sluices are then opened to allow the remainder of the water to flow into the basin where the process of ebb generation begins. An example of a barrage which can operate using dual generation is the La Rance Tidal Barrage in France; it consists of 24 bulb turbines which are rated at 10 MW each. Construction of the barrage was completed in 1966, and it is now the second largest worldwide after the recent development of the Sihwa Power Plant. The Rance Barrage is able to generate on both the flood and ebb tide and can also use the installed turbines to pump water in either direction [19].

Double-basin barrages. This type of barrage is essentially a standard single-basin barrage with a second basin constructed within. The method of operation within the first basin would most likely be ebb generation; a portion of power generated from the first basin would then be used to pump water into the second basin. The purpose of having the second basin is to create the ability to generate power at any time without having the restrictions of the tidal flow; this would allow for the second basin to begin generation when electricity demand is highest. Currently no barrages have been constructed using this method.

4.5.2 Tidal Lagoons

Tidal lagoons work on the same principle as tidal barrages and have been proposed as a possible alternative. A lagoon is defined as a body of shallow water separated from the sea through a low sandbank or coral reef. Tidal lagoons are artificial impoundments built up of large boulders, rubble and sand which act as a basin that captures water on the flood tide and releases it through turbines on the ebb tide. Tidal lagoons are designed to be constructed on shallow offshore tidal flats. The main attraction to tidal lagoons is that they offer generation capacities similar to barrages but do not have the same environmental effects. The extent of the environmental impact will depend on the site and construction design.

The Swansea Bay Tidal Lagoon has recently been given the go-ahead by the UK government, provided the final three challenges are met. These challenges centre around environmental concerns regarding possible silting and sand dredging, as well as the impact on fishes; an updated validation report to convince the government that the technology is viable; and agreement on the government subsidy that would need to be paid for the power generated. Assuming these challenges are met, the Swansea Bay Tidal Lagoon will be the first human-made energy generating lagoon with a 320 MW installed capacity that can generate power through both flood and ebb to give 14 hours of clean renewable power every day for the next 120 years. At the time of writing the Swansea Bay lagoon project aims to start construction on site in 2018; it is estimated that the construction will take 4 years with the first power generated in year 3. It is expected that the final decision will be made by 2016 and that the lagoon construction will begin shortly afterwards to ensure it would be operational in 2019. The proposed lagoon will form a 9.5 km long loop which runs up to 4 km out to sea. It will incorporate 16 low-head bulb 7 m diameter turbines. Once operational, it is expected that this lagoon will provide sufficient power for 155 000 homes, which is more than the requirements for Swansea and equates to 11% of the annual Welsh domestic use.

4.5.3 Other

Another method of capturing the energy from the tidal range is through a concept known as tidal delay which is an innovative idea being developed by Woodshed Technologies. The concept relies on the delaying effect that costal features such as peninsulas and isthmuses have on the tidal range either side of these features. The water level can take up to several hours to reach the same height on both sides of such costal landforms which act as natural barrages, therefore there is energy potential formed. The tidal delay concept captures this energy by channelling the water from the higher level to the lower through the use of pipes which incorporate turbines. These pipes can travel underground or along the top of the landform when a siphoning technique is used [20].

4.6 Tidal Stream

In contrast to barrages and tidal lagoons, tidal stream turbines (TSTs) use the kinetic energy of the tide directly, and unlike the impoundment schemes, TSTs allow the water to pass through and around them and do not require the storage of water. Therefore local tide velocity, turbulence, bathymetry, water column velocity profile and depth, seabed mounting, local shipping requirements and concerns associated with marine fishes and mammals are just some of the key issues facing the successful development of tidal stream devices. Much of the fundamental technology associated with TSTs, however, is derived from the wind industry which in some respects circumvents much of the early developments phases, such as blade profile testing and the basics of the ensuing hydrodynamics. The medium in which TSTs operate, however, produces higher structural loading, when compared with air driven turbines, with the addition of biological fouling from marine life, possible impact from marine creatures and flotsam, increased material corrosion from salts and the possibility of blade cavitation at shallower water depths [21]. As a result the design criterion for a TST requires a high degree of robustness with a limited maintenance schedule to reduce both operational cost and embodied CO_2 emissions from increased material usage and site to shore transport.

TSTs are also fully submerged below the water level and thus do not offer a visual obstruction to the seascape, though some designs will include a supporting structures such as the stanchion which will breach the water surface, for example the SeaGen Turbines (Figure 4.2a). They can be seabed mounted, for example via a pile-driven stanchion, or floated to a desired depth in the water column using buoyancy. There are a number of devices currently under development that can be used to extract energy from local energy fluxes. Most fall into two general categories: the horizontal- and vertical-axis turbines. Others include venturi devices that can be used to concentrate the flow, and oscillating hydrofoils that move up and down through the water column generating electricity via the pumping of hydraulic fluids [22].

There are numerous publications and websites that discuss the various aspects of these technologies and therefore they will not be discussed in detail here. The independent report by the Sustainable Development Commission covers the relative performance of differing types of marine energy conversion technologies in comprehensive detail [22].

By placing such devices at locations where the flow is restricted, such as through channels, between islands and around peninsulas, the tidal currents that occur have the potential to result in high energy extraction. Examples of these include bays, fjords and harbours. This type of flow possesses a significant energy resource which could make a valuable contribution to either a local or national energy supply. Global areas of potential interest include the straits of Messina between Italy and Sicily and concentrated flows between the Indonesian islands. Others include Long Island Sound and New York Harbour, New York, USA [23]. However, this is by no means an exhaustive list of potential sites; along the European coastline around 106 sites have been identified for electricity generation [24]. The UK in particular has a significant domestic tidal stream resource, representing around half of the European resource and around 10–15% of the known global resource [25]. However, before the year 2006 measurable energy contributions from tidal stream simply do not exist, indicating that the technology has yet to reach the supply radar and is still in the development stage. However, the motivation for the technology is clear: of the ~360 TWh of electricity demand in the UK [26], the tidal stream resource has the potential to generate nearly 16 TWh, which is approximately 4% of the UK electricity demand [27]. However, this figure is slightly reduced by the 'significant impact factor' (SIF) as presented by Black and Veatch [28]. The resulting figure of 12 TWh per year represents the UK tidal stream resource that could be economically exploited if the technology were to be fully developed and deployed. Interestingly it also represents around 71 % of the energy that could be extracted from the Cardiff-Western Barrage.

Although in its infancy, to date UK tidal stream technology has resulted in a number of installed full-scale devices. Marine Current Turbines (MCT) introduced the world's first offshore TST, the Seaflow, which was built into the seabed 1.5 km offshore from Lynmouth, Devon, at a total cost of £3.2 million. It comprises an 11 m diameter twin-bladed turbine and is capable of producing 300 kW of electricity at a tidal flow of about $2.8 \, m \, s^{-1}$ (5.5 knots) [29]. With around £4.27 million worth of grant from the Department of Trade and Industry, MCT has also developed the more recent 1.2 MW SeaGen project at Strangford Lough off the coast of Northern Island (Figure 4.2a). MCT has also commenced studies for a small array of 12 turbines off Foreland Point, North Devon. The Dublin-based OpenHydro Group Ltd has recently been selected to partner with

(a)

SeaGen MCT twin turbines near Strangford Lough [30]

(b)

OpenHydro concept with seabed mounting [31]

Figure 4.2 Selection of tidal turbines technologies.

EDF Energies Nouvelles for a 14 MW project off the Normandy coast, with each turbine having a power output of 2 MW (Figure 4.2b).

4.6.1 Available Resources

Tidal currents with a velocity less than $1\,\text{m s}^{-1}$ have been deemed as economically non-viable in terms of power generation [32]; the ideal peak tidal velocity should be $2-3\,\text{m s}^{-1}$ [33]. This is because the power density that will be generated from a velocity of $1\,\text{m s}^{-1}$ is only $500\,\text{W m}^{-2}$ [34]. In fact, the Carbon Trust [35] has suggested that a mean annualised power density of $1.5\,\text{kW m}^{-2}$ in a depth greater than $15\,\text{m}$ is required for the economic viability of most projects. For example, based on these criteria and the cost of energy, Black and Veatch ranked suitable sites around the UK as detailed in Table 4.1.

Table 4.1 Ranked UK possible sites for tidal stream energy (adapted from the Carbon Trust [36]).

Ranking	Site	Area km²	Mean sea level m	Cost of energy p/kWh	Type of tidal current
1	Kyle Rhea	1	34	21	TS
2	Strangford Lough	3	35	25	HC
3	Pentland Firth Shallow	20	30	24	HC
4	Westray Firth	22	27	35	TS
5	Race of Alderney	68	31	26	TS
6	Pentland Firth Deep	240	62	17	HC
7	Blue Mull Sound	2	35	21	TS
8	Mull Of Kintyre	9	116	23	TS
9	Ramsey Island	18	42	25	TS
10	Islay / Mull of Oa	128	37	25	TS
11	Rathlin Island	4	102	25	TS
12	East Rathlin Sound	2	45	27	TS
13	Carmel Head	50	38	29	TS
14	Yell Sound – West Channel	2	30	32	TS
15	Bristol Channel – Minehead	17	30	34	RES
16	West Islay	93	31	36	TS
17	Big Russel	1	36	36	TS
18	North Of N. Ronaldsay Firth	11	39	38	TS
19	Bristol Channel – Mackenzie	4	22	40	RES
20	Isle of Wight	21	29	57	TS
21	Uwchmynydd	1	29	57	TS
22	Mull of Galloway	9	29	59	TS
23	Barry Bristol Channel	9	26	60	RES
24	West Casquets	16	23	61	TS
25	Portland Bill	1	29	61	TS
26	East Casquets	61	22	62	TS
27	North East Jersey	35	21	72	TS
28	N. Ronaldsay Firth	17	17	74	TS
29	South Jersey	60	20	75	TS
30	South Minquiers (Jersey)	11	16	90	TS

TS, tidal streaming; HC, hydraulic current; RES, resonant system.

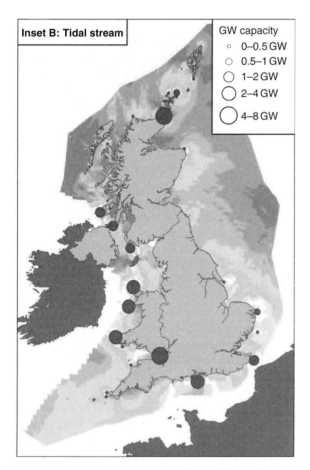

Figure 4.3 Tidal stream resources around the UK [36].

As can be seen from Table 4.1, the tidal stream resources occur at discrete sites around the UK. It has been estimated that the theoretical UK resource for tidal stream is 95 TWh per year (32 GW) [37]. Figure 4.3 shows the distribution of resources around the UK.

Given the tidal stream resource data published by the Crown Estate [36], the actual energy availability at an identified site is very much reliant on bathymetry and the local upstream tidal velocity profile. These features will have a major influence on the performance of individual turbines, whether sited individually or in arrays. Moreover, wake interaction between turbines will further influence bathymetric positioning and turbine spacing. Although this is extremely site-specific it will have a significant effect on performance and, once fully developed, on the UK annual tidal stream resource estimate of 12 TWh. The location of the turbine through the water column can be influenced by local shipping requirements and/or the local bathymetry within the vicinity of the turbine or turbine arrays. In the case where the turbine occupies a significant proportion of the water column, the upstream velocity profile and inherent turbulence characteristics could have the potential to introduce asymmetric loading across the turbine diameter, which not only affects the turbine's ability to extract energy effectively, but also can have a detrimental influence on components through mean stress/strain fatigue. The resulting areas of main concern would be the blades, blade to hub connection and main drive components such as the driveshaft, bearings and seals.

Long-term performance and reliability are a key issue if the TST technology is to be expanded to the implementation of arrays at all the potential UK sites. Since tidal stream technology is at an early stage it is unclear what the actual nature of site-specific resource is and/or what the final availability for the UK will be. Early indications from prototypes, such as those previously discussed, indicate that the devices are meeting or exceeding the predicted energy capture [38]. Given that very little site-specific performance data is available for tidal stream devices, localised site-specific data is still needed to understand the performance of a given turbine design and how key operational parameters might change under varying conditions. In a detailed study on the implementation of TSTs within the races around the Channel Islands, Bahaj and Myers [39, 40] showed that with an appropriately positioned array of turbines an energy yield in excess of 7.4 TWh could be realised, which was stated to be equivalent to 2% of the UK requirements for the year 2000. It was also observed that although the energy resource was totally predictable the power production was observed to be uneven. The latter point elucidates the complexity of individual sites as knowledge of tidal patterns grows. Further work on the effects of bathymetry and water column velocity profile is therefore required to help build a fundamental picture on the performance characteristics of a given design when compared to an idealised operational environment. Used in a scientific manner, the comparisons made could be used to further adjust the current resource estimates and how the performance of the turbine is measured.

The mitigation of supply intermittency and generation redundancy was also shown to be possible with careful choice of six phase-locked tidal stream power installations [41]. Research indicated that by installing devices in the Severn Estuary, Menai Strait, Mersey, Clyde, Tyne and Humber with power capacities of 45, 20, 40, 20, 55 and 25 MW respectively, an overall rate of about 45 MW throughout the UK tidal cycle could be realised, with the potential to stabilise the total supply. This, though, would require large portions of these locations to be dedicated to power generation, which could potentially cause major disruption to local shipping lanes. This is especially the case for the Severn Estuary as large cargo vessels are commonplace with vessel drafts up to 14 m. With such large drafts the operational depth of tidal stream devices would therefore be severally restricted to deeper depths. This, however, places the device in the lower boundary of the water column which could potentially exhibit high rates of shear. Generally, it would be undesirable to locate a horizontal-axis tidal turbine (HATT) at a depth whereby its lower diameter occupies a portion of the water column that equates to around 25% of the overall depth, since this is the region of highest shear rate [42]. However, if a stable electrical contribution to the UK supply is to be realised then some form of compromise would be required either in local shipping or the operational size and/or depth for HATT placement. If the latter option were to be considered then the performance of a HATT within the lower 25% boundary would need to be investigated with the aim of establishing its effects on key performance characteristics such as power extraction, torque and axial thrust loads. To achieve a true representation of this a site measured velocity profile would be required. Moreover, with 63% of the total mean spring peak velocity (V_{msp}) in depths greater than 40 m [25], it may be the case that modular arrays could potentially occupy the lower 25% of the water column, placing the rotors in a high shear velocity field.

4.6.2 Turbine Characteristics

All turbines are based on the design of a hydrofoil that has been optimised to create maximum lift and minimum drag. Since blade design is so critical to the success of tidal energy, a brief summary of the key parameters of a hydrofoil is given in Figure 4.4. The angle of attack, α, is defined as the angle between the chord direction and the direction of the free stream, if the blade is not moving, or the relative velocity if the blade is rotating through moving water.

As with all submerged bodies, the flow past the hydrofoil is dependent on Reynolds number (Re), which represents the ratio of inertial effects to viscous effects. Therefore at low Re the viscous effects dominate, while at larger Re the inertial forces start to dominate and the fluid starts to separate from the hydrofoil at a point on the top surface near to the trailing edge, leading to separation behind it. The separation point and the location of the separation bubble are also dependent on the angle of attack. As the angle of attack increases, the pressure over the front portion of the hydrofoil decreases while the pressure over the rear portion increases. Consequently, the separation point moves forward towards the leading edge. The wake starts from the points of separation, therefore as the separation point moves forward, the wake widens.

The relative motion between the hydrofoil and the water gives rise to wall shear stresses, due to the viscous effects and normal stresses due to the pressure. The resultant of the shear stress and the pressure distribution gives rise to a resultant force (F_{res}) which acts at the centre of pressure, where the component normal to the direction of the relative velocity is known as lift (L) and the component along the direction of the relative velocity is known as drag (D); see Figure 4.5. For a TST, it should be noted that ω is the angular velocity of the turbine and r is the radial distance from the rotation axis of the hydrofoil on the blade.

Therefore as the angle of attack increases and the pressure distribution changes, the lift increases rapidly before reaching a maximum point and then decreases with a drop as the critical angle or stall angle is passed. As the angle of attack increases further, the upper surface flow becomes more and more fully separated and the hydrofoil produces less lift. Correspondingly, as the angle of attack increases, the levels of drag on the blade

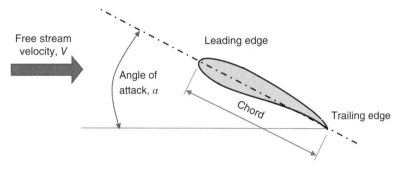

Figure 4.4 Basic definitions of a hydrofoil.

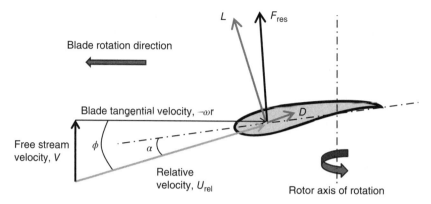

Figure 4.5 Profile of a hydrofoil and the resultant, lift and drag forces.

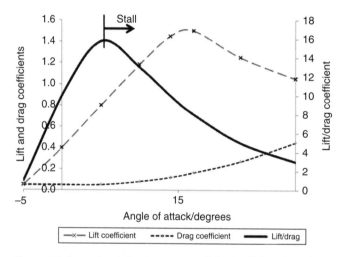

Figure 4.6 Example relation between coefficients of lift and drag and angle of attack.

rise slowly at first and then as the rate of increase in lift falls around the stall angle the levels or drag increases rapidly (Figure 4.6).

The lift and drag are often expressed non-dimensionally as the lift (C_L) and drag (C_D) coefficient respectively:

$$C_L = \frac{L}{\frac{1}{2}\rho A V^2}, \quad C_D = \frac{D}{\frac{1}{2}\rho A V^2}, \tag{4.2}$$

where A is the projected area when looking from an upstream direction. The variation of C_L and C_D as well as their ratio as a function of the angle of attack is therefore important in designing a hydrofoil section. As can be seen in Figure 4.6, there is an optimal angle of attack when the C_L/C_D ratio is at a maximum. As the angular velocity changes, the angle of attack also changes. In order to keep the angle of attack constant, a twist is normally introduced along the length of the blade such that the optimal angle of attack for the maximum C_L/C_D ratio is maintained along the whole blade span.

The forces on the blades can also be expressed as an axial load (F_A) and a tangential load (F_T), where

$$F_A = L\cos\Phi + D\sin\Phi \tag{4.3}$$

$$F_T = L\sin\Phi - D\cos\Phi \tag{4.4}$$

The tangential force creates a torque (T) about the rotational axis of the turbine from which the power generated by the rotor (P) can be determined, given that $P = T\omega$. The axial load, commonly known as the thrust load, imposes a structural load on the stanchion. Given that the turbine needs to be designed to generate as much power as possible, it is clear from Figure 4.6 that turbine designs that maximise lift and minimise drag will perform best.

As with all types of turbines, the torque, power and thrust load are normally non-dimensionalised to define the turbine characteristics. The power is non-dimensionalised against the available power as determined from an upstream flow of cross-sectional area as the swept area of the turbine, and is defined as the coefficient of power, C_p:

$$C_p = \frac{P}{\frac{1}{2}\rho A V^3} \tag{4.5}$$

There is an upper limit to C_p, or maximum theoretical efficiency, of 0.593. This is known as the Betz limit, which assumes a turbine in an open free flow and that all friction and swirl components are neglected. This measure is the ratio of the energy extracted by the turbine to the kinetic energy flux contained within a stream tube the size of the turbine swept area. The maximum energy extraction occurs when the ratio of the velocity downstream of the turbine to that of the upstream velocity is equal to 0.333. The full derivation of the Betz limit is covered in the literature elsewhere [43].

The torque coefficient, C_θ, is defined relative to the theoretical torque that can be produced from the upstream flow through an equivalent swept area and is given by

$$C_\theta = \frac{T}{\frac{1}{2}\rho A R V^2}, \tag{4.6}$$

where R is the turbine diameter. The thrust coefficient, C_T, is defined relative to the theoretical load from an upstream flow through an equivalent swept area, that is,

$$C_T = \frac{F_A}{\frac{1}{2}\rho A V^2}. \tag{4.7}$$

Finally, the angular velocity of the turbine is also normally non-dimensionalised in relation to the upstream velocity. This parameter is known as the tip speed ratio, λ:

$$\lambda = \frac{\omega R}{V}. \tag{4.8}$$

By plotting the turbine characteristics against the tip speed ratio (Figure 4.7), it can be seen that C_p, C_T and C_θ are independent of the size of the rotor or the upstream velocity.

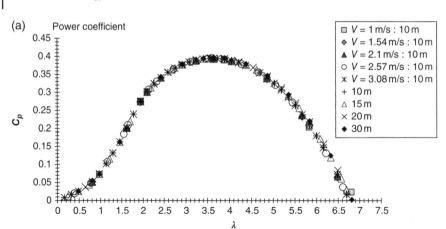

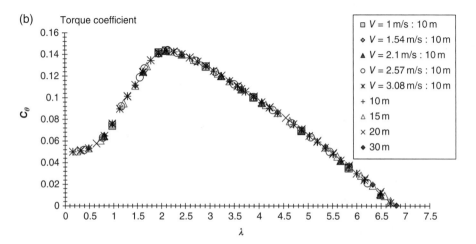

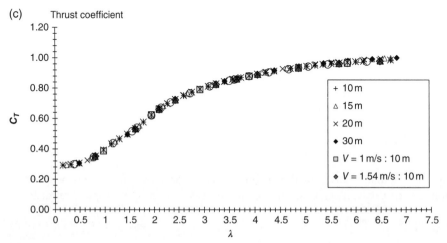

Figure 4.7 Turbine characteristics for different size turbines ($\phi = 10$–30 m) and different upstream velocities (1–3.08 m s^{-1}).

This only applies for Reynolds independent flows, and although the critical Reynolds number is dependent on the actual design, a number of researchers have shown that it is of the order of 10^6 [44, 45]. This can be seen in the set of results obtained for the Cardiff Tidal Turbine, shown in Figure 4.7. Although these results are based on a uniform upstream velocity, the same holds true when a profile flow is considered, as long as the upstream velocity is determined from the volumetric averaged velocity over the swept area.

Solidity (σ) is another important design parameter that must be considered. It is defined as the ratio of the total blade area to the swept area of the rotor. Solidity is therefore a function of the averaged chord length (c) and the number of blades (B) and was defined by Duquette and Visser [46] as

$$\sigma = \frac{Bc}{\pi R}. \tag{4.9}$$

The optimum number of blades is dependent on both the economics of blade construction and materials when compared to the energy increase with increasing blade number [47], [48]. It has been shown that the peak power increases with the number of blades and that the tip speed ratio at which peak power occurs decreases, corresponding with a decrease in the operating range for increasing number of blades. However there is a diminishing return with the increase in number of blades, as illustrated in Figure 4.8.

The pitch angle (β) is defined as the angle between the chord of the blade and the normal to the rotational axis of the turbine hub. It is important that the optimal pitch angle is defined, as the pitch angle has a direct effect on the turbine performance characteristics, as shown in Figure 4.9. If a turbine is designed to optimise power output, it can be seen from these curves that the optimal pitch angle is 6°. However, this pitch angle, while not leading to the highest values of C_T and C_θ, does not produce the lowest values either, indicating that there has to be a compromise between designing for peak power, for minimal thrust loads and for the required start-up torque. For example, the pitch angles of 9° and 12° give higher start-up torques, which would allow the turbine to operate in slower-moving water.

The power output from a turbine will vary with the free stream velocity. At very low velocities, the blades will not rotate since insufficient torque is generated until a critical velocity, the cut-in speed, is reached when the blades will start to rotate and power is generated. Under some circumstances, the turbine may require a small electrical input to overcome start-up torque to instigate rotation. As the tidal velocity increases further, the power rises. However, this is limited by the electrical generator such that a turbine has a maximum power that it can generate, known as the rated power output, which corresponds to the rated tidal velocity. If the tidal velocity increases above this rated velocity, then the design of the turbine must limit the power that can be generated. This is normally achieved by adjusting the blade angle (feathering the blade) to keep the power below the required threshold. However, a further threshold is reached when the forces on the turbine exceed the design specification and there is a risk of structural damage to the rotor. At this stage, the rotor has to be brought to a standstill and the corresponding tidal velocity is known as the cut-out speed (Figure 4.10).

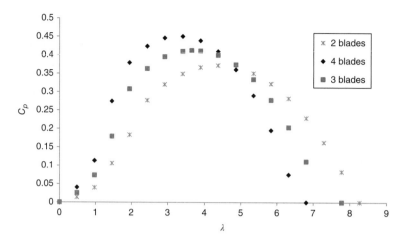

Figure 4.8 Variation in C_p with number of blades for the Cardiff Tidal Turbine.

(a) Power coefficient

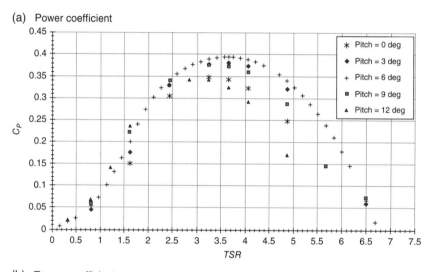

(b) Torque coefficient

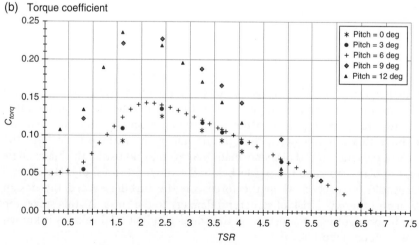

Figure 4.9 Effects of blade pitch angle on turbine characteristics for the Cardiff Tidal Turbine.

(c) Thrust coefficient

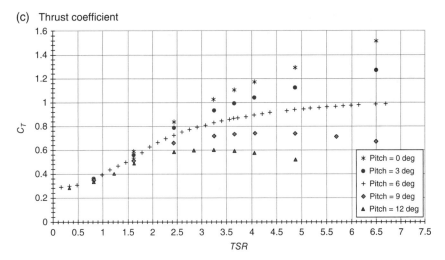

Figure 4.9 (Continued)

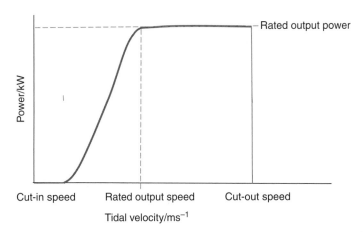

Figure 4.10 Rated turbine characteristics.

4.6.3 Cavitation

If the local pressure on the hydrofoil falls below the vapour pressure of the fluid, cavitation, that is, the formation of the vapour phase within the water, can form in the low-pressure area and move towards the blade's trailing edge where the bubbles that are formed suddenly collapse as the pressure increases. This changes the lift and drag characteristics of the blade and increases noise. The effect of the collapsing can also cause permanent damage to the structure of the blade. For a given blade design the blade pitch angle must be altered to avoid cavitation issues to ensure that the pressure distribution on the blade does not lead to cavitation.

The relative pressure throughout the flow can be defined by the pressure coefficient (C_{pr}), which for incompressible flows is given by

$$C_{pr} = \frac{p_L - p_o}{0.5\rho U^2}.$$
(4.10)

where p_L is the local pressure, at the point the coefficient is being calculated, and p_o is the free stream pressure. Therefore cavitation will occur when C_{pr} is less than the critical value, defined when the local pressure equals the vapour pressure, p_v.

Cavitation is normally characterised by the cavitation number (σ), which is a special form of the Euler number, and is defined as

$$\sigma = \frac{p_o - p_v}{0.5\rho U^2}.$$
(4.11)

Since σ is constant for a given depth and freestream velocity, a simple criterion based upon σ and C_{pr} can be defined such that cavitation will occur when $\sigma \leq -C_{pr}$. Therefore for a given angle of attack it is possible to predict if and where cavitation will occur [49].

4.6.3.1 Shaft Design

As previously discussed, tidal turbines extract power from tidal streams through the formation of hydrodynamic lift and drag forces, creating a torque and thrust load. The axial thrust loads on each blade give rise to a bending moment about the x and y direction from the rotational axis of the turbine. To determine the actual bending moment on the shaft, the resultant bending moment, $M = \sqrt{M_x^2 + M_y^2}$, is calculated. For any given flow condition, M varies with the angular position of the blades and whether there is a stanchion or not (Figure 4.11).

For a tethered design, that is, when there is no stanchion present, the resultant bending moment is negligible compared to when there is a stanchion present. In addition, the resultant bending moment and its associated direction depend on the clearance distance between the rotor and the stanchion, and whether the stanchion is upstream or downstream of the rotor [50]. The direction is given relative to the y direction, with positive defined in a clockwise sense.

The presence of such significant fluctuating loads is undesirable due to its inevitable impact on the wear of the turbine parts. Avoiding the transfer of these loads to the power take-off system will be important to avoid fatigue issues and fluctuations in the power generation. Bearings along the drive shaft of the turbine will also suffer significant fatigue issues from such loading, resulting in shorter maintenance periods and increasing the cost per kilowatt-hour on the turbine.

4.6.3.2 Whirling of Shafts

The centroidal axis of the shaft is designed to coincide with the axis of rotation. However, small imperfections during manufacture or asymmetric design details that are not balanced will lead to eccentricity between the centroidal axis of the shaft and the axis of rotation. As the turbine rotates and the angular velocity of the rotor increases, the inertia effects will increase the eccentricity.

This eccentricity causes the shaft to deflect, creating a lateral vibration. When the rotational speed coincides with the natural frequency of these lateral vibrations the

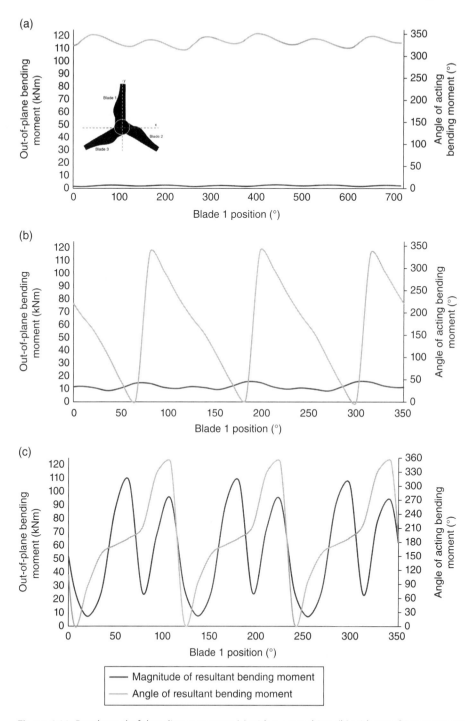

Figure 4.11 Resultant shaft bending moments: (a) with no stanchion; (b) with stanchion upstream; and (c) with stanchion downstream.

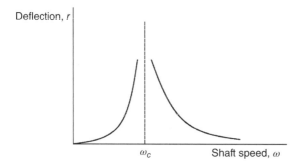

Figure 4.12 Whirl characteristics of a shaft.

shaft deflection increases significantly (Figure 4.12). The rotational speed of the shaft is known as the critical (whirling) speed and should be avoided in practice. The critical speed is dependent on the shaft dimensions, the loads acting on the shaft, the mass of the shaft and any associated parts, the stiffness of the shaft, the magnitude and location of the unbalance, the bearings used to support the shaft.

4.7 Types of Devices

A clear advantage of TSTs is that they can be sized to suit the requirements of the local environment (coastal restrictions, tidal flow, tidal range, seabed topography, etc.), and can be placed on either an individual or 'farm' configuration. As such, no large civil works are required and this method would therefore be less disruptive to wildlife, marine activity (and possibly the coastline) and would not present a significant barrier to water transport. It has been stated that the ideal site for a TST is within 1 km of the shoreline and at a depth of 20–30 m [52]. The ideal tidal speed is 2–3 m s^{-1} (4–6 knots). However, as the loading is due to the square of the tidal speed, higher speeds could lead to blade loading problems. The Pentland Firth, which will play a significant part in testing these figures, has local tidal velocities that can reach 4 m s^{-1}. Moreover, since around 53–58% of the UK tidal stream resource is estimated to come from the Pentland region [25, 52], designing marine devices to operate at tidal velocities of these magnitudes may be an absolute requirement if the full resource is to be taken advantage of.

Although there are several devices under development, most designs can be categorised based on whether they produce a rotational or linear motion, the direction of the rotational axis or linear motion and the inclusion of any flow acceleration mechanism. The main categories into which most devices fall are therefore horizontal-axis tidal turbines (HATTs), vertical-axis tidal turbines (VATTs), venturi effect devices and oscillating hydrofoils, though other novel designs such as 'kites' can be found.

4.7.1 The Horizontal-Axis Turbine

The main feature which identifies a HATT is that the rotational axis of the turbine is parallel to the tidal flow. There are many forms of HATTs, differing on how many turbine blades they have and how they are fixed.

MCT's SeaGen (Figure 4.2a) was installed in Strangford Lough, Northern Ireland, in 2008 as the first commercial-scale TST to generate in UK waters [53]. The device was a mono-piled structure with two 16 m diameter, twin-bladed rotors powering two 600 kW generators at a rated flow of 2.4 m s^{-1} [52]. The horizontal arm which carries the generators can be raised and lowered on the pile, allowing the turbines to be raised out of the water for maintenance or protection. Having proven the technology, MCT continues to develop it and has developed a 2 MW rated device which has a rotor diameter of 20 m. It is now focusing on the first tidal array in the UK and on jointly developing a floating tidal current turbine with Bluewater Energy Services BV and Minas Energy for installation in the Bay of Fundy, Canada.

The device from SMD, known as the TidEl, is a 1 : 10 scale prototype on which tests have been carried out. The device is built up of two contra-rotating rotors/generators attached together by means of a cross-beam. The whole unit is buoyant and kept fully submerged in waters more than 30 m deep using a flexible mooring system. This arrangement is believed to allow the device to align itself with the tidal flow. The full-scale prototype is planned to have a rated power of 1 MW [50].

Open Hydro have developed a unique turbine assembly called the Open Centre Turbine. As can be seen in Figure 4.13, the rotor consists of a circular arrangement of turbine blades which form a hollow centre which is designed to be friendlier to marine life. The Open Centre Turbine is a simple design which uses a direct drive generator system. A prototype was installed at the European Marine Energy Centre (EMEC) in Orkney in 2006; other devices have since been installed at the Bay of Fundy [54]. Tidal Generation Ltd (TGL) is another company which in 2010 deployed a 500 kW device, Deep Gen, at EMEC and has supplied over 100 MWh to the Scottish grid [55]. Another promising device is the Delta Stream from Tidal Energy Ltd which consists of three turbines mounted onto a structure; the prototype of a single 400 kW rated device was successfully deployed at Ramsey Sound in Pembrokeshire in 2015 [56]. The Open Centre Turbine, Deep Gen and also the Delta Stream are secured using a gravity-mounted structure.

Figure 4.13 OpenHydro's Open Centre Turbine [54].

Figure 4.14 CoRMat rotor concept [58].

Nautricity has developed a HATT which removes the use of the large structural components to support the turbine. CoRMat utilises two different sets of rotors: three blades at the front with four blades immediately behind (Figure 4.14). The two rotors are contra-rotating, such that the torque is divided between the rotors and balanced. The relative angular velocity drives a generator, without the need for a gearbox to obtain the required speeds. The resultant drag ensures that the device remains aligned to the flow. A second-generation 500 kW turbine was tested at EMEC in 2014 [57].

4.7.2 The Vertical-Axis Tidal Turbine

VATTs extract energy from the flow in a similar way to HATTs, but the rotational axis of the turbine is perpendicular to the tidal flow [59]. There are very few VATTs in development, compared to HATTs. The main advantage of a VATT is that it can operate regardless of the direction of tidal flow without loss of operational efficiency and without the need for any pitch or yaw mechanism to rotate the blades or rotor [60]. A VATT can have straight blades, reducing the design and manufacturing costs compared with the more complex HATT blades [61]. In addition, since the rotational velocity of VATTs tends to be lower, they produce less noise [62] and pose a reduced risk of collision [57] which may be beneficial for marine life. However, a reduction in noise may result in marine life coming closer to the device, which would counteract the reduced collision risk due to lower blade velocity. The disadvantages of VATTs include lower efficiency [55], with typical peak values of around 37–40% [54, 63], although Eriksson *et al.* [57], state that the lower efficiencies could be a result of more research based on horizontal turbines than on vertical turbines in the wind industry. In addition, the efficiency of VATTs means that the blades will have a limited period when they are providing suitable lift/drag characteristics. The rotational axis also means a blade will be subjected to the wake of a upstream blade. Other disadvantages of VATTs include

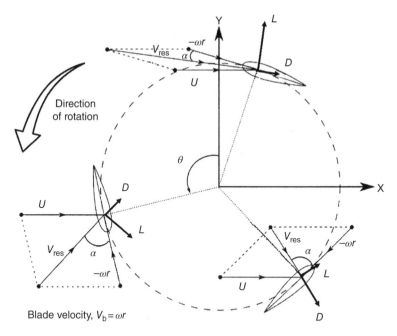

Figure 4.15 Lift and drag diagram for a three-bladed VATT.

the low starting torque, meaning that unlike most HATTs they may need a starting mechanism [58], and the torque ripple due to the changing angle of attack through the rotation cycle [57].

VATTs can be divided into two main groups depending on the blade design. The first type is a straight bladed VATT, for example the Kobold Turbine, and the second type has helically shaped blades, an example of which is the Gorlov Turbine).

VATTs are characterised using the same dimensionless groups as HATTs. However, for VATTs, the angle of attack changes as the blades rotate. Figure 4.15 shows the forces and relative velocities at three different blade rotational positions. This clearly shows that while the radial distance to a blade remains constant, the tangential force and hence the torque and power will vary with the angle of attack.

4.8 Oscillating Hydrofoils

These devices use an oscillating motion rather than a rotary motion and can consist of a hydrofoil which is mounted on a pivoting arm. The tidal flow over the hydrofoil results in the arm reciprocating back and forth in either the vertical or horizontal direction due to the hydrodynamic lift and drag forces on the wing.

One example of an oscillating hydrofoil is the Stingray developed by the Engineering Business Ltd [64]. The device is attached to the seabed and reciprocates in a vertical motion powering the pivoting arm, and hydraulic cylinders attached to the arm pressurise fluid which is used to power a hydraulic motor which turns a generator. A 150 kW device was tested in the Yell Sound, Shetland, in 2002 [65].

The bioSTREAM is another example of an oscillating hydrofoil [66]. This device oscillates in a horizontal motion; the hydrofoil resembles a fish tail and mimics its movement while in operation. The bioSTREAM is currently being developed by Bio Power Systems which has been extensively tank-testing a 1 : 15 scale model. The full-scale design will incorporate the company's O-Drive power take-off module which works through the use of hydraulics. The design allows the device to adapt to tidal flows in any direction [67].

4.9 Venturi Effect Devices

There are two general forms of venturi effect devices. The first is essentially a HATT consisting of a funnel-like duct which surrounds the turbine, concentrating the flow through the device and increasing the stream velocity. The second, less common form consists only of a venturi duct and operates utilising the pressure difference caused within the duct. As the flow passes through the throat of the duct the pressure of the fluid decreases; this drop in pressure can be used to draw a secondary fluid through a separate turbine and then into the flow.

An example of a device which uses the latter approach is the Spectral Marine Energy Converter [68]. This device replaces the air used in the Rochester Venturi with water. The flow through the device induces a pressure differential which causes a secondary, higher-velocity flow of water which is pulled through an ordinary axial turbine [69].

An example of a venturi ducted turbine is the Rotech Tidal Turbine [70]. The turbine is a bi-directional HATT which is encased in its own module; this allows the turbine to be raised to the surface for maintenance without having to retrieve the whole structure. The device is moored to the seabed and is fully submerged; power is transferred from the rotor to the generator through the use of a hydraulic system. The rotor turns a fixed-displacement hydraulic pump which then feeds two variable-displacement motors which are attached to the generator [71].

A second device which is similar in design is the Tidal Turbine from Alstom [72]. This device can be mounted either on the seabed or on a floating structure and also incorporates a venturi duct to increase flow through the turbine.

4.10 Other Devices

Some tidal stream devices use a unique technology or design and so cannot be placed into any of the previous categories.

A more recent design has been developed by engineers at Oxford University. The team claims that the transverse horizontal-axis water turbine (THAWT;, Figure 4.16) is more robust, efficient and cheaper to build and maintain than anything currently in operation. The device is essentially a cylindrical rotor that rotates around its horizontal axis, supported by two concrete posts at either end. Like the VATT, the THAWT can generate energy in both flow directions, with an estimated power generation of 12 MW from a 10 m diameter and 60 m long rotor [73] and with an efficiency significantly greater than the Lanchester–Betz limit [74].

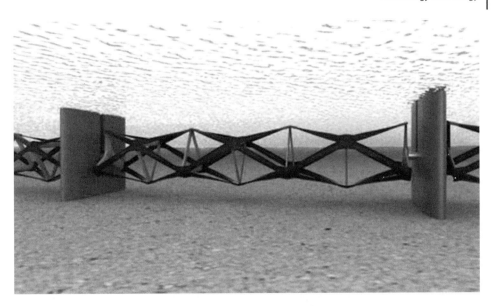

Figure 4.16 Oxford University's THAWT (*source:* Kepler Energy Ltd).

Figure 4.17 Deep Green device [75].

The Norwegian Flumill Power Tower consists of two helix screws mounted vertically and parallel to each other. The buoyant device is attached to the seabed using a flexible mooring system. The corkscrew turbines are directly linked to a permanent magnet generator and can generate in tidal flows as little as $0.5\,\mathrm{m\,s^{-1}}$ [52]. The company has deployed a 600 kW pilot device in the EMEC test facility [58].

Finally, the Deep Green developed by Minesto offers a unique approach since it is essentially a wing with a turbine attached to the lower half (Figure 4.17). The device is tethered to the seabed and moves through the water in a similar fashion to a kite; this is due to the hydrodynamic lift created by the flow of water over the wing. The movement through the water is controlled via a rudder and servo system. The Deep Green has the advantage of being able to operate in low-velocity flows as the device can move up to ten times the speed of the current; which creates sufficient flow through the turbine which is directly linked to a generator located in the nacelle [75].

4.11 Computational Fluid Dynamics

The hydrodynamic conditions around a tidal device are fundamental to its design and can be modelled using computational fluid dynamics (CFD). To predict the performance characteristics of a marine device, it is important that these hydrodynamic conditions, along with geometric features such as the rotor blades, nacelle and support structure, are modelled during the design phase. The data obtained from a CFD analysis can be used to predict kinetic energy extraction and structural forces when the device is subjected to rectilinear or profiled flow and, if appropriate, surface waves. Modelling surface waves becomes an important factor if their wavelength is large enough, relative to the depth of the turbine. The penetration depth of a surface wave is half its wavelength and therefore can interact with turbine blades and introduce significant asymmetric blade loading. In the case of wave devices, waves are the desired effect with energy extracted from the oscillatory motion. For a TST waves may pose a considerable risk via extreme loading and reduced energy extraction. Combined with flow and wave characteristics, the position of a tidal turbine in the water column can also be changed in CFD to investigate torque, power and axial load. Data relating to torque, power and loads can be generated quasi-statically or through a number of rotation or oscillation cycles; this latter option is of interest if a transient study is required where temporal changes in flow are a key factor. However, before such data are calculated from a CFD model there are three fundamental steps required. The first of these is to develop the device's geometry. The complexity of the geometry will depend on the parameters to be studied and will include decisions on whether the CFD model should include a support structure or not, such as a stanchion or seabed mounted frame for a TST as shown in Figure 4.18a.

To simplify the discussion, for this brief introduction, only the geometric requirements of a horizontal-axis TST will be discussed. However, similar procedures will be required for other device types. For a TST, it is vital that the support structure is considered early

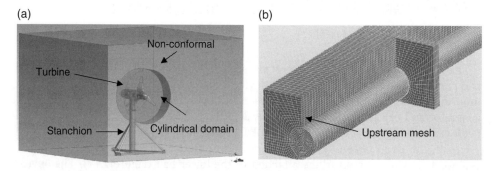

(a)   (b)

Figure 4.18 (a) Turbine geometry and (b) example mesh

Figure 4.19 Turbine upstream of stanchion.

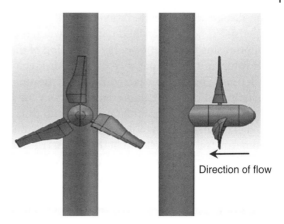

Direction of flow

Figure 4.20 Turbine downstream of stanchion.

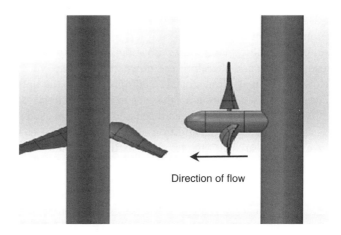

Direction of flow

on in the geometric design, as it will have a significant effect on the turbine's hydrodynamic performance. The key geometric features for a TST are the blade design, the distance between the turbine rotor and the support structure, and the projected area of the stanchion. The latter point is important if the turbine is to operate during flood and ebb tides and will have a significant effect on the turbine's hydrodynamic performance if a fixed or limited rotor pitch and nacelle yaw are used (Figures 4.19 and 4.20) [50].

Although the optimisation of both the kinetic energy extraction and load reduction is an iterative design sequence, it is important that the turbine's geometric features are reasonably approximated at the outset and represent the proposed design as closely as possible; this will give a good starting point but will depend on the experience of the designer. The design of the turbine blade can be optimised using CFD; here geometric features such as the blade profile, chord length, blade length, twist, angle of attack and number of blades are considered. The latter design criteria are applied following the selection of a profile from a NACA library (there are numerous validated profile sections to choose from), or from a profile designed using blade element momentum theory (BEMT), for example. BEMT will be briefly discussed later. For the purposes of this introduction, it will assumed that the blade profiles are known.

The design of the blade is again an iterative process and can be computationally expensive if not approached correctly when using CFD.

Optimising the kinetic energy extraction of the rotor blades is achieved by separating the rotor and hub from any other hydrodynamic interactions, such as the support structure. This somewhat contradicts the opening statement of this section about including all the structural components in the flow field; however, there are limits in mesh size and therefore the numerical run time which in turn is a function of the CPU used. The mesh size will be discussed later, but it is important to note here that the geometric features included in the geometric design phase will affect the mesh later and ultimately the solution and results. By separating the rotor and hub from the support structure the blade profile, chord length, blade length, twist, attack angle and number of blades can be theoretically optimised against the Betz limit, while minimising the cell count or mesh size. Once the kinetic energy extraction of the rotor is optimised, other geometric features such as the nacelle and support structure are then added and their influence on power output and loads calculated.

Using this methodology, both the rotor and its hydrodynamic relationship with its support structure are optimised. Using the CFD method to optimise the rotor and support structure has another advantage in that the flow field around the structures is also available and therefore the downstream wake and upstream effects are detailed. However, there is often a need to compromise on the optimal mesh density as detailed in the later discussion on mesh size and computational expense.

Two CFD models have been discussed so far, one used to optimise the blade and hub design and one with the addition of a support structure. These two models are used to investigate near-field effects, whereas the wake and upstream effects are considered far-field and in most cases would require a third CFD model. Far-field models require a higher cell or mesh density downstream of the rotor and support structure to resolve the local velocity and pressure gradients. Finer far-field meshes are crucial to capture features in the wake such as swirl and velocity deficit as the wake expands. Parameters such as these are important in the study of device arrays and the larger effects of arrays on local coastlines and sediment transfer.

So far, it has been established that for a TST there is a potential for at least three CFD models, excluding arrays and larger coastal effects. To create the CFD model a mesh is required to discretise these geometries. The discretised geometries are created in a number of volumes or domains. As an example, the tidal turbine CFD model is made of a rectangular domain to represent the tidal stream with velocity inlet and pressure outlet boundaries at either end, the set-up of these depending on the approach and methodology. An axial aligned cylindrical domain is created from which the turbine volume is subtracted, leaving a cavity in the shape of the rotor and hub.

The cylindrical domain is then subtracted from the rectangular domain to form a volume that can be numerically rotated, and can have a non-conformal interface between the sea and turbine volumes (Figure 4.21). The non-conformal interface is simply two faces that occupy the same location in space allowing the turbine volume to rotate since the surface elements at the interface do not share common nodes. This approach also allows transient studies to be run. If only the kinetic energy extraction is required a steady-state analysis neglecting the non-conformal interface is possible; however, if this is to be combined with wake studies, blade tower interaction and the recovery of the wake velocity to upstream conditions then a transient solution is necessary.

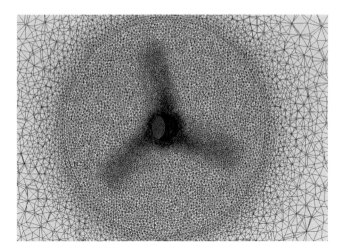

Figure 4.21 Tetrahedral meshed rotating domain and quadrilateral channel mesh with non-conformal interface.

The cylindrical and rectangular domains or volumes are meshed with 'mesh controls' to control cell density. The meshing method will be subject to the geometry and the meshing tools available. The cell count in the sea channel, for example, will depend on the purpose of the model, the hydrodynamic performance of the rotor or the near- or far-field wake characteristics, as previously discussed. The same approach is used for the other tidal devices by simply breaking down their geometry into a number of volumes and subtracting it from the volume of the main flow field. The details of these procedures are well beyond the level of this brief introduction and are not limited to one methodology, so will not be discussed further. The point for this introduction is to outline the process and to consider the purpose of the model so as not to introduce excessive cell count and needlessly increase the numerical runtime. Once a suitable mesh is established, at this stage via the experience of the analyst, it is necessary to conduct grid dependency checks, by increasing or decreasing the mesh density, for example on the blades or within the turbine domain or a portion of the rectangular domain that encompasses the far-field wake expansion downstream of the rotor. This process continues until performance or a monitored flow characteristic shows no change with mesh density. Once the mesh and cell count is established and is suitable for the model, boundary conditions are applied.

The velocity inlet and outlet boundaries have already been introduced, but a boundary to the side walls of the rectangular domain also needs to be applied. The sides and top walls can be represented by a frictionless surface so that a velocity profile is not formed during the analysis. For the floor or seabed of the rectangular domain a no-slip boundary constraint is used so that a near-wall boundary constraint is applied; this is typically called the 'law of the wall' and is well documented in the literature. This boundary condition allows a velocity profile to develop near the boundary. To limit the computational expense of CFD models, Reynolds averaged Navier–Stokes (RANS) equations are used to relate the Reynolds stresses to the mean velocity gradients. To close the RANS equations a turbulence model is applied. There are a number of turbulence models available in most commercial CFD codes, with the k–ε, k–ω, Reynolds stress and shear stress transport models being some of those typically used, although there are other turbulence models available [76]. RANS models have been shown to correlate well with measured data on kinetic energy extraction and the performance of the turbine [77]; however, RANS viscous models tend to be weaker in predicting wake characteristics such as velocity deficits. For wakes, large eddy simulation (LES) or hybrid RANS–LES models can be used to estimate the combined kinetic performance and wake features; for the latter computational expense is of major concern. The theory behind viscous models is ubiquitous in literature and therefore will not be discussed further here, but is mentioned as a point of note for the reader. In order to maintain an economical computational model, a free surface interaction between the water and air is not typically included; however, if the interaction with waves becomes a necessity then a free surface will be required. As previously discussed, surface waves penetrate the water column to a depth of half the surface wavelength and so have a potential to influence the turbine's performance.

4.11.1 Finite-Element Analysis and Fluid–Structure Interaction

As discussed, CFD is used to model the hydrodynamic performance of the tidal turbine and to establish the forces on its structure. Knowledge of these forces is essential in the design and development of the turbine and its components, such as the driveshaft, gearbox, bearings and seals. To investigate the effect of these forces on the structure, numerical procedures such as finite-element analysis (FEA) can be used. FEA uses a grid of elements, with each element's behaviour being characterised by a specific material stiffness that is has been established via material test procedures. Many of these test procedures follow international standards but can be bespoke for certain new or more challenging materials.

Most of the numerical challenges experienced when using FEA involve numerical accuracy and stability, with each of these being effected by mesh quality and density. The elements are connected in a grid via nodes that form a continuum where displacement, stress and strain can be calculated. If an accurate representation of the geometry is achieved then it is possible to reliably predict reaction forces, displacements, stress and strain since the shape of the blade is defined by the shape and location of the hydrofoil sections along the length of the blade and hence the lift and drag forces that result.

In the design of a tidal turbine the magnitude of the lift and drag forces are important to establish the kinetic energy extraction which is transferred into torque through the driveshaft. Not all the kinetic energy in the flow field, however, can be converted into shaft power, which is the product of the shaft torque and rotational speed of the driveshaft. The remaining drag forces from the blade, support structure and nacelle play an important role in establishing the structural integrity and safety limits. Like CFD, the numerical methods and procedures behind the development of the finite-element method are well documented [78, 79] and therefore will not be discussed in depth here. The main objective for this discussion is to outline its application to the design of a tidal turbine and how it can be used in conjunction with CFD to develop a first-order design. However, it is not limited to this portion of the development stage and can be used at more advanced stages. The basic principle of FEA is to discretise the turbine volume by dividing it into smaller parts or elements. In the CFD case this involves discretising the fluid domain, whereas in the FEA model the structure of the blade, stanchion and nacelle or internal components, such as the driveshaft and gears, are discretised and modelled with structural material properties.

These properties can be linear, plastic or hyperelastic where constitutive models are used to characterise the behaviour of the material. Material properties are typically characterised by the elastic modulus, Poisson's ratio and elastic/plastic stress strain data, which are derived from material test procedures. There is an added complexity to the characterisation of the turbine blades as their design may consist of composites where the directionality of the material lay-up must be specified along with its material properties.

FEA combined with CFD forms a powerful tool in the design and performance characteristics of TSTs to limit development costs and to experiment with new materials. The interaction between the hydrodynamic force of the fluid and its effect on a structure, in most cases, is a one-way effect and can be treated has a weak coupled system. A weak coupling here refers to the magnitude of deformation. If the deformation of the structure is significant then its new position will have an effect on the hydrodynamics which, for example, will reduce the drag forces which would allow the turbine rotor to deflect less. This has the potential to set up oscillatory motion and the risk of increased fatigue loading. By transferring load and displacement data between CFD and FEA, it is possible to account for this type of strongly coupled systems with fluid–structure interaction (FSI). Via system coupling, FSI techniques allow the hydrodynamic loads, calculated in the CFD analysis, to be transferred to the turbine blades in the FEA model. The loads are mapped on to the structural mesh of the turbine blade, and deformation occurs in relation to the material properties applied. The new node positions for the FEA mesh are then transferred, again via system coupling, to the CFD model where the new positions are mapped on the fluids mesh. This process continues until a stable solution or until the desired time step is reached. FSI modelling can be computationally expensive and will require high-performance computing in many cases, such as modelling a TST, but if there is a strong coupling (i.e. significant deformation in the structure) then FSI is will be a powerful tool in the design process and a requirement for reliable fatigue analysis.

4.11.2   Blade Element Momentum Theory

BEMT can be used to design or to analyse an existing rotor design performance. It uses two methodologies to establish the operational performance of the rotating blades. By combining lift (C_L) and drag coefficients (C_D) and a momentum balance, on r multiple rotating stream tubes, BEMT is used to study parameters such as the optimum blade pitch angle and associated tip speed ratio (λ). There are many documents that develop BEMT in great depth such as [80], and more specifically to TSTs in a 2015 paper by Masters *et al.* [81], therefore it will not be covered in detail here; however, a brief outline of the theory will be discussed in the context of a design tool for turbines.

There are two fundamental assumptions applied with the use of BEMT. The first of these is that there are no dynamic interactions between each of the turbine blades or indeed the blade sections, such as those imposed by wake vortices. The second is that the lift and drag forces on each of the HATT blades are determined from the lift and drag coefficients and can be found from standard profiles in published data, such as NACA. For a performance-based study using BEMT the power coefficient C_p can be estimated using an iterative approach to solve equations relating to the axial and rotational flow. Hydrodynamic loads can also be estimated and transferred to FEA. The limitation using BEMT is a lack of information on the flow field, but this can be improved via other approaches such as panel methods.

The development costs for TSTs are high and therefore numerical methods such as CFD, FEA and BEMT have a significant role to play in its continued development and future advancements. The main limitations are with computational expense and analyst experience. These smaller obstacles are not, however, insurmountable with effort and time.

4.12   Security, Installation and Maintenance

To allow the operation of the HATT there must be a means of fixing the turbine at some depth through the water column. The means by which the turbine is attached will greatly depend on the depth of the water and proximity to the nearest onshore service location [82]. Given the results from acoustic Doppler current profiler measurement the position of the rotational axis of the turbine will ideally be placed at a depth with minimum shear and higher velocity band. Any tidal energy device will need to be secured through one of the methods outlined below. The first two methods are suited to depths between 30 and 40 m. Beyond this depth it is likely that floating systems would be employed.

Gravity base. This is where the device is attached to a weighted structure which remains in place on the seabed through the effects of gravity. Some devices attach to pre-laid foundations usually constructed from concrete [83].

Pile. This is where large poles which are either driven or drilled into the seabed; devices can be mounted to either single or multiple piles, depending on the system. The main limiting factor with piles is the depth at which they can be installed: current wind farm installation techniques work to depths of around 50 m. However, the maximum theoretical depth is thought to be 100 m, though this raises the question whether this would be economically viable [84].

Flexible and rigid mooring. Mooring systems are generally used for floating or buoyant devices. Flexible mooring systems consist of a tether in the form of chains, cables or ropes which secure the device to the seabed by means of some form of anchor; this allows some movement of the device, mainly in rotation. This form of mooring is very common with devices which require alignment to interact with oncoming waves or tidal flows such as attenuators or buoyant HATTs. Rigid mooring systems hold the device in a fixed position and do not allow for any movement. Mooring systems can consist of multiple configurations, including single point, spread and dynamic. There are further subcategories but these are not covered in this chapter; further information is available via [85].

The first three seabed fixing methods are suited to depths between 30 m and 40 m. At depths greater than 50 m the application of a stanchion as a seabed fixing method becomes more difficult [86]. Another fixing method is via the last of those discussed, the buoyant tethered system. As suggested by [87], this system has the advantage of positioning the device higher in the water column without introducing excessive turning moments such as those imposed on a pile or stanchion fixed into the seabed. The tethered device could be attached to the seabed using sinks and floatation devices at the water surface. Alternatively ballast could be applied to the HATT assembly with neutral buoyancy, allowing the assembly float at a predefined depth. It has been suggested that using this type of methodology would have a number of potential advantages over the more conventional stanchion/pile design. One such advantage is the elimination of the torque moment transmitted to the support structure, thereby reducing the height from the seabed at which the device could potentially operate [88]. However, the method used to tether a floating HATT will need to limit the amount of yaw induced from reactive torque generated by the turbine blades during power extraction. In an attempt to overcome this problem Clarke *et al.* proposed the use of a contra-rotating turbine design where the reactive torque of the rear turbine corrects the yaw returning alignment and rectilinear flow [89]. It also allows a simple and economic mooring system for deep water applications. A further advantage of the system included the reduction of stable wake vertical elements in the wake of the HATT. The latter feature may have implications when considering the spacing of an array of tethered HATTs. Some of the problems associated with tethered systems were also covered by Snodin, who stated that a location for the turbine assembly itself but also for the transmission of power to the shore via the power cables [90]. Unlike seabed fixed devices, the power cables would have to be flexible to account for the induced movements.

When considering an array of HATTs the comparative costs between tethered and piled fixing methods were summarised in a 2001 study undertaken to investigate the installation of a 3–5 m diameter HATT in 20 m water depth at fixed navigation marks. Capital costs of £400 000 and £600 000 for a tethered and fixed installation were given, respectively [91]. From information reported by Previsic [92] and summarised by Clarke *et al.* [93] on the estimated breakdown of cost for a farm of pile-mounted tidal turbines, 33% of the estimated cost was in the structural steel elements alone and a further 16% for the turbine installation. Apart from the power conversion system at 35%, the stanchion mounting system made up a considerable proportion of the overall cost. This then builds an economic case for free-floating systems that can be tethered to the seabed. However, it is likely that for the early development stages of the technology the use of pile-driven stanchions will remain in depths between 20 m and 40 m. Black and Veatch [25] suggested that "UK technology development should be concentrated on devices that are suitable for sites of depth >40 m, with the highest focus on devices that are suitable for deep sites with high velocities". The position of the device through the water column, however, will still depend on the accompanying velocity profiles. As stated, pile-driven stanchions would not be suitable at these depths due to large turning moments as the turbine is positioned further up the water column. Companies such as TGL had already considered designs that operate at greater depths. These devices would have to be either floated and tethered at some defined depth or fixed closer to the seabed using a modular frame. Given the depth of operation, the latter option could potentially place the rotor in a high-velocity shear rate band. Given the high current velocities at depth, a case could be made for studying the performance and blade loading at these depths and shear rates. By taking a peak Vmsp of $5.5\,\mathrm{m\,s^{-1}}$ and applying the 1/7 power law at a depth from the seabed comparable with that of a turbine positioned at mid-depth of a 40 m location, where the peak velocities are of the order of $3.5\,\mathrm{m\,s^{-1}}$, it can be shown that the velocity at the rotational centre of the 60 m total depth turbine is still greater than that for the 40 m depth. For example, at a depth of 60 m and Vmsp = $5.5\,\mathrm{m\,s^{-1}}$ the velocity a rotational depth of 40 m gives

$$V_{\mathrm{deep}} = 5.5 \left(\frac{20}{60} \right)^{1/7} = 4.7 \mathrm{m\,s^{-1}}. \tag{4.12}$$

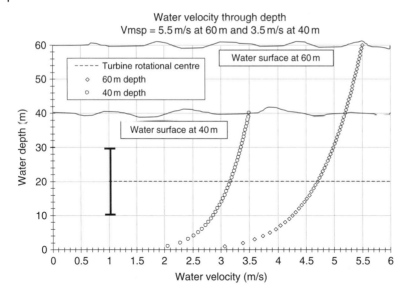

Figure 4.22 Velocity attenuation with depth using the 1/7 law.

At a depth of 40 m and a Vmsp = 3.5 m s^{-1} the velocity a rotational depth of 40 m gives

$$V_{\text{mid-depth}} = 3.5 \left(\frac{20}{40} \right)^{1/7} = 3.2 \, \text{m s}^{-1}. \tag{4.13}$$

Figure 4.22 shows the velocity profiles at the 60 m and 40 m depths with the location of the rotational axis of a 20 m turbine positioned 20 m above the seabed. The 40 m depth scenario matches the approximate conditions of such devices such as the SeaGen and SeaFlow projects.

If operational parameters are understood at these depths, power extraction under high-velocity shear within the lower 25% of the velocity profile still has a significant contribution to make to the overall resource. A number of developers such as OpenHydro have proposed deep-water designs that involve the use of modular frames consisting of a single or an array of turbines that could be sunk to the ocean floor. However, with velocity rates approaching 4.7 m s^{-1} at depths of 50 m and greater, issues associated with axial thrust loads are of greater importance due to the high inaccessibility. In the case of an array of devices the combined axial loads from the turbines and frame could be significant.

Although Black and Veatch were referring to power extraction higher in the water column when they stated that "There appears to be little merit in focusing UK technology development on devices that are only suitable for sites of depth <30 m," a case still may still exist for seabed fixed designs. The implications of high axial loads on a modular frame are covered in the study by Mason-Jones *et al.* when subjected to an upstream profiled velocity field and structural support [94].

The installation procedures of devices need to be well planned so they can be executed without any faults; poorly installed devices could cost millions of pounds. An example of this is when an offshore wind company found that 164 poorly installed turbines were moving on their foundations; they estimated the cost to be around £13 million to rectify the problem [95]. There is potential for the marine industry to gain valuable knowledge from the experiences of the wind industry.

In order for marine energy devices to be economically viable they require lengthy life spans. This then leads to the requirement of maintenance which can be a costly procedure for offshore devices especially, due to the remote locations and depths in which they may be situated. There are currently many proposed methods for maintenance procedures. Some devices are simply detached from their moorings and towed back to shore for offsite maintenance. This is slightly more challenging for submerged devices attached to the seabed; however, the Rotech Tidal Turbine has an effective method whereby the turbine and generation unit can be removed from the device in a cassette unit, leaving the main structure on the seabed [96]. Piled devices offer relatively easy access for maintenance as submerged devices can be raised out of the water. Such systems include the SeaGen, Open Centre Turbine and the Pulse Stream; however, the latter two have only been placed on piles for ease of access during testing. Barrages and lagoons would generally require minimal maintenance; there is, however, the possible need for regular dredging of the surrounding areas due to the silting effects which may occur.

4.13 Worked Examples

1 Suppose that the optimum rotational speed of a 10 m diameter tidal stream turbine in a flow velocity of $3\,\mathrm{m\,s^{-1}}$ is 22 rpm when the angle of attack at the blade tip is 8°.

 A What is the tip speed ratio, λ?

 B What is the tip pitch angle?

 C If the blade is 4 m long, what angle of blade twist is required to maintain the optimum angle of attack?

Solution:

a) $\omega = \dfrac{22}{60} \times 2\pi = 2.3\,\mathrm{rads^{-1}}$

 $\lambda = \dfrac{\omega r}{U} = \dfrac{2.3 \times 5}{3} = 3.833$

b) $U_{\mathrm{tang}} = \omega r = 2.3 \times 5 = 11.5\,\mathrm{ms^{-1}}$

 $\tan\phi = \dfrac{3}{11.5} = 0.26$

 $\phi = 14.62°$

 $\gamma + \theta = \phi$

 $\theta = 14.62° - 8° = 6.62°$

c) $U_{\mathrm{tang}} = \omega r = 2.3 \times 1 = 2.3\,\mathrm{ms^{-1}}$

$$\tan\phi = \frac{3}{2.3} = 1.3$$

$$\phi = 52.52°$$

$$\text{twist} = 52.52° - 14.62° = 37.9°$$

2 Suppose the net tangential force per blade on a 12 m diameter three-bladed tidal stream turbine rotor operating at a tip speed ratio of 4 in a flow velocity of $2.4\,\text{m s}^{-1}$ is 10.6 kN, and the force acts at a radial position of 5 m.

A What is the approximate C_p of the turbine? (Assume the density of water is $1000\,\text{kg m}^{-3}$.)

B Why is this value an approximation?

C If the solidity of the rotor is 0.2, what is the mean chord length of each blade?

Solution:

a) Available power $= \frac{1}{2}\rho A U^3 = 0.5 \times 1000 \times \pi \times 6^2 \times 2.4^3 = 781728.8\,\text{W}$

Torque per blade $= 10600 \times 5 = 53000\,\text{Nm}$

Total torque $= 53000 \times 3 = 159000\,\text{Nm}$

Power $= \text{Torque} \times \omega = 159000 \times 1.6 = 254400$

$$C_p = \frac{\text{Power}}{\text{Available power}} = \frac{254400}{781728.8} = 0.325$$

b) Drag losses on hub are neglected.

c) $\sigma = \dfrac{NC}{\pi r}$

$$C = \frac{\sigma \pi r}{N} = \frac{0.2 \times \pi \times 6}{3} = 1.26\,\text{m}$$

3 For the same rotor design, how much larger must a wind turbine be to produce the same rated power in a wind velocity of $10\,\text{m s}^{-1}$ than a tidal stream turbine in a water velocity of $3\,\text{m s}^{-1}$? (Assume density of air is $1.225\,\text{kg m}^{-3}$ and density of water is $1000\,\text{kg m}^{-3}$.)

Solution:

$$\frac{1}{2}\rho A_w U^3 = \frac{1}{2}\rho A_t U^3$$

$$0.5 \times 1.225 \times A_w \times 10^3 = 0.5 \times 1000 \times A_t \times 3^3$$

$$612.5\,A_w = 4500\,A_t$$

$$\frac{A_w}{A_t} = \frac{4500}{612.5} = 7.35$$

$$\frac{\pi r_w^2}{\pi r_t^2} = 7.35$$

$$\frac{r_w}{r_t} = \sqrt{7.35} = 2.71$$

4 What are the advantages and disadvantages of tidal stream energy when compared with other energy sources?

Solution:
Answers should include points regarding:

Advantages
- No emissions
- Renewable/infinite supply
- Predictable output
- No visual impact if completely submerged

Disadvantages
- Intermittent and variable output
- Expensive to install
- Difficult to maintain
- Harsh operating environment
- High and variable loads

5 Discuss some of the possible environmental impacts associated with tidal stream turbines.

Solution:
Answers should include points regarding:

- Injuries to diving birds and marine mammals from blade strike
- Rapid pressure changes can cause problems for fish by affecting swim bladders
- Disrupted sediment transfer
- Noise could attract/deter marine life and confuse animals that use echolocation.
- Energy extraction could affect far-field water quality by reducing mixing
- Change flow patterns could increase coastal erosion

6 Show that the Betz limit for a tidal stream turbine in a free flow is 59.3%. The following assumptions can be applied:
- No hub.
- Infinite number of blades.
- No drag.
- Uniform flow/thrust over disk.
- No swirl in wake.

Solution:

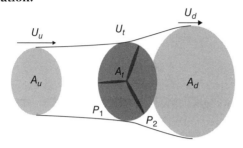

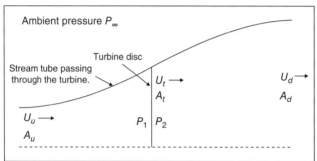

Continuity

$$A_u U_u = A_t U_t = A_d U_d$$

Momentum Equation

$$F = change\ in\ momentum$$

$$(P_1 - P_2) A_t = \rho A_u U_u (U_u - U_d)$$

Using Bernoulli's equation:

$$P_\infty + \frac{1}{2}\rho U_u^2 = P_1 + \frac{1}{2}\rho U_t^2$$

$$P_\infty + \frac{1}{2}\rho U_d^2 = P_2 + \frac{1}{2}\rho U_t^2$$

$$P_1 - P_2 = \frac{1}{2}\rho\left(U_u^2 - U_d^2\right)$$

$$= \rho \frac{A_u}{A_t} U_u (U_u - U_d)$$

$$= \rho U_t (U_u - U_d)$$

$$U_t = \frac{1}{2}(U_u + U_d)$$

$$Power = (P_1 - P_2) A_t U_t$$

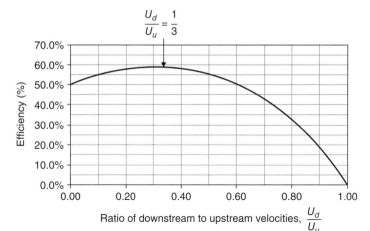

Variation of efficiency with the ratio of
downstream to upstream velocity

$$= A_t \frac{1}{2}\rho(U_u + U_d)(U_u - U_d)\frac{1}{2}(U_u + U_d)$$

$$= A_t \frac{1}{4}\rho(U_u^3 - U_d^3)$$

$$= A_t \frac{1}{4}\rho U_u^3\left(1 - \frac{U_d^3}{U_u^3}\right)$$

$$\frac{\text{Power}}{\frac{1}{2}\rho A U_u^3} = \frac{1}{2}\left(1 - \frac{U_d^3}{U_u^3}\right)$$

$$\eta = \frac{1}{2}\left(1 - \frac{U_d^2}{U_u^2}\right)\left(1 + \frac{U_d}{U_u}\right)$$

$$\eta = \frac{1}{2}\left(1 - \alpha^2\right)(1 + \alpha)$$

$$\frac{d\eta}{d\alpha} = \frac{1}{2}\frac{d}{d\alpha}\left[\left(1 - \alpha^2\right)(1 + \alpha)\right]$$

$$= \frac{1}{2}\left[\left(1 - \alpha^2\right) - 2\alpha(1 + \alpha)\right]$$

$$= \frac{1}{2}\left(1 - 3\alpha^2 - 2\alpha\right)$$

$$= \frac{1}{2}\left(1 - 3\alpha\right)(1 + \alpha)$$

2 solutions for $\dfrac{d\eta}{d\alpha} = 0$

Only valid one is $\alpha = \dfrac{1}{3}$

Substituting back in gives efficiency $= \dfrac{16}{27} = 59.3\%$

References

1 International Energy Agency. 2012. Global carbon dioxide emissions increase by 1.0 Gt in 2011 to record high. http://www.iea.org/newsroomandevents/news/2012/may/name,27216,en.html [accessed 20 May 2013].

2 European Commission. 2011. Communication from the Commission to the European Parliament and the Council Renewable Energy: Progressing towards the 2020 target. http://eurex.europa.eu/LexUriServ/LexUriServ.do?uri=COM:2011:0031:FIN:EN:HTML [accessed 18 March 2013].

3 Bahaj AS, Myers L. 2004. Analytical estimates of the energy yield potential from the Alderney Race (Channel Islands) using marine current energy converters. *Renewable Energy* 29(12), 1931–1945.

4 US Department of Energy. 2009. Report to Congress on the Potential Environmental Effects of Marine and Hydrokinetic Energy Technologies. US Department of Energy. http://www1.eere.energy.gov/water/pdfs/doe_eisa_633b.pdf [accessed July 2015].

5 SeaGen. 2011. Environmental Monitoring Programme. Edinburgh: Royal Haskoning Enhancing Society. http://www.marineturbines.com/sites/default/files/SeaGen-Environmental-Monitoring-Programme-Final-Report.pdf [accessed 2 March 2011].

6 Switzer T, Meggitt D. 2010. Review of literature and studies on electro magnetic fields (EMF) generated by undersea power cables and associated influence on marine organisms. Oceans 2010, Seattle, 20–23 September.

7 Open Hydro. 2012. Technology. http://www.openhydro.com/news/100908.html [accessed 4 February 2012].

8 Tidal Electric. 2010. Technology. http://tidalelectric.com/technology.shtml [accessed 27 February 2012].

9 Shi W, Park H-C, Baek J-H, Kim C-W, Kim Y-C, Shin H-K. 2012. Study on the marine growth effect on the dynamic response of offshore wind turbines. *International Journal of Precision Engineering and Manufacturing* 13(7), 1167–1176.

10 Pugh DT. 1987. *Tides, Surges and Mean Sea Level*. New York: John Wiley & Sons.

11 La Rance Tidal Power Plant. http://www.british-hydro.org/downloads/La%20Rance-BHA-Oct%202009.pdf [accessed 30 October 2015].

12 Tidal Power in South Korea. http://tidalpower.co.uk/south-korea[accessed 30 October 2015].

13 Lureg. 2012. Tidal Range Background Information. http://www.engineering.lancs.ac.uk/lureg/renewable_energy/tidal_range.php [accessed 3 Dec 2011].

14 Green Rhino Energy Ltd. 2013. http://www.greenrhinoenergy.com/renewable/marine/tidal_range.php [accessed July 2015]

15 Crown Estate. 2013. UK Wave and Tidal Key Resource areas Project; Technical Methodology Report 2013. http://www.thecrownestate.co.uk/media/5476/uk-wave-and-tidal-key-resource-areas-project.pdf [accessed August 2015]

16 Crown Estate. 2012. UK Wave and Tidal Key Resource Areas Project Summary Report http://www.thecrownestate.co.uk/media/5478/uk-wave-and-tidal-key-resource-areas-technological-report.pdf [accessed June 2015]

17 Nova Scotia Power. 2012. Annapolis Tidal Station. http://www.nspower.ca/en/home/environment/renewableenergy/tidal/annapolis.aspx [accessed 26 February 2012].

18 International Water Power. 2011. Tidal Turbine Development http://www.waterpowermagazine.com/story.asp?storyCode=2041292 [accessed 26 February 2012].

19 Wyre Tidal Energy. 2012. La Rance Barrage http://www.wyretidalenergy.com/tidal-barrage/la-rance-barrage [accessed 26 February 2012].

20 Woodshed Technologies. Tidal Delay. http://www.woodshedtechnologies.com.au/images/Tidal%20Delay(R).pdf [accessed 25 February 2012].

21 Douglas CA, Harrison GP, Chick JP. 2008. Life cycle assessment of the SeaGen marine current turbine. *Proceedings of the Institution of Mechanical Engineers, Part M: Journal of Engineering for the Maritime Environment* 222(1).

22 Sustainable Development Commission. 2007. Tidal Power in the UK Research Report 2: Tidal Technologies Overview http://www.sd-commission.org.uk/data/files/publications/TidalPowerUK2-Tidal_technologies_overview.pdf [accessed July 2015].

23 Pugh D. 2004. *Changing Sea Levels, Effects of Tides, Weather and Climate*. Cambridge: Cambridge University Press

24 Non Nuclear Energy – JOULE II. 1996. Wave Energy Project results. The Exploitation of tidal marine currents. EU Joule Contract JOU2-CT94-0355.

25 Black and Veatch. 2005. Tidal Stream Energy Resource and Technology Summary Report. Carbon Trust.

26 Department of Energy and Climate Change. 2015. Energy Trends Section 5: Electricity. https://www.gov.uk/government/statistics/electricity-section-5-energy-trends [accessed August 2015]

27 https://www.gov.uk/guidance/wave-and-tidal-energy-part-of-the-uks-energy-mix [accessed 30 October 2015].

28 Black and Veatch. 2005. Tidal Stream Energy Resource and Technology Summary Report. Carbon Trust.

29 DTI. 2005. Development, Installation and Testing of a Large-Scale Tidal Current Turbine. http://webarchive.nationalarchives.gov.uk/+/http://www.berr.gov.uk/files/file18130.pdf [accessed August 2015].

30 Siemens plc. 2012. Hat trick for SeaGen tidal current turbine. http://www.siemens.co.uk/en/news_press/index/news_archive/seagen-tidal-currant-turbine-milestone-hat-trick.htm [accessed August 2015]

31 Open Hydro. http://www.openhydro.com/images.html [accessed August 2015].

32 Couch SJ, Bryden I. 2006. Tidal current energy extraction: Hydrodynamic resource characteristics. *Proceedings of the Institution of Mechanical Engineers, Part M: Journal of Engineering for the Maritime Environment*, 220(4), 185–194.

33 Soares C. 2002. Tidal power: The next wave of electricity. *Pollution Engineering* 34(7),6–9.

34 Griffin D, Hemer M. 2010. Ocean Power for Australia – Waves, Tides And Ocean Currents. *Proceedings of Oceans 2010*. Piscataway, NJ: IEEE.

35 Carbon Trust. 2011. Accelerating marine energy. The potential for cost reduction – insights from the Carbon Trust Marine Energy Accelerator. http://www.carbontrust.com/media/5675/ctc797.pdf [accessed July 2015].

36 Carbon Trust. 2011. UK Tidal Current Resources and Economics. https://www.carbontrust.com/media/77264/ctc799_uk_tidal_current_resource_and_economics.pdf [accessed July 2015].

37 Crown Estate. 2012. UK Wave and Tidal Key Resource Areas Project. http://www.thecrownestate.co.uk/media/5476/uk-wave-and-tidal-key-resource-areas-project.pdf [accessed July 2015].

38 Kerr D. 2006. Marine energy. *Philosophical Transcripts of the Royal Society A* 365, 971–992

39 Bahaj AS, Myers L. 2004. Analytical estimates of the energy yield potential from the Alderney Race (Channel Islands) using marine current energy converters. *Renewable Energy* 29(12), 1931–1945

40 Myers L, Bahaj AS. 2005. Simulated electrical power potential harnessed by marine current turbine arrays in the Alderney Race. *Renewable Energy* 30(11), 1713–1731.

41 Hardisty J. 2009. *The Analysis of Tidal Stream Power*. Chichester: Wiley.

42 Bryden I, Naik S, Fraenkel P, Bullen CR. 1998. Matching tidal current plants to local flow conditions Energy Procedia 23(9).

43 Douglas JF, Gasiorek JM, Swaffield JA, Jack LB. 2005. *Fluid Mechanics* (5th edn). Harlow: Pearson/Prentice Hall.

44 Bachant P, Wosnik M. 2014. Reynolds number dependence of cross-flow turbine performance and near-wake characteristics. *Proceedings of the 2nd Marine Energy Technology Symposium 2014.*

45 Mason-Jones A, O'Doherty DM, Morris CE, O'Doherty T, Byrne CB, Prickett PW, Grosvenor RI, Owen I, Tedds S, Poole RJ. 2012. Non-dimensional scaling of tidal stream turbines. *Energy* 44 (1), 820–829.

46 Duquette M, Visser KD. 2003. Numerical implications of solidity and blade number on rotor performance of horizontal-axis wind turbines. *Journal of Solar Energy Engineering* 125(4), 425–432.

47 Hau E. 2006. *Wind Turbines: Fundamentals, Technologies, Application, Economics.* Berlin: Springer.

48 Morris CE, Mason-Jones A, O'Doherty DM, O'Doherty T. 2016. Evaluation of the swirl characteristics of a tidal stream turbine wake. *International Journal of Marine Energy* 14, 198–214.

49 Douglas JF, Gasiorek JM, Swaffield JA, Jack LB. 2005. *Fluid Mechanics* (5th edn). Harlow: Pearson/Prentice Hall.

50 Frost C, Morris C, Mason-Jones A, O'Doherty DM, O'Doherty T. 2015. The effect of tidal flow directionality on tidal turbine performance characteristics. *Renewable Energy* 78, 609–620.

51 Fujita Research. 2000. Wave and Tidal Power. www.fujitaresearch.com/reports/tidalpower.html [accessed30 October 2015].

52 Carbon Trust 2005. Variability of UL marine resources. https://www.carbontrust.com/media/174017/eci-variability-uk-marine-energy-resources.pdf [accessed August 2015].

53 Sea Generation Ltd. http://www.seageneration.co.uk [accessed 30 October 2015].

54 Open Hydro. 2012. Technology. http://www.openhydro.com/news/100908.html [accessed 4 February 2012].

55 Tidal Generation. 2010. Product Development. http://www.tidalgeneration.co.uk/products/ [accessed 26 March 2012].

56 Tidal Energy Limited. http://www.tidalenergyltd.com/ [accessed March 2015].

57 http://nautricity.com/ [accessed March 2015]

58 esru.strath.ac.uk [accessed October 2015]

59 Renewable UK. 2011. Wave and tidal energy in the UK. http://www.bea.com/pdf/ marine/Wave-Tidal_Energy_UK.pdf [accessed November 2012]

60 Eriksson S, Bernhoff H, and Leijon M. 2008. Evaluation of different turbine concepts for wind power. *Renewable and Sustainable Energy Reviews* 12(5), 1419–1434.

61 Khan MJ, Bhuyan G, Iqbal MT, Quaicoe JE. 2009. Hydrokinetic energy conversion systems and assessment of horizontal- and vertical-axis turbines for river and tidal application. A technology status review. *Applied Energy* 86, 1823–1835.

62 Riegler H. 2003. HAWT versus VAWT: Small VAWTs find a clear niche. *Refocus* 4(4) 44–46.

63 Han SH, Park JS, Lee KS. 2013. In-situ demonstration test for tidal current power generation with a helical turbine. 10th European Wave and Tidal Energy Conference, Aalborg.

64 The Engineering Business Ltd. 2003. Stingray Tidal Stream Device – Phase 3. Northumberland: DTI: http://www.inference.phy.cam.ac.uk/sustainable/refs/tide / StingrayPhase3r.pdf [accessed 29 February 2012].

65 The Engineering Business Ltd. 2003. Stingray Tidal Stream Device – Phase 3. Northumberland: DTI: http://www.inference.phy.cam.ac.uk/sustainable/refs/tide / StingrayPhase3r.pdf [accessed 29 February 2012].

66 BioPower Systems. 2011.BIoSTREAM. http://www.biopowersystems.com/biostream. html [accessed 22 February 2012].

67 Riegler H. 2003. HAWT versus VAWT: Small VAWTs find a clear niche. *Refocus* 4(4), 44–46.

68 Han SH, Park JS, Lee KS. 2013. In-situ demonstration test for tidal current power generation with a helical turbine. 10th European Wave and Tidal Energy Conference, Aalborg.

69 VerdErg. 2012. SMEC Technology, Basic Principles. http://www.verderg.com/index. php?option=com_content&view=article&id=95&Itemid=66 [accessed 25 February 2012].

70 Lunar Energy. 2010. RTT The Product. http://www.lunarenergy.co.uk/ productOverview.htm [accessed August 2015].

71 Salman SK *et al*. 2008. Integration of tidal power based-electrical plant into a grid [e-journal] 42 (6) Available through: Blackwell Science Synergy database [accessed 12 June 2005].

72 Clean Current Power Systems Inc. 2012. Tidal Turbines. http://www.cleancurrent.com/ tidal-turbines [accessed August 2015]

73 Jha A. 2008. Second generation tidal turbines promise cheaper power. The Guardian, 4 September. http://www.guardian.co.uk/environment/2008/sept/04/waveandtidalpower. renewableenergy [accessed 30 October 2015].

74 Cross-Flow Turbine Hydrodynamics. Tidal Energy Research Group, University of Oxford http://www.eng.ox.ac.uk/tidal/research/thawt [accessed 28 Nov 2016].

75 Minesto. 2012. Deep Green Technology. http://minesto.com/deepgreentechnology/ index.html [accessed 25 February 2012].

76 Versteeg HK, Malalasekera W. 2007. *An Introduction to Computational Fluid Dynamics, The Finite Volume Method* (2nd edn). Harlow: Pearson Education.

77 Mason-Jones A, O'Doherty DM, Morris CE, O'Doherty T, Byrne CB, Prickett PW, Grosvenor RI, Owen I, Tedds S, Poole RJ. 2012 Non-dimensional scaling of tidal stream turbines. *Energy*, 44 (1), 820–829.

78 Hughes TJR. 2000. *The Finite Element Method: Linear Static and Dynamic Finite Element Analysis*. Mineola, NY: Dover Publications.

79 Zienkiewicz OC, Taylor RL, Fox D. 2013. *The Finite Element Method for Solid and Structural Mechanics*. Oxford: Elsevier Science & Technology.

80 Leishman JG. 2006. *Principles of Helicopter Aerodynamics*. Cambridge: Cambridge University Press.

81 Masters I, Chapman JC, Willis M, Orme JAC. 2011. A robust blade element momentum theory model for tidal stream turbines including tip and hub loss corrections. *Journal of Marine Engineering &Technology*, 10(1), 25–35.

82 Snodin H. 2001. *Scotland's Renewable Resource. Volume II: Context* http://www.scotland.gov.uk/Resource/Doc/47176/0014635.pdf [accessed 4 Dec 2011].

83 EMEC. 2012. Tidal Devices. http://www.emec.org.uk/tidal_ devices.asp [accessed 14 February 2012].

84 University of Strathclyde. 2012. Technical Constraints. http://www.esru.strath.ac.uk/EandE/Web_sites/10-11/Tidal/tech.html [accessed 15 March 2012].

85 Harris RE *et al.* 2004. Mooring systems for wave energy converters: A review of design issues and choices [e-journal], 19, Available through: Zotero [accessed 13 Mar 2012].

86 Fraenkel PL. 2002. Power from marine currents. *Proceedings of the Institution of Mechanical Engineers, Part A: Journal of Power and Energy* 216(1), 1–14.

87 Clarke JA, Connor G, Grant AD, Johnstone CM, Ordonez-Sanches S. 2008. Contra-rotating marine current turbines: performance in field trials and power train developments. In *Renewable Energy 2008: 10th World Renewable Energy Congress* (pp. 1049–1054). London: Sovereign World.

88 University of Strathclyde. 2012. Technical Constraints. http://www.esru.strath.ac.uk/EandE/Web_sites/10-11/Tidal/tech.html [accessed 15 March 2012].

89 Clarke JA, Connor G, Grant AD, Johnstone CM, Ordonez-Sanches S. 2009. Contra-rotating marine current turbines: Single point tethered floating system - stability and performance. 8th EWTEC, Uppsala, Sweden.

90 Zienkiewicz OC, Taylor RL, Fox D. 2013. *The Finite Element Method for Solid and Structural Mechanics*. Oxford: Elsevier Science & Technology.

91 DTI. 2006. Tidal Lagoon power generation scheme in Swansea Bay. A report on behalf of the Department of Trade and Industry and the Welsh development Agency. AEA Technology Contract No 14709570 and 14709574. URN No 06/1051.

92 Previsic M, Polagye B, Bedard.2006. System Level design, performance, cost and economic assessment: San Francisco Tidal In-Stream Power Plant. EPRI report TP-006-SFCA.

93 Clarke JA, Connor G, Grant AD, Johnstone C, MacKenzie D. 2007. Development of a contra-rotating tidal current turbine and analysis of performance. In *7th European Wave and Tidal Energy Conference*. Instituto de Engenharia Mecanica.

94 Mason-Jones A, O'Doherty DM, Morris CE, O'Doherty T. (2012) Influence of a velocity profile & support structure on tidal stream turbine performance. *Renewable Energy*, 52, 23–30.

95 Renewable Energy World. 2010. Flaw hits hundreds of EU offshore wind turbines. http://www.renewableenergyworld.com/rea/news/article/2010/04/ flaw-hits-hundreds-of-EU-offshore-wind-turbines [accessed 26 March 2012].

96 Tidal Energy Limited [accessed March 2015].

5

Device Design

Lars Johanning[a], Sam D. Weller[b], Phillip R. Thies[c], Brian Holmes[d]
and John Griffiths[e]

[a] Professor of Ocean Technology, University of Exeter, Penryn Campus, UK
[b] Senior Research Fellow, University of Exeter, Penryn Campus, UK
[c] Senior Lecturer Renewable Energy, University of Exeter, Penryn Campus, UK
[d] MaREI Centre, Environmental Research Institute, University College Cork, Ringaskiddy, Ireland
[e] Associate, EMEC Ltd, Chair of UK National Committee on Wave & Tidal Standards

5.1 Standards and Certification in Marine Energy

This section describes the progress on standards and certification in the marine energy sector (wave and tidal stream). This has some precedents to draw on from the wind industry, but there are also substantial differences. The chapter indicates why standards and certification are required and summarises the progress to date and current activity with standards-making in an international context. The International Electrotechnical Committee (IEC) is the international body responsible for marine energy standards, and the organisation is briefly described, showing the relationship between standards and the certification process. The section concludes with a summarised overview of what is involved in type and project certification.

5.1.1 Why are Standards Needed?

Marine energy is a nascent industry with technologies that have significant innovation and that are in the early stages of proving from a reliability viewpoint. At the time of writing they are expensive technologies (in terms of levelised cost of energy, LCoE) and cost reduction, although studied extensively, cannot be realised without substantial deployment in numbers. Deployments have been slow, partly due to overcoming technical problems and partly due to the difficulties of raising funds to build and deploy devices. So, as many people have asked, why try to produce standards so early?

Standards are a sign of either mature industries or those seeking to become mature. It is necessary to have standards to open up world-wide markets, so it is vital that wave energy and tidal energy resources are evaluated consistently around the world. Equally, it is vital that the definition and measurement of performance of devices has a common basis internationally, and that an appropriate risk assessment is in place for the design.

Wave and Tidal Energy, First Edition. Edited by Deborah Greaves and Gregorio Iglesias.
© 2018 John Wiley & Sons Ltd. Published 2018 by John Wiley & Sons Ltd.

As the technologies demonstrate that they work and fulfil their design intent, it is necessary to raise funds to deploy devices to demonstrate reliability of performance and to start to run down the cost of energy produced. Raising such investment will require investors and insurers to begin to build confidence in the technologies, and that implies a need for certification systems which can give an objective and independent view of performance and cost. To carry out certification, i.e. verification that performance and reliability are suitable, there must be a yardstick to measure against; and this is where standards are needed.

Clearly there will not be detailed standards produced for all components and subsystems in the short term, so basic evaluation and measurement methods must figure early on. Guidance may be documented at fairly high level on matters such as design, but all detail must rely on relevant existing standards for components and subsystems. Even then experience shows that standards may not always apply as the conditions under which components need to operate may be different from those envisaged within existing standards. For example, many wave energy converters (WECs) are dynamic and in motion at all times. However, many of the components may have standards relating to items being fastened to a firm and motionless foundation, which may result in the standard not being met by the component(s) concerned.

So, standards are needed and must be developed in parallel with technologies, even though they may not be instantly applicable.

5.1.2   Wat has been done so far?

An international committee (designated TC/114) was set up in 2008 by the IEC[1] to manage the writing of standards documents for the wave, tidal and other water currents sector. Due to the early stage of the industry the standards are 'technical specifications' (TSs), which is a document type one level below that of a full standard. After a period of use, usually 3–4 years, the experience gathered and documented may be sufficient to allow a rewrite to upgrade to a standard.

Seven TSs have been published to date:

Terminology **(IEC TS 62600-1:2011).** The TS defines the terms relevant to ocean and marine renewable energy (MRE). For the purposes of this TS, sources of ocean and marine renewable energy are taken to include wave, tidal current, and other marine energy converters (MECs). The TS is intended to provide uniform terminology to facilitate communication between organisations and individuals in the MRE industry and those who interact with them. It is necessarily a dynamic document which will need to be updated to reflect terms used in other TSs.

Wave Energy Converters Power Performance Assessment **(IEC TS 62600-100:2012).** This provides a method for assessing the electrical power production performance of a WEC, based on the performance at a testing site. It provides a systematic method which includes: measurement of WEC power output in a range of sea states; WEC power matrix development; and an agreed framework for reporting the results of power and wave measurements.

Power Performance Assessment of Tidal Energy Converters **(IEC TS 62600-200:2013).** This is a systematic methodology for evaluating the power performance of tidal current energy converters (TECs) that produce electricity for utility-scale and localised

---

1 See http://www.iec.ch/index.htm (accessed May 2016).

grids, including: a definition of TEC rated power and rated water velocity; a methodology for the production of the power curves for the TECs in consideration; and a framework for the reporting of results.

***Wave Energy Resource Assessment* (IEC TS 62600-101:2015).** This establishes a system for estimating, analysing and reporting the wave energy resource at sites potentially suitable for the installation of WECs. This TS is to be applied at all stages of site assessment from initial investigations to detailed project design. In conjunction with IEC TS 62600-100 (WEC performance), it enables an estimate of the annual energy production of a WEC or WEC array to be calculated.

***Tidal Energy Resource Assessment* (IEC TS 62600-201:2015).** This establishes a system for analysing and reporting, through estimation or direct measurement, the theoretical tidal current energy resource in oceanic areas including estuaries (to the limit of tidal influence) that may be suitable for the installation of arrays of Tidal Energy Converters (TECs). It is intended to be applied at various stages of project lifecycle to provide suitably accurate estimates of the tidal resource to enable the array's projected annual energy production to be calculated at each TEC location in conjunction with IEC 62600-200.

***Assessment of Moorings Design* (IEC TS 62600-10:2015).** This provides uniform methodologies for the design and assessment of mooring systems for floating MECs. It is intended to be applied at various stages, from mooring system assessment to design, installation and maintenance of floating MEC plants. It is applicable to mooring systems for floating MEC units of any size or type in any open water conditions. The intent of this TS is to highlight the different requirements of MECs.

***Design Requirements for Marine Energy Systems* (IEC TS 62600-2:2016).** This TS provides general guidance on design of MECs, particularly focused on the structural aspects. The drafts have been the subject of a very large number of comments of which a significant number have not been fully resolved. Accordingly, a maintenance team will be established to process these comments and to reissue the TS within a two-year period. In the meantime the document will be useable with care and reference to the unresolved comments.

The task facing the industry is to attempt to use the above documents and provide feedback and comments on the experience gained to the ad hoc groups (AHGs) or maintenance teams formed to consider future editions of the documents. In many cases it is hoped to develop the TSs into standards at the next rewrite.

Furthermore, in October 2015 DNV GL published a standard dedicated to tidal energy: 'Tidal Turbines' (DNVGL-ST-0164). This standard is applicable to all types of support structures, fixed or floating, and all types of turbines. This standard is supported through a 'service specification' document also dedicated to tidal turbines, with the title 'Certification of Tidal Turbines and Arrays' (DNVGL-SE-0163) published in October 2015.

5.1.3 What is in hand?

There are eight international project teams working on the following documents to be published mostly during 2016–17:

***WEC Power Performance at Second Location* (IEC DTS 62600 -102 Ed 1.0).** This TS will describe the required methods and the required conditions to determine the power performance of a specific WEC 2 at a specific location, possibly at a different scale and with configuration changes to accommodate the new site conditions, based on a measured power performance for a reference WEC design and reference location. The TS will allow for assessment at Location 1 or Location 2 based on limited/incomplete data material, as long as this is accompanied by a validated numerical model and assessment of the uncertainty involved.

***Power Quality* (IEC TS 62600-30 Ed. 1.0).** This TS will aim to: identify power quality issues and conformity requirements (non-device-specific and non-prescriptive) of single-/three-phase, grid-connected/off-grid ocean wave and tidal power systems; establish adequate measures of nominal conditions and allowable deviations from these nominal conditions that may originate within the power source and/or electrical load; prescribe limits of voltage and frequency variations both during steady-state and transient conditions; identify the requirements for monitoring electrical characteristics of single-phase and poly-phase AC power systems; and establish the measurement methods, application techniques and result-interpretation guidelines.

***Early Stage Development & Scale Testing – Wave* (IEC/TS 62600-103 Ed. 1.0).** This TS will be based around the relevant sections of existing international documents addressing structured development schedules for WECs and will expand on the detailed tests necessary for the successful progression of a WEC from small-scale ($c.1 : 50$) proof of concept and optimisation, (stage 1), through the medium design scale ($c.1 : 15$, stage 2), to large-scale ($c.1 : 4$) system proving (stage 3).

***Scale testing of Tidal Stream Energy Systems* (IEC/TS 62600-202 Ed. 1.0).** This will provide specification of purpose of test both for single-turbine or infrastructure loading and for multi-turbine interactions. Specification of the environment for tidal stream sites is included, as well as the representation of flows for scale and purpose of tests, including tests in steady flows and under towing tests. The specification of flow profiles and turbulence, the flow and waves, together with the tidal forcing, is considered. Scale and purpose specifications also relate to suitable test facilities and instrumentation equipment for flow measurement and mechanical parameter measurement. Typically, the tests need to represent correctly the turbine characteristics for scale and purpose of the tests, including turbine subcomponents and support structures (farm subcomponents). Test procedures encompass quality control, use of the data acquisition system, calibration of the instrumentation, scheduling, analysis and reporting.

***Ocean Energy Thermal Conversion Design* (IEC/TS 62600-20 Ed. 1.0).** Ocean thermal energy conversion (OTEC) is a technique that uses the difference of temperature between deep seawater and shallow seawater and is beingresearched and developed in Europe, America and Japan. Although it has low efficiency, it can be developed for a semi-commercial capacity of 10 MW, and is close to the commercialisation stage for niche settings such as island communities in the regions between the tropics. The aim of this specification is to provide guidance on how to design the OTEC plant and to assess the structural safety of the plant.

***Acoustic Measurement Method* (IEC/TS 62600-40 Ed. 1.0).** The scope of this specification will encompass methods and instrumentation to characterise converter sound and ambient noise, as well as the presentation of this information for use by regulatory agencies, industry, and researchers. References are given to relevant calibration procedures for sound measurement devices, as are procedures for deploying measurement devices from different platforms (bottom fixed, compliantly moored, and drifting).

Criteria for spatial and temporal stationarity are given and specific guidance is provided for evaluating low-frequency sound (<100 Hz). This TS covers wave, current, and ocean thermal energy converters including single devices and arrays.

River – Power Assessment (**IEC/TS 62600-300 Ed. 1.0**). This TS will provide a systematic methodology for evaluating the power performance of river current energy converters (RECs) that produce electricity for utility scale and localised grids; a definition of REC rated power and rated water velocity; a methodology for the production of the power curves for the RECs in consideration; and a framework for the reporting of the results.

River Resource Assessment (**IEC/TS 62600-301 Ed. 1.0**). The objective of the TS will be to provide guidance on how to characterise a river reach to determine the factors that affect its hydrokinetic energy potential. These include deriving (or estimating) the historical velocity and turbulence distributions and how to combine that information with the power curve of the candidate REC. Unlike tidal zones, river current velocities and turbulence can fluctuate seasonally and annually, depending upon weather and climate related factors and the morphology of the river reach. Once the velocity distribution has been calculated, it can be applied to a specific REC's power curve to estimate the annual energy output and its potential statistical variance.

5.1.4 How is it Organised?

There are two sides to the organisation of the IEC: the Standardization Management Board (SMB) oversees standards-making, while the Conformity Assessment Board (CAB) has responsibility for certification systems. Figure 5.1 shows the organisation.

The standards-making and certification are kept separate to avoid conflicts of interest. In practice some of the same people are active in both areas and the committee system rules prevent them from getting conflicted. The technical committees (TCs), of which TC/114 is an example, are responsible to the SMB. The IEC provide support to

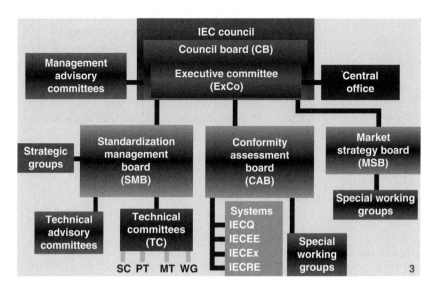

Figure 5.1 Organisation of IEC.

TCs by attendance at meetings and answering technical and procedural questions as needed. The SMB sets the rules for standards-making activity and arranges publication of documents.

On the certification side, IECRE[2] is the system for certification of renewable energies, namely wind, solar photovoltaic and marine energy. IECRE is described in more detail in Section 5.1.6.

The Market Strategy Board (MSB) consists of 15 members appointed from international corporations plus a number of IEC officers and identifies the major technological trends and market needs in the IEC's fields of activity. It sets strategies to maximise input from primary markets and establishes priorities for the technical and conformity assessment work of the IEC, improving the Commission's response to the needs of innovative and fast-moving markets. It may establish special working groups under the leadership of an MSB member to investigate certain subjects in depth or to develop a specialised document.

5.1.5 Standards-Making

The overall process for standards-making is shown diagrammatically in Figure 5.2. The TCs are responsible for overseeing the production of standards. They produce a strategic business plan[3] (SBP) which is normally set over a five-year time horizon and in which the priorities for standards to be produced are summarised against a background of the trends in technologies and the relevant markets.

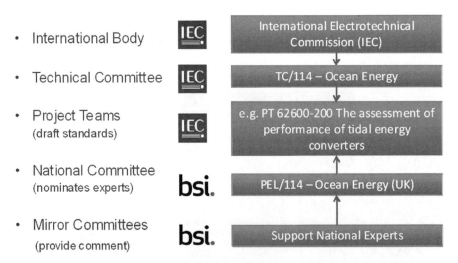

Figure 5.2 Summary of the process for standards-making.

2 See: http://www.iecre.org/ (accessed in May 2016).
3 See: http://www.iec.ch/dyn/www/f?p=103:30:0::::FSP_ORG_ID,FSP_LANG_ID:1316,25 (accessed May 2016). Reference to the SBP for TC/114 may be found in the documents listed.

The TC is responsible for setting up project teams (PTs), which are the international groups that write the standards. They have a convenor, and experts are nominated by member countries to join. In each country the experts are supported by the national committee (in the case of the UK, PEL/114 is the national committee). All decisions are voted upon, including convenors, new work items, committee drafts, draft standards or TSs, and approvals for publication. There must be a clear majority for any vote to be carried. Currently the USA holds the chairmanship of TC/114 and BSI holds the secretariat.

The TCs may also set up AHGs, which are working groups with a specific purpose and a limited life. The AHGs may be required to develop an idea for a standard where the concept and scope are not properly defined and work is needed to get a clear scope for a new standard. They may also be responsible for collecting information on experience of the use of a standard pending the revision of the document at the end of the 'stability period', which may be 3-4 years for a TS, or longer for a full standard. A maintenance team may then replace the AHG to rewrite the standard or upgrade the TS to a standard.

5.1.6   Certification Scheme: IECRE

The CAB organises the certification systems as shown in Figure 5.3. There are now four systems, as shown, one of which (IECRE) is the focus of attention in this subsection.

The IECRE system is headed by the Renewable Energy Management Committee (REMC). Reporting to this are three operational management committees (OMCs).

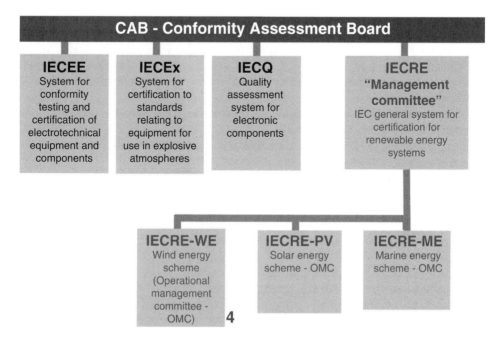

Figure 5.3 Organisation of certification activity.

One of these is dedicated to marine energy (ME-OMC). These committees set rules which are harmonised around a common rule set and are compatible with the overarching rules set by the REMC, but reflecting the necessary differences due to the technologies. ME-OMC has established a number of working groups to look at rules, scopes of certification and testing, as well as the range of documentation required for the certification process. The IECRE system has national bodies as members and the representation will, in various circumstances, include certification bodies (CBs) and test laboratories (TLs). These CBs and TLs have to be peer-reviewed in accordance with IEC policy; reviews are limited to adherence to IECRE procedures where an existing accreditation is held, such as ISO 17025 for the TLs.

5.1.7 Certification Process

5.1.7.1 Type Certification
The basis of certification is used to document the functional, safety, environmental and reliability targets of the device. It will also describe the operating conditions and design survival conditions for the device. The process is broken down into the elements shown in Figure 5.4.

Design Assessment
The design assessment uses a risk-based approach in order to qualify and verify the system in question. This is required to assess the innovative areas of the design where no codes or standards exist due to the emergent nature of marine energy converters technologies. This process is generally referred to as 'qualification' of the technology.

The design evaluation will include:

- establishment an overall plan for the certification. This is a continuing process and needs updating after each step using the available knowledge on the status of the qualification.
- screening of the technology based on identification of failure modes and their risk.
- design review and attendance at tests.
- assessment of maintenance, condition monitoring and possible modifications required to reduce the risk of failure.

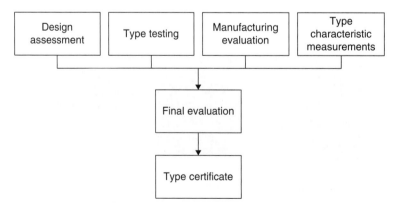

Figure 5.4 Outline diagram of a type certification process.

Type Testing

All tests carried out will be completed to the requirements of relevant international standards and will include any additional tests identified as critical and necessary during the failure mode identification and design review process. Standards will be clearly identified in test reports, along with the component being tested and the conditions for which the tests have been carried out. Deviations from the testing as specified by the international standard employed will be reported and an assessment of their impact on the test results will be made. Tests will be carried out to the satisfaction of the CB. Requirements for surveillance and use of an accredited TL may be defined on a case-by-case basis, depending upon the criticality and complexity of the test.

Manufacturing Evaluation

A manufacturing evaluation shall be carried out in order to ensure that the component or system is produced to the necessary specifications and quality as detailed in the design documentation. Manufacturers of materials, components and equipment will be approved according to criteria established by the CB, as applicable and defined in the respective standards or from the certification process. Manufacturers will demonstrate their capability to carry out fabrication of adequate quality in accordance with the relevant standards, and with any additional requirements based on criticality of processes, before construction is started. Any required quality control of materials, components and equipment shall be traceable and documented in writing. Quality control will be carried out by qualified personnel at facilities and with equipment suitable for that control. Welding of important structures, machinery installations and equipment will be carried out by qualified and approved welders to qualified and approved weld procedures, with approved welding consumables and at welding shops accepted by the CB. During fabrication and construction work, the CB will have safe access to the works at all reasonable times, insofar as the work affects certification.

Type Characteristic Measurements

The device performance will be tested under specified conditions to ensure that it conforms to the critical operating parameters defined in the certification basis. These tests will be performed under surveillance by the CB. The tests may include any or all of the following as relevant:

- power quality;
- acoustic noise;
- other systems.

Where any of these characteristics are defined as device requirements in the certification basis, measurements will be carried out as part of the type testing program to verify compliance. The measurement procedures will conform to requirements agreed with the CB. A test report will be produced describing the measurement conditions, instrumentation, calibration and analyses.

Final Evaluation

The final evaluation requires that a report be produced, which includes:

- verification that the required documentation is complete and whether or not the type testing program has confirmed that the requirements set out in the certification basis have been satisfied;

- review of all design documentation, including drawings, specifications, manufacturing, commissioning and installation procedures;
- verification that the design meets the supporting design calculations;
- a reference list of all supporting documentation relevant to the type certificate.

The CB will confirm that the report clearly identifies any safety critical items and that the functional requirements were demonstrably achieved.

Type Certificate

A type certificate can be issued once the CB has verified that there are no outstanding design issues to be resolved. The type certificate shall include:

- identification of the device;
- references to all relevant conformity statements, and final evaluation reports;
- device specifications as outlined in the certification basis, including environmental conditions, electrical network conditions and design lifetime;
- standards under which certification and testing have been carried out;
- conditions for validity of certificate.

5.1.7.2 Project Certification

A project certification may be based on a type certification and include all its elements (see Figure 5.5).

Project Design

Project certification will confirm for a specific site that energy converters meet requirements governed by site-specific external conditions and are in conformity with other requirements relevant to the site (soil and environmental conditions, mooring/anchoring, etc.). This includes: site assessment (soil, metocean and other environmental conditions); design assessment of site-specific built components: moorings and/or foundations and evaluation of installation (plan and quality and surveillance of the activity).

Type Certified Machine(s)

Type certificates must be provided for each separate manufacturer's machine installed on the site.

Operations and Maintenance Surveillance

This will include review of operations and maintenance (O&M) over the life of the project after an initial conformity statement has verified that installation, commissioning and

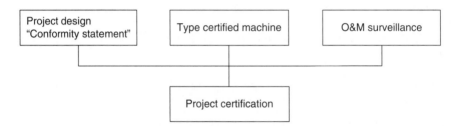

Figure 5.5 Project certification outline.

early operations are in line with the certification previously issued. Conformity will be demonstrated by:

- identification of the WEC or TEC;
- identification of the specific site;
- reference to the device(s) type certificate;
- reference to relevant documentation (e.g. O&M manual, O&M evaluation report);
- standard under which certification is being carried out;
- conditions for validity of the conformity statement.

Project Certification

The project certificate can be issued once the CB has verified that there are no outstanding issues relating to the deployment of the specific device at the site in question. The project certificate shall include:

- identification of the WEC or TEC;
- identification of the site for deployment;
- references to the relevant reports and conformity statements (including site assessment, installation, foundation and mooring design, O&M);
- standard under which the certification has been carried out;
- conditions of validity for the certificate;
- type certificate for the WEC or TEC, including device specification.

5.2 Reliability

Reliability testing is essential for any product development programme, especially if the development risks are high [1]. This is certainly the case for the MRE sector which is now emerging from a research and development phase to the deployment of full-scale prototypes and even small commercial projects. Reliability test programmes are crucial in order to prove that the design is reliable under the harsh marine operating environment and the dynamic loads experienced by most marine energy systems.

The necessity to engage in component testing at this stage of development is mainly due to:

- the need to reduce costly safety factors in the design of MECs;
- the current development of marine energy towards commercial deployments, necessitating reliable estimates of plant-performance indicators, i.e. reliability, availability and maintainability;
- the need on the part of both investors and insurance companies for assurances of reliability for safety and economic reasons – reliability test results are a key component of this assurance.

The implementation of proven technologies and components in the design of MECs is confronted by the different environmental and operational conditions which lead to uncertainty as to how previous knowledge should be applied. Component testing in truly representative conditions will allow the establishment of the necessary specific information about component performance and failures.

A component test facility to collect reliability data especially for WECs was proposed by Salter [2]. The suggested design of a floating raft included test beds to operate rams, seals, belts and cables at their operational use conditions, to conduct cavitation testing and to expose components to marine fouling conditions to assess the effectiveness of coatings. The initiative was not pursued further, principally due to the cost and practicality of such a platform.

Wolfram [3] suggested the application of accelerated testing for critical components to assess if the environment alters known failure rates. Any test results should be collected in a collaborative failure rate database.

The Carbon Trust [4] proposed both knowledge transfer from established offshore industries (e.g. engineering standards, risk assessment methods) and rigorous testing of components, subassemblies and device prototypes to mitigate technical risks. And indeed many companies have had their devices assessed in this way by consultants such as WS Atkins or by certifying authorities such as DNV GL or Lloyd's. Mueller and Wallace [5] called for extensive testing to improve reliability and establish a statistical database of component reliability in the marine renewable environment. The failure rate data currently available are scarce and often generic, making crude adjustments necessary that lead to necessarily conservative and highly uncertain results [6].

5.2.1 System Reliability Assessment

An operational availability threshold of 75% has been identified for MRE devices. In order for the sector to progress towards higher TRLs the reliability of components and subsystems must be demonstrated as this plays a key role in the overall availability of the device [3] as well as shaping efficient maintenance intervals [7]. Reliability assessment has the following objectives: to ensure that an acceptable level of system reliability can be achieved; to quantify life-cycle costs over the expected lifetime of the array (i.e. 25 years); to plan O&M strategies; and to identify design weaknesses for system improvement.

A widely used metric for assessing the multi-level reliability of systems (comprising subsystems and components) is the mean time to failure (MTTF) based on the reliability function of a component, subsystem or system:

$$MTTF(T) = \int_0^\infty R(t)dt = \int_0^\infty \left[\int_t^\infty f(\tau)d\tau \right] dt \int_0^\infty e^{-\lambda t} = \frac{1}{\lambda}. \tag{5.1}$$

This function describes the probability of continued reliability from a particular point in time and is based on a statistical probability density function (PDF) of reliability performance gathered from laboratory component testing, field analysis or numerical analysis. The most basic PDF is exponential, which is used to describe failures which occur during the 'useful' life of the component, illustrated as the central portion of the 'bathtub' curve in Figure 5.6. It is widely used as it gives a simple estimate of the MTTF as the inverse of the failure rate. Fundamentally, this approach assumes that components have a constant rate of random failures and have therefore not started to degrade or wear out with use. In reality the early life of a component is characterised by high (but decreasing) failures which then increase at the end of the component life. Alternative distributions are more suitable for describing failures before and after the

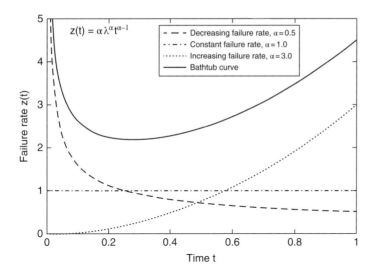

$$z(t) = \alpha \lambda^\alpha t^{\alpha-1}$$

- – – – Decreasing failure rate, α = 0.5
- – · – · Constant failure rate, α = 1.0
- ········· Increasing failure rate, α = 3.0
- ——— Bathtub curve

Figure 5.6 'Bathtub' curve failure rate behaviour, modelled with Weibull distribution with $\lambda = 1$ [8].

useful interval of component life; for example, the Weibull distribution tends to give a more refined representation of end-of-life failures [8].

The sensitivity of component failure rates to different applications is accounted for in bottom-up statistical methods in which influence factors are applied to base failure rates to account for variations in quality, environment and stress (e.g. temperature, use rate or load). Using this approach, reliability calculations can be carried out with a relatively small amount of information such as the type and number of components and operating and environmental conditions. While this method is straightforward to implement, it is highly reliant on the underlying data, and although influence factors are available in databases such as Mil-Hdbk-217 F [9] for electronic components or OREDA® database [10] produced by the oil and gas industry, the use of factors designed for other applications introduces uncertainty into the predicted failure rate.

A more detailed approach to assessing reliability is provided by physics of failure bottom-up methods. These are used to identify principal failure mechanisms (i.e. structural, mechanical, electrical, thermal or chemical) which contribute to component failure as well as their effect. The aim of this approach is to develop empirical relations between the dominant failure mechanisms and the predicted MTTF. Crucially, unlike the bottom-up statistical approach, this enables a greater depth of understanding regarding event or time-dependent failure mechanisms (e.g. the load-bearing capacity or degradation of components). Although empirical relations for particular components exist (e.g. [11]), as with bottom-up statistical methods it is possible that nuances of the application will not be fully accounted for. While both approaches introduced in this section have widely recognised limitations, they still provide a first-order estimate for reliability studies to identify critical subsystems and components.

As yet, a common failure database has not been established in the MRE industry, despite the development of initial reliability models. An endeavour similar to the offshore wind

SPARTA (System Performance, Availability and Reliability Trend Analysis)[4] project for the MRE sector is likely to occur in the near future. The absence of a database is due to a lack of design convergence within the sector (particularly for wave energy devices), the use of custom-made components and also the commercial confidentiality of designs. At present, developers must therefore rely on the adaptation of existing reliability assessment methods, as well as knowledge gained during earlier TRL stages, to predict operating conditions. Component reliability testing either in the field or laboratory environment plays a pivotal role in providing a means of validation for prediction methods.

5.2.2 Subsystem and Component Reliability

A generic MEC is typically modelled as series of subsystems (Figure 5.7) in the form of reliability block diagrams (RBDs). The series structure consists of independent components and the system reliability $R_S(t)$ can be calculated as the product of the component reliability survivor functions $R_i(t)$ for the number of components/subsystems in series $i = 1,..., n$:

$$R_S\left(t\right) = \prod_{i=1}^{n} R_i\left(t\right). \tag{5.2}$$

For a bottom-up statistical method it is required to assign subsystem reliabilities, with component failures assumed to follow an exponential distribution which is used to represent the 'useful' or main portion of component life [12]. The method is suited to different system hierarchies (such as parallel and series relationships, and where active redundancy is built into the system). One shortcoming of this approach is often that failures are assumed to occur independently and hence cascade failures are not considered. On an assembly level this is often an acceptable approach; however, in the case of mooring line components that are usually connected in series, for example, this is not the case, where the failure of one component usually results in loss of functionality of the entire line.

Often failure rates are specified as single values representing the mean of a set of analysed failure events. A single value does not provide information regarding the scatter of failures around the mean. In order to carry out confidence analysis if upper and lower failure rate bounds are known (i.e. exist in the database) the RBD analysis can be carried out using these values to provide a measure of confidence in the reliability estimate.

The consequence of a failure occurring will range from total loss of functionality of the component or subsystem to a negligible reduction in performance. Within the OREDA Handbook [10], failures are categorised into three groups: critical, degraded and incipient failures.

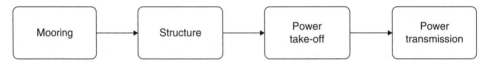

Figure 5.7 Reliability block diagram of generic floating marine energy converter.

---

4 https://ore.catapult.org.uk/sparta

5.2.3   Component Failure Rate Modelling and Prediction

Given the current availability of failure rate (λ) statistics within the MRE sector, the RBD method requires using failure rate statistics associated to each component. An RBD is shown as an example for an electrical system architecture for an array of WECs with mooring and foundation modules (Figure 5.8). As discussed above, failure rates are required from a database that allows several reliability statistics to be calculated recursively up through different hierarchical levels as outlined in the following subsections.

The reliability at a user-specified time[5] (t) is calculated at each hierarchical level. This metric ranges from 0 (not functional) to 1 (fully functional). Assuming that all components in the system have a constant failure rate (i.e. they do not degrade with time or wear out) and hence an exponential failure rate distribution is representative of component failures, the following equation is used at the component level to calculate reliability at time t:

$$R(t) = e^{-\lambda t}. \tag{5.3}$$

Working up through the hierarchical levels of the system, assembly, subsystem, device, string and finally array reliabilities are progressively calculated. The calculation scheme

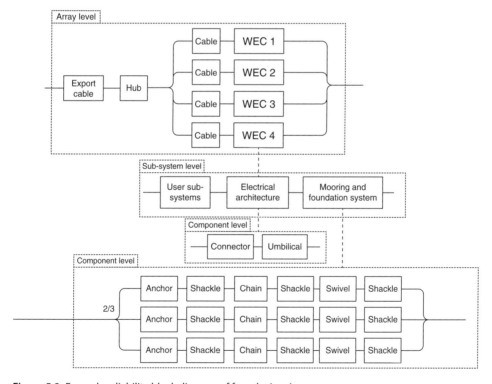

Figure 5.8 Example reliability block diagram of four devices in an array.

---

5 This could be the required lifetime of the array.

that can be used is dependent on the relationship between elements at the same hierarchy level. For example, using the layout shown in Figure 5.8, it can be seen that each mooring line is composed of a number of components connected in series. In this case reliability of each line (or assembly) is calculated as the product of n component reliabilities:

$$R_{\text{ser}}(t) = \prod_{i=1}^{n} e^{-\lambda_i t}. \tag{5.4}$$

This assumes that any parallel elements possess active redundancy (i.e. total failure occurs when all elements fail). In this case reliability is calculated differently from a series system, using the equation

$$R_{\text{par}}(t) = 1 - \prod_{i=1}^{n} (1 - R_i(t)). \tag{5.5}$$

A third case can also be analysed, where failure of part of a parallel system is tolerated due to the provision of redundancy. The mooring system in Figure 5.3 falls into this category, where failure of one mooring line is permitted as part of an accident limit state analysis [13]. This scenario is widely referred to as an 'm of n' system (where n is the number of mooring lines and $m = n - 1$) using the following approach:

$$R_{mn}(t) = \sum_{r=k}^{n} \binom{n}{i} R_i^r(t)(1 - R_i(t))^{n-r}. \tag{5.6}$$

Mean time to failure is then calculated as the reciprocal of component or (equivalent) assembly, subsystem, device, string or array failure rates, for example,

$$MTTF_{\text{array}} = \frac{1}{\lambda_{\text{array}}}. \tag{5.7}$$

The way in which equivalent failure rates are calculated for all hierarchical levels except the component level is dependent on the relationship between elements within the same level. For the mooring components within each mooring line in Figure 5.3, the equivalent failure rate of each line is calculated as the sum of individual component failure rates,

$$\lambda_{\text{sys}} = \sum_{i=1}^{n} \lambda_i, \tag{5.8}$$

where λ_{sys} is the system failure rate and the λ_i are the individual subsystems/components.

The method of calculating equivalent failure rates for parallel systems with active redundancy is more complex than for series systems. In this case the binomial expansion of the terms is as follows:

$$MTTF_{\text{array}} = \sum_{i=1}^{n} \frac{1}{\lambda_i} - \sum_{i=1}^{n-1} \sum_{j=i+1}^{n} \frac{1}{\lambda_i + \lambda_j} + \sum_{i=1}^{n-2} \sum_{j=i+1}^{n-1} \sum_{k=j+1}^{n} \frac{1}{\lambda_i + \lambda_j + \lambda_k} - \ldots + (-1)^{n+1} \frac{1}{\sum_{i=1}^{n} \lambda_i}. \tag{5.9}$$

Table 5.1 Probability of failure bands and corresponding frequencies.

Probability of failure/annum	<0.01	$0.01 \le p < 0.1$	$0.1 \le p < 1.0$	$1.0 \le p < 10.0$	$10.0 \le p < 50.0$	$p \ge 50.0$
f	0.0	1.0	2.0	3.0	4.0	5.0

This could be based on relevant standards (e.g. for mooring systems, DNV-OS-E301 specifies an annual probability of failure of 1×10^{-4} for ultimate limit state and consequence class 1 [13]). Alternatively, specific reliability targets could be used.

For further assessment, risk priority numbers (RPNs) can be used, which are numbers for a metric incorporating probability of failure and consequence severity. The use of RPNs can be approached using NREL's *MHK Technology Development Risk Management Framework* [14] and can be thought of as a simplified way of carrying out integrity management. The analysis is typically conducted at component level in order to be able to identify at-risk components readily. The probability of failure is first calculated for each component and each value is allocated a corresponding frequency (with values provided in Table 5.1). Once the corresponding frequencies have been assigned to each component, RPNs are calculated using the relevant severity level (*SL*) of the failure rate:

	Severity level	
	1	2
0	0	0
1	1	2
2	2	4
3	3	6
4	4	8
5	5	10
6	6	12
7	7	14
8	8	16
9	9	18
10	10	20

(Risk priority number)

Figure 5.9 Risk priority number colour codes.

$$RPN = SL \times f \tag{5.10}$$

where $SL = 1$ and 2 for non-critical and critical failure rates, respectively. For visual representation of these values colour codes are often applied, as illustrated in Figure 5.9.

5.2.4 Component Testing

Many important technologies are characterised in their early stages of development by frequent breakdowns, unanticipated failure modes, low reliability and high unavailability. This is true for the early motor cars, aeroplanes and computers. Unfortunately, it has also been true for many of the early MRE devices. Survivability and reliability of devices have been identified as major challenges for the MRE sector to emerge successfully from the research/testing/prototype stages towards economic and commercial deployment [15]. Consequently, reliability assessment and demonstration for MRE devices is essential. In particular, the characterisation of appropriate component failure rates is recognised as a key requirement for the deployment of commercial-scale MRE arrays. This is the case for new components along with non-marine components being used in a marine environment and for marine components being used under different operating conditions [16]. For both cases the available reliability information is often scarce and/

or not directly applicable. The challenge is to accurately establish failure modes and associated reliability data for the components and subsystems of marine renewable devices and, when reliability is unsatisfactory, to develop components that are fit for their purpose in a hostile marine environment.

Reliability testing and demonstration are an integral part of the overall reliability programme of product development. While reliability testing aims to reveal any design weaknesses and tries to establish if the component/equipment under test meets the operational requirements, the demonstration of reliability will also provide evidence that the component meets a specified reliability target under stated conditions [17].

Reliability testing is widely used in numerous industries to provide assurance that components and products are fit for purpose. The general requirements and procedures are described, for example, in British Standard 5760-4:2003 [18]. Electronic components have been systematically tested for decades, providing information about failure modes, MTTF and stressors [19]. Other industries that make extensive use of reliability testing are the automotive, aviation, offshore oil and gas, mining and astronautics industries. In all such cases the reliability of systems has to be assured before operational deployment/product launch, and long-term specifications (e.g. operational safety, fatigue life) have to be established with limited operational experience. A bibliographical list of reliability test applications was compiled by Dhillon [20]. Cardenas *et al.* [21] used accelerated testing to achieve a more rapid degradation process of umbilical cables in order to investigate the combined environmental effects of cyclic loading, marine environment and ultraviolet radiation.

A large variety of tests are described in the literature, but a broad distinction can be made in terms of the phase of product development. At an early stage the focus lies on accurate reliability prediction, whereas later stages are more concerned with reliability growth considerations. In general, the type of tests can be classified by their purpose [22]:

- Testing of full-scale structures – indicating fatigue-critical items of a structure, obtaining crack growth data to schedule service inspections, validation of damage tolerance requirements.
- Testing of specimens – obtaining data on crack growth and specimen life to support and optimise structural design.
- Comparative tests – investigation of design parameters and variables.
- Model validation testing – verifying model predictions, e.g. for fatigue life and crack growth.

The type of test can be further distinguished, depending on how accurate the field loads are replicated and to what extent they are accelerated [23]:

- field testing of the actual system under accelerated operating conditions;
- laboratory testing of actual system through physical simulation of field loads;
- virtual (computer-aided) simulation of system and field loads.

Raath [24] investigated the relationship between test type, acceleration and required safety factors. He argues that the test type determines the necessary safety factor. For example, in-service testing applies the actual loadings to the component, so lower safety factors can be applied with confidence, whereas a series of single-axis tests necessitate a higher safety factor to compensate for the simplified load assumptions. However, if tests are highly accelerated, they implicitly assume failure mechanisms are independent of cycle frequency, which is not always the case (e.g. corrosion fatigue).

The process of providing evidence of system/component reliability under operational conditions can generally be divided into four successive steps [25]:

- measuring of realistic load data;
- identification of representative loading regimes;
- testing of a (representative) sample on a laboratory test rig under the representative loads;
- root cause analysis and statistical evaluation of test results.

To generate a representative and meaningful test regime an iterative five-stage process has been applied in industrial applications [26–28];

1) Define an operating period and conditions (e.g. operating time, environmental conditions, wave climate) resulting in an expected operating profile.
2) Undertake field load measurements under operational conditions (or alternatively calculations/ assumptions).
3) Process the measured load samples to establish a 'load library' for different operating conditions.
4) Implement the operational profile for the planned test regime.
5) Finally, generate the test load sequence through the combination of operating profile and load library segments.

Steps 1 and 2 would represent data input related to a specific MEC device and site conditions. Ideally, the operating conditions and load data would be defined and measured directly in the field, e.g. the wave climate and loads experienced by the components of a prototype device or subsystem under full-scale sea conditions.

Steps 3 to 5 are based on established theoretical methods that can be applied and which will be discussed in the following. To close the loop for comprehensive component reliability testing a root-cause analysis of occurring failures would need to be implemented in addition.

5.3 Moorings and Anchors

5.3.1 Overview on Moorings and Anchors

The purpose of a mooring and foundation system is to provide offshore equipment with a means of station-keeping that is sufficiently robust to resist environmental loading (e.g. tidal, wind, wave, current and ice), impact and operational procedures. Although the station-keeping of vessels and offshore equipment has been carried out for centuries, MRE devices represent a relatively new field of application with specific requirements and challenges. The Wind, Wave and Tidal Energy report by Research Councils UK [29] identified the development of cost-effective MRE foundations and support structures for deep water as a 'high-level research challenge': 'moorings and seabed structures require design optimisation to improve durability and robustness and reduce costs, particularly for deep water tidal; and improved station-keeping technologies'.

To date a number of wave and tidal energy technologies have been trialled offshore to establish proof of concept. Of the concepts which have so far reached this development stage (Technology Readiness Levels 7–8) most are either single devices or small arrays (fewer than 10 devices). In order for MRE to reach commercial viability, large-scale deployments comprising many tens or hundreds of devices are required (Figure 5.10). The provision of robust and economical mooring and foundation systems for this number of devices over the lifetime of the array will be a significant challenge to the MRE industry. Previous published assessments of mooring or foundation options have assessed station-keeping options for generic devices (e.g. [31]), reflecting the state of the industry and variety of possible MRE device designs. Studies focused on particular technologies (e.g. [32]) provide valuable insight into the decision-making process of device developers.

MRE mooring systems can be divided into three categories: passive, active and reactive. The main function of a passive mooring system is to provide station-keeping only. These systems tend to be used for large floating platforms which support

(a)

(b)

Figure 5.10 Examples of MRE arrays: (a) Wave Star wave energy converter; (b) Sabella tidal turbine array [30].

(a)　　　　　　　　　　　　　　　　(b)

(c)　　　　　　　　　　　　　　　　(d)

Figure 5.11 Example WEC devices: (a) BlueTEC floating platform (image source: Bluewater); (b) Poseidon Floating Power Plant (image source: Floating Power Plant A/S); (c) two Pelamis P2 WECs; (d) CETO 5 and CETO 6 WECs [33].

multiple MRE devices (e.g. Figure 5.11). In addition to providing station-keeping, the response of an active mooring system has a significant influence on the dynamic response of the moored device, to the extent that both responses are closely coupled and hence affect the power output of the device. Many of the proposed WEC designs fit into this category (e.g. Figure 5.11). In the case of a reactive system the mooring is an integral part of the system, perhaps linking the floating part of a WEC to the power take-off (PTO) system, for example the CETO 5 device and the devices studied by Vicente et al. in [34].

5.3.2　Special Mooring Design Needs

While several variants exist (Figure 5.12); broadly there are two geometries relevant to MRE devices: catenary and taut-mooring systems. These types are widely used in the offshore industry, particularly for floating production storage and offloading (FPSO), floating production storage (FPS) facilities (Figure 5.13) as well as single-point mooring and reservoir and catenary anchor leg mooring (CALM) structures. Other categories of moorings include single anchor leg mooring, articulated loading column, and fixed tower mooring systems. In terms of device scale, geometry and mass, the CALM

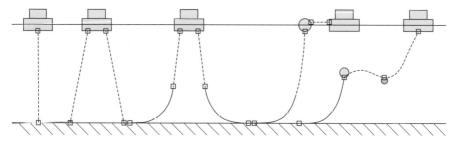

Figure 5.12 Schematic of possible mooring arrangements for a single MRE device: (from left) taut-moored systems with single and multiple lines; basic catenary system; catenary system with auxiliary surface buoy; and lazy wave system with subsea floater and sinker. The combined use of synthetic ropes and chains (dashed and solid lines, respectively) may be feasible.

Figure 5.13 FPSO with turret mooring system [35].

buoy [36] has perhaps the closest similarities with large buoy-like MRE devices, albeit is designed to operate in much deeper water depths (around 1200 m).

Catenary mooring systems comprise single or multiple lines with a catenary geometry which provide the necessary horizontal and vertical restoring forces to keep a device on station while allowing for changes in the water depth due to tidal variations. For MRE devices the compliance of a catenary system allows motions in several degrees of freedom for power generation. The horizontal compliance of a catenary mooring system can be increased by use of the 'lazy-wave' system which includes float and sinker components attached to the line. It is necessary at the design stage to determine what level of compliance is required to achieve the permissible magnitude of mooring tensions and device displacements. A system which is excessively compliant may allow large device motions in the desired degree(s) of freedom (i.e. heave for a WEC point absorber) but increase the risk of collision with adjacent devices or water users.

Although it is possible to use steel components (wires and chains) for the entire length of the line, alternative materials (i.e. synthetic ropes) tend to be used for the mid or upper sections of the line to reduce the cost and weight of mooring system. 'Rider' or 'ground' chains are used for the lower sections to provide tension to the line and transfer loads horizontally to the anchor or foundation.

Taut-mooring systems provide a much stiffer connection between the device and seabed, with compliance only provided by the axial properties of the mooring components, such as synthetic ropes. Ropes constructed from high-modulus polyethylene (HMPE) have been successfully used for platforms located in deep and ultra-deep water locations [37]. Because both horizontal and vertical restoring forces are provided by this type of mooring system, it is necessary to specify foundations and anchors which can operate under both loading directions (usually drag embedment type anchors are not suitable in this case). Unless a large mooring footprint is specified, the limited compliance of a taut-moored system may mean that the device becomes submerged during large-amplitude waves or in locations with high tidal ranges. Full or partial submersion of the device is not an issue for some designs (i.e. CETO 5) and may be a way of limiting device displacements in large-amplitude waves [38].

The proposed separation distance of MRE devices in arrays (tens of metres) means that consideration must be given to the siting of devices and the required level of mooring system compliance. Close separation distances have been suggested as a way of sharing electrical infrastructure and also to take advantage of the influence of hydrodynamic interactions on power production [40, 41]. This is necessary to reduce the risk of mooring line entanglement and device collisions and to allow suitable clearances between the devices for vessel access during installation, maintenance and decommissioning procedures. The separation distance specified in the DNV-OS-E301 Position Mooring guidelines [13] between offshore accommodation units and fixed equipment is necessarily large for the application, but not relevant for MRE devices which are typically unmanned during operation. An alternative and arguably more suitable approach suggested in the DNV-OS-J103 Design of Floating Wind Turbine Structures guideline [42] is to base the separation distance on maximum possible surge or sway displacements during normal operation and the possibility of mooring line failure (accident limit state, assuming that the mooring system has built-in redundancy). Shared mooring system infrastructure as shown, for example, in Figure 5.14 (i.e. common anchoring points and/or device interconnections), has been suggested as a means of reducing both capital costs and also the number and difficulty of installation/decommissioning operations [39, 43, 44]. Such benefits are clearly scalable to large arrays; however, with the exception of MRE devices attached to a common structure (e.g. Wave Star Energy's Wave Star system and MCT SeaGen), to date no arrays have been deployed comprising shared mooring or anchoring systems.

5.3.3 Mooring Design Simulation and Analysis

Numerical models provide designers and operators with a platform to simulate the conditions of an offshore situation without having to carry out difficult and expensive operations. In terms of mooring and foundation systems it is essential to undertake modelling of the system in order to understand how it will respond to environmental loading before

(a)   (b)

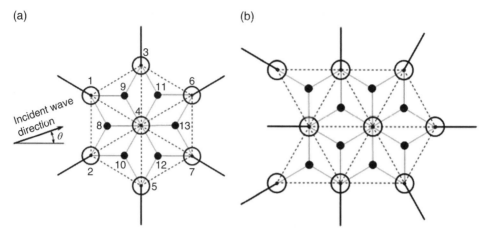

Figure 5.14 Schematic of proposed array layouts comprising (a) seven and (b) nine buoys with interconnecting lines (grey) and shared connection points (black dots) [39].

it is deployed and hence to ensure that the design is fit for purpose. Numerical models allow conditions which would be unrealistic to test using laboratory methods (e.g. testing how a particular mooring system would react to a storm of the severity that is likely to occur once in 100 years) to be simulated in a fast, safe and cost-effective way.

There are many commercially available tools on the market each with different capabilities, requirements and costs, from simple programs which can model just one part of a system through to computational fluid dynamics (CFD) simulations of arrays of devices (Figure 5.15). Many models can be run using standard personal computer equipment; however, greater computing power, such as provided by computer clusters, is often required for more complex models. For the most demanding models supercomputers may be needed to run the simulations.

Mooring and foundation numerical tools designed for the shipping, petroleum and offshore construction industries have many features that can be used in the modelling of MRE devices. It is possible to conduct detailed analysis of moored offshore structures using several commercially available tools. The body of the structure, mooring lines, risers and other components can be modelled, allowing static, quasi-static or dynamic analyses of a system to be carried out. The hydrostatic and hydrodynamic responses of the device can be simulated by software such as WAMIT, AQWA, WADAM and Sesam HydroD. Analysis of the mooring lines, chains and components can be performed by software including Orcaflex, Optimoor, Deeplines, DIODORE, ARIANE7, Sesam DeepC and AQWA Suite.

Although widely used in the design of offshore equipment, it is not possible to model all of the distinct features of MRE devices using existing mooring system software. Key subsystems such as PTO systems are not covered by the majority of currently available tools. 'Wave-to-wire' models such as WaveDyn by GL-Garrad Hassan and ACHIL-3D by Ecole Centrale de Nantes have been developed to simulate the dynamic response of WECs. WaveDyn is designed to be a standard software tool for the design of WECs and comprises modules to model the hydrodynamics of WEC devices, the PTO system, structural dynamics and the mooring system. Orcaflex is a

(a)

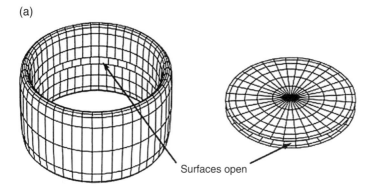

Surfaces open

(b)

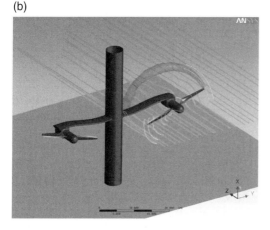

Figure 5.15 Example design tool applications: (a) an MRE device WAMIT mesh [45]; (b) CFD simulation of a tidal turbine using ANSYS (image used courtesy of ANSYS Inc).

widely used software package for a number of modelling applications. It can be used to undertake time domain analysis of different types of mooring systems, vessel objects and buoy objects (e.g. Figure 5.16). MARINTEK/SINTEF have developed a number of tools, such as Simo and Riflex, which can undertake detailed analysis of offshore vessels, structures and buoys. For example, the Sesam DeepC package which is similar to Orcaflex and AQWA can be used to conduct time domain analysis of coupled integrated moored structures as well as fixed bodies.

Some of the tools mentioned are standalone and suitable for simulating the

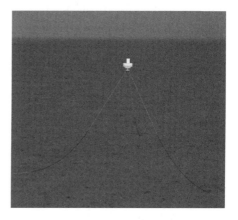

Figure 5.16 Visualisation of a simplified Orcaflex model of the South West Mooring Test Facility.

dynamic response of a moored body based on its physical attributes (geometry, dimensions, mass, inertia, etc.), mooring system, site characteristics and environmental

Table 5.2 Input parameters typically required by mooring system software.

Contact parameters	Body/vessel motion characteristics
Mooring line geometry	Environmental loading
Mooring line component properties	Site characteristics

Table 5.3 Typical frequency-dependent hydrodynamic parameters calculated by potential theory codes.

Excitation forces and phases (1st and 2nd order)	Added mass coefficients
Mean drift forces and moments	Radiation damping coefficients
Response amplitude operators	Pressure and fluid velocities

conditions. Other software packages require hydrodynamic parameters to be specified in order to function (e.g. Table 5.2). Hydrodynamic parameters can be quantified by physical testing and/or using numerical codes based on potential theory (e.g. Table 5.3).

Potential theory codes are used to solve the velocity potential around a defined geometry caused by the radiation and diffraction of an incident wave-field. Boundary element methods (BEMs) are used to integrate the flow-field over the immersed surface of the geometry. This modelling approach assumes that the fluid is ideal (inviscid, incompressible and irrotational) and that the first- and second-order linear wave forces resulting from small-amplitude waves lead to small device motions. Hence the variation of calculated hydrodynamic parameters with varying device position (i.e. draft or large displacements) is not accounted for. Commercially available tools include: WAMIT, AQUAPLUS, AQUADYN, HYDROSTAR, AQWA Diffraction and DIFFRAC.

These parameters are frequency-dependent and can be used to solve the equation of motion in the frequency domain, provided that the incident waves are harmonic. In order to account for irregular waves, superposition of linear parameters is used. It is possible to include a basic representation of the mooring system in some potential theory codes (i.e. with a user-defined stiffness matrix).

Time domain analysis is necessary for nonlinear system responses, where the equation of motion is solved at each time step. This allows coupled analysis to be carried out in which one or more subsystems are nonlinear (i.e. the PTO system, mooring system and the performance of devices in arrays). It is unclear how effective software packages such as Orcaflex are at accounting for the spatial variability of device response due to hydrodynamic interactions across devices arrays, even if hydrodynamic parameters are obtained from array-based BEM simulations. In terms of mooring system design such interactions should definitely be accounted for because device displacements and mooring loads are likely to differ between individual devices.

More complex hydrodynamics, wave breaking, sloshing, run-up, nonlinear storm waves, large amplitudes (i.e. resonant responses, viscous effects due to large velocities) require advanced methods as the flow is no longer irrotational. Viscous effects can be accounted for either in linear models (i.e. the addition of viscous drag or damping using the Morison equation, or nonlinear Froude–Krylov forces on the instantaneous immersed surface) or using CFD to solve the Reynolds-averaged Navier–Stokes equation of the flow. Smoothed particle hydrodynamics is a relatively new application of CFD; it is a mesh-free method of modelling the flow by dividing the fluid into discrete particles. This enables the prediction of variables such as velocity, direction of flow, pressure and energy and can be used to model fluids in complex situations where traditional grid-based CFD simulations cannot.

5.3.4 Materials for Marine Anchoring Systems

Examples of commonly used mooring components are shown in Figure 5.17. Economic considerations for typical components are reported in the EquiMar report 'Consideration of the cost implications for mooring MEC devices' [46]. Although these components provide bending flexibility along their length, the axial stiffness of steel components is considerably higher than that of alternative materials (Table 5.4). Ropes constructed from synthetic materials such as polyester, aramid, nylon and HMPE have been used successfully for the last two decades in the offshore industry for vessel mooring, towing and equipment station-keeping [47]. These materials have particular advantages

(a)   (b)   (c)

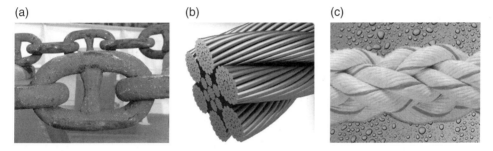

Figure 5.17 Examples of mooring components: (a) Studlink anchor chain (image source: https://pixabay.com/); (b) Bridon Diamond Blue wire rope (image source: www.bridon.com); (c) Lankhorst TIPTO®EIGHT 8-strand plaited rope (image source: http://www.lankhorstropes.com/).

Table 5.4 Selected properties of steel and several synthetic fibre materials [47].

Material	Density (g/cm³)	Melting/charring point (°C)	Moisture (%)[a]	Modulus (N/tex, GPa)	Tenacity (mN/tex)	Strength (MPa)	Break extension (%)
Steel	7.85	1600	0	20, 160	330	2600	2[b]
HMPE	0.97	150	0	100, 100	3500	3400	3.5
Aramid	1.45	500	1–7	60, 90	2000	2900	3.5
PET	1.38	258	<1	11, 15	820	1130	12
PP	0.91	165	0	7, 6	620	560	20
PA6[c]	1.14	218	5	7, 8[d]	840	960	20

a)   At 65% rh and 20 °C.
b)   Yield point of steel.
c)   PA6.6 has a higher melting point (258°C) than PA6.
d)   The modulus and strength of nylon are approximately 15% lower when wet.

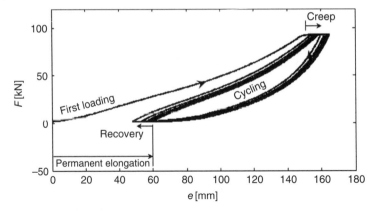

Figure 5.18 Load-extension behaviour of a new nylon mooring rope sample subjected to 10 cycles of bedding-in (tests reported in [50]).

compared to steel components, including low cost and mass (per unit length) and load-extension properties that can be harnessed to reduce peak loadings [48].

Unlike steel components, materials such as nylon and elastomers have nonlinear and time-dependent load-extension properties [49]. Changes to the compliance of these materials are possible over the lifetime of the component, and this should be factored into the design of the mooring system. For example, the initial loading of certain synthetic ropes results in permanent extension (Figure 5.18). As part of a maintenance plan, component inspection should be carried out (e.g. [51, 52]). Relevant procedures for the design and usage of mooring system components include DNV-OS-E301 Position Mooring [13], DNV-OS-E302 Offshore Mooring Chain [53], and DNV-OS-E304 Offshore Mooring Steel Wire Ropes [54].

5.4 Foundations

5.4.1 Introduction to Foundation Requirements

Foundations are used to provide a durable connection between the seafloor and the MRE device usually via a support structure (e.g. a seafloor-mounted tidal turbine). Foundation selection is generally dictated by the characteristics of the site, the operational and maintenance requirements of the device as well as the type of the support structure and the loads which it transfers to the foundations (e.g. Table 5.5). At present MRE foundation designs are broadly similar to those used in the existing offshore industries (Figure 5.19): gravity base structures (GBSs), piles (pipe and pin), shallow foundations (similar to GBSs with perimeter skirts and shear keys) and suction buckets.

Project economies tend to have a direct bearing on which technology is selected, as foundation systems are generally not off-the-shelf components with costs dependent on design and raw material variability (e.g. the commodity cost of steel [56]). The construction and installation of foundations broadly represents approximately 10% of the cost of energy of MRE projects [57], of a similar magnitude to offshore wind foundations [58].

Table 5.5 Compatibility matrix for MRE foundations based on support structure and seafloor type (soil suitability for pile and GBS foundations from [59] and suction caissons from [60]).

		Support structure					Seafloor type						
		Pile	Sheath	Telescopic	Shroud	Jacket	Soft clay/mud	Stiff clay	Sand	Boulders	Soft rock/coral	Hard glacial till	Hard rock
Foundation	Pile	✓	✓	✓	✓	✓	✓*	✓	✓		✓	✓	✓*
	Shallow	✓	✓	✓	✓	✓	✓	✓	✓	✓	✓	✓	✓
	GBS	✓	✓	✓	✓	✓	✓	✓	✓	✓	✓	✓	✓
	Suction bucket				✓	✓	✓	✓	✓				

✓* Other technology types may be preferable.

Figure 5.19 (a) Atlantis AR-1000 turbine with GBS (image source: Atlantis Resources Ltd); (b) quadrapod support structure for use with pin piles [55]; (c) schematic of a shallow foundation with shear keys; (d) suction bucket foundation (photo courtesy of Forewind).

Installation, maintenance and decommissioning requirements will also steer initial design decisions due to the variability of vessel and equipment day rates and site accessibility (e.g. weather windows [61]). A recent example of this is the choice of quadrapod support structure and multiple piles instead of a single monopile for the MCT device (Figure 5.19b; [55, 62]). Innovative designs such as shared infrastructure and towed-out structures can help to reduce both the capital expenditures and operational expenses of foundations for large-scale commercial arrays comprising hundreds or thousands of devices by reducing the number and complexity of offshore operations.

To-date most MRE projects have opted for conventional foundation technologies, drawing upon prior knowledge and certification practices developed by the offshore petroleum and wind industries [63–66]. Without similar long-term deployment experience, MRE developers are forced to adopt a conservative design approach to ensure that the specified foundation(s) and surrounding seafloor have sufficient capacity to withstand the effects of cyclic and infrequent loading, environmental exposure and changes to material properties over time. Uncertainty in the probability of these processes occurring necessitates the application of safety factors to design loads. Most of the currently available MRE guidelines generated by certification agencies (e.g. [66–69]) refer back to existing offshore guidelines (e.g. [70]) with modified consequence criteria to suit this new application. Additional non-certification agency MRE guidelines have been also developed (e.g. [71, 72]).

The selection and design of foundations are also greatly influenced by the expected environmental impact of potential technologies and can be a constraint of the site lease. Noise propagation during device installation is an issue which has received considerable attention [73, 74], due to the potential negative impacts on marine life. The installation of piles is a significant source of noise [75] requiring the use of mitigation techniques such as marine mammal observers and staged installation procedures. The environmental effects of MRE foundations are not yet fully understood due to the lack of long-term monitoring data and difficulty of quantifying effects. Studies have noted that the presence of human-made structures fixed to the seafloor is likely to alter local flow patterns (i.e. affecting sediment transport [76]) as well as being a possible migration barrier and collision risk for marine life. Not all impacts may be negative, though; for example, the no-catch zones of arrays could also provide habitat for certain species [77]).

5.4.2 Design Concepts for Sediment–Foundation Interactions

The foundation selection criteria introduced in the previous subsection focused on design criteria of the foundation and device. The site bathymetry, geophysical characteristics of the seafloor, its composition and geotechnical properties will significantly influence the selection and performance of a foundation. Informed decisions on foundation suitability and design therefore require detailed site data gained from site surveys and laboratory testing, the procedures of which have been standardised and are widely used (e.g. [78, 79]). Given that the foundations of large MRE device arrays are likely to be positioned over several square kilometres, it is possible that the lease area will incorporate several geological environments and bedform profiles. Because both the topology of the seafloor and its constitutive materials (and hence geotechnical properties) have a direct impact on foundation suitability ([59] and Table 5.5), certain sites may significantly constrain the foundation selection process. Perhaps surprisingly,

Table 5.6 Typical soil index properties and laboratory test methods.

	Parameter(s)	Procedure
General	Classification	By origin or grain size
Index property	Water content	ASTM D2216
	Grain size distribution	ASTM D422-63
	Atterberg limits (liquid and plastic limits and plasticity index)	ASTM D4318
	Carbonate and organic carbon content	ASTM D4373-02
	Dry unit weight	ASTM D2937
	Specific gravity	ASTM D854 or ASTM C127
Physical property	Undrained shear strength, over-consolidation ratio	Vane shear test or unconsolidated-undrained triaxial test
	Unconfined compressive strength	Unconsolidated-undrained triaxial compression test
	Effective stress parameters (Mohr–Coulomb failure envelope)	Consolidated-drained or consolidated-undrained triaxial compression test
	Compression and recompression indices, coefficients of consolidation, permeability and secondary compression, over-consolidation ratio	One-dimensional consolidation test

the possibility of geohazards or seafloor heterogeneity impacting MRE array design has yet to receive much attention, except for a few studies (e.g. [80]).

Due to the costs associated with sampling over large areas, surveys usually start with a search of published literature and data. For some regions coarse estimations of the seafloor soil type can be obtained from data in the public domain (e.g. the EMODNET platform [81]). In order to determine if, from a cost-versus-risk standpoint, a site is suitable for development, further information can be gained from acoustic sub-bottom profiling (to determine strata layers), limited (i.e. shallow) soil sampling, sidescan sonar surveying and even visual remotely operated vehicle surveys. Because the information gained from such a reconnaissance will be limited in detail, a full assessment will require further *in situ* testing (i.e. using vane or cone penetrometers) as well as grab or core sampling, particularly if the site is complex (i.e. if the proposed foundation will encounter significant variations in horizontal and/or vertical soil heterogeneity). Variability in physical soil composition (i.e. degree of uniformity) and state means that quantification of soil sample properties is recommended rather than relying exclusively on published tables of properties [59]. The choice of soil properties to investigate will depend on which soil types have been obtained from sampling and also the relevance of particular properties to the type of foundation technology which has been proposed (Table 5.6).

Most of the MRE performance studies conducted to date have focused on the influence of device position on flow pattern modification and loads (e.g. [82]). The impact of devices on seafloor materials has received less attention, although it has recently been demonstrated that far-field sediment transport can be altered many kilometres downstream from devices (e.g. [76]). Sediment scour and its effect on foundation integrity are a particular issue for offshore wind turbines installed in certain soils, and several mitigation methods are currently used (e.g. rock dumping). Tidal current turbines are usually positioned in high-energy sites and therefore scour will also need to be considered for foundations installed in seafloors susceptible to this process. While studies have already been conducted to investigate this, *in situ* measurements are required to validate the numerical techniques used (e.g. [83]).

The loads experienced by MRE foundations and surrounding geomaterials are dependent on the mode of device operation, environmental loading, in addition to the choice and design of foundation and seafloor composition [84]. In general terms, loads transferred to the seafloor via the foundation can be split into three categories: static loading (e.g. due to the deadweight of the structure); cyclic loading (caused by wave, wind and current loading); and infrequent loading (impulse loads caused by turbulence, breaking waves, wind gusts, impact and seismic events). Within these loading categories there are a number of time-scales; for example, a tidal stream device will experience loads due to water-level changes lasting several hours in addition to harmonics at turbine rotation and blade passing frequencies. Such effects are relatively well under-stood for offshore wind turbine monopiles with the specification of soft–soft, soft–stiff or stiff–stiff structures to avoid potentially damaging excitation frequencies [85]. As a result, the loading regimes experienced by MRE foundations and anchors are typically complex, and likely to vary in terms of mean load, load amplitude and load frequency as well as direction [84]. Such variability is further compounded for shared foundation points. Highly variable loading is likely to influence the properties and loading capacity of certain seafloor materials. In particular, transient cyclic loading can cause reductions in sediment strength if excess pore pressures develop in partially or undrained soils (perhaps due to soil anisotropy) and are not permitted to fully dissipate over successive load cycles [86]. Drained or partially drained soils can also experience volume changes resulting in a reduction in soil stiffness and strength. If multiple foundation points are in very close proximity then these sediment effects can be further compounded by the influence of neighbouring foundations experiencing cyclic loading. In extreme cases liquefaction [85] may occur in low-permeability sediments (including fine sands and coarse silts). While these effects have been studied for conventional offshore founda-tions subjected to variable loading (e.g. [87]), few studies have considered these load effects on soil–MRE foundation interactions [88].

5.4.3   Analysis Techniques for Seabed and Foundation Systems

The analysis of MRE foundations is typically an iterative process (see [59]) in which design parameters are altered (e.g. foundation geometry) until the loading capacity of the soil–foundation interface is sufficient compared with the predicted design loads and certain failure modes are avoided. Figure 5.20 illustrates several potential failure modes of offshore piles installed in rock. Other foundation failure modes include bearing capacity failure, overturning, horizontal sliding, slow or non-uniform dis-placements, scour and undermining. In addition, it is vital that the foundation is structurally fit for purpose in terms of holding capacity but also long-term durability in what is generally an extremely harsh operating environment. The analysis method used

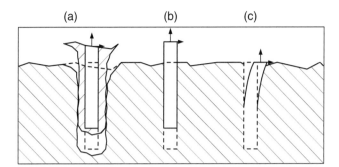

Figure 5.20 Typical failure modes of offshore piles installed in rock reproduced from [59]: uplift failure of (a) pile and rock mass and (b) grout-to-rock bond; (c) lateral bending failure of rock and pile.

Table 5.7 Soil–foundation interaction performance metrics based on [59, 60].

		Performance metrics
Foundation	Pile	Lateral load capacity, compressive bearing capacity (including end bearing and shaft resistance), uplift capacity
	Shallow	Compressive bearing capacity, lateral sliding resistance, eccentricity of loading, settlement (i.e. suction below foundation)
	GBS	Compressive bearing capacity, uplift capacity, lateral sliding resistance, eccentricity of loading, settlement (i.e. suction below foundation)
	Suction bucket	Compressive bearing capacity, uplift capacity, failure envelope, embedment due to self-weight and suction

and required performance metric depend on the foundation type but are broadly focused on vertical and horizontal bearing capacity, holding capacity, consolidation and installation considerations (i.e. settlement); see Table 5.7.

Simple analytical methods have been developed to quantify the metrics listed in Table 5.7 to a degree of accuracy that is usually acceptable for initial analysis of short- and long-term static and cyclic loading [59, 60]. Many of these approaches are based on empirical data of soil properties and soil–foundation performance derived from experimental tests (i.e. $p–y$ curves used in lateral pile analysis [78]). These methods tend to be used in piecewise fashion in order to check that the metrics listed above satisfy the required performance criteria. Although simple to implement and computationally inexpensive, this means that solutions are potentially suboptimal unless further design refinement is carried out.

Seafloor materials are inherently complex, and considerable variations of composition as well as physical (i.e. mechanical and hydrological) properties can occur both locally to a foundation and also across a site. The complexity of applied loads means that sophisticated analysis may be necessary to fully capture soil–foundation interactions and failure modes. Finite-element analysis (FEA) has been widely used to model 2D and 3D foundation structures and soil–foundation interactions. Because FEA techniques

Table 5.8 Constitutive soil models used in conventional foundation analysis.[a]

Model	Applicable soil type(s)	Attributes
Bounding surface plasticity	Clay	Isotropic and kinematic hardening. Captures cyclic behaviours including cyclic shakedown and strength degradation.
BWGG	Cohesive soils	1D static or dynamic response. Nonlinear effects including cyclic mobility, liquefaction, stress–strain loops, pore pressure build-up. Stiffness and strength degradation during cyclic loading.
Disturbed state concept	Cohesive (saturated) soils	Inelastic loading and unloading responses, drained and undrained behaviour and pore water pressure. Cyclic loading.
Drucker–Pager	Sand, clay	Examples of modelling plastic deformation with von Mises yield criterion.
Elastoplastic	Soft cohesive soils	Examples of von Mises and Tresca yield criterion being used. Depth dependent shear strength. Can be applied in some cases to cyclic loading problems.
Elastoplastic based on Cam–Clay model	Clay/sand, layered material	Applicable to a range of strain levels. Can model cyclic loading and liquefaction (sand/clay layering). Previously used in earthquake applications.
Generalised plasticity framework	Sand	Isotropic response under monotonic and cyclic loading (granular soils), nonlinear elastic and plastic soil responses. Liquefaction in loose sands. Memory effects and cyclic loading accounted for.

a)   When seafloor stability is an issue, the limit equilibrium method is used.

consider the domain as a number of small elements, they have a number of advantages over simplified methods, namely being able to account for complex geometries, linear and nonlinear soil and structure behaviour, fluid transport, soil heterogeneity and time-varying mechanisms [64]. Of particular importance is the inclusion of multiple material constitutive models that can be used to represent the behaviour of different soil types (Table 5.8) as well as user-defined models. The accuracy of these models will have a direct impact on predicted soil–foundation interactions and therefore care must be taken in selecting a suitable model as not all effects may be adequately represented (e.g. plastic deformation caused by cyclic loading).

At present it is not clear which of the constitutive models that have been developed can be applied successfully to MRE foundation scenarios, and hence further research is required in this area [84]. Model parameters are usually obtained from laboratory or *in situ* testing. For sites comprising a large number of soil types and/or variation in soil properties extensive (and potentially costly) sampling may be necessary in order to model accurately spatial variations in soil properties. Warranting consideration are the processing requirement and computational limits (i.e. number of degrees of freedom) of large-scale models comprising a great number of (possibly interacting) MRE foundations spread over a computational domain representative of the lease site. While most commercial codes are designed to run on commercial off-the-shelf workstations, complex models may need to be run on parallel computing systems to avoid prohibitive simulation times. Mesh refinement techniques, such as applying fine elements close to structures and increasingly coarse elements further away (e.g. [64]) or comparison of different element schemes (e.g. [86]), are often used to reduce the computational requirements of the model. However, caution must be exercised when specifying coarse elements to ensure that any interactions with far-field structures are adequately modelled; this is typically achieved by conducting a mesh sensitivity study. Although undoubtedly sophisticated in terms of capturing soil–structure interactions, full multi-physics coupling is beyond the scope of the commercial packages currently available. This would require CFD to capture the effect of wind, wave and current loading on MRE device structure responses, representation of the PTO system and its coupling with fluid flows and load coupling between the structure, mooring system (if applicable), foundation and surrounding soil. At the time of writing this is not possible for even a single device, and instead global load analysis with boundary conditions is carried out to uncouple these interactions.

References

1 O'Connor, P.D.T. (2002) *Practical Reliability Engineering*, 4th ed., Chichester: Wiley.
2 Salter, S.H. (2003) Proposals for a component and sub-assembly test platform to collect statistical reliability data for wave energy, *Proceedings of the Fourth European Wave Energy Conference*, Cork.
3 Wolfram, J. (2006) On assessing the reliability and availability of marine energy converters, *Proceedings of the Institution of Mechanical Engineers, Part O: Journal of Risk and Reliability*, 220, 55–68.
4 Callaghan, J. and Boud, R. (2006) Future marine energy. Results of the Marine Energy Challenge: Cost competitiveness and growth of wave and tidal stream energy. Technical report, Carbon Trust.
5 Mueller, M. and Wallace, R. (2008) Enabling science and technology for marine renewable energy. *Energy Policy* 36, 4376–4382.
6 Thies, P.R., Flinn, J. and Smith, G.H. (2009) Is it a showstopper? Reliability assessment and criticality analysis for wave energy converters. *Proceedings of the European Wave and Tidal Energy Conference (EWTEC)*, 7–10 September, Uppsala, Sweden, pp. 21–30.
7 Abdulla, K., Skelton, J., Doherty, K., O'Kane, P., Doherty, R. and Bryans, G. (2011) Statistical availability analysis of wave energy converters. *Proceedings of the 21st International Offshore and Polar Engineering Conference*, Maui, HI, USA.
8 Thies PR, Smith GH, Johanning L. (2012) Addressing failure rate uncertainties of marine energy converters. *Renewable Energy* 44, 359–367.
9 Department of Defence (1991) *Reliability Prediction of Electronic Equipment*. MIL-HDBK-217 F. Washington, DC: Department of Defense.
10 OREDA (2009) Offshore REliability Data. www.oreda.com [accessed: 10/07/2014].

11 Naval Surface Warfare Centre (2011) *Handbook of Reliability Prediction Methods for Mechanical Equipment*. CARDEROCK Division, West Bethesda, MD.

12 Weller, S.D, Thies, P.R., Gordelier, T., Davies, P. and Johanning, L. (2015) The role of accelerated testing in reliability prediction. *Proceedings of the 11th European Wave and Tidal Energy Conference*, Nantes, France.

13 Det Norske Veritas (2013) Position Mooring. Offshore Standard DNV-OS-E301.

14 D. Snowberg and J. Weber (2015) NREL's MHK Technology Development Risk Management Framework. http://www.nrel.gov/docs/fy15osti/63258.pdf

15 Department of Energy and Climate Change (2010) Marine Energy Action Plan.

16 Ricci, P., Villate, J.L., Scuotto, M., Zubiate, L., Davey, T., Smith, G.H., Smith, H., Huertas-Olivares, C., Neumann, F., Stallard, T., Bittencourt Ferreira, C., Flinn, J., Boehme, T., Grant, A., Johnstone, C., Retzler, C. and Soerensen, H.C. (2009) Equitable Testing and Evaluation of Marine Energy Extraction Devices in terms of Performance, Cost and Environmental Impact. [EquiMar] Deliverable D1.1, Global analysis of pre-normative research activities for marine energy.

17 Santhamma, K., Rajagopalan, S. and Ghose, M.K. (1988) Reliability testing and demonstration. *Reliability Engineering and System Safety* 23, 207–217.

18 British Standards Institute (2003) Reliability of systems, equipment and components. Guide to the specification of dependability requirements, BS 5760-4:2003.

19 Meuleau, C.A. (1965) High reliability testing and assurance for electronic components. *Microelectronics Reliability* 4, 163–177.

20 Dhillon, B.S. (1992) Reliability testing: Bibliography. *Microelectronics Reliability* 32, 1115–1135.

21 Cardenas, N.O., Machado, I.F. and Goncalves, E. (2007) Cyclic loading and marine environment effects on the properties of HDPE umbilical cables. *Journal of Material Science* 42, 6935–6941.

22 Schijve, J. (1985) The significance of flight-simulation fatigue tests. Technical report LR-466, Delft University of Technology.

23 Klyatis, L.M. and Klyatis, E.L. (2006) *Accelerated Quality and Reliability Solutions*. London: Elsevier.

24 Raath, A.D. (1997) A new time domain parametric dynamic system identification approach to multiaxial service load simulation testing in components. *International Journal of Fatigue* 19, 409–414.

25 Weltin, U. (2010) Reliability in Engineering Dynamics. Lecture notes, Reliability Engineering, TUHH, available at: https://www.tuhh.de/t3resources/mec/pdf/Scripte/Reliability_2010.pdf

26 Heuler, P. and Klätschke H. (2005) Generation and use of standardised load spectra and load-time histories. *International Journal of Fatigue* 27, 974–990.

27 Kam, J.C.P. (1992) Wave action standard history (wash) for fatigue testing offshore structures. *Applied Ocean Research* 14, 1–10.

28 Etube, L.S., Brennan, F.P. and Dover, W.D. (2001) Stochastic service load simulation for engineering structures. *Proceedings of the Royal Society A*, 457(2010), 1469–1483.

29 Research Councils UK Energy Programme Strategy Fellowship (2013) Energy Research and Training Prospectus: Report No 7 (Wind, Wave and Tidal Energy). https://workspace.imperial.ac.uk/rcukenergystrategy/Public/reports/Final%20Reports/RCUK%20ESF%20prospectus%20-%20WWT.pdf

30 Zhou, Z., Benbouzid, M., Charpentier, J.-F., Scuiller, F. and Tang, T. (2017) Developments in large marine current turbine technologies – A review. *Renewable and Sustainable Energy Reviews* 71, 852–858.

31 Harris, R.E., Johanning, L. and Wolfram, J. (2004) Mooring systems for wave energy converters: A review of design issues and choices. *Proceedings of the 3rd International Conference on Marine Renewable Energy*, Blyth, UK.

32 Vaitkunaite, E., Ibsen, L.B., Nielsen, B.N. and Molina, S.D. (2013) Comparison of foundation systems for wave energy converters Wavestar. *Proceedings of the 10th European Wave and Tidal Energy Conference*, Aalborg, Denmark.

33 Ding, B., Cazzolato, B.S., Arjomandi, M., Hardy, P. and Mills, B. (2016) Sea-state based maximum power point tracking damping control of a fully submerged oscillating buoy. *Ocean Engineering* 126, 299–312.

34 Vicente, P.C., Falcão, A.F.O. and Justino, P.A.P. (2013) Nonlinear dynamics of a tightly moored point-absorber wave energy converter. *Ocean Engineering* 59, 20–36.

35 Berntsen, P.I.B., Aamo, O.M. and Leira, B.J. (2009) Ensuring mooring line integrity by dynamic positioning: Controller design and experimental tests. *Automatica* 45, 1285–1290.

36 Cozijn, J.L. and Bunnik, T.H.J. (2004) Coupled mooring analysis for a deep water CALM buoy. *Proceedings of the 23rd International Conference on Offshore Mechanics and Arctic Engineering*, Vancouver, Canada.

37 da Costa Mattos, H.S. and Chimisso, F.E.G. (2011) Modelling creep tests in HMPE fibres used in ultra-deep-sea mooring ropes. *International Journal of Solids and Structures* 48(1), 144–152.

38 Stallard, T.J., Weller, S.D. and Stansby, P.K. (2009) Limiting heave response of a wave energy device by draft adjustment with upper surface immersion. *Applied Ocean Research* 31(4), 282–289.

39 Vicente, P.C., Falcão, A.F.O., Gato, L.M.C. and Justino, P.A.P (2009) Dynamics of arrays of floating point-absorber wave energy converters with inter-body and bottom slackmooring connections. *Applied Ocean Research* 31(4), 267–281.

40 Cruz, J., Sykes, R., Siddorn, P. and Taylor, R.E. (2010) Estimating the loads and energy yield of arrays of wave energy converters under realistic seas. *IET Renewable Power Generation* 4(6), 488–497.

41 Weller, S.D., Stallard, T.J. and Stansby, P.K. (2010) Experimental measurements of irregular wave interaction factors in closely spaced arrays. *IET Renewable Power Generation* 4(6), 628–637.

42 Det Norske Veritas (2013) Design of Floating Wind Turbine Structures. Offshore Standard DNV-OS-J103.

43 Young, P.A. (2011) The power of many? Linked wave energy point absorbers. Master's thesis, University of Otago.

44 Ricci, P., Rico, A., Ruiz-Minguela, P., Boscolo, F. and Villate, J.L. (2012) Design, modelling and analysis of an integrated mooring system for wave energy arrays. *Proceedings of the 4th International Conference on Ocean Energy*, Dublin, Ireland.

45 Payne, G.S., Taylor, J.R.M. Bruce, T. and Parkin, P. (2008) Assessment of boundary-element method for modelling a free-floating sloped wave energy device. Part 1: Numerical modelling. *Ocean Engineering* 35, 333–341.

46 Johanning, L. and Smith, G.H. (2009) Consideration of the cost implications for mooring MEC devices. EquiMar Deliverable 7.3.2. http://www.equimar.org/equimar-project-deliverables.html

47 Weller, S.D., Johanning, L., Davies, P. and Banfield, S.J. (2015) Synthetic mooring ropes for marine renewable energy applications. *Renewable Energy* 83, 1268–1278.

48 Ridge, I.M.L., Banfield, S.J. and Mackay, J. (2010) Nylon fibre rope moorings for wave energy converters. *Proceedings of the OCEANS 2010 Conference*, Seattle, USA.

49 Flory, J.F., Banfield, S.J. and Petruska, D.J. (2004) Defining, measuring, and calculating the properties of fibre rope deepwater mooring lines. *Proceedings of the 2004 Offshore Technology Conference*, Houston, USA. OTC 16151.

50 Weller, S.D., Davies, P., Vickers, A.W. and Johanning, L. (2014) Synthetic rope responses in the context of load history: Operational performance. *Ocean Engineering* 83, 111–124.

51 Det Norske Veritas (2005) Damage Assessment of Fibre Ropes for Offshore Mooring. Recommended Practice DNV-RP-E304.

52 Noble Denton Europe Ltd (2006) Floating production system JIP FPS mooring integrity. Research Report 444 prepared for the Health & Safety Executive.

53 Det Norske Veritas (2008) Offshore Mooring Chain. Offshore Standard DNV-OS-E302.

54 Det Norske Veritas (2009) Offshore Mooring Steel Wire Ropes. Offshore Standard DNV-OS-E304.

55 Renewable Energy Focus (2008) Africa, Asia and the Middle East. *Renewable Energy Focus* 9(3), 8.

56 B. van der Zwaan, R. Rivera-Tinoco, S. Lensink and P. van den Oosterkamp (2011) Cost reductions for offshore wind power: Exploring the balance between scaling, learning and R&D. *Renewable Energy* 41, 389–393.

57 Low Carbon Innovation Coordination Group (2012) Technology Innovation Needs Assessment (TINA), Marine Energy Summary Report.

58 R. Gross (2010) Great Expectations: The cost of offshore wind in UK waters – understanding the past and projecting the future. UKERC Technology and Policy Assessment.

59 D. Thompson and D. J. Beasley (2012) *Handbook of Marine Geotechnical Engineering* (SP-2209-OCN). Port Hueneme, CA: Engineering Service Center.

60 Randolph, M. F., Cassidy, M.J., Gourvenec, S. and Erbrich, C.T. (2005) The challenges of offshore geotechnical engineering. *Proceedings of the International Symposium on Soil Mechanics and Geotechnical Engineering*, Osaka, Japan, 2005.

61 Walker, R.T., van Nieuwkoop-McCall, J., Johanning, L. and Parkinson, R.J. (2013) Calculating weather windows: Application to transit, installation and the implications on deployment success. *Ocean Engineering* 68, 88–101.

62 Fraenkel, P.L. (2010) Development and testing of Marine Current Turbine's SeaGen 1.2 MW tidal stream device. *Proceedings of the 3rd International Conference on Ocean Energy*, Bilbao, Spain.

63 Carswell, W., Arwade, S.R., DeGroot, D.J. and Lackner, M.A. (2015) Soil–structure reliability of offshore wind turbine monopile foundations. *Wind Energy* 18, 484–498.

64 Andresen, L., Jostad, H.P., Andersen, K.H. and Skau, K. (2008) Finite element analyses in offshore foundation design. *Proceedings of the 12th International Conference of International Association for Computer Methods and Advances in Geomechanics (IACMAG)*, Goa, India.

65 Chen, J.-Y., Gilbert, R.B., Puskar, F.J. and Verret, S. (2013) Case study of offshore pile system failure in Hurricane Ike. *Journal of Geotechnical and Geoenvironmental Engineering* 139, 1699–1708.

66 Det Norske Veritas (2011) Design of Offshore Wind Turbine Structures. Offshore Standard DNV-OS-J101.

67 Det Norske Veritas (2008) Certification of Tidal and Wave Energy Converters. Offshore Service Specification DNV-OSS-312.

68 Det Norske Veritas (2005) Guidelines on Design and Operation of Wave Energy Converters. Report commissioned for the Carbon Trust.

69 International Electrotechnical Commission (2015) Marine energy – Wave, tidal and other water current converters – Part 10: The assessment of mooring system for marine energy converters (MECs), IEC/TS 62600-10 Ed. 1.0.

70 Det Norske Veritas (2011) Design of Offshore Steel Structures, General (LRFD Method). Offshore Standard DNV-OS-C101.

71 J. Gribble and S. Leather (2011) Offshore Geotechnical Investigations and Historic Environment Analysis: Guidance for the Renewable Energy Sector, Commissioned by COWRIE Ltd (project reference GEOARCH-09).

72 A. Rahim and R. F. Stevens (2013) Design procedures for marine renewable energy foundations. *Proceedings of the 1st Marine Energy Technology Symposium*, Washington DC, USA.

73 S. Patrício, A. Moura and T. Simas (2009) Wave energy and underwater noise: State of art and uncertainties. *Proceedings of OCEANS 2009 Europe*.

74 J. Garrett, M. Witt and L. Johanning (2013) Underwater sound levels at a wave energy device testing facility in Falmouth Bay, UK. *Proceedings of the Effects of Noise on Aquatic Life, 3rd International Conference*, Budapest, Hungary.

75 Mueller-Blenkle, C., McGregor, P.K., Gill, A.B., Andersson, M.H., Metcalfe, J., Bendall, V., Sigray, P., Wood, D.T. and Thomsen, F. (2010) Effects of Pile-driving Noise on the Behaviour of Marine Fish, COWRIE Ref: Fish 06-08, Technical Report.

76 Ahmadian, R., Falconer, R. and Bockelmann-Evans, B. (2012) Far-field modelling of the hydro-environmental impact of tidal stream turbines. *Renewable Energy* 38, 107–116.

77 Natural Environment Research Council (2013) Impacts on Fish and Shellfish Ecology. Wave and Tidal Consenting Position Paper Series.

78 American Petroleum Institute (2011) Geotechnical and Foundation Design Considerations. ANSI/API Recommended Practice 2GEO First Edition.

79 Det Norske Veritas (1992) Foundations, Classification Note DNV CN30.4.

80 Barrie, J.V. and Conway, J.W. (2014) Seabed characterization for the development of marine renewable energy on the Pacific margin of Canada. *Continental Shelf Research* 83, 45–53.

81 EMODNET (2015) Seabed habitats. http://www.emodnet-seabedhabitats.eu/default.aspx?page=1974.

82 Stallard, T., Feng, T. and Stansby, P.K. (2015) Experimental study of the mean wake of a tidal stream rotor in a shallow turbulent flow. *Journal of Fluids and Structures*, 54, 235–246.

83 Chen, L. and Lam, W. (2014) Methods for predicting seabed scour around marine current turbine. *Renewable & Sustainable Energy Reviews* 29, 683–692.

84 Heath, J.E., Jensen, R.P., Weller, S.D., Hardwick, J., Johanning, L. and Roberts, J.D. (2017) Applicability of available geotechnical approaches and geomaterial constitutive

models for foundation and anchor analysis of marine renewable energy arrays. *Renewable and Sustainable Energy Reviews* 72, 191–204.

85 Oka, F., Kodaka, T. and Kim, Y.S. (2004) A cyclic viscoelastic-viscoplastic constitutive model for clay and liquefaction analysis of multi-layered ground. *International Journal for Numerical and Analytical Methods in Geomechanics* 28, 131–179.

86 Cuéllar, P., Mira, P., Pastor, M., Fernández Merodo, J.A., Baeßler, M. and Rücker, W. (2014) A numerical model for the transient analysis of offshore foundations under cyclic loading. *Computers and Geotechnics* 59, 75–86.

87 Randolph, M.F. (2012) Offshore design approaches and model tests for sub-failure cyclic loading of foundations. In G.G. di Prisco and D.M. Wood (eds), *Mechanical Behaviour of Soils Under Environmentally-Induced Cyclic Loads* (pp. 441–480). Vienna: Springer.

88 Stevens, R.F., Soosainathan, L., Rahim, A., Saue, M., Gilbert, R.B., Senanayake, A.I., Gerkus, H., Rendon, E., Wang, S.T. and O'Connell, D.P. (2015) Design Procedures for Marine Renewable Energy Foundations,' in *Proceedings of the Offshore Technology Conference* (OTC-25960-MS), Houston, Texas.

6

Power Systems

Andrew R. Plummer[a], Andrew J. Hillis[b] and Carlos Perez-Collazo[c]

[a] *Professor of Machine Systems, Centre for Power Transmission and Motion Control, University of Bath, UK*
[b] *Senior Lecturer in Mechanical Engineering, Centre for Power Transmission and Motion Control, University of Bath, UK*
[c] *PRIMaRE Research Fellow, School of Engineering, University of Plymouth, UK*

6.1 Introduction to Power Take-Off Systems

6.1.1 Wave Energy PTO Systems

The power take-off (PTO) is a critical system within any wave energy converter (WEC). It converts the power available at the primary wave interface – often mechanical power associated with the movement of a float or paddle – into electrical power. The characteristics of the PTO are dictated by both its design and any control algorithm employed, and these will interact in a complex way with the primary wave interface to generate electrical power. Thus the relationship between the force (or torque) applied to the PTO and its resultant motion is critical to the ultimate power output, and the ideal relationship will vary with wave conditions. Crudely, if the PTO is too stiff, motion is small and so power capture is low; if the PTO is too compliant, force (or torque) is small and so power is again low. Although the conversion efficiency of the PTO itself is important, how the PTO affects the overall WEC behaviour is usually more important to the power generation.

Another challenge in PTO design is part-load efficiency. Within one wave period, instantaneous power input varies from zero to a maximum, and that maximum power varies by orders of magnitude in different sea states. Sizing all PTO components for peak power can lead to a large, costly system, and one which is very inefficient for small sea states. The expected variation in wave conditions at a particular installation site will dictate the best PTO size, one which can efficiently transmit power from waves of an energy level which occurs frequently, without having to shed power from the most energetic waves too often. This is sometimes achieved by having parallel power transmission paths which can be switched in and out depending on conditions. Having a PTO with an energy storage capability is one approach, as this can be used to smooth power variations so that components downstream of the energy store can be sized for the average power rather than the peak power, often meaning their capacity can be many times smaller. These components are then operating closer to their most efficient design condition.

Wave and Tidal Energy, First Edition. Edited by Deborah Greaves and Gregorio Iglesias.
© 2018 John Wiley & Sons Ltd. Published 2018 by John Wiley & Sons Ltd.

PTOs often have a role to play in protecting the WEC in rough seas, being able to reduce power capture to protect electrical generating equipment, or changing their characteristics to limit forces or motions at the primary wave interface.

PTO design and control is strongly influenced by the WEC concept used. The nature of the PTO required for common concepts is as follows:

- Point absorbers (heaving devices of small width compared to the wavelength, e.g. OPT Powerbuoy [1]), require a PTO accepting a linear motion input.
- Attenuators (these lie parallel to the predominant wave direction and 'ride' the waves, e.g. Pelamis) have rotary joints with a small range of motion, with power extraction usually through a linear translator bridging the joint (e.g. hydraulic cylinder).
- Terminators (devices perpendicular to the wave direction, appearing to block the wave; examples are Salter's Duck [2] and flap-type oscillating wave surge converters such as the Aquamarine Power Oyster [3]). These also have rotary axes, often with a small enough range of motion to allow power extraction through a linear translator bridging the joint (e.g. hydraulic cylinder).
- Oscillating water column (OWC) devices: in these air is forced through a turbine by varying water level in a chamber.
- Overtopping devices: these capture water from incident waves which is then passed through a turbine (e.g. the Wave Dragon [4])

Examples and a more detailed discussion of classification are given in Chapter 3.

Other than OWC or overtopping devices, WECs typically require PTOs accepting a limited range of motion, which could be rotary but is more commonly linear. Speeds are low compared to the relative speed of rotor and stator in conventional electrical machines, so some gearing-up is usually employed. High-pressure hydraulics is most commonly used to convert the low speed linear motion into high speed rotary motion to drive a generator, as this technology can handle high powers, is relatively light and compact, and is sufficiently robust to cope with harsh conditions and impact loading. Nevertheless, direct drive linear electrical generators have also been proposed, thus avoiding the complexity of the hydraulic transmission stage, but the low flux cutting speed results in a very large electrical machine. Mechanical transmissions, for example cable-winch systems, are a third option, and have the advantage of simplicity, but lack the controllability of hydraulic PTOs. The main options are illustrated in Figure 6.1, and a breakdown of the electrical power system options is shown in Figure 6.2.

6.1.2 Tidal Energy PTO Systems

Tidal energy conversion almost universally involves the use of rotating turbines to extract power. Tidal range generation requires the water to be trapped behind a barrage or in a lagoon on a rising tide, and then released through turbines as the tide drops. Alternatively, the turbines can be driven as a lagoon is flooded, or driven during both a rising and falling tide. In contrast to barrages and tidal lagoons, tidal stream turbines use the kinetic energy of the tide directly, and tidal stream turbines allow the water to pass both through and around them, not requiring the storage of water. As described in Chapter 4, a variety of rotating turbine designs, including both axial and cross-flow types, has been proposed for tidal energy conversion.

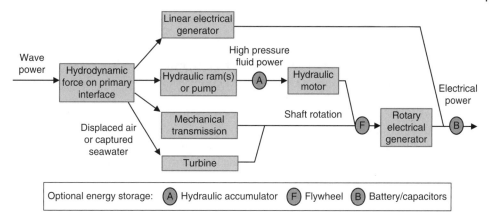

Figure 6.1 Alternative power take-off configurations.

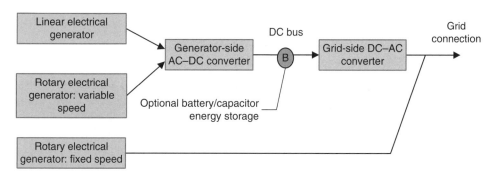

Figure 6.2 Alternative electrical power system configurations for variable or fixed speed generators.

Power take-off for tidal turbines is usually rather more straightforward than for WECs. A geared connection of the turbine to a rotary electrical generator is usually used. Thus tidal turbine PTOs have similarities to those used in wind turbines. For example, SeaGen has two horizontal-axis turbines with variable speed induction generators, of the doubly-fed induction generator type. The Kobold turbine is a vertical-axis turbine which drives a permanent magnet synchronous generator (PMSG) through a step-up gearbox. Clean Current is a ducted horizontal-axis turbine design with a direct drive (gearless), variable-speed PMSG. Openhydro also uses a direct drive generator, but this time integrated with the turbine, so that flow through the centre of the machine drives a bladed rotor, which turns within the generator stator built into the surrounding housing. One device which uses a different approach is the Rotech turbine from Lunar Energy. In this design the turbine drives the generator through a hydrostatic transmission, giving a continuously variable gear ratio and hence greater controllability.

The need to provide robust and durable support for the turbine shaft also influences PTO design. Bearing failure accounts for a significant share of downtime in wind turbines, and the same will be true for tidal turbines, as bearings must react both large static and transient hydrodynamic loads.

6.1.3. Chapter Outline

This chapter includes the following:

- An introduction to electrical generators, which transform mechanical to electrical power, both rotary and linear. The rotary generator technology described is applicable to be wave and tidal energy converters.
- Turbines for WECs are described, covering air-driven turbines which are applicable to OWC designs, and water-driven turbines which can be used in overtopping devices.
- An introduction to hydraulic power transmission technology.
- Hydraulic PTO design for WECs, including control for maximum energy capture.
- A brief description of alternative solutions using direct mechanical power transmission.
- Electrical infrastructure, describing components, frequency converters, array connection and including grid integration.

6.2 Electrical Generators

6.2.1 Linear Electrical Generators

A linear generator offers the possibility of directly converting movement of the primary wave interface (e.g. buoy) into electrical energy. During early wave power research this option was investigated, but it was concluded that these machines would be too heavy, inefficient and expensive [2]. However, new rare-earth permanent magnet materials and the reduced costs of frequency converter electronics mean that this possibility is now being reconsidered. The air gap speed between the rotor and stator in conventional rotary generators is high (upwards of 50 m/s), giving high rates of change of flux. Linear oscillatory motion from a WEC, however, may have a peak of only about 2 m/s, and an average of considerably less [5]. So this is even lower than the air gap speed for direct drive generators being developed for wind turbines (around 5–6 m/s).

A trial was conducted with a large linear permanent magnet generator PTO in the Archimedes Wave Swing submerged point absorber [6]. The generator had a 7 m stroke, 1 MN maximum force, and 2.2 m/s maximum velocity; the total moving mass was 400 tonnes. An average electrical power output of about 200 kW was predicted based on the trial results [7]. The basic concept of a linear generator for a point absorber is to have a translator on which magnets are mounted with alternating polarity directly coupled to a heaving buoy, with the stator containing windings, mounted in a relatively stationary structure (either connected to a drag plate, a large inertia, or fixed to the sea bed), as in Figure 6.3.

The design of electrical generators for direct drive WECs was examined by Mueller [5] by comparing the longitudinal flux permanent magnet machine with the transverse flux permanent magnet machine. He identified the transverse flux machine as having the best potential, owing to the design having higher power density and efficiency, compared to the longitudinal design. Despite the high shear stress offered by transverse flux machines (up to 200 kN/m^2 [8]), their topology requires structural support and they suffer from a low power factor requiring reactive power compensation [9].

Figure 6.3 A schematic of a linear electrical generator based on a permanent magnet generator.

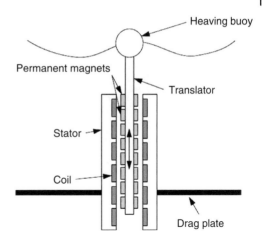

6.2.2 Rotary Electrical Generators

Traditional power stations use synchronous generators operated at a virtually constant speed, matching the alternating current (AC) generation frequency to that of the grid connection. However, unless the PTO has a local energy storage capability, the variability of many renewable energy sources translates into variable generator rotational speed, giving variable frequency generation from a synchronous machine. So this will require an electronic frequency converter for grid connection. Variability is not such a problem for tidal energy, in which the resource is predictable and more or less constant in velocity apart from some transition periods between ebb and flow, but the wave energy resource is intermittent in nature and slow in motion, making it more challenging effectively to harness and convert this energy resource into electricity.

Depending on the conversion system of the device and the characteristics of the electric grid used to transport the electricity from the offshore site to the onshore grid (see Section 6.8), four types of generators are commonly used for wave and tidal energy applications (Figure 6.4):

- wound rotor induction generator;
- squirrel cage induction generator;
- wound rotor synchronous generator;
- permanent magnet synchronous generator.

The *wound rotor induction generator* (WRIG) is a type of induction machine where by varying the resistance of the rotor windings, the torque-speed characteristic of the generator can be controlled. Changing the resistance of the rotor windings can be done by either connecting an external circuit to the rotor via slip rings and carbon brushes or by using a modified rotor, known as OptiSlip, where the resistors are mounted on the rotor itself together with an electronic control system that allows the resistance to be changed by an optical communication device mounted on the rotor shaft. These two types of WRIG allow the resistance of the rotor windings to be modified and so control the slip, which is the relationship between the frequency of AC electricity generation and the rotational speed. This means that the power output of the generator can be controlled. The range of the dynamic speed control depends on the size of the resistors,

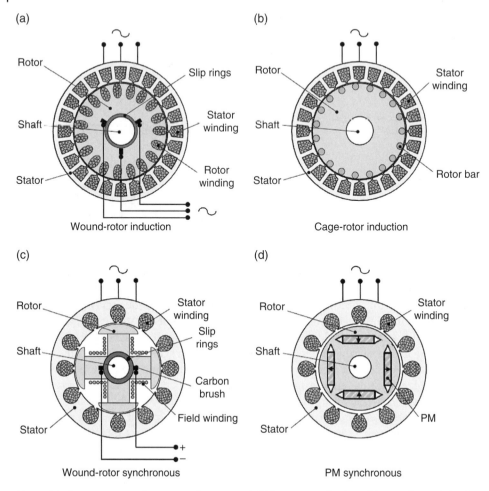

Figure 6.4 Common types of generators for wave and tidal energy applications: (a) wound rotor induction generator; (b) squirrel cage induction generator; (c) wound rotor synchronous generator; (d) permanent magnet synchronous generator [10]. Reproduced courtesy of The Electromagnetics Academy.

and typically the speed range is between 0% and 10% above the synchronous speed. The speed range can be increased by feeding the rotor windings by a frequency converter to perform slip power control. This type of WRIG is known as a doubly-fed induction generator (DFIG).

The *squirrel cage induction generator* (SCIG) is a type of induction machine where rotor windings have been replaced by a squirrel cage made of aluminium or copper bars. This type of machine has the advantage that it exhibits higher efficiency and higher power density than the WRIG because it uses the squirrel cage instead of the windings and does not need slip rings and carbon brushes. Furthermore, the SCIG is a robust and low-cost alternative, although it does not allow speed control and so is only suitable for fixed-speed devices.

The *wound rotor synchronous generator* (WRSG) is a type of synchronous machine where the rotor windings are fed with direct current (DC) which is provided from an

Table 6.1 Evaluation of common types of generators used for wave and tidal energy. A grade point system is used in which 1 is the worst and 5 is the best [10]. Reproduced courtesy of The Electromagnetics Academy.

	WRIG	SCIG	WRSG	PMSG
Efficiency	3	4	4	5
Power density	3	4	3	5
Power factor	4	3	5	5
Power conditioning	5	3	3	3
Robustness	4	5	4	4
Maturity	5	5	5	4
Cost	3	4	4	3
Power level	4	4	5	4
Maintenance	3	5	4	5
Total	34	37	37	38

external source via slip rings and carbon brushes. Compared to the SCIG, the WRSG has higher power generation efficiency, can be operated with a power factor at or close to unity, and is widely used in traditional power plants. As a drawback, the presence of slip rings and carbon brushes increases the maintenance costs. Furthermore, this type of machine requires a full-scale frequency converter to adjust the power output over a range of speeds. The speed of a synchronous machine is fixed by the number of poles in the rotor and the frequency of the stator.

The *permanent magnet synchronous generator* is a type of synchronous machine where the rotor DC windings have been replaced by permanent magnets. This type of machine has the advantage of no rotor copper loss coupled with high power density; however, its drawback is the high cost of the permanent magnets. Due to its high efficiency and power density, together with the cost reductions seen in permanent magnets recently, PMSGs are becoming more and more attractive, even though a full-scale frequency converter is needed to operate the machine at variable speed.

Chau *et al.* [10] compare these four electrical machines in different renewable energy applications, including wave and tidal stream energy. The comparison is made in terms of their efficiency, power density, power factor, power conditioning requirement, robustness, maturity, cost, power level and maintenance requirement (Table 6.1). The authors conclude that the most suitable generator for renewable energy applications is the PMSG. Similarly, O'Sullivan and Lewis [11] discuss these generator options with regard to suitability for an OWC application, by examining the advantages and disadvantages in terms of environmental, electrical and cost factors, and by using a time domain model. The favoured generators used in wind turbines (DFIG) driven via a gear box, and direct drive low-speed synchronous generators with dedicated power electronics are possible candidates for use in WECs. O'Sullivan and Lewis's study concludes that the synchronous generator is the preferred option due to its better energy yield, weight and controllability, despite the requirement for a full frequency converter between the generator and the grid to cope with rotary speed variations.

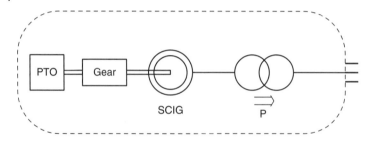

Figure 6.5 MRE device topology type A.

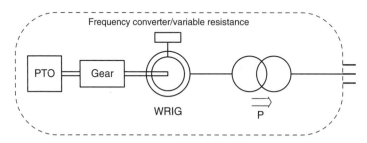

Figure 6.6 MRE device topology type B.

A common practice in the wind energy industry is to classify wind turbines into topology types A to D, according to their ability to control speed and the type of power electronics used [12]. A similar classification can be applied to marine renewable energy (MRE), including an additional type E suggested by [13], as defined below.

Type A: Fixed speed. This topology defines a fixed-speed wave or tidal energy device with an SCIG directly connected to the offshore grid (Figure 6.5). This type could be directly connected to the grid via a step-up transformer stage, previous to the connection point to the offshore grid, or positioned in a different part of the offshore grid (see Section 6.8). This configuration may require a gearbox in order to scale up the low speed from the PTO to the high speed required by the SCIG and a capacitor bank for reactive power compensation. A hydraulic PTO would typically output high-speed rotation so that a gearbox is not needed. The capacitor bank is required in order to provide the additional reactive power required by the SCIG to compensate the fluctuations in the velocity from the PTO and keep constant the voltage output at the grid connection point without reducing the quality of the electricity produced. The main advantage of this type is its low cost and simplicity. However, its drawbacks are the lack of speed control, stiff grid or dynamic reactive power compensation requirement, and the need to tolerate high mechanical stresses.

Type B: Limited variable speed. This topology includes wave and tidal energy devices using a WRIG with slip rings connected to a resistor or with the OptiSlip variant, where a limited control of the speed can be performed (Figure 6.6). Similar to type A, the generator can be directly connected to the grid via a step-up transformer or positioned at a different point in the offshore grid. The WRIG also may require a gearbox to increase the low velocity from the PTO to the high speed required for the machine to act as a generator. Even though the reactive power requirements of this

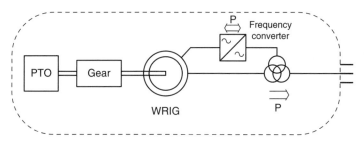

Figure 6.7 MRE device topology type C.

configuration are lower than for type A, a capacitor bank is still required to mini-mise the effects of the generator over the grid, and keep the voltage output constant. The main advantage of this configuration is that it allows limited variable speed control without the need for an expensive frequency converter unit. The drawbacks are that it only allows the speed to vary by up to approximately 10% above the synchronous speed, it requires a stiff grid, and if slip rings are used it has increased maintenance costs.

Type C: Variable speed with DFIG. This topology defines the first type with fully controlled variable speed configuration, and uses a DFIG. This configuration uses a WRIG as in type B; however, in this case the rotor windings are connected to a small-scale frequency converter through the slip rings (Figure 6.7). The small-scale frequency converter allows power to be fed either into the rotor or into the grid depending on grid and generator conditions. A step-up transformer with an inter-mediate voltage output is required for this configuration in order to feed the small-scale frequency converter and deliver the high voltage required by the grid. As with the previous two types, a gearbox may be required to increase the low speed from the PTO to that required by the generator. The frequency converter is usually rated at approximately 30% of the nominal generator power and it typically allows the synchronous speed to vary between −40% and +30%; however, this range depends on the size of the frequency converter used. The main advantages of this topology are that it allows a wide range of speed control with a small frequency converter, which has made this option the most common among wind turbines. Its drawbacks are the increased maintenance costs due to the slip rings and its vulnerability in case of grid faults.

Type D: Variable speed with single-phase AC/DC inverter. This topology is the second type with fully controlled variable speed configuration, where an SCIG or a PMSG are connected to a single-stage AC/DC frequency converter, delivering the power output directly to an offshore direct current grid (Figure 6.8). A gearbox may be required for the configuration using an SCIG as this is a high-speed machine, but direct drive PMSGs are possible, as a PMSG can be designed to be a low-speed machine. This topology is popular with wave and tidal energy developers due to the limited space available inside their devices. It also helps to minimise maintenance and reduce overall costs as it does not have slip rings or a step-up transformer, and only requires a single-stage full-scale frequency converter. This configuration uses an innovative DC offshore grid instead of the standard AC (see Section 6.8), and requires two conductors rather than three, thus reducing grid costs.

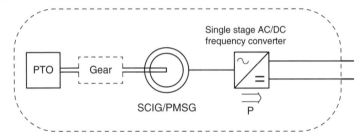

Figure 6.8 MRE device topology type D.

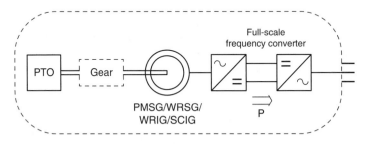

Figure 6.9 MRE device topology type E.

Type E: Variable speed with full-scale frequency converter. This topology has a fully controlled variable speed configuration, which can be assembled with any of the four types of generator (WRIG, SCIG, WRSG and PMSG) and coupled with a full-scale frequency converter, allowing full control of the PTO speed and guaranteed quality of the power output to the grid (Figure 6.9). The space limitation on wave and tidal devices means that the step-up transformer is usually positioned downstream in the offshore grid, combining the power from different devices. As with previous types, a gearbox is sometimes required. The advantage of this topology is that the frequency controller performs the speed regulation to compensate the reactive power and allows a smooth grid connection. However, the cost of the full-scale frequency converter increases considerably the cost of the whole unit.

6.3 Turbines for WEC Power Take-Off

6.3.1 General Considerations for WEC Turbines

Most turbine types can be driven by air or water and broadly can be defined as being impulse or reaction types. Impulse turbines redirect a high-speed working fluid jet in a bucket-shaped receiver and the fluid momentum is imparted to the rotor. Reaction turbines are immersed in the working fluid and foil-shaped blades generate radial and axial forces, with power generated from the radial component.

For WEC applications, the reciprocating flow requires either a rectification system or a turbine that is self-rectifying (i.e. provides unidirectional rotation in a bidirectional flow). The accepted philosophy in the field is that rectification systems (a network of non-return valves) add complexity and reduce reliability. Self-rectifying turbines are therefore the preferred option where possible, despite low efficiency, poor self-starting ability and higher noise compared to conventional turbines. This applies to reaction turbines only; impulse turbines require a rectification system.

Different designs are suitable for different head and flow ranges, and different designs operate with varying ranges of acceptable efficiency. The choice of turbine type will largely depend on the conditions for the specific WEC. Some turbine designs have variable geometry, which greatly extends their operating range, but at greater expense and reduced reliability.

The most important consideration for any WEC is the cost over its operating lifetime. Impulse type turbines tend to be mechanically simpler, for example not having variable geometry and not requiring pressure seals around the driveshaft. This makes them cheaper to construct and maintain, and more reliable than reaction type turbines. However, reaction turbines (particularly those with variable geometry) have a wider operating range and generally give a superior power to size ratio.

Generator and grid connection must also be considered. If synchronous generators are used then a near constant rotational speed must be achieved. If a low-speed turbine is used then gearing would be required, increasing cost and reducing efficiency. Achieving a constant speed is difficult in the variable conditions imposed by a WEC, though controlled variable geometry turbines can achieve this. If a frequency converter is used then the requirement for constant speed is removed, though overall efficiency may still be improved if it can be achieved.

The following sections describe some commonly used self-rectifying turbines. They are split into air- and water-driven types. In each case only the commonly used types for WEC applications are discussed; many other types of turbine exist and also many variations on the main operating principles.

6.3.2 Air-Driven Turbines

6.3.2.1 Wells Turbines

The Wells turbine was the first design to provide unidirectional motion in a reciprocating flow and is a type of reaction turbine. Aerofoil blades generate axial and tangential forces in an axial airflow; the tangential component results in rotation and can be used for power generation. Rectification is achieved by utilising symmetrical aerofoils perpendicular to the flow direction. The concept is illustrated in Figure 6.10.

Wells turbines exhibit poor starting performance and low efficiency compared to other types; they can also generate high noise levels as a result of high operating speeds. However, they are also simple and robust, making them the standard choice for OWC WECs. The basic Wells concept can be enhanced to improve efficiency, for example, guide vanes, twin contra-rotating turbines as used by LIMPET [14], and self-pitching blade control [15] have all shown improvements.

6.3.2.2 Impulse Turbines

In an impulse turbine, guide vanes direct air flow onto bucket shaped blades which redirect it to generate a tangential force. This is illustrated in Figure 6.11. The guide vanes may be fixed or pitched (self-pitch controlled or actively controlled) to improve efficiency. In general the mean efficiency of impulse turbines in an OWC system is higher than that for a Wells turbine [16]. Further benefits over Wells turbines include

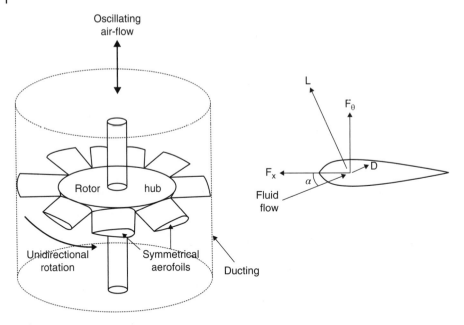

Figure 6.10 Operating principle of the Wells turbine [14].

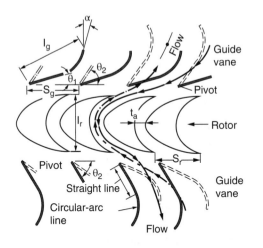

Figure 6.11 Operating principle of an air impulse turbine [16].

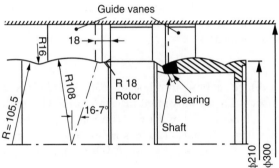

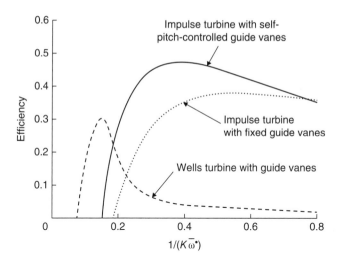

Figure 6.12 Performance comparison of Wells and impulse turbines [16].

improved self-starting, better efficiency over a wider range of flow conditions, and lower operating speeds and therefore lower noise. An application example is the NIOT plant in India [17].

6.3.2.3 Performance Comparison

Figure 6.12 compares the performance of Wells and impulse turbines in terms of energy conversion efficiency in irregular waves against a dimensionless flow coefficient. This suggests that impulse turbines are more efficient over a wider range of conditions than Wells turbines. It also illustrates the improvement that may be gained from using pitch-controlled guide vanes.

6.3.3 Water-Driven Turbines

Water-driven turbines suitable for use with WECs may be categorised as impulse or reaction types. An impulse turbine is driven by jets of water which are redirected to impart their momentum to the turbine, the most common example being the Pelton turbine. A reaction turbine is completely immersed in the flow of water and the blades generate a tangential force. The most common examples are the Kaplan and Francis turbines. Both impulse and reaction types have been applied in WECs.

6.3.3.1 Pelton Wheel

The operating principle is the same as for air-driven impulse turbines. Jets of water are directed onto buckets mounted around the edge of the wheel. The buckets redirect the jets which are decelerated as they impart their momentum to the wheel. Low-velocity water exits the buckets and is returned to an outlet in the casing. The operating principle is illustrated in Figure 6.13.

Two parallel sets of buckets may be used to split the water jets with the intention of reducing axial loading and thus increasing bearing life. Efficiency for Pelton wheels is generally good, and they are best suited to high-pressure applications. An example of Pelton wheel use for WECs is in the onshore power conversion system used with Oyster [19].

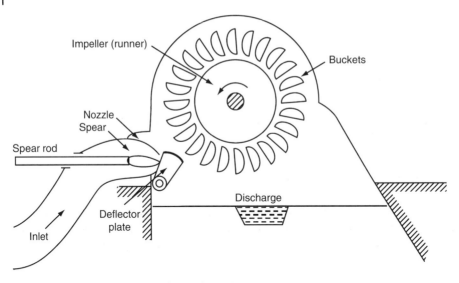

Figure 6.13 Operating principle of the Pelton Wheel. Source [18].

6.3.3.2 Kaplan Turbine

Kaplan turbines are axial flow reaction types and are widely used for low head applications. Inlet guide vanes direct water flow onto rotor blades which extract power from both kinetic energy in the flow and hydrostatic head. Typically the guide vanes are adjustable to regulate the flow rate, and the blade pitch is adjustable to improve efficiency over a wide range of flow and pressure conditions. The basic concept is illustrated in Figure 6.14. Kaplan turbines are very efficient compared to other forms, though the added complexity increases costs and reduces reliability. Kaplan turbines are particularly suited to overtopping devices, and are used in the Wave Dragon WEC [20].

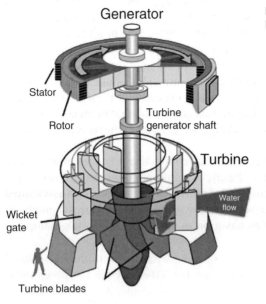

Figure 6.14 Operating principle of the Kaplan turbine.

Figure 6.15 Operating principle of the
Francis turbine [18].

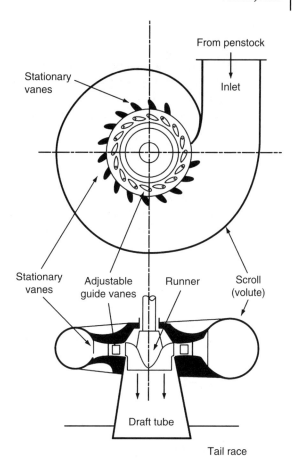

From penstock

Stationary
vanes

Inlet

Stationary
vanes

Adjustable
guide vanes

Runner

Scroll
(volute)

Draft tube

Tail race

6.3.3.3 Francis Turbine

The Francis turbine is a type of reaction turbine with very similar operating principles to the Kaplan turbine. Unlike the Kaplan turbine, the blades are fixed, which reduces the comparative efficiency but also reduces cost and increases reliability. Francis turbines are very widely used in hydroelectric power stations and are suited to high head applications. The concept is illustrated in Figure 6.15.

6.3.3.4 Performance Comparison

The choice of turbine to achieve the best efficiency for a given application will depend on the head and flow characteristics. Figure 6.16 shows a comparison of the head versus flow operating regions for Pelton, Kaplan and Francis turbines. Pelton wheels operate best at high head (>20 m) and low flow (<1 m^3/s). Kaplan turbines operate best at low head (<10 m) and high flow (>1 m^3/s). However, both Pelton wheels and Kaplan turbines operate with reasonable efficiency over a wide range of head and flow conditions. Francis turbines lie between Pelton and Kaplan devices in terms of head/flow operating regions and are generally less efficient than the other types when operating away from their ideal conditions.

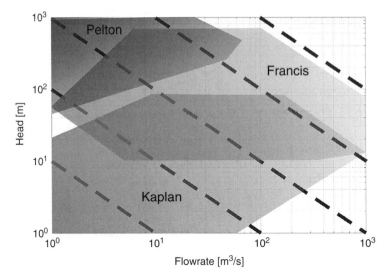

Figure 6.16 Comparison of operating regions for different turbines.

6.4 Hydraulic Power Transmission Systems

6.4.1 Introduction: Hydraulic Fluids and Circuits

In hydraulic PTOs power is transmitted by the flow of liquid under pressure. As power is given by the product of pressure and flow rate, hydraulic systems usually operate at high pressures to maximise the power transmission for a system of fixed size; pressures in the range 100–300 bar are typical. Alternative hydraulic fluids include the following:

- Mineral oil: by far the most common.
- Water, or water with additives: lacks the good lubrication properties of oil, and necessitates expensive components using stainless steel and ceramic materials. Additives can be used to inhibit corrosion, increase lubricity and reduce biological growth.
- Synthetic oil: expensive, but usually designed for specific demanding applications, such as a wide temperature range.
- Vegetable oil based fluids: biodegradable, but not as stable as mineral oils (e.g. oxidation starts at lower temperatures).

For marine energy applications, seawater seems like an attractive fluid, but lack of lubricity, corrosion, biological growth and dirt ingress are very serious problems.

Figure 6.17 shows an example hydraulic circuit. Standard CETOP symbols are used to represent the components (these are incorporated in International Standard ISO 1219, and British Standard BS 2917). Figure 6.18 explains the the symbols.

6.4.2 Hydraulic Pumps

Hydraulic systems use *positive displacement* pumps. These are pumps which are designed to produce a flow rate which is proportional to their speed of rotation, i.e. their ideal characteristic is

$$Q = D_p \omega_p, \tag{6.1}$$

Figure 6.17 Example hydraulic circuit.

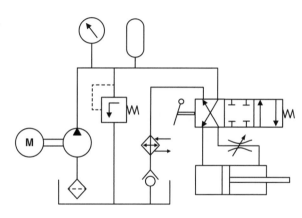

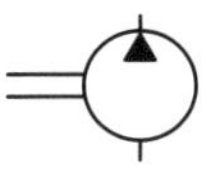	**Pump** This one is fixed displacement (i.e. constant volume flow rate for constant speed of rotation) and can only pump in one direction.	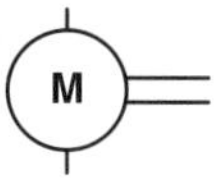	**Drive unit** The drive unit powers the pump. This one is an electric motor.
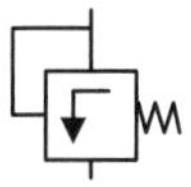	**Pressure relief valve** A valve which keeps the hydraulic fluid at the pump outlet at a constant pressure. Contains an orifice which opens when the pilot pressure (dashed line) overcomes the spring force.	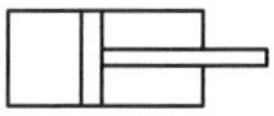	**Cylinder** Also known as a hydraulic jack, hydraulic ram or simply a hydraulic actuator. This one is powered in both directions, and is single-ended (i.e. piston rod at one end only).
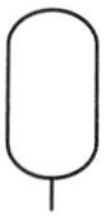	**Accumulator** Storage vessel for high-pressure fluid. The hydraulic equivalent of an electrical capacitor. Can 'smooth out' variable flow from a cylinder in a PTO.		**Directional control valve** Changes flow connections. This one is a three-position valve with a closed centre, and has manual left–right movement with spring centring.
	**Restrictor** Causes a restriction and hence a pressure drop in the hydraulic line. The oblique arrow indicates that the restriction is variable.		**Reservoir** Tank of hydraulic fluid.
	**Filter** For filtering out contaminants. A coarse filter is called a strainer.	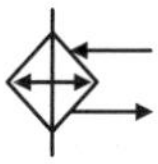	**Cooler** A heat exchanger: removes heat from hydraulic fluid by water cooling.
	**Pressure gauge** Measures gauge pressure of fluid (i.e. pressure above atmospheric).		**Check valve** Allows flow in one direction only –downwards in the orientation shown.

Figure 6.18 Hydraulic circuit symbol key.

where Q is the volume flow rate, D_p is the pump displacement (or capacity), and ω_p is the angular velocity of the pump drive shaft. Thus, based on the conservation of energy (i.e. fluid output power is equal to mechanical input power), it can be shown that the ideal torque required to drive the pump is

$$T_p = D_p \Delta P, \tag{6.2}$$

where ΔP is the pressure increase generated by the pump.

Due to leakage, a pump never quite delivers its theoretical maximum flow rate. Also, due to friction, the torque required to turn the pump is more than the ideal. Hence equations (6.1) and (6.2), which give the ideal characteristics, are in practice modified as

$$Q = v_p D_p \omega_p, \tag{6.3}$$

$$T_p = \frac{D_p \Delta P}{m_p}, \tag{6.4}$$

where v_p is called the volumetric efficiency and is defined as

$$v_p = \frac{\text{actual flow rate}}{\text{ideal flow rate}},$$

and m_p is called the mechanical efficiency and is defined as

$$m_p = \frac{\text{ideal torque}}{\text{actual torque}}.$$

Both v_p and m_p typically in the range 0.8–0.95. Their values can vary depending on the operating conditions of the pump. The overall efficiency of the pump is the product of volumetric and mechanical efficiencies:

$$\eta_p = v_p m_p. \tag{6.5}$$

6.4.2.1 Pump Design

There are three main pump design concepts used in high-pressure hydraulics, although there are many variants within these three families. Piston pumps can handle high powers, and are the most efficient, but they are also the most expensive. Gear and vane pumps are simpler machines but performance is not quite as good. A summary of pump designs and performance is given in Table 6.2. These are illustrated in Figure 6.19.

Piston pumps (and occasionally vane pumps) can be of the variable displacement type, i.e. it is possible to mechanically vary the displacement D_p during operation. The displacement change can either be manual or hydraulically or electronically controlled. In an 'over-centre' pump, the displacement can become negative, in which case the pumping direction is reversed. The term 'bidirectional pump' indicates a unit which is designed to pump in either direction, either by changing the sign of the displacement or by reversing the direction of shaft rotation. Symbols for these pump variants are shown in Figure 6.20.

Table 6.2 Pump types and characteristics.

Pump type	Variants	Comments	Specification (typical maximums)
Gear pump	External gear pump	Most common type of pump. Cheap and simple. Fixed displacement.	250 bar, 4000 rpm 100 kW 90% efficiency
	Internal gear pump	Similar to external gear pump; slightly more complex mechanical design and hence less common. However, it is quieter.	Similar to above
Piston pump	Axial piston pump	Can operate efficiently at very high pressure; most common design if variable displacement is required.	350 bar, 5000 rpm 2 MW 90% efficiency
	Radial piston pump	Generally only suitable for low shaft angular velocities.	250 bar, 1500 rpm 5 MW 90% efficiency
Vane pump		Variable displacement if the eccentricity can be changed.	150 bar, 3000 rpm 300 kW 85% efficiency.

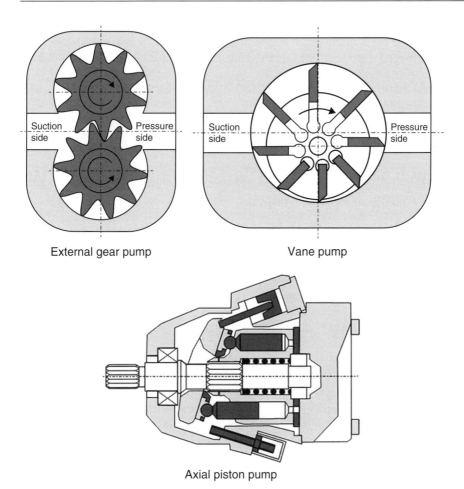

External gear pump

Vane pump

Axial piston pump

Figure 6.19 Common pump designs.

Variable displacement pump
Displacement D_p can be varied either manually or through some electromechanical or hydraulic control.

Bidirectional pump
Can pump in either direction, dependent on direction of rotation. Note: variable displacement versions are available; these can sometimes pump in either direction without changing the direction of rotation.

Figure 6.20 Symbols for pump variants.

6.4.3 Hydraulic Motors

Hydraulic motors convert hydraulic flow into continuous mechanical rotation. They are very similar in design to pumps, and some units are designed to be operated as either pumps or motors. Thus gear, piston and vane motors are available. Piston motors are often favoured due to their efficiency and their variable displacement capability. Radial piston motors are particularly suited to operation as low speed drives (e.g. winches, hub motors for off-road vehicles). The circuit symbol for a motor has an inward pointing triangle. Variants are shown in Figure 6.21.

The analysis of motors is very similar to that of pumps. The flow/speed and pressure/torque relationships are summarised below:

$$Q = \frac{D_m \omega_m}{v_m}, \tag{6.6}$$

$$T_m = m_m D_m \Delta P, \tag{6.7}$$

where the volumetric efficiency is defined as

$$v_m = \frac{\text{ideal flow rate}}{\text{actual flow rate}},$$

and the mechanical efficiency is defined as

$$m_m = \frac{\text{actual torque}}{\text{ideal torque}}.$$

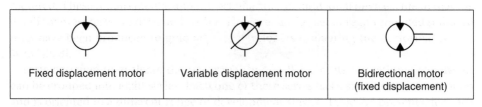

| Fixed displacement motor | Variable displacement motor | Bidirectional motor (fixed displacement) |

Figure 6.21 Symbols for motor variants.

Again v_m and m_m are typically in the range 0.8–0.95. The overall efficiency of the motor is the product of these two individual efficiencies:

$$\eta_m = v_m m_m.$$

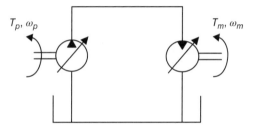

Figure 6.22 Variable ratio hydrostatic transmission.

6.4.4 Hydrostatic Transmissions

A hydrostatic transmission is a means of transmitting rotary mechanical power from one location to another, and possibly increasing torque or rotational speed in the process. Hence it is an alternative to mechanical transmission elements such as gears, belts or chains. Hydrostatic transmissions are beginning to be used for large wind turbines, and could have applications in tidal turbines. They are currently widely used in off-road vehicle transmissions.

A major advantage of hydrostatic transmissions over mechanical transmissions is the freedom to locate input and output shafts wherever required. If flexible hoses are used to join the hydraulic components then the relative position and orientation of input and output shafts can change during use. Another major advantage is that a continuously variable transmission ratio can easily be obtained by use of a variable displacement pump and/or a variable displacement motor (both are variable in Figure 6.22). Variable displacement units make the system more expensive, but high efficiencies can be maintained.

6.4.5 Hydraulic Actuators

Hydraulic cylinders are also known as jacks or rams. The most important dimensions of a cylinder are its bore (internal diameter) and stroke (difference between fully extended and retracted positions). Some variants are shown in Figure 6.23 with their standard

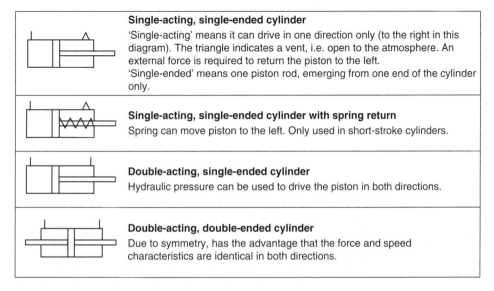

Figure 6.23 Hydraulic cylinders.

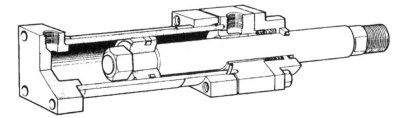

Figure 6.24 Detail of typical hydraulic cylinder.

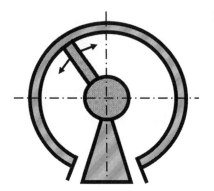

Figure 6.25 Rotary vane actuator.

symbols, and typical construction is illustrated in Figure 6.24. Rotary hydraulic actuators, of limited stroke (less than 360°) are also available. Figure 6.25 shows the principle. They are less common than linear cylinders as sealing around the moving vane is more difficult, leading to leakage and wear issues.

6.5 Hydraulic PTO Designs for WECs

Hydraulic transmissions are often favoured for the conversion of the low-speed linear wave motion to high-speed rotary motion to drive the generator due to their high power density, robustness and controllability. As yet, an industry standard design configuration or control strategy for hydraulic PTOs is far from being established. As sizes of trial WECs increase beyond peak powers of 1 MW, serious challenges in PTO design are being encountered.

There is a trade-off between the complexity (and efficiency) of the hydraulic system and the electrical generation system. Two possible approaches are:

- a simple hydraulic system with limited energy storage, driving a variable speed generator;
- a more complex hydraulic system, designed to store the variable supply of power from the wave, and release smoothly to a constant speed generator.

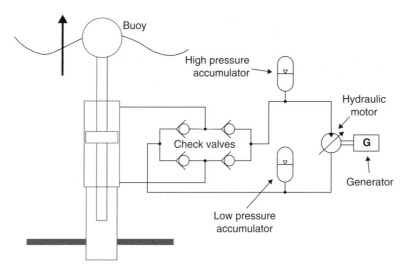

Figure 6.26 Simple hydraulic PTO.

Figure 6.27 Discrete-level force control PTO.

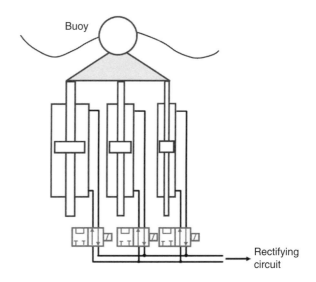

In both cases, control of force or torque on the primary wave interface (e.g. buoy) is required to extract maximum energy from the prevailing waves within the stroke and velocity limits of the PTO.

In the following, the example of a point absorber WEC will be used. For a point absorber, power is extracted from the vertical motion of a buoy relative to the seabed, a submerged reaction plate (drag plate), or a large reaction mass. Figure 6.26 shows a point absorber with a simple hydraulic PTO. Only limited control over the cylinder force is possible, which is achieved by managing the state of charge of the accumulators. An alternative design which allows force control is shown in Figure 6.27. A control strategy which switches the valves to vary the effective total piston area gives force

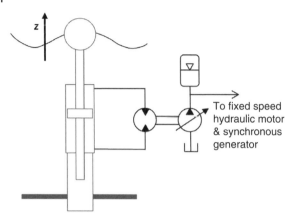

Figure 6.28 Variable displacement pump PTO.

To fixed speed
hydraulic motor
& synchronous
generator

control in discrete steps with a constant hydraulic pressure. However, frequent switching may ultimately give reliability problems, and switching can cause large force or motion transients [21].

A further option is to use a hydraulic motor and pump, or hydraulic transformer, as shown in Figure 6.28. With an over-centre pump, varying pump displacement can be used to give control of force as well as to provide unidirectional flow. However, losses in such a system (particularly part load losses) have been shown to be potentially quite high [22].

6.6   Direct Mechanical Power Take-Off

Hydraulic PTOs provide a relatively robust and controllable means of converting the slow linear (or limited angle rotary) motion of a typical WEC to the fast rotary motion required by conventional electrical generators. Nevertheless, mechanical means of achieving this conversion are also feasible, for example rack and pinion systems, ball screws, or winch systems. In particular the latter, with a steel cable winding on a drum, has been suggested as a simple way to achieve the required conversion. A simple system would have no power smoothing due to the lack of an energy storage system, and with a large gear ratio the reflected inertia of the electrical generator would be significant and so influence the WEC behaviour. However, there are ways in which the basic mechanical concept can be improved, such as by adding energy storage using flywheels as in the case described below.

The Manchester Bobber is an example WEC for which a mechanical PTO was proposed. The WEC design consists of a grid of buoys suspended beneath a platform fixed over the sea surface, thus forming a point absorber array. Each buoy is connected to one end of a cable, which runs around a pulley, and is connected to a counterweight at the other end, as shown in Figure 6.29. Each pulley drives a generator through a ratchet and a gearbox, and a flywheel is mounted on the same shaft as the generator. Whenever the pulley speed attempts to exceed the geared-down generator speed during buoy descent, the ratchet engages and the generator/flywheel is accelerated. As this is a relatively simple design it may exhibit high reliability. Also the vulnerable mechanical and electrical components are housed in a protected environment above sea level, which makes for ease of accessibility for maintenance. However, one disadvantage is that impact loading as the ratchet engages may damage the transmission. Gearbox failures which are a

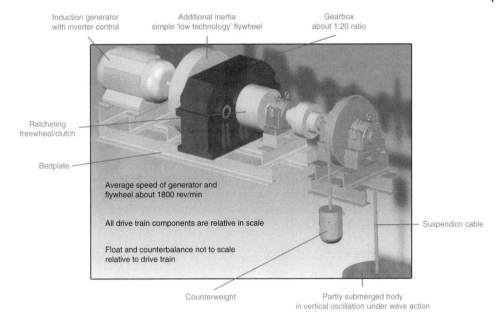

Figure 6.29 Manchester Bobber PTO (courtesy Renolds Gears).

problem in wind turbines may also be an issue here. Also there is limited controllability: the behaviour is highly dependent on the relative buoy and flywheel velocities which are present at any particular time.

6.7 Control for Maximum Energy Capture

A variety of control concepts for maximising energy capture have been studied, and the two most prominent ones are described here. Both relate to matching the characteristics of the WEC to the predominant frequency of the incident waves so that it behaves as though at resonance. A point absorber, such as that shown in Figure 6.30, is used for analysis. A more detailed analysis of the dynamics of such a device, including an example calculation of its resonant period, is given in Section 3.7. Understanding the practicalities of these control approaches, particularly the effect of PTO limitations and losses, is still at an early stage.

6.7.1 Reactive Control

The vertical buoy dynamics can be described using a linear hydrodynamic model:

$$M\frac{d^2z}{dt^2} + C\frac{dz}{dt} + Kz = f_e - f_p,$$

or

$$\left(Ms^2 + Cs + K\right)z = f_e - f_p, \tag{6.8}$$

where M is the buoy mass and added hydrodynamic mass, C is a linearised radiation resistance/drag term, K is the buoyancy stiffness, and s is the differential operator. The buoy upward displacement is z, the upward wave excitation force acting on the buoy is f_e, and the downward PTO force is f_p.

Assume the PTO force can be controlled to be a linear function of buoy position. The following relationship is sufficient to demonstrate reactive control [23, 24]:

$$f_p = \left(C_p s + K_p\right) z. \tag{6.9}$$

Thus the PTO behaves like a damper in parallel with a spring (with either positive or negative spring constant). The damper force component, in phase with velocity, relates to resistive (absorbed) power, and the spring force component, lagging the velocity by 90°, gives reactive power which averages to zero over one cycle. From equations (6.8) and (6.9), the buoy velocity is

$$v = \frac{s}{Ms^2 + \left(C + C_p\right)s + \left(K + K_p\right)} f_e. \tag{6.10}$$

A method for choosing C_p and K_p is required. Consider a regular (sinusoidal) wave of frequency ω. In the frequency domain, the amplitude of velocity (V) and excitation force (F_e) are related by

$$V = \frac{1}{\sqrt{(C + C_p)^2 + \left(\dfrac{K + K_p - M\omega^2}{\omega}\right)^2}} F_e, \tag{6.11}$$

or

$$V = \frac{1}{\sqrt{(1 + r_c)^2 + \left(\dfrac{K + K_p - M\omega^2}{C\omega}\right)^2}} \frac{F_e}{C}, \tag{6.12}$$

where $C_p = r_c C$ (thus PTO damping is chosen as a multiple r_c of hydrodynamic damping). The average power absorbed from the buoy by the PTO is the averaged product of the damping force $C_p v$ and velocity v, which for a sinusoid is

$$P_{\mathrm{av}} = \frac{1}{2} r_c C V^2, \tag{6.13}$$

so

$$P_{\mathrm{av}} = \frac{4 r_c}{(1 + r_c)^2 + \left(\dfrac{K + K_p - M\omega^2}{C\omega}\right)^2} \left(\frac{F_e^2}{8C}\right). \tag{6.14}$$

It can be shown that to maximise the power from the buoy (i.e. maximise P_{av}):

$$K_p = -K + M\omega^2,$$ (6.15)

$$r_c = 1.$$ (6.16)

K_p (which is usually negative) alters the natural frequency of the WEC to match the wave frequency, i.e. the buoy is in resonance. The 'reactive term' – the second term in the denominator of equation (6.14) – disappears because the inertia and stiffness forces cancel.

However, this choice for K_p and r_c will often give too large a motion amplitude in heavy seas (from equation (6.12)). To reduce the amplitude K_p or r_c can be increased, but it is clear from equation (6.14) that the latter is best as r_c also appears in the numerator and power will not be so greatly affected. Note that this control strategy requires an estimation of the dominant wave frequency.

6.7.2 Latching Control

Reactive control requires relatively high-resolution PTO force control, and also an oversized PTO to handle large reactive powers. A simpler alternative is latching control, applicable to devices with resonant frequencies higher than the wave frequency, which is the normal situation.

Phase control by latching was first introduced by Budal and Falnes in 1980 [25]. It consists of locking the buoy in position at the instant when its velocity is zero and releasing it after a certain delay such that the wave force is in phase with the body velocity (Figure 6.30). The latching duration effectively increases the resonant period of the device to match the frequency of the wave, and this tends to maximise the buoy motion amplitude. The release time of the body represents the control variable and studies have been undertaken to determine the best way to calculate this; prediction of the future wave excitation force is necessary to determine the optimum value [26]. The method gives significant power capture improvement in regular and irregular waves, with release timing strategies to maximise buoy amplitude or keep buoy velocity and wave excitation force in phase giving similar improvements [27].

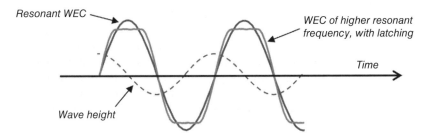

Figure 6.30 Latching control displacements.

6.7.3 Specific Hydraulic PTO Studies

6.7.3.1 Force Control

The PTO in the Pelamis WEC is described by Henderson [27]. A pair of single-ended cylinders mounted across each segment joint pumps fluid, via control manifolds, into high-pressure accumulators. These accumulators provide energy storage and in turn provide smoothed flow to hydraulic motors which drive grid-connected generators. Thus the generators are rated for the mean incident power.

Pelamis uses reactive control. However, the device is designed to have a resonant frequency close to the incident wave frequency, so the reactive power can be kept small. The PTO can control the moment about each joint to four discrete levels in each direction by switching different actuator chambers into the high-pressure line. Four equispaced levels can be achieved using 3 : 1 area ratio cylinders, as shown in Figure 6.31.

Measured PTO efficiencies in a test rig are reported to be around 90% for mechanical to fluid power conversion, and predicted at over 80% in a range of in-service conditions (with greater reactive power) [28]. Pressure drops in valves, manifolds and pipes are the main losses, with seal friction accounting for an estimated 5%. As suggested in the paper, the efficiency of a conventional variable hydrostatic transmission (or transformer) used to convert a constant high pressure source to variable cylinder force, in a range of operating conditions, is likely to be much less than 60%.

The idea of electronically switched cylinder chambers for WEC control originated in the 1970s in Salter's group in Edinburgh, with the need for a very efficient high power (>1 MW) rotary PTO for the Duck [2]. This led to the Digital Displacement® approach latterly developed by Artemis Intelligent Power, in which electronically controlled poppet valves were provided for each pumping chamber in a piston pump [29–31]. These pumps are of the radial piston type.

Reactive control is also the focus of [32], in which the Wavestar device – an array of point absorbers – is modelled. In this study, each absorber drives an over-centre variable displacement swashplate motor such that the output shaft can be controlled to always rotate the generator in the same direction (Figure 6.32). PTO force is controlled by a combination of motor displacement and generator torque control. Maximising motor displacement, and thus relying on generator control much of the time, was found to be the most efficient control strategy. When the power direction is reversed,

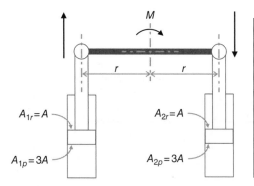

Connection to high(P_s) or tank (0) pressure				Moment, M (multiple of $P_s Ar$)
A_{1p}	A_{2r}	A_{1r}	A_{2p}	
P_s	P_s	0	0	4
P_s	0	0	0	3
P_s	0	P_s	0	2
0	P_s	0	0	1
0	0	0	0	0
0	0	P_s	0	−1
0	P_s	0	P_s	−2
0	0	0	P_s	−3
0	0	P_s	P_s	−4

Figure 6.31 Pelamis moment control concept.

i.e. the generator becomes a motor, its stored kinetic energy reduced the need for electrical power input. Simulation results predicted overall conversion efficiencies (buoy input to electrical output) of 52–68% for a range of wave conditions, which the authors considered inadequate. Part of the challenge with this design is the need to handle large peak powers as there is no hydraulic accumulation for smoothing.

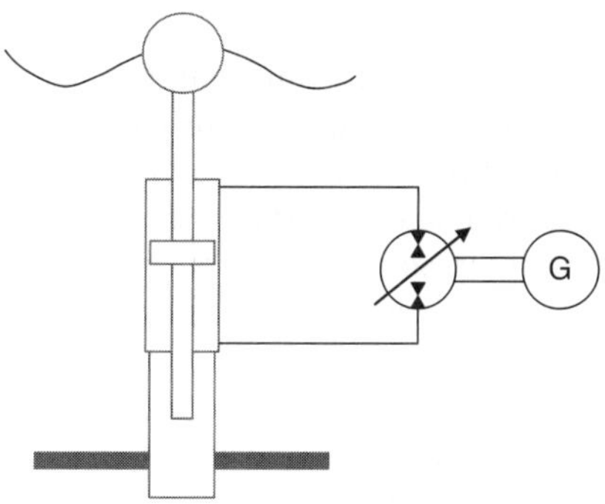

6.7.3.2 Resistive PTOs

A number of researchers have considered, in simulation, the simple rectifying/smoothing circuit of Figure 6.26 [33–39]. In this circuit, the piston is hydraulically restrained unless the force is high enough to overcome the accumulator pressure. This gives a latching effect, and also a resistance force akin to Coulomb friction, as shown in Figure 6.33. Note that the latching force may be much smaller than the optimum, but power capture can still be higher than just a viscous damper for a device operating below its resonant frequency [33, 37–39]. For the SEAREV point absorber, results indicate that two design parameters of the PTO (high pressure level and generator rated power) have a large influence on the power generated and these parameters can be

Figure 6.32 Force control by varying motor displacement and generator torque.

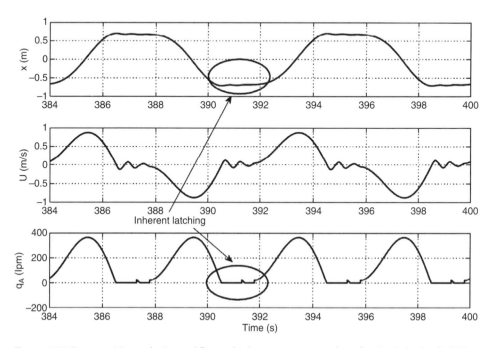

Figure 6.33 Buoy position, velocity, and flow to high-pressure accumulator for simple hydraulic PTO of Figure 6.26 [33].

optimised to sea state. It has also been shown that parameters in a PTO can be optimised for a given wave condition in [34]. Some controllability might be possible by altering motor displacement to reach a different accumulator state of charge and hence pressure, to match characteristics to different wave periods [33].

Design concepts for the hydraulic PTO of the Wavebob point absorber are discussed in [39]. Using a hydraulic transformer, in an arrangement similar to Figure 6.28, is considered, but dismissed due to the need for a motor rated to peak powers, and reliability concerns associated with its intermittent operation. Instead, a rectifying circuit charging a high-pressure accumulator, as in Figure 6.26, is proposed, but with a check valve before the accumulator, and an extra flow path from the cylinder through a second hydraulic motor. Both motors are variable displacement, and drive the same generator. The extra motor allows the generator to be driven in small sea states where the wave force is below the equivalent accumulator force. Additional flexibility is provided by having two pump/generator modules of different size, and either or both can be operative at any time. Thus a reasonable degree of versatility in damping force control is possible, but reactive control is not proposed.

Three PTO designs are simulated for a terminator WEC in [40]. A constant pressure system (Figure 6.26) is compared with linear damping (PTO force proportional to velocity) implemented by removing the accumulation and varying hydraulic motor displacement and generator torque with constant generator speed. The terminator simulated has no inherent stiffness and thus no resonance, so the latching effect of the constant pressure system is of no benefit, and linear damping is found to be best. Connecting four terminators to a single centralised, onshore, hydraulic motor plus generator is also studied. The terminators can be positioned to operate out of phase with one another. Linear damping at the individual terminators can no longer be achieved, but controlling pressure to be proportional to total flow into the central motor gives good power capture.

6.7.3.3 System Modelling

Until recently, there has been very little published research on detailed dynamic predictions of complete system behaviour, incorporating both hydrodynamic and PTO modelling. Many control studies have assumed the PTO can provide a user-defined mechanical impedance, and PTO efficiency is not considered. Some of the first papers to include a reasonable PTO model are by Falcão [35, 36]. Equations for a heaving buoy attached to a simple rectifying hydraulic PTO (as in Figure 6.26) are presented, but there are significant simplifications in the PTO model, such as lossless components. A more detailed hydraulic PTO unit is presented in [39], but critical characteristics, like hydraulic motor losses, are represented just by a constant efficiency value. Convincing system simulation results, incorporating sophisticated PTO characteristics, have only been published since 2010 [32–34, 38].

Whole-system simulation is important to predict absorbed power. The PTO cannot be optimised in isolation, for changing a PTO design to increase its own efficiency will alter the wave–buoy interaction force and motion, which may reduce absorbed power. Accurate PTO simulation is also necessary to enable the best control method to be selected. Losses and non-ideal characteristics such as friction, leakage, compressibility, inertia, and valve/pipe pressure drops may mean that accurate force control is not achieved open loop. Simulating such characteristics can also show that larger PTOs, designed to handle reactive power, have larger losses and so the advantage of reactive control is lost.

6.8 Electrical Infrastructure and Grid Integration

6.8.1 Electrical Infrastructure Components

The power harnessed from waves and tides by the PTO, once converted into electricity by the electrical generator, needs to be transported from the offshore location where the MRE device is located to the onshore grid connection point. Hence, the main aim of offshore subsea electrical arrays is to deliver this electricity in an acceptable form and at an affordable price. Furthermore, an electrical array has to be able to provide an auxiliary power connection to each one of the devices and act as a communication network between the devices and the main control unit.

To understand the different types of subsea electrical arrays (see Section 6.8.2) and optimise for a particular MRE prototype or project, it is crucial to understand how each component of the grid fits into the whole system. The various components that make up a subsea electrical system are defined next.

6.8.1.1 Transmission Cable Systems

Transmission cable systems are composed of the cables themselves, the protection or armouring and other accessories required to connect the cables to other elements of the subsea electrical array or to safely cross over other subsea cables. The main purpose of these systems is to link the offshore power grid to the shore, providing a power transport cable as well as a communication network. The technology in use by the subsea cable industry is well advanced thanks to the offshore wind industry expansion in recent years. A submarine power cable can range in diameter from 70 mm to over 210 mm and can be one of two main types, according to the type of current: high-voltage alternating current (HVAC) power cable or high-voltage direct current (HVDC) power cable. The selection criteria to decide which specific type of cable to use and its diameter are strongly dependent on factors such as route length, seabed morphology, voltage, current capacity and grid synchronisation. HVAC is the type of submarine power cable commonly selected for MRE projects due to some of the technical constraints and the proximity to shore of the projects.

Protection or armouring systems for submarine cables include cable burial, armour casing, concrete mattress, rock dumping and horizontal direct drilling (HDD). Cable burial is widely used in the offshore wind industry as a fast and cheap solution; however, cables can only be buried in a soft seabed which is often not possible at MRE sites. When the cable cannot be buried deep enough or needs to lie on the seabed due to the existence of a rocky or mixed surface, the cable can be enclosed in a metallic flexible armour case before installation. Other solutions can be to cover the cable using a concrete mattress, which has been used at EMEC [40], or dump rocks on top of the cable, as has been done at the Wave Hub [41]. If the site is close to shore and none of the discussed methods can be used, due to either the conditions of the seabed or a strong trawling activity, HDD can also be used; however, this is most expensive. HDD consists of routeing the cable under the seabed in a hole drilled from the shoreline. HDD has been used at Strangford Lough [42] and is the selected preferred method for the MeyGen project [43].

6.8.1.2 Dynamic Umbilical Cable

Dynamic umbilical cables are used to connect the transmission cable or any other type of underwater system with a floating MRE device or platform. This type of cable is widely used in the oil and gas industry at depths well in excess of those expected for MRE. However, such cables have been designed for a completely different sector with different standards and requirements, and using them for MRE devices of 1–2 MW size may not be economical. Furthermore, wave and tidal devices will stress dynamic umbilical cables under a different type and range of forces from those for which they have been designed.

Dynamic umbilical cables are often used together with bend restrictors and bend stiffeners, which are devices used to prevent the umbilical cable from being stressed beyond its design limits. Bend stiffeners are positioned at the connection ends of the umbilical cable, where the cable goes into the device or junction box. Bend restrictors are placed along the umbilical cable to prevent the cable from bending more sharply than its minimum bending radius. Restrictors will allow the cable to move freely up to the point when they will lock into a safety position preventing further bending. Buoyant restrictors can also be used to keep the umbilical cable at a certain depth, to reduce the weight of the cable pulling on the device [13].

6.8.1.3 Subsea Connectors

Subsea connectors are used to connect two parts of the subsea grid, and are required for the high-voltage power grid, the low-voltage auxiliary power system and communication networks. Subsea connectors are designed to tolerate the corrosive action of the seawater and marine life, and must operate at high pressures. Subsea connectors can be grouped in two classes: wet mate connectors (which can be connected under water) and dry mate connectors (for which the connection needs to be made onshore or on a vessel).

Dry mate connectors are a mature piece of technology, able to handle a wide range of currents and voltages at an affordable price. To proceed with the connection or disconnection of a dry connector, this needs to be lifted out of the water. This requires a considerable amount of free cable on both sides of the connector (approximately two or three times the water depth on each side).

Wet mate connectors are novel solutions which makes them expensive. They are still limited to low voltage ranges (6.6–11 kV); however, connectors that will operate at 36 kV are at the early stage of development. Wet connectors can be connected directly to the device or by using a stab plate arrangement. The direct connection will need a remotely operated vehicle or a diver to physically perform the connection or disconnection operation, while the stab plate option will be integrated in the foundation of the device, which is fixed to the seabed, then a removable part of the device will be connected to the foundation and at the same time the connectors in the stab plate will be connected. This is the system used by Alstom and Atlantis tidal devices.

Multi-way connectors allow more than one connection, usually three-way or four-way. Multiple connections are required for radial arrays (see Section 6.8.2). A four-way connector will allow each device to have an input and output from the previous and following devices on the radial array; this configuration increases the quality of the generated power. However, this type of device is expensive and requires additional space for the extra connection, which is sometimes not possible.

6.8.1.4 Frequency Converters

Frequency converters are power electronic systems that convert the variable frequency power output of an asynchronous generator to the fixed frequency required to supply the electricity to the onshore grid. Converters are usually divided into two different parts. The first is the machine or generator part, also known as converter (Figure 6.34a) which converts the variable frequency AC output from the generator to a DC signal. Then the grid side part (Figure 6.34b) converts the DC signal into an AC signal at the grid fixed frequency; this part is also known as inverter. All MRE devices need a frequency converter at some point in the electrical system. The position of this converter is one of the main factors to consider at the design phase of a submarine electrical array.

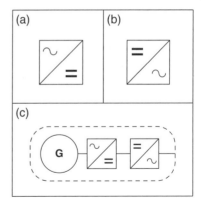

Figure 6.34 Symbols for different types of converter: (a) the generator-side converter; (b) the grid-side converter or inverter; and (c) the back to back frequency converter and the variable frequency generator.

6.8.1.5 Transformers

Transformers are used to raise or reduce the voltage level at certain points of a power grid. Usually at the generation side voltage needs to be raised from the medium voltage levels at the generator output (0.4–11 kV) to high or transport voltage levels (33 kV or 132 kV) in order to reduce the transmission power losses, as these losses are inversely proportional to the square of the voltage. It is worth noting that usually electrical grids over 132 kV are considered as transport grids and these are operated and managed by a grid operator company. Subsea transformers are also commercially available and have been used by the oil and gas industry for the last two decades. These transformers could be easily adapted for use as part of a subsea electrical array for MRE.

6.8.1.6 Connection Hubs

A connection hub or hub arrangement is a node on the network that shapes the whole grid, acting as the backbone of a large electrical array. Connection hubs are compatible with devices with both on-board and remote converter units. According to the type of MRE device used, connection hubs can be classified as: aggregation only; aggregation and transformation; and conversion, aggregation and transformation. Aggregation-only hubs can be used to combine together the power output from an array of fixed frequency devices, with on-board converters. Aggregation and transformation hubs combine the power output from a number of devices and then step up the voltage to 33 kV or 132 kV to transmit it in a more efficient way to the shore. Conversion, aggregation and transformation hubs are used when the MRE devices in the farm use variable frequency generators. In this configuration, the conversion units are integrated together at the hub. This configuration has the possibility of using one or more grid-side medium voltage converters to fix the frequency output. Figure 6.35 shows the scheme for the three possible configurations.

According to where the connection hub is located, it can be classified as: a surface piercing hub; a floating hub; or a subsea hub. Surface piercing hubs are located on a

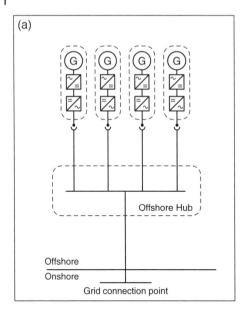
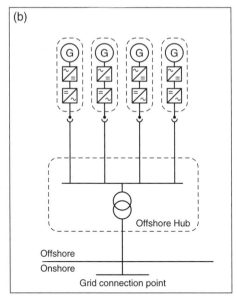
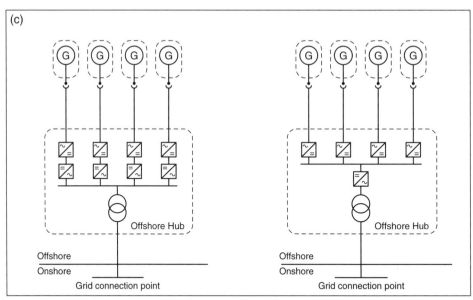

Figure 6.35 Different types of offshore connection hubs, (a) aggregation-only hub; (b) aggregation and transformation hub; and (c) conversion, aggregation and transformation hub, for the two possible configurations with one (left-hand side) and multiple grid-side medium voltage converters (right-hand side). Image based on [13].

platform situated on top of a structure which is fixed to the seabed, usually a jacket frame or a monopile foundation. This is the standard approach used by the offshore wind industry, which makes this solution the most advanced one at the moment and the one presenting lowest risk. Floating hubs could represent a cheaper solution in the future, and are especially attractive for floating MRE devices; however, the possible commercial solutions for floating hubs are still at an early development stage and it is still not clear that this type of solution can make it through the planning and consent phases. Finally, subsea hubs are pieces of hardware that are available from many manufacturers (e.g., MacArtney, JDR, Hydro Group, Alstom). All subsea hubs are of the aggregation-only type, and so only allow collection of the power output from a number of fixed frequency MRE devices. An example of a subsea hub is the one developed by JDR deployed at the Wave Hub site.

6.8.2 Offshore Electrical Arrays

When designing the grid topology for an MRE park there are a number of constraints that should be considered. These are driven by the choice of device and the size and location of the project. The distance from the park location to the onshore connection point to the main land grid is one of the critical points when designing a grid topology, as losses increase with the distance and current MRE devices generate at only 6.6 kV or 11 kV. For larger distances, offshore hubs with transformers would be required to step up the transmission voltage. The number of devices in the park would impact significantly on the grid topology, so for initial parks with a few devices, a grid topology with directly connected devices is viable, but this would be prohibitively expensive for parks with a larger number of devices, where a star cluster or radial topologies would suit better. Finally, the type of device selected for the park will have a clear impact on the grid topology, not only for the voltage range and its power output, but also depending on whether it hosts the converter unit on-board or not. Variable frequency devices would mean that converters would need to be installed downstream on the grid either on an offshore hub or onshore. The three possible grid topologies (directly connected devices, star cluster configuration and radial configuration) are defined next.

6.8.2.1 Directly Connected Devices

In this type of electrical array, MRE devices are connected directly to shore with their own individual cables. This alternative is suitable where each device has a variable frequency generator and the converter of each device is located onshore; however, it can also be used for devices with on-board converters (Figure 6.36a). This is a solution applicable for initial arrays of MRE devices with small numbers of units deployed close to shore; however, as the number of devices and the distance from shore increases this solution becomes expensive and the losses in the transmission cables become large. This array topology is the one selected for the MeyGen project [43].

6.8.2.2 Star Cluster Configuration

In a star cluster configuration, MRE devices are grouped around hubs which collect individually the power output from each device then export it to the shore by means of an individual transport cable. This array topology can be adapted to any type of generator (fixed or variable frequency) and to any type of hub. For variable frequency generators hubs will host the converters for each turbine then export the power at medium voltage to shore. This option will require the hubs to be close to the devices in order to reduce transport losses. Therefore, a surface piercing cluster will be a good solution, as a larger platform can be obtained for low price. For fixed frequency generators hubs will

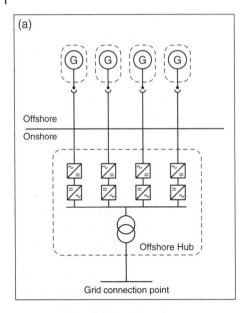

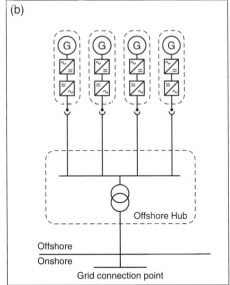

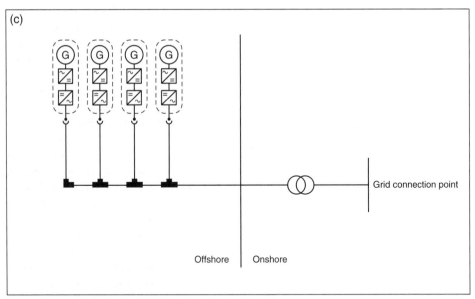

Figure 6.36 Array types for offshore electrical grids: (a) directly connected devices; (b) star cluster array; and (c) radial array. Image based on [13].

aggregate only the power from each device and so submarine or floating hubs can be used, as these are relatively cheap option compared with a surface pierced one. An example of a star cluster grid topology is the Wave Hub [41] where a submarine hub aggregates the power output of each one of its four berths then transmits it to the grid connection point onshore. Figure 6.36b shows an example star cluster configuration for the offshore grid topology; however, any of the examples shown for the different hub types in Figure 6.35 are also valid examples.

6.8.2.3 Radial Configuration

In a radial configuration, the power output from each MRE device is collected on a central ring which is formed by the transmission cable, and each device is connected to the main ring by a three-way connector (Figure 6.36c). This grid topology configuration is a variation of the one most extensively used for offshore wind parks, where each wind turbine has two connection points, one to connect to the next turbine in the array and the other to connect to the previous turbine in the array, with an internal ring which allows the unit to connect to the grid or just keep the continuity of the string. However, this configuration is not possible for current MRE devices as they do not have enough space to accommodate the two connections and the internal ring. Radial grid topologies are only possible with fixed frequency generators, where the converters are mounted on-board the MRE devices.

The main limitation for this grid topology is the number of MRE devices that can be connected in each string; as most recent MRE devices operate at 6.6 kV or 11 kV, the number will be limited to 6 or 10 devices respectively for 1 MW per string. This limitation would require several strings to be connected at hubs where the current would be increased and then transmitted to shore. Another possible solution would be to step up the voltage of each individual device or of a reduced group of devices by means of subsea transformers, which would deliver the power to the radial grid at 20 kV or 33 kW, increasing the capacity of the string. These limitations show that even though radial grid topology configurations are possible for MRE devices, these will require further development and are not a short-term option. They are an alternative to be considered for future larger arrays.

6.8.3 Grid Integration and Power Quality

All facilities generating electricity and supplying it to the grid, and especially those facilities generating electricity from renewable energy sources, need to ensure the reliability and quality of the supply. In particular, MRE needs to harness a natural resource which is inherently variable. In order to ensure the proper operation of the general grid and an MRE power generation facility, there are two main aspects to consider: firstly, the grid integration conditions to which the supplied power needs to adhere; and secondly, the quality of the produced power. Is it not the aim of this chapter do an intensive review of these two points and only the main concepts will be tackled. More details about grid integration and power quality of MRE can be found in [44, 45].

6.8.3.1 Grid Integration

In order to ensure a smooth operation of power generation facilities working with renewable energy sources, and especially those harnessing variable resources, a set of grid codes and distribution codes for electrical transmission and distribution systems are in place. Grid codes which have been defined for wind power plants under normal and grid disturbance conditions are of special relevance for the MRE sector. From these codes there are four main parameters to consider when looking at the grid integration: these are the frequency (f), voltage (V), reactive power (Q), and the fault ride through (FRT) behaviour.

As seen in previous sections, the majority of MRE devices use a variable frequency asynchronous generator, which requires the presence of a frequency converter at some point in the electrical infrastructure. Most commercial frequency converters are already adapted to the diverse national grid codes. Thus, by ensuring the compliance of the grid codes by the selected frequency converters, only the FRT test would be a main requirement to ensure compliance requirements for the MRE generation facility.

The nominal grid frequency (f) in Europe is 50 Hz, allowing a frequency operation range which varies depending on the country and the voltage level (Germany has the widest range in Europe, going from 46.5 Hz to 53.5 Hz). The allowable voltage range, under normal operation conditions, differs between countries; however, typically when a power generation facility is under continuous operation, the voltage at the connection point is expected to be within ±10% of the nominal grid voltage. Reactive power (Q) regulations, which relate to the phase relationship required between voltage and current waveforms, depend on the actual (real) power, or the connection point voltage or both. Reactive power is usually quantified by the power factor, which is the ratio of the real power to the apparent power (which is the power value obtained if voltage and current were assumed to be in phase). Reactive power requirements vary considerably between countries, depending on transmission voltage ranges and even on whether it is a high or low energy demand period for the grid. However, a power factor higher than 0.95 at the grid connection point is normally required [44].

FRT, sometimes called low voltage ride through, is the capability of an electric generator to stay connected in short periods of lower electric network voltage. A fault such as a short circuit on a network would cause a voltage dip and it is important than generators do not fail in this condition. FRT requirements for a power plant differ considerably between countries and in general depend on parameters such as generation capacity, connection point voltage and voltage dip duration. The more restrictive FRT test curves for some European countries are presented in Figure 6.37. Further details about the requirements for the FRT test at different European countries can be found in [44]. Testing the FRT characteristics of an MRE park or device represents the greatest

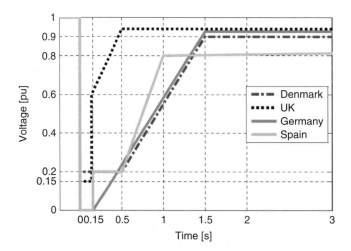

Figure 6.37 Low voltage ride-through test curves for different European countries [44].

challenge when following the grid codes. For these tests, MRE devices should be decoupled from the grid in order to avoid interactions between this and other devices in the same grid. Portable test platforms, mounted on containers or truck trailers, have been developed for wind or solar farms, which could be also suitable to be used for MRE parks.

6.8.3.2 Power Quality

Defining the quality of the power supplied to the grid by an MRE power plant requires monitoring a wide set of indicators, apart from the ones already defined in the grid codes, and seen for the grid integration. For MRE devices the International Electrotechnical Commission (IEC) currently has several standards under development, which will describe the required technologies, measurement procedures, etc. IEC 62600, 'Marine energy – Wave, tidal and other water current converters', defines the root standard for MRE devices. Most of the standards defined under the MRE root are still under development; however, it is expected that they will be similar to the ones already defined for wind turbines. IEC 61400-21 defines the relevant parameters to characterise the power quality of a wind turbine; these can be found in [45, p. 80].

One of the main actions to improve the power quality of MRE parks is to minimise the fluctuations in the power output, which can be achieved for example by using improved control strategies, or by incorporating mechanical, hydraulic or electrical energy storage units. From the offshore wind experience, it has been learned that the positioning of devices in an array relative to the incident resource has a strong influence on the quality of the power delivered by the individual devices [44]. Hence, results and lessons learned from the first MRE parks will be crucial in order to define strategies to increase the quality of the power output of future MRE parks.

6.9 Summary of Challenges for PTO Design and Development

Wave and tidal PTO design and control is an immature field. Suitable design solutions for tidal turbine PTOs can be based on more well-established approaches, and design practice mirroring wind turbines is possible. The best approach to achieve power take-off for any particular WEC design is usually less clear. However, most PTOs for large-scale reciprocating WECs under development are hydraulic. Only very recently have there been serious attempts, with the help of simulation, to analyse and optimise PTO design and control, properly accounting for hydrodynamic, hydraulic, mechanical and electrical characteristics. Nevertheless, rigorous experimental validation of models and results for many types of WEC at large scale is still almost non-existent. Some of the challenges facing WEC PTO designers are as follows:

- Ensuring extremely high robustness and reliability, despite extreme environmental condition.
- Maximising power absorption with highly variable seas (long periods with small seas states, and short periods with large waves and massively increased input power).
- Balancing the benefits of achieving ideal PTO behaviour (e.g. reactive control) with the consequences for PTO design (e.g. larger, more expensive PTO with higher losses).

- Determining the benefits of energy storage (typically hydraulic accumulation) to allow sizing motors/generators for mean rather than peak powers, improving efficiency and reliability through constant-speed running, but making buoy force control more difficult.
- Improving part-load efficiency of components (e.g. variable displacement hydraulic motors).
- Considering implications of arrays (farms) of WECs working together.
- Developing controllers which automatically adapt to wave conditions.
- Sourcing electrical power system componentry, perhaps requiring adaptation from systems developed for the offshore wind or oil and gas industries.

References

1 Ocean Power Technologies Inc. (2014). OPT web page. Available: http://www. oceanpowertechnologies.com/ (accessed 22/6/2016)

2 S. H. Salter, J. R. M. Taylor and N. J. Caldwell (2002) Power conversion mechanisms for wave energy. *Proceedings of the Institution of Mechanical Engineers, Part M: Journal of Engineering for the Maritime Environment* 216, 1–27.

3 Aquamarine Power Ltd. (2015). Aquamarine Power web page. Available: http://www. aquamarinepower.com (accessed 22/6/2016).

4 Wave Dragon AS. (2005). Wave Dragon web page. Available: http://www.wavedragon. net/index.php (accessed 22/6/2016).

5 M. A. Mueller (2002) Electrical generators for direct drive wave energy converters. *IEE Proceedings – Generation, Transmission and Distribution* 149, 446–456.

6 H. Polinder, M. E. C. Damen and F. Gardner (2004), Linear PM generator system for wave energy conversion in the AWS. *IEEE Transactions on Energy Conversion* 19, 583–589.

7 H. Polinder, M. E. C. Damen and F. Gardner (2005) Design, modelling and test results of the AWS PM linear generator. *European Transactions on Electrical Power* 15, 245–256.

8 N. Iwabuchi, A. Kawahara, T. Kume, T. Kabashima and N. Nagasaka (1994) A novel high-torque reluctance motor with rare-earth magnet. *IEEE Transactions on Industry Applications* 30, 609–614.

9 M. R. Harris, G. H. Pajooman and S. M. Abu-Sharkh (1997) The problem of power factor in VRPM (transverse-flux) machines. In *Eighth International Conference on Electrical Machines and Drives* (Conference Publication No. 444, pp. 386–390). London: Institution of Electrical Engineers.

10 K. T. Chau, W. L. Li and C. H. T. Lee (2012) Challenges and opportunities of electric machines for renewable energy (invited paper). *Progress in Electromagnetics Research B* 42, 45–74.

11 D. O'Sullivan and T. Lewis (2008) Electrical machine options in offshore floating wave energy converter turbo-generators, In *Proceedings of the Tenth World Renewable Energy Congress (WREC X)*, Glasgow, UK (pp. 1102–1107).

12 L. H. Hansen, P. H. Madsen, F. Blaabjerg, H. C. Christensen, U. Lindhard and K. Eskildsen (2001) Generators and power electronics technology for wind turbines. In *IECON '01. The 27th Annual Conference of the IEEE Industrial Electronics Society* (vol. 3, pp. 2000–2005). Piscataway, NJ: IEEE.

13 ORE Catapult (2015) Marine Energy Electrical Architecture. Report 3: Optimum Electrical Array Architectures, PN000083-LRT-008.

14 M. Folley, R. Curran and T. Whittaker (2006) Comparison of LIMPET contra-rotating wells turbine with theoretical and model test predictions. *Ocean Engineering* 33, 1056–1069.

15 M. Inoue, K. Kaneko, T. Setoguchi and H. Hamakawa (1989) Air turbine with self-pitch-controlled blades for wave power generator (estimation of performance by model testing). *JSME International Journal. Ser. 2, Fluids Engineering, Heat Transfer, Power, Combustion, Thermophysical Properties* 32, 19–24.

16 T. Setoguchi, S. Santhakumar, H. Maeda, M. Takao and K. Kaneko (2001) A review of impulse turbines for wave energy conversion. *Renewable Energy* 23, 261–292.

17 S. Santhakumar, V. Jayashankar, M. A. Atmanand, A. G. Pathak, M. Ravindran, T. Setoguchi *et al.* (1998) Performance of an impulse turbine based wave energy plant. *Eighth International Offshore and Polar Engineering Conference.*

18 J. F. Douglas, J. M. Gasiorek, J. A. Swaffield and L. B. Jack (2011) *Fluid Meachanics* (6th edn.). Harlow: Prentice Hall.

19 L. Cameron, R. Doherty, A. Henry, K. Doherty, J. Van't Hoff, D. Kaye, *et al.* (2010) Design of the next generation of the Oyster wave energy converter. In *Proceedings of the 3rd International Conference on Ocean Energy* (pp. 1–12).

20 P. Frigaard, J. P. Kofoed and W. Knapp (2004) Wave Dragon: Wave power plant using low-head turbines. *In Proceedings of Hidroenergia 04: International Conference and Exhibition on Small Hydropower*, Falkenberg, Sweden (p. 8).

21 U. A. Korde (1999) Efficient primary energy conversion in irregular waves. *Ocean Engineering* 26, 625–651.

22 K. Budal and J. Falnes (1975) A resonant point absorber of ocean-wave power. *Nature* 256, 478–479.

23 K. Budal and J. Falnes (1980) Interacting point absorbers with controlled motion. In B. M. Count (ed.), *Power from Sea Waves* (pp. 381–399).

24 A. Babarit and A. H. Clément (2006) Optimal latching control of a wave energy device in regular and irregular waves. *Applied Ocean Research* 28, 77–91.

25 A. Babarit, G. Duclos and A. H. Clément (2004) Comparison of latching control strategies for a heaving wave energy device in random sea. *Applied Ocean Research* 26, 227–238.

26 J. Falnes (2002) *Ocean Waves and Oscillating Systems: Linear Interactions Including Wave-Energy Extraction.* Cambridge: Cambridge University Press.

27 A. R. Henderson (2006) Design, simulation, and testing of a novel hydraulic power take-off system for the Pelamis wave energy converter. *Renewable Energy* 31, 271–283.

28 M. Ehsan, W. H. S. Rampen and S. H. Salter (1997) Modeling of digital-displacement pump-motors and their application as hydraulic drives for nonuniform loads. *Journal of Dynamic Systems, Measurement, and Control*, 122, 210–215.

29 W. H. S. Rampen (1992) The digital displacement hydraulic piston pump. PhD thesis, University of Edinburgh.

30 R. H. Hansen, T. O. Andersen and H. C. Pedersen (2011) Model based design of efficient power take-off systems for wave energy converters. In *Proceedings of the 12th Scandinavian International Conference on Fluid Power (SICFP 2011)*, Tampere, Finland (pp. 18–20).

31 A. F. de O. Falcão (2007), Modelling and control of oscillating-body wave energy converters with hydraulic power take-off and gas accumulator. *Ocean Engineering* 34, 2021–2032.

32 A. F. de O. Falcão (2008), Phase control through load control of oscillating-body wave energy converters with hydraulic PTO system. *Ocean Engineering* 35, 358–366.

33 C. J. Cargo, A. R. Plummer, A. J. Hillis,and M. Schlotter (2011) Optimal design of a realistic hydraulic power take-off in irregular waves. In *Proceedings of the 9th European Wave and Tidal Energy Conference (EWTEC)*, Southampton, UK, 2011.

34 C. J. Cargo, A. R. Plummer and A. J. Hillis (2011) Hydraulic power transmission and control for wave energy converters. In *Proceedings of the 52nd National Fluid Power Conference*, Las Vegas.

35 C. J. Cargo, A. R. Plummer, A. J. Hillis and M. Schlotter (2011) Determination of optimal parameters for a hydraulic power take-off unit of a wave energy converter in regular waves. *Proceedings of the Institution of Mechanical Engineers, Part A: Journal of Power and Energy* 226, 98–111.

36 M. Livingston and A. R. Plummer (2010) The design, simulation and control of a wave energy converter power take off. In *Proceedings of the 7th International Fluid Power Conference (IFK)*.

37 C. Josset, A. Babarit and A. H. Clément (2007) A wave-to-wire model of the SEAREV wave energy converter. *Proceedings of the Institution of Mechanical Engineers, Part M: Journal of Engineering for the Maritime Environment*, 221, 81–93.

38 K. Schlemmer, F. Fuchshumer, N. Böhmer, R. Costello and C. Villegas (2011) Design and control of a hydraulic power take-off for an axi-symmetric heaving point absorber. In *Proceedings of the 9th European Wave and Tidal Energy Conference (EWTEC)*, Southampton, UK.

39 Y. Kamizuru and H. Murrenhoff (2011) Improved control strategy for hydrostatic transmissions in wave power plants. In *Proceedings of the 12th Scandinavian International Conference on Fluid Power (SICFP 2011)*, Tampere, Finland.

40 European Marine Energy Centre (2015). *EMEC web page.* Available: http://www.emec.org.uk/ (accessed 8/10/2015).

41 Wave Hub Ltd. (2014). *Wave Hub web page.* Available: http://www.wavehub.co.uk/ (accessed 5/5/2014).

42 Marine Current Turbines Ltd. (2015). *Marine Current Turbines web page.* Available: http://www.marineturbines.com/ (accessed 16/10/2015).

43 MeyGen Ltd. (2013). *MeyGen web page.* Available: http://www.meygen.com/ (accessed 18/3/2016).

44 J. Giebhardt, P. Kracht, C. Dick and F. Salcedo (2014) Deliverable 4.3: Report on grid integration and power quality testing. Marine Renewables Infrastructure Network.

45 T. Ackermann (2005), *Wind Power in Power Systems.* Chichester: John Wiley & Sons.

7

Physical Modelling

Martyn Hann[a] and Carlos Perez-Collazo[b]

[a] *Lecturer in Coastal Engineering, School of Engineering, University of Plymouth, UK*
[b] *PRIMaRE Research Fellow, School of Engineering, University of Plymouth, UK*

7.1   Introduction

Physical modelling is a crucial tool for the development of marine renewable energy (MRE) devices and is recommended in most development guidelines and laboratory testing standards [1–3]. This is especially true at early stages of development, such as for wave energy converters (WECs) with a technology readiness level (TRL) between 1 and 5. Testing model devices at scale in the controlled environment of a flume or basin has many advantages. These include the proof (or otherwise) of novel design concepts, the ability to test in systematically changing conditions and the ability to test in conditions which have low occurrence probabilities (i.e. extreme events). Physical modellers not only are able to collect detailed quantitative measurements of a device's performance but also can qualitatively observe their device interact with wave and currents in a way not possible when deployed at sea. If done correctly these measurements and observations can lead to evolution of device designs and concepts and reduce the chance of costly failure if and when devices are eventually deployed at sea.

Physical model testing is a common engineering tool which has a long history. For example, the systematic scale testing of hydraulic models goes back to the nineteenth century. In physical modelling, certain laws of similarity and common standard practices are followed and several reference books are available that discuss these in detail [4, 5].

This chapter presents the basic concepts and common practices which should be considered when performing physical modelling of an MRE device. The focus is on the physical modelling of a device's interaction with wave and current loading. The testing of individual components, such as the PTO, are not considered here. The chapter is structured as follows. In Section 7.2 physical modelling is framed into the development process of an MRE device, and the main steps which need to be considered when planning a test campaign are presented. Section 7.3 presents the similitude laws and scaling concepts. Section 7.4 goes into model design and construction techniques, including the different techniques used to represent the power take-off (PTO) of the device. Section 7.5 defines the different techniques commonly used to represent real fixings or moorings at a model scale. Section 7.6 discusses instrumentation commonly

Wave and Tidal Energy, First Edition. Edited by Deborah Greaves and Gregorio Iglesias.
© 2018 John Wiley & Sons Ltd. Published 2018 by John Wiley & Sons Ltd.

used for physical modelling. Section 7.7 presents the different techniques used to calibrate scale models. Section 7.8 defines the standard approaches which are usually followed to model the marine environment in the laboratory. Section 7.9 presents the different test facilities commonly used for physical modelling of MRE devices. Finally, Section 7.10 goes through the recommended set of tests used at the different stages of development of an MRE device.

7.2 Device Development and Test Planning

There are diverse development and evaluation protocols for MRE devices, which are based on the traditional engineering methods related to TRLs first introduced by the National Aeronautics and Space Administration (NASA). These protocols define five different development stages or phases which cover the nine TRLs. Scale physical modelling is commonly used to develop devices at stages 1 and 2 and to test some components at larger scale in stage 3, covering TRLs 1–5. For a more detailed discussion on MRE design and development, see Chapter 5. The TRL of the MRE device being tested is crucial when planning a physical modelling test campaign as the requirements to test a TRL 1 device are different from those of a TRL 4 device. An appreciation of where a device is in its development, including the results of previous studies and the requirements for future development, are crucial to the planning of a successful and useful physical modelling campaign. Table 7.1 shows the different technical considerations to take into account when planning a physical modelling campaign for an MRE device. Figure 7.1 shows images of the same MRE device being tested at different development stages.

In Figure 7.2 a typical pathway undergone by an experimental campaign is laid out. This chapter addresses each step in this diagram as a means to examine the physical modelling process and to provide guidance on how to successfully test scaled MRE devices. Many of the first steps are covered during the planning of the physical modelling. The proper planning of experimental work is crucial to ensure that the maximum benefit is achieved from the available tank time, while reducing the risk of costly (and embarrassing) mistakes occurring. Two of the most important criteria to consider during the planning stage of a physical model test are: to identify what are the expected outcomes from the physical modelling as guided by the current development status of the MRE device; and to decide upon the suitable scale at which to conduct the tests.

The required experimental outcomes should be considered when choosing a model scale at the planning stage. For example, an experimental campaign for a proof of concept requires a basic understanding of the associated physical processes. However, a model validation study requires a precise representation of all physical processes occurring within the prototype. These two experimental campaigns have very different outcomes and therefore can be conducted at very different scales. This is discussed in greater detail in the next section.

7.3 Scaling and Similitude

Nearly all physical modelling of MRE devices is conducted at scale. Testing at a larger scale (closer to prototype size) is generally advantageous in reducing scale effects. The choice of model scale is generally a compromise between the cost of the project/model,

Table 7.1 Technical output of TRLs 1–5 and the three first development phases of an MRE, based on [1].

Development	Phase 1 Model validation			Phase 2 Model design	Phase 3 Process model	
	Concept	Performance	Optimisation	Validation	Subsystems validation	Sea trials
TRL	1	2	3	4	5	6
Objectives	Concept validation / Performance variables / Response amplitude operators / PTO and mooring characteristics / Diffraction and radiation		Real sea performance / Dir. sensitivity / Deployment seaways / Response amplitude operators / Seakeeping	Verify phase 1 / Active control / Seakeeping / Failure modes	Wave-wire performance / Mooring forces / Survival seakeeping	Full fabrication / True PTO and electrical generator
Model	Idealised with quick change options / Simulated PTO (0 to ∞ damping range) / Standard mooring and mass distribution		Distributed mass / Minimal drag / Design dynamics	Final design (internal view) / Mooring layout	Advanced PTO simulation / Special materials	
Model scale		$\lambda = 1:25\text{–}100$		$\lambda = 1:10\text{–}25$	$\lambda = 1:10\text{–}15$	$\lambda = 1:3\text{–}10$
Facility		Flume or basin		Basin	Basin	Nursery site
Measures			Motions, forces, waves, mooring loads, PTO performance		Subsystem performance	Control strategy / Electrical supply
Specials	Degrees of freedom (heave, pitch only) / 2D solo and multihull	Short-crested seas / Angled waves as required	Storm seas / Focused waves	Realistic PTO / Bench test PTO and generator	Survivability conditions / Performance of components	Marine corrosion and growth / Permissions
Mathematical methods	Hydrodynamic, numerical frequency domain to solve the model / Undamped linear equations of motion		Finite waves / Applied damping / Multi-frequency	Time domain response model and control strategy / Complete model of the external structure, PTO, mooring and anchorage system		Economic and businesses plan

Figure 7.1 The Wavestar model at different stages of development. Top left: an early-stage 1 : 40 model being tested in the laboratory (TRL 2). Top right: model with an advanced PTO being tested in the laboratory (TRL 4–5). Bottom left: model of a large single float being tested in the laboratory with real-time control PTO and force measurements on bearings and floater surface (TRL 5). Bottom right: full device model with 38 floats being tested at the Nissum Bredning nursery test centre from 2006 until 2011 (TRL 6). Pictures courtesy of Wavestar AS [6].

the limitations of the laboratory or other equipment, the requirements for similitude and the required outcomes from the tests. The four major factors that influence the scale selection for an MRE model test are model development stage, model construction, tank blockage and generation capabilities.

As Hughes [4, p. 51] says: 'The basis of all physical modelling is the idea that the model behaves in a manner similar to the prototype it is intended to emulate. Thus a properly validated physical model can be used to predict the prototype under a specified set of conditions.'

In a successfully scaled experiment all the relevant parameters which affect a model's reactions and responses are in proportion between the experimental model and the full-scale prototype, and all those factors that are not in proportion have a negligible effect. Scale factors are used to measure the correspondence between a relevant parameter in the model and the same parameter in the prototype:

$$\lambda_X = \frac{X_p}{X_m} = \frac{\text{Value of } X \text{ in prototype}}{\text{Value of } X \text{ in model}}. \tag{7.1}$$

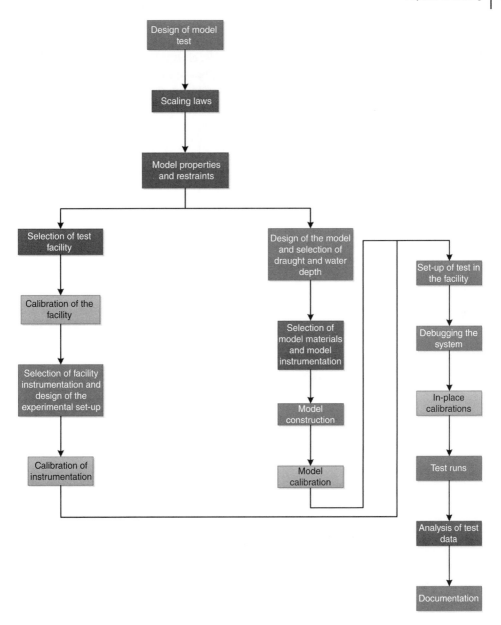

Figure 7.2 Typical pathway of a physical model test.

To relate the responses of a scaled model with the expected ones at the prototype scale the modelling laws or principles of similitude are considered. For the study of fluid-related phenomena, there are three basic principles of similitude that must be taken into account: geometric similitude, kinematic similitude and dynamic similitude.

When applying the three principles of similitude to an MRE problem there are six relevant non-dimensional numbers which should be considered: the *Froude, Reynolds,*

Cauchy, *Weber*, *Euler* and *Strouhal* numbers. The relevance of each of these non-dimensional numbers is dependent on the device and situation being tested. The first four are the ratio of the inertial forces to the gravity, viscous, elastic and surface tension forces, respectively; the Euler number is the ratio of the pressure forces to the inertial forces; and the Strouhal number is the ratio of the temporal inertia force to the convective inertia forces. These six numbers are given respectively by

$$Frd = \frac{\text{Inertia forces}}{\text{Gravity forces}} = \frac{\rho V^2 L^2}{\rho g L^3} = \frac{V^2}{gL}, \tag{7.2}$$

$$Re = \frac{\text{Inertia forces}}{\text{Viscous forces}} = \frac{\rho V^2 L^2}{\mu VL} = \frac{VL}{\nu} = \frac{\rho LV}{\mu}, \tag{7.3}$$

$$Ca = \frac{\text{Inertia forces}}{\text{Elastic forces}} = \frac{\rho V^2 L^2}{EL^2} = \frac{\rho V^2}{E}, \tag{7.4}$$

$$We = \frac{\text{Inertia forces}}{\text{Surface tension forces}} = \frac{\rho V^2 L^2}{\sigma L} = \frac{\rho V^2 L}{\sigma}, \tag{7.5}$$

$$Eu = \frac{\text{Pressure forces}}{\text{Inertial forces}} = \frac{p L^2}{\rho V^2 L^2} = \frac{p}{\rho V^2}, \tag{7.6}$$

$$St = \frac{\text{Temporal inertia forces}}{\text{Convective inertia forces}} = \frac{\left(\rho L^3\right)\left(V/t\right)}{\left(\rho L^3\right)\left(V^2/L\right)} = \frac{L}{Vt} = \frac{\omega L}{V}, \tag{7.7}$$

where V is the velocity, L the length of the device, g the gravitational acceleration, ρ the density of the fluid, ν the kinematic viscosity of the fluid, μ the dynamic viscosity of the fluid, E the modulus of elasticity of the fluid or material, σ the surface tension coefficient, p the fluid pressure and ω the frequency of oscillation of the fluid.

The Froude, Reynolds, Cauchy, Weber, Euler and Strouhal criteria require each of the respective dimensional numbers to be the same at model and prototype scale. For complete similitude all criteria should be met; this is, however, not possible. For example, it is not possible to meet the Froude and Reynolds criteria without increasing g and/or decreasing ν. This is not feasible when testing MRE devices at scale and therefore a decision over which criterion to uphold must be made.

The majority of WEC tests are scaled according to the Froude criterion. It is also an important criterion to consider when investigating the impact of waves on tidal energy devices. The Reynolds similitude criterion gains relevance for modelling flows where viscous forces predominate (e.g. modelling tidal turbines). The Cauchy similitude criterion is of special relevance in the design of models where the elastic forces are important (e.g. in MRE models this criterion is applicable for modelling elastic taught mooring lines and elastic membranes). The Weber similitude criterion may become important when testing models at small scale, when thin films of a fluid are studied and when air bubbles are entrained in the water column (e.g. in MRE models the Weber criterion should be considered when testing models at a small scale when the surface tension force of the water free surface is significant compared to the inertial forces from the model). The Euler similitude criterion is usually considered when the pressure forces have a dominant effect over the inertial forces (e.g. for MRE models this criterion is applicable when investigating the cavitation occurring on a tidal turbine). The Strouhal similitude criterion is considered of special relevance for the scaling of unsteady oscillating flows (e.g. in MRE models the Strouhal criterion is used to scale the turbulence intensity and eddy size in current flows, which are of special relevance for tidal stream energy devices). The scaling ratios for Froude and Reynolds criteria between the model and prototype for the most common physical variables can be found in Table 7.2.

7.3.1   Scaling MRE Devices

In most MRE devices the forces associated with elastic compression are relatively small and thus can be neglected. This means that the choice of scaling law depends on whether gravity or viscous forces are dominant for the device being tested. Although dependent on the individual case, generally the Froude criterion is applied to WECs, where gravity is the dominant force and the Reynolds criterion is applied to tidal energy devices, where viscous forces are dominant.

Table 7.2 Geometric similitude scaling ratios for Froude and Reynolds criteria between model and prototype.

Variable	Dimension	Froude	Reynolds
Geometry			
Length	$[L]$	λ	λ
Area	$[L^2]$	λ^2	λ^2
Volume	$[L^3]$	λ^3	λ^3
Rotation	$[L^0]$	–	–
Kinematics			
Time	$[T]$	$\lambda^{1/2}$	λ^2
Velocity	$[LT^{-1}]$	$\lambda^{1/2}$	λ^{-1}
Acceleration	$[LT^{-2}]$	–	λ^{-3}
Volume flow	$[L^3T^{-1}]$	$\lambda^{5/2}$	λ
Dynamics			
Mass	$[M]$	λ^3	λ^3
Force	$[MLT^{-2}]$	λ^3	–
Pressure, stress	$[ML^{-1}T^{-2}]$	λ	λ^{-2}
Energy, work	$[ML^2T^{-2}]$	λ^4	λ
Power	$[ML^2T^{-2}]$	$\lambda^{7/2}$	λ^{-1}

Once the Froude or Reynolds criterion is selected, it is applied together with the geometric similitude through the scaling ratios from Table 7.2. These ratios are used to directly obtain the results for a full-scale prototype from the experimental physical model. For example, using a 1 : 30 scale model of a WEC where the Froude criterion was followed, the power extracted would be multiplied by a factor of $30^{3.5}$ (147,885) to obtain a prediction of prototype performance.

Scale effects are the differences between the prototype and model response due to being unable to simulate all relevant forces at the chosen scale. For example, scale effects will arise if viscous forces have a significant impact when testing is conducted under Froude scaling. Care should be taken of scale effects arising from surface tension forces at very small model scales. This can be checked with use of the Weber number. Laboratory effects are differences between prototype and model response due to laboratory limitations, such as reflections from tank walls [7].

One of the crucial elements of the development of an MRE device is the comparison of the results obtained between the models tested at each one of the development stages. This comparison presents a number of advantages: for example, it verifies the smaller, previous model results; validates the applications of the similitude law to MRE device development; highlights possible weakness in the physical modelling testing methodology; checks the relevance of different physical properties and processes on the device behaviour and performance; and reduces the uncertainty in the results [2].

7.3.2 Common Problems Scaling MRE Devices

It is sometimes necessary to break the geometric similitude between model and prototype. Examples of this situation include the use of distorted models and situations where both Froude and Reynolds criteria rule over different parts of the model.

Distorted models are sometimes required where the different dimensions of the full-scale system differ by orders of magnitude. For example, when modelling a section of coastline, horizontal lengths may be tens of kilometres long, while water depth may by tens of metres deep. To model this system in a laboratory requires a horizontal scale small enough to fit the model into the available space, and a vertical scale factor large enough to avoid significant surface tension effects. Different vertical and horizontal length scales are therefore applied. More details about distorted models can be found in [4].

Oscillating water column (OWC) type WECs are an example of where Froude and Reynolds criteria are relevant to different parts of the model. The wave action on the OWC chamber is dependent on gravity and therefore requires the Froude criterion. The air flowing in and out of the chamber through a turbine or orifice and driven by water motion inside the chamber is dominated by viscous forces and therefore requires the Reynolds criterion. Various approaches to deal with this may be followed. In the first a multi-domain approach is adopted, where the part of the model where the Reynold criterion rules apply is geometrically distorted [8, 9]. In the second the same scale factor is applied to the whole model and the resulting scale effects accepted as it is believed that modifying the geometry to follow the multi-domain approach will have bigger effects. This is the case for floating OWC where increasing the air volume of the chamber will modify the dynamics of the buoy [10, 11], or when the model is at a very preliminary development phase (proof of concept).

7.4 Model Design and Construction

Physical models need to be designed so that they reproduce as well as possible the behaviour of a full-scale prototype at the scale that has been chosen. In nearly all cases this requires the external dimensions of the model to be geometrically scaled, along with the mass distribution. The modelling of PTO systems is a particular issue. The PTO systems used in MRE devices are discussed in Chapter 6. Unfortunately PTO dynamics and losses do not scale particularly well, especially as model scale decreases. Therefore, instead of using a model of the prototype PTO, an alternative system is often used which provides the correctly scaled damping characteristics to the model's response while allowing the capture power to be measured. Recommendations for model material selection and design are made below, before some of the alterative model PTOs are described.

7.4.1 Material Choice and Model Design

Models tested at small to medium scale are rarely manufactured from the same materials as proposed for the full-scale device. This is usually due to cost, manufacturing ease and a wish in some instances to correctly scale the prototype material properties. The choice of material is easiest when the device, or device component, is non-deformable under the expected hydrodynamic load. In this case any material that can be used in the reproduction of the correct geometry and mass distribution, survive submersion in water for the duration of the experimental campaign and withstand the predicted loads without deformation can be considered. When material deformation is predicted under the applied load at prototype scale it is important to select a material that models this deformation as well as possible at model scale. This can be especially important for tidal devices, where blade deformation affects both performance and loading.

Sheet steel is often used due to its low cost and ease of manufacture. Coating is required to prevent or reduce corrosion of the steel. Various paints are available for this kind of protection. Galvanising is also an option, although the resulting surface finish will be rougher then before and finer geometric details can be lost [12]. Stainless steel offers a non-corrosive alternative, but is a lot more expensive and difficult to machine. Aluminium is an attractive lightweight and non-corrosive alternative, although more expensive than mild steel and should be avoided if chlorine is in the water. Care should also be taken to avoid galvanic corrosion when using metal in models, which can occur when two different metals are in contact with water [12].

Many non-metallic materials are also used in the construction of MRE models, including wood, PVC, acrylic and high-density closed cell foam (such as Divinycell [12]). Fibre reinforced plastic is commonly used in the construction of ship models [13] and in MRE testing. Depending on the device to be tested, certain components may need to be manufactured to a higher level of precision then others. For example, the leading edge, trailing edge and tip of tidal blades require high precision [14], with Doman *et al.* [15] reporting model blade dimension machining tolerances of 0.0005 m. Materials used for such high-precision blade construction include hard anodised aluminium [16, p. 56] and Telene^{TM} DCPD resin [17]. High precision is also required when conducting array tests. Small differences in model devices make it harder to measure possible array interactions.

Reproduction of the scaled geometry is all that is required for static structures. However, for moving components, and especially for floating bodies, correctly reproducing the mass distribution is essential. Water can be used as a quick and adjustable ballasting option, but is unsuitable for moving bodies due to sloshing. Sand is another cheap option, although due to its ease of movement it is not suitable if large motions are predicted (splitting the model into smaller compartments reduces this problem). If space is limited then metal shot, such as lead or steel, provides a denser option. These forms of ballast are limited in the control they provide over vertical mass distribution. Options to add this control include designing the ability to add ballast at different vertical positions within the model and using ballast rods held at different vertical position with foam [12]. Whichever ballasting option is chosen, it is essential to measure the resulting location of the model's centre of mass and its moments of inertia (Section 7.7), and design the ability to make adjustments to this if needed.

When designing larger-scale models it is important to consider laboratory access and deployment issues. Consultation should be conducted with the chosen test facility regarding how the model will be deployed. This often requires lifting eyes for crane attachment being designed into the model.

7.4.2   Power Take-off

7.4.2.1   Orifice Plate

Orifice plates are often used during physical model tests of OWCs and other wave energy devices which at full scale use an air turbine as a PTO. An orifice plate simply consists of a flat plate with a circular hole in the middle placed in a straight tube through which the oscillating air flow generated by the wave energy device passes (e.g. Figure 7.3). Orifice plates have a quadratic relationship between pressure and mass flow rate, and as such they are often used to model impulse turbines [18, 19]. The level of PTO damping

Figure 7.3 Orifice plate used in an OWC model; image shows an orifice plate mounted on the top lid of an OWC model, the wave gauge used to measure the free surface oscillation inside the OWC chamber and the four tubes used to measure the internal air pressure [24].

can be controlled by changing the diameter of the orifice (or equivalently, changing the number of orifices in the system).

To calculate the pneumatic power (P) extracted by a wave energy device using an orifice plate the pressure drop across the plate (Δp) and the volumetric flow rate of air (q) through the orifice need to be known:

$$P(t) = \Delta p(t) q(t). \tag{7.8}$$

Measurements of the pressure drop can be conducted relatively easily with the use of a differential pressure transducers. If the orifice is open to the atmosphere (e.g. during an OWC test) the total inner pressure within the chamber can be measured and the pressure drop calculated from simultaneous measurements of atmospheric pressure.

Measurement of the unsteady and bidirectional volume flow is more complicated. Kooverji [20], based on Hayward [21] and Figliola and Beasley [22], identifies four classes of instruments that are capable of measuring bidirectional turbulent flow: pressure differential meters, pitot-static probes, thermal anemometers and vane anemometers. However, in many cases the direct measurement of bidirectional air flow at the required accuracy is impractical and alternative techniques to estimate the flow rate thought the orifice plate are used. For example, when testing an OWC the air flow rate is calculated from the measured velocity of the free surface within the inner chamber (assuming conservation of mass and that the air is incompressible). Alternatively, a calibration of the orifice plate can be conducted to determine its characteristic curve (details are given in [23]). This enables calculation of pneumatic power from the pressure measurements only.

7.4.2.2 Porous Media

Porous media such as some felts or air filters can be used to model an air turbine in scaled testing. Such media are used due to the linear relationship between flow rate (q) and the differential pressure (Δp) as air passes through them. This linear damping models certain bidirectional turbines such as the Wells turbine, which are commonly used in OWCs [25, 26]. As with the orifice plate, the pneumatic power is calculated from equation (7.8).

7.4.2.3 Capillary Tubes

Capillary tubes can also be used to replace an air turbine in model testing of wave energy devices. Oscillating air flow generated by the device is forced through a large number of capillary tubes. These tubes provide a linear, and predictable, impedance to the airflow and therefore act as a PTO. By adjusting the number of capillary tubes through which the air flows, the PTO impedance is adjusted (see [27] and [28, pp. 131–132]). Instantaneous power can be calculated from differential pressure and volume flow rate measurements using equation (7.8). However, as long as the flow can be assumed to be laminar and incompressible, volume flow rate (and therefore power) can also be calculated directly from the differential pressure measurement across the tubes using the Poiseuille–Hagen equation:

$$\Delta p = \frac{128 v_a \rho_a L_p Q}{N \pi d_p^4}, \tag{7.9}$$

where L_p and d_p are respectively the length and diameter of an individual capillary tube, N is the number of open capillary tubes, and ν_a and ρ_a are respectively the kinematic viscosity and density of air. Alterations to this formulation can be made if air incompressibility cannot be assumed, as detailed in [25, pp. 136–140].

7.4.2.4   Tidal Turbines and Rotating Shaft WEC

Most tidal stream turbines, as well as some WECs, generate useful power by rotating a shaft connected to a generator. 'Free-running tests' are sometimes conducted, where the mechanical power to electrical power conversion process is also scaled [14]. However, drive-train friction losses do not scale with the model scale and hence these tests are generally of only limited used. Instead models tests are usually carried out with either torque control via brakes applying resistive load to the rotating shaft or by rotational speed control. Speed control is usually performed by connecting the rotating shaft to a motor-gearbox assembly [14, 17]. This can be used to adjust the tip speed ratio (TSR) and also drive the turbine to measure performance curves at different inflow speeds [15]. Some dynamometers, which measure the torque of a rotating shaft by applying a resistance to the rotation (Section 7.6.5), are also used directly as a PTO.

Electrical motors can also be used as a model PTO for WECs. For example, [29] used a motor as a PTO on scaled tests of a flap-type WEC. The motor provided a breaking torque proportional to the flap's measured angular velocity, with the motor amplifier adjusted to obtain optimum damping to maximise power capture.

Power is calculated by multiplying the torque and rotational speed of the PTO system.

7.4.2.5   Dampers and Brakes

Various types of dampers and brakes have been used to model PTOs at scale by directly applying damping to a devices motion. These include 'off-the-shelf' pneumatic and hydraulic cylinders (e.g. Figure 7.4) used as linear actuators [30, 31]. These nearly always have the option to adjust the damping factor. Rotary viscous dashpots can also be used as linear dampers, again with the damping rates often being adjustable [32].

Direct frictional brakes are sometimes used in small-scale models. For example, Folley *et al.* [33] applied a coulomb friction brake to the shaft of a model oscillating wave surge device. The level of damping is varied by adjusting the force applied by the brake, and a near constant torque opposing the paddle rotation is provided. Vantorre *et al.* [34] also used a frictional brake, consisting of a wheel attached to a pulley which is rotated by a device. Felt brakes are applied to the wheel to damp rotation. Göteman *et al.* [35] modelled the linear PTO used in the SeaBased wave energy device with frictional brakes acting on a translator.

More sophisticated model PTOs have been developed to impose complicated variations in PTO damping and control strategies. Henriques *et al.* [36] describes one such system, which uses eddy current brakes to impose PTO damping, and has the potential to apply 'non-linear PTO damping characteristic curves.'

7.4.2.6   Bilge Pumps and Flow Meters

Overtopping type WECs use low head water turbines as PTO, extracting the potential energy from water collected in a water reservoir which is higher than the mean water level. The usual approach to measure extracted power is by measuring the flow rate of the water leaving the reservoir (the discharge flow rate, q_d), based on which the energy

Figure 7.4 Wavestar model using a hydraulic actuator as PTO being tested at the COAST laboratory. Image courtesy of Wavestar AS [6].

output of the prototype device can be calculated. Two options are available to extract water from the reservoir, thereby simulating the low head turbine.

In the first approach bilge pumps are used to pump water in the reservoirs out of the model through a hose (Figure 7.5). The use of pumps requires a control system to turn pumps on and off, depending on reservoir water levels. An estimation of the overtopping flow rate into the reservoirs can then be made based on how long the pump has been operating. The combination of pump and the control system substitutes the PTO for the purposes of calculating the power that would be generated by a real PTO in the prototype. This technique requires a calibration procedure in order to determine the relationship between the mean discharge flow rate, the different levels of water in the reservoir and the amount of pumping time. Figure 7.6 shows an example of a bilge pump PTO system being calibrated.

Figure 7.5 WaveCat being tested [37, 38].

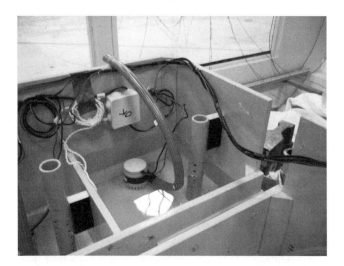

Figure 7.6 WaveCat's Bilge pump PTO and control system being calibrated [37, 38].

The second approach to removing water from the reservoirs is by using flow meters and valves. The flow meter is positioned in a duct that connects the bottom of the reservoir to an outlet below the mean water level. A valve positioned between the reservoir and the flow meter is used to control the discharge times, and the flow rate is measured directly at the flow meter.

The power output of the overtopping device (P) can be obtained from

$$P(t) = \Delta\big(gq_0(t)\big) \cdot \Delta H(t), \tag{7.10}$$

where g is the gravitational acceleration, q_0 is the discharge flow and ΔH the water head between the reservoir and the mean still water surface.

7.5   Fixing and Mooring

The mounting of model devices which are designed to be rigidly fixed to the seabed, shoreline or other rigid structure is a relatively simple task in the laboratory. However, for structures designed to be moored the proper scaling of mooring properties is essential to fully reproduce the impact these have on the device's dynamic response to environmental loads. The measurement of scaled mooring loads also provides useful information regarding the survivability of a device and for comparison with numerical models of mooring response. Different mooring options for floating MRE devices are discussed in detail in [39]. Here catenary and taut moorings are considered.

7.5.1   Catenary Mooring

Catenary moorings are one of the most common type of mooring solutions adopted in the traditional maritime sector due to their low cost and robustness and have been widely adopted for MRE devices. Four points need to be considered when modelling a catenary mooring: (i) geometric similitude, i.e. the catenary length and rope diameter or chain size needs to be scaled accordingly to the model scale factor; (ii) dynamic similitude, i.e. catenary moorings work via the dynamic response of the submerged catenary, which makes the weight per unit of length of the mooring a critical element to correctly scale; (iii) if the rope used for model mooring lines is neutrally buoyant it will need to be ballasted and the total weight of the scaled catenary (rope plus added weights) used when considering weight per unit length, and (iv) the breaking load of the selected model mooring rope or chain should be higher than the equivalent breaking load at prototype scale. If facility dimensions prevent geometric similitude from being achieved then experimental techniques exist where truncated mooring systems are designed to achieve approximate dynamic similitude while reducing the horizontal extent of the mooring system.

7.5.2   Taut Mooring

The scaling of taut moorings involves different challenges compared to catenary moorings. The Cauchy scale criterion is applied to find as close as possible a representative mooring line materials that correctly scales in terms of geometry, elasticity and density [4, pp. 223–226]. Often an as affective approach is to use non-elastic model mooring lines connected to linear tension springs which provide the correctly scaled spring constant and initial mooring tension [40]. However, failure to correctly model the geometry of the full-scale mooring line can result in the hydrodynamic damping of the line being incorrectly scaled, which can have effects on the low-frequency response of floating bodies [41, p. 1038].

When using springs to model taut moorings care should be taken that the springs' elastic limit is large enough to accommodate the expected mooring expansion. It is generally good practice to measure the spring constant before and after testing to check for any changes in spring property. A combination of springs can also be used to model nonlinear elastic behaviour in full-scale moorings.

7.5.3 Fixed Guides

Physical modelling of floating MRE devices should generally be conducted with moorings that model those proposed at full scale. However, there are cases when using a system that restrains a model to move in only specified degrees of freedom is useful, for example during proof of concept tests at early TRL and when trying to reduce complexity for comparison or calibration of numerical models. Linear guide rails provide a relatively simple way of restraining floating devices to either heave, or heave and pitch only [34, 42]; see Figure 7.7. More complicated combinations of unrestrained and restrained degrees of freedom are sometimes possible with specially designed towing posts.

7.6 Instrumentation

A wide range of instruments are used for the physical modelling of MRE devices. The choice of instrumentation depends on a number of factors, including what properties need to be measured, the scale of the test and instrument availability. It is important to consider which instrumentation will be used during the initial planning stage. Depending on scale, the size and mass of an instrument can have a significant impact on a model's physical properties when mounted to it. These impacts need to be mitigated against in the

Figure 7.7 SQ3 device restrained to heave with linear guide rails in Plymouth University's COAST laboratory [24].

model design. Similarly, the impact on drag and added mass of any instrument mounted underwater should be considered, along with the influence that instrumentation has on the flow and waves interacting with the model device. For example, an underwater camera (and associated support structure) mounted directly in front of a tidal turbine will influence the flow field the turbine experiences.

When choosing instrumentation it is also important to have an idea of the range of measured values expected to occur. The appropriate instrument can then be selected. An instrument with too small a range will become saturated, while an instrument with a range that is significantly too large will likely be less precise than a more appropriate choice. Here the more commonly used instrumentation used in physically modelling MRE devices is discussed.

7.6.1 Water Surface Elevation

The measurement of water surface elevation is used to measure the wave field around a model, to measure water heights in overtopping basins and to measure surface elevation within OWCs and other wave energy devices which generate an oscillating air–water interface as part of their PTO. Instruments for measuring water surface elevation include resistance wave gauges, capacitance wave gauges, acoustic wave gauges, ultrasonic gauges, pressure readings and camera systems. All these systems, with the exception of certain camera systems, provide point measurements.

7.6.1.1 Resistance Wave Gauge

Resistance (or conductance) wave gauges are the most widely used in physical modelling. Resistance wave gauges consist of two rods (self-supporting) or wires (with corresponding tensioning structure) of conductive material (usually stainless steel), which are isolated from each other. A current is sent around the two rods/wires, with the water surface completing the circuit. A change in water level changes the physical length of the circuit and therefore results in a change in voltage. Resistance wave gauges are normally calibrated by being moved vertically up and down in know intervals in water. In certain circumstances the bottom of the gauges may be fixed to the tank floor to increase their stability. In this case water level is changed during calibration.

It is important that the gauges are calibrated in the same water in which they will be used as their calibration depends on water conductivity. Due to this dependence it is recommended that gauges are calibrated once a day, especially if water properties (such as temperature or the levels of dissolved solids such as chloride) are expected to change. Resistive wave gauges are not suitable for use in water that might become contaminated (such as by oil). Modern electronic wave gauges experience far less calibration drift then previous generations of gauges, and therefore if water properties do remain constant, the length of time between recalibration can be extended.

It is important to mount wave gauges so that their locations relative to models are known as accurately as possible, while also limiting movement and vibrations during operation. Gauge vibration can be reduced by using gauges of a minimum acceptable length, thereby limiting the lever arm about the mounting point (e.g. not using 1 m length rods when measuring maximum wave heights of 0.1 m). The orientation of the two resistive rods to the waves can also have a significant impact. Generally gauges are set up so that both rods are in line with the incoming wave crests (assuming long-crested waves). This allows for greater accuracy in gauge location. However, resistance gauges can also be positioned at right angles to the wave crests, with the measured surface elevation assumed to occur between the two rods. This approach has the potential to reduce vortex induced vibrations. Movement of resistive wave probes can also be reduced by using wires attached to the tank floor and tensioned from above. However, this increases the complexity of both set-up and calibration.

7.6.1.2 Capacitance Wave Gauge

Capacitance wave gauges are used in similar experimental situations to resistance wave gauges and are mounted and calibrated in the same way. They consist of a single insulated wire and supporting frame. The wire insulation acts as a dielectric and the wire and surrounding water are conductors. Variation in wire immersion generates a change in capacitance which corresponds to a change in water level via calibration. Capacitance gauges have the advantage of reducing flow restriction compared to the two wires of a resistance gauge (although this depends on the size of the support frame). They require that the insulation layer is constant along the length of the gauge and are susceptible to breaches in the insulation, potentially making them less robust then their resistance counterparts.

7.6.1.3 Others

Various other techniques for measuring water surface elevation are discussed by Lawrence *et al.* [43] and Bourdier *et al.* [44]. These include wave slope optical wave gauges, acoustic point measurement, pressure transducers and the use of velocity measurements to infer wave surface elevation. Ultrasonic distance sensors are a particularly useful non-intrusive alterative. For example, Perez and Iglesias [45] used three non-contact ultrasonic water level gauges to measure the varying water level within an OWC wave energy converter, while Christensen *et al.* [46] demonstrated their use for measuring waves at sea.

7.6.1.4 Measuring Wave Reflection

When modelling a WEC in the laboratory, waves are reflected by beaches, the model itself and other bodies or boundaries present during the test. The reflected waves interact with the incident waves from the wave maker, modifying the wave field to which the model is exposed. Notwithstanding the fact that the majority of modern laboratories have the capability to detect and absorb these reflected waves, thereby minimising their effect over the wave field (see Section 7.10.2), it is common standard procedure to perform a reflection analysis.

Analysis of the reflected waves can be carried out with and without the model in place to examine the influence of tank reflections compared to combined tank and model reflections. The analysis without the model in place will often be conducted as part of the tank calibration. Measurements with the model in place increase understanding of how the model interacts with the wave field by quantifying the reflected and transmitted waves. This is basic to understanding how the device interacts with the near and far wave field, as seen in Chapters 2, 3 and 9.

There are several methods available to analyse the reflected waves, obtain the reflection coefficients and reconstruct the incident and reflected waves from recorded measurements. Additionally, the transmitted wave can be determined by the direct measurement of the surface elevation at the rear of the model, either using a single wave probe or an array of probes to measure spatial distribution.

All methods used to quantify the reflected wave are based on positioning a number of wave gauges to simultaneously measure the water free surface elevation in a line or in a 2D array, usually facing the incident wave. These methods can be grouped according to the number of probes needed to apply the method: (i) the two-probe group includes methods such as those proposed by Goda and Suzuki [47] and by Frigaard and Brorsen [48]; (ii) the three-probe group such as the method proposed by Mansard and Funke [49] which was revised by Baquerizo *et al.* [50]; and (iii) other types of methods, such as that proposed by Zelt and Skjelbreia [51] which is valid for an arbitrary number of probes positioned in line.

Once the reconstructed incident and reflected waves have been obtained from the probe data, the reflection coefficient (K_R) and the transmission coefficient (K_T) can be defined. For regular waves,

$$K_R = \frac{H_r}{H_i}, \tag{7.11}$$

$$K_T = \frac{H_t}{H_i}, \tag{7.12}$$

where H_r, H_t and H_i are the reflected, transmitted and incident wave heights, respectively. For irregular waves,

$$m_{0j} = \int_{f_{min}}^{f_{max}} S_j(f) df, \tag{7.13}$$

where m_{0j} is the j zeroth-order moment, S_j is the j power spectral density, and j refers to the incident (I), reflected (R) or transited (T) waves. Then

$$K_R = \sqrt{\frac{m_{0R}}{m_{0I}}}, \tag{7.14}$$

$$K_T = \sqrt{\frac{m_{0T}}{m_{0I}}}. \tag{7.15}$$

Once K_R and K_T have been obtained, the absorption or loss coefficient[1] (K_A) can be calculated from [52]

$$K_R^2 + K_T^2 + K_A^2 = 1. \tag{7.16}$$

---

1 Note that this coefficient groups all the energy that is not reflected or transmitted. It is called the absorption coefficient as traditionally in civil engineering this term represents the energy absorbed by a breakwater or coastal structure. However, when talking about wave energy conversion, this term quantifies the energy absorbed by the WEC as well as energy loses due to, for example, friction or radiation phenomena.

7.6.1.5 Directional Wave Spectrum Analysis

The reflection analysis techniques presented previously characterise long-crested wave fields. However, reflection analysis on a short-crested sea state (also known as a multi-directional or 3D wave fields) requires a directional wave spectrum analysis. Several measurement techniques exist to measure the directional wave spectrum *in situ* or in the laboratory [53]. These can be grouped into single-point systems, wave gauge arrays, and remote-sensing systems.

For laboratory applications wave gauge arrays are the most widely used technique. They are composed of several instruments mounted at various locations over a fixed frame [54]. This technique can be used just with wave gauges or by combining wave gauges, current meters and pressure sensors [55–57]. Special attention must be given not only to the number of probes but also to their geometrical spacing. Young [58] shows that, for optimum performance, an array should have as many non-redundant spatial lags between probes as possible.

Once the directional wave field has been measured, a directional wave spectrum analysis method needs to be followed to characterise the directional wave spectrum. Benoit *et al.* [54] present an overview of the different methods available to analyse ocean waves. Among these methods, the maximum likelihood method [59], the Bayesian directional mthod [56] and the maximum entropy method [60] and their derivatives are the most extensively used approaches to analyse directional wave spectra in the laboratory.

7.6.2 Fluid Velocity

The accurate measurement of fluid velocity and turbulence effects is essential during the physical modelling of tidal energy devices. It can also be required when testing certain WECs, either as part of the measurement of power or when examining the impact of currents on a moored device. As with surface measurements there are various techniques to measure fluid velocity, including acoustic Doppler velocimeters (ADVs), pitot-static tubes, turbine flow meters, laser Doppler velocimetry (LDV), particle image velocimetry (PIV) and particle tracking velocimetry, hot-wire and hot-film anemometers and electromagnetic flow meters.

7.6.2.1 Pitot-static Tube

Pitot-static tubes can measure point velocity in both fluid and gas flow. They consist of two concentric tubes placed opposing the flow. The inner tube has an opening facing the flow and measures stagnation pressure. The outer tube has an opening perpendicular to the flow to measure static pressure. The difference between these two gives a measure of velocity. Pitot-static tubes are cheap and simple to use. However, they should not be used in unsteady, accelerating or fluctuating flow. Furthermore, they only measure the component of velocity in the direction they are pointing so can only be used in situations where this direction is constant. They are susceptible to becoming blocked if sediment is present in the water.

7.6.2.2 Turbine Flow Meters

Turbine flow meters are also intrusive point measurements which provide one-directional velocity measurement. As the name suggests, they consist of a turbine, the rotation of which is used to measure fluid velocity. They are a rugged, relatively cheap and simple-to-use instrument which responds more quickly than pitot-static tubes to changes in speed. Although highly accurate in steady high-speed flows, they are less effective at lower speeds and require homogeneous flow.

7.6.2.3 Acoustic Doppler Velocimeters

Acoustic Doppler velocimeters measure fluid velocity in the three directional axis components and are particularly useful for turbulence measurements, slow and low energy flows and rapidly varying flows [61]. They are an intrusive instrument, in that they need to be placed within the water column, although the volume of water sampled is remote. ADVs consist of an acoustic transmitter and acoustic receivers. Acoustic waves are transmitted, scattered by particles moving in the flow and then detected by the receivers. The three axis velocity components within the remote sample volume are calculated using the shift in frequency of the scattered acoustic waves and the orientation of the receivers. For ADVs to work effectively in the laboratory, seeding usually needs to be added to the water to give an appropriate amount of scattering. Rusello [61] provides greater detail on the operating principle and data quality checks of ADVs. Two measures of data quality are usually available: signal-to-noise ratio (SNR) and beam correlation. Both need to be at an acceptable level for measurements to be reliable. An approximate rule of thumb is for the beam correlation to be above 70% and the SNR to exceed 15. More detailed quality control and filtering methods for ADV measurements are examined by Cea *et al.* [62]. ADVs have been used to measure wake effects of model tidal turbines, for example [63, 64].

7.6.2.4 Laser Doppler Velociemeters

Laser Doppler velocimeters, also known as laser Doppler anemometers, use a laser to take point measurements of velocity. It is a non-intrusive instrument which can measure all three directional velocity components at a high sample rate. A laser beam is split, and then the two halves focused on the probe volume where they generate an interference pattern. This volume is kept as small as possible to increase the brightness of the light reflecting from particles in the water and to ensure the interference pattern is as consistent as possible. Particles moving in the flow generate reflections from the interference pattern. Velocity is then calculated from the resulting Doppler shift. LDVs are used to characterise flows in tidal and wave experiments (e.g. Gaurier *et al.* [17]) and for measuring wake effects from tidal turbines [65].

7.6.2.5 Particle Image Velocimetry

Particle image velocimetry provides an instantaneous velocity vector map in a cross-section of the flow. Laser pulses are expanded into a sheet of light. A synchronised high-speed camera captures illuminated particles in the water. The timing between pulses is set so that pulses are long enough to capture displacement of particles and short enough that the majority of particles do not leave the light sheet. A stereoscopic arrangement of two cameras allows velocity to be measured in three dimensions. Examples of PIV used in the physical modelling of MRE devices include [65, 66].

7.6.2.6 Hot-Wire and Hot-Film Anemometers

Hot-wire and hot-film anemometers consist of either a conducting wire or film heated by an electrical current. Fluid flow acts to cool the conducing element, the resistance of which is temperature-dependent. Depending on the particular system, flow is measured via calibration and the associated electronics maintaining either the voltage, electrical current or temperature of the element [67]. Frequent calibration is required. Velocity measurements are made over a small volume and have a high sampling frequency and therefore can be used for quantifying turbulence [67]. Sensors with two or three individual anemometers allow the measurement of two or three velocity components. Hot-wire and hot film anemometers are fragile compared to other instruments and there is often a more suitable instrumentation choice when testing MRE devices.

7.6.3 Pressure and Force Measurements

Pressure transducers measure pressure, usually via the mechanical deflection of a strain-gauged diaphragm. There is a wide range of available transducers designed with different installation options. Three different kinds of pressure measurement are possible. Absolute pressure is measured relative to zero pressure (i.e. a vacuum). Gauge pressure is measured relative to a constant reference point (usually atmospheric pressure). Differential pressure measures the difference between two points in a system, with the absolute or gauge pressure at both these points potentially changing (i.e. either side of an orifice plate).

Pressure transducers are generally factory-calibrated and instruments should be returned for recalibration on a semi-regular basis. However, if considered necessary, there are several easy techniques to check calibrations when carrying out physical modelling. Transducers designed to measure water pressure can simply be submerged to various depths, with the resulting calibration calculated using the resulting hydrostatic pressure. Air pressure transducers can be calibrated by fitting them into an enclosed pipe or volume attached to a monometer and then pressurising this system to various values.

Load cells can be used for measuring force both in tension and compression. This is usually along one direction, although three-axis load cells are also available. A common use in MRE device testing is the measurement of forces in mooring lines and connections. The majority of load cells use a similar arrangement of strain gauges in a Wheatstone bridge configuration as pressure transducers. In-house calibration of a load cell can be conducted by fixing the load cell at one end and hanging a series of weights from the other. The working range of load cells can vary from 5 N to 50 MN [68]. It is therefore important to have an idea of the range of forces expected when planning an experiment.

Strain gauges are also used directly to measure strain on tidal blades. Examples of this include [17, 69].

7.6.4 Body Motion

By their nature most MRE devices involve part or all of the device being in motion. There are various instruments which can be used to measure this motion, the suitability of which depends on the modelling being conducted. Some common versions include wire potentiometers, accelerometers and optical tracking systems.

A wire potentiometer (or string pot) is a relatively cheap device for measuring displacement in one direction. It consists of a spring-loaded spool of cable. The spool is fixed to a non-moving base out of the water and the cable end attached to the body. As the body moves, cable is pulled in and out of the spool, causing a rotation which is measured and translated into linear displacement via calibration. String pots are used when measuring motion restrained to a single translation (e.g. a floating body fixed to heave only with guide rails). However, a string pot should not be used if more than one translational degree of freedom is present. The range of the expected motion also needs to be known so that the string both remains in tension at all times and is not pulled past its maximum length.

Device motion can be recorded by attaching accelerometers onto the device. One- and three-axis accelerometers are available, which measure acceleration relative to the local coordinate system of the instrument. These are relatively cheap devices to use, but their attachment to a model adds an additional mass which should be accounted for in the model design. Displacement is determined by integrating the measured acceleration twice with respect to time. This requires a relatively noise-free signal, as SNR reduces during integration.

Optical tracking systems allow the motion of a body (or several bodies) to be remotely tracked. A set of infrared cameras is used to track reflective targets which are attached to the model. Qualysis® is an example of an optical tracking system, which is able to track targets on multiple bodies both above and below the water (Figure 7.8). With the correct target layout the six degrees of freedom of the bodies' motion, with reference to either a global or local frame of reference, are directly output (e.g. [40]). Except for very small models, the mass of the targets is generally insignificant (although this is not always the case if support structures are required to put the targets in the cameras' field of vision). Calibration of optical tracking systems is very sensitive to camera movement,

Figure 7.8 Qualysis® cameras used to track targets attached to a floating body [24].

and therefore care is needed to ensure cameras are not disturbed once they have been set up to cover the required field of view.

7.6.5 Torque

Measurement of torque is necessary to calculate power for devices which connect to a PTO via a rotating shaft (e.g. tidal turbines). Dynamometers are commonly used to measure torque. A resistance is applied to the rotating shaft and torque measured by measuring the resulting force and knowing the torque arm (the distance between the force measurement and the centre of the shaft) [68]. Some dynamometers can also measure thrust [70]. Torque sensors are also available which can measure static and dynamic torque.

7.6.6 Measurement Error and Repeatability

When conducting any form of physical experiment, it is important to have an understanding of how reliable the measurements are. This is where the concepts of accuracy and precision become relevant. Figure 7.9 demonstrates these two principles. A precise measurement is one that experiences very little scatter between repeat measurements; however, these repeat measurements may not necessary be recording the 'actual' value. An accurate measurement will (possibly with several repeats) record on average the 'actual' value; however, individual measurements may be scattered around this 'actual' value.

It is always recommended that several repeat experiments are made of representative experimental runs to assess the precision of measurements. The precision of measurements

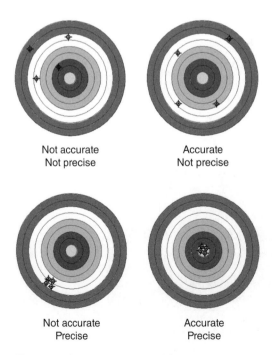

Not accurate
Not precise

Accurate
Not precise

Not accurate
Precise

Accurate
Precise

Figure 7.9 Accuracy versus precision.

depends on random experimental errors (such as from electronic noise affecting the instruments). Random errors generally have a normal distribution, and therefore the mean of several repeat measurements will give a good estimate of the measured quantity. However, in most wave basin and flume experiments the repeatability of wave and current generation and the repeatability of the device's response to external loading is likely to have a larger influence than random measurement errors. A measurement of precision can be made by calculating the standard deviation of the repeat measurements.

The accuracy of an experiment is far harder to assess. A loss of accuracy is caused by systematic measurement errors, such as changes in gauge calibration during the experiment. For some instruments, such as wave gauges, the calibration can change in response to environmental conditions. In these situations calibration should be checked daily. Failing this, instrumentation calibration at the start and end of the experimental campaign should at least give some indication as to whether this source of systematic error has occurred.

The majority of variables that are measured within experiments are continuous in nature. However, all instruments have finite precision, and therefore the continuous variables are assigned to discrete 'bins'. The size of the bins determines the measurement error. For example, a meter rule with millimetre intervals has a measurement error of ± 0.5 mm. This means that a measurement of 3.11 m is actually 3.11 ± 0.005 m. The error in a value calculated from this measurement can be estimated by multiplying the measurement error with the partial differentiation of the calculated value with respect to the measured value. For example, if the power of a regular wave is being calculated from a measurement of wave height with a measurement error of ΔH, then the error in power (ΔP) is given by

$$\Delta P = \frac{\partial P}{\partial H} \Delta H = 2 \frac{\rho g^2}{32\pi} TH\Delta H. \tag{7.17}$$

7.6.7 Common Problems

Noise affects all instruments to different extents. It is important to check before carrying out any of the experimental plan that noise levels are at an acceptable level. This should be done for all instruments at the lowest wave amplitude or current that is planned [12]. Although the impact of noise can be reduced to a certain extent in post-processing, it is best practice to reduce levels as much as possible during data collection. Various approaches can be used. These include grounding all instruments to a single point, preferably the same as the power supply. The use of shielded cable can also help, where the shield acts as a Faraday cage, as will low-pass filters. In some circumstances amplification of the signal can also help. It is recommended that for best resolution the maximum expected signals reach two-thirds the maximum range of the data logger [12]. Regular checks of noise levels should be made during the experiment. Increasing noise levels can indicate that an instrument is starting to fail.

Drift in instrument calibration make any experimental results unreliable. The likelihood of this occurring depends on the type and age of the instrument. It is recommended that for any instrument which requires user calibration, the calibration process is conducted at least at the start and end of the experiment to check for drift.

Thermal shock can occur when an instrument suddenly experiences a change in temperature. When testing MRE devices this can occur if a sensor passes from air to water. Guaraglia and Pousa [71, p. 414] show an example of thermal shock occurring in an instrument measuring water level.

7.7   Model Calibration

Scaled models of MRE devices need to be calibrated in order to ensure they are in similitude with the prototype. For components that do not move in response to hydrodynamic loading (e.g. a fixed OWC or the substructure of a tidal energy device) only geometric similitude is necessary. However, for floating or mobile MRE devices all three similitudes (geometric, kinematic and dynamic) are required. Therefore the correct modelling of mass distribution and associated moments is fundamental to ensuring that model and prototype have comparable responses.

The relevant physical properties that define the kinematic and dynamic response of the model are centre of buoyancy, centre of gravity, mass moment of inertia, water plane area, metacentre, metacentric height, and added mass and added inertia. In order to measure and adjust the physical properties of the constructed model a number of tests can be carried out, which can be classified as follows.

7.7.1   Dry Tests

Following model construction, the mass properties of the model (centre of gravity and mass moment of inertia) need to be measured and corrected when necessary. There is a wide set of standard balance and suspension tests to measure these properties. Of these, the bifilar suspension method (Figure 7.10) has the advantage that it allows the

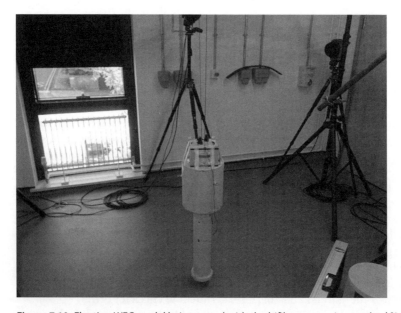

Figure 7.10 Floating WEC model being tested with the bifilar suspension method [24].

measurement of all properties at the same time with a quick and easy set-up. Hinrichsen [72] presents a simple bifilar method where the motions are monitored with a simple accelerometer or distant meter. However, the use of a six-degrees-of-freedom motion capture system (such as Qualisys®) is recommended if available.

Centre of mass is measured by performing an incline test [72, p. 24]. The model is suspended on the bifilar pendulum, aligning the plane of the bifilar pendulum to one of the principal axes of the model (XX for the longitudinal axis, YY for the transversal axis and ZZ for the vertical axis). With the model stationary, different weights are hung from one of the model extremes (within the model plane of interest). Figure 7.11 presents an example of a test measuring the vertical distance to the centre of gravity (l_2), which can be obtained from

$$l_2 = \frac{m_d}{M}\left(\frac{L_S}{\tan\theta_S} - b_S\right), \qquad (7.18)$$

where m_g is the mass hung at the distance L_S and a depth b_S from the pivot point P and θ_S is the inclination of the model in response to the mass. The procedure is repeated for the other two principal model axes to fully locate the centre of mass.

The XX and YY moments of inertia can be measured by performing a pitch–surge double pendulum test [72, pp. 17–19]. The model is suspended on the bifilar pendulum, aligning the plane of the bifilar pendulum to the YY principal axis of the model (see Figure 7.11 (right)). Two different modes of motion are excited in the model by rotating it to two different angles θ_1 and θ_2 and then releasing it to freely oscillate. Model motion

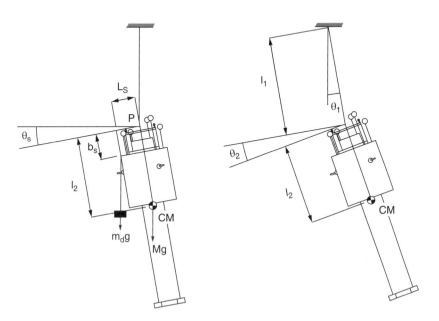

Figure 7.11 The model inclined at pitch angle θ_S by a mass m_d hung at a distance L_S from the pivot point P (left) and the model being excited in two different modes of motion (pitch and surge) by applying two different angles θ_1 and θ_2 (right).

is measured and the periods of oscillation of both modes of motion (T_1 and T_2) found using fast Fourier transform analysis. The moment of inertia can be calculated by applying the following equations:

$$I_{YY} = mk_{YY},\tag{7.19}$$

$$k_{YY} = \frac{g}{4\pi^4 T_S^2}\sqrt{T_1^2 T_2^2 \left(T_S^2 - T_1^2\right)\left(T_2^2 - T_S^2\right)},\tag{7.20}$$

$$T_S = 2\pi\sqrt{\frac{l_1}{g}},\tag{7.21}$$

where m is the mass of the model, k_{YY} the Pitch gyradius of the model, g the gravitational acceleration, T_S the sway period of oscillation and l_1 the length of the pendulum rope from the suspension point to the pivot point P. The XX moment of inertia can be obtained by following the same procedure but aligning the plane of the bifilar pendulum with the XX principal axis of the model.

The ZZ moment of inertia can be measured using a yaw oscillation test [72, pp. 7–9]. The model is suspended on the bifilar pendulum, aligning the plane of the bifilar pendulum to the YY principal axis of the model. The model is displaced and then released to generate motion around its Z-axis. Model sway is measured and the period of oscillation in yaw (T_Y) extracted. The moment of inertia is then calculated by applying the equations

$$I_{ZZ} = mk_{ZZ},\tag{7.22}$$

$$k_{YY} = \frac{T_Y}{T_S}d,\tag{7.23}$$

where m and k_{ZZ} are the mass and yaw gyradius of the model and d is the distance between one of the bifilar pendulum ropes to the vertical axis of the model.

When applying any of the bifilar pendulum methods it is good practice to keep the pendulum length to a minimum, in order to reduce errors due to the elasticity of the rope.

7.7.2 Wet Tests

7.7.2.1 Static

Wet tests are used to determine the physical properties of the model not measurable using the dry rig. The first set of wet tests measure the model's metacentric height ($\overline{GM}$). An appropriate moving mass is positioned at, or vertically above, the centre of gravity of the model. The mass is then moved to different positions along the longitudinal or transverse axis of the model and measurements taken of the model's inclination angle. The metacentric height can be directly obtained from these measurements [2, pp. 39–40].

Once the metacentric height has been obtained, and using the mass properties obtained from the dry tests, the main natural periods can be calculated. These are the resonance period in heave (T_H), pitch (T_θ) and roll (T_ϕ), which are respectively given by

$$T_H = 2\pi \sqrt{\frac{m+m_a}{\rho g A_{wp}}}, \qquad (7.24)$$

$$T_\theta = 2\pi \sqrt{\frac{I_{YY}+a_{YY}}{\rho g \nabla \overline{GM}_L}}, \qquad (7.25)$$

$$T_\phi = 2\pi \sqrt{\frac{I_{XX}+a_{XX}}{\rho g \nabla \overline{GM}}}, \qquad (7.26)$$

where m is the mass of the floating body, m_a the added mass of the floating body, ρ is the density of water, g the gravitational acceleration, ∇ is the volume of displacement of the body, A_{wp} the water plane area, I_{YY} and I_{XX} the moments of inertia over the Y- and X-axis respectively, a_{YY} and a_{XX} the added moments of inertia over the Y- and X-axis respectively and $\overline{GM}$ and $\overline{GM}_L$ the transversal and lateral metacentric heights, respectively.

7.7.2.2 Free Oscillation

Free oscillation or decay tests have two main purposes: to check dynamically the natural periods calculated in the static wet test; and to determine the hydrodynamic coefficients of the model at these natural periods. The hydrodynamic coefficients are the added mass (a), hydrodynamic damping (b) and hydrodynamic stiffness (c), and occur in the model's equation of motion. The linear equations of motion for a model (mass m) experiencing non-driven translational displacement (z) in still water, and for a non-driven angular displacement (Φ), are respectively given by

$$(m+a)\cdot\ddot{z}+b\cdot\dot{z}+c\cdot z=0, \qquad (7.27)$$

$$(m+a)\cdot\ddot{\Phi}+b\cdot\dot{\Phi}c\cdot\Phi=0. \qquad (7.28)$$

Free decay tests of an MRE model in still water can be carried out only for heave, pitch and roll motions (assuming no moorings are present), as these motions are the only ones that have a hydrodynamic restoring force or moment. They are usually conducted with the model in its experimental set-up configuration (i.e. moored), at its testing position and with the appropriate water depth. The model is then forced to an initial displacement, along the main axis, and then is allowed to return naturally to its equilibrium position, while the model motions are recorded. Figure 7.12 shows the result of a free oscillation test in heave of an OWC type WEC scale model. The natural period in heave (T_H) is measured as the period between nodes (peaks) of the recorded motion. The added mass and the hydrodynamic damping coefficient are determined from the exponential decay curve (the curve that the oscillation peaks trace, the dotted line in Figure 7.12) and the following equations:

$$a = \frac{c}{\omega_0^2} - m, \qquad (7.29)$$

$$b = \frac{2kc}{\omega_0}, \qquad (7.30)$$

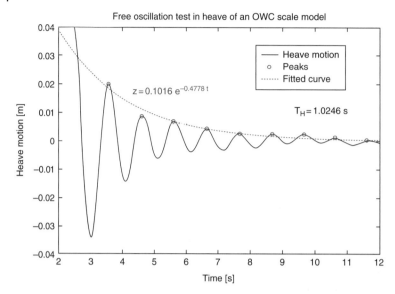

Figure 7.12 Free oscillation test in heave of an OWC buoy scale model.

where $\omega_0 = 2\pi / T_H$ is the natural frequency of oscillation, c is the stiffness coefficient and k the decay coefficient. The stiffness coefficient is known from the geometry of the model and is defined for heave (c_H), pitch (c_θ) and roll ($c_\varnothing$) respectively by

$$c_H = \rho g A_{wp}, \tag{7.31}$$

$$c_\theta = \rho g \nabla \overline{GM}_L, \tag{7.32}$$

$$c_\Phi = \rho g \nabla \overline{GM}. \tag{7.33}$$

Finally, k is measured from the figure by

$$k = \frac{1}{2\pi} \ln\left(\frac{z(t)}{z(t+T_H)}\right), \tag{7.34}$$

where $z(t)$ is the value of the fitted curve at a selected node and $z(t+T_H)$ is the value of the fitted curve at the consecutive node [73].

7.7.2.3 Forced Oscillation

Forced oscillation tests in still water are used to determine the relationship between both added mass and hydrodynamic damping and the frequency of oscillation. These are obtained from the loads excited by the forced oscillation as well as the coupling coefficients between motions. With the appropriate test configuration forced oscillation tests can be carried out for any of the linear or angular motions, but usually only the three motions with hydrodynamic restoring forces are studied.

Tests are carried out with the model attached to a test rig consisting of a system of one or more linear actuators which apply linear or angular motions to the model at different frequencies. Tests can be conducted with and without device moorings present. Testing with moorings allows analysis of second-order motions caused by the mooring restoring forces. This information is of relevance for the mooring design at later stages of development.

Figure 7.13 shows a schematic representation of the set-up of a forced oscillation test in heave of an OWC Masuda type WEC model. Other examples for angular motions can be found in [73]. The linear actuator at the top of the figure creates a vertical motion with an amplitude (z_a) and frequency (ω). By varying driving frequency the relationship between both added mass (a) and damping (b) with frequency is found.

During the experiment vertical forces (F_a) and displacements of the buoy (z) are measured. The linear equation of motion of the model undergoing forced heave oscillation is

$$(m+a)\cdot \ddot{z}+b\cdot \dot{z}+c\cdot z = F_a \sin\left(\omega t + \varepsilon_{F_z}\right),\tag{7.35}$$

where the subscribed buoy motion is $z(t) = z_a \sin(\omega t)$ and ε_{F_z} is the phase difference between the measured force and displacement.

The hydrodynamic stiffness coefficient is directly obtained from the geometry (equation (7.27)) and the added mass (a) and hydrodynamic damping coefficient (b) respectively from

$$a = \frac{c - \dfrac{F_a}{z_a}\cos\left(\varepsilon_{F_z}\right)}{\omega^2} - m,\tag{7.36}$$

$$b = \frac{\dfrac{F_a}{z_a}\sin\left(\varepsilon_{F_z}\right)}{\omega}.\tag{7.37}$$

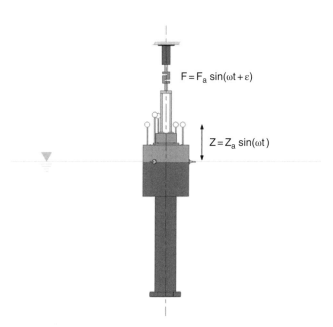

Figure 7.13 Forced oscillation test in heave for an OWC Masuda buoy.

7.7.3 Calibration of Tidal Turbine Models

The main physical properties that define the kinematic and dynamic response of a tidal turbine model are the centre of gravity of the rotor, the moment of inertia of the rotor, the drive train friction, the thrust at the turbine driveshaft, and the drag coefficient. A number of tests can be carried out to measure and adjust these physical properties in a scale model [74].

The centre of gravity and the different moments of inertia of the rotor (or other turbine components) can be measured using the bifilar pendulum method (Section 7.7.1).

The thrust characteristic curve of the turbine drive shaft can be measured by applying a series of known torques at different positions on the rotor and measuring the resulting force transmitted to the driveshaft. This is done by positioning a series of static weights at different radial distances along the rotor and measuring the force transmitted to the driveshaft with strain gauges.

The drive train friction is obtained by rotating the turbine rotor in a dry test. This can be done using a motor connected to the driveshaft, which is often also the generator or brake PTO of the model (Section 7.4.2). Friction is calculated by applying a range of rotational speeds to the rotor and recording the resulting angular velocities and torque.

The last physical property to determine is the drag coefficient, which is measured directly from the drag forces acting over the main body of the turbine by means of a load cell connected to the test rig. Note that there are two different drag forces acting over the main body of the turbine. The first is from the rotor, due to the interaction of the fluid velocity and the blades. The second is caused by the interaction between the main body of the turbine and the fluid velocity. In order to be able to measure the drag forces at the rotor it is necessary to test the main body of the turbine without the rotor in place, under the same set of fluid velocities as will be used to test the full model later.

The manufacturing process of the blades of a tidal rotor is complex, which makes it difficult to adjust them once they are built. The International Towing Tank Conference (ITTC) [74] has produced guidelines for the acceptable manufactured tolerances for model propellers. It is recommended that these are also followed when manufacturing model tidal rotors.

The kinematic viscosity of water is an order of magnitude higher than that of air and more variable with temperature. This means that for low Reynolds numbers the characteristic curves of NACA profiles (power and drag coefficients plotted against relative velocity or TSR) become very temperature-dependent (v is 1.267×10^{-6} m^2 s^{-1} at 10°C and 9.7937×10^{-7} m^2 s^{-1} at 20°C), as can be appreciated in Figure 7.14. This means that for lower relative velocities (turbine fluid) the resulting drag and power coefficients can vary significantly. For this reason, it is advisable to test at the highest Reynolds number possible and to regularly measure water temperature during calibration and experimental tests to check for changes. Testing at a higher Reynolds number (in excess of 5×10^5 [76]) also has the advantage that performance characteristics such as power, torque and thrust coefficients become independent of the Reynolds number, significantly simplifying the scaling process.

7.8 Modelling the Environment

Wave basins and flumes can generate regular, irregular and focused waves. The choice of which wave type is most suitable if generally based on the objectives of the experimental campaign. If used correctly, irregular waves most closely represent 'real' ocean conditions.

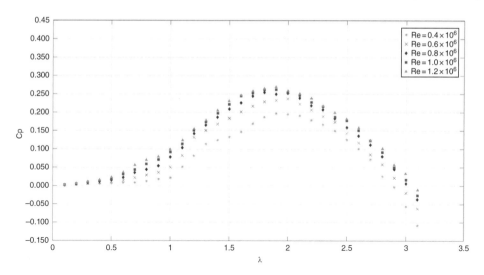

Figure 7.14 Variation of the power coefficient (C_P) with tip speed ratio (λ) for different Reynolds numbers for vertical axis turbine tidal model [75].

However the use of regular and focused waves can provide valuable information. Generation of current is required for the physical modelling of tidal energy devices and when examining the impact that current has on the performance of wave energy devices. Careful mapping, and possible manipulation, or turbulence and flow profile characteristics is generally required.

7.8.1 Regular Waves

Regular waves have a single frequency, amplitude and propagation direction. As such they can be considered the 'simplest' wave type used in physical modelling. They are most often used during initial proof of concept testing during early development stages. Regular waves allow device response at individual excitation frequencies to be explored. This can be particularly interesting when testing resonating wave energy devices. By testing at a range of frequency intervals the size and bandwidth of the resonance response can be identified. Parametric studies on the impact of wave amplitude and wave steepness can also be conducted using regular waves.

The use of regular waves provides a useful data set for calibration and/or validation of analytical and numerical models. However, performance data obtained only from regular wave measurements can sometimes overestimate the performance a prototype would demonstrate in the ocean. Irregular wave tests should be conducted when wishing to make accurate performance predictions.

Various analytical theories are available to calculate the properties of regular waves. Figure 7.15 outlines the wave steepness and water depth regimes in which these theories are applicable. Linear (or Airy) wave theory is the 'simplest' of these theories and is often used as a first-order approximation outside its applicable range. A surface profile given by

$$\eta(x,t) = a\cos(kx - wt) \tag{7.38}$$

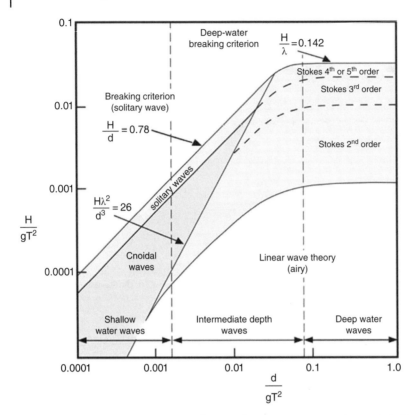

Figure 7.15 Validity of several theories for periodic water waves, according to Le Méhauté [77].

is assumed, where a is wave surface amplitude, and this results in the dispersion relationship relating the wave's angular frequency (ω) and wave number (k),

$$\omega^2 = gk\tanh(kh),$$ (7.39)

where h is water depth and g is gravitational acceleration. The mean wave-energy density per unit horizontal area E,

$$E = \frac{1}{2}\rho g a^2,$$ (7.40)

and the group velocity c_g,

$$c_g = \frac{\partial \omega}{\partial k} = \frac{1}{2}\sqrt{\frac{g}{k}\tanh(kh)}\left(1 + kh\frac{1-\tanh^2(kh)}{\tanh(kh)}\right),$$ (7.41)

can be used to calculate the average power of a regular wave per unit crest width,

$$P_w = E c_g.$$ (7.42)

When $h > 0.5\lambda$ the deep water approximation can be applied ($\tanh(kh) \to 1$). Then equation (7.42) reduces to

$$P_W = \frac{\rho g^2}{32\pi} TH^2. \tag{7.43}$$

7.8.2 Irregular Waves

Irregular waves reproduce the properties of most full-scale sea states more accurately than regular waves. Long-crested irregular waves are formed in the laboratory using a sum of regular waves with different frequencies, amplitudes and randomised phase (ϕ):

$$\eta(x,t) = \sum_{i=1}^{N} a_i \cos(k_i x - \omega_i t + \phi_i). \tag{7.44}$$

The range of frequencies in equation (7.44) is dependent on the generation capacity of the wave paddles (where f_{max} is the maximum frequency that can be generated and f_{min} is the minimum), while the number of components (N) is dependent on the repeat time (RT),

$$N = (f_{max} - f_{min})RT. \tag{7.45}$$

The repeat time is the length of time before the generated wave signal is reproduced and determines the number of wave components (N) and therefore the interval between frequency components.

The amplitudes of the individual components can be calculate using amplitude wave spectrum. Various analytical expressions exist that define standardised spectrum. These include the Pierson–Moskowitz spectrum, representing fully developed seas, the JONSWAP spectrum, representing fetch-limited conditions, and the Bretschneider spectrum, which accounts for duration and fetch limitation empirically. Using these expressions, a sea state with specific zero-crossing or peak period and significant wave height can be generated. These three spectrums are all single-peak. For multimodal sea states the Ochi–Hubble spectrum can be used to better represent certain real-world conditions. Alternatively, site-specific sea measurements will produce a measured spectrum which can be reproduced at scale in a laboratory.

When testing with irregular waves it should be remembered that there is no unique surface profile that corresponds to a specific spectrum. This is not usually an issue when testing a device's performance. However, when examining a device's survivability different time series with the same spectral content will contain different combinations of large waves, as shown in Figure 7.16, resulting in differences in the measured extreme response.

Irregular waves can alternatively be generated by conducting a Fourier analysis on a previously measured time series to give the components required in equation (7.44). The use of wave spectrum is generally considered preferable to this approach as obtained results can be associated with a generalised sea state. However, this approach might be useful when a specific wave time series has a particular significance, for example if trying to reproduce a failure event.

Generation of short-crested irregular sea states is possible in most three-dimensional wave basins. This allows the influence of sea state directionality to be investigated. A cosine-squared distribution approximates the spreading of waves relatively well.

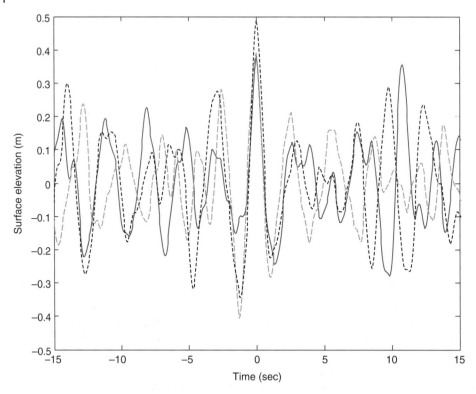

Figure 7.16 Three extracts of surface profiles generated for a 1 : 30 scale representation of a three-hour-long sea condition with a once in 100 years return period at Wave Hub. Time series have been shifted so that the largest wave in each profile occurs at time $t = 0$.

The wave spectrum $(S(\omega))$ is multiplied by the cosine-squared distribution to produce the directional spectrum $(S(\omega,\psi))$,

$$S(\omega,\psi) = \frac{2}{\pi}\cos^2(\psi - \psi_0)S(\omega), \tag{7.46}$$

where ψ_0 is the primary angle of propagation and ψ is the angle of propagation considered. The average power of an irregular wave per unit crest width can be calculated as

$$P_w = \rho g \sum_{i=1}^{N} S_i . C_{gi} . \Delta f, \tag{7.47}$$

where S_i is the spectral density, C_{gi} the group velocity of the ith wave component, and Δf the frequency interval between different wave components. In deep water $(h > 0.5\lambda)$ wave power per unit width is given by

$$P_w = \frac{\rho g^2}{64\pi} H_{m0}^2 T_e, \tag{7.48}$$

where H_{m0} is significant wave height and T_e is wave energy period, both of which can be calculated from the spectral moments of the measured wave spectrum.

7.8.3 Focused Waves

The third category of waves that can be generated in most wave tanks are focused waves. Focused waves are a concentration of wave energy at a specified time and location in the tank which can be used to investigate the response of an MRE device to an extreme wave. Various techniques are employed in the generation of focused waves. A common approach is frequency focusing. A wave is generated using the same approach as irregular waves (equation (7.44)), but with the phase relationship between different components adjusted so that all components constructively superimpose at a specific location and time.

A specific type of focused wave group was introduced by Tromans *et al.* [78] and is called NewWave. For a given sea state NewWave theory produces the average shape of the highest wave with a specified exceedance probability. Figure 7.17 shows an example time series of a NewWave focused group.

This technique is a valuable way of generating a standalone extreme wave. However, the extreme loading on MRE devices that move in response to wave action depends not only on the amplitude and shape of the largest wave, but also on the device's motion due to previous wave interactions [40]. One approach to systematically investigate this is the use of embedded (or constrained) focused waves [79]. This approach generates a deterministic focused wave within an irregular wave time series, allowing the impact of different irregular time series on the response to an extreme event to be examined.

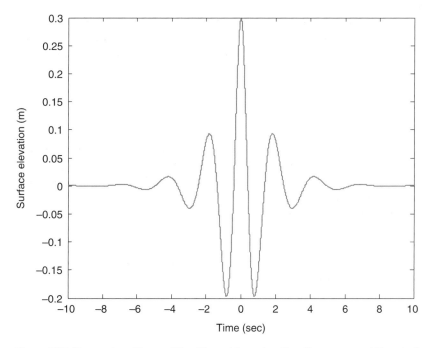

Figure 7.17 Time series of focused NewWave at focus location. Focus occurs at time $t = 0$, with a crest amplitude of 0.3 m. Generated using a Pierson–Moskowitz spectrum with a peak frequency of 0.5 Hz.

7.8.4 Flow

There are several relevant parameters to characterise water flow generated in the laboratory. Of these, average flow at individual locations is the simplest to obtain and understand. However, it generally does not provide sufficient detail. Turbulence intensity has been identified as having a significant impact on both tidal turbine performance and wake recovery. Total turbulence intensity (I) is defined by

$$i = \frac{\sqrt{\frac{1}{3}\left(\overline{u_x'^2} + \overline{u_y'^2} + \overline{u_z'^2}\right)}}{U} = \frac{\sqrt{\frac{2}{3}k}}{U} \tag{7.49}$$

where $\overline{u_i'^2}$ is the mean of the square of the velocity fluctuations and U is mean velocity. The mean velocity used to calculate turbulence intensity varies depending on the situation (mean of a point measurement, mean of the whole flow, etc.), and the value used should be clearly identified when reporting turbulence intensity values. Often U is defined by

$$U = \sqrt{U_x^2 + U_y^2 + U_z^2}, \tag{7.50}$$

where U_i is the mean velocity magnitude, in the longitudinal, lateral and vertical ($i = x$, y, z) directions respectively. The velocity fluctuations for each specific measurement (j) are defined by

$$u_{i,j}' = u_{i,j} - U_i. \tag{7.51}$$

Turbulent kinetic energy, k, is defined as

$$k = \frac{1}{2}\left(\overline{u_x'^2} + \overline{u_y'^2} + \overline{u_z'^2}\right) \tag{7.52}$$

The turbulence intensity (I_i) along a specific direction can also be defined:

$$I_i = \frac{\sqrt{\overline{u_i'^2}}}{U} \tag{7.53}$$

Turbulence length scales are another important characteristic of turbulence. The integral length scale, L, can be considered to be the size of the eddies containing the greatest proportion of turbulent energy [80]. Blackmore *et al.* [80] estimate integral turbulence length scale as

$$L \approx \tau U, \tag{7.54}$$

where τ is the integral timescale,

$$\tau = \int_0^\infty \frac{R(s)}{R(0)} ds, \tag{7.55}$$

and $R(s)$ is the autocovariance (assuming a statistically stationary flow), as given by u_i'

$$R(s) = \left\langle u_i'(t)u_i'(t+s)\right\rangle, \tag{7.56}$$

where *t* is time and *s* is time lag. Turbulence length scale has been found to have significant effects on the drag on tidal turbines (Blackmore *et al.* [80]). To represent full-scale conditions, turbulence intensity and eddy size should be scaled to preserve non-dimensional parameters: L/R and $u'/\Omega R$ [14, p. 10], where R is turbine radius and Ω is turbine rotational speed.

Vertical flow profile and velocity shear can impact the performance and loading of tidal turbine blades, especially for devices whose cross-section span a significant vertical cross-section. It is therefore important for the vertical flow profile to be measured upstream of a model device. It is good practice to check the horizontal flow profile as well.

7.9 Test Facilities

7.9.1 Wave Generation and Absorption

Most wave generation systems used in the physical modelling of MRE devices can be fitted into two broad categories: flap paddles and pistons paddles. Flap paddles generate waves via an angular rotation about the bottom of the paddle, which is located either on the bottom of the tank or on a ledge above the tank floor. These are designed to generate deep water waves, with orbital particle motions which decay towards zero motion on the tank floor.

Piston paddles are best used when generating shallow water waves. The paddles remain vertical and are driven to produce oscillating horizontal displacement. This is used to generate the elliptical particle motion observed in shallow water waves, including the significant horizontal fluid motion on the tank floor. Piston paddle generation systems are ideally suited for testing bottom-mounted wave energy devices.

The first analysis of wave generation theory was published by Biesel and Suquet [81]. Since then wave generation theory and the resulting wave control software, for both flap and piston paddles, have developed to increasingly improve the quality of generated waves. This includes second-order wave generation and active wave absorption. Wave generation needs to be as accurate and repeatable as possible. Paddle motion creates unwanted evanescent waves which decay with distance and can be minimised by trying to match the paddle motion to the water motion [82, p. 149]. It is also the case that, especially at higher frequencies, several wavelengths are required for a 'natural' wave to develop from the motion of water generated by the paddles. A general rule of thumb is that at least two wavelengths should be left between the wave generator and the model being tested. However, if in doubt waves should be measured at the model's location before installation to check this.

The necessity of pre-installation wave measurements depends on both the facility in which tests are conducted and the nature of the tests. Each wave paddle will have a transfer function which relates paddle displacement to wave amplitude. Most facilities start with theoretically derived transfer functions, which are then refined through tank calibration experiments. These calibrations are often only valid over a predefined test region and care should be taken when testing larger models, or when conducting array tests, that go outside this 'calibrated region'. Similarly, a lack of spatial homogeneity in the generated wave field can be an issue when testing devices that move significantly (e.g. a floating device with a compliant mooring), as the device could end up exposed to wave properties different than what has been specified.

The simplest form of waves to generate are those travelling in a perpendicular direction to the front face of the wave paddles. This form of wave is created in wave flumes where two-dimensional experiments are conducted with blockage ratios close to 100%. Long-crested waves are also produced in wider tanks with reduced blockage effects. The generation of angular waves is possible in wider tanks using a number of wave paddles which can be independently controlled. A straight line of wave paddles can generate waves travelling at an angle to the perpendicular of the paddle surface by applying a phase difference to the motion of each panel. In wider rectangular tanks this approach can be extended using the Dalrymple method, where waves are reflected from the tank walls to generate an angled wave elsewhere along the tanks length. The ability to control individual paddle motion means that directional spreading can be applied to generate directional sea states (Section 7.8.2). This is made easier using wave paddles which are distributed in a curved, or even circular, arrangement, although such arrangements make generating long-crested waves much harder.

The generation of a high-quality wave field requires the presence of wave absorption within the tank facility to minimise reflections from tank sides and models adversely affecting the generated wave field. Both passive and active wave absorption systems are used in modern test facilities. Passive wave absorption generally consists of either a linear slope or a parabolic beach or (generally in smaller flumes) triangular sections of porous foam. Wave-absorbing beaches induce wave breaking and the resulting energy dissipation. Various designs are made porous to channel some of the water from the breaking wave under the beach back into the basin, thereby reducing the impact on incoming waves. Additional surface roughness is added to beaches to induce breaking and further energy dissipation.

The reflection coefficient from a beach will depend on the incoming wave frequency, amplitude and angle, and therefore beaches cannot be optimised for all wave conditions. This also makes it difficult to fully characterise the reflection characteristics of a basin, especially if a variable floor depth is present. It has been estimated that the length of a beach needs to be at least half the wavelength of the incident wave to achieve 90% absorption [82, p. 155].

Active wave absorption is where wave generation paddles are able to absorb incoming waves, while continuing to generate the required wave field. This is a useful feature of nearly all modern basins. Energy reflected back from the model and passive wave absorber is removed, thereby limiting secondary reflection from the paddles and allowing the required wave field to be generated for longer. This both increases the time over which experiments can be successfully run and reduces the time spent waiting for the water to settle between experiments. Unlike passive wave absorbers, active wave absorbers can also have a limited effect on the cross-waves that can build up in rectangular facilities when the wave frequencies matches the tanks resonance frequencies.

Active wave absorption systems fit into two broad categories. In the first the wave elevation is measured close to the paddles and the measurement feds into the piston control system. The second approach, which is more common in newer facilities, uses a force measurement on the pistons themselves as the feedback signal. Some basin facilities deploy wave paddles on multiple sides of the tank as opposed to using passive wave absorbers. This has the advantage of increasing the usable space within the facility.

However, as with passive wave absorbers, active wave absorbers work better at certain frequencies than others, and as such a combination of passive and active absorption systems can be effective.

7.9.2 Basin and Flume Flow

Flow generation is possible in many flumes and basins, usually using recirculating hydraulic multi-pump systems. These provide the option of testing scaled tidal energy devices in the same configuration as they would appear at full scale, for example mounted on the seabed or supported from a moored structure. It is important to characterise the flow around the test section that the device will be placed in, including downstream of the device if wake effects are to be measured. This characterisation includes measuring the flow profile (the variation of velocity through the water column and potentially across the tank), the turbulence intensity and turbulent kinetic energy of the flow. This characterisation is often made with ADV or LDV measurements (Section 7.2).

Within the majority of facilities it is difficult to influence the current profile, although within flumes a series of tubes with decreasing spacing down the water column can be used to generate a profile through local blockage [83]. A certain amount of control is also possible over the level of turbulence. Flow straighteners can be added or removed to decrease or increase turbulence, respectively. For example, Mycek *et al.* [84] studied tidal turbine behaviour at an ambient turbulence intensity of 3% and 15% respectively with this technique. Alternatively, Myers *et al.* [85] describe a system consisting of two grids aligned orthogonally to the flow which are driven in a horizontal and vertical direction respectively to generate relatively persistent turbulent flow structures.

In many facilities it is possible to model the impacts of bathymetry and surface roughness on the performance of tidal energy devices. This can be useful when examining the impact of site-specific features creating additional turbulence and unexpected changes in flow characteristics [14].

Some facilities are capable of generating waves and current flow simultaneously. In flumes this is clearly limited to waves travelling either with or against currents. However, some 3D basins, such as Plymouth University's COAST Laboratory Ocean Basin or the FloWave TT Basin, can generate a wide range of waves and currents from different relative directions. Such facilities allow the impact of waves on performance and loading of tidal energy devices, and likewise the impact of currents on the performance and loading of wave energy devices, to be investigated. In such experiments, especially when currents are not in line with the wave propagation direction, additional care is sometimes needed to ensure that instruments such as wave probes are not vibrating in response to the application of loads from different directions.

7.9.3 Towing Tanks

Towing tanks, although originally designed for ship testing, are often used as an alternative to current generation in flumes and basins for the testing of tidal turbines. Examples of these facilities include those at Southampton and Strathclyde University in the UK and INSEAN in Italy. The turbine is fixed to a carriage and towed through stationary water, to simulate water flowing past a fixed turbine. For a turbine designed to be mounted from a floating platform this installation is similar to the prototype design. It represents a greater deviation for seabed-mounted systems; however, towing tanks can still be used to collect valuable data for control and component testing.

Towing tidal turbines through stationary water does not realistically reproduce full-scale turbulence and velocity profiles. Turbulence can be generated by placing a grid 'upstream' in front of the tidal turbine. By varying the mesh properties turbulence intensity can be partially controlled [14]. Most towing tanks are equipped with wave generating capabilities, providing the capability to investigate the impact long-crested waves have on tidal turbine performance and load.

7.9.4 Blockage Effects

When generating waves or currents at an experimental facility tank blockage effects need to be considered when designing the test campaign. In wave generating facilities blockage effects are due to the interaction between the incident wave field, the model and the walls of the tank. In current generation facilities blockage effects occur when the proportion of the tank's cross-section that the model takes up is large enough for contraction effects to occur, and flow interactions with the tank wall and floor start to significantly impact the model behaviour.

For wave generation facilities, blockage is considered to be significant when the transverse dimension of the model is of the same order of magnitude as the width of the tank. A common rule of thumb is that the ratio of model width to tank width should be at least 5 : 1 to limit the impact of blockage [5, pp. 128–129]. For example, if testing a vertical cylinder the tank walls should be at least 2.5 diameters from the centre lines of the cylinder.

For current generation facilities, Elsaesser *et al.* [14] suggests a blockage ratio (the ratio of cross-section of the model to the cross-section of the channel) of less than 10%. If this is impractical then blockage correction techniques exist to 'correct' experimental results. Cavagnaro and Polagye [86] compare the performance of three of these techniques, finding them to reduce the variation between performance curves for different levels of blockage.

7.10 Recommended Tests

7.10.1 Standard Tests for Wave Energy

Designing the tests to be carried out during the physical modelling of a WEC is not a trivial task. As previously discussed (Section 7.2), knowledge of previous study outcomes, the device's development requirements and the required test outcomes are all required. These requirements, and the peculiarities of individual devices, mean that the details of most tests are designed *ad hoc*. However a wide set of recommended standard tests have been produced to guide device developers evaluating the performance of a WEC [2, 3].

Tests designed to evaluate the performance of a WEC at its operational conditions can be grouped into eight series. Each one of these series has a specific set of outcomes and is oriented to a different range of development stages. Table 7.3 presents, for each

Table 7.3 Standard tests for wave energy devices.

Test series	TRL	Facility	2D, 3D	Test duration
Series A: Linear regular waves	1–4	flume, basin	2D	50–100 waves
Series B: Nonlinear regular waves	3–5	flume, basin	2D	(300 if resonance)
Series C: Long-crested irregular waves	1–5	flume, basin	2D	1 h full scale or
Series D: Spectral shape	2–5	flume–basin	2D, 3D	(>500 waves)
Series E: Directional long-crested waves	2–5	basin	3D	
Series F: Short-crested waves	2–5	basin	3D	
Series G: Combined waves and ocean currents	2–5	flume, basin	2D, 3D	test specific
Series R: Repeatability	1–5	flume, basin	2D, 3D	

test series, the range of TRLs for which the series is applicable, the type of facility in which the experiments can be carried out, if the waves need to be reproduced in 2D or 3D and the recommended duration for the tests. Each one of these test series is defined in more detail below.

7.10.1.1 Series A: Linear Regular Waves

Linear regular wave tests are one of the most valuable engineering design tools to evaluate a WEC design at its initial stages of development (i.e. TRLs 1–4). The tests are used to measure device performance and characteristic indicators, including the power matrix, the capture width or capture width coefficient, the response amplitude operator (RAO) for each motion, the PTO damping, reflected and transmitted waves, wave forces acting on the device, and the motions of the mooring lines and the forces acting on them.

The power matrix (Figure 7.18) characterises the power output of the device under a wide range of wave height and period conditions. The capture width (C_W) and capture width ratio (C_{WR}) are two indicators of the efficiency of the device at a specific wave frequency, respectively defined by

$$C_W = \frac{P_m}{P_w}, \tag{7.57}$$

$$C_{WR} = \frac{P_m}{P_w l}, \tag{7.58}$$

where P_W is the incident wave power (equation (7.42)), P_m the mean power output of the device and l the characteristic dimension of the device.

The RAO is a widely used ratio to study the relationship between different input and output (or response) parameters of a linear system. RAOs are usually used to analyse the response of a device in the different degrees of freedom (surge, sway, heave, roll, pitch and yaw) relative to incident wave amplitude. However, RAOs can also be used to study the relationship between other parameter and wave amplitude, such as horizontal

	Regular waves power matrix [kW] –1:1 scale masuda buoy, for H< 5.5 m											
	Wave period T [s]											
Wave height H [m]	6	7	8	9	10	11	12	13	16	18	20	22
0.50	0.137	3.773	2.325	1.126	0.626	0.594	1.010	0.469	0.000	0.000	0.000	0.000
1.50	3.302	27.593	38.127	20.306	10.818	9.676	6.232	3.472	0.106	0.029	0.028	0.038
2.50	10.396	61.793	110.247	69.637	32.960	24.838	13.800	8.126	0.218	0.098	0.083	0.101
3.50	14.181	101.288	193.511	143.898	65.691	44.269	23.297	14.254	1.400	0.298	0.152	0.276
4.50	26.123	153.131	296.310	236.387	101.410	68.080	33.873	21.296	2.956	0.555	0.301	0.687
5.50	40.492	223.575	298.653	326.293	146.654	95.622	44.860	29.968	5.752	1.228	0.547	1.210

Figure 7.18 Regular waves power matrix of an OWC Masuda buoy model, 1 : 50 scale [38].

or vertical velocities at a fixed depth, pressure, etc. The RAO is defined as a non-dimensional operator [87],

$$RAO_i(f) = \frac{1}{l^k}\sqrt{\frac{m_{0i}}{m_{0W}}}, \qquad (7.59)$$

where m_{0i} is the zeroth-order moment of the motion of the body displacement in oscillation mode i (i.e. surge, sway, heave, roll, pitch and yaw), m_{0w} is the zeroth-order moment of the incoming waves, l is the characteristic dimension of the WEC (typically its width or diameter) and k is an exponent which is 0 for the translation modes and 1 for the rotation modes. The respective zeroth-order moments are defined respectively for the motions or the incoming waves from equation (7.13).

A transfer function (or RAO spectrum) is typically composed of individual values of the RAO obtained from a regular wave test series. This spectral response can then be multiplied directly with any regular or irregular wave spectrum to obtain the response of the WEC under a wide set of possible conditions. Figure 7.19 shows an example of the heave and pitch RAOs of an OWC Masuda buoy scale model.

Analysing the influence that PTO damping has on the efficiency of a device and how this is affected by wave properties can lead to significant improvement in energy extraction. López *et al.* [88] presents a method to select the optimum turbine-induced damping for an OWC using a set of regular and irregular wave tests.

A reflected and transmitted wave analysis (see Section 7.6.1) allows the characterisation of the incident wave that reaches the model, the reflected wave from the model and the transmitted wave at the lee of the model. This analysis is relevant for understanding the impacts of the WEC on the near and far fields.

The wave forces acting on the device can be measured directly with pressure transducers mounted over the surface of the model. The distribution of forces acting over the surface of the model can then be studied [89].

The motions of the mooring lines and the forces acting on them can be measured using load cells at different positions along the mooring line and by tracking the motions of the model in six degrees of freedom.

The standard approach when testing a WEC in regular waves is to use between 50 and 100 waves in each test. However, if resonance effects are expected or are observed it is

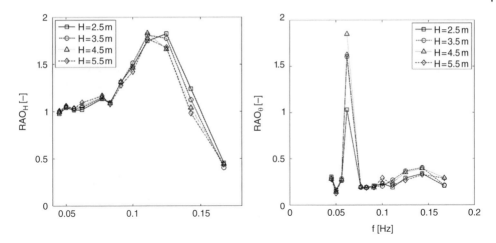

Figure 7.19 Regular wave heave RAO (left) and pitch RAO (right) of an OWC Masuda buoy model, 1 : 50 scale [38].

good practice to extend the length of the tests to at least 300 waves. Short time series, especially for lower wave heights, do not allow the resonance phenomena to develop enough to be measured accurately.

7.10.1.2 Series B: Nonlinear Regular Waves

The nonlinear regular wave test series are of special relevance for devices deployed in water depths that are not considered deep ($h < \frac{1}{2}\lambda$) or are expected to be exposed to waves with low periods and high wave heights (e.g. flap type devices). Figure 7.15 defines the regions where the different wave theories are valid. These tests are not intended to be carried out at the initial stages of device development, where assumption of linear wave theory is acceptable. However, at later development stages (TRL levels 3–5) a nonlinear regular wave test series can be considered to evaluate the performance of the model under a more realistic environment.

The recommended duration of this test series is the same as for series A: 50–100 waves, increasing to at least 300 waves if resonance phenomena are expected or observed.

7.10.1.3 Series C: Long-crested Irregular Waves

Long-crested irregular waves are used in the three initial development stages of a device (TRLs 1–5) to evaluate performance under real sea conditions. A choice of spectral shape is required, with two classic options available. The first choice is to use the Bretschneider generic spectra, which is a site-independent approach. The second is to use a JONSWAP or Pierson–Moskowitz spectrum, based on data from a particular test site of interest (e.g. a wave energy test centre such as Wave Hub). Alternatively, a more complicated spectrum (e.g. the Ochi–Hubble spectrum) or a site–specific spectrum (based on measurements) is sometimes used.

For devices at higher TRLs (3–5) it is advisable to test the full scatter diagram of expected sea states, or at least the part that covers the operative range of conditions for the device. From these results the annual mean values of energy production at a selected location can be estimated.

At lower TRLs (1–2) it is often enough to just select some representative sea states at which to test. Options for how to make this choice include: selecting the one or two most occurring sea states at each significant wave height; defining a sea state with the average peak period (weighted with the occurrence of each peak period of each sea state) for each significant wave height; or defining different sea states by varying the peak period for a fixed wave height [3].

The evaluation of device performance in long-crested irregular waves can be done by using the same seven indicators already defined for regular waves, with the power matrix bins defined by significant wave height and a representative wave period, such as energy period. The equations defined previously for the different indicators are still valid for irregular waves, with the exception of the RAO. For irregular waves the RAO is defined as

$$RAO_i(f) = \frac{1}{l^k} \sqrt{\frac{S_i(f)}{S_W(f)}}, \tag{7.60}$$

where S_i is the spectral power density in oscillation mode i (i.e. surge, sway, heave, roll, pitch and yaw), S_W is the spectral power density of the incoming waves, l is the characteristic dimension of the WEC (typically its width or diameter) and k is an exponent which is 0 for the translation modes and 1 for the rotation modes. Unlike with regular waves, where a single value of RAO was obtained for each test, for irregular waves a complete spectral RAO is obtained from a single irregular wave test.

The standard approach when testing a WEC in irregular waves is to test for the equivalent scaled time of 1 hour at prototype scale. However, in order to ensure that the selected spectrum is fully represented, it is good practice to check that the number of waves contained in the time series is above 500 [90]. If this is not the case it is recommended that the test duration is extended. It is also good practice to partially repeat the irregular time series, in order to avoid the transient effects occurring in the tank due to interactions between the incident wave field, the walls, wave absorber and model. This can be done by repeating the initial 20% of the time series at the end of it, then selecting the final part of the data to analyse.

7.10.1.4   Series D: Spectral Shape
In order to evaluate the influence of the spectral shape on the performance of the device, it is recommended to test some or all the sea states tested for series C under one or more different spectral shapes. For example, if a JONSWAP spectrum was selected to test the model at series C, a Pierson–Moskowitz or Bretschneider spectrum can be selected for series D and the performance of the device compared. These types of tests are advised for later TRLs (2–5), as they help determine the device's dependence on the wave field properties.

7.10.1.5   Series E: Directional Long-crested Waves
A directional long-crested wave test series is recommended for none axisymmetrical devices to evaluate their performance under different wave directions. This can be done by either directly generating directional waves or turning the device to different angles relative to the incident waves. This test series is recommended during the performance and optimisation phases of the device development (TRLs 2–3) as well as for later TRLs (4–5) when a dependence on wave direction has been identified.

7.10.1.6 Series F: Short-crested Waves

The aim of a short-crested wave test series is to evaluate how sensitive the device is to short-crested seas compared to long-crested ones. It is recommended to test some or all the sea states tested at series C under one or more spreading coefficient values. This test series is recommended for TRL 2–3 during the performance and optimisation phases of the device development, along with later TRLs (4–5) for those devices where a change in performance has been identified. The same approach to test duration as detailed in series C is recommended.

7.10.1.7 Series G: Combined Waves and Ocean Currents

Tests combining any of the previous wave series with currents may need to be considered during the development of a wave energy device if the currents at the selected deployment site are considered to be significant. Ocean currents are likely to increase loads on the moorings and hull of a device, as well as modifying the wave's steepness. When designing the duration of the test series, the same approach as for the waves tested without the current should be followed.

7.10.1.8 Series R: Repeatability

Repeatability tests are aimed to measure the accuracy of the experimental set-up and should be carried out during all experimental campaigns. To check the repeatability a series of representative tests are selected and repeated several times throughout the duration of the experiments.

7.10.2 Survivability Tests for Wave Energy

Survivability is as critical for a WEC design as power performance and should be considered from the very beginning of the design. It is therefore recommended that tests are conducted to start establishing an understanding of how the device behaves in extreme conditions from TRL 2 onwards. However, the procedure to test survivability of WECs is not as well as established as the procedure to evaluate performance in more benign conditions. In many cases the set of conditions that produce the largest loading and resulting stress on a structure is often unique to a particular device and is not necessarily caused by the largest wave. Different device failure modes further complicate the decision on what conditions are important for failure. Determining these conditions is not an easy task and is unlikely to be achieved through physical modelling alone.

Regardless of these points, conducting model tests in extreme conditions can still provide valuable data if correctly done. It can be especially useful in examining the important nonlinear effects that occur in more energetic and steep wave climates and which are still difficult to accurately model in most numerical modelling tools. One of the first issues to be considered is the magnitude of the 'storm' that is to be modelled. This is usually defined in terms of return period (e.g. 1 in 100 year wave conditions) or exceedance probability (1% chance of being exceeded in a year). The choice of return period is largely dependent on the design objectives of the device developer. The wave conditions and spectral shape relating to this return period then need to be established. This will be site-specific, and indeed most survivability testing is conducted with reference to a particular planned deployment location.

By their very nature survivability wave conditions are larger than operating conditions. It is therefore a greater challenge to reproduce them in wave tank facilities. Correct reproduction of the wave field may require testing at a smaller scale then is required for performance testing. Consideration should be made regarding the increased scale effects when analysing results. Special consideration should be taken if breaking wave impacts involving air entrapment are considered to be important, as large scale-effects are introduced as model scales decreases. Many WECs also have survivability control strategies as part of their design. It is clearly important to model these strategies as accurately as possible.

Generally, and as with performance modelling, the greater the number of parameters that can be measured during the survivability testing the better. This assumes that deployment of instrumentation does not significantly influence the model's behaviour and that additional installation and calibration times do not limit the number and duration of tests conducted. For floating structures measurement of mooring loads and device motion is essential. For all structures impact pressures can be recorded, and for fixed structures the loading on the support structure should be measured. Depending on the device's survival strategy, loading of the PTO may also be required.

Table 7.4 presents suggested survivability test series that can be conducted using the extreme wave conditions identified. Each one of these test series is discussed in more detail below.

Initial survivability testing is conducted using long-crested waves (series H). As discussed in Section 7.8.2, there is no unique irregular wave time series for a particular wave energy spectrum. Therefore if the extreme response or failure condition is believed to be due to a particular combination of loading values then multiple experiments may be required.

Even an axisymmetric point absorber will have a level of asymmetry in its mooring arrangement. Therefore investigating the influence of wave direction can be important (series I). In series J a directional spreading function is applied to the sea state to generate a short-crested sea. As with performance testing, the impact of ocean current on the models wave response may be of interest, depending on the proposed site (series K). Finally, conducting repeat experiments of representative experimental cases is again highly recommended.

There is often interest in examining how a device behaves in the event of the failure of a specific component. Physical modelling is well suited to this form of investigation, assuming that only the required component is made to fail.

Table 7.4 Survivability tests for wave energy devices.

Test series	TRL	Facility	2D, 3D	Test duration
Series H: Long-crested	2-5	flume, basin	2D, 3D	
Series I: Long-crested and directional	3-5	flume, basin	2D, 3D	
Series J: Short-crested	3-5	basin	3D	3 hrs (full scale)
Series K: Combined wave and ocean current	3-5	basin	3D	
Series R: Repeatability	2-5	flume, basin	2D, 3D	

7.10.3   Standard Tests for Tidal Energy

Designing tests to aid the development of a tidal energy device, as with wave energy devices, is not a trivial task. Previous study outcomes, the device's development requirements and the required test outcomes all influence the programme. Although tidal energy device designs tend to vary less than with WECs, the individual peculiarities of a device mean that the details of most tests are designed *ad hoc*. Often specialised tests will be conducted to examine specific aspects of a device, such as the performance of a particular blade design. In this section recommendations are made for tests often conducted on scale models of a whole device.

7.10.3.1   Performance

Various coefficients exist to characterise the performance of a horizontal axis tidal turbine. These include the turbine power coefficient, C_p, the ratio between power P extracted by a turbine and the maximum power available from the incoming flow through the turbine area S. The turbine power coefficient is given by

$$C_p = \frac{P}{\frac{1}{2}\rho S U_\infty^3} = \frac{Q\omega}{\frac{1}{2}\rho \pi R^2 U_\infty^3} = \frac{Q}{\frac{1}{2}\rho \pi R^3 U_\infty^2} TSR, \tag{7.61}$$

where Q is turbine torque, R is the turbine radius, ρ is the water density, ω is the rotational speed, TSR is the tip speed ratio and U_∞ is the streamwise flow velocity. The torque coefficient, C_Q, is the ratio between the turbine torque and the theoretical torque,

$$C_Q = \frac{P}{\frac{1}{2}\rho R S U_\infty^3} = \frac{Q}{\frac{1}{2}\rho \pi R^3 U_\infty^2} = \frac{C_p}{TSR}, \tag{7.62}$$

while the thrust coefficient, C_T, is the ratio between the drag force, T, acting on the turbine and the kinetic energy of the incoming flow through the turbine cross-sectional area,

$$C_T = \frac{T}{\frac{1}{2}\rho S U_\infty^2}. \tag{7.63}$$

Experiments examining the performance of horizontal-axis blades examine how these ratios change with tip speed ratio, as defined by

$$TSR = \frac{\omega R}{U_\infty}. \tag{7.64}$$

This ratio is varied by changing the turbine rotational speed or by changing the streamwise velocity. Often tests are conducted at several fixed flow velocities while the turbines rotational speed is varied to give the required TSR range. This generates several performance curves which can be compared to examine repeatability and the effect of testing at different Reynolds numbers. When testing at higher Reynolds numbers (in excess of 5×10^5 [76]), the power, torque and thrust coefficients become independent of Reynolds number, simplifying the scaling process. The radius-based Reynolds number is defined as

$$Re_\infty = \frac{U_\infty R}{\nu}, \tag{7.65}$$

where ν is the fluid kinematic viscosity.

Similar coefficients are defined for vertical-axis turbines and oscillating hydrofoil devices. In these instances the cross-sectional area used, S, is the frontal area of the turbine exposed to the flow, for example the turbine diameter multiplied by the height of model in the case of vertical-axis devices.

Options for varying turbulence properties within laboratory flow were discussed in Section 7.9.2. Investigating the influence of turbulence on performance and loading is strongly recommended. Depending on the planned location site, the influence of flow direction relative to the device and both vertical and horizontal flow profiles across the rotor should also be systematically examined.

7.10.3.2 Wave Interactions

It is important to examine the effect that waves have on a tidal energy device. For devices mounted from floating platforms the importance of examining the effect of waves on performance is clear. However non-floating devices are also affected by waves, especially in terms of fatigue loading [17]. A test programme similar to that for wave energy (Section 7.10.2) is advised, although the specific tests conducted should be informed by how and where a prototype device is to be deployed.

7.10.3.3 Wake

Nearly all tidal energy devices are intended to be deployed in an array. In this case wake recovery behind the device is important. Measurements of wake velocity profiles behind model devices are generally conducted using ADV, LDV or PIV systems. It has been shown that wake recovery is strongly dependent on turbulence intensity within the flow [91]. It is therefore again important to either properly model turbulence with respect to a proposed deployment site, or examine the influence of turbulence intensity on measurements of wake recovery. Wake recovery can be defined in terms of the velocity deficit (U_{def}) with respect to the free-stream flow speed [65]:

$$U_{def} = 1 - \frac{U_c}{U_\infty}, \tag{7.66}$$

where U_c is the wake velocity along the rotor centreline (for a horizontal-axis turbine).

7.10.3.4 Survivability

Chen and Lam [92] reviewed the factors influencing tidal turbine survivability, identifying extreme weather, seabed scour, fatigue failure, corrosion/erosion and marine fouling to be significant influences. Of these factors physical modelling tends to be used to investigate the loading and moments experienced by the blades and drive trains in both operational and extreme conditions. Time-varying and cyclic loads are of particular importance, as this is likely to result in accelerated fatigue to the rotor and blades [93]. Galloway *et al.* [93] have used physical modelling to measure cyclic loading due to the presence of waves and a misalignment of the rotor with the current. Similarly, Gaurier *et al.* [17] have shown that wave–current interaction causes large additional loading when compared to when testing with currents alone, concluding that fatigue performance is dominated by wave contributions. Milne *et al.* [94] conclude that a characterisation of unsteady hydrodynamic loads imparted by waves and turbulence is essential for the prediction of both the fatigue life and the ultimate loads experienced by tidal turbine blades. When modelling tidal stream devices for survivability it is considered critical to consider the influence of waves on blade and drive train loading, alongside variations in current directionality, levels of turbulence and the existence of velocity gradients across the blades.

Loading on the device as a whole is often measured using a six-axis load cell in the mounting of the device within the test facility. The survivability of tidal stream devices incorporating moored floating platforms should be assessed using a test program similar to that discussed in Section 7.10.2 for testing floating wave energy devices.

References

1 Hydraulics Marine Research Centre and Marine Institute of Ireland (2003) Ocean Energy: Development & Evaluation Protocol. Part 1: Wave Power. HMRC, September.

2 B. Holmes (2009) *Tank Testing of Wave Energy Conversion Systems*. Orkney: European Marine Energy Centre.

3 K. Nielsen (2003) Annex II Report 2003. Development of recommended practices for testing and evaluating ocean energy systems. IEA – Ocean Energy Systems, Technical report.

4 S. A. Hughes (1993) *Physical Models and Laboratory Techniques in Coastal Engineering*. Singapore: World Scientic.

5 S. K. Chakrabarti (1994) *Offshore Structure Modeling*. Singapore: World Scientific.

6 Wave Star AS (2012). Wave Star Energy web page. Available: http://wavestarenergy.com/ (accessed 5/11/2015).

7 H. Bailey (2009) The effect of a nonlinear power take off on a wave energy converter. PhD thesis, University of Edinburgh.

8 J. W. Weber and G. P. Thomas (2001) An investigation into the importance of the air chamber design of an oscillating water column wave enerrgy device. In *Proceedings of the Eleventh International Offshore and Polar Engineering Conference (ISOPE)*, Stavanger, Norway, pp. 581–588.

9 A. J. N. A. Sarmento, L. M. C. Gato and A. F. de O. Falcão (1990) Turbine-controlled wave energy absorption by oscillating water column devices. *Ocean Engineering* 17, 481–497.

10 R. P. F. Gomes, J. C. C. Henriques, L. M. C. Gato and A. F. de O. Falcão (2015) Testing of a small-scale model of a heaving floating OWC in a wave channel and comparison with numerical results. In C. Guedes Soares (ed.), *Renewable Energies Offshore*, pp. 445–454. Boca Raton, FL: CRC Press.

11 R. P. F. Gomes, J. C. C. Henriques, L. M. C. Gato and A. F. de O. Falcao (2015) Wave channel tests of a slack-moored floating oscillating water column in regular waves. In *Proceedings of the Eleventh European Wave and Tidal Energy Conference (EWTEC)*, Nantes, France, p. 9.

12 G. Payne (2008) Guidance for the experimental tank testing of wave energy converters. Technical report, University of Edinburgh.

13 ITTC (2002) Recommended procedures: Ship models ITTC Report 7.5-01-01-01. International Towing Tank Conference.

14 B. Elsaesser, P. Schmitt, C. Boake, A. Lidderdale, F. Salvatore, G. Germain *et al.* (2013) Tidal Measurement Best Practice Manual, Deliverable 2.7. Marine Renewables Infrastructure Network, Work Package 2 of the Marinet Project.

15 D. A. Doman, R. E. Murray, M. J. Pegg, K. Gracie, C. M. Johnstone and T. Nevalainen (2015) Tow-tank testing of a 1/20th scale horizontal axis tidal turbine with uncertainty analysis. *International Journal of Marine Energy* 11, 105–119.

16 P. W. Galloway (2013) Performance quantification of tidal turbines subjected to dynamic loading. PhD thesis, Faculty of Engineering and the Environment, University of Southampton.

17 B. Gaurier, P. Davies, A. Deuff and G. Germain (2013) Flume tank characterization of marine current turbine blade behaviour under current and wave loading. *Renewable Energy* 59, 1–12.

18 M. T. Morris-Thomas, R. J. Irvin and K. P. Thiagarajan (2007) An investigation into the hydrodynamic efficiency of an oscillating water column. *Journal of Offshore Mechanics and Arctic Engineering* 129, 273–278.

19 K. Thiruvenkatasamy and S. Neelamani (1997) On the efficiency of wave energy caissons in array. *Applied Ocean Research* 19, 61–72, 2.

20 B. Kooverji (2014) Pneumatic power measurement of an oscillating water column converter. Master's thesis, Faculty of Engineering, Stellenbosch University.

21 A. Hayward (1979) *Flowmeters: A Basic Guide and Source-Book for Users*. London: Macmillan, 1979.

22 R. S. Figliola and D. E. Beasley (1991) *Theory and Design for Mechanical Measurements*. New York: John Wiley & Sons.

23 ISO (1991) *Measurement of Fluid Flow by Means of Pressure Differential Devices*, ISO 5167.

24 COAST Laboratory (2015) COAST Laboratory image repository, Plymouth University.

25 A. El Marjani, F. Castro Ruiz, M. A. Rodriguez and M. T. Parra Santos (2008) Numerical modelling in wave energy conversion systems. *Energy* 33(8), 1246–1253.

26 A. J. N. A. Sarmento (1993) Model-test optimization of an OWC wave power plant, *International Journal of Offshore and Polar Engineering* 3, 66–72.

27 J. R. Chaplin, V. Heller, F. J. M. Farley, G. E. Hearn and R. C. T. Rainey (2012) Laboratory testing the Anaconda. *Philosophical Transactions of the Royal Society of London A*, 370, 403–424.

28 M. R. Hann (2013) A numerical and experimental study of a multi-cell fabric distensible wave energy converter. PhD thesis, Faculty of Engineering and the Environment, University of Southampton.

29 M. Folley, T. J. T. Whittaker and A. Henry (2007) The effect of water depth on the performance of a small surging wave energy converter. *Ocean Engineering* 34(8), 1265–1274.

30 P. Stansby, E. C. Moreno, T. Stallard and A. Maggi (2015) Three-float broad-band resonant line absorber with surge for wave energy conversion. *Renewable Energy* 78, 132–140.

31 P. Stansby, E. C. Moreno and T. Stallard (2015) Capture width of the three-float multi-mode multi-resonance broadband wave energy line absorber M4 from laboratory studies with irregular waves of different spectral shape and directional spread. *Journal of Ocean Engineering and Marine Energy* 1, 287–298.

32 F. Flocard and T. D. Finnigan (2012) Increasing power capture of a wave energy device by inertia adjustment. *Applied Ocean Research* 34, 126–134.

33 M. Folley, T. Whittaker and M. Osterried (2004) The oscillating wave surge converter. Paper presented to the Fourteenth International Offshore and Polar Engineering Conference, 23–28 May, Toulon, France.

34 M. Vantorre, R. Banasiak and R. Verhoeven (2004) Modelling of hydraulic performance and wave energy extraction by a point absorber in heave. *Applied Ocean Research* 26, 61–72.

35 M. Göteman, J. Engström, M. Eriksson, M. Hann, E. Ransley, D. Greaves *et al.* (2015) Wave loads on a point-absorbing wave energy device in extreme waves. *Journal of Ocean and Wind Energy* 2, 176–181.

36 J. C. C. Henriques, M. F. P. Lopes, M. C. Lopes, L. M. C. Gato and A. Dente (2011) Design and testing of a non-linear power take-off simulator for a bottom-hinged plate wave energy converter. *Ocean Engineering* 38, 1331–1337.

37 H. Fernandez, G. Iglesias, R. Carballo, A. Castro, J. A. Fraguela, F. Taveira-Pinto *et al.* (2012) The new wave energy converter WaveCat: Concept and laboratory tests. *Marine Structures* 29, 58–70.

38 COAST Research Group (2014) Coastal, Ocean and Sediment Transport Research Group (COAST) web page. https://collaborate.plymouth.ac.uk/sites/cerg/Pages/COAST.aspx (accessed 5/3/2014).

39 S. Weller, L. Johanning and P. Davies (2013) Best practice report – mooring of floating marine renewable energy devices. *MERiFIC Project Deliverable Report* 3.5.3.

40 M. Hann, D. Greaves and A. Raby (2015) Snatch loading of a single taut moored floating wave energy converter due to focussed wave groups. *Ocean Engineering* 96, 258–271.

41 S. K. Chakrabarti (2005) Physical Modelling of Offshore Structures. In S. K. Chakrabarti (ed.), *Handbook of Offshore Engineering*, pp. 1001–1054. London: Elsevier.

42 S. Beatty, B. Buckham and P. Wild (2013) Experimental comparison of self-reacting point absorber WEC designs. In *Proceedings of the Tenth European Wave and Tidal Conference (EWTEC)*, Aalborg, Denmark.

43 J. Lawrence, B. Holmes, I. Bryden, D. Magagna, Y. Torre Encirso, J. M. Rousset *et al.* (2012) D2.1 Wave instrumentation database Rev. 05. Work Package 2 of the Marinet Project.

44 S. Bourdier, K. Dampney, H. Fernandez, G. Lopez and J. B. Richon (2014) Non-intrusive wave field measurement Rev. 4. Work Package 4 of the Marinet Project.

45 C. Perez and G. Iglesias (2012) Physical modelling of an offshore OWC wave energy converter mounted on a windmill monopile foundation, In *Proceedings of the International Conference on the Application of Physical Modelling to Port and Coastal Protection (COASTLAB)*, Ghent, Belgium.

46 K. H. Christensen, J. Röhrs, B. Ward, I. Fer, G. Broström, Ø. Saetra *et al.* (2013) Surface wave measurements using a ship-mounted ultrasonic altimeter. *Methods in Oceanography* 6, 1–15.

47 Y. Goda and T. Suzuki (1976) Estimation of incident and reflected waves in random wave experiments. *Proceedings of the 15th International Conference on Coastal Engineering*, pp. 828–845.

48 P. Frigaard and M. Brorsen (1995) A time-domain method for separating incident and reflected irregular waves. *Coastal Engineering* 24(3–4), 205–215,.

49 E. P. Mansard and E. R. Funke (1980) The measurement of incident and reflected spectra using a least squares method. In *Proceedings of the 17th International Conference on Coastal Engineering.*

50 A. Baquerizo, M. A. Losada and J. M. Smith (1998) Wave reflection from beaches: A predictive model. *Journal of Coastal Research* 14, 291–298.

51 J. A. Zelt and J. E. Skjelbreia (1992) Estimating incident and reflected waves fields using an arbitrary number of wave gauges. *Proceedings of the 23rd International Conference on Coastal Engineering.*

52 E. B. Thornton and R. J. Calhoun (1972) Spectral resolution of breakwater reflected waves. *Journal of the Waterways, Harbors and Coastal Engineering Division* 98, 443–460.

53 K. Horikawa (1988) *Nearshore Dynamics and Coastal Processes: Theory, Measurement and Predictive Models.* Tokyo: University of Tokyo Press.

54 M. Benoit, P. Frigaard and H. A. Schäffer (1997) Analysing multidirectional wave spectra: A tentative classification of available methods. In *IAHR Seminar Multidirectional Waves and their Interaction with Structures*, San Francisco, pp. 131–158.

55 N. Hashimoto (1997) Analysis of the directional wave spectrum from field data. In P. Liu (ed.), *Advances in Coastal and Ocean Engineering.* vol. 3, pp. 103–144. Singapore: World Scientific.

56 N. Hashimoto, K. Kobune and Y. Kameyama (1987) Estimation of directional spectrum using the Bayesian approach, and its application to field data analysis. Report of the Port and Harbour Research Institute, Yokosuka, Japan.

57 M. Benoit and C. Teisson (1994) Laboratory comparison of directional wave measurement systems and analysis techniques. *Proceedings of the 24th International Conference on Coastal Engineering.*

58 I. R. Young (1994) On the measurement of directional wave spectra. *Applied Ocean Research* 16, 283–294.

59 J. Capon (1969) High-resolution frequency-wavenumber spectrum analysis. *Proceedings of the IEEE* 57, 1408–1418.

60 T. E. Barnard (1969) Analytical studies of techniques for the computation of high-resolution wavenumber spectra. Dallas Science Services Division, Texas Instruments Inc.

61 P. J. Rusello (2009) A practical primer for pulse coherent instruments. Nortek Technical Note No. TN-027.

62 L. Cea, J. Puertas and L. Pena (2007) Velocity measurements on highly turbulent free surface flow using ADV. *Experiments in Fluids* 42, 333–348.

63 T. Stallard, R. Collings, T. Feng and J. Whelan (2013) Interactions between tidal turbine wakes: Experimental study of a group of three-bladed rotors. *Philosophical Transactions of the Royal Society of London A* 371(1985).

64 S. C. Tedds, I. Owen and R. J. Poole (2014) Near-wake characteristics of a model horizontal axis tidal stream turbine. *Renewable Energy*, 63, 222–235.

65 F. Maganga, G. Germain, J. King, G. Pinon and E. Rivoalen (2010) Experimental characterisation of flow effects on marine current turbine behaviour and on its wake properties. *IET Renewable Power Generation* 4(6), 498–509.

66 I. López, A. Castro and G. Iglesias (2015) Hydrodynamic performance of an oscillating water column wave energy converter by means of particle imaging velocimetry. *Energy* 83, 89–103.

67 C. G. Lomas (1986) *Fundamentals of Hot Wire Anemometry*. Cambridge: Cambridge University Press.

68 A. Têtu, P. Frigaard, J. P. Kofoed, M. F. P. Lopes, A. Iyer, S. Bourdier *et al.* (2013) D2.5 Report on instrumentation best practice. Work Package 2 of the Marinet Project.

69 P. Davies, G. Germain, B. Gaurier, A. Boisseau and D. Perreux (2013) Evaluation of the durability of composite tidal turbine blades. *Philosophical Transactions of the Royal Society of London A* 371(1985).

70 A. S. Bahaj, A. F. Molland, J. R. Chaplin and W. M. J. Batten (2007) Power and thrust measurements of marine current turbines under various hydrodynamic flow conditions in a cavitation tunnel and a towing tank. *Renewable Energy* 32, 407–426.

71 D. O. Guaraglia and J. L. Pousa (2014) *Introduction to Modern Instrumentation: For Hydraulics and Environmental Sciences*. Warsaw: Walter de Gruyter.

72 P. F. Hinrichsen (2014) Bifilar suspension measurement of boat inertia parameters. *Journal of Sailboat Technology* 1, 1–37.

73 W. W. Massie and J. M. Journée (2001) Offshore hydromechanics. Delft University of Technology.

74 G. S. deBree (2012) Testing and modeling of high solidity cross-flow tidal turbines. Master's thesis, Mechanical Engineering, University of Maine.

75 P. Bachant and M. Wosnik (2016) Effects of Reynolds number on the energy conversion and near-wake dynamics of a high solidity vertical-axis cross-flow turbine. *Energies*, 9, 73.

76 A. Mason-Jones, D. M. O'Doherty, C. E. Morris, T. O'Doherty, C. B. Byrne, P. W. Prickett *et al.* (2012) Non-dimensional scaling of tidal stream turbines. *Energy* 44, 820–829.

77 B. Le Méhauté (1976) *Introduction to Hydrodynamics and Water Waves*. New York: Springer.

78 P. S. Tromans, A. R. Anaturk and P. Hagemeijer (1991) A new model for the kinematics of large ocean waves-application as a design wave. Paper presented to the First International Offshore and Polar Engineering Conference, 11–16 August, Edinburgh.

79 M. J. Cassidy (1999) Non-linear analysis of jack-up structures subjected to random waves. DPhil. thesis, University of Oxford.

80 T. Blackmore, W. M. J. Batten, G. U. Müller and A. S. Bahaj (2013) Influence of turbulence on the drag of solid discs and turbine simulators in a water current. *Experiments in Fluids* 55, 1637.

81 F. Biesel and F. Suquet (1951) Les appareils générateurs de houle en laboratoire. *La Houille Blanche* 2, 723–737.

82 J. Cruz (2008) *Ocean Wave Energy: Current Status and Future Perspectives*. Berlin: Springer.

83 S. Tedds (2014) Scale model testing of tidal stream turbines: Wake characterisation in realistic flow conditions. PhD thesis, University of Liverpool.

84 P. Mycek, B. Gaurier, G. Germain, G. Pinon, and E. Rivoalen (2014) Experimental study of the turbulence intensity effects on marine current turbines behaviour. Part I: One single turbine. *Renewable Energy* 66, 729–746.

85 L. E. Myers, K. D. Shah, and P. W. Galloway (2013) Design, commissioning and performance of a device to vary the turbulence in a recirculating flume. In *Proceedings of the Tenth European Wave and Tidal Energy Conference (EWTEC)*, Aalborg, Denmark.

86 R. Cavagnaro and B. Polagye (2014) An evaluation of blockage corrections for a helical cross-flow turbine. In *Proceedings of the Third Oxford Tidal Energy Workshop*.

87 M. F. P. Lopes (2011) Experimental development of offshore wave energy converters. PhD thesis, Instituto Superior Técnico, Lisbon.

88 I. López, B. Pereiras, F. Castro and G. Iglesias (2016) Holistic performance analysis and turbine-induced damping for an OWC wave energy converter. *Renewable Energy* 85, 1155–1163.

89 M. M. Jakobsen, S. Beatty, G. Iglesias and M. M. Kramer (2016) Characterization of loads on a hemispherical point absorber wave energy converter. *International Journal of Marine Energy*, 13, 1–15.

90 Y. Goda (2010) *Random Seas and Design of Maritime Structures*. Singapore: World Scientific.

91 L. E. Myers and A. S. Bahaj (2010) Experimental analysis of the flow field around horizontal axis tidal turbines by use of scale mesh disk rotor simulators. *Ocean Engineering* 37, 218–227.

92 L. Chen and W. H. Lam (2015) A review of survivability and remedial actions of tidal current turbines. *Renewable and Sustainable Energy Reviews* 43, 891–900.

93 P. W. Galloway, L. E. Myers and A. S. Bahaj (2014) Quantifying wave and yaw effects on a scale tidal stream turbine. *Renewable Energy* 63, 297–307.

94 I. A. Milne, A. H. Day, R. N. Sharma and R. G. J. Flay (2013) Blade loads on tidal turbines in planar oscillatory flow. *Ocean Engineering* 60, 163–174.

8

Numerical Modelling

Thomas Vyzikas[a] and Deborah Greaves[b]

[a] *Associate Researcher, School of Engineering, University of Plymouth, UK*
[b] *Professor of Ocean Engineering, School of Engineering, University of Plymouth, UK*

8.1   Introduction

The term 'modelling' refers to the representation of a physical system, commonly undertaken by means of physical or numerical models. Physical modelling refers to experimental representation of a natural phenomenon usually at a smaller scale in a laboratory, while numerical modelling is the mathematical representation of the physical processes included in a natural phenomenon, based on certain assumptions. The purpose of modelling is the prediction of the behaviour of the system in response to a change in its conditions. Often modelling is used for examining the response of the system in extreme conditions, which occur rarely in nature, such as fatigue and failure modes, storms and tsunamis, and for estimating its long-term behaviour, such as the shoreline evolution over the years or the annual energy production of a wave farm.

Traditionally, the modelling of natural phenomena has been performed with physical models, due to the complexity of the processes occurring and the lack of advanced computation tools. However, physical modelling is often limited to scaled models of natural systems based on the laws of dynamic similarity [1] and might introduce important uncertainties in the results of the model for certain types of problems. On the other hand, the increasing processing power and emergence of sophisticated numerical tools have led to an increase in the application of numerical models, which, generally speaking, are not subject to scaling effects. Nowadays, numerical models are widely applied for a large variety of problems, thanks to their versatility and range of options offered to users. Despite their success, numerical models should always be employed with care and often hand-in-hand with physical model data. The strengths and weaknesses of numerical models are discussed here in order to highlight the potential advantages and disadvantages of numerical modelling compared to physical modelling [2].

One of the strong points of numerical modelling is the existence of large validation databases, which include at least the basic cases and provide confidence in the use of the numerical codes for further applications, although for new applications further validation may be required. The main reason why numerical modelling became such a popular tool is its low cost compared to physical model testing, since costly facilities are not required. Different configurations can easily be tested and simulations can be

Wave and Tidal Energy, First Edition. Edited by Deborah Greaves and Gregorio Iglesias.
© 2018 John Wiley & Sons Ltd. Published 2018 by John Wiley & Sons Ltd.

performed even a long time after the initial study, provided that the software is still supported. The human resources required to perform a numerical study are minimal and the user can even operate the model remotely. Another practical aspect is the use of non-intrusive methods for extracting the data, which is rarely the case for physical models, and the high density of flow field data, which offers an integrated view of the problem and is almost impossible with physical modelling.

On the downside, most of the numerical models rely on parameterisations of the physical processes and assumptions related to the conditions of the problem, which restrict the range of applicability of the models to certain cases. Outside this range, the numerical results might be inaccurate or completely unrealistic. The numerical result is also subject to the quality of the computational domain, e.g. number of grid points and treatment of complex geometries, and the temporal discretisation, which can make the results grid- and time-step-dependent.

Numerical modelling has gained a crucial role in the design process and is established as a rapidly developing sector. A major weaknesses has been, in some cases, its high computational cost, which limits long simulations of large domains with complex solvers. This has begun to be mitigated through the increasing processing power of personal computers and the use of high-performance computer (HPC) facilities. Present-day numerical models offer optimised parallelising algorithms for efficient computations. New horizons in numerical modelling can open up with the use of open-source software, which promotes the sharing of knowledge and valuable contributions from user communities. Opportunities also arise through extensive validation of the models with detailed experimental results obtained with advanced measurement techniques, both in the laboratory and in the field, in order to gain reliability and confidence for using the codes.

As demonstrated, the development of numerical models is strongly bound with advances in physical modelling. Complex physical problems should be approached by means of both modelling techniques using the most appropriate tools for each case. A new modelling methodology has been proposed recently, namely composite modelling, which refers to the integrated and balanced use of physical and numerical modelling for mitigating the weaknesses and exploiting the strengths of the two classic modelling methods [2].

The design of marine renewable energy (MRE) projects combines the use of physical models, numerical models, field studies and sea demonstration of prototype devices. MRE projects require multidisciplinary approaches, since they incorporate classic offshore and coastal engineering studies with the problem of energy generation, storage and distribution. Possible coastal and environmental impacts make MRE projects very challenging. At the same time, there are many different types of MRE devices, which hinders the adoption of general practices and modelling techniques. For instance, the modelling of wave energy converters (WECs) is essentially similar to other marine structures, while tidal and offshore wind energy have their roots to the onshore wind turbine sector [3]. The different design aspects of an MRE project are depicted in Figure 8.1 in a simplified way, since most of these aspects are interconnected. For example, the design of moorings is associated with the survivability of a device, but in combination with the power take-off (PTO) mechanism, moorings control the hydrodynamic response of the device and its performance in arrays. Subsequently, the performance is associated with the resource characteristics for optimising the device.

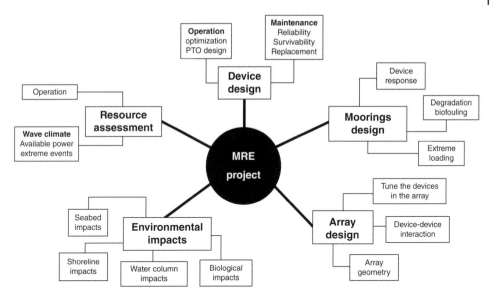

Figure 8.1 Technical aspects of an MRE project [4].

Biofouling on the mooring or on the device itself alters the response of the mooring, which affects the performance of the device and affects the vortex-induced vibrations that may cause disturbance in the water column, with possible environmental effects.

The role of numerical modelling in an MRE project is vital at all stages, especially when the project is at technology readiness level (TRL) 4–5 or higher [5]. However, numerical modelling can start much earlier at the stage of the proof of the operational concept of a device. As shown in Figure 8.1, the complexity of the problem can include a wide range of numerical modelling techniques related to the different processes. The hydrodynamics of the water flow, such as currents, waves, tides and their interaction with the structure, play a significant role from an engineering perspective and are always an essential part of the entire modelling process. Therefore, in this chapter, the main focus is the numerical modelling of water flows [6] and the presentation of a large variety of numerical modelling techniques, including well-established and recently developed methods and practices in industry and research.

This chapter starts with a description of the basic equations of fluid mechanics and continues with specific numerical models developed for the simulation of water flows in oceanic and coastal waters. Building on this fundamental knowledge, one can combine different models to simulate complex problems, such as sediment dynamics processes, fluid–structure interaction problems and wave-to-wire problems. Then a description of commonly used open-source software for MRE-related simulations is presented, in order to suggest freely available and validated numerical tools to potential modellers. In the final section of the chapter, tables on the applicability of numerical models for simulating different physical processes are presented, with a focus on MRE projects. In this sense, these tables can be used as guidelines and decision-making tools, according to the physical problem, available expertise and computational resources, and the required level of accuracy.

8.2 Review of Hydrodynamics

8.2.1 The Primitive Equations of Fluid Mechanics

The basic equations that govern the fluid motion are mathematical statements of the conservation laws of physics. There are various formulations of these equations depending on the variables used to describe the flow [7]. If the fluid is considered as a continuum and the analysis neglects the molecular level, the fluid flow can be described by macroscopic variables, such as velocity U, pressure p, density ρ and temperature T and their spatial and temporal derivatives [8].

The conservation laws of physics that constitute the primitive equations of fluid mechanics are: the conservation of mass or continuity equation; the conservation of momentum or Newton's second law; and the conservation of energy or first law of thermodynamics. These three conservation laws are commonly linked by the equation of state for achieving thermodynamic equilibrium, comprising the fundamental equations of fluid mechanics.

The detailed derivation of the equations and the different forms that they can take can be found in the extensive relevant literature. The variations of the equations arise from the different considerations of the fluid continuum. Here, a brief description of the governing equations is presented with reference to a stationary fluid element or particle, which is considered as an infinitesimally small part of the fluid, whose macroscopic properties are not influenced by individual molecules. Practically, the fluid element is a very small control volume in three-dimensional space that contains a suitable number of molecules and the variables of the flow can be examined [8].

8.2.1.1 Mass Conservation

The mass conservation or continuity equation states that the rate of increase of mass in a fluid element equals the net rate of mass flow into the fluid element. In other words, there is no production or destruction of mass of fluid in the fluid element.

Therefore, since the volume of the fluid element is constant, the rate of change of the mass inside the fluid element can only be a result of changes in the fluid density ρ and the flux in the element. The rate of increase of mass due to inflow or outflow into and out of the fluid element is expressed as the product of the fluid density, area of each face of the fluid element and the velocity component of the fluid normal to each face. Assuming that the area of the faces is constant, the flow is a product of density and velocity. Consequently, the continuity equation can be expressed in a longhand and compact notation respectively as

$$\frac{\partial \rho}{\partial t} + \frac{\partial \rho u}{\partial x} + \frac{\partial \rho v}{\partial y} + \frac{\partial \rho w}{\partial z} = 0, \tag{8.1a}$$

$$\frac{\partial \rho}{\partial t} + \nabla \cdot \left(\rho \vec{V} \right) = 0, \tag{8.1b}$$

where u, v, w refer to the velocity components in the x, y, z direction respectively and $\vec{V}$ to the velocity vector.

Equations (8.1a) and (8.1b) can describe the conservation of mass for any unsteady compressible or incompressible fluid flow. For many common applications, the assumption of incompressibility is considered, which suggests that the density of the fluid is constant. This assumption reduces the continuity equation to the incompressible mass conservation equation,

$$\nabla \cdot \vec{V} = 0. \tag{8.2}$$

8.2.1.2 Momentum conservation

According to Newton's second law of motion, the sum of the forces acting on a fluid particle is equal to the rate of change of its momentum. This is commonly expressed in a mathematical form as $\sum F = \dfrac{d(mU)}{dt}$, where mU is the acceleration of the object.

The forces acting on a fluid particle can be divided into surface forces and body forces. Surface forces are imposed on the surface of the fluid element by the surrounding fluid and refer to pressure and viscous forces. Pressure, p, results in normal stress distribution on the element faces, while viscous stresses result in shear and normal stresses on the faces, which are denoted by τ_{ij}. The suffixes i and j in τ_{ij} indicate that the stress component acts in the j-direction on a surface normal to the i-direction. The resultant surface force on the fluid element in the x-direction is given by

$$\frac{\partial(-p + \tau_{xx})}{\partial x} + \frac{\partial \tau_{yx}}{\partial y} + \frac{\partial \tau_{zx}}{\partial z}. \tag{8.3}$$

Similar expressions can be written for the other directions.

Body forces are imposed on the centre of mass of the fluid element without requiring direct contact with other bodies. Common body forces are gravity, centrifugal force and Coriolis force [8]. In practice, body forces are included as source terms in the equations, while surface forces are included as distinguished terms S_{M_j}, with j being the direction of the force.

The rate of change of the x-, y- and z-component of the momentum is denoted by $\rho \dfrac{Du}{Dt}$, $\rho \dfrac{Dv}{Dt}$ and $\rho \dfrac{Dw}{Dt}$ respectively, and it is equal to the sum of the body and surface forces in the corresponding directions. This yields the momentum conservation equations in the three directions [9]:

$$\rho \frac{Du}{Dt} = -\frac{\partial p}{\partial x} + \frac{\partial \tau_{xx}}{\partial x} + \frac{\partial \tau_{yx}}{\partial y} + \frac{\partial \tau_{zx}}{\partial z} + S_{M_x}, \tag{8.4a}$$

$$\rho \frac{Dv}{Dt} = -\frac{\partial p}{\partial y} + \frac{\partial \tau_{xy}}{\partial x} + \frac{\partial \tau_{yy}}{\partial y} + \frac{\partial \tau_{zy}}{\partial z} + S_{M_y}, \tag{8.4b}$$

$$\rho \frac{Dw}{Dt} = -\frac{\partial p}{\partial z} + \frac{\partial \tau_{xz}}{\partial x} + \frac{\partial \tau_{yz}}{\partial y} + \frac{\partial \tau_{zz}}{\partial z} + S_{M_z}. \tag{8.4c}$$

It should be noted that equations (8.4) are presented in a non-conservative form. The momentum equations are called the Navier–Stokes equations (NSE) in honour of M. Navier and G. Stokes, who derived them independently two centuries ago [9].

8.2.1.3 Energy Conservation

The first law of thermodynamics refers to the conservation of energy, which for a fluid element states that 'the rate of change of energy in a fluid element is equal to the sum of the net flux of heat added to the fluid element plus the net rate of work done by external forces on the fluid element' [9].

The net rate of *heat transfer* to the fluid particle can be calculated by the heat flux vector $\vec{q}$, which represents the rate of heat added to the fluid element per unit volume due to heat flow across its faces:

$$\frac{\partial q_x}{\partial x} + \frac{\partial q_y}{\partial y} + \frac{\partial q_z}{\partial z} = \nabla \cdot \vec{q}. \tag{8.5}$$

According to Fourier's law of heat conduction, heat flux is related to the local gradients of temperature via the thermal conductivity κ:

$$q_x = -\kappa \frac{\partial T}{\partial x}, \quad q_y = -\kappa \frac{\partial T}{\partial y}, \quad q_z = -\kappa \frac{\partial T}{\partial z} \quad \text{or} \quad \vec{q} = -\kappa \nabla T, \tag{8.6}$$

where T is the temperature and q_x, q_y, q_z the heat fluxes in the corresponding directions. Consequently, the total rate of heat transferred to the fluid element is given by combining equations (8.5) and (8.6), yielding

$$\nabla \cdot \vec{q} = -\nabla \cdot \left(\kappa \nabla T \right). \tag{8.7}$$

The total rate of *work* done per unit volume on the fluid particle is calculated as the sum of the net rate of work of all the surface forces and body forces in each direction. For the case of surface forces, the rate of change of work per unit volume can be expressed, after some algebra, as [8]

$$-\nabla \cdot \left(p\vec{V} \right) + \frac{\partial \left(u\tau_{xx} \right)}{\partial x} + \frac{\partial \left(u\tau_{yx} \right)}{\partial y} + \frac{\partial \left(u\tau_{zx} \right)}{\partial z} + \frac{\partial \left(v\tau_{xy} \right)}{\partial x} + \frac{\partial \left(u\tau_{yy} \right)}{\partial y} + \frac{\partial \left(u\tau_{zy} \right)}{\partial z}$$
$$+ \frac{\partial \left(w\tau_{xz} \right)}{\partial x} + \frac{\partial \left(w\tau_{yz} \right)}{\partial y} + \frac{\partial \left(w\tau_{zz} \right)}{\partial z}. \tag{8.8}$$

The work done by body forces can be included in the energy equation as a source term, namely S_E, which is defined as the rate of change of work per unit volume per unit time due to body forces. As the fluid element moves in a force field, it stores potential energy. For instance, if gravity is the acting body force, the fluid element stores gravitational potential energy [8].

Summing up the work done by heat, surface and body forces yields the rate of change of a fluid element per unit volume:

$$\rho \frac{DE}{Dt} = -\nabla \cdot \left(p\vec{V} \right) + \frac{\partial \left(u\tau_{xx} \right)}{\partial x} + \frac{\partial \left(u\tau_{yx} \right)}{\partial y} + \frac{\partial \left(u\tau_{zx} \right)}{\partial z} + \frac{\partial \left(v\tau_{xy} \right)}{\partial x} + \frac{\partial \left(u\tau_{yy} \right)}{\partial y}$$
$$+ \frac{\partial \left(u\tau_{zy} \right)}{\partial z} + \frac{\partial \left(w\tau_{xz} \right)}{\partial x} + \frac{\partial \left(w\tau_{yz} \right)}{\partial y} + \frac{\partial \left(w\tau_{zz} \right)}{\partial z} + \nabla \cdot \left(\kappa \nabla T \right) + S_E, \tag{8.9}$$

where E is the total energy given as a sum of the internal energy i, kinetic energy $\frac{1}{2}(u^2 + v^2 + w^2)$ and potential energy.

Equation (8.9) can be used to derive the internal energy equation or the kinetic energy equation [8], which are useful forms of the energy equation for practical applications. Further simplifications allow the expression of the internal energy equation in terms of specific heat c and temperature T, by considering $i = cT$ for incompressible flows. Moreover, the equation for enthalpy for compressible flows can be derived from equation (8.9) by employing the definition of the total enthalpy h_0:

$$h_0 = i + p/\rho + \frac{1}{2}\left(u^2 + v^2 + w^2\right) = E + p/\rho. \tag{8.10}$$

8.2.1.4   Equations of State

As mentioned, the motion of a fluid can be described by the mass, momentum and energy conservation equations that have as unknowns the thermodynamic variables of pressure p, density ρ, internal energy i and temperature T. These variables are linked together through the equations of state.

A necessary assumption for employing the equations of state is to consider the fluid in thermodynamic equilibrium. This assumption is valid if the changes in the transport properties of the fluid occur in such time scales that allow the fluid to thermodynamically adjust itself to the new conditions. This assumption is only violated for very fast flows, for example shock waves. Therefore, for most common applications the fluid can adjust its thermodynamic state almost instantaneously and remain in thermodynamic equilibrium [8].

The equations of state can be expressed in terms of two variables only. If ρ and T are used as independent variables, the other two variables can be calculated as a function of them: $p = p(\rho,T)$ and $i = i(\rho,T)$. For perfect gases the exact expression for the equation of state is given by

$$p = \rho R T \quad \text{and} \quad i = C_v T, \tag{8.11}$$

where R is the universal gas constant and C_v is the specific heat capacity.

The equations of state are useful for compressible flows, in order to provide the link between the mass and momentum conservation equations on the one hand and the energy equation on the other hand. This interconnection between the equations is important, because in compressible flows variations of density might occur, due to changes in pressure and temperature. In incompressible flows, though, there are no density variations and therefore there is no need to use the equations of state. In these cases, the flow can be described adequately by the mass and momentum conservation equations. The energy conservation equation is not required unless heat transfer takes place in the problem [8].

8.2.2   The Navier–Stokes Equations

The momentum conservation equations are known as the Navier–Stokes equations, and here they will be presented in their conservative form, complementing equations (8.4) [9].

Using the definition of the substantial derivative, the rate of change of the momentum in the x-direction can be written as

$$\rho\frac{Du}{Dt} = \rho\frac{\partial u}{\partial t} + \rho\vec{V}\cdot\nabla u. \tag{8.12}$$

Employing the properties of the partial derivatives, $\rho\dfrac{\partial u}{\partial t}$ can be expanded as follows:

$$\frac{\partial(\rho u)}{\partial t} = \rho\frac{\partial u}{\partial t} + u\frac{\partial \rho}{\partial t} \;\Rightarrow\; \rho\frac{\partial u}{\partial t} = \frac{\partial(\rho u)}{\partial t} - u\frac{\partial \rho}{\partial t}. \tag{8.13}$$

Additionally, the product of velocity vector and the divergence of the x-velocity of equation (8.12) can be rewritten as

$$\nabla\cdot(\rho u\vec{V}) = u\nabla\cdot(\rho\vec{V}) + (\rho\vec{V})\cdot\nabla u \;\Rightarrow\; \rho\vec{V}\cdot\nabla u = \nabla\cdot(\rho u\vec{V}) - u\nabla\cdot(\rho\vec{V}). \tag{8.14}$$

Substituting equations (8.13) and (8.14) into equation (8.12) yields an expression which contains the continuity equation (8.1a) and can be further simplified to give

$$\rho\frac{Du}{Dt} = \frac{\partial(\rho u)}{\partial t} - u\frac{\partial\rho}{\partial t} + \nabla\cdot(\rho u\vec{V}) - u\nabla\cdot(\rho\vec{V}) \Rightarrow \rho\frac{Du}{Dt} = \frac{\partial(\rho u)}{\partial t} + \nabla\cdot(\rho u\vec{V}). \tag{8.15}$$

The same mathematical formulations can be derived for the y- and z-directions and, when substituted into equation (8.4), yield the conservation form of the NSE:

$$\frac{\partial(\rho u)}{\partial t} + \nabla\cdot(\rho u\vec{V}) = -\frac{\partial p}{\partial x} + \frac{\partial\tau_{xx}}{\partial x} + \frac{\partial\tau_{yx}}{\partial y} + \frac{\partial\tau_{zx}}{\partial z} + S_{M_x}, \tag{8.16a}$$

$$\frac{\partial(\rho v)}{\partial t} + \nabla\cdot(\rho v\vec{V}) = -\frac{\partial p}{\partial y} + \frac{\partial\tau_{xy}}{\partial x} + \frac{\partial\tau_{yy}}{\partial y} + \frac{\partial\tau_{zy}}{\partial z} + S_{M_y}, \tag{8.16b}$$

$$\frac{\partial(\rho w)}{\partial t} + \nabla\cdot(\rho w\vec{V}) = -\frac{\partial p}{\partial z} + \frac{\partial\tau_{xz}}{\partial x} + \frac{\partial\tau_{yz}}{\partial y} + \frac{\partial\tau_{zz}}{\partial z} + S_{M_z}. \tag{8.16c}$$

The essential element for solving equations (8.16) is the definition of the normal stresses (τ_{ii}) and shear stresses ($\tau_{ij}, i \neq j$). Newton suggested that the stress in a fluid is proportional to the rate of strain, which is expressed by the velocity gradient. When this assumption is valid, the fluid is called Newtonian. For many common applications, including water wave mechanics, the fluid can be considered Newtonian and the viscous stresses can be given by the constitutive equations, as suggested by Stokes in 1845 [9]:

$$\tau_{xx} = \lambda\nabla\cdot\vec{V} + 2\mu\frac{\partial u}{\partial x}, \quad \tau_{yy} = \lambda\nabla\cdot\vec{V} + 2\mu\frac{\partial v}{\partial y}, \quad \tau_{zz} = \lambda\nabla\cdot\vec{V} + 2\mu\frac{\partial w}{\partial z}$$
$$\tau_{xy} = \tau_{yx} = \mu\left(\frac{\partial v}{\partial x} + \frac{\partial u}{\partial y}\right), \tau_{xz} = \tau_{zx} = \mu\left(\frac{\partial u}{\partial z} + \frac{\partial w}{\partial x}\right), \tau_{yz} = \tau_{zy} = \mu\left(\frac{\partial w}{\partial y} + \frac{\partial v}{\partial z}\right), \tag{8.17}$$

where λ and μ are two constants, namely the bulk viscosity coefficient and the dynamic viscosity coefficient, respectively. μ relates the stresses to the linear deformations and λ

relates the stresses to the volumetric deformation. Little is known about λ, because its effect is small and if the fluid is incompressible ($\nabla \cdot \vec{V} = 0$), the effect of λ is not accounted for in the equations of the stresses [8]. A good approximation for common applications assumes $\lambda = -\frac{2}{3}\mu$ [10].

The so-called 'complete' NSE for Newtonian fluids can be obtained by substituting the viscous stresses (equation (8.17)) into the conservation momentum relation (equation (8.16)). The NSE are considered the fundamental equations of fluid mechanics that can describe any fluid flow.

Even though, historically speaking, the NSE refer to the momentum conservation equations, in modern computational fluid dynamics (CFD) terminology, Navier–Stokes solvers refer to the full system of the equations that describe the fluid motion (mass, momentum, energy conservation and state equations). Thus, in practice 'Navier–Stokes' simulations imply solution of viscous flows using all the equations of fluid motion [9]. However, solving the full NSE and accounting for all the scales of turbulence requires very complex numerical models. As a result, additional simplifications are considered, which are discussed in the next paragraphs.

In this sense, it is useful to derive a general form of partial differential equations (PDEs) of any fluid variable Φ that can be reduced to any of the conservation equations with appropriate modifications [8]. The generalised form of the transport equation is expressed in its conservative form as

$$\frac{\partial(\rho\Phi)}{\partial t} + \nabla \cdot \left(\rho\Phi\vec{V}\right) = \nabla \cdot (\Gamma\nabla\Phi) + S_\Phi, \tag{8.18}$$

where Γ is the diffusion coefficient and $S\Phi$ refers to the source terms. This form is particularly useful for numerical applications. In words, the transport equation means that the sum of the rate of increase of a fluid variable in a fluid element and the rate of outflow of this variable from the fluid element is equal to the rate of change of the variable, due to diffusion and to the sources. The first term of the left-hand side of equation (8.18) is the transient term and the second is the convection term. The first and second terms on the right-hand side are called the diffusion and source terms, respectively.

8.2.3 Modelling of Turbulence

The behaviour of a fluid flow can be identified by the way fluid particles move. As was demonstrated experimentally by O. Reynolds in 1883 [11], there are two types of flows, namely laminar and turbulent. The laminar regime is characterised by relatively low fluid velocities and fluid particles moving in an orderly way and retaining their relative positions during their propagation. Once the flow velocity increases beyond a certain point, the fluid particles stop moving in an orderly manner and the velocity and pressure of the flow fluctuate. This type of flow is called turbulent and is characterised by chaotic behaviour. The flow regime is determined by the Reynolds number, which is the ratio of the inertial forces that cause disturbance to the fluid particles, to the viscous forces that restrain them to move in streamlines [12].

Turbulent flows are highly unsteady and exhibit a three-dimensional behaviour, which includes an element of randomness. They are associated with the generation of vortices that span a broad range of length scales and cause vorticity in the flow. Turbulence also causes diffusion and mixing of conserved quantities in the flow. Energy dissipation is another consequence of turbulence, due to the transformation of energy of the fluid into internal energy or heat [13]. These characteristics make the theoretical and numerical study of turbulence a complicated task. Turbulence is also related to many important engineering parameters, such as frictional drag, flow separation, transition from laminar to turbulent flow, thickness of boundary layers, extension of secondary flows and spreading of jets and wakes. Thus, in many real-life applications that include the study of fluid–structure interaction or rapid flow variations, turbulence may play a significant role.

The computational cost of solving with the NSE for all the spatial and temporal scales of turbulence would be extremely high for practical applications. However, if the physics of the problem allows for assumptions regarding the type of flow to be made, the NSE can be simplified and achieve similar numerical results with a minimal computational cost. Such simplifications can include, for example, the removal of viscosity or vorticity, which reduces the NSE to the Euler or potential flow (PF) equations, respectively.

If the problem requires modelling of turbulence, appropriate mathematical models have to be considered. These models have been developed specifically for certain scales of turbulence. So far, there is not a generally valid and efficient universal mathematical model for turbulence, due to the complexity of the problem [14]. Three main mathematical models for turbulence will be presented here: the Reynolds-averaged Navier–Stokes (RANS) equations with appropriate turbulence closure models; the large-eddy simulation (LES); and direct numerical simulation (DNS) [8].

8.2.3.1 RANS Equations

The Reynolds-averaged Navier–Stokes model is the most popular method of solving the NSE for CFD applications in industry and research, because of its efficiency and relatively low computational cost. In this section the RANS equations for incompressible flows are presented, describing the essence of 'averaging'. For the full derivation of the equations and for compressible flows, the reader should refer to the relevant literature [8].

For better understanding of the physical phenomenon of turbulent flow, a control volume can be considered with fluid flow parallel to its faces, which causes generation of eddies. These eddies cause transfer of momentum between the control volume and the external fluid, which results in acceleration of slow-moving layers and deceleration of fast-moving layers. The resulting changes in momentum imply the existence of additional turbulent shear stresses, namely the Reynolds stresses.

A typical measurement of the velocity in a turbulent flow is depicted in Figure 8.2. It can be seen that the velocity is random, but consists of a steady mean value U and a fluctuating value $u'(t)$. This approach of splitting the variable into a mean and fluctuating component is called Reynolds decomposition. The same approach can be applied for all the velocity components and the pressure of the fluid:

$$\vec{v} = \vec{V} + \vec{v}', \quad u = U + u', \quad v = V + v', \quad w = W + w', \quad p = P + p'. \tag{8.19}$$

RANS modelling is based on the time-averaging of the governing equations, and its description includes the continuity and the momentum equation in three dimensions.

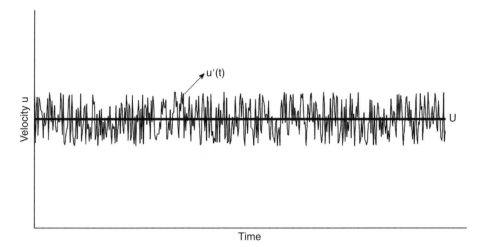

Figure 8.2 Typical measurement of the *x*-component of the velocity in turbulent flows [8].

For the case of incompressible flows, after considering that $\overline{\nabla u} = \nabla \vec{V}$, the continuity equation reads

$$\nabla \cdot \vec{V} = 0. \tag{8.20}$$

The corresponding momentum equations are:

$$\frac{\partial U}{\partial t} + \nabla \cdot \left(U \vec{V} \right) = -\frac{1}{\rho} \frac{\partial P}{\partial x} + \nu \nabla \cdot \left(\nabla (U) \right) + \frac{1}{\rho} \left[\frac{\partial \left(-\rho \overline{u'^2} \right)}{\partial x} + \frac{\partial \left(-\rho \overline{u'v'} \right)}{\partial y} + \frac{\partial \left(-\rho \overline{u'w'} \right)}{\partial z} \right],$$

$$\tag{8.21a}$$

$$\frac{\partial V}{\partial t} + \nabla \cdot \left(V \vec{V} \right) = -\frac{1}{\rho} \frac{\partial P}{\partial y} + \nu \nabla \cdot \left(\nabla (V) \right) + \frac{1}{\rho} \left[\frac{\partial \left(-\rho \overline{u'v'} \right)}{\partial x} + \frac{\partial \left(-\rho \overline{v'^2} \right)}{\partial y} + \frac{\partial \left(-\rho \overline{v'w'} \right)}{\partial z} \right],$$

$$\tag{8.21b}$$

$$\frac{\partial W}{\partial t} + \nabla \cdot \left(W \vec{V} \right) = -\frac{1}{\rho} \frac{\partial P}{\partial z} + \nu \nabla \cdot \left(\nabla (W) \right) + \frac{1}{\rho} \left[\frac{\partial \left(-\rho \overline{u'w'} \right)}{\partial x} + \frac{\partial \left(-\rho \overline{v'w'} \right)}{\partial y} + \frac{\partial \left(-\rho \overline{w'^2} \right)}{\partial z} \right].$$

$$\tag{8.21c}$$

Equations (8.20) and (8.21) together constitute the RANS equations, which contain the Reynolds stresses in longhand form. For better comprehension, the normal and the shear Reynolds stresses are presented here as well:

$$\tau_{xx} = -\rho \overline{u'^2}, \quad \tau_{yy} = -\rho \overline{v'^2}, \quad \tau_{zz} = -\rho \overline{w'^2},$$

$$\tau_{xy} = \tau_{yx} = -\rho \overline{u'v'}, \quad \tau_{xz} = \tau_{zx} = -\rho \overline{v'w'}, \quad \tau_{yz} = \tau_{zy} = -\rho \overline{v'w'}. \tag{8.22}$$

The time averaging eliminates the effect of the unsteadiness of the flow, by employing the mean value. The instantaneous fluctuations of the flow are expressed through the Reynolds stresses, which contain the turbulent variables of the flow. The calculation of the flow properties requires additional equations for the mean flow variables. The equations used for the closure of the system of RANS equations are called turbulence models, which are mathematical models that include relations for the unknown quantities [8]. The widespread use of RANS modelling in industry and research has led to the emergence of four main categories of turbulence models: algebraic, also known as zero-equation models; one-equation models; two-equation models; and stress transport models. Each category includes various models that differ in the implementations and parameterisations that they employ.

The selection of the most appropriate turbulence model is not a trivial task, since it requires good knowledge of the physics involved in the problem and possible parametric analysis [14]. Here, three of the most popular turbulence models for RANS equations are briefly presented, including the $\kappa-\epsilon$ model, the $\kappa-\omega$ model and the Reynolds stress model, which are appropriate for general-purpose CFD modelling. For additional details of these models, the reader should refer to one of the classic CFD books [8, 13].

8.2.3.2 The $\kappa-\epsilon$ Model

The $\kappa-\epsilon$ is a two-equation turbulence model, which accounts for production and destruction of turbulence and for the transport of turbulent properties by convection and diffusion. κ refers to the turbulent kinetic energy and ϵ to the rate of dissipation of turbulent kinetic energy per unit mass by the action of viscosity on the smallest eddies [14]. This turbulence model assumes that the turbulent viscosity μ_t is isotropic, which means that the ratio between Reynolds stress and mean rate of deformation is the same in all directions. The isotropic μ_t assumption can fail for complex flow studies, leading to inaccuracies [8].

The standard $\kappa-\epsilon$ model [15] comprises of two PDEs, one for κ and one for ϵ, which are based on the best understanding obtained so far of the processes associated with these variables. κ and ϵ are used to define the velocity scale θ and the length scale l, which are representative properties of the large-scale turbulence [8]:

$$\theta = \kappa^{1/2} \quad \text{and} \quad l = \frac{\kappa^{3/2}}{\epsilon}. \tag{8.23}$$

The transport equations used in the standard $\kappa-\epsilon$ read:

$$\frac{\partial(\rho\kappa)}{\partial t} + \nabla \cdot \left(\rho\kappa\vec{V}\right) = \nabla \cdot \left(\frac{\mu_t}{\sigma_\kappa}\nabla\kappa\right) + 2\mu_t S_{ij} \cdot S_{ij} - \rho\epsilon, \tag{8.24a}$$

$$\frac{\partial(\rho\epsilon)}{\partial t} + \nabla \cdot \left(\rho\epsilon\vec{V}\right) = \nabla \cdot \left(\frac{\mu_t}{\sigma_\epsilon}\nabla\epsilon\right) + C_{1\epsilon}\frac{\epsilon}{\kappa}2\mu_t S_{ij} \cdot S_{ij} - C_{2\epsilon}\rho\frac{\epsilon^2}{\kappa}, \tag{8.24b}$$

where S_{ij} are the rates of deformation of a fluid element and σ_κ and $\sigma\epsilon$ are the Prandtl numbers connecting the diffusivities of κ and ϵ to the eddy viscosity μ_t, which is obtained by dimensional analysis as $\mu_t = C_\mu \rho \theta l = \rho C_\mu \dfrac{\kappa^2}{\epsilon}$, with C_μ being a dimensionless constant [8]. In words, the transport equations say that:

Rate of change of κ and ϵ	+	Transport of κ or ϵ by convection	=	Transport of κ or ϵ by diffusion	+	Rate of production of κ and ϵ	−	Rate of destruction of κ and ϵ

The transport equations include five adjustable constants: C_μ, σ_κ, σ_ϵ, $C_{1\epsilon}$ and $C_{2\epsilon}$. The values for these constants have been obtained by comprehensive data fitting to the standard $\kappa - \epsilon$ model for a wide range of turbulent flows (Table 8.1).

Reynolds stresses can be computed using the Boussinesq relationship,

$$\tau_{ij} = -\rho \overline{u_i' u_j'} = \mu_t \left(\frac{\partial U_i}{\partial x_j} + \frac{\partial U_j}{\partial x_i} \right) - \frac{2}{3}\rho\kappa\delta_{ij} = 2\mu_t S_{ij} - \frac{2}{3}\rho\kappa\delta_{ij}, \tag{8.25}$$

where δ_{ij} is the Kronecker delta.

The standard $\kappa - \epsilon$ model discussed here requires the specification of detailed boundary conditions for the variables κ and ϵ, as shown in Table 8.2. These conditions can be obtained from measurement of the variables, but in practice, values are not available for many specific applications and CFD modellers have to refer to the literature and explore the sensitivity of the results to the inlet distributions [8].

A broadly used modification of the $\kappa - \epsilon$ model is the RNG $\kappa - \epsilon$, which is based on statistical mechanics approaches that have been extended to eddy viscosity models. RNG stands for 'renormalisation group', which is a method of treating the small scales of motion and effectively replacing them by means of their effects on large-scale motions and modifying viscosity [8].

8.2.3.3 The $\kappa - \omega$ model

The $\kappa - \omega$ turbulence model is another widely used two-equation model, which considers the turbulent viscosity μ_t as isotropic, similarly to the $\kappa - \epsilon$ model. The model is derived from substitution of the rate of dissipation of turbulent kinetic energy ϵ by the turbulence frequency $\omega = \epsilon/\kappa$ in the expression of the length scale l (see equation (8.23)).

Table 8.1 Values of $\kappa - \epsilon$ model coefficients [8].

$C_\mu = 0.09$	$\sigma_\kappa = 1.00$	$\sigma_\epsilon = 1.30$	$C_{1\epsilon} = 1.44$	$C_{2\epsilon} = 1.92$

Table 8.2 Boundary conditions for the $\kappa - \epsilon$ model [8].

Inlet	Distributions of κ and ϵ must be given
Outlet, symmetry axis	$\partial\kappa/\partial n = 0$ and $\partial\epsilon/\partial n = 0$
Free stream	κ and ϵ must be given or $\partial\kappa/\partial n = 0$ and $\partial\epsilon/\partial n = 0$
Solid walls	Approach depends on Reynolds number

Table 8.3 Values of $\kappa - \omega$ model coefficients [8].

$C_\kappa = 2.0$	$\sigma_\omega = 2.0$	$\gamma_1 = 0.533$	$\beta_1 = 0.075$	$\beta^* = 0.09$

Table 8.4 Boundary conditions for the $\kappa - \omega$ model [8].

Inlet	Distributions of κ and ω must be given
Outlet	$\partial \kappa / \partial n = 0$ and $\partial \omega / \partial n = 0$
Free stream	$\kappa \rightarrow 0$ and $\omega \rightarrow 0$, with $\kappa \neq 0$
Solid walls	Not dependent on Reynolds number

The classic $\kappa - \omega$ model was proposed by Wilcox [16] by assuming that $l = \sqrt{\kappa}/\omega$, which leads to the expression of the eddy viscosity as $\mu_t = \rho \kappa / \omega$.

At high Reynolds numbers, the transport equations for κ and ω read as follows:

$$\frac{\partial(\rho \kappa)}{\partial t} + \nabla \cdot \left(\rho \kappa \vec{V} \right) = \nabla \cdot \left[\left(\mu + \frac{\mu_t}{\sigma_\kappa} \right) \nabla \kappa \right] + P_\kappa - \beta^* \rho \kappa \omega, \tag{8.26a}$$

$$\frac{\partial(\rho \omega)}{\partial t} + \nabla \cdot \left(\rho \omega \vec{V} \right) = \nabla \cdot \left[\left(\mu + \frac{\mu_t}{\sigma_\omega} \right) \nabla (\omega) \right] + \gamma_1 \left(2 \rho S_{ij} \cdot S_{ij} - \frac{2}{3} \rho \omega \frac{\partial U_i}{\partial x_j} \delta_{ij} \right) - \beta_1 \rho \omega^2, \tag{8.26b}$$

where μ is the molecular viscosity; β_1, β^*, γ_1 are adjustable constants; and P_κ is the rate of production of turbulent kinetic energy, expressed as $P_\kappa = 2 \mu_t S_{ij} \cdot S_{ij} - \frac{2}{3} \rho \kappa \frac{\partial U_i}{\partial x_j} \delta_{ij}$.

The Reynolds stresses τ_{ij} are computed with the Boussinesq expression in the same way as in the $\kappa - \epsilon$ model (equation (8.25)). The values of the constants and coefficients in equations (8.26) are shown in Table 8.3.

The advantage of the $\kappa - \omega$ model is that it does not require wall-damping functions in low Reynolds numbers for the integration to the wall, which made it popular in many industrial applications. At the wall, the value of the turbulent kinetic energy κ is set to zero, while the frequency ω tends to infinity. Alternatively, a very large value can be given to ω at the wall or, as was proposed by Wilcox, a hyperbolic variation, $\omega_P = 6\nu/\left(\beta_1 y_P^2 \right)$, can be used at the near-wall grid point. In practice, though, it has been shown that the results are not very sensitive to the precise value of ω.

The boundary conditions need to be specified as shown in Table 8.4. The most challenging condition is that of ω in the free stream when both the turbulent kinetic energy and turbulent frequency tend to zero, since if $\omega = 0$, then μ_t is indeterminate. To circumvent this a very small value of ω can be specified, instead of setting it to zero, which might have a strong impact on the results, especially for applications of external aerodynamics and aerospace simulations [8].

There are also other variations of the $\kappa - \omega$ model, such as the SST (shear stress transport) $\kappa - \omega$ model developed by Menter [17], where $\epsilon = \kappa \omega$. This model reduces to the $\kappa - \epsilon$ model away from the wall and has the $\kappa - \omega$ function for the treatment of the near-wall region. The SST $\kappa - \omega$ has the advantage of being less sensitive to the free stream values, but might not be as accurate as the classic $\kappa - \omega$ near walls [8].

8.2.3.4 The Reynolds Stress Model

As discussed, the two-equation models presented above are very useful for many applications, but they are subject to the underlying assumption of isotropic turbulent viscosity μ_t. This might lead to unrealistic results for strain fields that arise from complex geometries, swirls and body forces, such as buoyancy. For such cases, the employment of a more rigorous relationship between the stresses and the strains is necessary, which allows for the calculation of the exact values of the Reynolds stresses and their directional effects.

The simplest way to form a turbulence model for classic flows is by solving seven additional PDEs for the six stress components and for ϵ, in order to obtain the length scale. This approach is called Reynolds stress equation model (RSM), also known as the second-order or second-momentum closure model [8]. This solution methodology was suggested by Launder *et al.* [15], who derived the equations using the following expression for the Reynolds stresses: $R_{ij} = -\dfrac{\tau_{ij}}{\rho} = \overline{u_i' u_j'}$. However, it is more appropriate to refer to these stresses as kinematic Reynolds stresses. In this light, the transport equation of the RSM model reads

$$\frac{DR_{ij}}{Dt} = \frac{\partial R_{ij}}{\partial t} + C_{ij} = P_{ij} + D_{ij} - \epsilon_{ij} + \Pi_{ij} + \Omega_{ij}, \tag{8.27}$$

where C_{ij}, P_{ij}, D_{ij}, ϵ_{ij}, Π_{ij} and Ω_{ij} are the convective, production, diffusion, dissipation, pressure–strain and rotational terms, respectively.

The full mathematical description and interpretation of the terms of equation (8.27) can be found in [8]. If analysed, the transport equation describes nine PDEs, six of which are independent: $u_1'^2, u_2'^2, u_3'^2, \overline{u_1' u_2'}, \overline{u_1' u_3'}, \overline{u_2' u_3'}$ and $\overline{u_2' u_1'} = \overline{u_1' u_2'}, \overline{u_3' u_1'} = \overline{u_1' u_3'}, \overline{u_3' u_2'} = \overline{u_2' u_3'}$.

An important relative advantage of the RSM compared to two-equation models is that the definition of free stream conditions is not required, since they are found from the exact calculation of the Reynolds stresses. For RSM modelling only the initial and boundary conditions must be provided, as listed in Table 8.5. These are similar to boundary conditions for elliptic flows, as discussed later in Section 8.2.4.

The RSM method for solving the RANS equations is considered a very advanced tool capable of simulating complex and non-equilibrium flows. Despite being complex, since it considers seven additional equations, the RSM is the simplest way to describe accurately all the mean flow properties and the Reynolds stresses without having to adjust all the coefficients for each different case, as is necessary with

Table 8.5 Boundary conditions for elliptic flows solved with RSM [8].

Inlet	Distributions of R_{ij} and ϵ must be given
Outlet, symmetry axis	$\partial R_{ij}/\partial n = 0$ and $\partial \epsilon /\partial n = 0$
Free stream	$R_{ij} = 0$ and $\epsilon = 0$ must be given or $\partial \kappa /\partial n = 0$ and $\partial \epsilon /\partial n = 0$
Solid walls	Use wall functions relating R_{ij} to either κ or u_τ^2, where $u_\tau^2 = \overline{u_{i'} u_{j'}}$

two-equation models. On the downside, solving additional equations increases the computational cost of the model and requires specification of extra parameters. At the same time, convergence issues related to the coupling of the mean velocity and the turbulent stress fields through source terms might cause problems in the application of the RSM. The performance of the RSM is subject to the ϵ-equation, which might result in poor behaviour similar to the $\kappa - \epsilon$ mode for certain complex cases [8]. These drawbacks have delayed the adoption of RSM in industrial engineering applications [14].

8.2.3.5 Large Eddy Simulation

A general remark from the discussion of RANS models is that a general-purpose turbulence model for RANS modelling is yet to be developed. The principal reason for this lies in the different behaviour of the small and the large eddies in the flow, which in RANS models is described by a single turbulence model. In a turbulent flow of high Reynolds number, small eddies are isotropic and appear to exhibit a universal behaviour. In contrast, the behaviour of the large eddies is characterised by anisotropy. Large eddies are mostly subject to the geometry of the domain, the boundary conditions and the body forces, and they determine the main flow characteristics by extracting energy from the flow. In cases where different scales of eddies are present, a single treatment of the turbulence with time-averaged variables is not adequate and special effort should be made to capture their different behaviour.

LES considers a time-dependent simulation for the larger eddies and employs a compact model based on the universal behaviour of the smaller eddies. As implied, a main difference between RANS and LES modelling is that the latter is not time-averaged. The treatment of the different scales of eddies in LES is achieved by employing a spatial filter for separating larger and smaller eddies. This filter is based on a function with a specified cut-off width and distinguishes the eddies that have a length scale greater than the cut-off width, when applied on the time-dependent flow equations. Once the filter is applied and the smaller eddies are separated from the larger, the information related to the former is filtered out. Only the larger eddies are properly resolved, while the smaller unresolved eddies and their interaction with the resolved flow are taken into account by a sub-grid-scale (SGS) stresses model [8].

The filtering function for distinguishing the smaller from the larger eddies in LES simulations has the general form

$$\overline{\phi(\mathbf{x},t)} \equiv \iiint G(\mathbf{x},\mathbf{x}',\Delta)\phi(\mathbf{x}',t)dx_1'dx_2'dx_3' \quad \text{or} \quad \overline{\phi} = G \star \phi, \tag{8.28}$$

where $\overline{\phi(x,t)}$ is the filtered function, $\phi(x,t)$ the original function, x the position vector and Δ the cut-off width. There is a range of filtering functions, such as top-hat or box filter, Gaussian filter and spectral cut-off function. It should be noted that LES filtering is an integration in three-dimensional space, in contrast to RANS modelling, where the equivalent filtering refers to the time-averaging integration [8].

The governing equations for fluid flow consist of the mass and the momentum conservation equations (see equations (8.1b) and (8.16)). If the flow is considered incompressible ($\nabla \cdot \vec{V} = 0$), the source terms in the momentum equations become zero. Further simplification can be achieved by considering a filtering function which is independent of the position. A uniform filter, such as $G(x,x') = G(x - x')$, which is linear, allows for swapping the order of filtering and differentiation with respect to time, as well as with respect to space [8]. After these implementations, the governing equations read

$$\frac{\partial \rho}{\partial t} + \nabla \cdot \left(\rho \overline{\vec{V}} \right) = 0, \tag{8.29}$$

$$\frac{\partial (\rho \overline{u})}{\partial t} + \nabla \cdot \left(\rho \overline{u \vec{V}} \right) = -\frac{\partial \overline{p}}{\partial x} + \mu \nabla \cdot (\nabla \overline{u}), \tag{8.30a}$$

$$\frac{\partial (\rho \overline{v})}{\partial t} + \nabla \cdot \left(\rho \overline{v \vec{V}} \right) = -\frac{\partial \overline{p}}{\partial y} + \mu \nabla \cdot (\nabla \overline{v}), \tag{8.30b}$$

$$\frac{\partial (\rho \overline{w})}{\partial t} + \nabla \cdot \left(\rho \overline{w \vec{V}} \right) = -\frac{\partial \overline{p}}{\partial z} + \mu \nabla \cdot (\nabla \overline{w}), \tag{8.30c}$$

where μ is the constant dynamic viscosity and the overbar indicates a filtered variable.

Equations (8.30) can be further simplified by analysing the second term on the left-hand side, which represents the convection, using the mathematical identity for divergence,

$$\nabla \cdot \left(\rho \overline{\phi u} \right) = \nabla \cdot \left(\rho \overline{\phi} \overline{u} \right) + \nabla \cdot \left(\rho \overline{\phi} u \right) - \nabla \cdot \left(\rho \overline{\phi} \overline{u} \right). \tag{8.31}$$

This facilitates the computations, since the filtered variables ($\overline{\phi}$) are readily available from the selected model.

The set of equations for LES has analogies with RANS models, but it can be shown that there is a substantial difference in the calculation of the SGS stresses τ_{ij}. LES models contain further contributions that are determined by the decomposition of the flow variables based on the cut-off width for the eddies [8].

LES modelling can provide good accuracy for the flow variables and the Reynolds stresses while at the same time remaining general, without requiring specific parameterising of the turbulence models, like in RANS modelling. A unique aspect of LES is that during the post-processing the flow information is synthetic, consisting of the mean flow and the statistics of the resolved fluctuations. LES models are computationally expensive compared to RANS models and might result in approximately twice the computational cost of RSM. Despite the high computational cost, which caused a delay in industrial applications with LES models, the inherent unsteadiness of the flow can be properly described, making LES a valuable tool for the simulation of complex flows [8].

8.2.3.6 Direct Numerical Simulation

As already discussed, a flow can be described adequately by the primitive equations of fluid mechanics. If the flow is incompressible, the continuity and the NSE comprise a closed set of equations with four unknowns, u, v, w and p, which can be calculated without any temporal or spatial filtering, like those used in RANS and LES modelling respectively. This approach is referred to as DNS modelling and it can be used to solve the flow equations, using sufficiently fine spatial and temporal resolution. DNS allows for the development of a transient solution for the whole range of turbulent eddies and the fastest fluctuations in the flow without employing any filtering [8].

The term 'direct numerical simulation' was coined by Orszag [18] for simulations that solve the governing equations without a turbulence model. The main advantage of the DNS approach is the precise description of the flow structures that it can provide, accounting even for instantaneous fluctuations. The results of DNS simulations can be used to compensate for the limitations of the instrumentation in physical model testing, by offering information about the turbulence structures in the flow that cannot be probed and visualised properly with the present technologies. In that sense, DNS modelling can be used to evaluate the performance of new measuring experimental techniques and to validate other numerical models. Additionally, DNS simulations can contribute to the calibration and validation of other parameterised turbulence models. For that scope, databases with results from DNS simulations have started to become available. DNS modelling is also very useful for fundamental turbulence research regarding flows that are hard to reproduce experimentally, such as shear-free boundary layers developing on walls at rest with respect to the free stream and effects of initial conditions on the development of self-similar turbulent wakes [19]. In coastal engineering, DNS can contribute to the study of complex phenomena, such as wave breaking [20].

The biggest drawback of the DNS method is the high computational cost, which emerges from the fact that a very high temporal and spatial resolution is required in the numerical model for resolving all the length and time scales of eddies in a turbulent flow. The computational cost of DNS is proportional to the Reynolds number and the ratio between the largest and the smallest eddies in the flow. For instance, the minimum number of computational cells required for a flow of a moderate Reynolds number of $Re = 10^4$ and a length eddy ratio of $Re^{3/4}$ is $Re = 10^3$ cells in each coordinate direction, resulting in 10^9 cells for a three-dimensional simulation. Additionally, the time step has to be sufficiently low, which further increases the computational effort. Consequently, DNS modelling is infeasible for high Re, since practical engineering applications today can only employ a few million cells. However, the advances in processing power and in efficient parallelised computational methods can make some DNS applications realistic in the future. Nevertheless, the contribution of DNS to fundamental research is invaluable [8].

8.2.3.7 Potential Flow

So far in this chapter, the derivation of complicated numerical approaches for simulating turbulence has been presented. However, there are flows with negligible turbulence, which can be described by simpler and more efficient mathematical models that do not account for turbulence and its effects. One commonly used approach is potential flow (PF) theory.

There are several assumptions that need to be valid in order to use a PF model. The first is that the effect of viscosity is negligible and the fluid behaves like an ideal fluid. In other words, the shear stresses acting in the tangential directions on fluid elements, which are a result of only the viscous forces, can be ignored. The result is that no rotation of the fluid particles takes place and no vorticity is created or destroyed, and thus PF theory can be applied for irrotational flows. Rotationality can be an issue for fluid–structure interaction studies, where boundary layers are created near the walls of the structure, resulting in important viscous effects. However, as the distance from the object increases, these effects diminish and the flow gradually takes its initial form. Therefore, PF models can be applied for fluid–structure interaction studies of initially irrotational flows, providing good results sufficiently far away from the structures [7]. In the case of laminar flows, PF models can even be applied in the vicinity of the structure for analysing external flows over solid surfaces.

The mathematical model of PF theory considers a velocity potential function ϕ, the gradient of which is used to express the velocity field of the flow as $\vec{V} = \nabla\phi$. According to vector calculus the curl of a gradient is zero: $\nabla \times \nabla\phi = 0$. Using these two equations, it can be demonstrated that $\nabla \times \vec{V} = 0$, which proves that the curl of the velocity field (i.e. vorticity) is zero. Assuming that the flow is incompressible and using the relationship $\vec{V} = \nabla\phi$, the continuity equation (8.2) can be rewritten:

$$\nabla \cdot \nabla\phi = 0 \quad \Rightarrow \quad \nabla^2\phi = 0, \tag{8.32}$$

where ∇^2 is the Laplace operator, also denoted by Δ.

Equation (8.32) is known as the Laplace equation. When combined with the Bernoulli equation, which is used to find the relationship between the potential function ϕ and the pressure p, the calculation of the velocity and the pressure becomes possible in space and time [7]. The Bernoulli equation shown below is the equivalent of momentum equation (8.4):

$$-\frac{\partial\phi}{\partial t} + \frac{1}{2}\left[\left(\frac{\partial\phi}{\partial x}\right)^2 + \left(\frac{\partial\phi}{\partial y}\right)^2 + \left(\frac{\partial\phi}{\partial z}\right)^2\right] + \frac{p}{\rho} + gz = C(t), \tag{8.33}$$

where $C(t)$ is a time-varying integration constant determined by the boundary conditions and g is the gravitational acceleration, which is considered to act only in the vertical direction.

PF theory can be applied for both incompressible and compressible flows. In its incompressible formulation, as presented here, it can be used to simulate the propagation and transformation of water waves, provided of course that the flow is not rotational. Therefore, PF theory should not be employed for breaking waves, overturning flows and interaction of waves with small bodies. With regard to compressible flows, PF theory is applicable for sub-, trans- and super-sonic flows at arbitrary angles of attack. In any case, it is not applicable for the flow description near walls with boundary layers [21]. It should be noted that PF simulations are considerably faster than RANS, LES and DNS that incorporate the modelling of turbulence.

8.2.4 Classification of Physical Behaviours

The numerical techniques for solving the governing equations of the fluid motion require the employment of appropriate boundary and initial conditions. A fundamental step is to classify the physical behaviour of the problems and select the most appropriate form of the PDEs. There are two categories of physical behaviours: equilibrium problems, solved with the elliptic equations, and marching problems, solved with parabolic or hyperbolic equations. Equilibrium problems describe steady-state situations, while marching problems refer to propagation problems, including unsteady and transient phenomena, such as transient heat transfer, unsteady flows and water waves. Nonetheless, there are cases of steady-state problems that can be described by marching equations by considering propagation in time-like coordinates [8].

8.2.4.1 Elliptic Equations

Elliptic equations are employed for the description of equilibrium problems, which reach a steady-state situation after the application of the boundary conditions. Some examples are the steady-state temperature distribution in a solid body, the stress distribution of a solid body under a load and many steady fluid problems, such as a low *Re* flow in a duct.

The basic elliptic equation is the Laplace equation (8.32), which is applicable for irrotational flows and steady-state conductive heat transfer. For incompressible flows in two dimensions the longhand form of the Laplace equation reads

$$\frac{\partial^2 \phi}{\partial x^2} + \frac{\partial^2 \phi}{\partial y^2} = 0. \tag{8.34}$$

Elliptic equations offer a unique solution for a variable ϕ in the whole domain and are called boundary-value problems, since they require definition of the variables on all boundaries. A distinguishing characteristic of the elliptic equations is that a change in the variable anywhere in the domain influences the solution everywhere else, since the value of the variable at a specific location is influenced by all the neighbouring values. In practice, this implies that a disturbance signal in the computational domain travels in all directions, provided that the employed numerical schemes connect the computational nodes adequately. As a result, the solutions obtained by elliptic equations are always smooth, even in the cases of discontinuous boundary conditions. This feature of elliptic equations is a considerable advantage for the formulation of a numerical model [8].

8.2.4.2 Parabolic Equations

Parabolic equations are used for the description of marching problems that include propagation in time. Therefore, the solution is time-dependent, as can be seen in the equation

$$\frac{\partial \phi}{\partial t} = \alpha \frac{\partial^2 \phi}{\partial x^2}. \tag{8.35}$$

The distinguishing characteristic of parabolic equations is that they include a significant amount of diffusion, which is useful for unsteady viscous flows and unsteady heat problems.

Similarly to elliptic problems, parabolic problems require the definition of conditions of a variable ϕ at all boundaries of the domain for all times $t > 0$, but they also require the initial conditions at $t = 0$. The latter characterises them as initial boundary-value problems. Any disturbance of the variable in the interior of the domain can only have an effect at later times. In parabolic problems, the solution propagates in time and diffuses in space. Eventually, after a sufficient number of time steps, the solution in the interior of the domain becomes smooth and reaches a steady state for an infinitely long time.

This can be observed by considering $\frac{\partial \phi}{\partial t} = 0$ in equation (8.35) and effectively obtaining a steady-state problem. The diffusivity included in the parabolic equations guaranties smooth solution in the interior of the domain, even for the case of discontinuous boundary conditions. Similarly to elliptic problems, this behaviour facilitates numerical modelling applications [8].

8.2.4.3 Hyperbolic Equations

Hyperbolic equations are used for the description of time-marching problems inside a computational domain. These equations are mainly used for oscillation problems without accounting for energy dissipation. The basic form of the hyperbolic equations presented below is also known as the wave equation,

$$\frac{\partial^2 \phi}{\partial t^2} = c^2 \frac{\partial^2 \phi}{\partial x^2},$$
(8.36)

where c is a constant that represents the wave speed.

Equation (8.36) requires the specification of initial conditions of a variable ϕ and one boundary condition for all $t > 0$, making hyperbolic problems initial boundary-value problems. The distinguishing characteristic of hyperbolic equations compared to elliptic and parabolic equations is that the propagation of the disturbance in the computational domain is controlled by the constant wave speed c. On the other hand, elliptic and parabolic models consider infinite propagation speed of the solution. As a consequence, the disturbance in hyperbolic problems has a limited region of influence, while in elliptic and parabolic problems it influences the entire domain. Moreover, the absence of damping makes the solution periodic in time, as there is no mechanism to reduce the solution to certain constant values. Another effect of the undamped character of the hyperbolic equations is that any discontinuity in the initial conditions remains in the solution without being smoothed out. These discontinuities can be described by means of Fourier analysis with linear summation of an adequate number of wave equations in order to form a single PDE [8].

For some common fluid flows, the classification is presented in a generic form in Table 8.6. It can be seen that the governing equations of fluid motion are elliptic for steady flows and parabolic for unsteady flows. In the case of inviscid flows, the absence of viscous high-order terms makes the equations hyperbolic. In thin shear layers, such as boundary layers, jets, mixing layers, wakes and fully developed duct flows, the governing equations include a diffusion term, which classifies them as parabolic [8].

Table 8.6 Classification of physical behaviours for low Mach numbers [8].

	Steady flow	Unsteady flow
Viscous flow	Elliptic	Parabolic
Inviscid flow	Elliptic	Hyperbolic
Thin shear layers	Parabolic	Parabolic

8.3 Numerical Modelling Techniques

8.3.1 Introduction

The description of physical phenomena is achieved by mathematical representation of the physical processes in a form of governing equations, such as ordinary differential equations (ODEs), PDEs and integration equations. Commonly, these mathematical forms include several assumptions and simplifications of the actual physical problem. Additional boundary conditions and initial conditions are selected to describe the evolution of the variables at specific places in the domain and the initial state of the solution, respectively. For most realistic problems, analytical mathematical solutions are not available, at least not to an acceptable degree of accuracy. In these cases, generic numerical modelling techniques should be applied in order to approximate the solution in time and space [22].

Numerical modelling is divided into three distinct parts, pre-processing, solution and post-processing, which are discussed in Sections 8.3.2–8.3.4. In short, the pre-processing part includes definitions of the appropriate mathematical model via selection of the governing equations and all the other actions required to commence the numerical simulation, such as the definition of the boundary conditions, initial conditions and computational mesh. In numerical modelling the domain is divided into discrete components, where the governing equations are solved. Everywhere else, the solution is calculated through interpolation between these discrete locations. The process of dividing up the domain is called discretisation and results in a number of computational nodes, which should sufficiently represent the continuous domain. The computational nodes comprise a mesh or grid with moving or stationary points and in more modern techniques a 'cloud' of freely moving particles. The computational nodes are related with some type of connectivity that determines their relative behaviour and controls the overall evolution of the solution based on the solution of the governing equations at the computational nodes. The accuracy of numerical simulations depends on the density of the computational nodes and the quality of the computational geometry.

Once the domain is discretised, the solution process requires the numerical discretisation of the governing equations. This is achieved by transforming the PDEs or integral equations into equivalent algebraic or ODEs, which can be solved by the computer using specific numerical algorithms. According to the nature of the physical problem (i.e. transient or steady state, parabolic, hyperbolic or elliptic), there are numerical schemes for calculating the algebraic equations in time and space, taking into account the solution at previous time steps and at neighbouring computational nodes.

The post-processing part takes place after the solution part for better interpretation of the results. It may include mathematical and statistical analysis of the raw numerical results and visualisation of the results with informative graphs. Post-processing is as important as the other parts of numerical modelling, since it allows engineers and scientists to assess the outcomes of the simulations.

To sum up, the required actions and steps in order to perform a numerical simulation of a physical problem include the following steps:

- Definition of governing equations
- Selection of proper boundary and initial conditions
- Domain discretisation into computational nodes

- Numerical discretisation of the governing equations
- Numerical schemes for solving the resultant discretised equations
- Post-processing and interpretation of the numerical results.

8.3.2 Pre-Processing

Pre-processing includes all the necessary preliminary actions and decisions before executing the numerical simulation. The two main parts of pre-processing refer to the definition of the problem and the selection of the boundary and initial conditions.

8.3.2.1 Definition of the Problem

The definition of the problem includes the selection of the geometry of the computational domain, the physical phenomena to be simulated and the numerical methods and schemes that will be employed for the discretisation of the equations and the numerical solution.

The definition of the geometry is a critical part of the modelling process, because it defines the location where the governing equations will be solved. It should be representative of the physical geometry of the problem, accurately designed and efficiently selected, so that it covers in detail only the necessary parts of the physical domain. The next step is to discretise the geometry into a number of computational nodes. This process is referred to as mesh generation for traditional numerical applications. In recently developed meshless solvers it refers to the number of particles selected. The mesh should be carefully designed especially in areas of interest, such as changes in the geometry or areas with high flow fluctuations. The number of computational nodes determines the accuracy of the numerical simulation, but it also affects the computational cost. The computational mesh can be 'structured', if there is a constant distribution of computational nodes in each axis, or 'unstructured', if the nodes are arbitrarily distributed in the domain [23]. Structured meshes can be uniform or non-uniform with areas of local refinement. The mesh can also be 'stationary' or 'dynamic', which allows it to adapt to the flow conditions or to any movement of simulated bodies. In mesh-free techniques, the number and distribution of particles have to be defined in the interior of the computational geometry. Usually, the mesh design is one of the most time-consuming processes of a numerical simulation, requiring many human hours to achieve an efficient mesh design. It was estimated that over 50% of modellers' time in industrial applications is dedicated to design of the computational mesh [8].

The definition of the physical phenomena is done through the selection of the governing equations. Most commonly, simplifications and assumptions have to be considered, in order to build the system of equations and facilitate the solution process. Physical properties of the medium (e.g. water), and the other materials comprising any structures in the domain have to be defined before the commencement of the solution process [22]. Density, viscosity, elasticity and other material properties are usually inserted into the model before the solution process.

8.3.2.2 Boundary and Initial Conditions

As already mentioned, the solution in the interior of the domain is specified by the governing equations, while at the borders of the domain and on the embedded structures in the domain, boundary conditions must be applied. These conditions determine the values of the variables or the relationship between different variables on the surrounding and the internal boundaries of the computational domain. The boundary conditions can be steady over time or have a pre-specified transient behaviour. For a fluid flow simulation, the boundary conditions of the velocity u, pressure p and temperature T should be determined.

Two very common boundary conditions are the Dirichlet or first-type boundary condition and Neumann or second-type boundary condition, which are referred to as the fixed boundary condition and the partial differential (gradient) boundary condition, respectively. There are many other mixed-type boundary conditions derived from these two primitive types [24]. Many numerical models give the option of defining symmetrical and cyclic boundary conditions in order to decrease the computational cost and exploit the symmetry of the problem. Another boundary condition that is particularly useful for simulations of water waves is the Sommerfeld boundary condition, often referred to as the free radiation or open boundary condition. The Sommerfeld condition allows free propagation of reflected waves from the coast to the open ocean without noticeable local disturbances.

Initial conditions define the variables' values in the domain and on the boundaries at the beginning of the simulations, $t = 0$. In practice, if the field values of the variables are known in advance, they can be applied as the initial state in the domain. Otherwise, only an assumption can be made for the initial conditions, which is usually taken as an equilibrium case.

The correct definition of the boundary conditions is crucial for realistic simulations and accurate initial conditions can make the simulation more efficient, as it can converge to the final state more quickly.

8.3.3 Discretisation Methods: Solution

8.3.3.1 Finite Difference Method

The finite difference method (FDM) is one of the oldest and simplest ways to solve PDEs numerically and is believed to have been introduced by Euler in the eighteenth century [13]. According to the FDM formulation, the time and space continuum have to be discretised into a finite number of discrete grid points, and then the PDEs that describe the physical problem are approximated at every grid point [7]. The core of the method is based on the fact that the value of a variable at a given point is calculated by a certain number of neighbouring points. Only one equation is solved per grid point with only one unknown variable, which is the same unknown variable at all the neighbouring cells. This technique allows the differential equations to become algebraic and be solved by computer models.

Usually, an approximation of the derivatives of the PDEs can be obtained by means of Taylor series expansion or polynomial fitting with respect to the coordinates of the grid point. In practice, a truncation is applied to the high-order terms of the approximation, which induces the so-called truncation error. The accuracy of the FDM can therefore be first-order, second-order or higher depending on the omitted terms of the approximation. Additionally, the values of the variables can be computed at locations other than the grid points, by interpolation [7].

Apart from accuracy, the FDM should be characterised by consistency, convergence and stability. Consistency requires the authentic representation of the PDE by the FDM for capturing the physical behaviour of the problem. For that, the FDM schemes have to be at least first-order accurate and the truncation errors should tend to zero for increasing spatial and temporal resolution. Convergence refers to the ability of the FDM schemes to approach the true solution of the problem. Provided that the PDEs and the boundary conditions are correct, continuous refinement of the mesh and decrease of the time step should lead to this true solution. Stability refers to the emergence and augmentation of errors in the numerical solution of transient problems that can 'blow up' the simulation, if the errors become larger than the actual computed values of the variables. Stability depends on the formulation of the numerical schemes employed. Explicit numerical schemes, which calculate the values of the solution based only on the previous time step and not the neighbouring cells at the examined time step, can be unstable. A way to overcome this issue is by defining stability criteria, such as the Courant number [25] that restricts the numerical solution to propagate over no more than one grid point in one time step [7]. On the other hand, implicit numerical schemes use iterations between two time states. In general, these schemes are harder to implement, but provided that they are formulated correctly, they do not suffer from instabilities.

The formulation of FDM schemes depends on the selection of the neighbouring points for the calculation of the approximation of the PDEs. This characterises the FDM and it affects the evolution of the solution in the numerical domain. Some of the commonly used schemes are the 'backward difference', 'forward difference' and 'central difference' that respectively use the backward node $(i-1)$, the forward node $(i+1)$ and a combination of both to approximate the value at node i [7].

The FDM is applicable only when the domain comprises a grid. In theory, this method can be applied on any grid, but it is mainly used for structured grids, because the regularity of structured grids facilitates the implementation of the FDM and makes it work very efficiently. It is also easy to employ high-order schemes on regular grids. On the other hand, the employment of the FDM at unstructured grids or complex geometries might raise important issues, making this method impractical for many practical engineering applications. Another disadvantage of the FDM is that the conservation of the variables in the equations is not necessarily guaranteed [13].

8.3.3.2 Finite Volume Method

The finite volume method (FVM) is a discretisation method for solving PDEs by transforming them into algebraic equations around a control volume. Control volumes are subdivisions of the computational domain that do not overlap. Each computational volume is represented by a computational cell, which has a node at the centre, where the variables are computed, and vertices and edges that connect it with the neighbouring cells.

The FVM and FDM have inherent similarities, since the former can be considered as an integral formulation of the differential form of the latter over the control volume around the nodal point. As such, the FVM and FDM have equivalent expressions for the numerical schemes that they employ. Consequently, the numerical accuracy and stability of the FVM can be assessed by the same criteria used in the FDM, such as the examination of the Taylor series expansion [7].

In the solution process, interpolation methods are employed for calculating the values of the variables at the surfaces of the control volume based on the values at the nodes and the adjacent cells. In staggered grids, the scalar variable Φ is defined on the cell centre, while the fluxes F are defined on the cell boundaries. Within the control volume, the conservation law is satisfied:

$$\frac{\partial}{\partial t} \int_V \Phi + \int_S \mathbf{F} \cdot \mathbf{n} dS = 0, \tag{8.37}$$

where V is the control volume, S is the surface around the control volume and n is the outward-pointing normal vector on the surface. Considering that the fluxes between adjacent cells are shared and thus are the same between the cells, the conservation law can be applied from a local level to a global level. As a result, the FVM formulation ensures the conservation of the variable both mathematically and numerically [7].

The FVM is a very popular approach for engineering applications, because the approximated terms of the PDEs on the cell have physical analogues. This makes FVM schemes comprehensive and easier to program [13]. From a practical point of view, another considerable advantage of the FVM is the ability to accommodate any type of grid, making it applicable for domains of high complexity. From a mathematical perspective, the formulation of the FVM makes it conservative by definition, since the control volumes that share a boundary have the same surface integrals, which describe the same convective and diffusive fluxes. On the downside, the development of high-order accuracy numerical schemes in three dimensions is more difficult than in the FDM, because the FVM requires three levels of approximation, namely interpolation, differentiation and integration.

In computational fluid flows, the FVM is a broadly applied method able to simulate highly nonlinear processes and distorted fluid surfaces, such as wave breaking. Its success lies in the unrivalled advantage of the FVM for accommodating multi-phase flows and simulating the interaction between air and water in free surface flows. The FVM has been adopted by many CFD software packages for practical applications, since it can handle unstructured and self-adjustable meshes for freely moving objects as well [26].

8.3.3.3   Finite Element Method

Similarly to the FDM and FVM, the finite element method (FEM) is a technique for descretising PDEs used in numerical simulations. Historically, the FEM was used initially for structural and solid mechanics. In the FEM, the computational domain is subdivided into a number of elements that are related to computational nodes. The number of nodes is always less than the number of elements, because each node can be shared by various elements. The discrete governing equations are constructed from contributions to the element level, which are later recombined. The characteristic distinguishing FEM from the other methods is that it employs variational methods in order to minimise the error between the approximated numerical solution and the true solution [27].

The FEM variational methods for solving PDEs assume that the solution has a prescribed form and belongs to a function space. The function space is built by varying functions that usually have polynomial forms. The varying functions connect the nodal points, which can be the vertices, mid-side points, mid-element points, etc. of the elements. There are also shape or interpolation functions, which are used to approximate the true solution of a variable $\Phi(x,t)$ in the entire domain as the linear summation of M nodal values of $\Phi(t)$ multiplied by the shape function $S(x)$ at each nodal point j and time t:

$$\Phi(x,t) = \sum_{j=1}^{M} S_j(x)\Phi_j(t). \tag{8.38}$$

As discussed, the numerical and the true solution of the PDEs are not the same irrespective of the discretisation methods. They differ by a residual ε, which is the error between the true solution of the PDE and the computed value. In order to minimise the residuals in the global solution, the FEM employs weighting functions W_k, which are built in such a way as to drive the weighted integral of the residuals to zero:

$$\int_V W_k \varepsilon \, dV = 0. \tag{8.39}$$

Different weighting functions have been developed, such as the collocation method, the least-squares methods and the Galerkin method, with the latter being the most popular method resulting in accurate results at the expense of great computational effort [7].

The FEM formulation eliminates all the spatial derivatives from the PDEs, and therefore differential-type boundary conditions for transient problems and algebraic type boundary conditions for steady-state problems can be considered. That makes the FEM very efficient, since it can incorporate higher-order accuracy and it is flexible in implementing different boundary conditions. The shape functions and the integral formulations of the FEM on the boundary conditions give the method a strong and rigorous mathematical basis [28]. The FEM is also known for its ability to handle complex geometries, including structured or unstructured grids consisting of triangular or quadrilateral elements.

Comparing with the previous methods for discretising the PDEs, the FDM and the approximate the governing equations in a differential and in integral form respectively, while the FEM directly approximates the PDEs using shape or approximation functions. Despite the fact that the FEM is fundamentally different from the FDM, the two methods have an inherent similarity referring to the forward-time-difference scheme, which is used to express the spatial derivatives at the next time step. In general, though, there is no clear relationship between FEM and FDM when complex shape functions are employed. A considerable limitation of FEM is the absence of well-established techniques for determining the numerical accuracy and stability of the computations, which restricts the expansion of FEM in CFD applications, where long time integration is often required. In contrast, an important advantage of the FEM and the FVM for realistic engineering applications is the efficient handling of complex geometries [7].

8.3.3.4 Spectral Method

The spectral method (SM) belongs to the generic group of methods of weighted residuals, which evolved from the traditional Galerkin methods. At core, these methods make use of trial and test functions, also known as approximation and weight functions, respectively [29]. In a similar way to the FEM, the SM is based on the Galerkin method. The difference between the two methods is that the FEM is characterised by the use of local low-order polynomials as test and trial function in subdomains called finite elements, while the SM uses the Galerkin method in combination with a Fourier representation [30]. It should be noted that in the SM the unknown coefficients of the approximation formulations are identified globally and not at the local nodes [31].

What also distinguishes the SM from similar discretisation methods is the selection of trial functions, since in the SM these functions comprise infinitely differentiable global functions. The selection of the test functions determines the spectral schemes. The most commonly used schemes are the Galerkin, the collocation or pseudospectral and the Tau version of the SM [29]. In general, the SM is considered to be of high accuracy with a relatively low computational cost [7]. However, the accuracy of SM is related to the number of grid points in combination with the test and trial functions employed [30]. Therefore, under certain conditions, the SM can yield larger errors than other discretisation methods [13].

As mentioned, the FEM and SM have inherent similarities in the way they are formulated and the functions used to approximate the solution. Comparing equation (8.38) for the FEM and the equation

$$\Phi(x,t) = \sum_{K=0}^{M} \alpha_K(t)\Phi_K(x), \tag{8.40}$$

for the SM, it can be seen that the formulations are equivalent, with K referring to the modal number in the wavenumber space in the SM and j referring to the nodal number of the finite elements in physical space in the FEM. In the SM, the function $\Phi_K(x)$ is continuous, representing the wave modes for different wavenumbers, and $\alpha_K(t)$ is an unknown coefficient associated with the wave mode K. Once the functions are selected, the accuracy of the SM depends only on the total number of wave modes M included in the summation [7].

Commonly, the test function $\Phi_K(x)$ is expressed in terms of simple periodic functions, such as discrete Fourier series for evaluating the spatial derivatives ($\Phi_K(x) = e^{iKx}$). When periodic functions are considered, care should be taken to eliminate aliasing, which is a common numerical error that causes different signals to appear the same, due to poor sampling. The SM can be easily generalised to include high-order derivatives, and, at the same time, techniques like the fast Fourier transform can be used to decrease the computational cost significantly, while maintaining the same level of accuracy [13].

As discussed, the SM has several advantages in comparison with other discretisation methods. However, its application is limited to specific conditions, such as uniform computational grids and periodic functions. A considerable drawback of the SM for applied engineering is the fact that it can handle only relatively simple computational domains [7]. For this reason, the SM is less suited than the FEM and FVM to general-purpose CFD simulations. Nevertheless, the SM can be very useful for certain types of simulation, such as wave propagation and turbulence [13].

8.3.3.5 Boundary Element Method

The boundary element method (BEM) is a discretisation method to solve PDEs by employing boundary integral equations (BIEs). The distinguishing characteristic of the BEM is the use of the exact solution of the differential equation in the computational domain and a parameterised solution on the boundary [32]. The history of the BEM can be traced to the end of the nineteenth century, with important developments taking place between 1960 and 1980 and the term BEM coined in 1977. A rich bibliography is available in books by Brebbia [33, 34]. Originally, the BEM was introduced for steady boundary-value problems described by the Laplace equation, but it was later extended to applications of potential flows, groundwater flows, fracture mechanics, and viscous fluid flows [7]. The BEM is increasingly used in structural mechanics thanks to its ability to handle material and geometric nonlinearities in the simulations. A common feature in all BEM approaches is the employment of fundamental solutions that are free in space and controlled by a point source [35].

BEM solutions are possible only for certain types of PDE, when the exact solution is known, and based on the mathematical identity between volume and surface integration. The BEM solution process requires the discretisation of the domain by a series of N linear elements with N nodes, where N boundary conditions are applied to specify a variable ϕ, its gradient or their relationship. As a result, the number of unknowns is reduced, guaranteeing a unique solution of ϕ and its gradient on the boundary. In practice this means that the BEM reduces the dimension of the problem by 1, converting 2D problems into 1D and 3D problems into 2D. However, this reduction does not necessarily translate into computational efficiency, because the resulting matrix is dense [7]. The solution of the PDEs in the domain has to be determined by means of boundary potentials, and provided that the boundary conditions are known, the solution can be calculated anywhere in the domain. Commonly used domain solutions for classical boundary value problems are Green's third identity for potential theory, Betti's formula for elasticity theory, and the Stratton–Chu formula for electrodynamics [32]. The numerical solution of the BIE on the boundary requires the employment of functions that are dependent on a finite set of parameters. These functions can be global, such as spherical harmonics or polynomials, or trial functions, similarly to the FEM. The approximate solution is based on linear equations related to the set of parameters through various methods, such as the collocation method, the Galerkin method and the least-squares method. The resulting linear system of equations is relatively straightforward to solve [32].

The BEM incorporates several advantages, the most important being the simple discretisation that needs to be defined only on the boundary. Especially for 2D simulations, where the boundary is represented by a curve, the input data and storage methods can be very simple. Additionally, the BEM is very efficient for simulating problems with unbounded domains, which essentially allows for simulations with infinite or semi-infinite domains. Thus, from the modeller's point of view, the BEM can be very convenient, because it minimises the amount of effort required to build the computational mesh. For practical applications, the BEM can accommodate adjustable meshes and symmetrical boundaries in a simplified way [35]. Moreover, the method is ideal for applications where the relevant data are given by boundary values and not in the interior of the domain, and it exhibits high-order accuracy and convergence [32].

On the other hand, the biggest drawback of the BEM is the need to know explicitly the fundamental solution for the PDEs, which is possible only for linear PDEs with constant or specifically varying coefficients, but not for problems with inhomogeneity or nonlinear PDEs. As a consequence, the BEM cannot be considered as a general solution for PDEs and it is certainly less versatile than the FEM, FVM and FDM [7]. The evaluation of the choice of the BIE and the approximation methods usually requires considerable mathematical analysis that can be time-consuming. Another issue with BEM arises for non-smooth boundaries with corners and edges and for locations where the boundary conditions are discontinuous. On these locations, singularities can appear that require special numerical treatment [32, 35].

In comparison with other discretisation methods, the BEM is most closely related to the FEM, because both numerical techniques are based on the method of the weighted residuals. However, they differ in the discretisation of the domain. The fact that the BEM does not require discretisation of the interior of the domain gives it a clear advantage when simulating concentration problems or problems with open boundaries. On the other hand, the FEM does not require knowledge of the exact solution, which makes it more applicable for practical problems [35]. Despite the fact that the popularity of the BEM has increased significantly in recent years, the FEM is still the most commonly used discretisation method, having more than six times as many related publications as the BEM [24].

8.3.3.6 Meshless Methods

Meshless methods have been developed in the last few decades in order to overcome some problems of the mesh-based techniques that are mainly related to the problem geometry and boundary conditions. Meshless techniques can be readily applied to complex geometries governed by a system of PDEs that employs arbitrarily distributed particles without requiring any mesh for connecting these particles [7]. In practice, though, many of the meshless methods employ a background mesh for various reasons associated with integration of system matrices and for improving the overall stability and accuracy of the simulation. In some cases, the predefinition of the particles' properties requires a background mesh as well [36]. There are various mesh-free techniques developed in the last three decades that differ by means of function approximation and implementation process. The most prominent techniques are smoothed particle hydrodynamics (SPH), the finite point method, the diffuse element method, the element-free Galerkin method, the reproduced kernel particle method, the HP-cloud method, the free mesh method, the meshless local Petrov–Galerkin method, the point interpolation method, and the mesh-free weak-strong form [22].

In contrast to classic mesh-based techniques for continuum mechanics solvers that might have issues with simulating problems with large deformations, phase transformations, discontinuities, breakage of material into fragments and crack growth, mesh-free techniques can be more flexible and easily applied. Additionally, the use of mesh-free methods minimises the concerns about the mesh quality and the human effort required to build and evaluate the mesh [36]. The computational cost of re-meshing and mapping the variables' fields after every iteration is not a concern in meshless techniques, but there is no guarantee that a mesh-free simulation will be computationally efficient [7]. For more details about meshless methods, there are dedicated books that discuss the advances in the field in greater depth [22, 37].

In the solution process of a meshless method the particles can be distributed in a random or regular way and they can be defined as stationary or moving. A certain number of particles is necessary in every simulation to represent the physical problem adequately. Each particle node has a domain of influence with radius r, as seen in Figure 8.3, which determines the local connectivity with the other particles. The

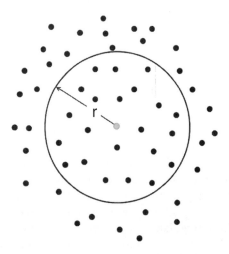

Figure 8.3 Illustration of the domain of influence of the central particle in meshless numerical approaches.

collective behaviour of the neighbouring particles can be used to approximate the function and/or its derivatives. The definition of the domain of influence is crucial for the reproduction of the physical phenomena, since it has to be large enough to represent accurately the local derivatives with many particles, but small enough to be locally representative [7].

We focus here on the SPH method, since it has been applied successfully for many fluid flow simulations, including breaking waves. SPH is a kernel-based method and was introduced in 1977 [38, 39] to simulate phenomena in astrophysics with very large computational domains. The conventional SPH method might suffer from instabilities near boundaries, but appropriate numerical damping techniques can account for that. A similar method, namely the moving particle semi-implicit (MPS) method, was also developed with the aim of improving the computational efficiency when the list of neighbouring particles is generated [40].

As a meshless technique, SPH does not require a grid for calculating the spatial derivatives of the equations of the fluid flow, which are instead solved for every particle. An interpolation formula based on a kernel estimation technique is employed to calculate the interactions between the particles and reveal the collective behaviour of the fluid [41]. Consequently, SPH is strongly bonded with statistical concepts. From a mathematical point of view, the SPH particles are considered as interpolation points, while from a physical point of view they represent parts of the material. Issues might arise in the application of the SPH method due to differences in the particle number density and the probability density functions. These issues can be resolved through appropriate treatment of the momentum equations [42].

The first step in the solution process of the SPH method is to discretise the computational domain with a set of particles and impose the boundary conditions. A spatial distance, known as the 'smoothing length', is given to every particle, over which the particle's properties are 'smoothed' by a kernel function. This function defines how the properties of the particles are summed in the range of the kernel. The value of a quantity is found through interpolation using the integral interpolant [42]

$$A_I = \int A(r') W(r - r', h) dr',$$ (8.41)

where the function W is the kernel and dr' is a differential volume element.

For numerical applications, the integral interpolant is approximated by a summation interpolant over the mass elements:

$$A_s(r) = \sum_b m_b \frac{A_b}{\rho_b} W(r - r_b, h),$$ (8.42)

where b denotes the particle label, m_b is the mass of the particle, and r_b and ρ_b the position and the density of the particle respectively.

The SPH method has inherited all the advantages of mesh-free techniques compared to traditional mesh-based numerical modelling, as discussed in detail by [36]. Complementary advantages, particularly useful for hydrodynamic problems, include the use of particles with different properties for representing different materials, such as sediment or floating bodies; savings in computational resources in terms of central processing unit (CPU) time and memory, thanks to the fact that only the regions of interest are discretised by a large number of particles; and multi-physics simulations, thanks to the commonalities between SPH and molecular dynamics [42].

On the downside, the SPH method has several limitations and issues, mainly referring to the stability, accuracy and convergence properties of the simulations, especially when large deformations and impulsive loads occur that cause high distortion to the particles [22]. The computational cost of the simulations might also be unrealistic for some practical engineering problems, when many particles are required. Other errors are associated with the integral interpolant and the summation interpolant [42]. Further issues might arise due to the weak compressibility of the simulated fluid, especially when the flow velocity has the same order of magnitude as the speed of sound [43].

8.3.3.7 Lattice Boltzmann Method

The lattice Boltzmann method (LBM) unlike the previous discretisation methods, simulates the fluid flow not by approximating the PDEs that describe it, but by solving the fundamental mass and momentum conservation equations at a mesoscopic level. The collective behaviour of this level can then be used to reproduce the macroscopic properties of real fluid flows [44]. The method is based on the lattice gas cellular automaton (LGCA), a technique that emerged in the 1980s for solving flow problems using discrete lattices, and it corresponds to a simplified molecular model operating with discrete space, time and velocity variables. The LBM can be categorised as a problem-based discrete formulation method [7] and is considered as a self-standing research field for fluid flow modelling.

To better understand the operating principle of the LBM, the levels of fluid motion should be considered. There are three levels that describe the fluid motion: the microscopic or molecular level, where the motion is reversible; the mesoscopic or kinetic level, where the motion is irreversible and can be described by the Boltzmann approximation; and the macroscopic level, where the approximations of continuum mechanics are valid. The foundations of the LBM lie in the theory that the macroscopic features of a fluid flow can be replicated by appropriate Boltzmann models, which are based on probabilistic assumptions [7]. A necessary assumption for moving from the Boltzmann level to the continuum level, described by the NSE, is that the mesoscopic movement should be much smaller than the macroscopic flow variations [45].

The solution process of the LBM requires the definition of the lattice and its properties. For demonstration purposes, a regular hexagonal symmetric lattice can be considered, where each lattice site is connected with the neighbouring lattice sites by six vectors, as seen in Figure 8.4. All the particles located in the lattice should have unit mass [44]. According to the LGCA technique, each node can host 0 or 1 particle that can move only along the six prescribed directions in a process called propagation. During a time cycle of the simulation, the particles can only move to a nearest neighbouring node, with shorter or longer movements being forbidden [44]. When two or more particles arrive simultaneously at the same node from different directions, they collide according to suitable collision laws that account for the conservation

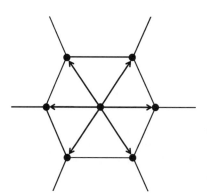

Figure 8.4 Hexagonal lattice.

of mass, momentum and energy [7]. This behavioural model of the particles is an idealised representation of a real gas, in which molecules can move freely in any direction. The unparalleled advantage of the LGCA representation is the simple coding for numerical models that it offers, which can be based on a single binary digit per lattice site and its direction. This intrinsic feature allows for excellent parallelisation in the solution process as well [46].

The advance introduced in the LGCA solution process by the LBM is that the Boolean particle number in a lattice direction is represented by an ensemble average, namely a density distribution function. Essentially, this neglects the individual motion of single particles and averages over larger numbers of particles. The particles' distribution function can be formed as the kinetic equation

$$f_i\left(x+c_i\Delta x, t+\Delta t\right) = f_i\left(x,t\right) + \Omega_i\left(f_i\left(x,t\right)\right), \quad i = 1,2,\ldots,M, \tag{8.43}$$

where f_i denotes the particle distribution function in the ith direction, Ω_i is the collision function, Δx and Δt the space and time intervals, c_i is the local velocity and M is the number of discrete directions of the particle velocities. Equation (8.43) has similarities with the forward difference scheme of the FDM, but without having knowledge of the PDE that the function represents. When Δx and Δt become sufficiently small, the lattice Boltzmann equation reduces to familiar PDEs for mass and momentum conservation equations [7].

The biggest advantage of the LBM is its simplicity in representing complex macroscopic phenomena with simple mesoscopic conservation laws and thus avoiding deriving specific complicated equations. The kinetic equation for the movement of the particles is linear, in contrast to the nonlinear expressions of velocities at the macroscopic level. Additionally, the pressure of the fluid can be readily calculated from the equation of state, which overcomes further issues that might occur with pressure–velocity coupling. Also, the development of simplified kinetic equations at Boltzmann (mesoscopic) avoids the complications in the modelling of molecular dynamics [46].

Applications of the LBM can cover single- and multi-phase fluid flows with irregular boundaries and complex phenomena, such as wave breaking, sediment transport and flow through porous media [7]. For the latter application, the LBM can be used to validate empirical laws, since it represents a numerical analogue of the physical problem [45].

8.3.4 Post-Processing

Post-processing refers to the last step of numerical modelling, performed after the numerical computation is complete, and it is a crucial part of the numerical process, since it allows for visualisation and interpretation of the results produced by the numerical model. Usually, these results comprise huge amounts of data stored in matrices that contain the calculated values of the variables at different output times and locations in the computational domain. It is nearly impossible to assess these results, unless they are transformed into graphical representations, such as plots of spatial distribution of the variables or time-histories of variables at specific locations. Post-processing brings the computational results closer to human comprehension with informative and illustrative figures.

In general, post-processing tools belong to the wider category of computer-aided engineering (CAE), which refers to the employment of computer software for facilitating engineering analysis. Similarly, there are pre-processing computer programs that are considered CAE software. The increasing availability of processing power and storage has led to the employment of more advanced solvers, complex mesh structures and the ability to handle of millions of computational nodes per numerical simulation. Simultaneously, post-processing software has advanced as well, providing effective tools for visualising complex geometries, 2D and 3D surface plots, particle tracking, colour PostScript output, animation for dynamic result display, etc. [8].

Being an essential part of the numerical modelling procedure, post-processing software is readily available as part of distribution package of numerical models. Commonly, post-processing tools are specifically designed for certain numerical models, in order to read their data in the most efficient way. In many cases, post-processing software is equipped with a graphical user interface (GUI), allowing for ease of use.

8.3.5 Best Practice in Numerical Modelling

8.3.5.1 Errors and Uncertainties

As discussed so far in this chapter, numerical modelling techniques have advanced significantly in the last few decades and become the norm in engineering applications. The available computation tools allow for studies of very complicated flow problems, and the capabilities of numerical models are constantly expanding. Despite the sophistication and efficiency of the numerical techniques employed, all numerical models are just a representation of the actual physical problems and their results are always subject to the underlying simplifications, approximations and assumptions. Consequently, all numerical models are associated with uncertainties and errors and should be used with an awareness of their capabilities and deficiencies.

The term 'uncertainties' refers to deficiencies caused by lack of knowledge regarding the physical processes of the phenomena examined and use of inadequate models, parameters or solution techniques. The effects of these uncertainties cannot be readily removed from the final solution. As a result, the uncertainties are case-specific and rely on the modeller's knowledge and experience and the availability of information regarding the problem. On the other hand, the term 'errors' refers to recognisable defects of the numerical models, due to the limitations induced by the methods employed in the modelling process. In this sense, the errors can be identified and minimised or taken into account in the final outcome. The errors in numerical modelling are classified here in seven categories; however, a universally accepted classification is not yet available [14].

The first category of errors refers to the selection of the model to represent the physics. In cases where some phenomena are not represented adequately or are omitted completely from the numerical model, important deviations may arise between the simulated and real solutions. For example, the modelling of turbulent flows is subject to the selection of the turbulence model from a range of different models that might produce significantly different results. In short, this category of errors comprises the inevitable consequences of solving the 'wrong' equations.

The second category refers to discretisation or numerical errors, which arise from the fact that the exact solution in numerical modelling is commonly calculated at specific locations only, namely grid nodes or particles, and not in the entire domain. At any other location, the solution is represented by interpolated values between the computational nodes. Therefore, these errors are the difference between the interpolated and the real solution at an arbitrary point in the domain. Theoretically, discretisation errors can be minimised by increasing the density of the computational nodes and by employing high-order numerical schemes for the interpolation, which can make the simulations computationally inefficient.

The third category of errors is associated with the errors induced by the iterations and convergence in the numerical simulation. As discussed in previous sections, the governing equations in a numerical model are usually solved iteratively, starting from some initial conditions and gradually finding the solution as time steps increase. The boundary conditions should always be satisfied, while the solution propagates in the domain according to some tolerance criterion, such as the Courant number. In most cases, restricting the tolerance criteria and increasing the computational nodes lead to convergence of the result to a final solution, which is usually referred to as a grid-independent or converged solution. However, there are simulations where convergence cannot be reached and the solution fluctuates between some limits, which is usually the case for turbulent flows.

Computers have a certain memory capacity for storing numbers and therefore the values used in the simulations are limited to a certain number of digits. The difference between the exact mathematical value and the rounded, approximated one stored in the computer's memory is called 'rounding' error. Most numerical codes used nowadays carry out 'double-precision' calculations, which corresponds to a floating-point representation with almost 16 decimal digits. 'Single-precision' and 'quadruple-precision' correspond to about 7 and 34 decimal digits, respectively. For more information, the reader should refer to the IEEE 754 standard for floating-point arithmetic [47]. The level of precision used in the computer codes might seem to cause no significant errors, however, it should be taken into account that during simulations millions of iterations take place and the cumulative effect of the rounding errors might cause instabilities in the simulation or unrealistic results.

A common type of error that is hard to predict or avoid at any stage of the modelling process is 'user error'. This error is caused by human mistakes or carelessness, such as mistyping or inserting the wrong parameters. If these mistakes do not cause a fatal error in the numerical algorithm, the model will calculate the solution, but of course the result will be unrealistic. Usually, more experienced users tend to make fewer errors or to find them and quickly fix them. Human carelessness is not forgiven by computer codes and if mistakes are made in the input, the output will be useless.

Another source of error, which might result in blocking the simulation or producing unnatural results, is associated with the numerical code itself. The numerical models used in engineering practice are usually complex and consist of many algorithms with hundreds or thousands of lines of code. Code errors include programming errors, inconsistencies in the algorithm, misuse of some variables, compiler errors and incompatibilities with hardware, and they are often referred to as 'bugs'. The contribution of the user community (particularly for open-source code) in discovering the code errors though testing the code with various applications and reporting them to the software developers is vital.

The last category of errors is associated with inaccuracies caused by lack of information regarding the physical problem. The source of these errors is uncertainties regarding the knowledge of the precise geometry, boundary conditions and flow characteristics. A small difference in these conditions might cause large deviations in the final results.

8.3.5.2 Recommendations and Guidelines

For some categories of error and uncertainty there are several actions that can be taken in order to minimise the problem or assess its impact on the overall result. These recommendations and best practice guidelines are presented in [14] for CFD solvers, but they can be generalised for any kind of numerical model. Similarly to physical modelling, the best results are achieved by experienced modellers with a good understanding of the physics and the processes involved in the problem.

A common source of error, related to convergence, can be addressed with appropriate monitoring of the residual during the simulation. Best practice recommends addressing convergence purely in terms of the levels of the residual error. The residuals should be dimensionless and checked both at a global level and at characteristic locations in the computational domain. Global residuals can be calculated for global balances, such as the conservation of mass, momentum, turbulent kinetic energy and mass flow balance. Local residuals can be addressed by local comparisons of the simulated results with target values of the parameters of the computation and might yield areas with poor mesh design or unrealistic velocities caused by incorrect boundary conditions. Moreover, the quality of the results can be examined not only from the residuals, but also from their rate of change with increasing number of iterations. For practical engineering applications, a converged solution might be computationally expensive and impractical. In such cases, the modeller should perform sensitivity analyses and be aware of the induced discretisation errors.

The discretisation errors in the simulation may be divided into spatial and temporal errors. Spatial errors can be assessed by modifying the resolution of the grid or the number of particles. Refinement and coarsening studies, commonly referred to as grid convergence studies, should be used to evaluate the quality of the results produced. High-order numerical schemes should also be used for spatial discretisation, thanks to the high accuracy that they offer compared to first-order schemes, which are more robust but not efficient for convergence problems. Temporal errors are induced by the magnitude of the time step and the selection of the numerical schemes for time discretisation. Similarly to spatial discretisation, time convergence studies should be performed to find the most appropriate time step, and high-order numerical schemes should be used for temporal discretisation. Spatial and temporal discretisation together determine the stability of the simulation, which can be controlled through appropriate conditions, such as the Courant number. The overall accuracy of time-dependent simulations is determined by the lowest-order component of the discretisation.

Round-off errors can readily be minimised if the accuracy of the representation of real numbers is increased. As mentioned, most numerical models are programmed for 64-bit platforms, corresponding to double precision. This should be adequate for the majority of applications, but for simulations that require even higher accuracy, some models have started offering quadruple precision.

The mesh is a vital element of numerical simulation. Its design determines the quality of the results and the computational efficiency of the simulation. The mesh quality should be checked while building the mesh with mesh-controlling algorithms, which are usually offered with mesh generators. Close to points of interest, such as fluid interfaces, structures, deformation in geometries caused by moving bodies, hydrodynamic, mechanical or thermal loads, the mesh resolution should be carefully selected. For simulations that include structures or have a complex geometry, CAD software should be employed in the pre-processing stage to facilitate the mesh design. The resulting mesh can be then imported into the numerical model for the simulation. There are many potential pitfalls in the transition from the CAD software to the numerical model, including the use of the wrong coordinate system or scale, different dimension units and even importing corrupted files. It is advisable to visually check the mesh before starting the simulation and during the solution process, if possible.

User errors can affect the quality of the simulation at various stages. If the modelling process is considered as a 'solution strategy' or 'modelling approach' of the physical problem, user errors can start from the preliminary stage of understanding the underlying physical processes and selecting the most appropriate model for the scope. Therefore, it is vital to spend some time understanding the physics of the problem. At this very first stage, it is well worth conducting a literature review of similar studies in order to be aware of the commonly adopted practices in the field. In some cases, the literature might not cover specific numerical applications, and validation of the model with experimental results might be necessary. If relevant experimental results are not readily available, the performance of any new or modified numerical model should be evaluated at least with some benchmark studies. When there are considerable uncertainties regarding vital elements of the problem, such as boundary, initial conditions and physical geometry, sensitivity analysis should be performed to give an estimation of the induced errors.

The numerical model should be chosen according to the available computational resources and level of desired accuracy. At the outset, the objectives and requirements of the simulation should be clear so that an efficient numerical simulation can be designed.

8.4 Numerical Modelling of Water Waves

In this section, the numerical models for simulating water waves are presented. The models fall into two categories, depth-resolving and depth-averaged, depending on their ability to calculate the flow variables throughout the water column. The models are classified according to the equations solved, in descending order of complexity, following the assumptions made on the governing equations.

8.4.1 Depth-Resolving Models

8.4.1.1 CFD/NSE Solvers

CFD is a specific field of fluid mechanics that employs numerical modelling to address fluid flows problems, including heat transfer and chemical reactions. In this sense, the term CFD can cover any kind of numerical modelling technique for fluid flow simulations. However, in practice, the term is used to refer mainly to computer codes that solve the fully nonlinear NSE in the three dimensions with the inclusion of appropriate turbulence closure models. In theory, this set of equations can handle very complex phenomena almost without limitations [7]. Therefore, CFD codes are very powerful numerical tools covering a broad field of applications from aerodynamics and turbo-machinery to marine and biochemical engineering. In industry and research, CFD codes are considered cutting-edge robust numerical models for gaining insight into physical phenomena, but their use is complex and computationally expensive, often requiring HPC facilities and well-trained staff.

CFD methods emerged in the 1960s for aerospace applications, but they became more popular in the 1990s, when HPCs became available and powerful personal computers became affordable. The latter triggered the expansion of CFD into many engineering applications in industry, and many codes were developed. The cost of commercial CFD software can be high, depending on the specifications of the numerical model and the parallelisation permitted [8]. On the other hand, there are many open-source freely distributed codes, but they are usually more challenging to use, as they usually come without GUI, documentation or customer support. Nevertheless, provided that the user has adequate programming skills and ways to verify the methodologies and validate the results of the open-source software, such a numerical model can be used for many cases at no cost and have expanded functionality, if the source code is modified. A list of the available free and commercial CFD codes can be found on the CFD-wiki webpage [48].

Nowadays, CFD models are commonly distributed in the form of integrated packages, with pre-processing, solution and post-processing capabilities, as discussed in Section 8.3. The solver is the actual computation tool of the model for finding the solution of the selected equations at every computational node in the domain at every time step. Each code uses different approaches for solving the NSE, different numerical schemes and different tools for pre- and post-processing. Model selection depends on the particular tools that the software offers, but also on the level of validation of the code, the available documentation, the size of the user community and the support and services offered. A thorough literature review is always recommended, since the codes are constantly updated with new capabilities.

With regard to wave modelling, CFD codes can simulate highly nonlinear problems, such as wave breaking, surf-zone effects, wave–current interactions and wave–structure interactions. The structures can have six degrees of freedom, undertake large translations and be considered as rigid or flexible bodies [7].

8.4.1.2 Potential Flow Models

PF theory is commonly used for applications with negligible turbulence, as discussed in Section 8.2.3. PF models are derived from the NSE using the Laplace equation approximations, and commonly the discretisation is performed with BEM using Green's theorem for surface integration [7]. Models of this type are the second most capable of simulating complex flows after CFD models and their computational cost is lower than that of CFD models, but still considerable.

Potential theory is applicable where the flow is irrotational and there are no boundary layers. It can be used for incompressible or compressible flows that are stationary or non-stationary. The area of applicability covers flows around aircraft, groundwater flow, acoustics, water waves, and electro-osmotic flows [49]. Most common applications refer to flow around aerofoils.

With regard to the modelling of water waves, PF theory can simulate highly nonlinear waves propagating in any water depth, nonlinear wave transformation over complex topographies, linear wave diffraction over large bodies and wave forces on large structures. The limitations of the method arise from the irrotationality assumption, which restricts PF models from simulating breaking waves, green water effects and wave interactions with small structures [7].

8.4.1.3 Hydrostatic Pressure Models

For specific types of flows, such as ocean circulation, the horizontal length scales are often of greater importance than the vertical scales. In some cases, though, information about the vertical circulation is essential, despite the fact that it is weak. For such cases, the problem cannot be reduced to two dimensions, but solving the full 3D NSE is computationally inefficient or even impossible. Instead, an assumption on the pressure distribution can be made to reduce the complexity of the NSE and simplify the solution process to a great degree, resulting in the derivation of hydrostatic primitive equations [7]. These are directly derived from the NSE, if the vertical momentum equation is reduced to an approximation of hydrostatic balance. As in the solution process of the full NSE, all the velocity components are obtained directly from the momentum equations [50]. When considering the hydrostatic approximation, though, the pressure is computed as hydrostatic, while the vertical component of the velocity is calculated through the vertical momentum equation. In this case, this momentum equation becomes just a diagnostic relation for the hydrostatic pressure. Consequently, the pressure difference between two points in the domain is $p_b = p_a + \int_z^\eta \rho g dz$. The 3D hydrostatic equations consisting of the continuity and momentum equations with the pressure assumptions can be found in the literature [51].

8.4.2 Depth-Averaged Models

8.4.2.1 Shallow Water Equations

The shallow water equations (SWE) were developed for the modelling of water waves in the shallow water regime, where the movement of the water particles can be considered basically horizontal and the vertical accelerations are negligible [52]. The SWE are derived from depth-averaging of the NSE, assuming that the horizontal length scale is much greater than the vertical. As a result, the horizontal velocity profile can be treated as uniform in depth, with very small vertical velocity components.

Mathematically, the solution of the continuity equation shows that the vertical velocity components are small and the solution of the momentum equation shows that the vertical pressure gradients are hydrostatic. The SWE consist of the mass conservation equation

$$\frac{\partial \eta}{\partial t} + \frac{\partial (\eta u)}{\partial x} + \frac{\partial (\eta v)}{\partial y} = 0, \tag{8.44a}$$

and the momentum conservation equations

$$\frac{\partial (\eta u)}{\partial t} + \frac{\partial}{\partial x}\left(\eta u^2 + \frac{1}{2} g \eta^2\right) + \frac{\partial (\eta u v)}{\partial y} = 0, \tag{8.44b}$$

$$\frac{\partial (\eta v)}{\partial t} + \frac{\partial (\eta u v)}{\partial x} + \frac{\partial}{\partial y}\left(\eta v^2 + \frac{1}{2} g \eta^2\right) = 0, \tag{8.44c}$$

which are derived for negligible Coriolis, frictional and viscous forces [53]. Here η is the total fluid column height; the 2D vector (u, v) is the fluid's horizontal velocity in the x- and y-direction, averaged across the vertical column; and g is the acceleration due to gravity.

As mentioned, the SWE have many applications in cases where the horizontal length scales are much greater than the vertical. Such applications include the modelling of tides, tsunamis, storm surges and river flows. SWE are not suitable for the deep water regime or when vertical phenomena are important. Different versions of the SWE have been developed in order to expand the applicability of SWE models. In their original form, the SWE are considered to be simpler than Boussinesq models (see below), because the flow is calculated as uniform and wave-dispersion effects are not taken into account for the wave propagation [7].

8.4.2.2 Boussinesq Equations

The Boussinesq equations were developed in order to describe the propagation of waves in intermediate water depth. In this transitional water regime, the water particle motion cannot be treated as horizontal, while linear theory does not hold as waves become steeper due to the shoaling effect. According to the Boussinesq approximation, the vertical velocity distribution is not calculated by the nonlinear balance equations. Instead, it varies almost linearly over depth, while the horizontal component of the velocity is assumed to be constant in the water column.

The original Boussinesq equations [54] were derived for a horizontal seabed by substituting the velocity potential function into the nonlinear dynamic and kinematic surface boundary conditions. For a non-horizontal seabed the one-dimensional Boussinesq equations read [55]

$$\frac{\partial \eta}{\partial t} + \frac{\partial}{\partial x}\left[(d+\eta)\bar{u}_x\right] = 0, \tag{8.45a}$$

$$\frac{\partial \bar{u}_x}{\partial t} + \bar{u}_x \frac{\partial \bar{u}_x}{\partial x} + g\frac{\partial \eta}{\partial x} = \frac{1}{2}d\frac{\partial^3\left(d\bar{u}_x\right)}{\partial t \partial x^2} - \frac{1}{6}d^2\frac{\partial^3\left(d\bar{u}_x\right)}{\partial t \partial x^2}, \tag{8.45b}$$

where $\bar{u}_x$ is the vertically averaged horizontal velocity. Essentially, these equations are the SWE with corrections for the vertical acceleration, seen on the right-hand side. The third-order derivatives are the result of the Laplace equation forcing the vertical velocity of the velocity potential function to be expressed in terms of the horizontal velocity distribution. These equations can be readily expanded into two horizontal dimensions [52].

There are many variations of the Boussinesq equations. Those as derived by Peregrine [55] refer to weakly dispersive and weakly nonlinear formulations, limited to the modelling of very long waves. The depth-averaged velocity expression makes this model invalid for simulating short waves in the coastal zone and for propagating waves in deep water, since the averaged velocity assumption breaks down for depths greater than one-fifth of the wavelength [6]. Madsen *et al.* [56] and Nwogu [57] extended the Boussinesq equations by employing a linear operator and adjusting the velocity profile respectively, in order to obtain better dispersive properties. Despite the improved dispersion relation, Boussinesq equations are limited to the simulation of weak wave nonlinearity, being valid only for the transitional regions from intermediate to shallow water regime, before wave breaking occurs [7]. The evolution of the Boussinesq equations through the contributions of many researchers and their engineering applications are discussed in the recent work of Brocchini [58].

The Boussinesq equations were originally used for the simulation of long waves, like seiches, but the inclusion of energy dissipation terms due to wave breaking and eddy viscosity can extend the use of Boussinesq equations to challenging engineering applications beyond the wave breaking limit [7]. In practice, Boussinesq-type equations are able to simulate wave propagation in coastal regions and handle surf zone hydrodynamics [59], such as wave run-up. Moreover, the Boussinesq model does not have the assumption of periodicity in the flow, making it applicable for the modelling of impulsive motions, such as solitary waves, tsunamis, landslide-induced waves and unsteady undulation in open channels [7].

Care should be taken in the numerical solution of the Boussinesq equations in order to prevent corruption of the solution due to truncation errors arising from the differencing of the leading-order wave equation terms. These errors should never be allowed to become comparable to the dispersive terms [60].

8.4.2.3 Mild-Slope Equation

The original mild-slope equation (MSE) was derived to describe the propagation of waves on slowly varying seabed, with the assumption that waves were linear. The MSE was proposed originally by Eckart in 1952 [61] and revisited by Berkhoff in 1972 [62]. It was developed to account for wave diffraction and refraction over large regions [7]. There are three different formulations of the MSE depending on the nature of the problem. The hyperbolic MSE is used for time-dependent wave fields, the elliptic MSE for steady-state wave fields and the parabolic MSE for steady-state wave fields with a primary direction for the wave propagation [7].

Here the elliptic form of the MSE is presented, as firstly introduced by Berkhoff [62], yielding the steady state under the assumption of linear waves on a mild slope:

$$\nabla \left(cc_g \nabla \hat{\eta} \right) + \kappa_w^2 cc_g \hat{\eta} = 0, \tag{8.46}$$

where c is the wave celerity, c_g the group velocity, $\hat{\eta}$ the complex amplitude of the water surface elevation and κ_w the wavenumber [63].

The success of the elliptic MSE in coastal engineering applications lies in the correct representation of the vertical velocity profile based on the linear wave theory. The elliptic MSE is valid from deep to shallow water, reducing to the linear SWE for long waves in shallow waters and to the Helmholtz equation for short waves propagating in deep waters. However, the elliptic form of MSE requires specific boundary conditions, which poses limitations for some simulations. For example, when large areas are simulated in the coastal zone, it is not a trivial matter to specify boundary conditions for the shoreline, especially without knowing the wave breaking location. To tackle these issues, the parabolic form of the MSE was developed, which does not require a definition of the boundary conditions *a priori* [6].

The MSE is a popular tool for coastal engineering applications with monochromatic and irregular waves, from deep to shallow waters. MSE models are commonly applied to simulate wave field changes near harbours and coasts and in particular the waves' interaction with cliffs, seawalls and breakwaters. However, they are rarely used for wave run-up and simulations in the swash zone. The original form of the MSE has inherited limitations, being only applicable for linear waves, propagation over mild and imperme-able bottom geometry without accounting for energy dissipation. The inclusion of non-linear waves, steep bottom, energy dissipation and porous seabeds has led to a more advanced form of the MSE, namely the modified MSE (MMSE). The MMSE can describe the combined effects of diffraction and refraction of water waves, model the propagation and transformation over varying bathymetries and simulate random wave fields. Despite the modifications included, the MSE cannot account for the nonlinear wave–wave and wave–current interaction effects and simulation beyond the breaking point, when new wave components are generated [7].

8.4.2.4 Spectral Models

Spectral models represent a different approach to the modelling of waves than the NSE-based models presented so far. Here the wave propagation is described through the energy balance equations. Spectral models do not consider the waves as individual components, but as an energy spectrum [7]. Therefore, their basic operating principle lies in the fact that they do not resolve the phases of the components and thus are called phase-averaged models. This type of wave modelling is computationally cheap and is commonly applied for large-scale problems, with grid cells varying from tens of metres to tens of kilometres.

The mathematical formulation of spectral models is based on the conservation of energy of the wave spectrum. In the absence of currents the energy of the spectrum is a conserved quantity, but in the presence of currents the energy is not conserved due the effect of currents on the mean momentum transfer of the waves. However, the wave action, which is the energy divided by the relative radian frequency $(A \equiv E/\sigma)$, is conserved [64]. The basic formulation of the wave action equation used in spectral models reads [52]

$$\frac{\partial N}{\partial t} + \frac{\partial c_{g,x} N}{\partial x} + \frac{\partial c_{g,y} N}{\partial y} + \frac{\partial c_\theta N}{\partial \theta} + \frac{\partial c_\sigma N}{\partial \sigma} = \frac{S}{\sigma}, \tag{8.47}$$

where N and S are functions of $(\sigma, \theta; x, y, t)$, representing the action density spectrum and the source terms in the action balance equation respectively. $c_{g,x}$ and $c_{g,y}$ refer to the propagation velocities in x- and y-space respectively, which account for the shoaling effect. c_θ is the propagation velocity in θ-space, accounting for depth-induced and cur-rent-induced refraction. c_σ is the propagation velocity in σ-space, accounting for shift in the relative frequency due to the effect of depth and current variations.

Spectral models are ideal tools for applications in the ocean. In particular, they can be used for hindcasting of the wave climate, when coupled with atmospheric models. Moreover, spectral models can be used for nearshore applications, being able to simu-late the effects of varying topography and currents on the propagation of the waves [7]. Additionally, the source terms in the wave action equation can be used to replicate the effect of various physical phenomena. Depending on the complexity of the mathemati-cal model, the source terms can include wind–wave interactions, wave–wave interac-tions, such as nonlinear interactions and triad–wave interactions, energy dissipation due to white-capping or depth-induced breaking, wave–ice interactions, etc. [64].

Spectral models are usually characterised as first-, second- or third-generation models depending on the treatment of the nonlinear interactions terms in the source terms of the wave action equation. Briefly, in first-generation models, the nonlinear interactions are calculated implicitly through other source terms, while in second-generation models they are calculated with parametric methods using observed spectra. Third-generation models, on the other hand, calculate the nonlinear energy transfers explicitly, without imposing spectral shapes or energy levels. Most of the commonly used spectral models nowadays are third-generation, which makes them more versatile, but at the same time more computationally expensive than first- and second-generation models [65].

The limitations of spectral models arise from phase-averaging and the underlying assumption of the linear theory for wave propagation. Only the source terms try to compensate for these limitations by including nonlinear effects. Consequently, spectral models can be used in the far field for simulating waves in the deep and intermediate water regime, but in coastal waters, where the modelling of the wave propagation becomes more complicated, the spectral approach has several drawbacks. For instance, the absence of diffraction in the equations restricts the area of modelling to at least a few wavelengths away from obstacles. In some cases, the source terms attempt to provide some representation of wave diffraction, which is a phase-related process, and in general it cannot be tolerated by spectral models. Also, linear theory does not hold for shallow waters and cannot account for wave breaking or energy dissipation from friction [66]. The strong triad–wave interactions that occur in finite depth and shallow water cannot be described adequately by empirical formulations of the source terms, especially for complex wave trains with multimodal spectra. The challenge for spectral models is to develop a triad–wave interaction module that can address these issues in the near future [67].

8.5 Commonly Used Open-Source Software

In this section various models are presented that cover all the available numerical techniques for solving the governing equations of fluid mechanics discussed earlier in this chapter. Additionally, some basic tools for the design of MRE devices are included to allow for more specific studies. A quick search in the literature reveals a large number of numerical tools [3], making the selection of the most representative models and their grouping not a trivial task. The main selection criteria for the numerical models presented are the availability of a public licence for the software, the number of applications and scientific publications for each, the support offered by the developers and the size of the user community. The models are categorised based on the governing equations that they employ. A similar description of numerical models including commercial software as well is available in a recent report on MRE applications [4]. More general descriptions of numerical models that cover a wider range of applications can be found in various reports [14, 23].

8.5.1 CFD

8.5.1.1 OpenFOAM

OpenFOAM (Open source Field Operation and Manipulation) is an extensive software package that has been widely used for industrial and academic applications. It is a freely distributed open-source CFD Toolbox for solving continuous mechanics problems.

OpenFOAM was initially developed in the late 1980s at Imperial College, London, and it was later rewritten in C++, incorporating the advantages of object-oriented programming. The ESI group has distributed and maintained OpenFOAM since 2012, the latest version being v3.0 [68]. The OpenFOAM Foundation is the copyright holder of the code, which is available to users under the GNU General Public Licence. A community-driven release is also available through the FOAM-Extend project [69]. OpenFOAM can be compiled in Linux operating systems, but a Windows version is also under development.

OpenFOAM is a general platform that includes solvers for complex fluid flows, chemical reactions, heat transfer, solid dynamics and electromagnetics. Users can also benefit from the modularity of C++ to add new solvers or extend the existing ones and from the user-friendly syntax of the PDEs in the source code [70]. The simulations can be parallelised to a theoretically unlimited number of processors. A large number of utilities and tutorials are offered together with some basic documentation for beginners [71]. A book was also published recently to try to cover the needs of new users [72]. OpenFOAM might lack a GUI, but the growing user community makes it a very popular toolbox, which is constantly updated. Post-processing is commonly performed with the third-party open-source software Paraview.

Two recently developed libraries, IHFOAM [73] and waves2Foam [74], extended the capabilities of OpenFOAM to simulating wave generation and absorption. Active and passive wave generation techniques, irregular waves, directional spectra, porous medium flows and wave–structure interaction modules became available through these libraries.

MRE applications in OpenFOAM include simulations of oscillating water columns (OWCs) [75–77], floating bodies [78], tidal turbines [79], offshore wind foundations [80] and extreme waves at MRE deployment sites [81, 82].

8.5.1.2 REEF3D

REEF3D is a newly developed open-source CFD package distributed under the terms of the GNU General Public Licence. The development of the model was undertaken at the Norwegian University of Science and Technology (NTNU Trondheim) and the source code can be downloaded from the website [83], together with documentation and tutorials. What distinguishes REEF3D from other CFD packages is that it is focused on hydraulic and marine applications. REEF3D is written in C++, reaping the benefits of object-oriented programming, such as the ease of handling the source code and the convenient implementation of new features. The simulations can be parallelised using the MPI method, which allows compilation both on shared-memory machines and distributed architecture HPCs. The code can be compiled on Mac OS X and Windows platforms.

The physics of the simulated problem is resolved by the RANS equations, which are derived for multi-phase fluid flows. The free surface is captured using the Eulerian level-set method [84], which can offer a sharp interface without requiring reconstruction processing. The current distribution of REEF3D offers a wide variety of turbulence models, velocity–pressure coupling techniques and numerical schemes for the spatial and temporal integration. In particular, the conservative fifth-order scheme for the discretisation of the convection employed in REEF3D can prevent the non-physical damping of the waves. Mesh generation can be performed with DiveMesh, which is a toolbox freely distributed alongside REEF3D [85]. Complex geometries and structures can be included in the computational domain using the immersed boundary method.

REEF3D can also handle Cartesian topographic files for realistic bathymetries. Suspended and bed sediment transport can be included in the simulations. Multiple wave generation and absorption methods are readily available for numerical wave tanks together with complementary tools for measuring the flow properties and forces on structures [85]. For more sophisticated post-processing, the freely available third-party software Paraview can be used.

Despite being a new CFD tool with a small user community, REEF3D has already been employed in practical coastal engineering studies, such as wave–structure interactions [86], breaking waves [87, 88], scouring around cylinders [89] and modelling OWCs [90, 91].

8.5.2 Smoothed Particle Hydrodynamics

8.5.2.1 SPHysics and DualSPHysics
SPHysics and DualSPHysics are SPH numerical codes for simulating free surface flows. These models are products of international collaborative work involving Johns Hopkins University (USA), the University of Vigo (Spain) and the University of Manchester (UK). SPHysics and DualSPHysics are open-source models, which are freely available from the official website with good documentation [92] and pre- and post-processing tools [93].

The first version, SPHysics, was released in 2007 and written in FORTRAN90. The second version, DualSPHysics, was written in C++ and Compute Unified Device Architecture (CUDA) in order to be able to handle simulations on CPUs and graphics processing units (GPUs), respectively [92]. Using a GPU has shown important improvements to the computational efficiency of numerical simulations, allowing for applications with large domains for realistic engineering problems. The source code can be compiled in both Windows and Linux platforms.

The SPH modelling of water waves is a relatively new technique and experience is still being gained in the pitfalls and limitations of the method. Most of the drawbacks of DualSPHysics are inherited from the mathematical formulation of the governing equations. Therefore, any errors are usually associated with the integral interpolant, the summation interpolant and the distortion of the particles [42].

Several test cases confirmed the ability of DualSPHysics to deal with free surface flows in two and three dimensions. These tests refer to the classic dam break simulation; waves generated by a paddle in a beach; a tsunami generated by a sliding wedge; floating objects in waves; floating bodies with 2D periodicity and a focused wave group approaching a trapezoid [43]. Wave breaking, green water overtopping on decks and wave–structure interaction were also modelled with DualSPhysics [94]. Other studies include tsunami generation and loading on typical structures [95], estimation of sea waves on coastal structures [96] and modelling of breakwaters with detailed armour units [97]. With regard to MRE applications, DualSPHysics has been validated against experimental data for fixed OWCs [93].

8.5.3 Potential Flow

8.5.3.1 QALE-FEM
QALE-FEM (Quasi Arbitrary Lagrangian-Eulerian Finite Element Method) was first developed in 2005 by Ma and Yan [98]. It is based on fully nonlinear potential theory using a quasi-Lagrangian–Eulerian FEM approach for simulating water waves. At the

moment, there is no official release of the software, but QALE-FEM is used as an advanced research code for free surface flow problems.

QALE-FEM differs from other models in the coupling of two different discretisation techniques for the fluid flow: the classic Eulerian methods, which are based on fixed meshes, where the fluid moves relative to the mesh; and the meshless Lagrangian formulations that allow for sharp interfaces, since the material particles are represented from different nodes. Eulerian techniques can replicate large distortions of the interface between two materials, provided that the mesh is fine enough, but they require great computational effort and have difficulties when simulating moving objects. On the other hand, Lagrangian techniques might be diffusive. The Arbitrary Lagrangian–Eulerian (ALE) formulation exploits the merits of both techniques, minimising the computational cost of regenerating the mesh at each time step. The complex mesh is generated only once at the beginning of the simulation and used at all the other time steps following the movement of the fluid and structures. The employment of the FEM instead of other discretisation methods, such as the BEM, further decreases the computational cost. However, a drawback of the FEM is that an unstructured mesh requires re-meshing at each time step, which can be overcome when the QALE technique is employed [98].

Initially, QALE-FEM was created to simulate complex wave–structure interactions or three-dimensional overturning waves over uneven seabeds [99]. Other applications of QALE-FEM involves the generation and propagation of freak waves with a fully nonlinear solver [100]. In particular, the model was also used to replicate the extreme wave that occurred near the Draupner platform in the North Sea through the interaction of two crossing wave groups [101]. With regard to wave–structure interactions, QALE-FEM was employed to examine wave interactions with six-degrees-of-freedom freely floating structures [102] or with moored floating bodies [103]. The advances of the QALE-FEM approach in the simulation of the responses of floating bodies and overturning waves are gathered together in a book chapter written by the developers of the model [104].

8.5.4 Hydrostatic Models

8.5.4.1 POM

POM (Princeton Ocean Model) is a community general ocean circulation numerical model, which was developed at Princeton University by G. Mellor and A. Blumberg in 1977. The model is freely distributed from the website, together with documentation and visualisation tools [105].

The numerical model focuses on water circulation in areas such as semi-enclosed seas, open seas and global oceans. It calculates a wide range of wave properties including current flow, temperature and salinity. POM is distributed in different modules that can be used separately depending on the processes simulated. All these subroutines are described in the manual [106]. POM is a σ-coordinate model with the vertical coordinate scaled on the water column depth. The horizontal discretisation is performed with curvilinear orthogonal coordinates and an Arakawa C differencing scheme. The time differencing is done both explicitly and implicitly, which allows for finer resolution at the free surface and bottom. The model has an internal and external time stepping controlled by the Courant condition and the wave celerity. Full thermodynamics were also implemented in POM for permitting relevant simulations.

Based on POM, the PROFS (Princeton Regional Ocean Forecasting System) was developed to perform high-resolution, accurate model simulations in coastal areas and semi-enclosed seas, such as the Gulf of Mexico, the Caribbean Sea, the Santa Barbara Channel and Santa Monica Bay. PROFS can model realistic ocean environments utilising nested grid techniques, data assimilations and high-resolution atmospheric models [107]. POM has also been used to investigate atmosphere–ocean coupled processes that might occur over the Yellow and East China Sea shelves in winter [108]. With regard to MRE studies, POM was used to examine the effects of tidal energy extraction in flow patterns [109].

8.5.4.2 COHERENS

COHERENS (COupled Hydrodynamical Ecological model for REgioNal Shelf seas) is a three-dimensional multi-purpose operational numerical model used for simulating coastal and shelf seas, estuaries, lakes and reservoirs. The model is open-source and freely available under the GNU General Public Licence [110]. COHERENS was initially developed in the 1990s by several European institutions and the Management Unit of the North Sea Mathematical Models and the Scheldt estuary (MUMM, now OD Nature) [111]. The latest version is COHERENS V2 [110].

The model is written in FORTRAN90 and uses the MPI library for parallel computations. It can be compiled in Linux operating systems, but is easy to implement across other platforms as well. COHERENS is structured in a modular way, based on a hydrodynamic model enriched with side modules for the biological and sediment transport features. An Arakawa C grid formulation is used in Cartesian or spherical coordinates. The horizontal grid can be rectangular or curvilinear, while the vertical one uses σ-coordinates. The model can be run in 1D, 2D and 3D mode. Sub-grid nesting is also allowed for greater accuracy in specific areas.

At the beginning, the main focus of COHERENS was to simulate environmental conditions in the North Sea. The test cases refer to the simulation of the seasonal cycle of temperature and the microbiological and resuspension processes in the central part of the North Sea [112]. There are also many case studies, which include waves, tides, salinity and temperature examination, but they focus mainly on the morphology and sedimentary processes [110]. COHERENS has not yet been used in MRE studies, but among the publications mentioned one can find interesting work related to tides.

8.5.4.3 Delft3D

Delft3D is a software suite developed by Deltares for 3D computations in coastal, river and estuarine areas. The software is divided into modules and since 2011 the flow, morphology and wave modules have been open-source and freely available [113]. The model can be compiled in Linux and Windows machines and includes a GUI. The source code is written in FORTRAN [114].

The core of Delft3D is based on the FLOW module, which is a program used for 2D (depth-averaged) or 3D hydrodynamic and transport simulations. It can calculate non-steady flows and transport phenomena caused by tidal and meteorological forcing. Rectilinear, spherical or curvilinear boundary fitted grids can be employed. The vertical discretisation for 3D simulations is performed with the σ-coordinate system or the Z-layer approach. The morphological changes are calculated through the MOR module, accounting for both suspended and bed total sediment transport.

The MOR module is coupled with the WAVE module in a dynamic way, so that currents and waves act as the driving forces for the sediment transport and morphological changes affect the evolution of the flow and the propagation of waves accordingly [113]. The WAVE module is based on the third-generation wave model SWAN (see Section 8.5.8). It can simulate the generation of wind waves and the evolution of sea waves, accounting for the transformation of waves in coastal waters and the effects of nonlinear wave–wave interactions and dissipation [115].

The applications of Delft3D cover a wide area in hydraulic and environmental research, including the modelling of tides and storm surges; stratified and density-driven flows; river flow simulations; simulations in deep lakes and reservoirs; simulation of tsunamis; hydraulic jumps; bores and flood waves; freshwater river discharges in bays; salt intrusion; thermal stratification; cooling water intakes and waste water outlets; transport of dissolved material and pollutants; sediment transport and morphology; tidal currents along open coast; wave-driven currents (rip currents); and non-hydrostatic flows. Scouring effects around structures have also been studied with Delft3D [115], as well as storm impacts on beaches [116].

Recent MRE studies with Delft3D have involved wave energy resource assessment [117] and tidal stream energy resource characterisation in coastal areas in the Philippines [118] and Spain [119]. Different ways of appropriate numerical modelling of tidal energy extraction have been analysed using Delft3D as well [120]. Additionally, the hydrodynamic behaviour of horizontal-axis tidal turbines has been modelled with Delft3D, and validation with experimental data demonstrated the capabilities of Delft3D in simulating the wakes and energy extraction by the turbines [121]. Impacts on the morphology in Pentland Firth, Scotland, from tidal energy extraction have also been studied with Delft3D [122].

8.5.4.4 TELEMAC-MASCARET

TELEMAC-MASCARET was initially developed by the Laboratoire National d'Hydraulique et d'Environnement (LNHE), which is a department of Electricité de France [123]. The model is written in FORTRAN and can be compiled in Windows, UNIX and Linux platforms. TELEMAC-MASCARET is freely available open-source software that can be downloaded from the website [124]. GUIs are provided for the convenience of users.

TELEMAC-MASCARET is an integrated suite of solvers for numerical simulations in the field of free surface flows, including currents, waves, transport of tracers and sedimentology. The model uses high-capacity algorithms based on the FEM. Space discretisation is achieved with unstructured grids of triangular elements. This grid design allows further refinement in specific areas of interest. The code is divided into modules that are connected through a single shared library for exchanging information during the simulation [124]: The available modules are:

- SISYPHE: Sediment transport and morphodynamic processes
- DREDGESIM: Modelling and predicting dredging operations
- ARTEMIS: Wave propagation nearshore or agitation in harbours from long waves
- MASCARET: One-dimensional free surface flows based on Saint-Venant equations
- TELEMAC-2D: Simulation of free surface flows in horizontal space
- TELEMAC-3D: Hydrodynamic modelling using SWE or non-hydrostatic formulation
- TOMAWAC: Wave propagation in coastal areas with FEM and wave action equations.

Some examples of the applications of TELEMAC-MASCARET are given on the website [124], such as the quantification of incoming, outgoing and total flows of the Saint-Chamas hydraulic power plant in the Berre lagoon (France) [125]. The software can be used for monitoring of environmental properties, such as water temperature or salinity and vertically heterogeneous flows. MRE applications of TELEMAC-MASCARET refer to tidal energy resource assessment in various UK sites [126] and to the evaluation of the impact of a tidal array in the sediment transport at Ramsey Sound, UK [127]. The effects of waves on currents for tidal resources assessment have been also studied using TELEMAC in the region of Anglesey in North Wales [128].

8.5.5 Shallow Water Equations

8.5.5.1 SHYFEM

SHYFEM (Shallow water HYdrodynamic Finite Element Model) was developed at ISMAR-CNR (Institute of Marine Sciences – National Research Council, Venice) and is based on the SWE formulation. SHYFEM is an open-source and freely available numerical model under the GPL licence [129]. The main part of the code is composed in FORTRAN77, but there are parts written in C, FORTRAN90, Perl and Shell scripts [130].

SHYFEM solves the hydrodynamic equations of water flow in lagoons, coastal seas, estuaries and lakes using finite elements, which are superior to finite differences in accounting for complex bathymetries. The 3D shallow water hydrodynamic model in SHYFEM is coupled with a wind–wave model. This uses a semi-explicit algorithm for the integration in time, combining the advantages of both the explicit and the implicit numerical schemes. The bottom friction is formulated by the Strickler method, allowing the friction coefficient to vary with the depth. Sub-grid processes are simulated with the employment of the General Ocean Turbulence Model, a turbulence closure model. The advection and diffusion of tracers can be simulated by Eulerian or Langragian transport modules. At the open boundaries, the water level is calculated by the Dirichlet condition [130].

SHYFEM has been applied to lagoons, estuaries, coastal and open seas, lakes and also for operational forecasting. For these cases, the model was used to examine the hydrodynamics, the wind and tidal mixing, the stratification, the temperature of the water and other environmental properties. Another application of SHYFEM is in the modelling of storm surge on the Italian coasts, providing crucial information for areas under flooding risk, such as Venice [129]. So far, SHYFEM has not been used in MRE studies; nevertheless it can be considered as a useful tool for modelling the hydrodynamics in tidal lagoons and environmental impacts in coastal areas.

8.5.5.2 SWASH

SWASH (Simulating WAves till SHore) is an open-source freely available flow solver that was released relatively recently by Delft University of Technology. The source code is written in FORTRAN and it uses similar parameters and format to the SWAN model (see Section 8.5.8) [131]. It can be compiled in Windows, Linux and Mac OS X operating systems. The distribution includes various test cases for new users [132].

The mathematical formulation of SWASH is based on the nonlinear SWE with a non-hydrostatic pressure assumption. It can account for a range of physical processes, such as wave propagation frequency dispersion, shoaling, refraction, diffraction, nonlinear wave–wave interactions, wave breaking, moving shorelines, bottom friction, wave–current interactions, turbulent effects and wind effects on waves [131]. Therefore, it can be applied to water depths varying from deep water to the swash region. Thanks to its computational efficiency, SWASH is commonly used in operational engineering problems, at scales greater than CFD. However, it has been also used at the laboratory scale for validation regarding the propagation of water waves [133].

SWASH's computational efficiency is achieved by the vertical discretisation in layers, which allows it to operate as a depth-averaged model (single-layer mode) or depth-resolving model (multi-layer mode). Additionally, an innovating edge-based compact difference scheme (Keller–Box) makes SWASH very efficient in simulating accurately the free water surface, even with very coarse vertical discretisation. This is thanks to the fact that Keller–Box allows for a zero pressure boundary condition on the free surface. Consequently, the numerical formulation of SWASH makes the SWE model able to simulate wave propagation in deep water as well and to achieve comparable results to Boussinesq models in intermediate waters [134]. In comparison with CFD models that require very fine vertical resolution, SWASH can achieve high accuracy with a small number of layers. In practical engineering applications, two layers are commonly employed [135].

SWASH has been validated for a series of experimental studies, including wave propagation over submerged obstacles [133] and wave interaction with coastal structures [134]. The results of SWASH have also been compared with results of SPH models for wave overtopping [136] and with those of RANS models for extreme wave generation [137].

8.5.6 Boussinesq Models

8.5.6.1 FUNWAVE

FUNWAVE is a phase-resolving, time-stepping Boussinesq model developed by Shi *et al.* [138] at the University of Delaware for the modelling of ocean surface wave propagation in the nearshore environment. The first release of the model was announced in 2012 and a revised version was published in 2013. The code is written in FORTRAN and can be compiled in Linux and Mac OS X operating systems. Parallel simulations are possible with the use of the MPI protocol. The model is open-source and freely distributed under the GNU General Public Licence [139].

FUNWAVE is based on the SWE for non-dispersive linear wave propagation, extended in order to include the lowest-order effects of nonlinearity and frequency dispersion. Original Boussinesq equations are invalid for intermediate and deep water, due to the assumption of weak frequency dispersion. On the other hand, extended Boussinesq models that include energy dissipation, due to wave breaking and eddy viscosity, can account for wave propagation in the coastal zone and replicate the surf zone hydrodynamics [59]. With regard to the numerical formulations of FUNWAVE, the latest version of the model, FUNWAVE-TVD (Total Variation Diminishing), has incorporated additional numerical features, such as the MUSCL-TVD finite volume scheme, together with adaptive Runge–Kutta time stepping, a shock capturing wave breaking scheme and a wetting–drying moving boundary condition with the HLL construction method [139].

A common issue in numerical modelling with Boussinesq equations is potential corruption of the solution caused by truncation errors. In FUNWAVE, all errors involved in the solution of the underlying nonlinear equations are reduced to fourth order in grid spacing and time-step size. A numerical filter is also employed for handling the super-harmonic waves. Other issues refer to the noise caused by the implementation of the boundary conditions, the staggered grid and the eddy viscosity formulation. All these issues seem to have been resolved in the latest version of the model.

Among other applications, Boussinesq-type equations have been used for the simulation of surf-zone phenomena, wave breaking, sub-grid mixing and nearshore circulation [140]. In particular, FUNWAVE was employed for the simulation of the wave-induced vortices near a breakwater [141], tsunami propagation [142] and long wave penetration in a port [143]. The TVD version of the model was developed mainly for simulating tsunami waves at regional and coastal scale, and coastal inundation and wave propagation at basin scale [144]. With regard to MRE applications, FUNWAVE has been used together with other models to study the energy shadowing effect in the lee of a wave farm [145].

8.5.6.2 COULWAVE

COULWAVE is a phase-resolving Boussinesq-type model, originally developed by Patrick Lynett at Cornell University. The code is open-source and was first released in 2008. COULWAVE is written in FORTRAN90 and can be parallelised using the MPI method. It can be compiled in Windows and Linux operating systems and used together with MATLAB post-processing tools, which are provided in the main distribution [146].

The free water surface modelling in COULWAVE is achieved by solving the depth integrated long wave equations, such as the nonlinear SWE and weakly dispersive Boussinesq-type equations. The spatial derivatives are discretised using a fourth-order finite difference scheme and the time integration is performed with a fourth-order iterative predictor–corrector scheme. Weak turbulent effects can be replicated with the implementation of a finite volume scheme [147]. Simulations of wind waves can be realised thanks to the incorporation of the two-layer [148] and multi-layer Boussinesq theory [149].

COULWAVE has the ability to simulate very large domains containing tens of millions of grid points and can handle irregular bathymetries. Initially, COULWAVE was developed for the simulation of tsunamis generated by landslides, tsunami propagation, nearshore evolution and inundation [150]. Classic validation cases and analytic description of the governing equations can be found in the users' manual [146]. The model has been also used to simulate wave run-up [151] and wave breaking, rip and long-shore currents, wave–wave interactions, and transformation of waves in coastal waters [147].

8.5.7 Mild-Slope Equation

8.5.7.1 REFDIF

REFDIF is a phase-resolving parabolic numerical model based on the MSE formulation. The model was developed by Kirby and Dalrymple to simulate combined wave refraction and diffraction problems and it is now freely distributed from the University of Delaware under the GNU licence. REFDIF is written in FORTRAN. A GUI is also available for the Windows installation [152].

REFDIF has a long history of development. Its predecessor was developed to bridge the gap between the nonlinear diffraction models and the linear MSE. At that stage, it could only account for weak currents. Later, a strong current formulation was developed that could handle larger angles of wave incidence. The latest developments have made REFDIF essentially a weakly nonlinear combined refraction and diffraction model. In more detail, the mathematical formulation of REFDIF is based on Stokes expansion, including a third-order correction for the wave phase speed. However, this is not a complete third-order theory, as the wave height is calculated by second-order theory [153]. The model includes also various energy dissipation mechanisms, due to bottom friction, porosity, wave breaking and turbulence. In the simulations, the wave climate can be represented as monochromatic wave trains or directional irregular waves.

There are several inherent assumptions in REFDIF that should be taken into account when designing the simulations. Since the mathematical formulation of the model is based on the MSE; the sea bottom variations should occur in distances greater than the wavelength. Accurate prediction of wave shoaling is performed only for slopes milder than 1:3. Additionally, since the model is based on a Stokes perturbation expansion, its validity is restricted to applications with Stokes waves only. Moreover, the wave direction should be confined to a sector of ±70° relative to the principal wave direction, due to the limitations of the parabolic approximation [153].

The model has been used in multiple applications to study wave transformation over varying seabed topographies, around islands and over currents. REFDIF can also be used to examine wave interactions with harbours and breakwaters. A long list of related articles can be found on the website [152].

8.5.8 Spectral Models

8.5.8.1 WAVEWATCH-III

WAVEWATCH-III is a full spectral third-generation wind–wave model, based on the principles of the WAM model [65]. The model is developed and distributed by the National Oceanic and Atmospheric Administration (NOAA) and National Centers for Environmental Prediction. The code is written in FORTRAN and can be compiled in Linux and Unix platforms. Parallel simulations are possible through the OpenMP and MPI protocols. WAVEWATCH is open-source and freely available under specific licensing.

The first version of the model (WAVEWATCH) was developed at TU Delft by Tolman in 1989 [154], and the second version (WAVEWATCH-II) was developed by NASA's Goddard Space Flight Center in 1992 [155]. The latest version, WAVEWATCH-III, has incorporated many improvements, turning it into an integrated wave modelling framework. The model includes various options and packages for the modelling process that can be selected at the compilation stage [155]. The good documentation and the structure of the model allow users to include their own developments in the source code [156].

WAVEWATCH-III solves the random phase spectral action density balance equation for directional spectra. It is possible to model large areas and long time scales, since the medium physical properties, namely the water depth and current properties, as well as the wave field, vary in time and space in scales much larger than a single wave. The effects of refraction and straining of the wave field, due to temporal and spatial variations of the mean water depth and of the mean current, can be readily modelled by

WAVEWATCH-III. The wave propagation in the model is linear and all nonlinear effects are included in the wave action equation as source terms [155]. The output of WAVEWATCH-III comprises wave spectra at specific locations that can be further post-processed using the GrADS graphics package.

Version 3.14 of WAVEWATCH-III has important improvements in the governing equations, model structure, numerical schemes and physical parameterisations. Especially in the surf zone, new source terms and wet–dry grid method implementations ensure better replication of shallow water phenomena. Spectral partitioning is also available as a post-processing option [155]. However, at this stage of development, WAVEWATCH-III is not able to simulate the strong three-wave interactions that occur in finite depth and shallow waters. For that situation, simplified empirical calculations are employed that usually result in large errors. The aim is to minimise these errors in the near future by implementing a more sophisticated three-wave interaction module [67].

With regard to the applications of the model, WAVEWATCH-III is used for operational wave forecasting at NOAA, with a grid of 50 km resolution with nested regional domains [155]. WAVEWATCH-III is commonly used to provide the open ocean boundary conditions for other numerical models that can simulate local effects at the location of deployment of MRE devices [157, 158]. The model has been tested in the Pacific Ocean for the Wave Information Study (WIS) using nested grids and wind field data from Oceanweather Inc. [159]. WAVEWATCH-III was also employed to study waves generated by cyclones using the new feature of 'mosaic' grids. An example is hurricane Lili in the Gulf of Mexico in October 2002 [160]. Other applications refer to oil spill transport in the Black Sea region with a real-time modelling system including weather, currents and wind-waves [161]. WAVEWATCH-III is employed by PREVIMER for coastal observation and forecasting. For MRE applications, WAVEWATCH-III was employed as a hindcast model for sea states around the French coast using the HOMERE database [161]. The model is also a good tool for wave energy resource assessment studies, such as those on the west coast of Vancouver Island, British Columbia, Canada [162], Uruguay [163] and the East and South China sea for wind-wave farms [164].

8.5.8.2 SWAN

SWAN (Simulating WAves Nearshore) is a third-generation wave model developed at TU Delft. It was based on the WAM model [65] with additional formulations for the shallow water regime. Some important contributions were carried out by the Waves in Shallow Water Environment (WISE) group, which was formed by several members of the Wave Model Development and Implementation (WAMDI) group, who focused their attention on coastal areas. The model is open-source and freely available under the GNU General Public Licence [165]. SWAN is written in FORTRAN90 and can be compiled in Windows, Linux and Mac OS X operating systems. SWAN can operate in parallel mode using the OpenMP and MPI libraries [166].

SWAN was primarily developed for obtaining realistic estimates of wave parameters in coastal areas, lakes and estuaries for given wind, bottom and current conditions. The model is based on the wave action balance equation with incorporated source and sink terms. These terms can be used to replicate bottom friction, wave breaking, white-capping, triad and quadruple wave–wave interactions. Directional and non-directional spectra can be computed as well. Several spectral and time-dependent outputs are available, such as the significant wave height, the peak or mean period and the direction of energy transport. SWAN can be employed at any scale relevant to wind-generated surface gravity waves; however, for ocean scale applications WAM and WAVEWATCH-III are more efficient than SWAN, the latter specialising in the transition from ocean scales to coastal scales [166]. The model is designed to allow coupling with WAM and WAVEWATCH-III by means of nesting inside coarse meshes, preferably on grids with spherical coordinates.

On the other hand, there are several limitations of SWAN requiring awareness while modelling. Diffraction is usually challenging for spectral models, and in the case of SWAN it is treated in a restricted sense. The model should be used in areas where variations in wave height are large within a horizontal scale of at least a few wavelengths. For cases with arbitrary geophysical conditions, a phase-decoupling approach is adopted, in order to represent diffraction and achieve the same qualitative behaviour of spatial redistribution and changes in wave direction. Moreover, SWAN cannot account for wave-induced current, which, if required, should be provided as an input to the model using, for example, an appropriate circulation model. The modelling of triad and four wave–wave interactions is performed with the lumped triad approximation and discrete interaction approximation (DIA), respectively. These approximations depend on the width of the directional distribution of the wave spectrum. For that scope, SWAN was tuned according to observations made in narrow wave flumes (long-crested waves), which seem to provide reasonable results for many applications. Despite that, DIA is considered a poor approximation for long-crested waves, because it depends on the frequency resolution as well [166].

SWAN has been employed successfully in many applications worldwide. The model was used to simulate a realistic tropical storm, after being validated for the refraction of monochromatic waves propagating from deep to shallow water and for wave generation from a wind field of fixed fetch and unlimited duration [167]. In another study, SWAN was used for calculations regarding erosion, shoreline protection and coastal management for the Bangkhuntien shoreline (Bangkok) over a period of 23 years, after being verified with wave buoy measurements [168]. The sediment transport patterns and their implications for topographic changes, due to waves and currents in Ystad Bay, Sweden, have also been studied with SWAN and EBED models [169]. The literature includes many other applications, such as energy dissipation due to vegetation, which is now part of SWAN, as a module named SWAN-VEG [170]. MRE applications with SWAN focus mainly on resource assessment studies for waves. The model has been applied to estimate the wave energy resource at various locations around the Spanish coast [171–173] and for the region of North Cornwall and the Isles of Scilly [174]. A review of different methods for assessing the wave energy potential using wave buoy measurements and SWAN simulations has been carried out for the west coast of Canada as well [175]. SWAN has also been applied for the wave modelling in a tidal energy resource assessment study performed with the Regional Ocean Modelling System tidal model, accounting for oblique propagation of waves in relation to the currents [176]. An interesting application refers to coastal impacts from the presence of a wave farm [177], which was focused on Perranporth beach in the UK [178]. It was surmised that the positioning of that wave farm could influence the sediment transport at the beach behind it and potentially prevent erosion [179].

8.5.9 Models for structural design and other tools

8.5.9.1 WAFO

WAFO (Wave Analysis for Fatigue and Oceanography) is a toolbox developed in MATLAB for the statistical analysis and simulation of random waves and loads [180]. It is freely distributed under the GNU General Public Licence. WAFO has a wide range of users worldwide in academic institutions and industry. The first version of WAFO was released in 2000 by Lund University, after combining the Wave Analysis Toolbox (WAT) with the Fatigue Analysis Toolbox (FAT) [181].

The structure and applications of WAFO are covered in the users' manual [182]. The toolbox can be used to simulate stochastic waves and sea states including linear and nonlinear Gaussian waves or other commonly used spectra, such as JONSWAP and Pierson–Moskowitz. It also includes directional wave spectra and joint wave height, steepness and period distributions. WAFO's statistical tools include extreme wave analysis, multivariate Gaussian probabilities, hypothesis tests, kernel density estimation and random number generators. Tools for plotting are available, together with scripts for comparing the results with benchmark studies from the literature [181].

Fatigue load analysis and rain-flow cycles can be performed in WAFO using random load models, provided that the load is given in the form of a time series of stress or strain measurements, power spectrum in the frequency domain or rain-flow matrix. More accurate predictions include Markov models for calculating the load cycle characteristics. A detailed description of the load, fatigue and damage estimation is provided in the users' manual [182].

There are several applications of WAFO, including probabilistic analysis of dangerous sea states [183], reproduction of wave spectra for PF simulations with WECs [184], studies of the variability of the wave energy resource [185] and decay tests for mooring systems of WECs [186]. In another study the rain-flow cycle analysis of WAFO was employed to calculate mooring loading cycles in order to estimate the damage to moorings [187].

8.5.9.2 SDWED

The Structural Design of Wave Energy Devices project (SDWED) is an international project led by Aalborg University and financially supported by the Danish Council for Strategic Research (2010–2014). During the project multiple tools were created for the investigation of hydrodynamics, moorings, PTOs, wave-to-wire integration and reliability of WECs. SDWED may not comprise an individual software tool, but it consists of many libraries that are freely available from the project's website [188].

In particular, the DTU Motion Simulator is a MATLAB utility for simulating the motion of a floating structure using weakly nonlinear equations of motion, which focuses on hydrodynamic complexity. A linear time-domain model for the hydrodynamics of arrays was also developed, which is able to handle six degrees of freedom for WECs. This program is coded in MATLAB and requires input from the WAMIT model for arrays of WECs. OCEANWAVE3D-SDWED2D is a C library that employs a high-order difference PF model to calculate linearised hydrodynamic coefficients of floating bodies. WAVEC-W2W is another MATLAB utility for solving the equations of motion for WECs requiring data from WAMIT and WAFO. This model has already been used for OWC modelling. Additional applications in MATLAB include a spectral based fatigue analysis model for WECs according to the DNV and ABS offshore standards. Special tools using MATLAB/Simulink have been developed for hydraulic energy storage systems and for wave-to-wire applications as well [188].

Applications and outcomes of SDWED can be found on the website of the project, including relevant PhD theses, project deliverables and scientific publications [188]. A large amount of the published work refers to the analysis of point absorbers and especially to the control system of the devices [189, 190] and dynamic responses [191].

8.5.9.3 Marine Systems Simulator

Marine Systems Simulator (MSS) is a set of MATLAB/Simulink libraries for the simulation of marine systems, which is freely available under the GNU General Public Licence [192]. It consists of three main libraries, MSS GNC, MSS HYDRO and MSS FDI Toolbox. MSS GNC is a time-domain navigation and control modelling toolbox, containing special libraries for vessel modes, manoeuvering trials, dynamic simulations of marine vessels, ships, rigs and underwater vehicles. MSS HYDRO is designed to read output files of hydrodynamic models, such as ShipX (Veres) and WAMIT, and process them for real-time simulations of marine vessels with six degrees of freedom in first- and second-order wave loads. MSS FDI Toolbox is a separate program for calculating the radiation forces and fluid memory effects on marine structures, such as marine crafts and WECs [193].

The applications of MSS mainly focus on seakeeping modelling in time and frequency domains, but its capabilities with regard to the dynamic modelling of marine structures can be expanded to MRE systems. For example, MSS has been used to study the collocation of WECs on floating wind turbines for controlling the motion of the system and generating additional energy from the waves [194]. A point absorber developed by 'CorPower Ocean' has also been studied with MSS FSI for the hydrodynamic loading from waves [195].

8.5.9.4 WEC-Sim

WEC-Sim (Wave Energy Converter SIMulator) is an open-source numerical tool for the simulation of WECs, which is freely distributed from GitHub under the Apache Licence. The code is developed in MATLAB/Simulink, as a product of joint work by the National Renewable Energy Laboratory (NREL) and Sandia National Laboratories (SNL) with funding from the US Department of Energy's Wind and Water Power Technologies Office [196]. The code can be used to simulate multi-body dynamics of devices represented as rigid bodies with PTO mechanisms and mooring systems. WEC-Sim solves the governing equations of motion for six-degrees-of-freedom WECs in the time domain. The model can be used to predict the power performance of WECs and for the design optimisation of the devices. These are achieved with the radiation and diffraction methods obtained from a frequency-domain BEM code, which uses linear coefficients for the dynamics of the system in the time domain. The latest version (8.2) includes nonlinear Froude–Krylov hydrodynamics and hydrostatics, state space radiation, wave directionality and importing of non-dimensionalised data from the pre-processing code BEMIO [196].

WEC-Sim has been employed in a large number of publications listed on the website [196]. Most focus on the validation of the model [197] and the inclusion of new features in the code, such as the simulation of PTO systems [198]. In a more applied study,

WEC-Sim has been used to model different floating oscillating surge WECs and assess the power generation performance of these devices [199].

8.5.9.5 NEMOH

NEMOH is a BEM code for calculating wave loads on offshore structures, which was developed by the Laboratoire de Recherche en Hydrodynamique, Energétique et Environnement Atmosphérique (LHEEA) at Ecole Centrale de Nantes. The code is open-source and freely available under the Apache 2 Licence. NEMOH is written in FORTRAN90 and has a MATLAB wrapper for operating in MATLAB GUI. It can be compiled in Windows and Linux operating systems [200].

The model is dedicated to the computation of first-order wave loads on offshore structures, such as added mass, radiation damping and diffraction forces, the estimation of the dynamic response of floating structures and the assessment of WEC performance. NEMOH has an advanced BEM formulation, which decouples the resolution of the linear free surface boundary value problem (BVP) from the body boundary conditions. Essentially this allows simulations with flexible structures, hydroelasticity, generalised modes and unconventional degrees of freedom. The code consists of a pre-processor for the definition of the mesh and body conditions, a solver for calculating the BVP, pressure field, hydrodynamic coefficients and wave elevation, and a post-processor for calculating the response amplitude operators and plotting the results. NEMOH is able to import and export meshes from WAMIT and to use Salome-Meca software for generating the meshes. Moreover, NEMOH's results can be used by Orcaflex for calculations regarding the moorings [200].

The MRE applications of NEMOH include the calculation of the structural forces on WECs, for instance on the floaters of Wavestar [201]. NEMOH has also been used to calculate the wave forces and hydrodynamic coefficients of floating wind turbines and to examine the response of the structure due to wind loading [202]. In a similar study, the numerical results of NEMOH have been compared with experimental results on the hydrodynamic excitation of floating wind turbines, showing good agreement for the primary parts of the structure (i.e. the column of the turbine) [203].

8.5.9.6 MoorDyn

MoorDyn is a dynamic mooring system code developed by Matthew Hall at the University of Maine for floating wind turbine research. The software is open-source and freely available under the GPL v3 Licence. MoorDyn is written in C++ and has been coupled with other programs, such as MATLAB, Simulink and FAST v7 [204]. MoorDyn can be used to calculate mooring loading due to damping forces, weight and buoyancy forces, hydrodynamic forces from Morison equation and vertical spring-damper forces from contact with the seabed. The mathematical formulation is based on a lumped-mass mooring line approach for discretising the cable dynamics over the mooring line. The model can account for arbitrary line interconnections, clump weights and floats, different line properties and internal axial stiffness [205].

The software was released very recently and there is no extensive literature on its applications. The validation of MoorDyn was performed against experimental results for moorings of the DeepCwind semisubmersible floating wind turbine [206]. MoorDyn was also employed for the calculation of the dynamic mooring loads of ISWEC (Inertial Sea Wave Energy Converter), which is a gyroscopic floating device, providing satisfactory agreement with the experimental results [207].

8.6 Applicability of Numerical Models in MRE

The MRE sector is still in its infancy and more experience, especially in prototype demonstrations, is required in order to foster the emergence of detailed protocols and guidelines regarding all the stages of the design process, including the application of numerical models for MRE projects. Nevertheless, important work has been done even at this early stage, mainly in small-scale physical testing and numerical modelling of almost all the aspects of MRE projects. Apart from the few guidelines for MRE projects, independent reports can be found on general development and evaluation standards [208] and more regionally focused MRE planning and development projects, such as that of the Wave Hub, which is a site for demonstration and testing of MRE devices in the South West of England [209].

The availability of numerical models for marine applications is ever increasing, with the appearance of new models or with the expansion of the capabilities of existing models by including new features. Additionally, most of the models tend to evolve into general modelling platforms that include modules switched on or off depending on the specific needs of the simulation. Therefore, potential users should take advice from manuals and relevant recent studies. Useful information can be found in general guidelines and reports on different mathematical and numerical approaches, which can be used by new modellers for selecting the most appropriate tools [23]. This section endeavours to present an assessment of the applicability of the various mathematical models and numerical techniques for the numerical simulation of coastal, ocean and especially MRE engineering problems.

As discussed in the introduction, the design process of MRE projects is a complex problem, which is even more challenging and demanding than classic offshore and coastal engineering studies, since it incorporates energy generation and optimisation of devices' responses to dynamic sea conditions as well. Apart from the design aspects presented in Figure 8.1, such as device optimisation and PTO design, extreme loading and fatigue, installation of foundations and moorings, grid connection, and the deployment, maintenance and monitoring of the system, MRE projects also include the study of socioeconomic impacts by their interaction with other industries (e.g. fisheries and tourism), and they are often subject to oil prices or the availability of other renewable energy sources. Numerical modelling can be used to examine these aspects as well, but here the main focus is on the technical aspects of the project. The role of numerical modelling has become crucial for all stages of MRE projects spanning from the proof of concept to the optimisation of full-scale models, especially due the high cost of physical model testing.

Arguably, a significant proportion of MRE numerical studies are concerned with resource assessment, mainly because numerical models are the most appropriate tools for large-scale studies and hindcasting. A suite of protocols on wave and tidal resource characterisation became available through the Equimar project, including guidelines for the most appropriate numerical modelling techniques, assessment of extreme conditions and identification of the constraints on the development of MRE projects at specific sites [210]. Additional intercomparisons among numerical models were also published by the Equimar project, referring to SWAN, TOMAWAC and MIKE 21 for wave resource assessment [211] and to TELEMAC-2D and MIKE21 for tidal resource assessment [212]. Similar reports were produced by other consortia as well, such as MERiFIC [4] and SuperGen Marine [213], on the applicability of various modelling approaches for MRE

Table 8.7 Physical processes according to the depth of the sea.

Physical process	Water regime			
	Deep	Intermediate	Shallow	Basin/harbour
Diffraction	*	*	**	****
Depth refraction/shoaling	*	***	****	***
Current refraction	*	**	***	*
Quad wave interactions	****	****	**	*
Triad wave interactions	*	**	***	**
Atmospheric	****	****	**	*
White-capping	****	****	**	*
Depth induced breaking	*	**	****	*
Bottom friction	*	****	**	*

* Negligible;
** moderately important;
*** very important;
**** essential.

projects. More general guidelines concerning the overall design aspects of wave energy conversion were also published by the European Marine Energy Centre (EMEC) [214].

In the remainder of this section, tables are presented highlighting the applicability of numerical models for different physical processes and modelling aspects related to MRE studies. It is emphasised that these tables are general and indicative, and in any case, apart from the basic equations solved by the available models, additional parameterisations can be used to reproduce more complex physical problems. Table 8.7 presents the importance of different physical processes related to wave propagation according to the depth of the sea [215]. The different weighting of these physical processes can give a good indication of what approximations can be considered in the mathematical formulation in order to choose the most appropriate numerical model for the case examined. This table can be useful especially for resource characterisation studies with regional numerical models, where physical processes act as source terms in the governing equations and can be examined independently.

The next step after identifying the dominant physical processes in the physical problem is to select the mathematical model of the governing equations. Table 8.8 [7] relates the different versions of the governing equations discussed in Section 8.4 and the various numerical techniques presented in Section 8.3 with the physical processes examined. This table includes more physical processes than Table 8.7, especially referring to processes in the vicinity of a marine structure and the problem of wave–structure interaction. A similar table for the suitability of numerical models with regard to the temporal and spatial characteristics of the simulation, the hydrodynamic analysis of the problem and the description of the free water surface can be found in SuperGen Marine guides [213].

As shown in Table 8.8, the applicability of the numerical model depends on the equations solved, as well as the different formulations and numerical methods employed.

Table 8.8 Model applicability based on the governing equations for various physical processes.

Physical Process Num. Model	Wave diffraction	Wave refraction		Wave dispersion		Wave–wave interactions		Wave–current interactions		Wave breaking		Wave–structure interaction			Turbulence effects	
		Depth	Current	Linear	Highly nonlinear	Quad	Triad	Weak	Strong	Depth	White capping	Run-up	Over topping	Green water	Weak	Strong
Navier–Stokes (CFD)	****	****	****	****	****	****	****	****	****	****	***	****	****	****	****	****
Hydrostatic			***	**	*	*	*	**	*	*	*				**	
Potential flow	***	****	****	****	***	***	***	****	***	*	*	***	*		*	
Shallow water equations	****		***	**	*	*	***	***	**	**		***	**		***	
Boussinesq	***	****	***	****	***	***	***	***	**	***		***	**		***	
Mild slope equations			**		*	*	**	***		*					*	
Spectral models	*		***			***	***			**	***				**	

* Not recommended;
** Conditionally suitable;
*** Moderately suitable;
**** Highly suitable;
Absence of a star denotes that the model is inapplicable.

Table 8.9 Efficiency of the numerical models.

	Required skill	Computational cost	Solver	
			Complexity	Stability
Navier–Stokes (CFD)	High	****	Complex	Possibly unstable
Hydrostatic	Medium	**	Simple	Stable
Potential flow	Low–high[1]	***	Simple[3]	Stable[4]
Shallow water equations	Medium	*	Simple	Stable
Boussineq	Medium	**	Simple	Possibly unstable
Mild slope equations	Low	**	Simple	Stable
Spectral models	Low–high[2]	*	Simple	Stable

1) There are four types of potential flow models which require different level of skills: linear BEM, semi-analytical techniques, time-domain formulation, and nonlinear BEM, requiring low, high, medium, and high skills, respectively.
2) Low corresponds to 'supra-grid' models and high corresponds to 'sub-grid' spectral models.
3) Potential flow models with nonlinear BEM are characterised as complex.
4) Time-domain potential flow models can be unstable.
* Cheap,
** considerable;
*** high;
**** very expensive.

It is clear that Navier–Stokes/CFD models are capable of describing all the physical processes accurately, while simpler models might be insufficient for solving complex flows with significant turbulence. The trade-off for accuracy is the high computational cost and the level of user expertise required for operating such complex models. These aspects are presented in Table 8.9, together with the relative complexity and stability of every model. This table is a combination of Lin's [7] and Folley's [216] summary tables. In general, more sophisticated numerical models tend to be unstable and hard to operate. All these aspects make complex models less suitable for operational engineering purposes that often include large-scale modelling and long simulations. However, as the processing computer power increases, at least the computational cost will have a lower weighting factor in the decision-making process.

The final step for choosing a numerical model for an MRE study is to find the most suitable mathematical model depending on the complexity of the physical process from Table 8.8 and the available resources and expertise from Table 8.9. This can be achieved by dividing the main aspects of an MRE project into device design and array design groups and examining the suitability of the different versions of the governing equations for each aspect separately [4], as presented in Table 8.10. This table attempts to provide guidance for choosing the most suitable numerical model for different aspects of MRE projects. In brief, it shows that CFD models are more suitable at device scale and spectral models at array scale. The device–device interaction, operation and annual energy production (AEP) were first presented in Folley's work [216], where they are referred as localised effects, dynamic control and AEP, respectively. The strong correlation between Tables 8.8 and 8.10 is based on the fact that MRE studies are a subcategory of flow–structure interaction studies.

Table 8.10 Suitability of numerical models for different aspects of MRE projects [4].

Application Num. Model	Device–design		Moorings		Array Design Resource	Device-Device Interaction	Array Geometry Optimization	Annual Energy Production	Environmental Impacts
	Operation	Maintenance	Response	Extreme					
Navier-Stokes (CFD)	***		***	****	*	****	*	**	**
Hydrostatic	*	**	**	**	***	***	***	**	****
Potential flow	* to ****¹	***	****	**	*	***	****	**	*
Shallow water equations	*	*	*	*	***	**	**	***	
Boussinesq	*	*	**	*	***	**	**	***	***
Mild slope equations	*	*	*	*	***	**	**	***	***
Spectral models	*	*	*	*	****	*	***	***	****

* Not recommended,
**conditionally suitable,
***suitable,
****accurate and efficient.
¹ Rating based on method, e.g.
*linear BEM,
**semi-analytical,
***time-domain formulation,
****nonlinear BEM

In this section, a methodology for selecting the most suitable mathematical model for the different problems of MRE projects was presented in a relatively general way. Selecting the most suitable software requires examination of additional aspects as well, such as licence availability, expertise, support and even personal preference. The verdict of the chapter is to use the most appropriate numerical tool for each case, being aware of the limitations induced by the mathematical models and numerical techniques.

References

1 Payne G (2008) *Guidance for the experimental tank testing of wave energy converters. SuperGen Marine*, University of Edinburgh. Available from: http://www.supergen-marine. org.uk/drupal/files/ reports/WEC{_}tank{_}testing.pdf.

2 Sutherland J, Barfuss SL (2011) *Composite modelling: Combing physical and numerical models. In: 34th IAHR World Congress.* Brisbane, Australia.

3 Day AH, Babarit A, Fontaine A, He YP, Kraskowski M, Murai M, *et al.* (2015) Hydrodynamic modelling of marine renewable energy devices: A state of the art review. *Ocean Engineering.* 108, 46–69. Available from: http://www.sciencedirect.com/science/ article/pii/S0029801815002280.

4 Vyzikas T (2014) *Application of numerical models and codes – Task 3.4.4 MERiFIC project.* University of Plymouth. Available from: http://archimer.ifremer.fr/ doc/00324/43550/43111.pdf.

5 Weber J (2012) *WEC Technology Readiness and Performance Matrix – finding the best research technology development trajectory.* International Conference on Ocean Energy.

6 Liu PLF, Losada IJ (2002) *Wave propagation modeling in coastal engineering. Journal of Hydraulic Research.* 40(3), 229–240.

7 Lin P (2008) *Numerical Modeling of Water Waves.* London: Taylor & Francis.

8 Versteeg HK, Malalaskekera W (2007) *An Introduction to Computational Fluid Dynamics: The Finite Volume Method.* 2nd edn. Pearson Prentice Hall.

9 Anderson JD (1995) *Computational Fluid Dynamics: The Basics with Applications.* Singapore: McGraw-Hill International Editions.

10 Schlichting H (1979) *Boundary-Layer Theory.* 7th edn. New York: McGraw-Hill.

11 Reynolds O (1883) *An experimental investigation of the circumstances which determine whether the motion of water shall be direct or sinuous, and of the law of resistance in parallel channels.* Philosophical Transactions of the Royal Society of London. 174, 935–982.

12 Douglas JF, Gasiorek JM, Swaffield JA, Jack LB (2005) *Fluid Mechanics.* 5th edn. Harlow: Pearson Education.

13 Ferziger JH, Perić M (2002) *Computational Methods for Fluid Dynamics.* Berlin: Springer.

14 MARNET CFD (2002) Best Practice Guidelines for Marine Applications of Computational Fluid Dynamics.

15 Launder BE, Spalding DB (1974) *The numerical computation of turbulent flows. Computer Methods in Applied Mechanics and Engineering.* 3(2), 269–289.

16 Wilcox DC (1998) Turbulence Modeling for CFD (Second eqn).

17 Menter FR (1994) Two-equation eddy-viscosity turbulence models for engineering applications. *AIAA Journal.* 32(8), 1598–1605. Available from: http://arc.aiaa.org/ doi/abs/10.2514/3.12149.

18 Orszag SA (1970) Analytical theories of turbulence. *Journal of Fluid Mechanics.* 41(02), 363. Available from: http://journals.cambridge.org/abstract{_}S0022112070000642.

19 Moin P, Mahesh K (1998) Direct numerical simulation: A tool in turbulence research. *Annual Review of Fluid Mechanics.* 30(1), 539–578. Available from: http://www.annualreviews.org/doi/abs/10.1146/annurev.fluid.30.1.539.

20 Hendrickson KL (2005) Navier-Stokes simulations of steep breaking water waves with a coupled air-water interface. Massachusetts Institute of Technology. Available from: http://dspace.mit.edu/handle/1721.1/33559.

21 Anderson JD (2002) *Modern Compressible Flow*. Boston: McGraw-Hill.

22 Liu GR, Liu MB (2003) *Smoothed Particle Hydrodynamics*. Singapore: World Scientific.

23 McCabe AP (2004) *An appraisal of a range of fluid modelling software*. Supergen Marine Workpackage 2.

24 Cheng AHD, Cheng DT (2005) Heritage and early history of the boundary element method. *Engineering Analysis with Boundary Elements.* 29(3), 268–302.

25 Courant R, Friedrichs K, Lewy H (1967) On the partial difference equations of mathematical physics. *IBM Journal of Research and Development.* 11(2), 215–234. Available from: http://ieeexplore.ieee.org/lpdocs/epic03/wrapper.htm?arnumber=5391985.

26 Greaves D (2010) *Application of the finite volume method to the simulation of nonlinear water waves*. In: Advances in Numerical Simulation of Nonlinear Water Waves. Hackensack, NJ: World Scientific. p. 357.

27 Lewis RW, Nithiarasu P, Seetharamu KN (2004) *Fundamentals of the Finite Element Method for Heat and Fluid Flow*. Hoboken, NJ: Wiley.

28 Wendt J (2008) *Computational Fluid Dynamics: An Introduction*. Springer Science & Business Media.

29 Canuto C, Hussaini MY, Quarteroni AM, A T, Zang J (1988) *Spectral Methods in Fluid Dynamics*. Springer Science & Business Media. Available from: https://books.google.com/books?id=c2XmCAAAQBAJ{&}pgis=1.

30 Fletcher CAJ (1984) *Computational Galerkin Methods*. Springer Science & Business Media. Available from: https://books.google.com/books?id=mav1CAAAQBAJ{&}pgis=1.

31 Fletcher CA (1990) *Computaional Techniques for Fluid Dynamics 1: Fundamental and General Techniques*. 2nd edn. Springer-Verlag.

32 Costabel M (1987) *Principles of boundary element methods*. Computer Physics Reports. 6, 175–251.

33 Brebbia CA (1983) *Progress in Boundary Element Methods*, Volume 2. New York: Springer-Verlag. Available from: https://books.google.com/books?id=95TkBwAAQBAJ{&}pgis=1.

34 Brebbia CA (1981) *Progress in Boundary Element Methods*, Vol. 1. New York: Pentech Press.

35 Gaul L, Kögl M, Wagner M (2003) *Boundary Element Methods for Engineers and Scientists: An Introductory Course with Advanced Topics*. Springer Science & Business Media. Available from: https://books.google.com/books?id=RZnsCAAAQBAJ{&}pgis=1.

36 Liu GR (2009) *Meshfree Methods: Moving beyond the Finite Element Method*. CRC Press.

37 Li S, Liu WK (2004) *Meshfree Particle Methods*. Springer.

38 Gingold RA, Monaghan JJ (1977) Smoothed particle hydrodynamics – Theory and application to non-spherical stars. *Monthly Notices of the Royal Astronomical Society.* 181, 375–389. Available from: http://adsabs.harvard.edu/abs/1977MNRAS.181.375G.

39 Lucy LB (1977) A numerical approach to the testing of the fission hypothesis. *Astronomical Journal.* 82, 1013.

40 Koshizuka S, Nobe A, Oka Y (1998) Numerical analysis of breaking waves using the moving particle semiimplicit method. *International Journal of Numerical Methods in Fluids.* 26(7). Available from: http://adsabs.harvard.edu/abs/1998IJNMF.26.751K.

41 Monaghan JJ (1992) Smoothed particle hydrodynamics. *Annual Review of Astronomy and Astrophysics.* 30, 543–574.

42 Monaghan JJ (2005) Smoothed particle hydrodynamics. *Reports on Progress in Physics.* 68(8), 1703–1759.

43 Gómez-Gesteira M, Rogers BD, Dalrymple RA, Crespo, A J C Narayanaswamy M (2010) User guide for the SPHysics code v2.0.

44 Succi S (2001) *The Lattice Boltzmann Equation: For Fluid Dynamics and Beyond.* Oxford Science Publications.

45 Qian Y, Succi S, Orszag S (1995) *Recent Advances in Lattice Boltzmann Computing.* World Scientific.

46 Chen S, Doolen G (1998) *Lattice Boltzmann method for fluid flows.* Annual Review of Fluid Mechanics. Available from: http://www.annualreviews.org/doi/abs/10.1146/annurev.fluid.30.1.329.

47 IEEE (2008). 754-2008 - IEEE Standard for Floating-Point Arithmetic. IEEE Xplore Digital Library. p. 1–70.

48 CFD-Online (2013) CFD Online Wiki. Available from: http://www.cfd-online.com/Wiki/Main{_}Page.

49 Kirby BJ (2010) *Micro- and Nanoscale Fluid Mechanics: Transport in Microfluidic Devices.* Cambridge University Press.

50 Marshall J, Hill C, Perelman L, Adcroft A (1997) Hydrostatic, quasi-hydrostatic, and nonhydrostatic ocean modeling. *Journal of Geophysical Research.* 102(C3), 5733.

51 Roelvink D, Reniers A (2012) A guide to modeling coastal morphology. Available from: http://www.vliz.be/imis/imis.php?module=ref{&}amp;refid=212305.

52 Holthuijsen LH (2007) *Waves in Oceanic and Coastal Waters.* Cambridge: Cambridge University Press.

53 Moler C (2011) Shallow water equations. In: Experiments with MATLAB. p. 241–246. Available from: https://uk.mathworks.com/moler/exm/book.pdf.

54 Boussinesq J (1872) Théorie des ondes et des remous qui se propagent le long d'un canal rectangulaire horizontal, en communiquant au liquide contenu dans ce canal des vitesses sensiblement pareilles de la surface au fond. *Journal de Mathématiques Pures et Appliquées.* p. 55–108. Available from: https://eudml.org/doc/234248.

55 Peregrine DH (1967) Long waves on a beach. *Journal of Fluid Mechanics.* 27(04), 815–827. Available from: http://journals.cambridge.org/abstract{_}S0022112067002605.

56 Madsen PA, Murray R, Sørensen OR (1991) A new form of the Boussinesq equations with improved linear dispersion characteristics. *Coastal Engineering.* 15(4), 371–388. Available from: http://www.sciencedirect.com/science/article/pii/037838399190017B.

57 Nwogu O (1993) Alternative form of Boussinesq equations for nearshore wave propagation. *Journal of Waterway, Port, Coastal, and Ocean Engineering.* 119(6), 618–638. Available from: http://ascelibrary.org/doi/10.1061/{%}28ASCE{%}290733-950X{%}281993{%}29119{%}3A6{%}28618{%}29.

58 Brocchini M (2013) A reasoned overview on Boussinesq-type models: The interplay between physics, mathematics and numerics. Proceedings of the Royal Society 469, 20130496.

59 Kirby JT, Wei G, Chen Q, Kennedy AB, Dalrymple RA (1998) Funwave 1.0, fully nonlinear Boussinesq wave model documentation and user's manual. Available from: http://repository.tudelft.nl/view/hydro/uuid:d79bba08-8d35-47e2-b901-881c86985ce4/.

60 Wei G, Kirby JT (1995) Time-dependent numerical code for extended Boussinesq equations. *Journal of Waterway, Port, Coastal, and Ocean Engineering.* 121(5), 251–261. Available from: http://ascelibrary.org/doi/abs/10.1061/(ASCE)0733 9 50X(1995)121:5(251).

61 Eckart C (1952) The propagation of gravity waves from deep to shallow water. Gravity Waves Proceedings of the NBS Semicentennial Symposium on Gravity Waves held at the NBS on June 18-20. Available from: http://adsabs.harvard.edu/abs/1952grwa. conf.165E.

62 Berkhoff JCW (1972) Computation of combined refraction-diffraction. In: 13th International Conference on Coastal Engineering. p. 471–490.

63 Kubo Y, Kotake Y, Isobe M, Watanabe A (1992) Time-dependent mild slope equation for random waves. In: Coastal Engineering Proceedings 1(23).

64 Tolman HL (2009) User manual and system documentation of WAVEWATCH-IIITM version 3.14. Technical note. (276). Available from: http://polart.ncep.noaa.gov/mmab/papers/tn276/MMAB{_}276.pdf$\delimiter"026E30F$npapers2://publication/uuid/298F36C7-957 F-4D13-A6AB-ABE61B08BA6B.

65 Group W (1988) The WAM model – a third generation ocean wave prediction model. *Journal of Physical Oceanography.* 18(12), 1775–1810. Available from: http://journals. ametsoc.org/doi/abs/10.1175/1520-0485(1988)018{%}3C1775:TWMTGO{%}3E2. 0.CO;2.

66 Booij N, Ris RC, Holthuijsen LH (1999) A third-generation wave model for coastal regions: 1. Model description and validation. *Journal of Geophysical Research.* 104(C4), 7649.

67 Sheremet A, Kaihatu JM (2009) *Development of Numerical 3-Wave Interactions Module for Operational Wave Forecasts in Intermediate-Depth and Shallow Water.* Dept of Civil and Coastal Engineering, Florida Univ Gainesville.

68 OpenCFD_Ltd (2015) The open source CFD toolbox. Available from: http://www. openfoam.com/.

69 OpenFOAM-Extend (2015) The OpenFOAM Extend Project website. Available from: http://www.extend-project.de/.

70 Jasak H, Jemcov A, Tukovic Z (2007) *OpenFOAM: A C++ library for complex physics simulations.* In: International Workshop on Coupled Methods in Numerical Dynamics. Dubrovnik, Croatia.

71 OpenCFD (2012) OpenFOAM: The Open Source CFD Toolbox User Guide Version 2.1.1.

72 Marić T, Höpken J, Mooney K (2014) The OpenFOAM Technology Primer. Sourceflux.

73 Higuera P, Lara JL, Losada IJ (2013) Realistic wave generation and active wave absorption for Navier–Stokes models Application to OpenFOAM. *Coastal Engineering.* 71, 102–118.

74 Jacobsen NG, Fuhrman DR, Fredsøe J (2011) A wave generation toolbox for the open-source CFD library: OpenFoam. International Journal for Numerical Methods in Fluids. p. 601–629.

75 Iturrioz A, Guanche R, Lara JL, Vidal C, Losada IJ (2015) Validation of OpenFOAM for oscillating water column three-dimensional modeling. *Ocean Engineering*. 107, 222–236. Available from: http://linkinghub.elsevier.com/retrieve/pii/ S0029801815003649.

76 Vyzikas T, Deshoulières S, Giroux O, Barton M, Greaves D (2017) Numerical study of fixed Oscillating Water Column with RANS-type two-phase CFD model. *Renewable Energy*. 102, 294–305.

77 Simonetti I, Cappietti L, Safti HE, Manfrida G, Matthies H, Oumeraci H (2015) *The use of OpenFOAM as a virtual laboratory to simulate Oscillating Water Column wave energy converters*. In: VI International Conference on Computational Methods in Marine Engineering MARINE 2015. Rome, Italy.

78 Ransley E (2015) *Survivability of Wave Energy Converter and Mooring Coupled System using CFD*. University of Plymouth.

79 Gebreslassie MG, Tabor GR, Belmont MR (2013) Numerical simulation of a new type of cross flow tidal turbine using OpenFOAM – Part I: Calibration of energy extraction. *Renewable Energy*. 50, 994–1004. Available from: http://www.sciencedirect.com/ science/article/pii/S0960148112005356.

80 Paulsen BT, Bredmose H, Bingham HB, Schløer S (2013) *Steep wave loads from irregular waves on an offshore wind turbine foundation: Computation and experiment*. In: OMAE2013: Proceedings of the ASME 2013 32nd International Conference on Ocean, Offshore and Arctic Engineering. Nantes, France. Available from: http:// proceedings.asmedigitalcollection.asme.org/proceeding.aspx?doi=10.1115/ OMAE2013-10727.

81 Vyzikas T, Ransley E, Hann M, Magagna D, Simmonds D, Magar V, *et al.* (2013) *Integrated numerical modelling system for extreme wave events at the Wave Hub site*. In: ICE Conference: Coasts, Marine structures and Breakwaters.

82 Ransley E, Hann M, Greaves D, Raby AC, Simmonds D (2013) Numerical and physical modeling of extreme waves at Wave Hub. *Journal of Coastal Research*. 29(65), 1645–1650. Available from: http://ics2013.org/papers/Paper4475{_}rev.pdf.

83 REEF3Dweb (2015) REEF3D Home page. Available from: https://reef3d. wordpress.com/.

84 Osher S, Sethian JA (1988) Fronts propagating with curvature-dependent speed: Algorithms based on Hamilton- Jacobi formulations. *Journal of Computational Physics*. 79(1), 12–49. Available from: http://www.sciencedirect.com/science/article/ pii/0021999188900022.

85 Bihs H, Kamath A, Chella MA, Ahmad N, Aggarwal A (2015) REEF3D: User Guide.

86 Kamath A, Chella MA, Bihs H, Arntsen ØA (2015) CFD investigations of wave interaction with a pair of large tandem cylinders. *Ocean Engineering*. 108, 738–748. Available from: http://www.sciencedirect.com/science/article/pii/S0029801815004473.

87 Alagan Chella M, Bihs H, Myrhaug D, Muskulus M (2015) Breaking characteristics and geometric properties of spilling breakers over slopes. *Coastal Engineering* 95, 4–19. Available from: http://www.sciencedirect.com/science/article/pii/S037838391400177X.

88 Alagan Chella M, Bihs H, Myrhaug D (2015) Characteristics and profile asymmetry properties of waves breaking over an impermeable submerged reef. *Coastal Engineering*. 100, 26–36. Available from: http://www.sciencedirect.com/science/article/ pii/S037838391500054X.

89 Ahmad N, Bihs H, Kamath A, Arntsen ØA (2015) Three-dimensional CFD modeling of wave scour around side-byside and triangular arrangement of piles with REEF3D. *Procedia Engineering.* 116, 683–690. Available from: http://www.sciencedirect.com/science/article/pii/S187770581502010X.

90 Kamath A, Bihs H, Arntsen ØA (2015) Numerical modeling of power take-off damping in an Oscillating Water Column device. *International Journal of Marine Energy.* 10,1–16. Available from: http://dx.doi.org/10.1016/j.ijome.2015.01.001.

91 Kamath A, Bihs H, Arntsen ØA (2015) Numerical investigations of the hydrodynamics of an oscillating water column device. *Ocean Engineering.* 102, 40–50. Available from: http://www.sciencedirect.com/science/article/pii/S0029801815001249.

92 Crespo AJC, Dominguez JM, Gesteira MG, Barreiro A, Rogers BD (2011) User Guide for DualSPHysics code.

93 DualSPHysics (2015) DualSPHysics web page. Available from: http://www.dual.sphysics.org/.

94 Dalrymple RA, Rogers BD (2006) Numerical modeling of water waves with the SPH method. *Coastal Engineering.* 53(2–3), 141–147. Available from: http://www.sciencedirect.com/science/article/pii/S0378383905001304.

95 Cunningham LS, Rogers BD, Pringgana G (2015) Tsunami wave and structure interaction: An investigation with smoothed-particle hydrodynamics. Proceedings of the Institution of Civil Engineers – Engineering and Computational Mechanics. Available from: http://www.icevirtuallibrary.com/doi/full/10.1680/eacm.13.00028.

96 Altomare C, Crespo AJC, Domínguez JM, Gómez-Gesteira M, Suzuki T, Verwaest T (2015) Applicability of Smoothed Particle Hydrodynamics for estimation of sea wave impact on coastal structures. *Coastal Engineering.* 96, 1–12. Available from: http://www.sciencedirect.com/science/article/pii/S0378383914001999.

97 Altomare C, Crespo AJC, Rogers BD, Dominguez JM, Gironella X, Gómez-Gesteira M (2014) Numerical modelling of armour block sea breakwater with smoothed particle hydrodynamics. *Computers & Structures.* 130, 34–45. Available from: http://www.sciencedirect.com/science/article/pii/S0045794913002824.

98 Ma QW, Yan S (2006) Quasi ALE finite element method for nonlinear water waves. *Journal of Computational Physics.* 212(1), 52–72.

99 Yan S, Ma QW (2010) QALE-FEM for modelling 3D overturning waves. *International Journal for Numerical Methods in Fluids.* 63(6), 743–768.

100 Yan S, Ma QW (2008) *Nonlinear simulation of 3D freak waves using a fast numerical method.* International Society of Offshore and Polar Engineers. Available from: https://www.onepetro.org/conference-paper/ISOPE-I-08-204.

101 Adcock TAA, Taylor PH, Yan S, Ma QW, Janssen PAEM (2011) Did the Draupner wave occur in a crossing sea? Proceedings of the Royal Society A. 467(2134), 3004–3021.

102 Yan S, Ma Q, Cheng X (2012) Numerical investigations on transient behaviours of two 3-D freely floating structures by using a fully nonlinear method. *Journal of Marine Science and Application.* 11(1), 1–9. Available from: http://link.springer.com/10.1007/s11804-012-1099-0.

103 Ma QW, Yan S (2009) QALE-FEM for numerical modelling of non-linear interaction between 3D moored floating bodies and steep waves. *International Journal for Numerical Methods in Engineering.* 78(6), 713–756. Available from: http://doi.wiley.com/10.1002/nme.2505.

104 Ma QW, Yan S (2010) *QALE-FEM method and its applications to the simulation of free-responses of floating bodies and overturning waves.* In: QW, Ma (ed.), Advances in Numerical Simulation of Nonlinear Water Waves. World Scientific. p. 165–202.

105 POM (2013) The Princeton Ocean Model web page. Available from: http://www.ccpo. odu.edu/POMWEB/index.html.

106 Mellor GL (2002) Users guide for a three-dimensional, primitive equation, numerical ocean model. *Ocean Modelling.* 8544(June), 0710. Available from: http://www7.civil. kyushu-u.ac.jp/kankyo/adcp/2011/pom2k.pdf.

107 Oey L (2008) The Priceton Ocean Model. Available from: http://www.aos.princeton. edu/WWWPUBLIC/htdocs.pom/index.html.

108 Iwasaki S, Isobe A, Kako S (2014) Atmosphere-ocean coupled process along coastal areas of the Yellow and East China seas in winter. *Journal of Climate.* 27, 155–167.

109 Hasegawa D, Sheng J, Greenberg DA, Thompson KR (2011) Far-field effects of tidal energy extraction in the Minas Passage on tidal circulation in the Bay of Fundy and Gulf of Maine using a nested-grid coastal circulation model. *Ocean Dynamics.* (61), 1845–1868.

110 ODNE. COHERENS web page. Available from: http://odnature.naturalsciences.be/ coherens/.

111 Luyten PJ, Jones JE, Proctor R, Tabor A, Tett P, Wild-Allen K (1999) COHERENS—A coupled hydrodynamical-ecological model for regional and shelf seas: User Documentation, MUMM Report. Management Unit of the Mathematical Models of the North Sea.

112 Luyten P (2000) COHERENS Home Page. Available from: www.mumm.ac.be/{˜} patrick/mast/coherens.html.

113 Deltares (2015) Delft3D Open Source Community. Available from: http://oss.deltares. nl/web/delft3d/home.

114 Deltares (2011) User Manual, Delft3D-FLOW version 3.15.

115 Deltares (2015) Delft3D wave: 3D/2D modelling suite for integral water solutions. Available from: http://content.oss.deltares.nl/delft3d/manuals/Delft3D-WAVE{_} User{_}Manual.pdf.

116 Bergsma E (2012) Process-based modelling of the Maumusson inlet: Interaction between inlet, ebb-delta and adjacent shoreline. Delft University of Technology.

117 Magar V, Joshi CR, Williams BG, Conley D (2013) Wave energy resource characterisation for Guernsey Island (UK). In: EWTEC 2013 Proceedings.

118 Abundo M, Nerves A, Paringit E, Villanoy C (2012) A combined multi-site and multi-device decision support system for tidal in-stream energy. *Energy Procedia.* 14, 812–817. Available from: http://linkinghub.elsevier.com/retrieve/pii/ S1876610211044365.

119 Carballo R, Iglesias G, Castro A (2009) Numerical model evaluation of tidal stream energy resources in the Ría de Muros (NW Spain). *Renewable Energy.* 34(6), 1517–1524. Available from: http://dx.doi.org/10.1016/j.renene.2008.10.028.

120 Baston S, Waldman S, Side J (2014) Modelling energy extraction in tidal flows. TeraWatt Consortium.

121 Mungar S (2014) Hydrodynamics of horizontal-axis tidal current turbines. PhD thesis, TU Delft.

122 Chatzirodou A, Karunarathna H (2014) Impacts of tidal energy extraction on sea bed morphology. Proceedings of Coastal Engineering. 34.

123 TELEMAC Modeling system, (2013) Operating manual, Release 6.2. March.
124 TELEMAC-MASCARET (2013) Open Telemac-Mascaret home page. Available from: www.opentelemac.org.
125 Ricci S, Piacentini A, Weaver A, Ata R, Goutal N (2013) A variational data assimilation algorithm to estimate salinity in the Berre Lagoon with TELEMAC-3D. In: XXth TELEMAC-MASCARET User Conference. p. 19–24.
126 Bourban S, Liddiard M, Durand N, Cheeseman S, Baldock A (2013) *High resolution modelling of tidal resources, extraction and interactionss around the UK*. In: 1st Marine Energy Technology Symposium METS13. Washington, DC.
127 Haverson D, Bacon J, Smith H (2015) *Modelling hydrodynamic and morphodynamic effects at Ramsey Sound*. In: 4th Oxford Tidal Energy Workshopth Oxford Tidal Energy Workshop. Oxford.
128 Hashemi MR, Neill SP, Robins PE, Davies AG, Lewis MJ (2015) Effect of waves on the tidal energy resource at a planned tidal streamarray. *Renewable Energy*. 75, 626–639. Available from: http://www.sciencedirect.com/science/article/pii/S0960148114006569.
129 SHYFEM. S.HY.F.E.M (2013) Shallow water Hydrodynamic Finite Element Model. Available from: https://sites.google.com/site/shyfem/home.
130 Umgiesser G (2005) SHYFEM Finite Element Model for Coastal Seas. User Manual.
131 The Swash Team (2011) User Manual SWASH version 1.05.
132 SWASHweb. Swash webpage. Available from: http://swash.sourceforge.net/.
133 Zijlema M, Stelling GS (2005) Further experiences with computing non-hydrostatic free-surface flows involving water waves. *International Journal for Numerical Methods in Fluids*. 48(2), 169–197.
134 Zijlema M, Stelling GS, Smit PB (2011) SWASH: An operational public domain code for simulating wave fields and rapidly varied flows in coastal waters. *Coastal Engineering*. 58(10), 992–1012. Available from: http://dx.doi.org/10.1016/j.coastaleng.2011.05.015.
135 Stelling GS, Zijlema M (2003) An accurate and efficient finite-difference algorithm for non-hydrostatic free- surface flow with application to wave propagation. International *Journal for Numerical Methods in Fluids*. 23(May 2002), 1–23.
136 St-Germain P, Nistor I, Readshaw J, Lamont G (2014) *Numerical modeling of coastal dike overtopping using SPH and non-hydrostatic NLSW equations*. In: Coastal Engineering Proceedings.
137 Vyzikas T, Stagonas D, Buldakov E, Greaves D (2015) *Efficient numerical modelling of focused wave groups for freak wave generation*. In: ISOPE-2015: 25th International Offshore and Polar Engineering Conference. Kona, Big Island, Hawaii, USA.
138 Shi F, Kirby JT, Harris JC, Geiman JD, Grilli ST (2012) A high-order adaptive time-stepping TVD solver for Boussinesq modeling of breaking waves and coastal inundation. Ocean Modelling. 43-44, 36–51. Available from: http://dx.doi.org/10.1016/j.ocemod.2011.12.004.
139 Kirby JT (2014) FUNWAVE homepage. Available from: https://www.udel.edu/kirby/programs/funwave/funwave.html.
140 Chen Q, Kirby JT, Dalrymple RA, Kennedy AB, Thornton E, Shi F (2000) *Boussinesq modeling of waves and longshore currents under field conditions*. In: Coastal Engineering. p. 651–663.
141 Hommel D, Shi F, Kirby JT, Dalrymple RA, Chen Q (2000) *Modelling of a wave-induced vortex near a breakwater*. In: Coastal Engineering. ASCE. p. 2318–2330.

142 Watts P, Grilli ST, Kirby JT (2002) *Coupling 3D tsunami generation with Boussinesq tsunami propagation.* In: 27th European Geophysical Society General Assembly. Nice.

143 McComb PJ, Johnson DL, Beamsley BJ (2009) Numerical study of options to reduce swell and long wave penetration at Port Geraldton. Coasts and Ports 2009: In a Dynamic Environment. p. 490. Available from: http://search.informit.com.au/document Summary;dn=863683725499435;res=IELENG.

144 Shi F, Kirby JT, Tehranirad B, Harris JC, Grilli S (2012) FUNWAVE-TVD Fully Nonlinear Boussinesq Wave Model with TVD Solver: Documentation and User's Manual.

145 Monk K, Zou Q, Conley D (2011) *Numerical and analytical simulations of wave interference about a single row array of wave energy converters.* In: EWTEC 2011 Proceedings.

146 Lynett PJ, Liu PLF, Sitanggang KI, Kim DH (2008) Modeling wave generation, evolution, and interaction with depth-integrated, dispersive wave equations COULWAVE Code Manual Cornell University Long and Intermediate Wave Modeling Package v. 2.0.

147 ISEC (2008) COULWAVE home page. Available from: http://isec.nacse.org/models/coulwave{_}description.php.

148 Lynett P, Liu PLF (2004) A two-layer approach to wave modelling. *Proceedings of the Royal Society A.* 460(2049), 2637–2669. Available from: http://rspa.royalsocietypublishing.org/content/460/2049/2637.abstract.

149 Lynett PJ, Liu PLF (2004) Linear analysis of the multi-layer model. *Coastal Engineering.* 51(5–6), 439–454. Available from: http://www.sciencedirect.com/science/article/pii/S0378383904000560.

150 Liu PLF, Lynett P, Fernando H, Jaffe BE, Fritz H, Higman B, *et al.* (2005) Observations by the international tsunami survey team in Sri Lanka. *Science.* 308(5728), 1595. Available from: http://www.sciencemag.org/content/308/5728/1595.abstract.

151 Lynett P (2005) A numerical study of the run-up generated by three-dimensional landslides. *Journal of Geophysical Research.* 110(C3), C03006. Available from: http://doi.wiley.com/10.1029/2004JC002443.

152 Kirby JT (2014) REF/DIF home page. Available from: https://www.udel.edu/kirby/programs/refdif/refdif.html.

153 Kirby JT, Dalrymple RA, Shi F (2002) Combined Refraction/Diffraction Model REF/DIF 1 Version 3.0, Documentation and User's Manual.

154 Tolman HL (1989) The numerical model WAVEWATCH: a third generation model for the hindcasting of wind waves on tides in shelf seas. *Communications on Hydraulic and Geotechnical Engineering.* (89-2), 72.

155 NOAA (2013) Centre, National Weather Service – Environmental Modeling. Available from: http://polar.ncep.noaa.gov/:http://polar.ncep.noaa.gov/waves/wavewatch/wavewatch.shtml.

156 Tolman HL (2009) User manual and system documentation of WAVEWATCH-III version 3.14. Technical note. Available from: http://polart.ncep.noaa.gov/mmab/papers/tn276/MMAB{_}276.pdf$\delimiter"026E30F$npapers2://publication/uuid/298F36C7-957 F-4D13-A6AB-ABE61B08BA6B.

157 Cornett A, Toupin M, Baker S, Piche S, Nistor I (2014) *Appraisal of IEC Standards for Wave and Tidal Energy Resource Assessment.* In: International Conference on Ocean Energy, ICOE 2014. Halifax, Canada.

158 METOCEAN-SOLUTIONS-Ltd (2008) Marine energy resources. Ocean wave and tidal current resources in New Zealand.

159 Spindler DM, Tolman H (2008) Example of WAVEWATCH III for the NE Pacific. Environmental Modeling Center Marine Modeling and Analysis Branch.

160 Tolman HL, Alves JHGM (2005) Numerical modeling of wind waves generated by tropical cyclones using moving grids. *Ocean Modelling.* 9(4), 305–323.

161 Boudière E, Maisondieu C, Ardhuin F, Accensi M, Pineau-Guillou L, Lepesqueur J (2013) A suitable metocean hindcast database for the design of marine energy converters. *International Journal of Marine Energy.* 3-4, e40 e52. Available from: http://www.sciencedirect.com/science/article/pii/S2214166913000362.

162 Kim CK, Toft JE, Papenfus M, Verutes G, Guerry AD, Ruckelshaus MH, *et al.* (2012) Catching the right wave: evaluating wave energy resources and potential compatibility with existing marine and coastal uses. PloS One. (11), e47598. Available from: http://journals.plos.org/plosone/article?id=10.1371/journal.pone.0047598.

163 Alonso R, Solari S, Teixeira L (2015) Wave energy resource assessment in Uruguay. Energy. 93, 683–696. Available from: http://www.sciencedirect.com/science/article/pii/S0360544215011950.

164 Zheng C, Zhuang H, Li X, Li X (2011) Wind energy and wave energy resources assessment in the East China Sea and South China Sea. *Science China Technological Sciences.* 55(1), 163–173. Available from: http://link.springer.com/10.1007/s11431-011-4646-z.

165 SWAN (2013) SWAN home page Available from: http://swanmodel.sourceforge.net/.

166 The SWAN team (2011) Swan User Manual. SWAN Cycle III version 4072A.

167 Zhang D, Kiu H, Lin P (2003) *The application of SWAN to the simulation of a storm surge.* In: Proceedings of the International Conference on Estuaries and Coasts. Hangzhou, China: IAHR.

168 Ekphisutsuntorn P, Wongwises P, Chinnarasri C (2010) The application of Simulating Waves Nearshore model for wave height simulation at Bangkhuntien shoreline. *American Journal of Environmental Sciences.* 6(3), 299–307.

169 Sadabadi SA (2011) *Modeling wave, currents, and sediment transport in Ystad Bay.* Lund University.

170 de Oude R (2010) *Modelling wave attenuation by vegetation with SWAN-VEG: Model evaluation and application to the Noordwaard polder.* Univeristy of Twente.

171 Iglesias G, Carballo R (2009) Wave energy potential along the Death Coast (Spain). *Energy.* 34(11), 1963–1975. Available from: http://www.sciencedirect.com/science/article/pii/S0360544209003454.

172 Iglesias G, Carballo R (2010) Offshore and inshore wave energy assessment: Asturias (N Spain). *Energy.* 35(5), 1964–1972. Available from: http://www.sciencedirect.com/science/article/pii/S0360544210000137.

173 Iglesias G, Carballo R (2010) Wave energy and nearshore hot spots: The case of the SE Bay of Biscay. *Renewable Energy.* 35(11), 2490–2500. Available from: http://www.sciencedirect.com/science/article/pii/S0960148110001333.

174 Smith H, Maisondieu C (2014) Cross-Analysis of Resource Assessment Methods. Task 1.5 of WP3 from the MERiFIC Project. Available from: http://www.merific.eu/files/2012/06/D3-1-5{_}CrossAnalysis{_}final.pdf.

175 Robertson B, Bailey H, Clancy D, Ortiz J, Buckham B (2016) Influence of wave resource assessment methodology on wave energy production estimates. *Renewable Energy.* 86, 1145–1160. Available from: http://www.sciencedirect.com/science/article/pii/S0960148115302998.

176 Lewis MJ, Neill SP, Hashemi MR, Reza M (2014) Realistic wave conditions and their influence on quantifying the tidal stream energy resource. *Applied Energy.* 136, 495–508. Available from: http://www.sciencedirect.com/science/article/pii/S0306261914010095.

177 Abanades J, Greaves D, Iglesias G (2014) Coastal defence through wave farms. *Coastal Engineering.* 91, 299–307. Available from: http://www.sciencedirect.com/science/article/pii/S0378383914001306.

178 Abanades J, Greaves D, Iglesias G (2014) Wave farm impact on the beach profile: A case study. *Coastal Engineering.* 86, 36–44. Available from: http://www.sciencedirect.com/science/article/pii/S0378383914000179.

179 Abanades J, Greaves D, Iglesias G (2015) Coastal defence using wave farms: The role of farm-to-coast distance. *Renewable Energy.* 75, 572–582. Available from: http://www.sciencedirect.com/science/article/pii/S0960148114006752.

180 Brodtkorb PA, Johannesson P, Lindgren G, Rychlik I, Rydén J, Sjø E (2000) WAFO – a Matlab toolbox for analysis of random waves and loads. In: Proc. 10th Int. Offshore and Polar Eng. Conf. ISOPE. Seattle, USA. p. 343–350.

181 WAFO-group (2007) WAFO webpage. Available from: http://www.maths.lth.se/matstat/wafo/index.html.

182 WAFO-group (2011) *WAFO – a Matlab Toolbox for Analysis of Random Waves and Loads.* Lund University.

183 Brodtkorb PA (2004) *The probability of occurrence of dangerous wave situations at sea.* Norwegian University of Science and Technology.

184 McComb C, Lawson M (2013) *Combining multi-body dynamics and potential flow simulation methods to model a wave energy converter.* In: Proceedings of the 1st Marine Energy Technology Symposium METS13. Washington, DC.

185 Nolan G, Ringwood JV, Holmes B (2007) *Short term wave energy variability off the west coast of Ireland.* In: Proceedings of the 7th European Wave and Tidal Energy Conference. Porto, Portugal.

186 Harnois V, Weller SD, Johanning L, Thies PR, Le Boulluec M, Le Roux D, *et al.* (2015) Numerical model validation for mooring systems: Method and application for wave energy converters. *Renewable Energy.* 75, 869–887. Available from: http://www.sciencedirect.com/science/article/pii/S0960148114007010.

187 Thies PR, Johanning L (2012) *Lifecycle fatigue load spectrum estimation for mooring lines of floating marine energy converter.* In: Proceedings of ASME 2012 31st International Conference on Ocean, Offshore and Arctic Engineering OMAE 2012. Rio de Janeiro, Brazil.

188 Aalborg-University. SDWED webpage. Available from: http://www.sdwed.civil.aau.dk/.

189 Nielsen SRK, Zhou Q, Kramer MM, Basu B, Zhang Z (2013) Optimal control of nonlinear wave energy point converters. *Ocean Engineering.* 72, 176–187. Available from: http://www.sciencedirect.com/science/article/pii/S0029801813002758.

190 Sichani MT, Chen JB, Kramer MM, Nielsen SRK (2014) Constrained optimal stochastic control of non-linear wave energy point absorbers. *Applied Ocean Research.* 47, 255–269. Available from: http://www.sciencedirect.com/science/article/pii/S0141118714000546.

191 Zurkinden AS, Ferri F, Beatty S, Kofoed JP, Kramer MM (2014) Non-linear numerical modeling and experimental testing of a point absorber wave energy converter. *Ocean*

Engineering. 78, 11–21. Available from: http://www.sciencedirect.com/science/article/pii/S0029801813004393.

192 MSS (2010) Marine Systems Simulator. Available from: http://www.marinecontrol.org.

193 Perez T, Fossen TI (2009) A Matlab Toolbox for Parametric Identification of Radiation-Force Models of Ships and Offshore Structures. *Modeling, Identification and Control: A Norwegian Research Bulletin.* 30(1), 1–15. Available from: http://www.mic-journal.no/ABS/MIC-2009-2009-1.asp.

194 Borg M, Collu M, Brennan FP (2013) Use of a wave energy converter as a motion suppression device for floating wind turbines. *Energy Procedia.* 35, 223–233. Available from: http://www.sciencedirect.com/science/article/pii/S1876610213012617.

195 Bånkestad M (2013) Simulation and dynamic control of a wave energy converter [Master's thesis]. Kungliga Tekniska Högskolan (KTH).

196 WEC-Sim (2015) WEC-Sim webpage. Available from: http://wec-sim.github.io/WEC-Sim/index.html.

197 Lawson M, Yu YH, Ruehl K, Michelen C (2015) *Improving and validating the WEC-Sim wave energy converter code.* In: Proceedings of the 3rd Marine Energy Technology Symposium METS2015. Washington, DC.

198 So R, Casey S, Kanner S, Simmons A, Brekken TKA (2015) *PTO-Sim: Development of a power take off modeling tool for ocean wave energy conversion.* In: Proceedings of IEEE PES.

199 Yu YH, Li Y, Hallett K, Hotmisky C (2014) *Design and analysis for a floating oscillating surge wave energy converter.* In: Proceedings of OMAE 2014. San Francisco, CA.

200 NEMOH (2015) NEMOH webpage. Available from: http://lheea.ec-nantes.fr/doku.php/emo/nemoh/start.

201 Borgarino B, Babarit A, Singh J, Kramer MM (2012) *Assessing the structural forces in the Wavestar wave energy converter, using a coupled hydro-elastic methodology.* In: 4th International Conference on Ocean Energy, ICOE. Dublin, Ireland.

202 Antonutti R, Peyrard C, Johanning L, Incecik A, Ingram D (2014) An investigation of the effects of wind-induced inclination on floating wind turbine dynamics: heave plate excursion. *Ocean Engineering.* 91, 208–217. Available from: http://www.sciencedirect.com/science/article/pii/S0029801814003242.

203 Andersen M, Hindhede D, Lauridsen J (2015) Influence of model simplifications excitation force in surge for a floating foundation for offshore wind turbines. *Energies.* 8(4), 3212–3224. Available from: http://www.mdpi.com/1996-1073/8/4/3212/htm.

204 Hall M (2014) MoorDyn webpage. Available from: http://www.matt-hall.ca/software/moordyn/.

205 Hall M (2015) MoorDyn User's Guide. Department of Mechanical Engineering, University of Maine.

206 Hall M, Goupee A (2015) Validation of a lumped-mass mooring line model with DeepCwind semisubmersible model test data. *Ocean Engineering.* 104, 590–603. Available from: http://www.sciencedirect.com/science/article/pii/S0029801815002279.

207 Vissio G, Passione B, Hall M (2015) *Expanding ISWEC modelling with a lumped-mass mooring line model.* In: European Wave and Tidal Energy Conference EWTEC. Nantes, France.

208 HMRC (2003) OCEAN ENERGY: Development & Evaluation Protocol, Part 1: Wave Power.

209 Greaves D, Attrill M, Chadwick A, Conley D, Eccleston A, Hosegood P, *et al.* (2009) Marine renewable energy development – research, design, install. *Proceedings of the Institution of Civil Engineers: Maritime Engineering.* 162(4), 187–196. Available from: http://www.scopus.com/inward/record.url?eid=2-s2.0-77956410019{&} partnerID=40{&}md5=3cc1eb1047daab0afef16492d7f3b9cb.

210 Davey T, Venugopal V, Girard F, Smith H, Smith G, Lawrence J, *et al.* (2010) Equimar Deliverable D2.7: Protocols for wave and tidal resource assessment.

211 Venugopal V, Davey T, Girard F, Smith H, Caveleri L, Bertotti L, *et al.* (2011) Equimar Deliverable D2.4: Wave model intercomparison.

212 Herry C, Girard F, Lawrence J, Venugopal V, Davey T (2011) Equimar Deliverable D2.5: Tidal model intercomparison.

213 Topper MBR (2010) *Guidance for numerical modelling in wave and tidal energy. SuperGen Marine.* University of Edinburgh.

214 EMEC, Department of Energy and Climate Change (2009) Assessment of performance of wave energy conversion systems. Marine Renewable Energy Guides. Available from: www.emec.org.uk/standards/ assessment-of-performance-of-wave-energy-conversion-systems/.

215 Venugopal V, Davey T, Smith H, Smith G, Cavaleri L, Bertotti L, *et al.* (2010) Equimar Deliverable 2.3: Application of numerical models. Assessment.

216 Folley M, Babarit A, Child B, Forehand D, O'Boyle L, Silverthorne K, *et al.* (2012) A review of numerical modelling of wave energy converter arrays. Volume 7: Ocean Space Utilization; Ocean Renewable Energy. p. 535–545. Available from: http:// proceedings.asmedigitalcollection.asme.org/proceeding.aspx?doi=10.1115/ OMAE2012-83807.

9

Environmental Effects

*Gregorio Iglesias[a], Javier Abanades Tercero[b], Teresa Simas[c], Inês Machado[c]
and Erica Cruz[c*]*

[a] *Professor of Coastal Engineering, School of Engineering, University of Plymouth, UK*
[b] *Offshore Renewable Energy Consultant, TYPSA, Spain*
Associate Researcher, School of Engineering, University of Plymouth, UK
[c] *Senior Researcher, WavEC – Offshore Renewables, Lisboa, Portugal*

9.1 Introduction

Wave and tidal energy devices extract energy from the environment. This extraction is bound to have effects at different levels, from the physical to the biological, which are the subject of this chapter. The next sections deal with the physical effects of marine renewable energy (MRE) on the environment: the effects of wave farms on the wave field (Section 9.2) and coastal processes (Section 9.3), and those of tidal stream farms on the hydrodynamics and sediment transport processes (Section 9.4). In Section 9.5 we consider the impacts of MRE on the marine biota. Finally, Section 9.6 is dedicated to the environmental impact assessment that must form part of any MRE project.

9.2 Wave Farm Effects on the Wave Field

9.2.1 Wave Farm Effects: Positive or Negative?

Some of the effects of wave farms may in fact be positive. For instance, in the case of a coastline experiencing wave-induced erosion, the extraction by a wave farm of some of the incident wave energy may mitigate the problem. This opens up the possibility of using wave farms as a means not only to produce carbon-free energy but also to protect coastal stretches at risk from erosion, complementing, or even replacing, conventional coastal defence solutions.

A significant proportion of the world's coastline is at risk of erosion or flooding, and these risks will be exacerbated by climate change, notably through sea level rise and

---

* G. Iglesias and J. Abanades contributed Sections 9.1–9.4; T. Simas, I. Machado and E. Cruz contributed Sections 9.5 and 9.6.

Wave and Tidal Energy, First Edition. Edited by Deborah Greaves and Gregorio Iglesias.
© 2018 John Wiley & Sons Ltd. Published 2018 by John Wiley & Sons Ltd.

increased storminess (e.g. [1]). Many of the coastal regions at risk are subject to energetic wave climates. At the same time, MRE is called upon to play a fundamental role in reducing our carbon footprint and tackling climate change goals, as identified, for example, in the SET-Plan [2]. Wave farms, by harnessing wave energy, influence the hydrodynamics and morphodynamics of coastal regions, and may serve not only to generate carbon-free energy but also to mitigate coastal erosion. In this way MRE would tackle the impacts of climate change on coastal environments, on the one hand, and the need to secure and diversify the energy supply, on the other – two of the main challenges facing humankind today.

The importance of the management of coastal erosion and flood risks can hardly be overstated. In the UK, for instance, the central government has committed over £5 billion to coastal management since 2005. Global and UK mean sea level, which has been observed to be rising at about 3 mm per year [3], is projected to rise by 1 m or more by 2100 according to the Intergovernmental Panel on Climate Change, increasing coastal flood risk and erosion. Additionally it is anticipated that climate change will bring about increased frequency and intensity of storms [4, 5] with the corresponding increase in extreme storm surge water levels. These factors will act to progressively degrade the standard of protection currently afforded by existing coastal defences [6]. The intensity of coastal erosion is also set to accelerate. Moreover, increased storminess will exacerbate damage and degradation rates of coastal defences and thus their cost of maintenance.

As waves propagate through a wave farm, its wave energy converters (WECs) absorb part of their energy, with the result that the wave energy transmitted past the farm is lower than the incident wave energy [7, 8]. This reduction in wave energy results in a milder wave climate in the lee of the farm. In work conducted on Perranporth Beach (SW England) it was found that a wave farm would, in effect, modify coastal morpho-dynamics, leading to a reduction in storm-induced (short-term) erosion on the beach face of over 30% in some sections, and to a significant mitigation of medium-term erosion (over a six-month period) [9–12]. These results vindicate exploring the utility and application of wave farms as a means of countering one of the effects of climate change: coastal erosion [13]. Indeed, wave farms have a number of advantages relative to conventional coastal defence solutions (detached breakwaters, groynes, etc.), including a lower (visual) impact and a natural adaptability to sea-level rise. For these reasons, wave farms are poised to become, in addition to an important contributor to the energy mix, a valuable tool in coastal management.

9.2.2 Near-Field Effects

The primary objective of the majority of the studies of near-field effects undertaken so far was to determine the optimal array configuration in order to minimise negative interactions between WECs, with a view to maximising the overall performance of the wave farm. For this purpose, sensitivity analyses were carried out applying different numerical models; however, it is not straightforward (and perhaps not possible at all) to establish an optimum layout of general validity due to the influence of a number of elements specific to each farm: the wave climate, the bathymetry of the area and distance to the coast, and, last but not least, the type of WEC. In the following we present the most relevant studies to date.

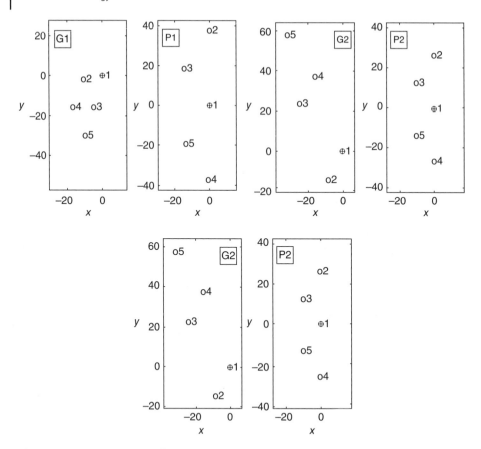

Figure 9.1 Optimum array configuration. G and P stand for the genetic algorithm and parabola intersection method, respectively. The number signifies the objective pursued: (1) to maximise power in an array of real-tuned devices, (2) to maximise power in an array of reactively tuned devices, and (3) to minimise power in an array of reactively tuned devices [14, 16].

The optimisation of the layout of a wave farm consisting of wave-activated bodies, with various array configurations under different wave conditions, was reported by [14, 15]. The arrays consisted of five floating devices, identical in geometry and power take-off characteristics. The interaction of each array with the wave field was determined by means of a numerical model that accounted for the incident, scattered and radiated waves, and for the motion of the devices. A genetic algorithm (G) and the parabola intersection (P) method were used to obtain the optimal configurations (Figure 9.1).

The interactions between devices in deep water as a function of their distance were studied by means of a boundary element model (BEM) based on linear potential flow theory under regular and irregular wave conditions [17]. Based on the results presented, deploying WECs in close proximity (under 100 m apart) was not recommended due to the negative interactions that would ensue. For greater distances, between 100 and 500 m, there are no general recommendations since wave interaction varied depending on the array configuration, particularly when the direction of the incident waves was not aligned with the array.

(a)   (b)

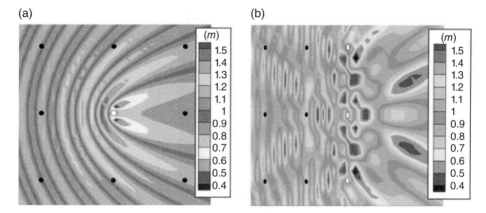

Figure 9.2 Wave amplitude around a set of (a) one or (b) three heaving cylinders in a nine-body array: (a) the wave amplitude around the white cylinder; (b) the wave amplitude around the three white cylinders in interaction [18].

The influence of the interactions between WECs on the overall annual energy production of the array was assessed; two WECs were considered, a heaving cylinder and a surging barge, arranged in arrays of nine with regular patterns, either square or triangular (staggered) grids (Figure 9.2) [18]. This study revealed that when WECs are very efficient, the square grid is not appropriate, particularly for short spacings between devices, as the masking or shadow effect is exacerbated. Triangle-based arrays may be the most suitable configuration, for they minimise destructive interactions between WECs.

Physical modelling has also been conducted to determine the interaction between converters [19], with large arrays of up to 25 heaving point-absorber WECs for a range of layouts and wave conditions. Although not all the configurations have been analysed so far, the results of the 5 × 5 WEC rectilinear array showed wave height attenuation in the lee of the farm. Figure 9.3 illustrates the modification of the wave field due to the devices for long-crested waves. While along the first rows of devices the wave height increased, in the lee of the devices wave heights were attenuated up to 18.10% at a distance of 30D, with D the diameter of the devices. The future analysis of the other array configurations will allow the determination of the best with a view to minimising the destructive interaction between devices.

The Wave Dragon WEC was implemented on a time-dependent mild-slope equation model (MILDwave) to establish the optimum wave farm configuration [20, 21]. To model the combined effects of reflection, transmission and absorption by a WEC, laboratory tests of a single device were conducted. The optimisation of the layout was carried out by means of the assessment of two key parameters: the distribution of the WECs and the spacing between WECs. Regarding the former, both staggered and aligned (also called square) configurations were studied in a nine-WEC array arranged in three lines. The effects on the wave height acting on the adjacent WECs and in the lee of the devices were represented by means of a disturbance coefficient, K_D, which represented the ratio between the disturbed significant wave height and the incident significant wave height. The reduction in the significant wave height (Figure 9.4) in front of the devices situated

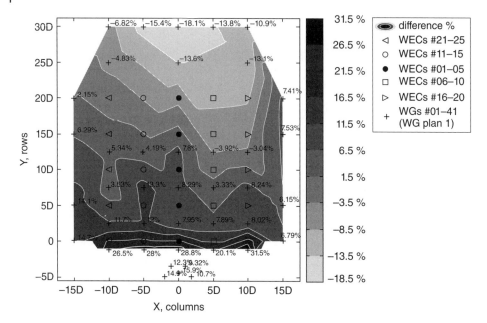

Figure 9.3 Perturbed wave field normalised by recorded undisturbed wave field for the 5 × 5 WEC array configuration [19]. The basin width (*X*, columns) and length (*Y*, rows) are expressed in number of WEC unit diameters, *D* = 0.315 m.

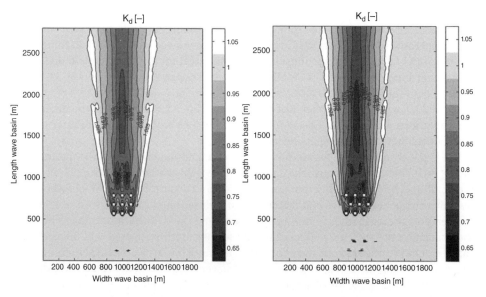

Figure 9.4 Calculated disturbance coefficient, K_D, for an aligned (left) and staggered (right) array formed by nine Wave Dragon WECs for irregular long-crested waves [21].

on the second and third rows is higher in the case of the aligned layout, and therefore the staggered formation presents better performance than the square grid. Whereas the capture-width ratio on the second and third row of devices was 45% and 35% for the staggered grid, both rows presented ratios of 30% in the case of the aligned layout [21]. However, this does not imply that the attenuation of the wave height in the lee of staggered grids is less intense than with aligned grids.

Considering the staggered array, the influence of the distance between devices was analysed. A five Wave Dragon farm was studied with three spacings: D, $2D$ and $3D$, with $D = 260$ m (i.e. the distance between the twin bows of the device). In general, for wave farms of overtopping converters, the greater the spacing between WECs, the lesser the interaction; however, the smaller the spacing, the cheaper the farm. The results showed that the layout with a distance between devices of $2D$ was the optimal farm layout (Figure 9.5), taking cost and spatial considerations into account.

In sum, although there is no optimum layout of general validity, a number of recommendations can be drawn from the near-field studies conducted so far. Spacings between devices below 100 m are inappropriate, especially for overtopping devices, in which the absorption of wave power by the WEC is the most relevant process. The works reviewed showed that avoiding the destructive interference between devices is essential in the design of the wave farm in order to not decrease their performance and therefore their economic viability. On the other hand, distances between WECs over 300–400 m are not recommended, for they would increase the costs of the wave farm significantly.

9.2.3 Far-Field Effects

This section is concerned with the effects of a wave farm on the wave conditions in its lee, in particular in nearshore areas. Given that waves are often the main driver of coastal processes (e.g. nearshore currents, sediment transport), any effects on the nearshore wave conditions will have some impact on these processes and, as discussed in the introduction, be reflected in the morphodynamic evolution of the coast, in some cases for the better (e.g. by mitigating erosion). The impacts on the morphodynamics will be dealt with in Section 9.3.

9.2.3.1 Introduction

In the following we present the relevant work to date, advancing in time from the first publications to the more recent ones.

The SWAN (Simulating WAves Nearshore) spectral wave model was used to investigate the effects of a wave farm in the Wave Hub project (UK) [22]. A fine grid was nested into a coarse computational grid with a resolution of 200 and 5000 m respectively. The wave field–wave farm interaction was studied by means of different notional energy absorption coefficients, leading to different energy transmission coefficients: (i) 0% (complete absorption), (ii) 40%, (iii) 70%, and (iv) 90%. The study analysed the reduction in wave height brought about by the Wave Hub along 300 km of shoreline in the SW coast of England under different wave conditions. The results showed that the reduction in the significant wave height at the coastline was around 5% in the case of a transmission coefficient of 40%, and considerably smaller in the cases with higher transmission coefficients, i.e. lower efficiencies (Figure 9.6). Notwithstanding the interest of these results, three aspects must be taken into account: the rather coarse computational

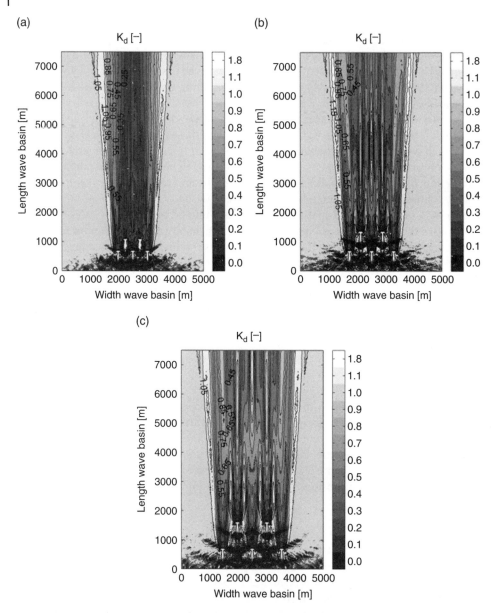

Figure 9.5 Calculated disturbance coefficient, K_D, of an array of nine Wave Dragons arranged in two rows with a spacing of (a) D, (b) $2D$ and (c) $3D$, with $D = 260$ m, for irregular long-crested waves [20].

grid (200 m) limited the spatial resolution and accuracy of the results; the wave farm was modelled as a single element, i.e. the individual WECs and their locations within the farm were not considered; and, perhaps most importantly, the values of the wave absorption coefficient were notional, i.e. they did not reflect the actual interaction of real WECs with the wave field. In the following we will see how these shortcomings were progressively addressed in subsequent works.

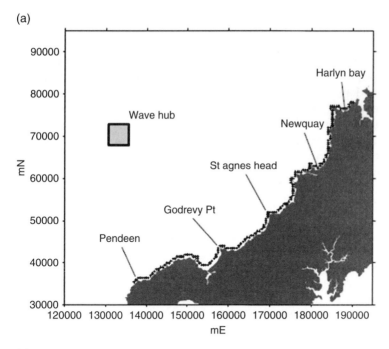

(a)

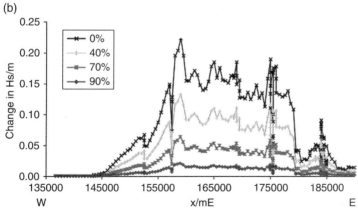

(b)

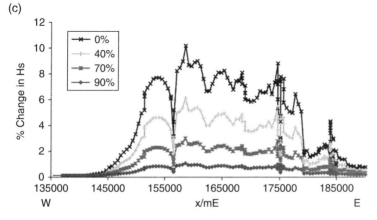

(c)

Figure 9.6 Map of the WaveHub area (left) and change in significant wave height in magnitude (a) and percentage (b) along the coastline for varying wave energy transmission percentages under a reference sea state with $H_{s0} = 3.3$ m, $T_m = 11$ s and direction = N.

The approach was upgraded [23] by implementing a wave absorption coefficient that varied as a function of the frequency, enabling the response function (Power Transfer Function, PTF) of the WECs to be accommodated, which defines the proportion of power extracted from the waves as a function of the wave frequency. The wave farm was represented in three configurations: a continuous 4 km barrier [22]; a series of shorter (100 m) barriers with a 200 m spacing between them; and two staggered rows of 100 m barriers with an inter- and intra-row spacing of 200 m and 500 m, respectively (Figure 9.7). The results were analysed at different distances from the wave farm – an example can be found in Figure 9.8. Two scenarios were considered: low-efficiency (narrow PTF) and high-efficiency (wide PTF) WECs. Low-efficiency WECs led to negligible reductions of the significant wave height in the lee of the farm, with values around 1% at a distance of 10 km from the farm. In the case of high-efficiency WECs the attenuation of significant wave height exceeded 5% at distances of 5 km behind the farm. The two major elements for improvement at this point were: the fact that wave absorption by the WECs was still modelled with notional values (not based on real data of wave–WEC interaction) and the low resolution of the computational grid.

A nonlinear Boussinesq wave model (MIKE 21) was used to assess the impact of a wave farm on the wave conditions off Orkney (Scotland) [24]. WECs were modelled as solid and porous structures. As solid structures, the devices mainly reflected waves (as would a vertical breakwater) and did not allow any wave energy absorption or transmission through them. As porous structures, the modification of the porosity values enabled different degrees of reflection, absorption and transmission to be represented. The wave farm consisted of five devices, with dimensions of 160 m (length) and 10 m (width) and a spacing between devices of 160 m. Figure 9.9 shows the comparison between the baseline (no farm) and wave farm scenarios. The 'non-absorbent' devices (solid structure, porosity $n = 0$) clearly led to unrealistic results, with reductions of wave height of up to 70%. The devices with a porosity coefficient of 0.7 significantly reduced the wave height in the lee of the farm, over 10% at the shoreline. The so-called disturbance coefficients ratio is presented in Figure 9.10 for different porosities.

The REFDIF parabolic mild-slope wave model was used to carry out a sensitivity analysis of different array configurations of Pelamis WECs in order to analyse their impact on the wave conditions along the west coast of Portugal [25]. The main difference from the previous works was the approach to modelling the energy absorption by the WECs: instead of using notional values, estimated coefficients were provided by the developer. The REFDIF model was adapted to include the effects of the WECs, but the model could only be run under monochromatic wave conditions. Five wave farm configurations were tested (Figure 9.11), named A–E. In configurations A–C the WECs were laid out along a single row perpendicular to the wave propagation direction. In configurations D and E the WECs were arranged on three and two staggered rows, respectively. Values of up to 23% of wave energy extraction were found. Configuration D led to the greatest reduction in significant wave height in the lee of the farm (approximately 20%) but the impacts extended a greater distance along the coast (along the 10 m contour) in the case of configuration E (Figure 9.12). This work revealed that the attenuation of wave height in the lee of the farm can indeed be significant.

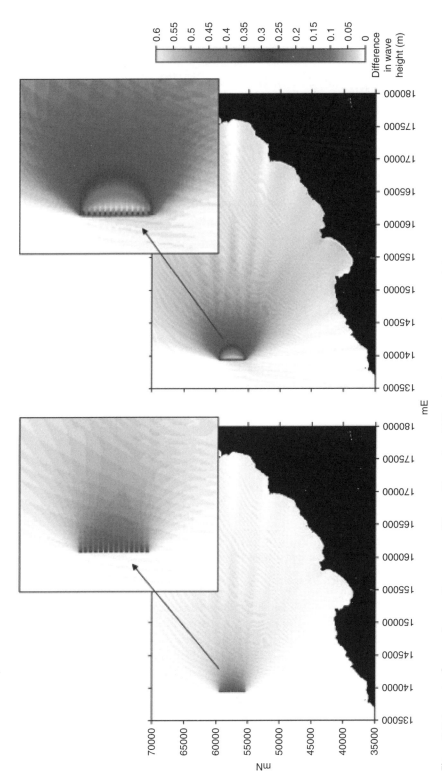

Figure 9.7 Impacts of a wave farm on the wave conditions with a series of 100 m barriers arranged in a row (left) and two staggered rows of 100 m barriers [23].

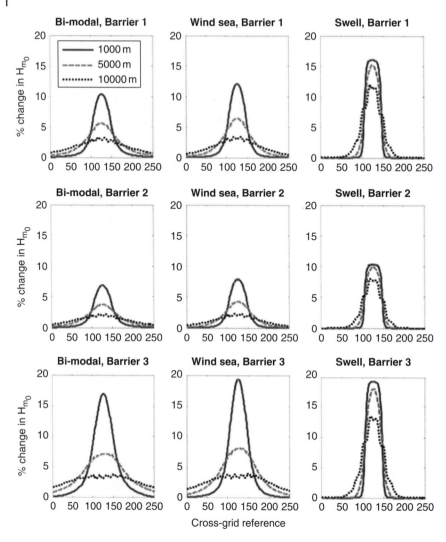

Figure 9.8 Results along the 25 km grid transect in the lee of the barrier representing the wave farm, at 100 m intervals [23].

9.2.3.2 Effects on Nearshore Wave Conditions Based on Laboratory Experiments of Wave–WEC Interaction

Thanks to the previous works two fundamental areas for improvement became clear: first, the importance of using data obtained from laboratory tests of model WECs, or from WECs at sea, to accurately represent wave–WEC interaction; and second, the need for high-resolution computational grids to define the position of the WECs in the farm, and to resolve the individual wakes of the devices, their interaction with one another, and their far-field effects. In this context, new work was carried out to investigate the impact of wave farms on nearshore wave conditions based on experimental data of wave–WEC interaction; more specifically, physical model tests of an overtopping WEC (WaveCat [26]) were carried out to obtain real (rather than notional) values

(a)

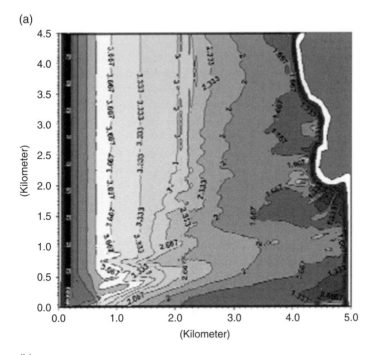

(b)

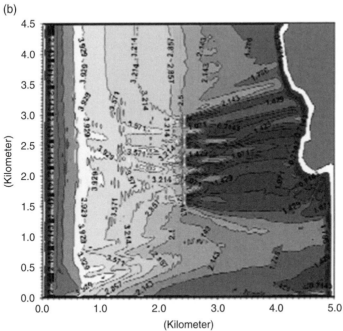

Figure 9.9 Modification of the significant wave height in the following scenarios: (a) no structure, (b) solid structure and (c) structure with porosity $n = 0.7$ [24].

(c)

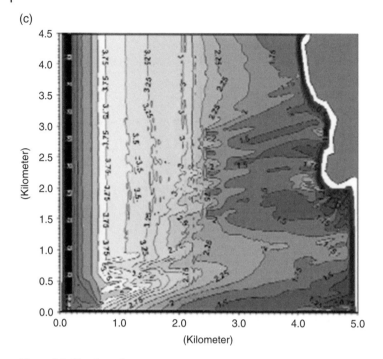

Figure 9.9 (Continued)

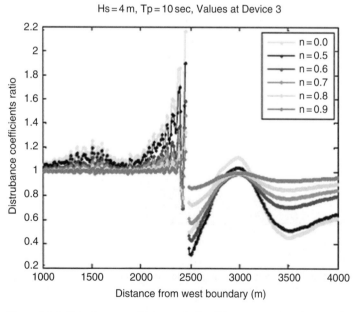

Figure 9.10 Disturbance coefficient ratio for different porosity values (*n*) along the line from offshore to the coast crossing device no. 3 [24].

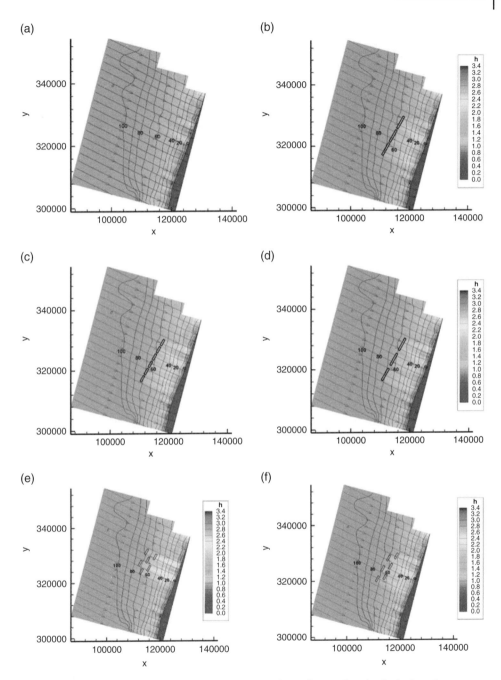

Figure 9.11 Values of significant wave height (colour scale) and wave direction in the baseline scenario (a) and for the different wave farm configurations ((b)–(f), for configurations A–E, respectively) under the January offshore wave condition: $H_s = 2.9$ m, $T_e = 11.1$ s, direction = NW [25].

(a)

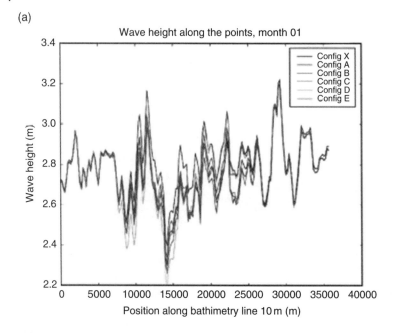

(b)

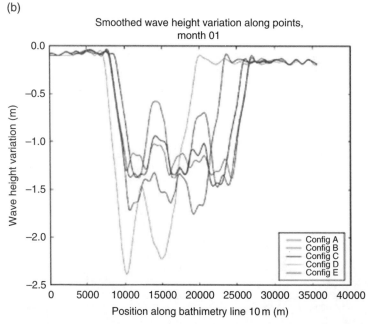

Figure 9.12 Wave height (left) and variation in the wave height (right) along the 10 m contour for the following scenarios: without (configuration X) and with wave farms (configurations A–E) [25].

of the transmission coefficient under characteristic sea states off Spain's Death Coast (Costa da Morte in the vernacular) (Figure 9.13) – one of the most promising areas for wave energy exploitation [27, 28]. The values of the transmission coefficient thus obtained were input into a high-resolution spectral wave model to assess the impact of different configurations of a wave farm, deployed at different distances from the coast [29]. This work was focused on the impact on nearshore wave power and significant wave height. In subsequent work the impact on the nearshore wave period and the directional spreading were investigated in the case of a wave farm for an island [30].

The WaveCat (Figure 9.14) is an overtopping WEC consisting of two hulls, like a catamaran – from which it takes its name. Unlike in the case of a conventional catamaran, however, the two hulls are not parallel but converge towards the stern, forming a wedge in plan view. With a single-point mooring system, WaveCat swings like a weathervane so that its bows always face the incoming waves, which propagate into the wedge. As a wave crest advances between the two hulls, wave height is enhanced by the tapering channel until, eventually, the inner hull sides are overtopped. Overtopping water is collected in reservoirs on each hull, at different levels – all above the mean sea level. The difference in elevation with respect to the exterior (sea) surface level is used to drive turbines connected to generators. Freeboard and draft, as well as the wedge angle between the hulls, can be varied according to the sea state.

WaveCat was selected for studying the impacts of a wave farm on the coast for its advantages. Unlike other overtopping WECs, in which wave crests impinge perpendicularly onto a ramp and overtopping occurs abruptly, in the case of WaveCat crests act obliquely on the hull side (Figure 9.14), so that overtopping occurs gradually along a length of hull. This has two main benefits: the motions of the WEC under waves merely shift the initial point of overtopping along the hull, without significantly affecting the overtopping volumes; and structural loadings are considerably lower. For survivability the wedge can be closed, i.e. the two hulls can be brought together so that no waves propagate between them, in which case WaveCat resembles a monohull or conventional ship. Maintenance costs are expected to be low thanks to the absence of submerged machinery and the fact that WaveCat can be towed to a shipyard in its closed configuration and repaired using existing facilities.

WaveCat models at a 1 : 30 scale were tested in wave tanks at the University of Porto (Portugal) and, very recently, at the University of Plymouth (UK) (Figure 9.14). Data from the Plymouth tests had not been analysed at the time of writing, so the following is based on the Porto tests. The model was built based on a prototype design with an overall length of 90 m. In total, 43 tests were performed in Porto, 25 with regular waves and 18 with irregular waves [7]. These tests were aimed not at optimising the device but at proving the concept. The wave conditions tested were defined based on the offshore wave climate of the Death Coast area (NW Spain), as determined from a hindcast study covering the period 1958–2001 (Figure 9.13) [7, 32, 33]. Different values of the angle between hulls (α = 30°, 45°, 60°, 90°) and reference freeboard (F = 0.04 m, 0.09 m, and 0.10 m, in model values) were tested. (We use a reference freeboard as the inner freeboard varies along the length of the hulls). The larger values of reference freeboard (F = 0.09 m and 0.10 m) were used to study the seakeeping performance of the device with no overtopping, whereas the smallest value (F = 0.04 m) was aimed at studying the device in operation (with overtopping). The experimental set-up included wave gauges in front of and behind the model in order to characterise wave field–WEC interaction.

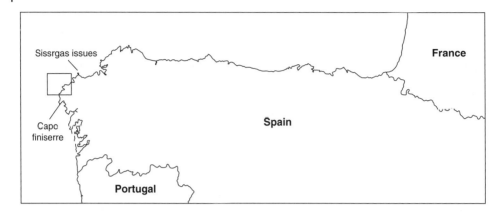

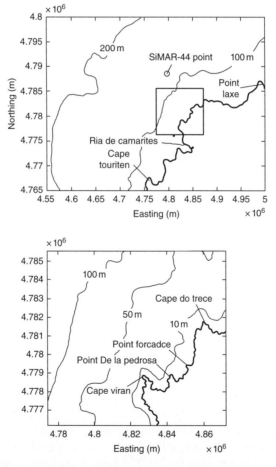

Figure 9.13 Map of the Costa da Morte, or Death Coast, in NW Spain, with the location of the SIMAR-44 grid node from which offshore wave data were obtained.

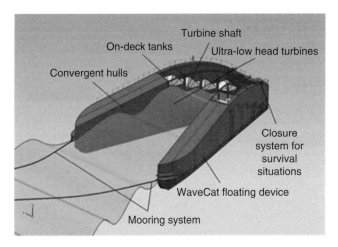

Figure 9.14 WaveCat schematic (above) [31] and model in the Ocean Basin of the University of Plymouth (below).

The transmission coefficient was calculated as

$$K_t = \frac{H_{st}}{H_{si}},$$
(9.1)

where H_{st} and H_{si} are the transmitted and incident significant wave heights, respectively (Table 9.1). The transmission coefficient did not present a large variability, remaining in the range $K_t = 0.78 \pm 0.03$ for all wave conditions with the exception of one outlier (test 5). The best overall results were obtained for an angle between hulls (α) of 45° (tests 3 and 4); consequently, the values of the transmission coefficient used to study the effects of the farm on the coast were $K_t = 0.776$ and $K_t = 0.756$, corresponding to tests 3 and 4, respectively; these values signify that 60.2% and 57.2% of the incident wave energy, respectively, is transmitted past a single WEC.

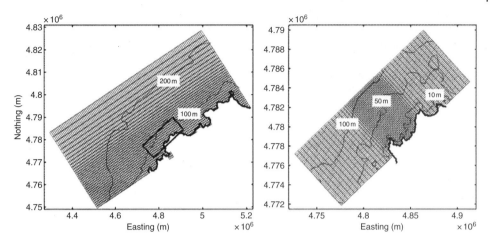

Figure 9.15 Coarse (left) and fine (right) grids for the SWAN model. For clarity, only one in three and one in nine coordinate lines is shown of the coarse and fine grid, respectively.

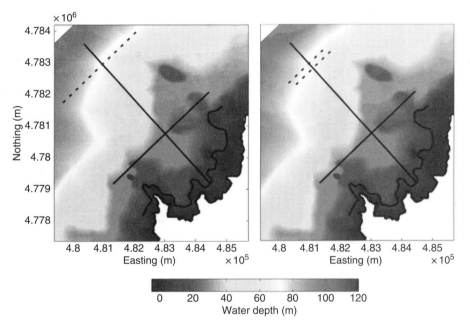

Figure 9.16 Configurations I (left) and II (right) superimposed on the bathymetry. The 10 m contour and a line at a distance of 3000 m from the wave farm and parallel to it are also delineated for later use.

distance between the twin bows of the device – was considered. For both configurations and the baseline scenario (without the wave farm) a number of wave conditions were tested, with different values of significant wave height, energy period and mean direction, representative of energetic and mean wave conditions in the area of interest (Table 9.2).

Table 9.2 Case studies, wave conditions and wave farm configurations (H_s = significant wave height; T_e = energy period; θ = wave direction; CONF = configuration (I, II or BL, baseline)).

Denomination	H_s (m)	T_e (s)	θ	NR
CS 1	3	10.8	NW	BL
CS 2	3	10.8	NW	I
CS 3	3	10.8	NW	II
CS 4	2.5	9.9	NW	BL
CS 5	2.5	9.9	NW	I
CS 6	2.5	9.9	NW	II
CS 7	3	10.8	WNW	BL
CS 8	3	10.8	WNW	I
CS 9	3	10.8	WNW	II
CS 10	2.5	9.9	WNW	BL
CS 11	2.5	9.9	WNW	I
CS 12	2.5	9.9	WNW	II

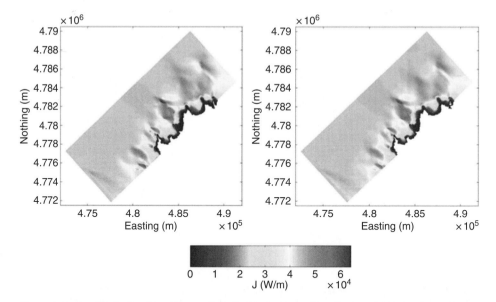

Figure 9.17 Spatial distribution of the undisturbed wave power (baseline) under energetic wave conditions (case study CS 1) as computed with the fine (left) and coarse (right) model grids, with grid spacings of 18 m and 36 m, respectively [7].

Figure 9.17 shows the spatial distribution of wave power without the wave farm for energetic wave conditions (case study CS 1). Wave power is fairly constant in deep water but, as waves approach the coast, their interaction with the irregular bathymetry leads to spatial variations in wave power, with nearshore areas of wave energy concentration

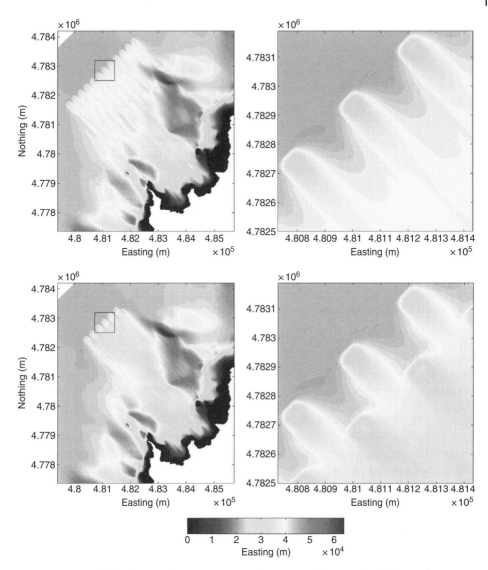

Figure 9.18 Spatial distribution of wave power in configurations I (above) and II (below), with H_s = 3m, T_e = 10.8 s and direction = NW (case studies CS2 and CS3, respectively) [7].

(also known as nearshore hotspots [34]). For these energetic wave conditions, the wave field with the wave farm is plotted in Figure 9.18. The impact of the wave farm on the wave conditions in its lee is apparent, and particularly relevant in the case of the farm formed by two rows of devices (configuration II).

The impact on wave power and significant wave height along the straight line through the mid-section of the farm delineated in Figure 9.16 is plotted in Figure 9.19. Points (1) and (2) denote the position of the first and second row of WECs. The greatest impact on wave power was found to occur 50 m behind the first row of WECs, with an attenuation of approximately 15 kW m^{-1} in energetic wave conditions and approximately 10 kW m^{-1}

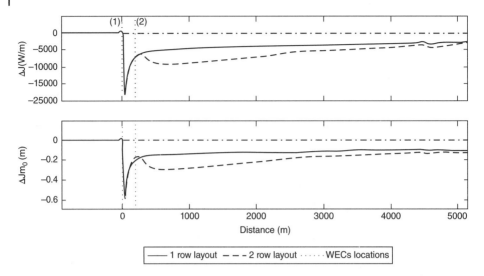

Figure 9.19 Wave farm impact on wave power (ΔJ) and significant wave height (ΔHm_0) along the line passing through the centre of the farm under energetic wave conditions from NW (case studies CS1, CS2 and CS3). Points (1) and (2) denote the position of the first and second row of WECs.

in mean wave conditions. In the case of the one-row array (configuration I) the abrupt drop in wave power immediately behind the WECs is essentially recovered over a short distance, approximately 200 m, through wave diffraction from the sides of the farm, after which the remainder of the wave power deficit is recovered more gradually. With configuration II, a second, less abrupt drop in wave power occurs due to the second row of devices, and the recovery is also more gradual. Regarding the influence of the direction, similar patterns were observed under NW and WNW waves, with only a slight difference – wave power and height recover within a somewhat smaller distance under WNW waves in the case of the two-row array.

For a better visualisation of the impacts of the wave farm on the wave patterns, the significant wave height and wave power are represented in Figure 9.20 along a line 3000 m behind the last WEC row and parallel to it (depicted in Figure 9.16). The DAPs (direct array projections) of the one- and two-row wave farms are denoted by the vertical lines (1), (4) and (2), (3), respectively. The reduction in wave power exceeds 4 kW m^{-1} under energetic wave conditions from the NW (CS2 and CS3). The difference between the one- and two-row configurations is relatively small: in section (2)–(3) the impact is slightly higher with the two-row layout, whereas in section (1)–(2) it is somewhat greater with the one-row layout. In section (3)–(4) the configuration that produces the greater impact changes between the W and E sides experiences a similar impact in both cases. Under WNW wave conditions the area with the largest impact is displaced according to the incident wave direction, and the reduction of wave power is approximately 6 kW m^{-1} under energetic wave conditions (CS8 and CS9).

In summary, the results indicate that the main impact of a wave farm on the wave conditions occurs immediately in the lee of the farm, in the form of an acute power deficit relative to the baseline scenario – a direct consequence of the absorption of wave power by the farm. From the area of maximum impact towards the coast, the power

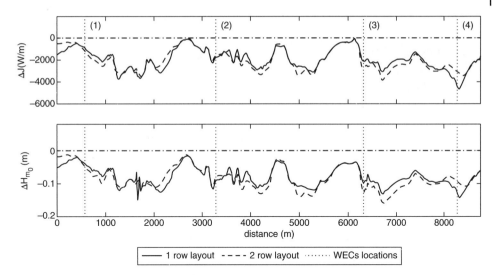

Figure 9.20 Wave farm impact on wave power (ΔJ) and significant wave height (ΔHm_0) along a 3000 m distance contour in the lee of the wave farm under energetic wave conditions from NW (case studies CS1, CS2 and CS3). Vertical lines (1) and (4) denote the DAP of the one-row farm, and vertical lines (2) and (3) the DAP of the two-row farm.

deficit reduces as wave energy is diffracted from the sides into the wake, i.e. wave power increases gradually as the distance from the farm increases. On this basis the impact of a wave farm on the nearshore wave conditions may be expected to vary with its distance from the shoreline [10, 31] – and this is the focus of the following section.

9.2.3.3 Influence of Farm-to-Coast Distance

The influence of the wave farm distance from the shoreline on its effects on the nearshore wave climate was analysed [31] by comparing three cases, with the wave farm located at reference distances of 2 km, 4 km and 6 km from the 10 m water depth contour along the same section of coastline, in water depths of 30, 50 and 80 m, respectively (Figure 9.21).

Wave farm–wave field interaction was modelled by means of SWAN using the same high-resolution grids as in the previous section, which allowed the WECs in the farm to be delineated and their impacts on the wave field to be modelled with accuracy. Regarding the wave conditions, two cases were selected: summer and winter (Table 9.3), obtained from the analysis of the Cape Villano offshore wave buoy (with data from 1998 to 2013) [10].

Figure 9.22 illustrates the wave patterns under winter conditions without and with the farm. Areas of wave power concentration (nearshore hotspots [34]) occur as a result of wave shoaling and refraction over the irregular bathymetry. In the baseline (no farm) scenario, wave power at these hotspots reached values as high as 80 kW m^{-1} and 20 kW m^{-1} in the winter and summer cases, respectively. In the wave farm scenarios, a significant modification in the wave patterns occurred. In the case of the locations farthest from the shoreline (CS 4) and intermediate (CS 3), all the WECs in the array are exposed to similar wave power values; however, in the case of the nearest location (CS 2) substantial differences in the levels of incident wave power appear across the array as a result of the nearshore hotspot to the NE of the farm and, more generally, of the irregular

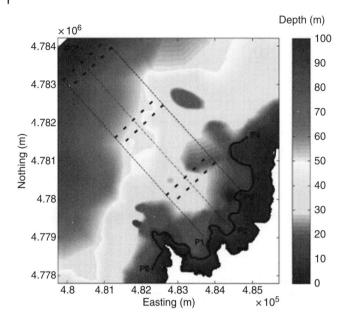

Figure 9.21 The three wave farm locations considered, at distances to the coast of $d = 2$ km (CS2, CS6), $d = 4$ km (CS3, CS7), and $d = 6$ km (CS4, CS8). The 10 m contour is depicted by a black line, and the Direct nearshore Array Projection (DAP) extends from points P1 to P3.

Table 9.3 Parameters of the case studies (d = farm-to-coast distance (– means no wave farm, i.e. the baseline case); h = reference water depth; H_s = significant wave height; T_e = energy period; θ = mean wave direction).

CS	Wave farm location	Wave case	d (km)	h (m)	H_s (m)	T_e (s)	θ (°)
1	– (Baseline)	Winter	–	–	3.3	12	307
2	Near	Winter	2	30	3.3	12	307
3	Intermediate	Winter	4	50	3.3	12	307
4	Far	Winter	6	80	3.3	12	307
5	– (Baseline)	Summer	–	–	1.8	9.5	306
6	Near	Summer	2	30	1.8	9.5	306
7	Intermediate	Summer	4	50	1.8	9.5	306
8	Far	Summer	6	80	1.8	9.5	306

wave power patterns that ensue from wave propagation over an irregular bathymetry. Overall wave power values are lower as the farm-to-coast distance decreases, as a result of wave friction with the seabed and other dissipative processes which lead to a gradual decline in wave power as waves propagate towards the shoreline. Generally speaking, the closer the wave farm to the coastline, the lower the overall incident wave power, and the greater the differences in incident wave power between the individual WECs – especially if the bathymetry is irregular.

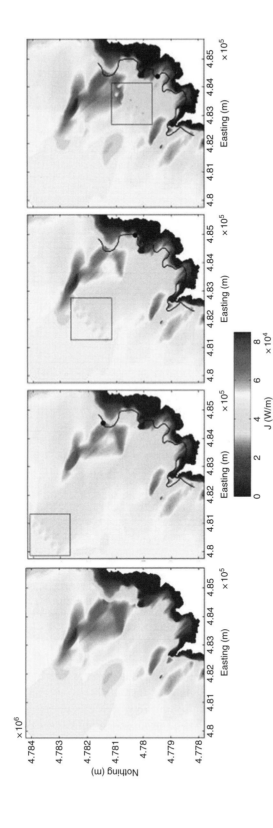

Figure 9.22 Wave power pattern in the baseline scenario (far left) and with the farm located at distances of 6 km (left), 4 km (middle) and 2 km (right) under winter conditions. The wave farm area is delineated by a box [29].

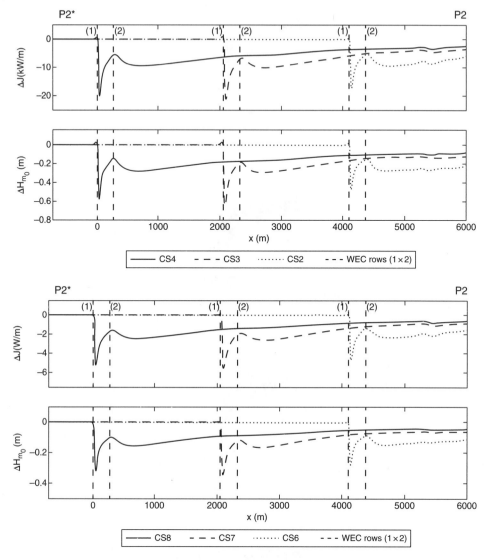

Figure 9.23 Variation in wave power (ΔJ) and significant wave height (ΔH_{m_0}) along the centreline of the wave farm (P2*–P2 in Figure 9.21) relative to the values incident on the farm in winter (above – CS2, CS3 and CS4) and summer (below – CS6, CS7 and CS8). For each case the position of the first and second rows of WECs is indicated by markers (1) and (2), respectively.

The wave power deficit and significant wave height deficit caused by the wave farm (Figure 9.23) peak immediately in the lee of the first row of WECs. By the time the waves reach the second row of devices these substantial deficits have been recovered to a great extent through wave diffraction. The second row of WECs increases the deficit again, albeit to a lesser extent. After this wave power and significant wave height gradually recover towards the coast, so that 500 m behind the farm the wave power deficit is approximately 20% in all cases.

9.2.3.4 Nearshore Impact Indicators

To better comprehend the impacts of the wave farm on the nearshore wave conditions, a number of *ad hoc* indicators were developed [29]. (In Section 9.3 additional impact indicators will be defined to characterise the impact of a wave farm on a sandy coastline.) We define the DAP as the direct array projection on the 10 m contour following the mean deep water wave direction – in other words, the 'shadow' of the farm. The first indicator is the nearshore impact (*NI*), i.e. the difference between the wave power with and without the wave farm at a generic point of the reference (10 m) contour. The second indicator is the maximum absolute nearshore impact (NI_{max}), defined as the maximum absolute value of *NI*. The point of maximum nearshore impact (MNI), where NI_{max} occurs, is naturally of interest. Figure 9.24 shows the values of *NI* and the variation in the significant wave height (ΔH_{m_0}) along the reference contour. If the farm-to-coast distance is sufficiently small, then the MNI point will be within the DAP. When the distance is larger, however, the MNI point can be outside the DAP, such as in case studies CS3 ($d = 4$ km) and CS4 ($d = 6$ km). Finally, the relative nearshore impact (*RNI*, Figure 9.25) measures the impact of the wave farm on the nearshore wave power as a percentage of the baseline wave power. While the *RNI* values in the DAP for the wave farm located at distances of 4 and 6 km were approximately 5% and 10%, respectively, they exceeded 20% with the farm at a distance of 2 km from the shoreline [29].

9.3 Wave Farm Effects on Coastal Processes

9.3.1 Introduction

Having assessed the impact of a wave farm on the nearshore wave conditions and its variation with the farm-to-coast distance, the next step is the investigation of its effects on coastal processes. This investigation can form the basis on which to apply wave farms to coastal protection.

The impact of a wave farm on the beach profile was analysed by means of the cross-shore profile model UNIBEST-TC [36]. The model was applied to a range of simulations, each with a duration of 6 months, in an area of SW Wales suited to the exploitation of the wave resource. It was found that, under certain conditions, a wave farm can lead to enhanced sand bar formation. This is in line with the findings of further work on the effects of wave farms on the beach profile, which is reported in the next section.

The impacts of a wave farm consisting of wave-activated bodies (DEXA) were reported in [37]. The farm was located a mere 650 m in front of a beach that is maintained periodically by artificial nourishment, in the region of Milano Marittima (Italy). Experimental tests of the DEXA WEC were carried out in order to determine the wave transmission coefficient (K_T) as a function of the length of the device and the length of the incident wave. The variations in the longshore sediment transport were assessed using the CERC formula [38]. For the wave conditions analysed, the wave transmission coefficient was approximately 0.8, and this resulted in a very substantial decrease in sediment transport, 43%.

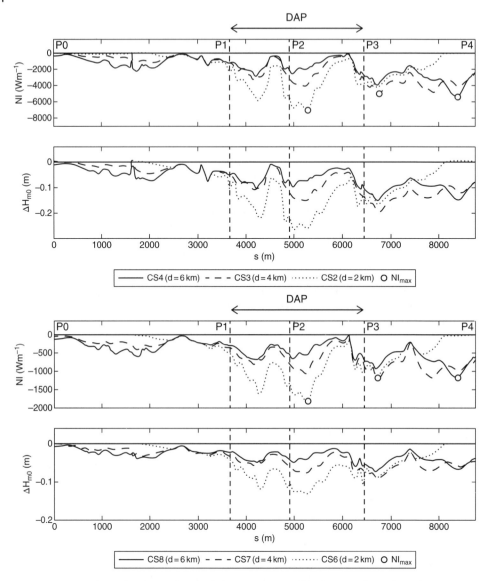

Figure 9.24 *NI* and significant wave height deficit, ΔH_{m_0}, caused by the wave farm along the 10 m contour in winter (left – CS1, 2, 3, 4) and summer (right – CS5, 6, 7, 8) conditions. *NI*$_{max}$ values are depicted by circles.

A wave farm consisting of DEXA WECs off the coast of Santander (Spain) was investigated on the basis of laboratory tests of a single device (at a 1 : 30 scale) and a three-WEC array (at a 1 : 60 scale), combined with numerical modelling with MIKE 21 [39]. The results showed that the transmission coefficient of a single device was approximately 0.8, while in the case of the array the transmission coefficients in the lee of the device located in the second row varied from 0.6 to 0.9 as a function of the wave

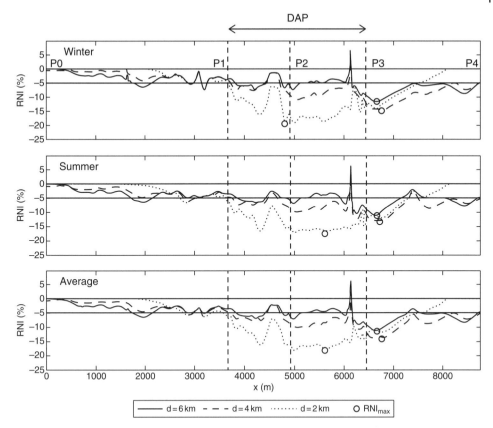

Figure 9.25 RNI in winter, summer and average (of winter and summer) for the three values of farm-to-coast distance. The points where the *absolute* value of RNI has a local maximum (RNI_{max}), indicative of maximum impact, are indicated by circles, and the line corresponding to RNI = −5% is also plotted.

conditions. Based on these coefficients, the numerical modelling of the wave farm impacts on the nearshore wave conditions showed a reduction in the wave height of up to 20% in some sections of the coastline (Figure 9.26) of wave energy.

The impacts of a wave farm consisting of 10 wave-activated bodies (OPT PB150 heaving buoys) were investigated by means of the wave propagation model OLUCA-SP, which solves the parabolic version of the mild slope equation, coupled with the sediment transport model EROS-SPm [35]. The farm was located in 50 m of water off Santoña (N Spain). Wave height attenuation in the lee of the farm (Figure 9.27) was found to be practically negligible, both directly behind the farm and along the coastline in its lee. Consequently, alterations to sediment transport patterns were negligible.

WECs of the floating, overtopping type – very different therefore from the wave-activated bodies considered in the previous paragraphs – were considered on the basis of a 13 Wave Dragon array by [40]. Transmission values were obtained in laboratory tests of a single Wave Dragon WEC. The impact of the wave farm on the wave conditions was significant, with reductions in the wave power of up to 55% (Figure 9.28).

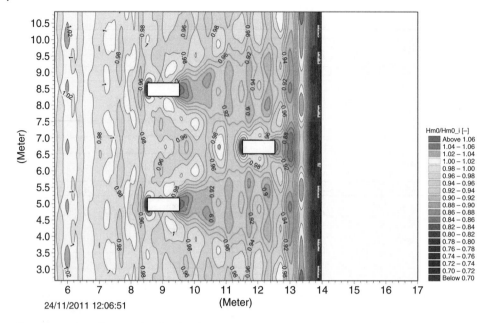

24/11/2011 12:06:51

Figure 9.26 Ratio between local and incoming (target) wave height in presence of the wave farm. The white area represents the emerged beach extension [39].

In another study [8] four wave farms consisting of different WECs were considered, specifically: three Wave Dragons (overtopping) arranged in two rows, eight Blow-Jets (hybrid system, whose main principle is overtopping) arranged in one row, 45 DEXAs (wave-activated body) arranged in five rows, and 18 Seabreath (offshore OWCs) arranged in two rows. The first two wave farms were located in the Bay of Santander (Spain) and the other two off Las Glorias Beach (Mexico). The impacts on the wave conditions were analysed by means of a 2D elliptic modified mild-slope model, WAPOQP, and the effects on the coast by means of a longshore sediment transport formula [41]. Depending on the type of WEC, the wave farms were located at different farm-to-coast distances, and consequently, water depths. The results for the Bay of Santander (Figures 9.29 and 9.30) show the effects of the wave farm on the wave patterns as a function of the WEC, with the Wave Dragons leading to a greater impact in the lee of the farm. Concerning sediment transport, the beach studied was Somo (on the right in Figure 9.29). Using the rates of longshore sediment transport, an index was developed to compare the impact between the baseline and wave farm scenarios, x_p/x_u, which indicates the accretion (positive) or erosion (negative) induced by the wave farm. The Wave Dragon and the DEXA WECs offered the greatest degree of coastal protection for the configurations and locations considered.

9.3.2 Effects on the Beach Profile

The above work outlined the possibility of applying wave farms to coastal protection on the basis of wave propagation by means of numerical modelling and, in some cases, longshore sediment transport formulae. As a continuation of this work it is necessary to consider the impacts of a wave farm on the beach profile and, more generally, the

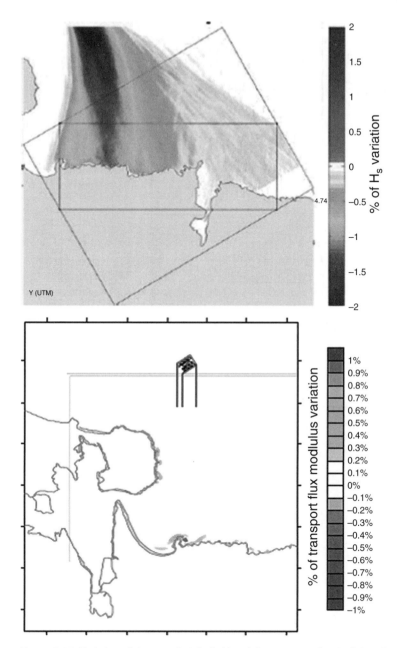

Figure 9.27 Variation of the wave height (left) and the transport flux (right) under the following wave conditions: $H_S = 6$ m, $T_p = 15$ s and direction = NW [35].

3D morphodynamics of the coast. An aspect of particular interest is storm-induced erosion, which cannot be correctly addressed with longshore sediment transport formulae.

With this in view, the XBeach coastal processes model was coupled with the SWAN wave propagation model for a case study at Perranporth Beach (Figure 9.31) [9, 11, 12].

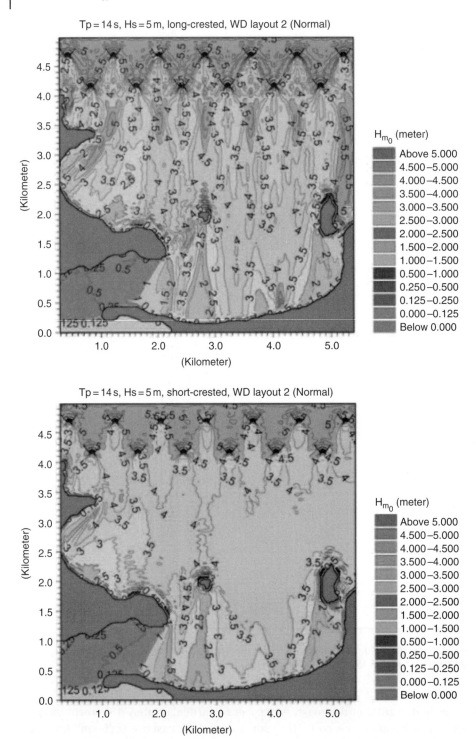

Figure 9.28 Effects of a wave farm consisting of 13 Wave Dragon devices on the wave conditions in Santander Bay, analysed by means of the disturbance coefficient, K_d, for long-crested (above) and short-crested (below) waves [40].

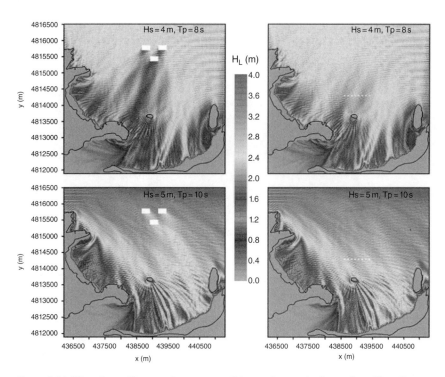

Figure 9.29 Wave farm effects on the wave conditions at Santander Bay: a three Wave Dragon wave farm (left) and an eight Blow-Jet wave farm (right). The wave conditions are indicated in the upper line [8].

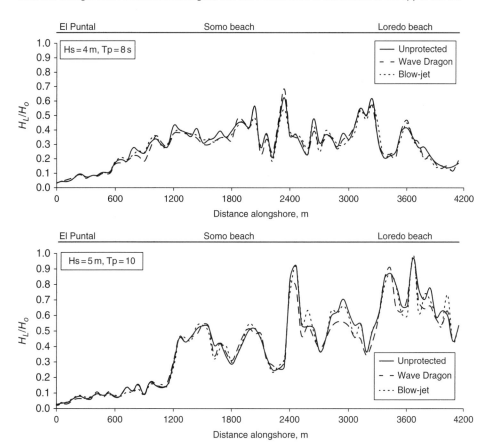

Figure 9.30 Wave height reduction nearshore for Santander Bay under different wave farm configurations [8].

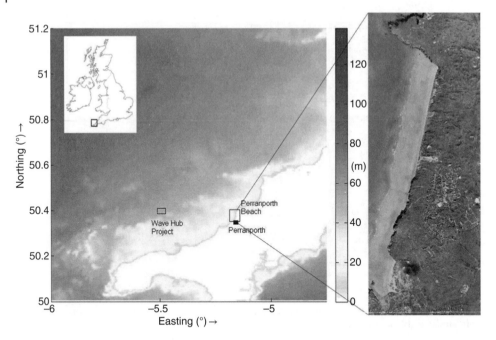

Figure 9.31 Location of Perranporth Beach and the Wave Hub in SW England (left; water depths in m), and aerial photo of the beach (right; courtesy of Coastal Channel Observatory).

A 4 km long sandy beach with a nearly flat intertidal area (tan β between 0.015 and 0.025), Perranporth has a submarine bar between −5 m and −10 m and a well-developed dune system (Figure 9.32). The selection of this location for a case study is motivated by two reasons: first, the interest of the area for wave energy development, as shown for example by the nearby Wave Hub, a grid-connected offshore facility for WEC testing [42, 43]; and second, the erosion experienced by the beach over the last years, and particularly under the extreme events that occurred during the winter 2013/2014 [44] (Figure 9.33). For these reasons, this section analyses the impact of wave farms on the beach profile in the medium term (6 months: 1 October 2007 to 1 April 2008) at Perranporth Beach using a suite of state-of-the art numerical models.

For the analysis of wave farm impacts, wave buoy data were used alongside hindcast (numerical modelling) data. Half-hourly data were obtained from the directional wave buoy off Perranporth, in approximately 10 m of water, operated by the Coastal Channel Observatory. The data from the wave buoy reflect the exposure of the area to heavy swells generated by the long Atlantic fetch, as well as to locally generated wind seas. The average values of significant wave height (H_s), peak period (T_p) and peak direction (θ_p) in the period covered by the wave buoy data (2006–2012) were 1.79 m, 10.36 s and 280°, respectively. The hindcast data were obtained from WaveWatch III, a third-generation offshore wave model run on global and regional (nested) grids, the latter with a resolution of 0.5° [45]. These values were prescribed at the outer boundaries of the wave model. In addition to these hindcast wave data, three-hourly wind data obtained from the Global Forecast System weather model were used to drive the wave model. The mean wind speed at a height of 10 m above the sea surface was $u_{10} = 9.5$ m s^{-1} during

Figure 9.32 Computer-modelled bathymetry at Perranporth Beach, including profiles P1, P2 and P3 (water depth in metres).

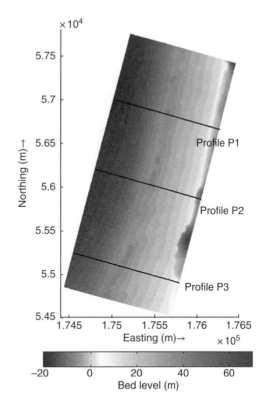

Figure 9.33 Damage at Perranporth Beach after the storms in winter 2013/2014 (courtesy of West Briton).

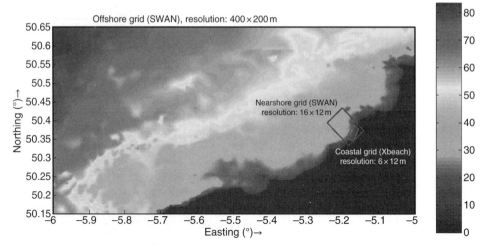

Figure 9.34 Boundaries of the computational grids used by the wave propagation and coastal processes models (SWAN and XBeach, respectively; water depths in metres).

the study period; the strongest winds, from the NW, had u_{10} values over 20 m s^{-1}. These data were used to force a suite of state-of-the-art numerical models – a coastal wave propagation model (SWAN) and a coastal processes model (XBeach) – in the first joint application of these models for these purposes.

To obtain high-resolution results in the coastal wave model – a requirement for an accurate coupling between the two models [7] – two computational grids were defined : a large-scale grid covering approximately 100 km × 50 km with a resolution of 400 m × 200 m; and a small-scale (nested) grid focused on Perranporth Beach, covering an area of approximately 15 km × 15 km with a resolution of 20 m × 20 m (Figure 9.34). The offshore and nearshore bathymetric information – obtained from the Digimap data centre and the Coastal Channel Observatory, respectively – was interpolated onto these grids. The wave farm considered consisted of 11 WaveCat WECs arranged in two rows (Figure 9.35) – using the optimum layout presented in the previous section, with a spacing between devices of 90 m. Thanks to the fine resolution of the nested grid, the individual WECs in the farm could be demarcated and their individual wakes modelled with accuracy.

The results of the coastal wave model were the input of the coastal processes model, XBeach. The spectral parameters obtained from the nearshore wave propagation model (the root-mean-square wave height, H_{rms}, mean absolute wave period, T_{m01}, mean wave direction, θ_m, and directional spreading coefficient, s) are used to construct time series of wave amplitudes, including wave groups; of relevance in beach behaviour under erosive conditions [46], wave groups are applied on the outer boundary of profiles P1 and P2 (Figure 9.33) to study their evolution.

9.3.2.1 Coastal Impact Indicators

Finally, a set of *ad hoc* impact indicators were defined to analyse the results and, on this basis, establish the impacts of the wave farm on coastal morphodynamics [11]: bed level impact (BLI), beach face eroded area (FEA), and non-dimensional erosion reduction

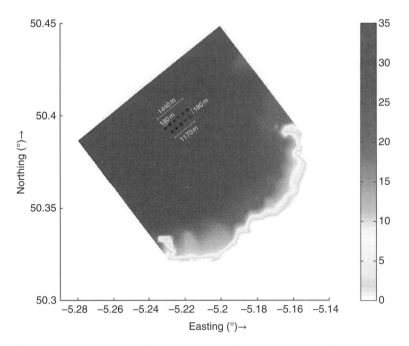

Figure 9.35 The array of WECs, or wave farm, off Perranporth (water depths in metres).

(NER). The BLI, with SI units of metres, represents the change in bed level caused by the wave farm, calculated as

$$BLI(x,y) = \zeta_f(x,y) - \zeta_b(x,y),$$ (9.3)

where $\zeta_f(x,y)$ and $\zeta_b(x,y)$ are the seabed level in the presence or absence of the farm, respectively, at a generic point of the beach. The y-axis is aligned with the general coastline orientation, and the x-axis is positive away from the sea. $BLI > 0$ ($BLI < 0$) signifies that the wave farm leads to a higher (lower) seabed level relative to the baseline – no farm – scenario, respectively.

The second coastal impact indicator, the beach FEA, is a profile function that quantifies the storm-induced erosion in the beach face, and has SI units of square metres. Unlike the BLI, which compared the farm and baseline (no farm) scenarios, the FEA index is defined separately for both scenarios, baseline (FEA_b) and wave farm (FEA_f):

$$FEA_b(y) = \int_{x_1}^{x_{max}} \left[\zeta_0(x,y) - \zeta_b(x,y) \right] dx,$$ (9.4)

$$FEA_f(y) = \int_{x_1}^{x_{max}} \left[\zeta_0(x,y) - \zeta_f(x,y) \right] dx,$$ (9.5)

where $\zeta_0(x,y)$ is the initial bed level at the point of coordinates (x,y), and x_1 and x_{max} are the values of the x-coordinate at the seaward end of the beach face and landward end of the profile, respectively.

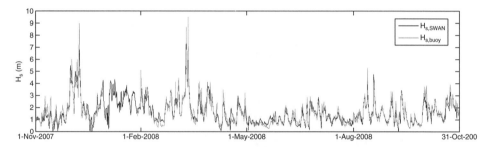

Figure 9.36 Time series of modelled ($H_{s, \text{SWAN}}$) versus observed ($H_{s, \text{buoy}}$) significant wave height.

The third coastal impact indicator, also a profile function, is the non-dimensional erosion reduction,

$$NER(y) = 1 - \left(x_{\max} - x_1\right)^{-1} \int_{x_1}^{x_{\max}} \left[\zeta_0(x,y) - \zeta_f(x,y)\right]\left[\zeta_0(x,y) - \zeta_b(x,y)\right]^{-1} dx, \qquad (9.6)$$

which quantifies the change in the eroded area of a generic profile (y) caused by the wave farm as a fraction of the total eroded area of the same profile; $NER > 0$ and $NER < 0$ signify a reduction and increase in the eroded area, respectively.

Having defined these indicators, the wave model was successfuly validated against the wave buoy data from November 2007 to October 2008 with correlation coefficient (R) and root mean square error of 0.97 and 0.38 m, respectively (Figure 9.36). Upon validation, the numerical model was applied to compare the wave patterns with and without the wave farm, and to establish the wave conditions that were to be the input to the coastal processes model. The reduction in the wave height in the lee of the farm for storm conditions (deep water wave conditions, $H_{s0} = 10.01$ m, $T_p = 15.12$ s, $\theta_p = 296.38$ °; Figure 9.37) shows an attenuation of the significant wave height over 30% in the direct wake of the farm, and a less pronounced nearshore reduction owing to the wave energy diffracted from the sides of the farm. The nearshore wave height deficit is particularly apparent in the northern section of the beach.

Three months into the simulation period it may be observed that the evolution of the profiles is characterised by erosion on the beach face – the section exposed to wave uprush – and deposition of the eroded sediment on lower sections of the profile (Figure 9.37). This pattern occurs both with and without the wave farm; however, the erosion with the wave farm is substantially weaker than without it (Figure 9.38). Furthermore, the extraction of wave energy by the WECs does not merely mitigate the erosion in the beach face (Figure 9.39); it also protects the landward section of the profile by shifting the landward extreme of the eroded area over 10 m towards the sea. This finding is of particular relevance in cases such as Perranporth, with buildings at risk due to their proximity of their foundations to the eroded sections of the profile (Figure 9.32).

The impact of the wave farm on the beach profile was analysed through the impact parameters defined above. The *BLI* for three points in time – 1 month (M1), 3 months (M3) and 6 months (M6) into the study period – shows a significant reduction in the erosion on the beach face and the submarine bar ($x \sim 600$ m; Figure 9.40). Given that the bar

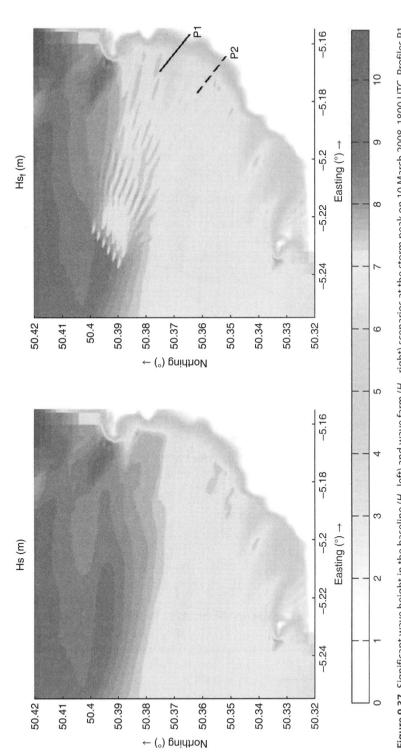

Figure 9.37 Significant wave height in the baseline (H_s, left) and wave farm (H_{sf}, right) scenarios at the storm peak on 10 March 2008, 1800 UTC. Profiles P1 and P2 are delineated for reference.

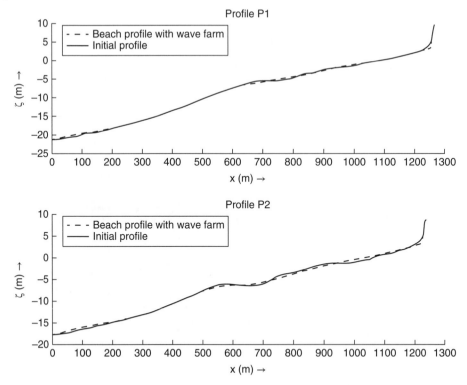

Figure 9.38 Profiles P1 and P2: initial (1 November 2007, 0000 UTC) and after 3 months (M3) with the wave farm.

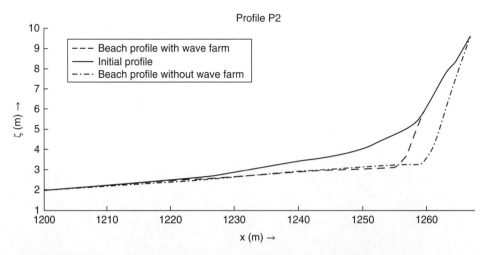

Figure 9.39 Beach face section of profile P2 at the beginning of the simulated period (1 Nov 2007, 0000 UTC) and after three months (M3) with and without the wave farm [22 Jan 2008, 15:47 UTC].

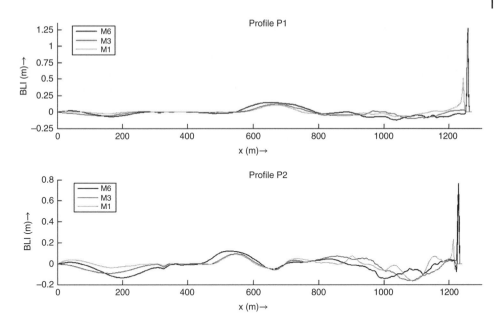

Figure 9.40 BLI values along profiles P1 and P2 at different points in time: 1 month (M1), 3 months (M3) and 6 months (M6) into the study period.

Table 9.4 *FEA* and *NER* factors for profiles P1 and P2 at different points in time: 1 month (M1), 3 months (M3) and 6 months (M6) into the study period.

	M1			M3			M6		
Profile	FEA_b (m^2)	FEA_f (m^2)	NER (%)	FEA_b (m^2)	FEA_f (m^2)	NER (%)	FEA_b (m^2)	FEA_f (m^2)	NER (%)
P1	20.53	14.11	31.27	16.30	10.42	36.07	23.85	18.66	21.76
P2	15.69	12.91	*17.72*	21.31	16.85	*20.93*	25.53	21.42	*16.10*

is essential to the natural capability of the coast for responding to storms, its reinforcement thanks to the wave farm clearly increases the resilience of the system. This effect strengthens over time, with BLI values soaring in particular over the submarine bar.

To better quantify the effects of the beach farm on the beach, the FEA and NER indicators were computed (Table 9.4) 1 month (M1), 3 months (M3) and 6 months (M6) into the period of simulation. The first two points in time (M1, M3) were associated to storm conditions, whereas the last point (M6) was associated with milder conditions. Comparing the two profiles, the effectiveness of the wave farm is more apparent in the northern section of the beach (profile P1). The comparison between FEA_b and FEA_f values, and the values of NER reflect the effectiveness of the wave farm in reducing storm-induced erosion, and in maintaining this reduction over the medium term (M6 values; Table 9.4). The relatively smaller reduction in proportional terms (NER) after 6 months is due to the fact that the later part of this period is not particularly stormy (March and April) – with fewer storms and less erosion, the reduction in storm-induced erosion is less significant globally. Nevertheless, NER values after 6 months are still relevant.

The results in Table 9.4 show a significant reduction in erosion over the medium term (6 months), which seems to indicate that the wave farm does afford some degree of coastal protection. In the following section we focus on the short-term effects of wave farms, and in particular on their ability to mitigate storm-induced erosion.

9.3.3 Mitigation of Storm-Induced Erosion

Having studied the medium-term effects of a wave farm on the beach profile, this section focuses on its short-term effects during a storm, from 5 December 2007 0000 UTC to 10 December 2007 0600 UTC [11]. The mean values of significant wave height, peak period and peak direction were H_s = 4.2 m, T_p = 12.1 s and θ_p = 295°, respectively. As in the previous section, based on the results of the SWAN wave propagation model, the XBeach coastal processes model was applied. The impact was quantified by means of the indicators defined above.

The reduction of the erosion was particularly apparent on the dune, with values of BLI over 4 m (Figure 9.41) – the outcome of wave energy extraction by the wave farm. Similarly, the submarine bar experienced reduced erosion, particularly in water depths between 5 m and 10 m and in the middle area of the beach, where the BLI parameter reached 0.5 m. The material eroded from the dune was deposited primarily on lower sections of the profile, between the bar and the dune, leading to negative values of BLI, of the order of –0.5 m.

The values of the beach FEA confirmed the effect of the wave farm in reducing erosion (Figure 9.42). The most drastic erosion occurred at the southernmost end of the beach ($y \sim 0$ m) and in its northern section, the reason being that the former is not backed by the dune system, and the latter is exposed to larger waves. In the mid-south area of the beach (500 m < y < 1500 m) the *NER* indicator exhibited large variability owing to the small values of baseline erosion (small values of the denominator in the expression for NER). The wave farm proved to be more effective in countering erosion in the northern area of the beach, in the direct lee of the WECs, as confirmed by the NER values – over 30% along more than 1000 m of beach (Figure 9.43). Another indicator, the cumulative eroded area, was also defined and applied to characterise the effects of the wave farm on sediment transport patterns in more detail, but cannot be reported here for reasons of space; the interested reader may consult [11].

9.3.4 Influence of Farm-to-Coast Distance

Finally, in view of the role of the distance of the wave farm from the shoreline on its impacts on the nearshore wave conditions, it is important to consider the role of the distance in the effects on coastal morphodynamics [10]. For this purpose we consider, following the approach presented in Section 9.2, a wave farm deployed at distances of 2 km, 4 km and 6 km from the 10 m contour at Perranporth Beach – corresponding to water depths of approximately 25 m, 30 m and 35 m, respectively (Figure 9.44). Following [47], and based on the analysis of the offshore wave climate in the area, we prescribe two wave conditions at the outer (ocean) boundary of the offshore grid (Table 9.5).

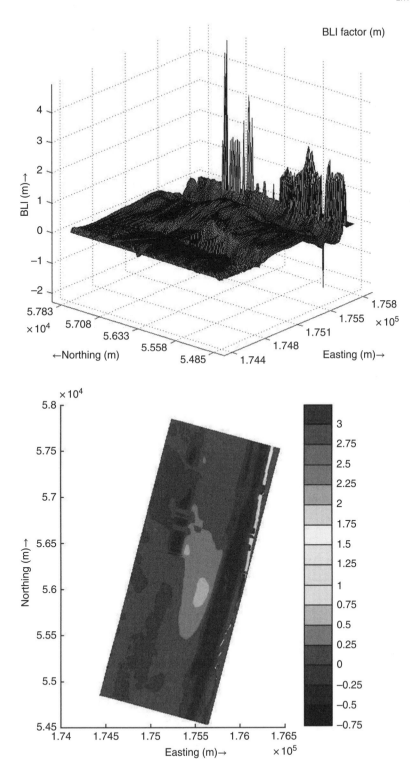

Figure 9.41 Bed level impact indicator after the storm, at the end of the study period.

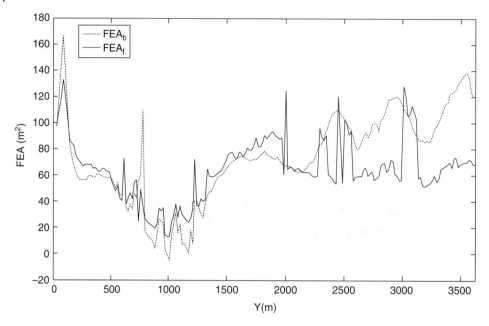

Figure 9.42 Beach face eroded area in the baseline (*FEA_b*) and wave farm (*FEA_f*) scenarios after the storm.

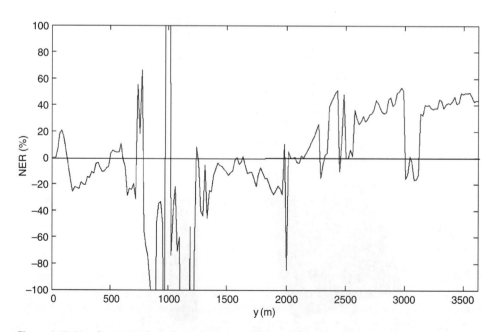

Figure 9.43 Nondimensional erosion reduction on the beach face after the storm.

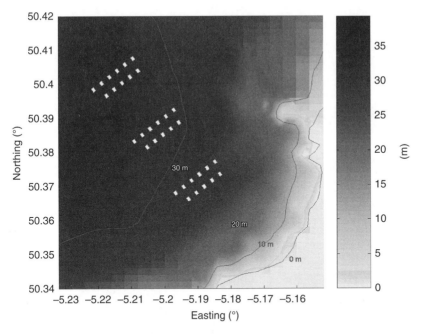

Figure 9.44 The three locations considered for the wave farm, at distances of 2 km, 4 km and 6 km from the reference (10 m water depth) contour.

Table 9.5 Wave conditions: significant wave height (H_s), peak period (T_p), direction (θ) and directional spreading (s), and wave transmission coefficient (K_t).

Case study	H_s (m)	T_p (s)	θ (°)	s (°)	K_t
CS1	3	12	315 (NW)	26.50	0.76
CS2	3.5	11	315 (NW)	26.34	0.78

The impacts of the wave farm on the nearshore significant wave height (H_s) for the three wave farm locations are shown in Figure 9.45 alongside the baseline values (without the wave farm). It is apparent that the reduction in the significant wave height in the lee of the farm varies according to its distance from the shoreline. The maximum reduction generally occurs in the lee of the second row of WECs, with significant wave height deficits of up to 50%. Moving towards the shoreline, this reduction decreases owing to wave diffraction. The relevance of the farm-to-coast distance is clear, in the first place, with respect to the nearshore wave conditions. The farthest farm alters the wave conditions along more than 7 km of shoreline with a relatively small significant wave height deficit along the reference (10 m) contour (average deficit of approximately 5%). In contrast, the nearest wave farm affects a shorter section of coastline (approximately 4 km) with a more pronounced significant wave height deficit (well over 10%). Similar patterns are observed with regard to the significant wave height along the outer boundary contour of XBeach (line AA′ in Figure 9.45; see Figure 9.46). The farm at 2 km affects mainly the

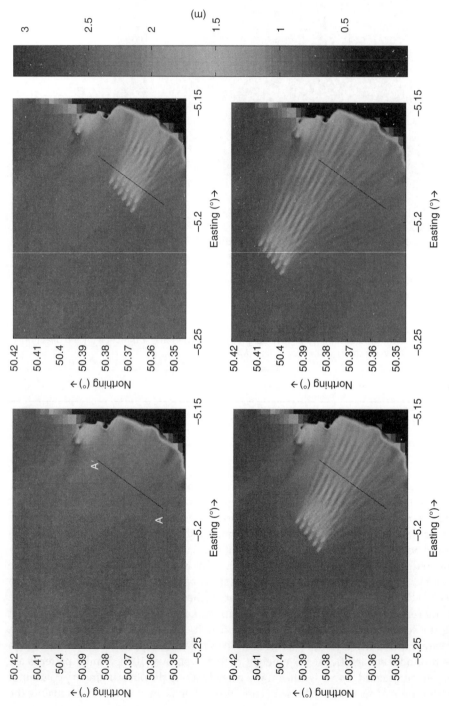

Figure 9.45 Significant wave height (in metres) in the baseline scenario and in the presence of the farm at distances of 2 km, 4 km and 6 km from the reference (10 m water depth) contour in CS1 (clockwise from above left). Line AA′ defines the outer boundary contour for the XBeach coastal processes model.

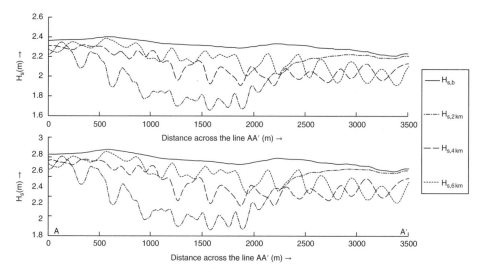

Figure 9.46 Significant wave height (in metres) in the baseline (no farm) scenario ($H_{s,b}$) and with the farm at a distance of 2 km ($H_{s,2km}$), 4 km ($H_{s,4km}$) and 6 km ($H_{s,6km}$) from the reference (10 m) contour across the AA′ line in case studies CS1 (top) and CS2 (bottom).

central section of the beach, and considerably less the northern section, while the farms at 4 km and 6 km affect a greater section of coast to a lesser degree.

Using the results of the coastal wave model as input to the coastal processes model, the modifications in the morphodynamics brought about by the wave farm were obtained. The main effects occurred on the beach face, with reductions in erosion leading to BLI values of up to 1.5 m (Figure 9.47). Lower (submarine) sections of the beach accreted with the materials eroded from the beach face. Comparing the different farm-to-coast distances, the wave farm at a distance of 2 km was more effective in mitigating erosion.

We have seen that the main effect of the wave farm occurs on the beach face. This effect can be quantified by means of the FEA impact indicator (Figure 9.48). The largest values occur in the south section of the beach, which, unlike the rest of the beach, is not backed by a well-developed dune system. Generally, erosion is greater without the farm than with it, especially in the middle and northern area of the beach (for values of the y-coordinate over 1250 m).

We now examine the role of farm-to-coast distance on beach face erosion by means of the NER indicator (Figure 9.49). Its values fluctuate considerably along the beach, but it is clear that in the central area (y-coordinate between 1200 m and 2000 m) they are consistently larger ($NER > 20\%$) in the case of the nearest farm, indicating a greater effectiveness in countering erosion. The results for the wave farms at distances of 4 km and 6 km do not present significant differences; for instance, for the central and northern sections of the beach ($y > 1250$ m) NER values for the wave farm at 4 km are only slightly greater (13%) than those for the farm at 6 km (11%). In terms of the average reduction in the erosion on the beach face along the beach, the nearest wave farm (2 km) offers a greater protection ($NER \sim 15\%$) than do the wave farms at 4 km and 6 km ($NER \sim 10\%$). Further details are reported in [10].

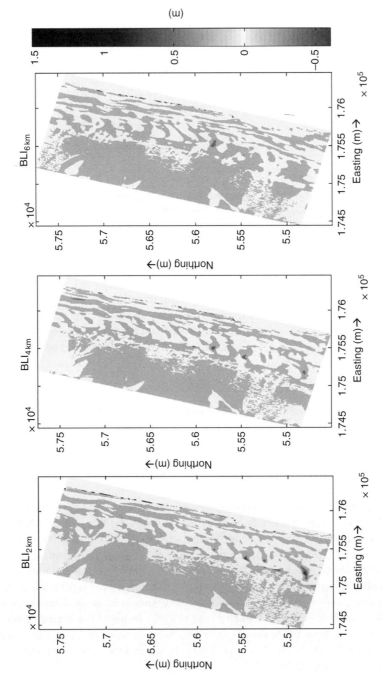

Figure 9.47 Bed level impact with the wave farm at a distance of 2 km (BLI_{2km}), 4 km (BLI_{4km}) and 6 km (BLI_{6km}) from the reference contour after the storm in case study CS1.

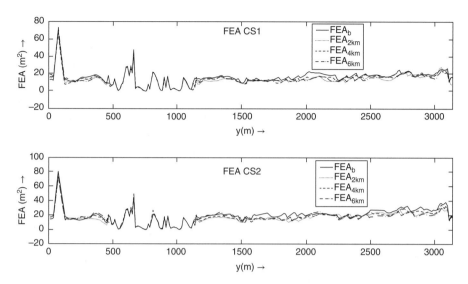

Figure 9.48 Beach face eroded area in the following scenarios: baseline (*FEA$_b$*) and with the wave farm at a distance of 2 km (*FEA$_{2km}$*), 4 km (*FEA$_{4km}$*) and 6 km (*FEA$_{6km}$*) at the end of the storm in case studies CS1 (top) and CS2 (bottom). The *y*-coordinate is the alongshore coordinate, with *y* increasing towards the north of the beach.

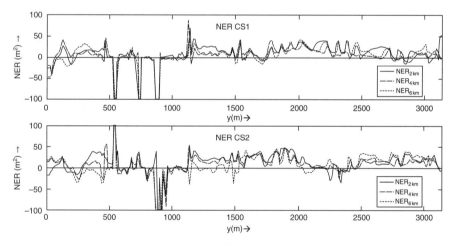

Figure 9.49 Non-dimensional erosion reduction on the beach face with the wave farm at distances of 2 km (*NER$_{2km}$*), 4 km (*NER$_{4km}$*) and 6 km (*NER$_{6km}$*) at the end of the storm in case studies CS1 (top) and CS2 (bottom). The *y*-coordinate is the alongshore coordinate, with *y* increasing towards the north of the beach.

In summary, the selection of the location for a wave farm is not trivial, and we have seen that the degree of coastal protection afforded by a wave farm and the manner in which this protection is distributed alongshore vary significantly with its distance from the coast.

9.3.5 Future Lines of Research and Development

In Sections 9.2 and 9.3 we summarised the state of the art concerning the impacts of wave farms on the wave field and on coastal morphodynamics, respectively, and in Section 9.1

we mentioned the possibility of applying wave energy extraction to coastal protection. However, prior to deploying wave farms for this purpose, it is necessary to investigate certain aspects which are not yet fully understood. The first concerns the detailed nature of wave energy transmission across the wave farm: as energy is extracted the incident wave energy is reduced and can be partially redirected due to diffraction, radiation and reflection by the wave farm. These interactions govern the magnitude and extent of shadow zones, regions of reduced wave energy and height, in the lee of wave farms. The second aspect concerns the long-term effects of wave farms on coastal morphodynamics; the analysis must transcend the short-term (storm) scale to deal with the multidecadal response of coastal systems to the extraction of wave energy by wave farms.

As regards the first aspect, wave energy transmission through a wave farm depends on the sea state parameters and the characteristics of the WECs forming the farm, which not only absorb part of the incident wave energy but, importantly, do so in a selective manner regarding the distribution of the energy across frequencies and directions. In other words, the wave farm acts as a complex bandpass filter, modifying the frequency and directional characteristics of the incident wave field. At present this process is poorly understood. The interaction between the wave field and the WEC array has been characterised through 'notional' wave transmission coefficients (not based on experimental data, e.g. [25]) and, in a small number of cases only, through wave transmission coefficients based on laboratory results, covering a limited range of wave conditions or WEC types [20, 31, 32]. There is a knowledge gap concerning relevant WEC types and, in particular, extreme wave conditions, under which WECs typically cease to operate and adopt so-called survival strategies to reduce the chances of mechanical or mooring failure, or damage. These mechanisms usually include disengaging the power take-off, changing the mooring orientation or pulling the device underwater. As a result, the effectiveness of the wave farm in absorbing the incoming wave energy is compromised to a greater or lesser extent – depending on the type of WEC and the survival strategy – and, more generally, the characteristics of the wave farm as a bandpass filter are modified. In any case, in this survival mode the wave farm still interacts with, and modifies, the incident wave field – and there is a conspicuous lack of data on this critical aspect.

9.4 Tidal Stream Farm Effects on Hydrodynamics and Sedimentary Processes

In this section we summarise the work carried out to date on hydrodynamics and sediment transport. The effects on the marine biota are dealt with in the next section, combined with those of wave farms.

The effects of tidal stream farms on hydrodynamics are typically analysed by means of numerical models of the hydrodynamics of the area in question based on the Navier–Stokes equations (Chapter 2). Ideally, the models should be calibrated and validated with field data of tidal levels and, wherever possible, flow velocities. The turbines are usually accounted for as a momentum sink [48–50], although other procedures may be used; in the near future it will be possible to combine a computational fluid dynamics model of the tidal farm (the near field) with a coastal hydrodynamics model, such as that described in Chapter 2, for the estuary or wider area (the far field). When implementing a coastal hydrodynamics model to investigate the impacts of tidal stream turbines it is important to use a high-resolution computational grid, so that the turbines can be properly represented [48]. If the area in question is not well mixed, as is often the case in estuaries [51], the numerical model should be 3D [52, 53]. In other cases, a 2DH model may be used [54].

A series of studies on the implementation of tidal stream energy in the estuaries in Galicia (NW Spain) showed that the tidal stream farms did have a non-negligible effect on flow velocities [48, 52–54]. The modifications occurred primarily in the area of the tidal farm and its vicinity. The extraction of power from the flow by the tidal turbines resulted in reduced flow rates in the tidal farm area and increased flow rates in adjacent areas, as tidal flow was diverted by the partial blockage effect induced by the tidal farm; in other words, the tidal flow patterns in the estuary adapted to accommodate the partial blockage. This adaptation modifies the power density in the flow, which must be taken into account when assessing the tidal resource. In a study in the Ribadeo estuary (Spain) it was found that the modifications in the magnitude of flow velocities were of the order of 10%, up to 15% at some points, relative to the baseline values. These changes reduced the available power density at the farm relative to the baseline (no farm) scenario by 21% and 12% for the two levels of power extraction considered [48]. In another study, the effects of a tidal farm on the transient and residual flow in the Ortigueira estuary were analysed [52]. Similar conclusions were found in respect to the effects on the flow velocity (transient circulation). As regards the residual circulation, it was found that the tidal farm did affect its magnitude but not the general 3D residual flow patterns in the estuary (a positive estuarine circulation). Naturally, these modifications of residual flow can have an effect on water quality [56], biological productivity, sediment transport, etc.

Regarding the effects of tidal stream turbines on sediment dynamics, a number of studies have recently been carried out. It was found that a tidal farm in the Alderney Race (English Channel) could lead to a significant change in the maintenance of headland sand banks [57]. Another study focused on the effects on sedimentary processes in an area of the Irish Sea with strong tidal asymmetry [58]; the results showed that energy extraction by the turbines led to a reduction in the overall magnitude of bed level change. In another study [59] in the area, the impact of tidal stream farms on sediment processes was found to be within the range of natural variability and, therefore, not detrimental to the local environment. In the Pentland Firth it was shown that the energy extraction by tidal turbines could have implications for the existing sandbanks and, more generally, the morphodynamic regime [60]. It is certainly early days in the study of the environmental effects of tidal stream farms, and future work is necessary.

9.5 Marine Biota

The study of biodiversity on both global and regional scales has mostly been concerned with the terrestrial environment despite the fact that the marine environment occupies around 70% of the Earth's surface. To date it is known that of the already identified 8.7 million species, 2.2 million species live in the ocean [61]. However, due to its longer evolutionary history, the ocean contains a higher diversity of genera and phyla than the land [62]. Nevertheless, because so much of the ocean is difficult to assess, the extent of marine biodiversity has been, and continues to be, a slow and difficult task leading to a great variation on the species number estimates. Marine species that are relatively easily monitored are those restricted to nearshore habitats, especially if they are sedentary or attached (e.g. seagrasses and corals) and those that spend time at the sea surface or on land (e.g. marine mammals and seabirds).

Coastal areas are commonly defined as the interface or transition areas between land and sea, where large number of biogeochemical processes occur. The peculiar characteristic of coastal environments is their dynamic nature which results from the transfer of matter, energy and living organisms between land and sea systems, under the influence of primary driving forces that include short-term weather, long-term climate, secular changes in sea level and tides. The high nutrient flows from the land and also from ocean upwelling, which bring nutrient-rich water to the surface (when the wind direction moves surface water offshore), promote high biological productivities in several coastal areas, supporting a rich biological diversity and frequently a valuable assortment of natural resources and providing a number of environmental goods and services [63, 64].

Due to this high productivity and also to their proximity to land, coastal areas are also those where most economic activities take place. In fact, besides the highest natural resource, the proximity to the coast is one of the most important criteria governing MRE site selection with regard to its economic feasibility, influencing operation and maintenance costs and logistics. The marine biota description considered herein concerns the coastal area up to 200 metres.

9.5.1 Marine Biota Habitats and Components

Marine habitats can be divided into coastal and open ocean habitats. Coastal habitats extend from as far as the tide comes on the shoreline out to the edge of the continental shelf (around 200 m depth). Most marine life is found in coastal habitats, even though the shelf area occupies only 7% of the total ocean area. Open ocean habitats are found in the deep ocean beyond the edge of the continental shelf. The ocean can also be divided vertically into the pelagic (water column) and benthic (seabed and immediate associated water layer) zones.

Depending on how deep the sea is, the pelagic zone can extend over up to five horizontal layers in the ocean (Figure 9.50). Up to 200 m depth the epipelagic zone corresponds to the water layer where most of the sunlight is available for photosynthesis and thus where nearly all the oceanic primary production occurs. A benthic zone exists for each pelagic area, and in the continental shelf their features are strongly dependent on the type of sediment (e.g. rocky, sandy, coarse).

Several different pelagic and benthic habitats exist in the continental shelf and associated water column layers, which are defined by physico-chemical (abiotic) and biological (biotic) factors. In fact, a habitat is made up of abiotic factors such as type of sediments and their chemical conditions (e.g. pH), range of water temperature and availability of light, as well as biotic factors such as the availability of food and the presence of predators. The availability of light is one of the most important factors influencing zonation of the water column and thus biodiversity distribution. In the top few to 200 metres of the water column, namely the euphotic zone, there is enough light (during the day) for the primary producers (e.g. phytoplankton and macroalgae) to photosynthesise. This top layer of the open ocean water is therefore a very different environment than that found below.

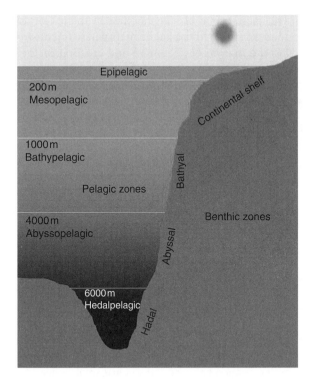

Figure 9.50 Pelagic and benthic division of the oceans.

Beyond a depth of 1000 m (most of the world's ocean volume) the ocean is effectively lightless, with a few small exceptions, such as bioluminescence produced by bacteria found in the light organs of deep sea biota.

Sunlight reaching the top layers of the ocean also transmits heat and thus the surface layers are the warmest and hence have lower density than the cold water beneath. Globally, the temperature of deep waters is relatively uniform at just a few degrees Celsius, in contrast with shallower water temperatures that fluctuate according to latitude and season. The rapid temperature change from the ocean top layers towards the seabed, named the thermocline, changes from place to place but is in general 10°C warmer in the tropics than in temperate regions, where it is permanent rather than seasonal. Moving away from the polar regions, the global ocean is stratified into a warm upper layer, a rapid cooling zone, and a lower cold zone. In the polar regions surface water is near freezing point at −1.85°C in the winter and just positive in the summer [65]. Exceptions to these areas are upwelling zones, distributed along coastal temperate and tropical regions, characterised by well mixed water columns where cold and nutrient-rich deep water reaches the surface.

Other gradients of salinity (halocline) and oxygen (pycnocline) may also lead to water layer stratification, although these are mainly local or regional instead of global (e.g. estuaries, and coastal lagoons). Salinity of seawater is typically around 35 psu (practical units of salinity) and only small changes occur both with latitude and longitude variation: slightly greater salinity values are expected in the tropics due to the increased evaporation of surface water and lower close to continents and in the Arctic due to the influence of fresh water runoff.

As the place where the organisms live, habitats may also be modified by them. Corals, kelp, mangroves and seagrasses are good examples of coastal habitats where organisms act as ecosystem engineers, reshaping the marine environment and creating further habitat for other organisms.

Coastal areas frequently contain critical terrestrial and aquatic habitats, particularly in the tropics. Such habitats together comprise unique coastal ecosystems, support a rich biological diversity and frequently contain a valuable assortment of natural resources. Examples of such habitats are estuarine areas, coral reefs, coastal mangrove forests and other wetlands, tidal flats and seagrass beds, which also provide essential nursery and feeding areas for many coastal and oceanic aquatic species.

Inhabitants of the pelagic zone can be divided into two categories based on their locomotor ability: plankton, organisms that are unable to counteract the influence of currents and thus drift passively in the horizontal plane; and nekton, organisms that can swim, overcoming the ocean currents. Some of the planktonic organisms are able to move vertically, adjusting their depth during dawn and dusk (diel vertical migrations). Planktonic organisms such as copepods and chaetognaths complete their entire life cycle as plankton (holoplankton), and others such as fish larvae and barnacle larvae spend only part of their life cycle as plankton (meroplankton). Plankton is particularly vulnerable to predation, but organisms with a planktic phase have more opportunities for dispersal and colonisation [66].

The high biological productivity of continental shelves fuels abundant plankton and fish fauna that are a key food resource for populations of top predators (e.g. seabirds and marine mammals) as well as major global fisheries. Seabed habitats of the continental shelves are bounded by the depth to which biota can burrow into the sediment (usually no more than 2 metres within the sediment) and the extent to which they penetrate the waters above (30–50 m for some of the largest kelp genera such as *Macrocystis*). In some cases, habitats are made of key biota such as in kelp forests, mussel or oyster reefs, maerl beds, corals and sponges. Due to their high biodiversity and ecological relevance to the sustainability of the marine ecosystem as whole, these habitats are usually strongly protected for conservation purposes and thus place constraints on the extent and range from habitat that can be utilised. Due to lack of or limited mobility benthic organisms living in these habitats are also more vulnerable to environmental changes and frequently utilised to help in the examination of the ecological quality of the local and regional marine environment [65].

9.5.2 Dynamics of Marine Biota: Ecological Processes

Marine organisms play crucial roles in many biogeochemical processes that sustain the biosphere, providing a variety of products (goods) and functions (services) which are essential to human well-being, including the production of food and natural substances, the assimilation of waste and regulation of the world's climate [67]. Primary production, or the conversion of inorganic carbon into biomass using light to spark the reaction (photosynthesis), is the starting point for all other life in the marine system. It is also the fundamental process to understand the ocean and its role as a sink of increasing levels of atmospheric carbon dioxide. As on land, marine photosynthetic organisms constitute most of the food source on which marine ecosystems are based. Seaweeds, seagrasses and microscopic algae and bacteria (phytoplankton) are the main primary producers in the oceans, which can be very productive. Coastal phytoplankton has an annual production that reaches 1 kg of carbon per square metre. The production per square metre of seaweeds and seagrasses is equal to, or in many cases greater than, that of terrestrial plant-based systems. Macroalgae dominated by the genus *Postelsia* have annual productivity rates of up to 14 kg of carbon per square metre, contrasting with those for mature rainforests and intensive alfalfa crop production of 1–2 kg of carbon per square metre [65].

Although many factors interact to determine net primary production in the ocean, light availability and quality is the dominant factor that determines the rate and extent of the photosynthetic activity. In waters, heavily laden with sediments or particulate matter, light penetration is much lower than in clear waters, although generally no sunlight penetrates below 1000 m.

In addition to carbon, oxygen and hydrogen, the supply of other inorganic nutrients (mainly nitrogen and phosphorus) – or simply nutrients, as they are usually referred to in aquatic ecology – is another factor limiting photosynthesis at sea. Waters that have low concentrations of such essential nutrients for algal growth are known as oligotrophic and thus are regions with low primary productivity, such as open ocean waters. Waters where those nutrients exist in high concentrations, such as coastal areas with high river runoff, are called eutrophic, with a high primary production rate and thus supporting very productive communities of organisms. The process of eutrophication can develop in eutrophic and very eutrophic (hypertrophic) areas, being defined as an increase in the rate of supply of organic matter to an ecosystem resulting in an increase of algal growth which sometimes is excessive and leads to adverse effects to the whole ecosystem. Mesotrophic waters sustain intermediate levels of primary production.

The complement to photosynthesis is respiration or simply the biological oxidation of the organic matter produced by autotrophic organisms (photosynthetic organisms) and heterotrophic organisms (those that get their carbon from already synthesised organic matter). Organic matter decomposition is another important process that continues the sustainable biological life cycle in the marine environment. Decomposition results in the production of carbon (carbon dioxide), nitrogen (ammonium) and phosphorus (phosphate), bringing the inorganic nutrients back to the marine system. Thus, this process is also called remineralisation and is dominated by micro-organisms, notably bacteria.

9.5.3   Sensitivity of Marine Habitats and Species

The major causes of biological impoverishment and loss of marine biodiversity are fishing and by-catch; hunting of marine mammals, birds and turtles; toxic chemicals, oil and nutrient pollution; laying of submarine cables; marine shipping; and the increased ultraviolet radiation due to ozone layer depletion. In the future, global climate change is also predicted to have a major impact in the marine environment.

Human activities within the marine environment give rise to a number of pressures on seabed habitats, including loss of habitat, habitat change and physical damage of the habitats the species that depending on it [68]. Habitat damage or loss can be caused by changes in the nutrient status and cycling, loss of food supplies, erosion or reduced sediment supply, changes in the sea level, inundation and increased exposure to natural disturbances [69], and as a result can be considered one of the main reasons for biodiversity and species abundance decrease. Furthermore, many marine species depend upon broad dissemination during motile life stages, and short-lived species in particular must be replenished by means of this dispersal and efficient recruitment. If dispersal routes or migration are interrupted by lethal environmental conditions, populations and ranges of species may be reduced. Nutrient and toxic chemical pollution are invariably associated with a reduction in biodiversity. Species that can adapt to or thrive under conditions stressful to most living organisms can dominate the biological community, thus changing the entire nature and function of the ecosystem. This may lead to an even greater loss of species from the system.

Not only human-induced activities place stress in the marine environment, natural variability can also change habitats and communities by altering dominant oceanographic conditions and the physical and chemical properties of seawater. These changes can also affect the distribution and extent of habitat and the structure of biological communities. As a result of the combined effects of human activities and changing environmental conditions, ecosystems have undergone a major structural shift in the last 40 years [70].

Within the human-induced activities fishing is widely recognised as one of the main causes of habitat and communities' change and loss. Fisheries have given rise to a new category of species depletion: commercial extinction. Fish and shellfish populations are depleted to the point that it is no longer economically feasible to fish them. While not extinct, these species are certainly no longer playing their traditional roles in their ecosystems, and some have been pushed to the brink of extinction. Fishing operations, such as trawling and dragging, destroy bottom habitats and impact species populations, and repetition of such activities delays or prevents recovery. Novel research studies have analysed global catch data to gain insights into the status of global fisheries, revealing an increase in collapsed stocks and a decline in new stocks [71]. Results show that species overexploitation and commercial depletion is worse than expected and that stock management should be stricter (Figure 9.51).

Large-scale oil and gas operations are also within the main non-fishing industrial activities taking place on the continental shelf and slope [73] and they have well-known implications on the perturbation of the coastal environment [74]. Oil and gas activities can affect marine habitats and communities by generating noise, pollution and

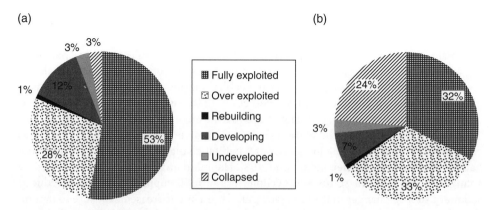

Figure 9.51 Comparison of the global catch data analysed by (a) FAO [72], and (b) recently published research analysis (adapted from [71]).

contamination, and disturbing the seabed [70]. Furthermore, industrial processes have led to bioaccumulation of contaminants [75], abnormal development of invertebrates, endocrine disruption, nutrient enrichment, toxic algal blooms and deoxygenation [69].

The assumption that certain geographic areas are more sensitive or vulnerable – no matter how these terms are defined – than their surroundings has been a central question in conservation and management for special habitats conservation and species preservation. The concepts of sensitivity and vulnerability have been commonly used as criteria (along with representativeness and uniqueness) in the identification of areas requiring special management or protection [76].

By definition, 'sensitivity' is the intolerance of a habitat, community or species to damage, or death, from an external factor. A habitat, community or species becomes 'vulnerable' to adverse effect(s) when the external factor is likely to happen [77]. These responses are potentially nonlinear and are likely to include interactions between stressors [76]. Also important is the definition of vulnerability, that is, the probability of a feature being exposed to a stressor to which it is sensitive [77, 78], combined in some way with the exposure (duration, magnitude, rate of change) to that stress.

Activities in the marine environment result in several pressures, which may result in an impact on environmental components that are sensitive to the pressure. Pressures can be physical, chemical or biological, and the same pressure can be caused by several different activities. Impacts are defined as the consequences of these pressures on components where a change occurs that is different from that expected under natural conditions. Different pressures can result in the same impact, for example, habitat loss and habitat structure changes can both result in the mortality of benthic invertebrates [68].

The criteria for identifying environmentally valuable areas for protection have been well established [79], and several methods have been developed to assess and quantify sensitivity and recoverability of marine species and biotopes [76, 77, 80]. More recently, research has been more focused on ecosystem sensitivity mapping as a decision support tool to help managers, stakeholders and public entities to perform marine spatial planning [80, 82].

Assessment of sensitive species is a qualitative process and must be made based on the best available scientific information and expertise [80]. Procedures may vary but in general they involve the review of available literature on the life history, distribution, environmental preferences and any effects of environmental perturbation on the chosen species. In the case of biotopes, information on the community ecology, structure of the biotope (or similar community) and its associated species must be collated. Key information for this process is considered relevant to understand the biology of a species or the ecology of a biotope, including those environmental factors important for the survival of the species and those key structural and key functional characteristics for the functioning of the biotope [81].

It may be that the marine ecosystem is too complex and there are too many species and biotopes to make universal prediction of cause and effect. Hence, there may be a need to use surrogates for the whole ecosystem, for instance, particular species, habitats or communities that are known to be responsive to change [81]. According to Tyler-Walters [77], it is important to be aware that not all species within a community affect its sensitivity to environmental change. High abundance, frequency or faithfulness of a species within a biotope are generally not good indicators of the contribution of a given species to the sensitivity of the biotope. Species that indicate the sensitivity of a biotope are identified as those species that significantly influence the ecology of that component community. The loss of one or more of these species would result in changes in the population of associated species and their interactions.

As far as concerns MRE projects, an evaluation of the sensitivity of the site (or site alternatives) with regard to both environmental and socioeconomic issues, should be carried out during the initial stage of an environmental impact assessment (EIA) process. The site selected for projects strongly influences both the potential environmental and socioeconomic impacts of the project, as each site will have its unique set of sensitivities. The environmental sensitivity of a site is generally associated with the identification of species, habitats and/or areas of marine natural heritage importance. Accurate characterisation of special natural features can largely support site selection [82]. The identification of spatial and temporal distribution of habitats and species (using both field work or existing data) can be made for each site as well as for its adjacent areas, taking into account the Red List species[1] [82] and international legislative documents that include selection criteria (e.g. EU Birds and Habitats Directives, Ramsar Convention, OSPAR guidelines (see Section 10.3), UNEP Convention on Biological Conservation) [79].

From the socioeconomic perspective, sensitive locations are those where conflicts may arise from the number and type of other uses. Where in place, strategic environmental assessment and/or marine spatial planning can assist MRE developers by in identifying the nature, location and extent of other existing uses to aid in the process of site selection and conflict management [81].

9.5.4 Marine Biota Observation and Experimentation

The general purpose of marine biota observation and experimentation during human activities at sea is to evaluate whether the anthropogenic disturbance changes the marine ecosystem, which groups are mainly affected and the extent or magnitude of such effects. In the case of ocean energy projects, the main objective is to identify interactions of the marine environment components with the devices or to assess the effects of modifications in the distribution of habitats and changes in species composition and populations' abundance. However, this is not a simple task due to the high diversity and dynamics of the marine biota. Furthermore, ocean energy projects are usually installed in highly energetic sites, presenting a considerable challenge for the observation and assessment of the marine biota.

Nowadays many techniques are available for the observation of the marine biota and the general trend is to use more advanced and autonomous technologies (e.g. acoustic devices, remotely operated vehicles), although traditional sampling techniques are still used for autonomous data collection validation.

9.5.4.1 Marine mammals

Marine mammals comprise two entirely aquatic orders, Cetacea and Sirenia, and a third order, Carnivora, which includes animals that spent at least a portion of their lives in the sea [83]. Considering the wave, currents and tidal resource distribution where ocean energy projects may be deployed, observation of and experimentation on marine

---

1 See http://www.iucnredlist.org/

mammals have been focused on species from the Cetacea order, such as dolphins and harbour porpoises, and on species from the Carnivora order, such as seals [84]. Due to their ecology, the assessment of such animals is mainly based on *in situ* visual observations and/or acoustic detection. This information is usually obtained from a fixed point or along transects [85, 86]. Fixed points are usually located on land in a vantage point (e.g. a headland) and are more suitable for monitoring areas and animals close to the shore, such as seals. Predetermined transects can be followed through vessels or aircraft using a distance methodology [87]. The main constraints of visual observations are the weather conditions, the sea state, and daylight hours [88].

Boat-based observations can be conducted using dedicated platforms or platforms of opportunity, such as vessels with regular lines. This technique is more suitable for observation of cetaceans such as dolphins and harbour porpoises. The main advantage is that additional and useful information can be collected such as the presence of calves and their behaviour. With regard to aerial surveys, these are more suitable for covering large areas in the same time window, although more expensive. Both aerial and boat-based surveys can be combined with bird surveys, reducing the costs of the monitoring programme.

Visual observations can be complemented with photo identification or mark–recapture techniques and passive acoustic detection. One of the assumption for transects is that all the animals are detected, but this is not always true since animals can be missed during diving. Passive acoustic monitoring complements visual observation through detection of sounds emitted by the animals, using hydrophones or C-PODs. In contrast to visual observations, the acoustic range of detection is less affected by meteorological conditions and therefore it can be used in higher sea states [89]. Acoustic equipment can also operate continuously, allowing detection of animals by day and night. Acoustic monitoring can be carried out at a fixed point or using towed hydrophone arrays [90]. Acoustic techniques can also be used in an active way.

Mark–recapture analysis using photographs is a useful tool to monitor movement patterns and population changes [86], but in areas where animals have a regular occurrence. This technique is only applicable to a few species, such as bottlenose dolphins [91, 92].

All the methods mentioned can be applied to monitor seals. Nonetheless, considering the differences in ecology and behaviour between pinnipeds and cetaceans, there are specific techniques better suited to the purpose of monitoring pinnipeds, such as aerial photographic counts and telemetry [93–95]. In addition to general information about their distribution and abundance at sea, the same information on land is required as well as their movements through the area of interest and connectivity with other areas and how the habitat use changes between seasons and in tidal sites with different states of the tide [96, 97]. The combination of photo identification and telemetry is useful to assess seasonal distribution and sea habitat usage [91].

Besides *in situ* techniques, demographic models can also be used to assess seal abundance. However, the application of these models requires information about survival rates and fecundity which may not always be available.

Although aerial surveys are more cost effective for large areas, considering the scale of wave and tidal projects, boat-based and land-based observations are likely to be chosen for site characterisation [96]. The main limitation of these types of observations can be topography and the oblique view since not all seals may be visible or closer seals may obscure other seals [96]. However, the data can be easily repeatable. Isolated land-based observations from a fixed point can only provide information about relative abundance and its change considering different factors, such as season, tide and time of the day. The limiting factor in get tingall this information will be the observation effort.

Seals at sea can be observed and counted from land using vantage points. However, telemetry techniques are more useful to understand the movement patterns of the animals, their behaviour and physiology. Additionally, telemetry studies can provide relevant information on how animals use a site which is particularly important on wave and tidal sites, for example to assess collision risks [96].

9.5.4.2 Seabirds

As for other groups, the basic information required to understand seabird dynamics in a given area is what species occur, their distribution and abundance and their type of occurrence (e.g. migrating species, resident species). This information is usually obtained through visual observations from land, boat or aerial surveys [98, 99]. The selection of the method will depend on the site characteristics and on the existing information. For projects sited near the coast (1.5 km from the shore) land-based surveys are more likely to be chosen, but aerial and boat surveys are more suitable to cover large areas as needed to understand the spatial coverage. Moreover, the use of transect sampling is a powerful tool to generate bird density surfaces [87, 98, 100]. Other information on bird behaviour is also needed to estimate and understand this group dynamics. Data obtained using different methods cannot be directly compared or easily combined [101] and therefore it is important to select common methods to study the same area. Radar techniques and telemetry tags are also very useful techniques to understand seabird dynamics.

9.5.4.3 Benthos

The distribution and abundance of benthic communities is irregular and therefore the aims and objectives of their study should be clearly defined at the outset in order to design a logical and coherent sampling plan [102]. The design of the monitoring plan should consider appropriate spatial and temporal scales to detect changes and differences. Another relevant aspect when studying benthic communities is the size of the sampling unit, which will be influenced by the objective of the study and the ecology of the animals.

The techniques for observation of the marine benthos can be split in two categories: destructive and non-destructive [103]. Destructive techniques require the removal of samples for subsequent analysis (e.g. grabbing or trawling), and are used to assess sedimentary habitats when they are unlikely to be damaged by the removal of samples. Non-destructive methods do not require removal of samples and are used *in situ* (e.g. remote imaging or direct diver observation).

Traditional methods, such as grabs, are the most common to establish the biological community composition of sedimentary habitats, mainly for sediments ranging from mud to medium sand. The use of grabs can provide a detailed description of the sampling station; however, its use is difficult in moderate current speeds or large swell sites [103]. Nowadays, alternative remote sensing techniques can be used to complement traditional techniques, such as acoustic mapping techniques or drop-down video/ photography.

Acoustic mapping can be used to detect seabed features, but ground truth is necessary using traditional techniques (sample collection). Drop-down video/photography and remotely operated vehicle technologies, apart from being used for ground truth of acoustic mapping, can also be used as monitoring tools. Drop-down video/photography and remotely operated vehicles (ROVs) are used to characterise a large area and to provide information on the species, habitat, biotope and substrata distribution [103, 104]. Additionally, ROVs may can provide quantitative data if they have a system of seabed area definition, such as a laser projection.

Benthos observation through sampling and image collection can be done by divers. However, the main limitations are the depth (ideally less than 30 m) and the diving period (ideally restricted to a few hours per day) [105]. Compared with other techniques, information obtained through divers is considered to be more accurate and thus best for obtaining quantitative data and/or good imaging documentation [103]. However, sampling by divers is only suitable when the area to sample is small and therefore can be observed in a short period.

9.5.4.4 Fish

The need for fish monitoring is usually focused on relevant species for commercial and recreational fisheries and species of conservation importance. Fish observation can be rather expensive, but compared to other groups it is easier to set up. A variety of methods can be used and combined, ranging from the use of traditional techniques to more advanced technology [106]. A common technique for fish monitoring is the use of telemetry tags, which is very useful for understanding the spatial dynamics of fish, although a large number of fish marked with tags are needed for assessments on a large spatio-temporal scale. However, there are some species that cannot be marked and therefore other methods are needed. Traditional methods include grabs, visual observation (using divers or video cameras) and gears. More advanced technologies include the use of side scan sonars and echo-sounders. Echo-sounders are useful to estimate fish abundance and complement other methods but are expensive unless the equipment is available on-board. As an alternative, trammel nets would appear to be most effective in detecting changes in abundance [107].

It is important that the method or combination of methods selected allows day and night (diel cycle) monitoring to cover the variation in vertical distribution of some fish species. A low-cost alternative are the single beam echo-sounders to estimate fish density and its variation with respect to seasonal, diel and tidal cycles. Acoustic cameras and dual-frequency identification sonars are another technology used in fish observation, although identification of species can be limited using such techniques [108]. To obtain information on fish species composition, trawling and diving are the best methodologies [109].

9.6 The Environmental Impact Assessment

The demonstration of the sustainable development of any project in the natural environment starts with the application of the EIA process to understand and evaluate its potential effects and implement mitigation measures to attenuate the adverse impacts. This assessment is typically used by the consenting or regulatory body to inform the consultation and decision-making process from the project concept to decommissioning. The environmental analysis should be considered a planning instrument, and thus it is desirable for it to form an integral part of the project development from the very beginning. The EIA is a formal approach reported in an environmental impact statement (EIS; or environmental statement), which compiles the information required to assess the effects of the project and which the applicant can reasonably be required to compile.

Although a project's EIA is based on site and project specificities, the comparison of EIA for ocean energy sites reveals some commonalities on the environmental components addressed as well as on impact evaluation. This is perhaps not surprising since the sites where technologies have been installed, mostly for testing, are of a similar size and require similar infrastructure. Besides the production of clean energy, other positive environmental impacts have been assigned to ocean energy deployments although not always considered for further evaluation during the EIA process. If marine renewable energy is going to contribute for a significant proportion of electricity, MRE parks are likely to become widespread along coastal regions and therefore their role in coastal management can be expected to be considerable. In such a scenario measures to promote sustainability and enhance the positive aspects of such parks can be expected to have a prominent role in their management.

9.6.1 The EIA process

As a stepwise approach, the EIA methodology requires continuous reappraisal and adjustment, as is shown by the feedback loops in Figure 9.52. The application of EIA to ocean energy projects has been specifically addressed in guidance documents, protocols and standards, providing information on EIA steps and requirements in terms of

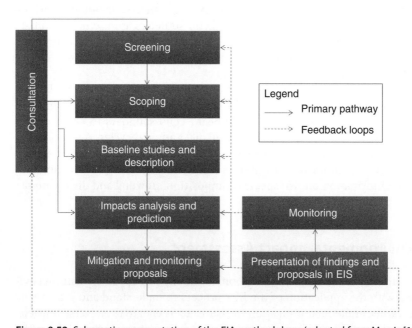

Figure 9.52 Schematic representation of the EIA methodology (adapted from Morris [114]).

information to be provided and suitable tools, methodologies and techniques that may be used at each step in order to improve quality and cost-efficiency of ocean energy EIAs [110–113]. These documents have been developed especially in Europe where more ocean energy projects have been installed at sea.

The ways in which EIAs are conducted differ by country, although they derive from the original framework set in USA by the US National Environmental Policy Act in 1969. Screening is the first step undertaken to determine whether an EIA is required. If needed, scoping is carried out to determine the content and extent of the matters that should be covered in the environmental information submitted to the competent authority. The next step is the baseline characterisation or the analysis of the existing environmental and socioeconomic conditions in the impact area. This characterisation is the basis for valid impact predictions as well as identification and implementation of effective mitigation and monitoring programmes. The impact analysis consists of three main levels of detail – identification, valuation and significance – and several techniques have been developed to carry out this exercise [114, 115]. With regard to MRE installations, there is typically a need to monitor any impacts that were assessed as either potentially of moderate significance or around which there was a reasonable degree of uncertainty, particularly where there are considerations with regard to other legislation on environmental conservation issues. This leads to the design and implementation of a monitoring programme with the ultimate objective of assessing the significance of impacts during installation, operation and decommissioning [116]. During the EIA process, appropriate levels of consultation (both statutory and public) should be undertaken to ensure interested parties or stakeholders are participating and contributing with concerns and opinions to the decision-making. Furthermore, the communication of the EIA results to decision-makers and the broader public is an important task which should be carefully prepared, avoiding jargon [117]. The outcomes of the EIA process are usually reflected in the licence or consent document in the form of specific terms and conditions to which the developer must comply.

Other complementary environmental assessment techniques such as the strategic environmental assessment (SEA), the environmental risk assessment (ERA) and the life cycle assessment (LCA) may be consulted or applied before and during the EIA process and their results further integrated into the EIS. The SEA (or programmatic environmental impact statement, as it is called in USA) is conducted to ensure that significant environmental impacts of certain plans and programmes are identified and considered during their preparation and before their adoption. The main objective is to help integrate the environment into the preparation and adoption of plans and programmes liable to have significant effects on the environment, by subjecting them to a prior environmental assessment at the planning stage. It aims to extend the principles of the EIA, carried out at the individual project level, to the decision-making at strategic level. It is intended that at this level the alternative approaches and their implications for the environment can more easily and appropriately be considered. Several SEAs on offshore renewable energy (including wind, wave and tidal) have been carried out in different countries (e.g. Canada, Ireland, UK, USA). The SEA has been recognised has an effective way to meet renewable energy demands while also safeguarding the seas through efficient negotiations and collaborations between regulators, statutory advisory bodies and developers [116].

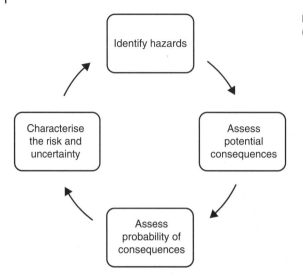

Figure 9.53 Risk analysis process (adapted from Gormley [120]).

Risk assessment or analysis is a well-established management tool for dealing with uncertainty. Risk assessment has only recently been extended to wider environmental considerations. 'Environmental risk assessment' is a generic term for a series of tools and techniques concerned with the structured gathering of available information about environmental risks and the formation of a judgement about them [118]. EIA and ERA are very similar concepts in that they have broadly the same goals, namely, to inform decision-makers on the frequency and magnitude of adverse environmental consequences. However a major additional aspect provided by ERA is that it provides the probability of a particular impact occurring [110]. ERA frameworks have already been proposed for marine renewable deployments [113, 119]. The ERA is a dynamic process involving four steps (Figure 9.53) and must be continuously revaluated in order to learning from new information as it comes to light. In that sense, monitoring activities provide a main source of information based on real data and represent an important tool to reduce uncertainty in the EIA process.

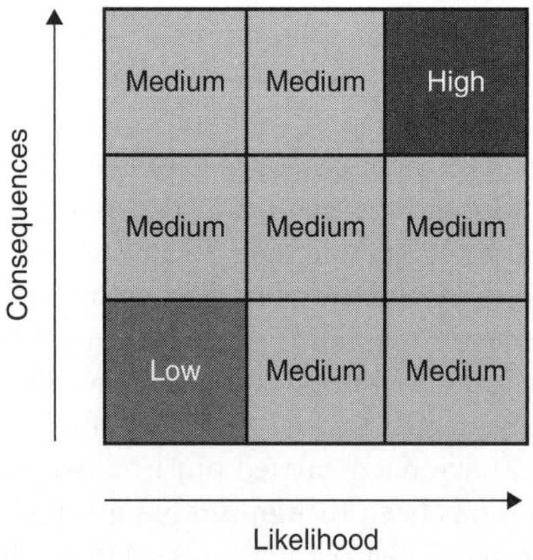

There are several techniques for hazard identification, but the objective is always to answer the question 'what can go wrong?' by recognising the threats that could lead to adverse effects. This step is usually based on information that already exists for the field of assessment or on the analysis of the same threats found for other technologies. Risk analysis involves the combination of the consequences and the likelihood of the threats identified during the hazard identification step, and the risk level is usually estimated using a risk matrix (Figure 9.54), which allows understanding the possible combinations of consequences and their likelihood. During the decision-making process risks

Figure 9.54 Risk matrix for general risk analysis.

classified as low level (low consequence and likelihood) are considered acceptable and only monitoring is recommended. In contrast, high risk levels (high consequences and likelihood) will require a strategy to manage the risk, which can involve, for example, the implementation of mitigation measures [120]. The medium risk level tends to correspond to risks that need special management to ensure that potential adverse effects are avoided or reduced. Risk level estimation can be done using qualitative, semi-quantitative or quantitative methods. Quantitative assessment requires calculations of the effects likelihood to produce a numerical risk score, while qualitative assessment is based on subjective scales and depends on the sensitivity of the team that is assessing risks.

Environmental effects of ocean energy projects are likely to be technology-, scale- and site-specific. Therefore, a lack of sufficient knowledge or experience in ocean energy projects compromises the correct quantification of the likelihood of adverse environmental effects. Another relevant aspect in risk analysis is the definition of appropriate indicators to assess the risk level of certain impacts. A range of such indicators by receptor (see Section 9.6.2) have been listed elsewhere [119].

The environmental risk analysis of ocean energy projects should also consider the risks that may arise from accidents during project phases and/or device malfunctions such as oil spills or vessel collision. For each risk identified a group of mitigation measures should also be listed to be implemented during the project development. The information delivered during the ERA process may inform the EIA process, i.e. the project risks may be translated into potential impacts. Further into the EIA process the extent, magnitude and significance of such impacts need to be confirmed by monitoring activities to adapt mitigation measures identified both in ERA and EIA.

LCA is used to estimate the cumulative environmental impacts resulting from the whole product life cycle, often including impacts ignored in traditional analyses (e.g. raw material extraction, transportation, maintenance process, final disposal, etc.). By including the impacts throughout the whole product life cycle, LCA enables a comprehensive view of the environmental aspects of the product or process and a more accurate picture of the true environmental trade-offs in product and process selection [110]. The LCA process is regulated by the ISO 14000 series. Examples of LCA application to different ocean energy technologies are available in the literature [121–123].

9.6.2 Identification of Stressors and Receptors

In the scoping phase of the EIA (see Section 9.5.1), the identification and definition of potential stressors and their corresponding natural receptors can help to identify and evaluate relevant impacts. The term *stressor* refers to all single characteristics or activities of an ocean energy project during construction, operation or decommissioning phases that may potentially lead to physical, chemical or biological changes in the marine environment. Examples of stressors are the introduction of residual chemicals, the increase of noise or even just the physical presence of the devices in the water.

The term *receptor* refers to elements of the ecosystem that may be affected by a stressor. The receptors are the different environmental features, usually operationally defined by an ecological entity (e.g. a single species or its indicator such as population size) [124, 125]. The connection of each stressor with each receptor is described as the effect (or impact) of the stressor on the receptor which may then be evaluated based on its duration, frequency, intensity and spatial coverage during the different phases of the project [69, 125].

It is important to note that, although stressors and receptors have already been identified for single ocean energy projects, when ocean energy devices are scaled up into farms new and cumulative environmental concerns are expected to arise in addition to those previously identified. All in all, MRE effects are dependent on the technology used, project dimension, array configuration and environmental specificities, such as sensitivity and vulnerability.

9.6.2.1 Ocean Energy Stressors

The main stressors to be considered in an MRE project vary among technology type (i.e. wave, tidal, currents). However, general stressors are identified listed for each project phase (as is shown in Table 9.6). Specific details on relevant stressors are given below.

Physical Disturbance   Different impacts may arise from the installation and presence of the devices in the water, such as the platform structure or its associated moorings or even from the increased traffic of vessels and ships in the areas. Physical impacts vary among the different phases of the project, but are expected to be higher during the construction and decommissioning phases as disturbance of the seabed is greater than during the operational phase.

Physical impacts can affect the biotic environment as they can change ecosystem functioning and benthic, fish and marine mammal communities as a consequence of altering the surrounding physical environment [126]. During construction and decommissioning, the introduction or removal of physical structures directly to the sea floor may affect sediment quality and the benthic species living within and on it. Effects may be caused by direct forces (e.g., sediment scouring) that lead to direct habitat losses. Changes in sediment dynamics can occur if the deployed structures alter the currents pattern [127]. During the installation/ decommissioning phase, the increase in local water turbidity arising from re-suspended sediments or by the smothering effects occurring from mobilised sediments can asphyxiate sedentary species [69]. On the other hand, in soft-bottom habitats, new installations represent new artificial hard

Table 9.6 Marine renewable energy environmental stressors occurring during the different phases of deployment.

Stressor/ phase	Construction	Operation	Decommissioning
Physical impacts	×	×	×
Reef effects		×	
Electromagnetic fields		×	
Heat energy		×	
Noise and vibration	×	×	×
Chemical impacts	×	×	×
Introduction of alien species		×	
Energy removal		×	

substrates that in the long term may provide settling grounds and new habitat for benthic invertebrates (e.g. bivalves, ascidians, etc. [128, 129]). As for pelagic species, one of the most straightforward effects is the risk of collision, for example, the collision of whales, fishes and turtles with ships or moving underwater and surface structures. In-water turbines, such as tidal energy devices, can lead to accidental 'blade strike', since the speed of the tip of some horizontal-axis rotors could be harmful for cetaceans, fishes, or diving birds [130]. Nevertheless, these turbines generally move at slow speeds and thus the likelihood of blade strike is low.

If the technology has physical structures above the water, the collision of birds, bats and migratory species is a major concern as moving parts of ocean energy devices can again lead to accidental collisions. These effects have been addressed by several authors [131, 132]. However, collision risk seems to be species-specific, as some species apparently avoid offshore structures. Collision is thought to depend mostly on the number, size and spacing of the devices, their moving parts, and whether the device is located along an organism's transit route [69].

Energy Removal During the operation phase energy is withdrawn from current or waves and this may have potential effects on both near- and far-field scales. Energy removal as well as blockage can locally hamper turbulence, stratification and vertical mixing of water masses (performed by wind and wave action) by installations covering a larger area of the water surface. These changes, in turn, can cause benthic sediment scouring leading to habitat changes. Alterations in mixing and stratification may affect the vertical movement and aggregation pattern of the pelagic flora and fauna, and their predators [125].

In addition, energy removal can affect the existing sediment dynamics and transport. If the device is near the coast it may affect beach replenishment and erosion or accretion, potentially affecting intertidal ecosystems and distribution of intertidal animals.

Reef Effects Once the construction of a device is completed, the resulting physical structure could be a positive environmental effect as it increases habitat heterogeneity. The introduction of new artificial hard substratum represents a colonisation opportunity for benthos, leading to an increase in biomass, in biodiversity, an opportunity for enhanced survival and growth, recycling of local nutrients, increase in refuge for juveniles of mobile species and the promotion of the ecosystem and the food web in the area [69, 133].

Habitats with greater physical heterogeneity have been shown to be functionally more important when compared with more homogeneous areas of the same extent [134]. Fisheries data show an increase in yields that is assumed to be the result of extensive invertebrate colonisation attracting fishes (artificial reef effects) [69, 109]. Installations are also thought to function as fish aggregation devices, since the shaded area below offshore structures such as platforms often provides shelter and feeding grounds where large aggregations of fishes can be found (e.g. [109, 135–138]). Physical structures may thus support local species diversity and even support the recovery of exploited fish stocks [137, 138]. The artificial reef effects appear to offer evidence of a positive impact on the area where the device is located. However, research has not yet determine if the reported biomass increase is provided by an additional source of recruits or if it is simply aggregating existing biomass from the surrounding areas. Species aggregation may create wider problems by recruiting species to the site that would normally replenish the species populations of existing adjacent habitats on a regular basis [69] or even by recruiting invasive species that can use the facility as a stepping-stone for colonising the natural environment.

The ocean energy infrastructures are also expected to cause disturbance resulting from shading effects as algae depend on natural light for photosynthesis, and shading or shadowing of water surfaces by installations may particularly affect microalgae and macroalgae production in the shaded area. Also, light might trigger vertical migration of zooplankton and their predators might be affected, as well as aggregation of fishes. The strength of these effects depends on the spatial and temporal dimension of shading, and on oceanographic features in the area.

Electromagnetic Fields   Submarine cables are involved in transmitting electrical energy produced by MRE devices to land substations (e.g. among devices, array to transformer, transformer to shore, etc.). The electrical currents transmitted in undersea cables induce electromagnetic fields, and these fields may affect marine organisms [139]. In particular, magneto-sensitive and electro-sensitive species, such as many fishes (particularly eels and elasmobranchs) and cetaceans, that often use electromagnetic fields for spatial location, large-scale orientation, feeding and mate finding, and these activities may be adversely affected by electromagnetic fields [125]. In addition, some species utilise electric fields behaviourally. Electro-sensitive species may be attracted or repelled by the electric fields, potentially resulting in congregation or dispersal depending on the extent of the electrical environment where multiple cable arrays exist.

Industry standard cables effectively shield against direct electric (E field) emissions but cannot completely shield against the magnetic component (B field). The leakage of B fields results in induced E fields adjacent to the cable independent of burial, as a result of magnetic properties [140]. Human-made electromagnetic fields are, among other factors, suspected to be a cause of stranding of cetaceans. Though little is known about the long-term effect of electromagnetic fields on marine organisms so far, the potential impact of this parameter needs to be considered. Research into the effects of ocean energy electric fields on sensitive species, particularly benthic ones, is required, especially when assessing the environmental impact at important feeding grounds or nursery areas.

Noise and Vibrations   MRE devices generate noise and vibrations, both above and under the water, through construction activities, moving rotors and vessel traffic. Acoustic signals can travel long distances underwater (longer than in air) and as sound is a mean of communication, navigation and echolocation used by vertebrates, such as fishes and marine mammals, they can be affected by noise pollution. Depending on its intensity, noise may affect animal behaviour and disrupt social interactions; it may damage hearing, and is suspected to be a cause of whale stranding [141–143]. Noise impacts are expected to be significantly higher during construction and decommissioning. These sources of noise could cause damage to the acoustic systems of species within 100 m of the source, and are expected to cause mobile organisms to avoid the area [144]. Any effects of the noise will depend on the sensitivity of the species present and their ability to get used to the noise. Effects will be reduced when the level of noise has decreased, i.e. following completion of the construction (or decommissioning) phases. To a lesser extent, though depending on the number of devices, noise disturbance during the operation phase may also be expected to affect a number of species (e.g. cetaceans, pinnipeds, teleost fishes and crustaceans) [69, 125].

Chemical Effects   Organic enrichment from natural and anthropogenic sources is a common disturbance in the marine environment. There are a multitude substances and materials that may harm the marine flora and fauna, among them are oil and oil-contaminated material (oily waste, oil filters), antifouling paint and other chemical pollutants, pharmaceuticals, and hazardous waste such as batteries. In the case of oil pollution, the lubricating or hydraulic oil leaking from ocean energy installations, as well as leakage of diesel oil from substations and vessels, has to be mitigated, but it is usually not an issue of major concern, due to the limited amount that is used. However, in the case of contamination and because most oil types float, swimming or floating organisms at the sea surface, such as seabirds, seals, sea otters, fishes, turtles, cetaceans, but also plankton, fish eggs and larvae, are the most affected [145]. Furthermore, chemical pollution may also affect water quality and organisms, for example inhibit phytoplankton growth [146].

Antifouling paint is commonly used on submerged structures, such as ships' hulls or devices, to prevent the surfaces from being heavily colonised by marine organisms such as barnacles, mussels and algae (biofouling) [125]. Antifouling components (e.g., copper, cuprous oxide, etc) are continuously leached and can accumulate in the food chain when concentrations are high, resulting in damage to target and non-target species. However, more research is needed to verify the impact of antifouling paint ingredients on marine life.

Introduction of Alien Species   Exotic species invasion due to shipping is a global problem arising from increasing trade and transport at sea. Alien species can be introduced in ships' ballast water or on ships' hulls (biofouling) and transported away from their native systems. Most invaders fail to establish, either because they do not survive the transit in dark and dirty ballast tanks, or due to unfavourable conditions at the point of release. However, there are successful marine invaders, including seaweed [147], gelatinous zooplankton [148], molluscs [149], and crabs [150] that may threaten native living communities after being released from the ballast tanks or attached to the ships' hulls.

When ocean energy devices are installed over long stretches of coast they create new hard bottoms that have the potential to be colonised and provide stepping-stones for settlement and spread invasive species [125, 136]. Invasive alien species may significantly affect habitat structure, native community composition, local biodiversity, food web structure and ultimately affect ecosystem services.

9.6.2.2   Environmental Receptors

Environmental receptors are the physicochemical and biological components of the ecosystem that, together with the socioeconomic aspects, can be affected by the deployment of an ocean energy project. Impacts in the physicochemical components may have effects in the overall biological receptors since the physicochemical environment is common to all marine organisms. Biological receptors can vary depending on the geographical area, and how they are affected depends on ecosystem sensitivity. Socioeconomic receptors involve human communities and their well-being. Effects on these receptors can be felt

Table 9.7 Examples of relevant environmental receptors that are usually considered during the EIA process of ocean energy projects. C: construction; O: Operation; D: Decommissioning.

Environment	Receptor	Project phase
Physicochemical	Water quality	C, O, D
	Water temperature	O
	Sediment dynamics	C, D
	Sediment quality	C, D
Biological	Benthic fauna and flora	C, O, D
	Pelagic fauna and flora	C, O, D
	Fish and turtles	C, O, D
	Birds and bats	C, O, D
	Marine mammals	C, O, D

during the entire project, although their magnitude may differ (Table 9.7). During a project's lifetime several stressors can affect the same receptor; however, their influence, intensity and magnitude are dynamic and can vary throughout the different stages of the project. For instance, benthos can be negatively affected by sediment resuspension during construction and cable burial, causing smothering of organisms; however, they can also be positively affected during the operation phase since the artificial structures increase the area for benthic colonisation.

Water Quality For proper functioning, ocean energy devices use oil and grease that can accidentally leak or spill into the marine environment and therefore degrade water quality. Accidental spillages are more likely to occur during construction and decommissioning, but can also take place during the operation phase. Water contamination can also occur from antifouling chemicals used in device coats or paint. This contamination usually occurs on a long-term basis [151]. Water quality might also be affected during construction works as a result of sediment removal and disturbance, increasing turbidity, remobilising contaminants from the sediments and increasing dissolved and particulate organic matter content in the water column [69].

Sediment Dynamics and Quality Ocean energy devices can work as obstacles for sediment. In addition, their operation removes energy from the system, and alterations in the hydrodynamic pattern may be expected. Consequently, long-term changes in the sediment dynamics may promote alterations in sediment deposition and erosion zones, contributing to the redefinition of seabed and coastline morphology. Impacts depend on the technology type, project scale (number of devices) and prevailing currents [125]. Local and temporary seabed disturbance, due to cable and mooring installation, occurs during device installation and decommissioning, and, as mentioned above, a potential effect on water quality can be expected through turbidity increase and remobilisation of contaminants from the sediments to the water column. These effects may temporarily affect benthic and pelagic flora and fauna.

Benthic Flora and Fauna   During the construction phase, several stressors can impact benthic habitats. Impacts such as increased turbidity, which may affect photosynthetic/ filter feeding organisms, increased suspended sediment and sediment deposition that can lead to benthic communities smothering, scouring and/or abrasion. In addition, the release of contaminants from dredged sediments or the contamination by spills or leakages can potentially have a harmful effect on benthic flora and fauna [124]. Habitat destruction can also occur locally in the short to medium term, as a consequence of excavation works or placement of moorings.

Species assemblages exhibit natural variation through space and time because of biotic interactions and natural environmental variability. Dredging-related disturbance has been shown to alter local species diversity and population density [152]. Nevertheless, the magnitude of the effects on the benthic community depends on the duration and intensity of the disturbance [153] and the resilience of the local infauna [154].

Research on human activities such as fishing or dredging has shown that after the disturbance ceases, recolonisation takes place over a period of months to years [155, 157]. Small opportunistic species, such as polychaetes and amphipods, are the quickest to colonise after physical disturbance, while epifaunal species assemblages are likely to take longer [144, 155]. Change may be rapid with soft substrata, and new habitat can be created if the conditions are suitable. On coarse and more stable substrata, change is likely to be slower [157].

Wave action and current dynamics have important implications for the welfare of benthic marine organisms. Wave action is a relevant factor influencing sediment movement and consequently the survivability of benthic animals. On the other hand, currents influence the bottom shear stress and as a consequence food availability for benthic communities and benthic secondary production [65]. During device operation, changes can occur in hydrodynamics that can potentially lead to changes in sediment type. The introduction of new habitat from foundation structures can have both positive and negative effects. It enhances marine growth by increasing hard substrate available for settlement (attachment of sessile hard bottom species, attraction of reef-dwellers) but it can also be available substrate for alien species to settle, grow and propagate to other areas.

Pelagic Flora and Fauna   As mentioned above, during construction and decommissioning works potential impacts can affect plankton species due to turbidity increase (reduced light penetration) and from the remobilisation of nutrients or pollutants from sediments. During commissioning and operation this group of organisms can also be exposed to residual chemicals (if spilled). Maintenance procedures may use cleaning solutions with high or low pH values and residual cleaning chemicals and products such as biocides may be harmful (e.g. antifouling paints).

Many pelagic species are vulnerable to noise disruption, particularly during spawning which may be disrupted during the construction phase. Placement of the device, mooring installation, scouring protection and cable laying operations may disrupt fish behaviour, particularly in relation to spawning and migration routes for diadromous fish and other migratory species. Studies show that a range of received sound pressure and particle motion levels will trigger behavioural responses in sole and cod (flat and round fish) [158]. Most species of fish are broadcast spawners, and so changes to the seabed and the placement of projects may not have severe long-term implications. However, disruption to the spawning periods of certain species should be avoided during the construction phase. Species of fish that deposit eggs on the sea floor are more likely to be affected by any activities that may disturb or displace them to areas of different sediment type. Fish aggregation effects are also expected to occur, but its impacts at the population level and its quantification has not been accounted yet. During the operation phase, effects on migrating species and the level of acoustic impacts are still to be understood [106]. To understand the interaction between fishes and ocean energy devices, tank experiments have been carried out. Although the extrapolation from such studies should be done with caution, the results provide information about impact collision [91, 159].

Marine Mammals   Changes in animal distribution, density and behaviour are most likely to occur during construction and decommissioning because of increased human activity in an area (increased noise, boat traffic, etc.), but this is likely to be a short-term impact. During operation, there are some risks related to strike and collision with, for example, turbine moving parts, which may result in injury or mortality. Tidal devices that are placed in narrow channels between land masses could form a barrier to organism movements or migration. Large expanses of ocean holding wave devices could have the same effect. The commercial scale of wave and tidal developments could also result in long-term displacement of animals. The potential effects of barrier effects and habitat displacement are likely to vary between species but are difficult to predict from small-scale commercial demonstrators [85]. Entanglement can also be of concern for devices installed with long cables or moorings.

Marine mammals are sensitive to sound since they use their highly evolved acoustic capabilities to communicate, navigate, find prey, escape predators and perceive their environment. Changes to the acoustic environment resulting from the increase of noise levels in an area can therefore interfere with the normal behavioural patterns of these animals. Marine mammal species can be classified into three functional hearing groups based on their auditory sensitivity: low (7 Hz to 22 kHz), medium (150 Hz to 160 kHz) and high frequency (200 Hz to 180 kHz). The level at which the animal receives the sound will depend on its frequency with relation to the species frequency sensitivity spectrum. Underwater, seals detect sound between 75Hz and 75 kHz, although harbour seals have been found to detect sound at 180 kHz if it is sufficiently intense. Seals' hearing in air is less acute, ranging up to 30 kHz. Potential impacts on marine mammals from noise associated with ocean energy developments can be divided into three categories: physical effects such as hearing damage as a direct result of noise produced; behavioural responses as a result of noise produced such as avoidance of an area; and indirect effects such as impacts on food availability [85].The effects of electromagnetic fields on marine mammals are poorly known; for species that rely on Earth's geomagnetic field, there is the potential for disorientation due to the magnetic fields emitted by power transmission cables, if they are large enough and/or discernible from background levels. Fundamental baseline data will be needed (mammal biology, species presence, absence and diversity, information on prey species) to understand projects' impacts and the cumulative effects as ocean energy reaches the commercial scale [125].

Marine Birds   Marine birds can potentially be affected by ocean energy projects including seabirds, seaducks, waterbirds and wildfowl. Seabirds feed and rest on marine waters, passing through the area, either on a daily basis or during migration. Such species include those listed above, as well as other wildfowl and passerines. The extent to which species are affected depends on the importance of the area for the species and their vulnerability. To understand the potential impacts of the ocean energy projects, it is necessary to assess the distribution of seabirds in the area and the changes of seabirds' populations on a seasonal basis [106]. Artificial reef effects are expected to occur underwater and as a consequence an increase in seabird prey is anticipated, what can increase collision risk [135, 160–164].

Sea Turtles Electromagnetic fields can alter these animals' ability to respond to naturally generated magnetic fields, decreasing their survivability, reproductive rate and migratory routes [69]. Underwater noise can also attract or deter sea turtles. Collision or entanglement risks are also expected specially if the projects include dynamic parts or a large area surface.

9.6.3 Impact Assessment Techniques and Mitigation Measures

Many potential effects are assigned to ocean energy projects, although to date evidence of their significance is theoretical since only single device demonstrations have been deployed at sea. However, impacts are expected to occur and these have been widely listed in several reports and peer-reviewed literature [165–168].

A number of approaches are used to determine the significance of environmental effects, which according to the BSI [112] follow a common iterative process consisting of four main steps: identification of environmental changes from the proposed activities and the receptors of interest that are likely to be affected; prediction of the impacts through the understanding of the nature of the environmental changes in terms of their exposure characteristics, the natural conditions of the system and the sensitivity of the specific receptors; evaluation of the vulnerability of the features to be affected as a basis for assessing the impact level and its significance; and management of significant impacts that need the implementation of mitigation measures and identification of the significance of residual impacts. The key significance levels consider in the third step may be classified as 'negligible' when the change does not have a discernible effect; 'minor' when effects tend to be discernible but tolerable and unlikely to require mitigation; 'moderate' when the changes are considered adverse and thus require mitigation; and 'major' when the effects are highest in magnitude and reflect the high vulnerability and importance of the receptor, and thus require mitigation. Those impacts that are classified as moderate or above are considered to be significant [112].

In general, the method most widely used by practitioners to assess the degree of significance of a predicted effect or impact is the application of a adapted version of the Leopold matrix [169]. This type of matrix shows cause–effect links between stressors and receptors, listing in the first column all project characteristics and activities (stressors) and on the top row the various environmental factors (receptors) to be considered. Each combination is then scored to indicate the magnitude and importance of the impact of each activity on each environmental factor and the two in combination are used to assess the significance of the impact. A simple matrix showing the main stressors and receptors to be considered in ocean energy projects is shown in Table 9.8.

It must be noted that the term 'significant' is often interpreted in different ways. In a statistical context, a 'significant change' means having a low probability of obtaining a test statistic at least as extreme as the one that was observed solely by chance, if the

Table 9.8 Ocean energy stressors and receptors during different project phases (adapted from Boehlert [171]).

Stressors	Receptors								
	Physical environment	Water quality	Sediment dynamics	Sediment quality	Benthic flora and fauna	Pelagic fauna and flora	Fishes and turtles	Birds and bats	Marine mammals
Construction phase									
Physical presence of devices	×				×	×	×	×	×
Chemical contamination		×			×	×	×	×	×
Underwater noise	×				×	×	×	×	×
Operation phase									
Physical presence of devices	×		×	×	×	×	×	×	×
Chemical contamination		×		×	×	×	×	×	×
Underwater noise	×				×	×	×		×
Electromagnetic fields					×	×	×		
Reef effects		×		×	×	×	×	×	
Energy removal	×	×	×	×	×	×	×		
Introduction of alien species					×	×	×	×	×
Decommissioning phase									
Physical presence of devices	×				×	×	×	×	×
Chemical contamination		×		×	×	×	×	×	×
Underwater noise	×				×	×	×	×	×

null hypothesis is true. In the context of EIAs, significance from screening to impact prediction, monitoring and mitigation has a different meaning, evolving during the process depending on the nature of related uncertainties [116]. However, when monitoring impacts a statistical interpretation of the meaning of significance may also be used, although due to the relatively short time frame in which monitoring is carried out, it is often difficult to distinguish any impact from background natural variability [85, 116, 163, 170].

Mitigation measures are formulated to avoid, minimise or reduce, remedy or compensate for the predicted adverse impacts of the project. Some of these measures may include: selection of alternative locations; modification of the methods and timing of construction; modification of design features; or minimisation of operational impacts (e.g. noise and collision). Different mitigation measures are needed according to the specific impacts on the environmental components and receptors. Measures are usually identified in the EIS report and include details on when and how to carry them out as well as the modifications they should undergo if post-project effects are different from those predicted. Where impacts of medium and high significance are identified, mitigation is usually proposed to reduce frequency, probability or extent of the impacts. A description of mitigation measures to be adopted during wave and tidal energy developments is available in several reports [172, 173].

9.6.4 Monitoring Potential Impacts

In order to understand the impacts of the project on each environmental receptor most studies employ a before–after control–impact (BACI) or gradient sampling design [85, 174]. Using this method, the disturbance caused by the development can be identified by monitoring two locations before and after the occurrence of the disturbance factor, if the impact location shows a different pattern after the disturbance compared with the control location [175, 176]. Although this method presents some constraints when the scale of the disturbance is unknown, this design may reduce the need to repeat studies [177].

To assess impacts in a systematic way, sampling designs and methods should be combined to answer the questions and purposes that led to the observation of the marine biota. Inadequate designs or methods can compromise the results and/or may not be appropriate for detecting effects. In addition to the receptor or process to be assessed, the spatial and temporal scale and resolution of the sampling methods are factors that may influence the selection of a specific methodology or a combination thereof. Therefore, it is important to tailor the experiment to fit the receptor dynamics and the effect that is to be monitored. The most relevant techniques used to monitor ocean energy effects and receptors are described below.

9.6.4.1 Benthos

As mentioned before the introduction of devices in high-energy environments may modify the local and regional water circulation patterns. The design and methodological requirements for benthic monitoring tend to be similar across technology types and project phases [103]. Before installation, information on local benthic habitats is usually scarce and therefore the main purpose of the monitoring activities is to characterise the area in terms of typical benthic communities that are expected to occur. During device operation, monitoring should focus on specific areas where impacts are likely to occur, using the same techniques as applied during baseline characterisation to allow 'before and after' deployment comparisons. In the SeaGen project installed in Strangford Lough (Northern Ireland), the modification of benthic species composition has been assessed with acoustic mapping using ground truthing and diver sampling. Broader benthic effects have been assessed in Orkney and Penthland Firth as well as in Paimpol-Bréhat (France) with drop-down video and grab and diver sampling [103].

9.6.4.2 Fish

The assessment of impacts of ocean energy on fish species focuses on the understanding of the population dynamics and habitat distribution at both spatial and temporal scales at the sites. When assessing environmental impacts of wave or tidal devices on fishes, it is important to consider the design of the device since it may influence the type of species attracted to them [178]. Acoustic cameras or dual-frequency identification sonars may be very useful in providing information on diel condition, turbine motion and fish size, although species identification can be limited using acoustic cameras [108]. Furthermore, models to estimate impacts (mainly collision and the likelihood of fish encounters with the device) have recently been developed [179]. Table 9.9 summarises the most common methods used for monitoring impacts of MRE projects on fish populations.

9.6.4.3 Marine Mammals

Currently, the main outputs for impact assessment come from single ocean energy demonstration projects. From the information collected so far it is assumed that the potential impacts of these devices on marine mammals are collision, entanglement, entrapment, acoustic disturbance, and displacement or habitat exclusion [84, 125]. Although collision is considered as a potential impact occurring both for wave and tidal projects, since both have underwater structures that may not be detected by the animals, the impact level is different due to the blade movements of tidal devices [84]. Detailed techniques for the monitoring of marine mammals with regard to ocean energy projects are available in the literature [84, 85] and summarised in Tables 9.9–9.11.

Table 9.9 Methods used to monitor MRE impacts on fish populations.

Objectives	Techniques	Examples
Fish abundance and density	Echo-sounders Single- or multi-beam sonars Diving Cameras	Open-hydro, Orkney [180] Wave Hub [181]
Fish behaviour	Acoustic cameras Ambient light video monitoring Acoustic tags (telemetry)	Roosevelt Island Tidal Energy tidal turbine, East River, New York [108] Wave Hub [181]
Influence of ecological factors	Acoustic cameras Single-beam sonars	

Table 9.10 Techniques for monitoring cetacean species in MRE parks.

Objective	Indicator	Technique	Examples
Assess disturbance or displacement	Changes in abundance (relative or absolute) Changes in distribution	Visual observations using vantage points (relative abundance) and transect surveys Passive acoustic monitoring (presence/absence) Photo identification	Horns Reef offshore wind farm [182] Nysted Wind Farm, Egmond aan Zee, wind farm [183] OpenHydro tidal turbine in Bay of Fundy [184] SeaGen device in Strangford Narrows
Interactions with devices	Collision, Entanglement Entrapment Presence of carcasses	Active sonar Underwater video/photography Monitoring of stranded animals and necropsy	SeaGen device in Strangford Narrows [185] OpenHydro

Table 9.11 Techniques used for pinniped monitoring.

Objective	Indicator	Technique	Examples
Assess disturbance or displacement	Changes in distribution and abundance	Visual observations Telemetry	Egmond aan Zee wind farm [94] Nysted and Rodsand II winfarms [95] Horns Reef wind farm [182, 188] SeaGen device in Strangford Narrows [185]
Assess interactions with devices	Collision Entanglement Carcasses	Underwater video Active acoustics Stranded animals monitoring and necropsy	SeaGen device in Strangford Narrows [185] OpenHydro, EMEC [191]

There is considerable uncertainty over ocean energy impacts on marine mammals and information from new pre-commercial and commercial projects in the coming years will be helpful in properly addressing the specific impacts of developments of this kind. Currently, the lack of knowledge about marine mammals is also related to their ecology. For example, little information is available on how marine mammals use tidal flows, and this is relevant to understand the potential for collision [186]. C-PODs have proved useful in looking for behavioural modifications of cetaceans, mainly porpoises, in the vicinity of devices [184, 187–190].Furthermore, the use of C-PODs is a cost-effective and non-invasive technique for collecting long-term data [85]. Although arrays of static hydrophones can be used to understand changes in the large-scale distribution and behaviour of animals they are not appropriate for obtaining fine-scale movement data to assess, for example, interactions between device and animals [91].

More recently, acoustic techniques are being considered to assess the impact of ocean energy projects at a smaller scale [91, 192]. For tidal devices the use of active acoustics is being tested and applied to detect the presence of animals, namely cetaceans and seals, to assess collision risk [185, 193]. Interactions between marine mammals and devices are also being studied through model applications. The most relevant example is the use of collision risk models to estimate this type of interaction with tidal turbines [130]. The use of models requires information on animal density, the velocity of the animals and the turbine blades, animal behavioural patterns and their reaction to the presence of devices which cannot be available for all sites. Behavioural responses by seals to marine renewable projects have been studied using telemetry and visual observations [93, 94, 188].

9.6.4.4 Seabirds

Compared to offshore wind, ocean energy devices are mainly submerged and have less above-water presence. Due to the lack of data, uncertainty about direct and indirect impacts of ocean energy devices is high but monitoring is focused on understanding the risk of collision, entanglement and/or entrapment especially for diving birds [163, 164, 181].

Seabirds are subject to different types of pressures and it is difficult to assess the influence of ocean energy projects on their populations. Studies should be site-specific and focus on species with high potential to be affected. According to Furness *et al.* [194] the vulnerability of seabirds to ocean energy projects may be evaluated with reference to vulnerability factors, which may be different depending on the type of the technology. For wave projects, the most relevant vulnerability factors are the risk of collision, exclusion from foraging habitat, benefit from roost platform, effects of fish attraction to device or biofouling, disturbance by structures and ship traffic, and habitat specialisation. Concerning tidal projects, vulnerability factors are the risk of drowning, diving depth, benthic foraging behaviour, the use of tidal races for foraging, feeding range, disturbance by ship traffic and habitat specialisation [194]. Traditional methods such as visual observations can provide this information, but radar observations or the use of telemetry tags can also be useful, for example to assess distribution and diving depths.

9.6.5 Adaptive Management

An adaptive approach involves exploring alternative ways to meet management objectives, predicting the outcomes of alternatives based on the current state of knowledge, implementing one or more of these alternatives, monitoring to learn about the impacts of management actions, and then using the results to update knowledge and adjust management actions. Adaptive management focuses on learning and adapting, through partnerships of managers, developers, scientists and other stakeholders learning together how to create and maintain sustainable resource systems [195].

The initial lack of information on new technologies such as ocean energy constrains the accurate assessment of environmental impacts. There is a need to learn from the experience of operating devices in order to validate the predicted environmental effects of an ocean energy project and adapt mitigation and/or monitoring strategies as knowledge improves [110]. The applicability of an adaptive management approach is described in the literature for general energy projects [196] and specifically for ocean energy projects [171, 173, 197].

The adaptive management process can be broken down into a few steps: identify the fields where environmental management decisions need to be made; make decisions based on the best available information; monitor the impacts of these decisions; adjust decisions based on monitoring results; repeat the monitoring and adjustment process. This cycle should be integrated into higher-level management reviews to ensure decisions achieve programmatic goals of protecting sensitive resources while allowing responsible energy development.

References

1 Wong, P.P., *et al.* (2014) Coastal systems and low-lying areas, in C.B. Field *et al.* (eds), *Climate Change 2014: Impacts, Adaptation, and Vulnerability. Part A: Global and Sectoral Aspects. Contribution of Working Group II to the Fifth Assessment Report of the Intergovernmental Panel of Climate Change*, Cambridge: Cambridge University Press.

2 European Commission (2007) *A European Strategic Energy Technology Plan (SET-Plan): Towards a low-carbon future*, Communication from the Commission to the Council, the European Parliament, the European Economic and Social Committee and the Committee of the Regions.

3 Woodworth, P., *et al.* (2009) Trends in UK mean sea level revisited. *Geophysical Journal International* 176(1), 19–30.

4 Chini, N., *et al.* (2010) The impact of sea level rise and climate change on inshore wave climate: A case study for East Anglia (UK). *Coastal Engineering* 57(11), 973–984.

5 Wadey, M., I. Haigh, and J. Brown (2014) A century of sea level data and the UK's 2013/14 storm surges: An assessment of extremes and clustering using the Newlyn tide gauge record. *Ocean Science Discussions* 11(4), 1995–2028.

6 Buijs, F., *et al.* (2007) *Performance and Reliability of Flood and Coastal Defences*. DEFRA.

7 Carballo, R. and G. Iglesias (2013) Wave farm impact based on realistic wave-WEC interaction. *Energy* 51, 216–229.

8 Mendoza, E., *et al.* (2014) Beach response to wave energy converter farms acting as coastal defence. *Coastal Engineering* 87, 97–111.

9 Abanades, J., D. Greaves, and G. Iglesias (2014) Wave farm impact on the beach profile: A case study. *Coastal Engineering* 86, 36–44.

10 Abanades, J., D. Greaves, and G. Iglesias (2015) Coastal defence using wave farms: The role of farm-to-coast distance. *Renewable Energy* 75, 572–582.

11 Abanades, J., D. Greaves, and G. Iglesias (2014) Coastal defence through wave farms. *Coastal Engineering* 91, 299–307.

12 Abanades, J., D. Greaves, and G. Iglesias (2015) Wave farm impact on beach modal state. *Marine Geology* 361, 126–135.

13 Iglesias, G. and J. Abanades (2015) Wave power – climate change mitigation and adaptation, in W.-Y. Chen, T. Suzuki, and M. Lackner (eds), *Handbook of Climate Change Mitigation and Adaptation*. New York: Springer.

14 Child, B.F.M. and V. Venugopal (2010) Optimal configurations of wave energy device arrays. *Ocean Engineering* 37(16), 1402–1417.

15 Child, B. and V. Venugopal (2007) Interaction of waves with an array of floating wave energy devices. *Proceedings of the 7th European Wave and Tidal Energy Conference, Porto, Portugal.*

16 Ransley, E. and D. Greaves (2012) Investigating interaction effects in an array of multi-mode wave energy converters. *27th International Workshop on Water Waves and Floating Bodies*, Copenhagen.

17 Babarit, A. (2010) Impact of long separating distances on the energy production of two interacting wave energy converters. *Ocean Engineering* 37(8–9), 718–729.

18 Borgarino, B., A. Babarit, and P. Ferrant (2012) Impact of wave interactions effects on energy absorption in large arrays of wave energy converters. *Ocean Engineering* 41, 79–88.

19 Stratigaki, V., *et al.* (2014) Wave basin experiments with large wave energy converter arrays to study interactions between the converters and effects on other users in the sea and the coastal area. *Energies* 2014. 7(2), 701–734.

20 Beels, C., *et al.* (2010) Application of the time-dependent mild-slope equations for the simulation of wake effects in the lee of a farm of Wave Dragon wave energy converters. *Renewable Energy* 35(8), 1644–1661.

21 Beels, C., *et al.* (2010) Numerical implementation and sensitivity analysis of a wave energy converter in a time-dependent mild-slope equation model. *Coastal Engineering* 57(5), 471–492.

22 Millar, D.L., H.C.M. Smith, and D.E. Reeve (2007) Modelling analysis of the sensitivity of shoreline change to a wave farm. *Ocean Engineering* 34(5–6), 884–901.

23 Smith, H.C.M., C. Pearce, and D.L. Millar (2012) Further analysis of change in nearshore wave climate due to an offshore wave farm: An enhanced case study for the Wave Hub site. *Renewable Energy* 40(1), 51–64.

24 Venugopal, V. and G. Smith (2007) Wave climate investigation for an array of wave power devices. *Proceedings of the 7th European Wave and Tidal Energy Conference, Porto, Portugal.*

25 Palha, A., *et al.* (2010) The impact of wave energy farms in the shoreline wave climate: Portuguese pilot zone case study using Pelamis energy wave devices. *Renewable Energy* 35(1), 62–77.

26 Fernandez, H., *et al.* (2012) The new wave energy converter WaveCat: Concept and laboratory tests. *Marine Structures* 29(1), 58–70.

27 Iglesias, G., *et al.* (2009) Wave energy potential in Galicia (NW Spain). *Renewable Energy* 34(11), 2323–2333.

28 Iglesias, G. and R. Carballo (2009) Wave energy potential along the Death Coast (Spain). *Energy* 34(11), 1963–1975.

29 Iglesias, G. and R. Carballo (2014) Wave farm impact: The role of farm-to-coast distance. *Renewable Energy* 69, 375–385.

30 Veigas, M., V. Ramos, and G. Iglesias, A wave farm for an island: Detailed effects on the nearshore wave climate. *Energy* 69, 801–812.

31 Fernandez, H., *et al.*, The new wave energy converter WaveCat: Concept and laboratory tests. *Marine Structures* 29(1), 58–70.

32 Pilar, P., C.G. Soares, and J.C. Carretero (2008) 44-year wave hindcast for the North East Atlantic European coast. *Hindcast of Dynamic Processes of the Ocean and Coastal Areas of Europe* 55(11), 861–871.

33 Guedes Soares, C. (2008) Hindcast of dynamic processes of the ocean and coastal areas of Europe. *Hindcast of Dynamic Processes of the Ocean and Coastal Areas of Europe,* 55(11), 825–826.

34 Iglesias, G. and R. Carballo (2010) Wave energy and nearshore hot spots: The case of the SE Bay of Biscay. *Renewable Energy* 35(11), 2490–2500.

35 Vidal, C., *et al.* (2007) Impact of Santoña WEC installation on the littoral processes. *Proceedings of the 7th European Wave and Tidal Energy Conference,* Porto, Portugal, 2007.

36 Neill, S. and G. Iglesias (2012) Impact of wave energy converter (WEC) array operation on nearshore processes. *Proceedings of 4th International Conference on Ocean Energy (ICOE),* Dublin.

37 Ruol, P., *et al.* (2011) Near-shore floating wave energy converters: Applications for coastal protection. *Proceedings of the International Conference of Coastal Engineering 2010,* Shanghai.

38 US Army Corps of Engineers (1984) *Shore Protection Manual.* Army Engineer Waterways Experiment Station, Vicksburg, MS.

39 Zanuttigh, B. and E. Angelelli (2013) Experimental investigation of floating wave energy converters for coastal protection purpose. *Coastal Engineering* 80, 148–159.

40 Nørgaard, J.Q.H. and T.L. Andersen (2015) *Investigation of Wave Height Reduction behind the Wave Dragon Wave Energy Converters and Application in Santander, Spain.* Department of Civil Engineering, Aalborg University, Aalborg.

41 Kamphuis, J.W. (1991) Alongshore sediment transport rate. *Journal of Waterway, Port, Coastal, and Ocean Engineering* 117(6), 624–640.

42 Reeve, D.E., *et al.* (2011) An investigation of the impacts of climate change on wave energy generation: The Wave Hub, Cornwall, UK. *Renewable Energy* 36(9), 2404–2413.

43 Gonzalez-Santamaria, R., Q.-P. Zou, and S. Pan (2013) Impacts of a wave farm on waves, currents and coastal morphology in South West England. *Estuaries and Coasts,* 38(s1), 159–172.

44 CISCAG (2011) *Shoreline Management Plan.* Cornwall and Isles of Scilly Coastal Advisory Group.

45 Tolman, H.L. (2002) *User manual and system documentation of WAVEWATCH-III version 2.22.*

46 Baldock, T.E., *et al.* (2011) Large-scale experiments on beach profile evolution and surf and swash zone sediment transport induced by long waves, wave groups and random waves. *Coastal Engineering* 58(2), 214–227.

47 Kenney, J. (2009) *SW Wave Hub metocean design basis.* SWRDA. Available at http://bit. ly/2xG4l0m.

48 Ramos, V., *et al.* (2013) Assessment of the impacts of tidal stream energy through high-resolution numerical modeling. *Energy* 61, 541–554.

49 Roc, T., D.C. Conley, and D. Greaves (2013) Methodology for tidal turbine representation in ocean circulation model. *Renewable Energy* 51, 448–464.

50 Ahmadian, R., R. Falconer, and B. Bockelmann-Evans, Far-field modelling of the hydro-environmental impact of tidal stream turbines. *Renewable Energy* 38(1), 107–116.

51 Dyer, K.E. (1997) *Estuaries: A Physical Introduction.* New York: John Wiley & Sons.

52 Sanchez, M., *et al.* (2014) Tidal stream energy impact on the transient and residual flow in an estuary: A 3D analysis. *Applied Energy* 116, 167–177.

53 Ramos, V., *et al.* (2014) Tidal stream energy impacts on estuarine circulation. *Energy Conversion and Management* 80, 137–149.

54 Carballo, R., G. Iglesias, and A. Castro (2009) Numerical model evaluation of tidal stream energy resources in the Ria de Muros (NW Spain). *Renewable Energy* 34(6), 1517–1524.

55 Sanchez, M., *et al.* (2014) Floating vs. bottom-fixed turbines for tidal stream energy: A comparative impact assessment. *Energy* 72, 691–701.

56 Kadiri, M., *et al.* (2012) A review of the potential water quality impacts of tidal renewable energy systems. *Renewable and Sustainable Energy Reviews* 16(1), 329–341.

57 Neill, S.P., J.R. Jordan, and S.J. Couch (2012) Impact of tidal energy converter (TEC) arrays on the dynamics of headland sand banks. *Renewable Energy* 37(1), 387–397.

58 Neill, S.P., *et al.* (2009) The impact of tidal stream turbines on large-scale sediment dynamics. *Renewable Energy* 34(12), 2803–2812.

59 Robins, P.E., S.P. Neill, and M.J. Lewis (2014) Impact of tidal-stream arrays in relation to the natural variability of sedimentary processes. *Renewable Energy* 72, 311–321.

60 Chatzirodou, A. and H. Karunarathna (2014) *Impacts of tidal energy extraction on sea bed morphology. Coastal Engineering Proceedings* 1(34), 33.

61 Mora, C., *et al.* (2011) How many species are there on Earth and in the ocean? *PLoS Biology* 9(8), e1001127.

62 Chapin, F.S.I., Matson, P.A., and Mooney H.A. (2002) *Principles of Terrestrial Ecosystem Ecology.* New York: Springer.

63 Scialabba, N.E. (1998) *Integrated Coastal Area Management and Agriculture, Forestry and Fisheries.* Rome: Environment and Natural Resources Service, FAO.

64 UNEP-WCMC (2011) *Marine and Coastal Ecosystem Services: Valuation Methods and their Application.* UNEP-WCMC Biodiversity Series No. 33 / UNEP Regional Seas Reports and Studies No. 188.

65 Kaiser, M.J., *et al.* (2005) *Marine Ecology – Processes, Systems and Impacts.* New York: Oxford University Press.

66 Pechenik, J.A. (1999) On the advantages and disadvantages of larval stages in benthic marine invertebrate life cycles. *Marine Ecology Progress Series* 177, 269–297.

67 Heip, C. (2014) Marine biodiversity.

68 Tillin, H., and Tyler-Walters, H. (2013) *Assessing the Sensitivity of Subtidal Sedimentary Habitats to Pressures Associated with Marine Activities. Phase 1 Report: Rationale and Proposed Ecological Groupings for Level 5 Biotopes against which Sensitivity Assessments would be best Undertaken.* Joint Nature Conservation Committee, Peterborough.

69 Gill, A.B. (2005) Offshore renewable energy: Ecological implications of generating electricity in the coastal zone. *Journal of Applied Ecology* 42(4), 605–615.

70 MacLean M, H. Breeze, J. Walmsley, and J. Corkum (eds) (2013) *State of the Scotian Shelf Report,* Canadian Technical Report of Fisheries and Acquatic Sciences 3074. Fisheries and Oceans Canada, Dartmouth, NS.

71 Froese, R., *et al.* (2012) What catch data can tell us about the status of global fisheries. *Marine Biology* 159(6), 1283–1292.

72 FAO (2012) *The State of World Fisheries and Aquaculture (2010).* Rome: FAO.

73 T. Worcester and M. Parker (2010) *Ecosystem Status and Trends Report for the Gulf of Maine and Scotian Shelf.* Fisheries and Oceans Canada.

74 Holdway, D.A. (2002) The acute and chronic effects of wastes associated with offshore oil and gas production on temperate and tropical marine ecological processes. *Marine Pollution Bulletin* 44(3), 185–203.

75 Matthiessen, P. and R.J. Law (2002) Contaminants and their effects on estuarine and coastal organisms in the United Kingdom in the late twentieth century. *Environmental Pollution* 120(3), 739–757.

76 Zacharias, M.A. and E.J. Gregr (2005) Sensitivity and vulnerability in marine environments: An approach to identifying vulnerable marine areas. *Conservation Biology* 19(1), 86–97.

77 Tyler-Walters, H., K. Hiscock, D. Lear, and A. Jackson (2001) *Identifying Species and Ecosystem Sensitivities*, Report to the Department for Environment, Food and Rural Affairs from the Marine Life Information Network (MarLIN). Marine Biological Association of the United Kingdom, Plymouth.

78 Holt, T.J., Jones, D.R., Hawkins, S.J., and Hartnoll, R.G. (1995) *The Sensitivity of Marine Communities to Man-Induced Change: A Scoping Report.* Countryside Council for Wales.

79 Derous, S., *et al.* (2007) A concept for biological valuation in the marine environment. *Oceanologia*, 49, 99–128.

80 Tyler-Walters, H. and A. Jackson (1999) *Assessing Seabed Species and Ecosystems Sensitivities. Rationale and user guide.* Report to English Nature, Scottish Natural Heritage and the Department of the Environment Transport and the Regions from the Marine Life Information Network (MarLIN). Marine Biological Association of the UK, Plymouth.

81 Hiscock, K. and H. Tyler-Walters (2006) Assessing the sensitivity of seabed species and biotopes – the Marine Life Information Network (MarLIN). *Hydrobiologia* 555(1), 309–320.

82 Simas T., M.A., R. Batty, B. Willson, J. Norris, M. Finn, D. Thompson, M. Lonergan, G. Veron, M. Paillard, and C. Abonnel (2010) *Equimar Deliverable 6.2.2. Scientific Guidelines on Environmental Assessment.* FP7 Project EQUIMAR Report

83 Evans, P., and Raga, J.A. (2001) *Marine Mammals: Biology and Conservation.* New York: Kluwer Academic/Plenum Publishers.

84 Sparling, C., *et al.* (2013) *Marine Mammal Impacts.* Wave and Tidal Consenting Position Paper Series. Natural Environment Research Council.

85 Macleod, K., *et al.* (2010) *Approaches to Marine Mammal Monitoring at Marine Renewable Energy Developments.* SMRU.

86 Evans, P.G.H., and P.S. Hammond (2004) Monitoring cetaceans in European waters. *Mammal Review* 34(1–2), 131–156.

87 Buckland, S.T., Anderson, D.R., Brurnham, K.P., Laake, J.L., Borchers, D.L. and Thomas, L. (2001) *Introduction to Distance Sampling. Estimating Abundance of Biological Populations.* Oxford: Oxford University Press.

88 Hammond, P.S., *et al.* (2002) Abundance of harbour porpoise and other cetaceans in the North Sea and adjacent waters. *Journal of Applied Ecology* 39(2), 361–376.

89 Gordon, J. and P.L. Tyack (2001) Sound and cetaceans, in P.G.H. Evans and J.A. Raga (eds), *Marine Mammals: Biology and Conservation.* New York: Kluwer Academic/Plenum Publishers

90 Leaper, R., Gillespie, D. and Papastavrou, V. (2000) Results of passive acoustic surveys for odontocetes in the Southern Ocean. *Journal of Cetacean Research and Management* 2(3), 187–196.

91 Thompson, D., Hall, A.J., Lonergan, M., McConnell, B. & Northridge, S. (2013) *Current Status of Knowledge of Effects of Offshore Renewable Energy Generation Devices on Marine Mammals and Research Requirements.*Sea Mammal Research Unit Report to Scottish Government.

92 Pesante, G., Evans, P.G.H., Baines, M.E. and McMath, M. (2008) *Abundance and Life History Parameters of Bottlenose Dolphin in Cardigan Bay: Monitoring 2005-2007.* CCW Marine Monitoring Report.

93 Edrén, S.M.C., *et al.* (2010) The effect of a large Danish offshore wind farm on harbor and gray seal haul-out behavior. *Marine Mammal Science* 26(3), 614–634.

94 Brasseur, S., Aarts, G., Meesters, E., van Polanen Petel, T., Dijkman, E., Cremer, J. & Reijnders, P. (2010) *Habitat Preferences of Harbour Seals in the Dutch Coastal Area: Analyses and Estimate of Effects of Offshore Wind Farms.* Wageningen IMARES.

95 McConnell, B., M. Lonergan, and R. Dietz (2012) *Interactions between Seals and Offshore Wind Farms.* Marine Estate Research Report. The Crown Estate.

96 Sparling, C., Grellier, K., Philpott, E., Macleod, K., and Wilson, J. (2011) *Guidance on Survey and Monitoring in Relation to Marine Renewables Deployments in Scotland. Volume 3. Seals.* Scottish Natural Heritage.

97 SMRU (2011) *Utilisation of Space by Grey and Harbour Seals in the Pentland Firth and Orkney Waters.* Scottish Natural Heritage Commissioned Report No. 441.

98 Camphuysen, K.C.J., Fox, A.D., Leopold, M.F., Petersen, I.K. (2004) *Towards Standardised Seabirds at Sea Census Techniques in Connection with Environmental Impact Assessments for Offshore Wind Farms in the UK. A Comparison of Ship and Aerial Sampling Methods for Marine Birds, and Their Applicability to Offshore Wind Farm Assessments.* Commissioned by COWRIE Ltd.

99 Maclean, I.M.D., Wright, L.J., Showler, D.A., Rehfisch, M.M. (2009) *A Review of Assessment Methodologies for Offshore Windfarms.* British Trust for Ornithology Report Commissioned by Cowrie Ltd.

100 Fox, A.D., *et al.* (2006) Information needs to support environmental impact assessment of the effects of European marine offshore wind farms on birds. *Ibis* 148(s1), 129–144.

101 Mellor, M., Maher, M. (2008) *Full Scale Trial of High Definition Video Survey for Offshore Windfarm Sites.* Report commissioned by COWRIE Ltd.

102 Underwood, A.J., and M.G. Chapman (2013) Design and analysis in benthic surveys in environmental sampling, in A. Eleftheriou (ed.), *Methods for the Study of Marine Benthos.* Chichester: John Wiley & Sons.

103 Saunders, G., Bedford, G.S., Trendall, J.R., and Sotheran, I. (2011) *Guidance on Survey and Monitoring in Relation to Marine Renewables Deployments in Scotland. Volume 5: Benthic Habitats.* Scottish Natural Heritage and Marine Scotland.

104 Carlier, A., Caisey, X., Jean-Dominique, G., Lejart, M., Derrien-Courtel, S., Catherine, E., Quimbert, E. and Soubigou, O. (2014) Monitoring benthic habitats and biodiversity at the tidal energy site of Paimpol-Breath (Brittany, France). *2nd International Conference on Environmental Interactions of Marine Renewable Energy Technologies*, Stornoway.

105 Smith, C.J. and H. Rumohr (2013) Imaging techniques, in A. Eleftheriou (ed.), *Methods for the Study of Marine Benthos.* Chichester: John Wiley & Sons.

106 Judd, A. (2012) *Guidelines for Data Acquisition to Support Marine Environmental Assessments for Offshore Renewable Energy Projects.* Centre for Environment, Fisheries & Aquaculture Science, Lowestoft.

107 Polunin N.V.C, H.J. Bloomfield, C.J. Sweeting, and D.T. McCandless (2009) *Developing Indicators of MPA Effectiveness: Finfish Abundance and Diversity in a Yorkshire Prohibited Trawl Area*. Marine and Fisheries Agency, London.

108 Zydlewski, G.B., Viehman, H., Staines, G., Shen, H. and McCleave, J. (2014) Fish interactions with marine renewable devices: Lessons learned, from ecological design to improving cost-effectiveness. *2nd International Conference on Environmental Interactions of Marine Renewable Energy Technologies*, Stornoway.

109 Wilhelmsson, D., T. Malm, and M.C. Öhman (2006) *The influence of offshore windpower on demersal fish*. ICES Journal of Marine Science: Journal du Conseil 63(5), 775–784.

110 Ingram, D., Smith, G., Bittencourt-Ferreira, C., and Smith, H. (2011) *Protocols for the Equitable Assessment of Marine Energy Converters*. Institute for Energy Systems, University of Edinburgh.

111 EMEC (2013) *Guidance for Developers at EMEC Grid-Connected Sites: Supporting Environmental Documentation.*, European Marine Energy Centre, Orkney.

112 BSI (2015) *Environmental Impact Assessment for Offshore Renewable Energy Projects – Guide*. British Standards Institution. London.

113 Marine Scotland (2012) *Survey, Deploy and Monitor Licensing Policy Guidance.*

114 Morris, P., and Therivel, R. (2005) *Methods of Environmental Impact Assessment*. London: Routledge.

115 Morgan, R.K. (1999) *Environmental Impact Assessment – A methodological procedure*. Dordrecht: Kluwer Academic Publishers.

116 Maclean, I.M.D., *et al.* (2014) Resolving issues with environmental impact assessment of marine renewable energy installations. *Frontiers in Marine Science*, 16 December.

117 Wood, G. (2008) Thresholds and criteria for evaluating and communicating impact significance in environmental statements: 'See no evil, hear no evil, speak no evil'? *Environmental Impact Assessment Review* 28(1), 22–38.

118 Brookes, A. (2009) Environmental risk assessment and risk management, in P. Morris and R. Therivel (eds), *Methods of Environmental Impact Assessment*. London: Routledge.

119 Isaacman, L., Daborn, G.R., Redden, A.M. (2012) *A Framework for Environmental Risk Assessment and Decision-Making for Tidal Energy Development in Canada*. Report to Fisheries and Oceans Canada and the Offshore Energy Research Association of Nova Scotia. Acadia Centre for Estuarine Research (ACER), Acadia University, Wolfville, NS.

120 Gormley, A.P., S., Rocks, S. (2011) *Guidelines for Environmental Risk Assessment and Management: Green Leaves III*. Report prepared by Defra and Cranfield University, UK.

121 Soerensen, H.C., and Russell, I. (2006) Life cycle assessment of the wave energy converter: Wave Dragon. *International Conference on Ocean Energy*, Bremerhaven.

122 Douglas, C.A., Harrison G.P., Chick, J.P. (2008) Life cycle assessment of the SeaGen marine current turbine. *Journal of Engineering for the Maritime Environment* 222(1), 1–12.

123 Miller, V.B., A.E. Landis, and L.A. Schaefer (2011) A benchmark for life cycle air emissions and life cycle impact assessment of hydrokinetic energy extraction using life cycle assessment. *Renewable Energy* 36(3), 1040–1046.

124 Latteman, S. (2010) *Development of an Environmental Impact Assessment and Decision Support System for Seawater Desalination Plants*. Leiden: CRC Press

125 Boehlert, G.W., and Gill, A.B. (2010) Environmental and ecological effects of ocean renewable energy development. *Oceanography* 23(2), 68–81.

126 Halcrow (2006) *Wave Hub Development Phase Coastal Processes Study Report*. Report produced for South West of England Regional Development Agency.

127 Wilding, T.A. (2014) Effects of man-made structures on sedimentary oxygenation: Extent, seasonality and implications for offshore renewables. *Marine Environmental Research* 97, 39–47.

128 Krone, R., *et al.* (2013) Mobile demersal megafauna at artificial structures in the German Bight – Likely effects of offshore wind farm development. *Estuarine, Coastal and Shelf Science* 125, 1–9.

129 Krone, R., *et al.* (2013) Epifauna dynamics at an offshore foundation – Implications of future wind power farming in the North Sea. *Marine Environmental Research* 85, 1–12.

130 Wilson, B.B., R. S., Daunt, F. and Carter, C. (2007) *Collision Risks between Marine Renewable Energy Devices and Mammals, Fish and Diving Birds*. Report to the Scottish Executive.

131 Larsen, J.K. and M. Guillemette (2007) Effects of wind turbines on flight behaviour of wintering common eiders: implications for habitat use and collision risk. *Journal of Applied Ecology* 44(3), 516–522.

132 Vanermen, N.B., R.; Stienen, E.; Courtens, W.; Onkelinx, T.; Van de walle, M.; Verstraete, H.; Vigin, L.; Degraer, S. (2013) Bird monitoring at the Belgian offshore wind farms: Results after five years of impact assessment, in S. Degraer, R. Brabant and B. Rumes (eds), *Environmental Impacts of Offshore Wind Farms in the Belgian part of the North Sea: Learning from the Past to Optimise Future Monitoring Programmes*. Royal Belgian Institute of Natural Sciences, Operational Directorate Natural Environment, Marine Ecology and Management Section, Brussels.

133 Bohnsack J.A., and D.L. Sutherland (1985) Artificial reef research: A review with recommendations for future priorities. *Bulletin of Marine Science* 37, 11–39.

134 Charbonnel, E., *et al.* (2002) Effects of increased habitat complexity on fish assemblages associated with large artificial reef units (French Mediterranean coast). *ICES Journal of Marine Science: Journal du Conseil* 59, S208–S213.

135 Langhamer, O. and D. Wilhelmsson (2009) Colonisation of fish and crabs of wave energy foundations and the effects of manufactured holes – A field experiment. *Marine Environmental Research* 68(4), 151–157.

136 Langhamer, O. (2012) Artificial reef effect in relation to offshore renewable energy conversion: State of the art. *Scientific World Journal*, 2012, 386713.

137 Reubens, J.T., *et al.* (2013) Aggregation at windmill artificial reefs: CPUE of Atlantic cod (*Gadus morhua*) and pouting (*Trisopterus luscus*) at different habitats in the Belgian part of the North Sea. *Fisheries Research* 139, 28–34.

138 Reubens, J.T., S. Degraer, and M. Vincx (2011) Aggregation and feeding behaviour of pouting (*Trisopterus luscus*) at wind turbines in the Belgian part of the North Sea. *Fisheries Research* 108(1), 223–227.

139 Öhman, M.C., P. Sigray, and H. Westerberg (2007) Offshore windmills and the effects of electromagnetic fields on fish. *AMBIO: A Journal of the Human Environment* 36(8), 630–633.

140 CMACS *et al.* (2003) *A Baseline Assessment of Electromagnetic Fields Generated by Offshore Windfarm Cables.* Commissioned by COWRIE Ltd.

141 Bradley, D.L., and Stern, R. (2008) *Underwater Sound and the Marine Mammal Acoustic Environment: A Guide to Fundamental Principles.* Prepared for the US Marine Mammal Commission.

142 Southall, B.L., *et al.* (2007) Marine mammal noise exposure criteria: Initial scientific recommendations (2007). *Aquatic Mammals* 33(4).

143 Haelters, J., L. Vigin, and S. Degraer (2012) The effect of pile driving on harbour porpoises in Belgian waters, in S. Degraer, R. Brabant and B. Rumes (eds), *Offshore Wind Farms in the Belgian Part of the North Sea: Heading for an Understanding of Environmental Impacts.* Royal Belgian Institute of Natural Sciences, Management Unit of the North Sea Mathematical Models, Marine Ecosystem Management Unit, Brussels.

144 Nedwell J.R., L.J., Howell D. (2004) *Assessment of Sub-sea Acoustic Noise and Vibration from Offshore Wind Turbines and its Impact on Marine Wildlife; Initial Measurements of Underwater Noise during Construction of Offshore Windfarms, and Comparison with Background Noise.* The Crown Estate, London.

145 Kingston, P.F. (2002) Long-term environmental impact of oil spills. *Spill Science & Technology Bulletin* 7(1–2), 53–61.

146 Walsh, G.E. (1978) Toxic effects of pollutants on plankton, in G.C. Butler (ed.), *Principles of Ecotoxicology.* New York: John Wiley & Sons.

147 Lowe, S., *et al.* (2004) *100 of the World's Worst Invasive Alien Species. A Selection from the Global Invasive Species Database.* Published by The Invasive Species Specialist Group (ISSG) a specialist group of the Species Survival Commission (SSC) of the World Conservation Union (IUCN).

148 Shiganova, T.A. (1998) Invasion of the Black Sea by the ctenophore *Mnemiopsis leidyi* and recent changes in pelagic community structure. *Fisheries Oceanography* 7(3–4), 305–310.

149 Probst, T.A. (2013) *Environmental Preferences, Impacts and Population Dynamics of the Invasive Screwshell Maoricolpus roseus.* University of Tasmania.

150 Grosholz, E.D., *et al.* (2000) The impacts of a nonindigenous marine predator in a California bay. *Ecology* 81(5), 1206–1224.

151 Champ, M.A. (2000) A review of organotin regulatory strategies, pending actions, related costs and benefits. *Science of the Total Environment* 258(1–2), 21–71.

152 Blyth, R.E., *et al.* (2004) Implications of a zoned fishery management system for marine benthic communities. *Journal of Applied Ecology* 41(5), 951–961.

153 van Dalfsen, J.A., *et al.* (2000) Differential response of macrozoobenthos to marine sand extraction in the North Sea and the Western Mediterranean. *ICES Journal of Marine Science: Journal du Conseil* 57(5), 1439–1445.

154 Drabsch, S.L., J.E. Tanner, and S.D. Connell (2001) Limited infaunal response to experimental trawling in previously untrawled areas. *ICES Journal of Marine Science: Journal du Conseil* 58(6), 1261–1271.

155 Harvey, M., D. Gauthier, and J. Munro (1998) Temporal changes in the composition and abundance of the macro-benthic invertebrate communities at dredged material disposal sites in the anse à Beaufils, baie des Chaleurs, eastern Canada. *Marine Pollution Bulletin* 36(1), 41–55.

156 Bradshaw, C., Veale, L.O., Hill, A.S. and Brand, A.R. (2000) The effects of scallop dredging on gravely seabed communities, in M.J. Kaiser and S.J. de Groot (eds),

The Effects of Fishing on Non-target Species and Habitats: Biological, Conservation and Socio-economic Issues. Oxford: Blackwell Science.

157 Kaiser, M.J., and B.E. Spencer (1996) The effects of beam-trawl disturbance on infaunal communities in different habitats. *Journal of Animal Ecology* 65(3), 348.

158 Mueller-Blenkle, C., McGregor, P., Gill, A., Andersson, M., Metcalfe, J., Bendall, V., Sigray, P., Wood, D. and Thomsen, F. (2010) *Effects of Pile-Driving Noise on the Behaviour of Marine Fish.* Lowestoft: Cefas on behalf of COWRIE Ltd.

159 Amaral, S., *et al.* (2011) *Evaluation of Fish Injury and Mortality Associated with Hydrokinetic Turbines.* Electric Power Research Institute, Palo Alto, CA.

160 Castro, J.J., J.A. Santiago, and A.T. Santana-Ortega (2002) A general theory on fish aggregation to floating objects: An alternative to the meeting point hypothesis. *Reviews in Fish Biology and Fisheries* 11(3), 255–277.

161 Løkkeborg, S., *et al.* (2002) Spatio-temporal variations in gillnet catch rates in the vicinity of North Sea oil platforms. *ICES Journal of Marine Science: Journal du Conseil* 59, S294–S299.

162 Baird, P.H. (1990) Concentrations of seabirds at oil-drilling rigs. *The Condor*, 92, 768–771.

163 Grecian, W.J., *et al.* (2010) Potential impacts of wave-powered marine renewable energy installations on marine birds. *Ibis* 152(4), 683–697.

164 Langton, R., I.M. Davies, and B.E. Scott (2011) Seabird conservation and tidal stream and wave power generation: Information needs for predicting and managing potential impacts. *Marine Policy* 35(5), 623–630.

165 Leeney, R.H., *et al.* (2014) Environmental Impact Assessments for wave energy developments: Learning from existing activities and informing future research priorities. *Ocean & Coastal Management*, 99, 14–22.

166 Lin, L., and Yu, H. (2002) Offshore wave energy generation devices: Impacts on ocean bio-environment. *Acta Ecologica Sinica* 32(3), 117–122.

167 Bonar, P.A., I.G. Bryden, and A.G. Borthwick (2015) Social and ecological impacts of marine energy development. *Renewable and Sustainable Energy Reviews* 47, 486–495.

168 Hastik, R., *et al.* (2015) Renewable energies and ecosystem service impacts. *Renewable and Sustainable Energy Reviews* 48, 608–623.

169 Leopold, L.B., *et al.* (1971) *A Procedure for Evaluating Environmental Impact.* Geological Survey Circular 645.

170 I. M. D. Maclean, *et al.* (2013) Evaluating the statistical power of detecting changes in the abundance of seabirds at sea. *Ibis* 155(1), 113–126.

171 Boehlert, G.W., G. R. McMurray, and C. E. Tortorici (2008) *Ecological Effects of Wave Energy in the Pacific Northwest.* US Department of Commerce.

172 Michel, J., Dunagan, H., Boring, C., Healy, E., Evans, W., Dean, J.M., McGillis, A. and Hain, J. (2007) *Worldwide Synthesis and Analysis of Existing Information Regarding Environmental Effects of Alternative Energy Uses on the Outer Continental Shelf.* Herndon, VA: US Department of the Interior, Minerals Management Service.

173 US Department of Energy (2009) *Report to Congress on the Potential Environmental Effects of Marine and Hydrokinetic Energy Technologies. Wind and Hydropower Technologies Program. Energy Efficiency and Renewable Energy.*

174 Andersen, K., *et al.* (2009) Environmental monitoring at the Maren wave power test site off the Island of Runde, western Norway: planning and design. *Proceedings of the 8th European Wave and Tidal Energy Conference*, Uppsala, Sweden.

175 Underwood, A.J. (1992) Beyond BACI: The detection of environmental impacts on populations in the real, but variable, world. *Journal of Experimental Marine Biology and Ecology* 161(2), 145–178.

176 Underwood, A.J. (1993) The mechanics of spatially replicated sampling programmes to detect environmental impacts in a variable world. *Australian Journal of Ecology* 18(1), 99–116.

177 Azzellino, A., *et al.* (2013) Marine renewable energies: Perspectives and implications for marine ecosystems. *Scientific World Journal* 2013, 547563.

178 Pickering, H. and D. Whitmarsh (1997) Artificial reefs and fisheries exploitation: A review of the 'attraction versus production' debate, the influence of design and its significance for policy. *Fisheries Research* 31(1), 39–59.

179 ABP Marine Environmental Research (2010) *Collision Risk of Fish with Wave and Tidal Devices.* Commissioned by RPS Group plc on behalf of the Welsh Assembly Government.

180 Williamson, B., and P. Blondel (2013) Multibeam imaging of the environment around marine renewable energy devices. *Proceedings of Meetings on Acoustics.* Acoustical Society of America.

181 Witt, M.J., *et al.* (2011) Assessing wave energy effects on biodiversity: The Wave Hub experience. *Philosophical Transactions of the Royal Society of London A* 370(1959), 502–529.

182 Tougaard, J., *et al.* (2009) Pile driving zone of responsiveness extends beyond 20 km for harbor porpoises (*Phocoena phocoena* (L.)). *Journal of the Acoustical Society of America* 126(1), 11–14.

183 Meike, S., *et al.* (2011) Harbour porpoises (*Phocoena phocoena*) and wind farms: A case study in the Dutch North Sea. *Environmental Research Letters* 6(2), 025102.

184 Tollit, D., Wood, J., Broome, J., Redden, A. (2011) *Detection of Marine Mammals and Effects Monitoring at the NSPI (OpenHydro) Turbine Site in the Minas Passage during 2010.* Acadia Centre for Estuarine Research, Acadia University, Wolfville, NS, Canada.

185 Keenan, G., *et al.* (2011) *SeaGen Environmental Monitoring Programme: Final Report.* Haskoning UK Ltd, Edinburgh.

186 Polagye, B., B. Van Cleve, A. Copping and K. Kirkendall (2011) Environmental effects of tidal energy development: Proceedings of a scientific workshop. *Tidal Energy Workshop.* Seattle, WA: US Department of Commerce (DOC).

187 Carstensen, J., O.D. Henriksen, and J. Teilmann (2006) Impacts of offshore wind farm construction on harbour porpoises: acoustic monitoring of echolocation activity using porpoise detectors (T-PODs). *Marine Ecology Progress Series* 321, 295–308.

188 Tougaard, J.T., and J. Teilmann (2006) *Rødsand 2 Offshore Wind Farm. Environmental Impact Assessment – Marine Mammals.* NERI Commissioned Report to DONG Energy.

189 Tougaard, J., P.T. Madsen, and M. Wahlberg (2008) Underwater noise from construction and operation of offshore wind farms. *Bioacoustics*, 17(1–3), 143–146.

190 Brandt, M.J., *et al.* (2011) Responses of harbour porpoises to pile driving at the Horns Rev II offshore wind farm in the Danish North Sea. *Marine Ecology Progress Series* 421, 205–216.

191 Joslin, J., B. Polagye, and S. Parker-Stetter (2012) Development of a stereo camera system for monitoring hydrokinetic turbines. *2012 Oceans.*

192 Wilson, B., S. Benjamins, and J. Elliott (2013) Using drifting passive echolocation loggers to study harbour porpoises in tidal-stream habitats. *Endangered Species Research* 22(2), 125–143.

193 Savidge, G., *et al.* (2014) Strangford Lough and the SeaGen tidal turbine, in *Marine Renewable Energy Technology and Environmental Interactions*. New York: Springer.

194 Furness, R.W., *et al.* (2012) Assessing the sensitivity of seabird populations to adverse effects from tidal stream turbines and wave energy devices. *ICES Journal of Marine Science: Journal du Conseil* 69(8), 1466–1479.

195 Murray, C., and Marmorek, D.R. (2004) Adaptive management: A science-based approach to managing ecosystems in the face of uncertainty. *Fifth International Conference on Science and Management of Protected Areas*. Victoria, BC: Science and Management of Protected Areas Association, Wolfville, Nova Scotia.

196 Williams, B.K., and E. D. Brown (2012) *Adaptive Management: The U.S. Department of the Interior Applications Guide*. Washington, DC: US Department of the Interior.

197 Oram, C., and C. Marriott (2010) Using adaptive management to resolve uncertainties for wave and tidal energy projects. *Oceanography* 23(2), 92–97.

10

Consenting and Legal Aspects

Anne Marie O'Hagan

Senior Research Fellow, MaREI Centre, Environmental Research Institute, University College Cork, Ringaskiddy, Ireland

10.1   Introduction

Wave and tidal energy (known collectively as ocean energy) developments represent a new use of the marine environment with great potential for growth. Internationally, and within the European Union, there are legal obligations on countries to address climate change by reducing greenhouse gas emissions and increasing the use of renewable energy. Such sources are, by definition, sustainable and infinite, hence they could have a major role to play in contributing to reductions in greenhouse gas emissions while also ensuring greater energy security. For a large proportion of countries across the globe, wave and tidal energy along with offshore wind (marine renewable energy, MRE) represent the only indigenous energy sources. Over time, development of these resources could result in greater energy independence as well as national and local economic development, particularly since the advancement of a promising industrial sector is likely to stimulate growth and innovation in the associated supply chain. The desire to develop a clean, secure and indigenous energy supply has instigated governments and other actors to publish dedicated marine renewable and/or ocean energy strategies mapping out the potential development path for the sector in their country. Such strategies represent an important policy context and can prompt further development and growth.

The regulatory system applicable to ocean energy derives from multiple sources. National, or domestic, legislation incorporates broader international obligations and, in the EU, also includes objectives all of which reflect the rights and duties of coastal states as recognised by international law. As an emerging industrial sector, wave and tidal energy projects currently tend to be subject to existing legislation and administrative procedures governing marine development, though in many countries this situation is evolving. Few countries have dedicated legislation or consenting procedures specifically designed for ocean energy development, though this is constantly evolving. It is usual for consenting to take account of aspects such as the occupation of sea space (seabed leasing), environmental impacts, terrestrial planning, grid/electrical connection and decommissioning [1]. The related requirements can vary according to location or, more

Wave and Tidal Energy, First Edition. Edited by Deborah Greaves and Gregorio Iglesias.
© 2018 John Wiley & Sons Ltd. Published 2018 by John Wiley & Sons Ltd.

correctly, to the maritime jurisdictional zone in which the project is to be located. Likewise, at different stages of project development, diverse regulatory requirements are applicable, replicating the planning, construction, operation and decommissioning realities of a project. Fundamentally, the legal aspects of a project involve identifying, managing, and mitigating risk. While each form of ocean energy may have a different origin, and involve different technologies for conversion, the regulatory considerations tend to be similar, with the exception of those relating to environmental impacts, given that different device types may potentially have different environmental effects.

Consenting is one of the most important, time-consuming and resource-intensive categories of legal considerations encountered by a project developer. It is also one of the most significant threats to the financial viability of a project because of the inherent 'regulatory risk' [2]. For these reasons it is essential that legal requirements are understood. In this chapter, relevant international, regional and EU legal instruments are described with a view to establishing whether current legal systems are sufficient to plan and manage ocean energy activities and any possible problems that may arise during the course of its growth. This chapter also presents an outline of the various aspects of consenting ocean energy projects namely occupation of sea space, terrestrial planning, environmental impacts, grid/electrical connection and decommissioning. An appropriate legal framework has the ability to facilitate the development of ocean energy, and examples of different consenting systems in a number of selected countries are summarised to demonstrate current practice. Given the development status of these technologies, ranging from research and development to the prototype stage and on to the pre-commercial stage, it is not possible to define what constitutes 'best practice' in terms of consenting. The chapter concludes with a section on the gaps and opportunities in the broader management framework, elucidating those elements of consenting and the legal system generally which could be adapted to enable development of this fledgling sector as well as forthcoming policy developments expected to change the way in which marine renewables may be governed and managed in future. It should be noted that the consenting and legal aspects of wave and tidal energy are under continuous and rapid development in line with the industry, although the information presented here is correct at the time of writing.

10.2 International Law

In contrast to development on land, where a state has sovereignty and is authorised to regulate all activities taking place on its territory, at sea the state's position is more restricted in accordance with the provisions of international law. The key legal framework for the oceans is the United Nations Law of the Sea Convention (LOSC). This prescribes the nature and extent of a coastal state's rights in offshore areas through definition of a number of maritime jurisdictional zones. Numerous other areas of international law are relevant to MRE including international development law, environmental law and civil aviation law.

10.2.1 United Nations Law of the Sea Convention

The LOSC is often heralded as a 'Constitution for the Oceans' as it comprehensively covers a multitude of marine issues. The LOSC organises marine space into a number of specific maritime jurisdictional zones, namely internal waters, territorial seas, the contiguous zone, the exclusive economic zone, archipelagic waters, the continental shelf, the high sea and the Area. Article 1 of the LOSC defines the 'Area' as the seabed and ocean floor and subsoil thereof, beyond the limits of national jurisdiction, so it is not considered further in this chapter. In principle, the LOSC prescribes the rights and obligations of coastal states and third states according to these jurisdictional zones. In this way the interests of individual states can be coordinated. For matters that cannot be governed solely through maritime jurisdictional zones the LOSC provides a legal framework for ensuring international cooperation on these issues, including straddling and highly migratory fish stocks, marine scientific research, marine pollution and protection of the environment generally.

Internal waters are those waters which lie landward of the baseline from which the territorial sea is measured [3]. Specifically, this means parts of the sea along the coast to the low water mark, waters enclosed by straight baselines, estuaries, ports and harbours. The seaward limit of internal waters is delimited by the baseline from which the territorial sea is measured. In internal waters, a coastal state exercises full sovereignty, meaning a state can make, apply and enforce its own laws, regulate any use and utilise any resource. This obviously extends to the use of the sea to produce energy. According to Article 3 of the LOSC, the territorial sea extends 12 nautical miles seaward from the baseline. Within this maritime jurisdictional zone, a coastal state also exercises territorial sovereignty in the water, the airspace over those waters and over the seabed and subsoil, but subject to the right of innocent passage which distinguishes the territorial sea from internal waters. The right to innocent passage reflects the freedom of navigation. The LOSC definition of 'passage' includes actual passage as well as stopping and anchoring in the territorial sea where this is necessary for reasons of distress or *force majeure* of that vessel or another vessel in those circumstances [4].

Passage is presumed to be innocent so long as it is 'not prejudicial to the peace, good order or security of the coastal State' [5]. The LOSC does, however, contain a list of activities that renders passage of a vessel non-innocent [6]. Obvious activities include practice with weapons and any threat or use of force against the sovereignty of the state, but non-innocent activities also incorporates fishing activities, research or survey activities and any other activity that does not have a direct bearing on passage. A coastal state is under an obligation not to hamper the innocent passage of foreign vessels through its territorial sea unless its passage is deemed non-innocent. Coastal states have the power to enact and enforce laws in the territorial sea relating generally to navigation, protection of cables and pipelines, fisheries, pollution, protection of the environment and prevention of pollution, marine scientific research and hydrographic surveys and the prevention of infringements of the customs, fiscal, immigration or sanitary laws of a coastal state [7].

To balance the right of innocent passage with the rights of a coastal state, the LOSC empowers a coastal state to designate sea lanes and traffic separation schemes in order to regulate the passage of ships [8]. When creating such lanes and schemes, a coastal state must take into account recommendations of the International Maritime Organization (IMO, the competent authority for such purposes), any channels routinely used for international navigation, the distinct features of particular ships and channels and the density of traffic [9]. Theoretically the construction and operation of ocean energy projects within the territorial sea has the potential to impinge upon sea lanes, traffic separation schemes and the right of innocent passage by ships from other jurisdictions. This has already happened in the English Channel as a result of safety concerns resulting from the construction and operation of the Wave Hub offshore testing facility in the waters off Cornwall. In that area, a recognised traffic separation scheme (TSS) operates off Land's End, between the UK mainland and the Isles of Scilly on the south-west coast of England. The UK regulatory agencies in association with the IMO amended the TSS, moving it 12 nautical miles to the north of the current boundaries and also amending the Inshore Traffic Zone to the east [10].

To date, all existing and planned wave and tidal energy projects are located in either internal waters or the territorial sea. Ongoing technological advances suggest that it is not beyond the realms of possibility that ocean energy projects could be constructed further offshore in future, once the technologies have reached technical and commercial maturity. This is confirmed by experiences in the offshore wind sector where there are already examples of planned and operational commercial scale projects in the Exclusive Economic Zone (EEZ) particularly off the coasts of Belgium, Denmark, Germany and the United Kingdom (England and Scotland). The EEZ refers to the sea space beyond and adjacent to the territorial sea extending 200 nautical miles from the baseline [11]. An EEZ must be claimed by a coastal state and it is subject to a dedicated legal regime established in Part V of the LOSC, where the rights and jurisdiction of the coastal state and the rights and freedoms of other states are prescribed. In the EEZ, a coastal state has sovereign rights for the purposes of 'exploring and exploiting, conserving and managing the natural resources, whether living or non-living, of the waters superjacent to the seabed and of the seabed and its subsoil, and with regard to other activities for the economic exploitation and exploration of the zone, such as *the production of energy from the water, currents and winds*' (emphasis added) [12].

It is clear from the preceding paragraph that coastal states have the right to produce energy from water, currents and wind in the EEZ. This right was included within the LOSC so as to enable coastal states to take advantage of future developments in technology. Realisation of this could involve construction of 'installations and structures', and, in relation to these, coastal states have jurisdiction (not sovereign rights) only, as provided for in Article 60 of the LOSC. This article permits coastal states to construct, authorise and regulate the construction, operation and use of installations and structures associated with exploring, exploiting, conserving and managing natural resources in the EEZ. Coastal states are required to provide a permanent means of warning of the presence of such installations and structures [13]. This provision also includes a requirement to remove any installations or structures that are abandoned or disused so as to safeguard navigation, taking into account any generally accepted international standards (discussed further in Section 10.5.1.4).

Coastal states may create 'reasonable safety zones' around installations and structures in their EEZ in order to guarantee safety of navigation and of the structures [14]. Safety zones must not exceed a distance of 500 metres, measured from each point of the structure's outer edge, unless generally accepted international standards permit otherwise [15]. The international body responsible for the creation of such standards is also the IMO. It has not adopted standards for safety zones around wave and tidal energy devices, though in 2008 both the United States of America and Brazil, in a joint proposal, advocated the drafting of comprehensive guidelines for safety zones around artificial islands, installations and structures larger than 500 metres in EEZs, to include offshore wind farms [16]. This was discussed in the IMO Sub-Committee on Safety of Navigation in 2010 where the USA stated that after more comprehensive consideration, it believed there was 'no demonstrated need, at present, for safety zones larger than 500 metres or the development of guidelines for such safety zones' [17]. The USA recommended that the Sub-Committee should focus the IMO's existing guidance on safety zones and on the available measures that, either individually or in combination with others, had demonstrated their effectiveness in providing for the safety both of navigation and other installations or structures in the EEZ. Other delegates maintained the opinion that the need for extension of safety zones beyond 500 metres might be necessary in the future, due to the unique nature of offshore installations, wind farms, aquaculture sites and energy exploitation activities. The matter was passed to the Ships' Routeing Working Group, which agreed with the latest US position, stating that there was at present no demonstrated need for larger safety zones and that continued consideration of this topic beyond 2010 was not necessary [18].

Article 76 of the LOSC defines the continental shelf of a coastal state as the area beyond the territorial sea comprising the seabed and subsoil of the submarine areas that extend beyond the territorial sea to a distance of 200 nautical miles from the baseline. In this maritime zone a coastal state exercises sovereign rights for the purpose of exploring the continental shelf and exploiting its natural resources [19]. 'Natural resources' in this context are mineral and other non-living resources of the seabed and subsoil together with living organisms belonging to sedentary species [20]. This definition would suggest that renewable energy could not be considered a 'natural resource' of the sea, though this has not yet been debated at UN level. Interestingly, all states are entitled to lay submarine cables and pipelines on the continental shelf, in accordance with the provisions of Article 79, which provides that any such cable-laying activity is subject to the consent of the coastal state and any conditions it deems necessary. Similarly, Article 81 provides that coastal states have the exclusive right to authorise and regulate drilling on the continental shelf for all purposes. Article 80 states that the provisions of Article 60 relating to artificial islands, installations and structures in the EEZ also apply *mutatis mutandis* to artificial islands, installations and structures on the continental shelf.

Beyond the 200-mile EEZ, are the high seas, sometimes referred to as 'areas beyond national jurisdiction'. According to Article 86, the term 'high seas' refers to all parts of the sea that are not included in the EEZ, the territorial sea or the internal waters of a state, or in the archipelagic waters of an archipelagic state. The high seas are open to all states, and freedoms of navigation, overflight, and the freedom to lay submarine cables and pipelines, subject to Part VI of LOSC, exist. There is also the freedom to construct artificial islands and other installations permitted under international law, freedom of fishing and scientific research, subject to legal requirements specified elsewhere in the LOSC. Where an EEZ has not been claimed by a state, the high seas regime is applicable and this could have important implications for the future development of wave and tidal technologies in this area. As it is not economically viable to develop ocean energy in the high seas at this time, it is not discussed further in this chapter. It should be noted, however, that in June 2015 the United Nations General Assembly agreed to develop a new, international, legally binding instrument under the LOSC on the conservation and sustainable use of marine biological diversity of areas beyond national jurisdiction [21]. A preparatory committee is progressing this work, including the role of area-based management tools and environmental impact assessment (EIA) [22].

Alongside the provisions relating to maritime jurisdictional zones, of fundamental importance when considering development in these areas is the general obligation of states to protect and preserve the marine environment under the LOSC [23]. This general obligation includes a requirement to take measures to prevent, reduce and control pollution of the marine environment from any source [24] and from the use of technologies under states' jurisdiction or control [25]. Coastal states are also required to monitor the risks or effects of pollution [26] and assess the potential effects of activities under their control which may cause substantial pollution of, or significant and harmful changes to, the marine environment [27]. Generally, these aspects are included in national consenting processes through mechanisms such as EIAs (see Section 10.4.4). The LOSC also covers marine scientific research in Part XIII, and the development and transfer of marine technology in Part XIV, which are also relevant to the development and exploitation of ocean energy. Following a Report of the Secretary-General on Oceans and Law of the Sea on marine renewables in April 2012 [28], the UN Open-ended Informal Consultative Process on Oceans and the Law of the Sea focused on MRE during its thirteenth meeting, in New York from 29 May to 1 June 2012 [29]. The latter reiterated that the LOSC was the appropriate legal framework for all activities in the oceans and seas, including MRE comprising offshore wind, wave and tidal. During that session a number of delegations highlighted the need to focus on the potential adverse environmental, social and cultural impacts of MRE [30]. Participants also advocated the need for a structured process to allocate ocean space for future development of MRE technologies and the role of marine spatial planning in managing competing demands for space and reducing conflicts [31].

10.2.2 United Nations Framework Convention on Climate Change

In 1992, countries joined the United Nations Framework Convention on Climate Change (UNFCCC) to work cooperatively on limiting average global temperature increases, and the resulting climate change, as well as what could be done to address the impacts of such effects. By 1995, countries realised that the provisions in the UNFCCC to deal with reductions of emissions were largely inadequate and consequently, two years later, the Kyoto Protocol was adopted. This places a legally binding obligation on developed countries to meet emission reduction targets. The Protocol's current commitment period runs to 2020. Article 4(b) of the Convention obliges contracting parties to formulate, implement, publish and regularly update national and, where appropriate, regional programmes containing measures to mitigate climate change by addressing anthropogenic emissions. While the Convention does not explicitly mention marine renewable energy or ocean energy, it is a key legal driver in the implementation of renewable energy, given its objective to reduce greenhouse gases. Parties are advised to promote and cooperate in the development, application and diffusion, including transfer, of technologies, practices and processes that control, reduce or prevent greenhouse gases in all relevant sectors, including energy [32].

Article 4 also contains a provision that developed countries, which are party to the Convention, take all practicable steps to promote, facilitate and finance, as appropriate, the transfer of, or access to, environmentally sound technologies and knowhow to other parties, particularly developing country parties, to enable them to implement the provisions of the Convention [33]. The actions necessary to achieve this objective are taken forward through the Technology Mechanism, agreed as part of the Cancún Agreement [34]. This decision recommends that it is necessary to accelerate action at different stages of the technology cycle, including research and development, demonstration, deployment, diffusion and transfer of technology in support of action on mitigation and adaptation [35]. Under the Kyoto Protocol a Clean Development Mechanism (CDM) was introduced to enable emissions reduction projects that generate certified emission reduction units to be traded in emissions trading schemes [36]. The overall purpose of the CDM is to promote clean development in developing countries. Essentially it can assist Annex I countries (developed countries) to meet part of their emission reduction commitments by buying certified emission reduction units from CDM emission reduction projects in developing countries. In an EU context, the objectives of the UNFCCC are taken forward through the various climate and energy policy packages as well as specific legislation such as the Renewable Energy Directive (2009/29/EC), discussed in Section 10.4.1.

10.2.3 United Nations Convention on Biological Diversity

The United Nations Convention on Biological Diversity (CBD) was opened for signature in 1992 with the UNFCCC. The CBD has three main goals: the conservation of biological diversity (or biodiversity); the sustainable use of its components; and the fair and equitable sharing of benefits arising from genetic resources. The CBD has a specific programme on marine and coastal biodiversity, which was strengthened significantly by the Jakarta Mandate in 1995. The programme identifies key operational objectives and priority activities within the five key programme elements, namely: implementation of integrated marine and coastal area management; marine and coastal living resources; marine and coastal protected areas; mariculture; and alien species [37]. The ecosystem approach, the precautionary principle, and the involvement of local and indigenous communities were identified by the parties as the basic principles for the implementation of the programme of work [38]. These are progressed through EU and national legislation worldwide.

In a similar fashion to the UNFCCC, the CBD does not mention marine renewable energy or ocean energy *per se*, but the CBD is fundamental to explaining why developers and regulators must consider the environmental impacts of development on the receiving environment. The CBD reiterates that under international law states have responsibilities to ensure that activities within their jurisdiction or control do not cause damage to the environment of other states or of areas beyond the limits of national jurisdiction. Article 14 of the CBD asserts that contracting parties must introduce appropriate procedures requiring environmental impact assessment of its proposed projects that are likely to have significant adverse effects on biological diversity with a view to avoiding or reducing such effects and allow for public participation in such procedures. This is advanced in the EU through both the Environmental Impact Assessment and Strategical Environmental Assessment Directives (Section 10.4.4) and Public Participation Directives (Section 10.4.5).

During implementation of the CBD and its provisions, other thematic areas have been the subject of new attention. Emerging issues identified by the Subsidiary Body on Scientific, Technical and Technological Advice (SBSTTA) include, for example, the impact of ocean noise on marine protected areas and to consider the scientific information on underwater noise and its impacts on marine and coastal biodiversity and habitats [39]. The Convention Secretariat has also considered marine spatial planning in the context of the CBD, given it is an approach that supports ecosystem-based management and has the potential to improve and enforce existing management frameworks, reduce the loss of ecosystem services, help address or avoid conflict on use of marine space, and establish conditions promoting economies of scale [40].

10.2.4 Other Sources of International Law Relevant to Ocean Energy

Consenting of ocean energy development is not neatly covered by a single legal instrument in any country, given that it encompasses many aspects of marine management and planning. Its regulation is therefore distributed across multiple legal instruments. Table 10.1 presents a non-exhaustive list of other relevant international legal instruments that may, implicitly or explicitly, impact upon the planning, development and management of ocean energy device deployments. As international law contains high-level objectives, the specifics of implementation are often detailed in domestic legislation or EU legislation, where applicable.

10.3 Regional Law

Regional law, in the context of consenting for renewable energy, refers to the Regional Seas Conventions as an intrinsic part of the UN Regional Seas Programme established in 1974. This programme was designed to enable neighbouring countries to work collaboratively to address issues of mutual concern and specific to those issues causing degradation of marine and coastal environments. The Regional Seas programmes are usually underpinned by a strong legal basis, coordinated and implemented by countries sharing a specific body of water. The issues addressed by the Regional Seas programmes are specific to the geographic water body concerned. Not all Regional Seas programmes consider energy, or indeed ocean energy. Accordingly, only the OSPAR Convention on the Protection of the Marine Environment of the North-East Atlantic is considered in this section.

The OSPAR Convention has its genesis in two other conventions, namely the Oslo Convention against dumping and the Paris Convention on pollution from land-based sources and the offshore industry. Both these were amalgamated to create the 1992 OSPAR Convention, which entered into force in 1998 [41]. OSPAR has been signed and ratified by the 15 governments of Belgium, Denmark, Finland, France, Germany, Iceland, Ireland, Luxembourg, the Netherlands, Norway, Portugal, Spain, Sweden, Switzerland and the United Kingdom. The EU is also a party to the Convention [42], and accordingly many OSPAR commitments are taken forward through EU law. OSPAR is implemented through strategies, decisions and recommendations that are binding on contracting parties. The original focus of OSPAR was the prevention and reduction of all sources of pollution in the north-east Atlantic, but this was expanded to include all human activities that might adversely affect the marine environment of the north-east Atlantic through the adoption of Annex V in 1998 [43]. Through its North-East Atlantic Environment Strategy [44], OSPAR works on the implementation of the ecosystem approach and five thematic strategies including biodiversity, offshore industry and hazardous substances. Legal drivers to reduce greenhouse gases and increase renewable energy have resulted in the OSPAR Commission adopting a new focus on offshore wind as this activity is increasing in the North Sea and Celtic Seas, both parts of the OSPAR area.

Table 10.1 International legal instruments with possible implications for Ocean Energy developments.

Instrument	Topic(s) with implication(s) for MRE	Example references
Convention on International Civil Aviation	Offshore wind turbine heights, location and lighting	International Civil Aviation Organisation (ICAO) Annex 14 standards, chapter 6.
	Implications for radar and aerial navigation	ICAO. (2009). European Guidance Material on Managing Building Restricted Areas, Technical Report, ICAO Eur. Doc. 015.
Convention on Wetlands (Ramsar)	Designation of wetlands of international importance	Article 2(4) of the Ramsar Convention.
	Guidance on how to consider wetlands in planning and operating energy infrastructure	Resolution XI.10 Wetlands and energy issues, 11th Meeting of the Conference of the Parties, Bucharest, Romania, 6–13 July 2012. Guidance for addressing the implications for wetlands of policies, plans and activities in the energy sector (2012).
Convention on the Conservation of Migratory Species of Wild Animals (Bonn)	Strict protection of species facing extinction involving the conservation or restoration of the places where they live, mitigation of obstacles to migration and control of other factors that might endanger them	Annex I
	Implications of renewable energy for migratory species	CMS Draft Resolution: Renewable Energy and Migratory Species (CMS and ASCOBANS), 8th Sept. 2014.
IMO Protocols and Resolutions	Ships routeing	General Provisions on Ships' Routeing, IMO Resolution A.572(14), 20 November 1985;
	Safety of navigation around offshore installations and structures	Safety Zones and Safety of Navigation Around Offshore Installations and Structures, IMO Resolution A.671(16), 19 October 1989;
	Decommissioning of offshore structures	Guidelines and Standards for the Removal of Offshore Installations and Structures on the Continental Shelf and in the Exclusive Economic Zone, IMO Resolution A.672(16), 19 October 1989.
International Convention for the Prevention of Pollution from Ships, 1973 (MARPOL), as amended	Prevention of pollution	Article 1. Generally applicable to 'ships' servicing energy installations.
Safety of Life at Sea Convention (SOLAS) 1974	Ships routeing	Chapter V, Safety of Navigation
	Exclusion zones	
Espoo Convention on Transboundary EIA 1991	Obligation on States to notify and consult each other on all major projects under consideration that are likely to have a significant adverse environmental impact across boundaries.	All Articles.

In 2004, the OSPAR Commission published a background document on potential advantages and disadvantages associated with the development of offshore wind farms [45]. A review of the state of knowledge on the environmental impacts associated with offshore wind farms [46] culminated in the publication of a dedicated guidance document on environmental considerations for offshore wind farm development [47]. The OSPAR Commission works on impacts relating to offshore wind farms, the placement of structures (other than oil and gas and wind farms), cables and pipelines, and artificial reefs, each of which are of potential significance to ocean energy deployments. OSPAR assesses these activities and, if necessary, develops programmes and measures to control them and to restore adversely affected marine areas. In 2010 the OSPAR Commission asked the International Council for the Exploration of the Sea (ICES) to provide it with advice on the environmental interactions of wave and tidal energy generation devices with respect to actual and potential adverse effects on specific species, communities and habitats; on specific ecological processes; and on the irreversibility or durability of these effects [48]. One of the objectives of the Biological Diversity and Ecosystems Strategy is to to keep the introduction of energy, including underwater noise, at levels that do not adversely affect the marine environment in the OSPAR maritime area [49]. This is supplemented by the European Commission through the Marine Strategy Framework Directive (Section 10.4.6).

The 2008 guidance on environmental considerations for offshore wind farm development includes decommissioning aspects [47]. This outlines the key international requirements covering decommissioning of structures generally but, in line with OSPAR's policy on waste disposal at sea, states that the removed components of a wind farm should generally be disposed of entirely on land. If a competent national authority decides that a component of the wind farm should remain on site, such as piles in the seabed or scour protection materials, the authority must ensure that such parts have no adverse impact on the environment, the safety of navigation and other uses of the sea [50]. The status of remaining parts should be monitored and, if necessary, appropriate measures should be taken. The guidance recognises that as offshore wind farms are still in the early stages of operation and are expected to operate for some decades, it is difficult to be certain about environmental impacts of their operation, but it is assumed that decommissioning will have effects similar to operation. The guidance recognises that removal techniques will evolve in coming years and that the offshore wind industry will learn from ongoing experiences with oil and gas installations [51].

10.4   EU Law and Policy

The EU has mandatory international obligations to meet both in terms of greenhouse gas reductions and conservation of marine and coastal biodiversity. To assist in achieving such obligations the EU has a vast array of legislation and policy instruments that will have some level of application to current and future development of ocean energy in European waters. It should be noted that the EU and its institutions have either exclusive, shared or supporting competences in relation to many areas. With respect to energy, for example, after the entry into force of the Treaty of Lisbon a specific legal basis for the field of energy was introduced with the creation of Article 194 of the Treaty on the Functioning of the EU. In practice this means that the EU is authorised to take measures at European level to ensure the functioning of the energy market and security of energy supply as well as promote energy efficiency and the interconnection of energy networks. Since the adoption of the Lisbon Treaty therefore, energy is one of the shared competences between the EU and member states and is subject to the principle of subsidiarity. In essence this means that the EU may only intervene if it is capable of acting more effectively than a member state itself.

The environment has been a topic of concern to the EU and its institutions since the 1970s. Historically the European Community was often criticised for putting economic development before environmental concerns. The introduction of the Single European Act in 1987 addressed these concerns in part by giving Community measures an explicit legal basis defining the objectives and guiding principles for action by the European Community relating to the environment. Further progress was made after the adoption of the EU Treaty in November 1993 through the addition of the concept of 'sustainable growth respecting the environment' to tasks by the Community on this theme. This included the introduction of the precautionary principle into the article on which environment policy is founded (Article 174) and gave actions on the environmental the status of a 'policy' in its own right except for matters such as town and country planning and land use, which remain an exclusive competence of individual member states. The latter partly explains why there cannot be an EU consenting system for marine or ocean energy. It is a member state competency, and consenting practices must incorporate overarching EU policy and legal objectives such as environmental assessment, public participation and conservation of biodiversity and water quality, to name but a few. Article 6 of the Treaty of Amsterdam resulted in the need for environmental protection requirements to be integrated into the definition and implementation of other policies as well as introducing the principle of integration so as to promote sustainable development.

The following subsections provide a succinct overview of key policies and legislative instruments that influence the national requirements placed on developers when undertaking a project in the marine environment.

10.4.1   EU energy Law and Policy

In 2010 the European Union launched its 2020 Energy Strategy (COM(2010)639) which sought to reduce its greenhouse gas emissions by at least 20%, increase the share of renewable energy to at least 20% of consumption, and achieve energy savings of 20% or more [52]. All EU countries were also required to achieve a 10% share of renewable energy in their transport sector. This was progressed through the adoption of specific legislation, namely the Renewable Energy Directive (2009/28/EC) which set binding national targets for increasing the share of renewable energy in the energy consumption of each member state. The Directive required member states to publish national renewable energy action plans (NREAPs) detailing how those targets are to be achieved. The plans include sectoral targets, the technology mix expected, the trajectory to be followed and any reforms necessary to overcome the barriers to developing renewable energy [53].

Fifteen of the 23 coastal EU member states specified national targets for offshore wind in their NREAPs to be reached by 2020. In contrast, only seven member states identified targets for wave and tidal energy, which were presented as a composite figure rather than a technology-specific target. The contributions expected from both offshore wind and ocean energy for the years 2015–2020 are presented in Table 10.2 [54]. While the overall renewable energy targets are legally binding, the path or mix that a member state

Table 10.2 Estimation of total contribution (installed capacity, gross electricity generation) expected from ocean (wave and tidal) and offshore wind to meet the binding 2020 targets and the indicative interim trajectory for the shares of energy from renewable resources in electricity for 2015–2020.

Member state	Source	2015 (MW)	2016 (MW)	2017 (MW)	2018 (MW)	2019 (MW)	2020 (MW)
Belgium	Tide, wave, ocean	n/a	n/a	n/a	n/a	n/a	n/a
	Offshore wind	n/a	n/a	n/a	n/a	n/a	n/a
Denmark	Tide, wave, ocean	0	0	0	0	0	0
	Offshore wind	1251	1277	1302	1328	1353	1339
France	Ocean current, wave, tidal	302	318	333	349	364	380
	Offshore wind	2667	3333	4000	4667	5333	6000
Germany	Tide, wave, ocean	0	0	0	0	0	0
	Offshore wind	3000	4100	5340	6722	8272	10000
Greece	Tide, wave, ocean	0	0	0	0	0	0
	Offshore wind	0	50	100	150	200	300
Ireland[*]	Tide, wave, ocean (a)	0	0	13	25	38	75
	(b)	0	0	125	225	352	500
	Offshore wind (a)	252	252	416	529	533	555
	(b)	539	827	1352	1802	2096	2408
Italy	Tide, wave, ocean	0	1	1	1	2	3
	Offshore wind	168	220	290	385	512	680
Portugal	Tides, waves, oceans	60	75	100	125	175	250
	Offshore wind	25	25	25	25	25	75
Spain	Tide, wave, ocean	0	10	30	50	75	100
	Offshore wind	150	500	1000	1500	2250	3000
Sweden	Tide, wave and ocean	0	0	0	0	0	0
	Offshore wind	129	140	150	161	171	182
The Netherlands	Tide, wave, ocean	0	0	0	0	0	0
	Offshore wind	1178	1978	2778	3578	4378	5178
United Kingdom	Tide, wave and ocean	0	200	400	700	1000	1300
	Offshore wind	5500	6810	8310	9800	11300	12990

Source: member state NREAPs
* Both a modelled and a non-modelled scenario are presented in the NREAP for Ireland. Figures (a) relate to the modelled scenario and (b) are from the non-modelled scenario. The non-modelled scenario is an 'export' scenario illustrating Ireland's potential to become a net exporter if the appropriate conditions (economic, technical and environmental) existed. It could be considered as aspirational.

selects to meet these targets is of its own choosing and not binding. The fact that some states have chosen to achieve their targets through the inclusion of ocean energy, however, could indicate a desire – and, more importantly, act as a driver at the national level – to progress ocean energy development.

Under the provisions of the Renewable Energy Directive, the European Commission is obliged to publish a progress report on implementation of the Directive and progress made every two years. The 2015 progress report finds that the EU is on track to meet its 20% renewable energy targets, with 25 member states expected to meet their 2013–2014 national targets [55]. This can be contrasted with the 2013 progress report which confirmed that there was a strong initial start in renewable energy growth under the Directive but concluded that, in light of the economic recession, delayed investment and ongoing administrative barriers, further efforts were needed to achieve the 2020 targets [56]. Specifically, the 2013 report found that progress in removing administrative barriers was 'still limited and slow', highlighting that sub-optimal planning and consenting processes raise the overall costs of renewable energy development [57]. The progress report for 2015, based on data submitted for 2013, found that only France and the Netherlands failed to meet their 2011–2012 targets, which is attributable to 'long procedures for permit granting (especially in the wind sector) coupled with technical barriers for wind and biomass in France' [58]. While the 2015 progress report recognises that the gap between planned and actually expected deployment rates is the highest for offshore wind and marine/ocean technologies, it is also stated that, from an analysis of member state progress reports for 2013, a 'large number of simplification measures' have been reported [59]. This includes the adoption of one-stop-shop approaches (Netherlands), merging of permits (Belgium), online information platforms (Portugal, Italy, Sweden, Ireland) and new time limits (UK) for decisions on renewable energy generally (not marine-specific). The 2017 progress report, based on data for 2015, found that the majority of member states are on track to reach their 2020 binding targets, but in relation to administrative procedures there was limited progress in relation to implementation of a one-stop-shop approach though many countries have introduced time limits for permitting [60].

The European Union, its institutions and member states have since agreed a new climate and energy package. The European Commission's communication on a policy framework for climate and energy, commonly referred to as the 2030 package (COM/2014/015 final), sets EU-wide targets and policy objectives for the period between 2020 and 2030 [61]. This includes a 40% cut in greenhouse gas emissions compared to 1990 levels and a new renewable energy target of at least 27% of final energy consumption in the EU by 2030 which is binding on the EU but not binding on the member states individually. This is a different approach than that which applies currently whereby the EU target was translated into national targets via the Renewable Energy Directive. The rationale for a change in approach is to allow member states to have the flexibility necessary to determine a low-carbon transition that reflects their specific conditions and needs, such as their capacity to produce renewable energy and their existing energy mix [61]. National approaches to achieving the new target will be incorporated into dedicated national plans that will set out the path to be followed to achieve the target with the principal aims of creating investor certainty, enhance coherence and progress towards meeting the objectives of the internal energy market and state-aid guidelines [61].

At supranational level, regional cooperation is also highlighted as having the potential to assist member states address common challenges and meet common objectives in a more cost-effective manner while simultaneously progressing market integration and averting market distortion [61]. Consultation with neighbouring member states is emphasised as a central component when preparing national plans. The Communication reiterates the need to provide regulatory certainty as early as possible for investors in low-carbon technologies in order to incentivise research, development and innovation in both new technologies and their associated supply chains. Reaction to the 2030 package proposed by the Commission has been mixed: the EU Climate Commissioner, Connie Hedegaard, has declared it ambitious and the European Council has broadly welcomed the package [62]. However, non-governmental organisations have criticised the package as falling far short of what the EU needs to do to fight against climate change [63], with Friends of the Earth, for example, calling for a higher target of 45% for the share of renewables in the European energy mix [64].

10.4.2 Integrated Maritime Policy and Blue Growth

The European Commission launched its first dedicated Integrated Maritime Policy (IMP) in 2007 [65]. This recognises that all matters relating to Europe's oceans and seas are interlinked and that sea-related policies must develop in a joined-up way if the desired economic results are to be achieved [65]. The IMP advocates an integrated maritime governance framework so that an integrated approach to planning and management can be applied at every level. Member states were invited to develop integrated maritime policies for their combined areas. To support efforts at integrated governance the IMP highlighted three key tools to support this action. These are maritime surveillance, to facilitate safe use of marine space; maritime spatial planning, to enable integrated planning; and access to data and information, viewed as critical to good decision-making. Since adoption of the IMP, member states have worked collaboratively to produce a number of sea-basin-specific strategies that are tailored to the specific needs of those areas. Currently there are sea-basin strategies for the Baltic Sea, Black Sea, Mediterranean Sea, Adriatic and Ionian Seas, North Sea, the Atlantic and the Arctic Ocean. The Action Plan for a Marine Strategy in the Atlantic Area (COM(2013) 279) has a number of objectives including the exploitation of renewable energy in the Atlantic area's marine and coastal environment [66].

In 2012, the Commission launched its Blue Growth (COM(2012) 494) agenda, which is a long-term strategy to support sustainable growth in the marine and maritime sectors and the maritime contribution to the Europe 2020 strategy for smart, sustainable and inclusive growth [67]. Blue Growth has five focus areas: aquaculture, coastal tourism, marine biotechnology, ocean energy and seabed mining. With respect to ocean energy, the EU is concentrating on this topic as it realises the potential future contribution the sector could make to reducing energy dependence, increasing energy security, reducing greenhouse gas emissions and creating jobs in peripheral coastal areas. As part of this work, in January 2014 the Commission published a specific communication on the action needed to deliver on the potential of ocean energy in European seas and oceans by 2020 and beyond (COM(2014) 8) [68]. This outlines the current status of the sector, existing support mechanisms and the challenges remaining for the sector. Technology costs, transmission grid and other infrastructure, revenue and grant support along with environmental impacts and consenting processes were identified as the key challenges to ocean energy development.

The Ocean Energy Communication concludes with a two-stage action plan to help the sector reach its full potential and build on research to date. The first phase of the plan (2014–2016) saw the creation of a dedicated Ocean Energy Forum, consisting of stakeholders from industry, academia and regulators, to formulate recommendations and possible solutions to the issues identified for the sector. The Forum is organised around three work-streams, including one on environment and consenting. This work-stream will focus on how to navigate processes relating to environmental assessment and requirements deriving from nature conservation legislation. The work of the Forum will culminate in a strategic roadmap for the ocean energy sector that will detail specific actions and timelines for their implementation [68]. The second phase (2017–2020) of the action plan aims to create a European industrial initiative on ocean energy. European industrial initiatives are public–private partnerships that bring together industry, researchers, member states and the Commission to set out and achieve clear and shared objectives over a specific time-frame and have proved successful in the past for wind energy. Sector specific guidance for the implementation of legislation, such as the Birds and Habitats Directives, will also separately be progressed during this phase [68].

10.4.3 Nature Conservation Legislation

Nature conservation legislation comprises two key directives, the Birds and Habitats Directives, which are fundamental to the protection of biodiversity in the EU. The Birds Directive (codified in Directive 2009/147/EC) provides protection for all of Europe's wild bird species, identifying 194 species and sub-species as particularly threatened and in need of special conservation measures. This is achieved through the designation of scientifically identified areas which are critical for the survival of the targeted species and are known as Special Protection Areas (SPA). The Directive bans activities that directly threaten birds and also creates rules relating to hunting. The Directive recognises that habitat loss and degradation are the most serious threats to the conservation of wild birds and consequently habitats of endangered birds and migratory species, listed in Annex I, are also protected through the provisions of the Directive. Many marine, coastal and wetland areas are critical to the survival of bird species, which is why an assessment of the species present at, or near, a project site is common in consenting processes across the EU.

The Habitats Directive follows a similar format to the Birds Directive. It seeks to deliver on the EU's commitments under the UN Convention on Biological Diversity. The main aim of the Directive is to promote the maintenance of biodiversity, taking account of economic, social, cultural and regional requirements. It covers a wide range of rare, threatened or endemic species, including around 450 animals and 500 plants. Approximately 230 rare and characteristic habitat types are also protected in their own right. The Directive provides for the designation of core sites for the protection of species and habitat types listed in Annexes I and II of the Directive. These are known as Special Areas of Conservation (SAC) and, together with SPA, form the Natura 2000 network of protected sites across Europe. Natura 2000 sites are selected on scientific grounds so as to ensure that the best areas in the EU are protected for the species and habitats of EU importance. Cumulatively, the overall objective of both directives is to ensure that the species and habitat types they protect are maintained or restored to a favourable conservation status throughout their natural range within the EU.

The site designation process involves both competent authorities at member state level and the European Commission. Firstly, the member state will send a list of sites for consideration to the Commission. The Commission will then select, with input from the European Environment Agency (EEA), member state officials and scientific experts, what are termed 'Sites of Community Importance' (SCIs). Once selected as an SCI, the site is part of the Natura 2000 network. Member states then have six years to designate them as SACs, and to introduce any additional management measures necessary to maintain or restore the species or habitats therein to favourable conservation status. In order to maintain this status, damaging activities that could significantly disturb the species or deteriorate the habitats for which the site is designated must be avoided in all Natura 2000 sites. Each designated Natura 2000 site has specific conservation objectives and measures set by authorities so as to maintain the species/habitats present. This in turn determines the type of management that is required to maintain and restore the site to a good state of conservation.

Articles 6(3) and 6(4) of the Habitats Directive contain a specific process for any plans or projects that are likely to have a significant effect on one or more Natura 2000 sites, either individually or in combination with other plans and projects. Under this process potentially damaging projects are subject to an Appropriate Assessment (AA) to determine the precise nature and extent of the potential impacts on the species and habitats of EU importance present at or near the site. The competent national authority must then decide whether or not to approve the plan or project. This decision can only be made after the authority has ascertained that the proposed plan or project will not adversely affect the integrity of that site. Unlike environmental assessment, the outcome is binding: if the AA finds that there will be adverse effects on the integrity of the site, the development cannot proceed. The only exception to this is if the plan or project is considered necessary for imperative reasons of overriding public interest (IROPI), there are no other alternatives, and all the necessary compensatory measures are in place to ensure that the overall coherence of the Natura 2000 network is protected.

In practice, EU member states operate a preliminary screening process for developments proposed in or near Natura 2000 sites. This focuses on whether a plan or project is directly connected to, or necessary for, the management of the site, and whether a plan or project, alone or in combination with other plans and projects, is likely to have significant effects on a Natura 2000 site in view of its conservation objectives. If the effects are estimated to be significant, potentially significant, or uncertain then the process must proceed to the next stage, the AA. This considers whether the plan or project, alone or in combination with other projects or plans, will have adverse effects on the integrity of a Natura 2000 site and includes any mitigation measures necessary to avoid, reduce or offset negative effects. As part of this stage, a developer will be required to submit a comprehensive ecological impact assessment of a plan or project in view of the site's conservation objectives. In the UK this is known as a Habitats Regulations Appraisal, and in Ireland it is called a Natura Impact Statement. The statement submitted enables the competent authority to carry out the AA.

The AA is usually an iterative process, so that improvements to the plan or project can be included so as to avoid adverse effects on the integrity of the Natura 2000 site. The competent authority will normally consider whether mitigation measures can be introduced or if other measures would reduce the effects to a non-significant level. Other alternatives to the plan or project must also be considered. Ultimately, the plan or project can only be approved if it has been demonstrated, beyond reasonable scientific doubt and on the basis of objective information, that there is no adverse effect on the integrity of the site. If these cannot be ruled out, then the competent authority must refuse consent or apply the derogation test of IROPI under Article 6(4). As the IROPI test is only applicable under very strict conditions and can only be passed in extraordinary circumstances, it is not discussed further here. Given the complexity of this subject, the European Commission has published a number of guidance documents, both general [69–73] and sector-specific [74–76] on this topic.

10.4.4 Environmental Assessment Legislation

Environmental assessment legislation incorporates the Environmental Impact Assessment Directive (2011/92/EU) and the Strategic Environmental Assessment Directive (2001/42/EC). The goal of both directives is to ensure that the environmental implications of decisions are taken into account before decisions are made. A central element in the impact assessment process is public participation as a result of international legal requirements but also to include the public in decision-making and hence strengthen the quality of decisions. The first European Environmental Impact Assessment Directive was adopted in 1985 (85/337/EEC); it was later amended three times to take account of the Espoo Convention on Transboundary Environmental Impact Assessment in 1997, to incorporate public participation in decision-making and access to justice in environmental matters into the EIA process in 2003, and to amend the lists of projects subject to EIA in 2009. The Directive and its amended versions were codified in 2011, which is the version currently in force (2011/92/EU), but this was amended in 2014 (2014/52/EU) to introduce a number of key changes that member states had to transpose into national legislation by 16 May 2017.

10.4.4.1 Environmental Impact Assessment Directive

The EIA Directive applies to a wide range of public and private projects, listed in Annexes I and II. It requires an assessment of the environmental impact (an EIA) of any project likely to have significant effects on the environment before permission can be granted. For projects listed in Annex I an EIA is mandatory as these projects are considered to have significant effects on the environment. Projects in this category include, for example, construction of overhead electrical power lines with a voltage of 220 kV or more and a length of more than 15 km. For projects listed in Annex II of the Directive, the competent authority in each member state has discretion to decide whether an EIA is necessary. This is done through a procedure known as 'screening', which determines the effects of a project on the basis of certain thresholds or criteria or on a case by case examination. When conducting a screening exercise the competent authority must take the criteria contained in Annex III into account. Projects listed in Annex II are, obviously, not in Annex I and include, for example, industrial installations for the production of electricity, installations for the harnessing of wind power for energy production (wind farms), and construction of harbours, port installations and marinas.

Wave and tidal energy farms are not mentioned in the Directive, but as they are installations for the production of electricity, they may be subject to an EIA so as to determine the effects of the development on the natural environment, species, biological and physical processes. The requirement to submit an EIA is generally dependent on the nature, size and location of the proposed development. Smaller-scale developments seem to have less rigorous EIA requirements in many countries (e.g. Mexico, Spain and

Portugal) [1]. In Spain, a reduced number of topics are considered in comparison to larger, full-scale projects, and the competent authority will examine these initially in order to determine whether a full EIA is necessary. In Scotland, the most recent guidance states that if a project is listed in Annex I or Annex II of the EIA Directive, it will require an EIA. Where, however, an EIA has been carried out under another consenting system relating to a project, Marine Scotland – Licensing and Operations Team (MS-LOT) may decide that this assessment is either entirely or partly sufficient and notify the developer accordingly [77].

Under the terms of the Directive, developers are required to submit comprehensive environmental data relating to both baseline conditions and possible environmental impacts of the development to the competent authority responsible for EIA in that jurisdiction. Article 3 specifies that the EIA must include a description of the aspects of the environment likely to be significantly affected by the proposed development, including:

1) human beings, fauna and flora;
2) soil, water, air, climate and the landscape;
3) material assets and the cultural heritage;
4) the interaction between the factors mentioned in the first, second and third indents.

The EIA must also include a description of the project, measures envisaged to avoid, reduce or remedy adverse environmental effects, the data required to identify and assess the main effects, the main alternatives identified and a non-technical summary of all of the preceding elements. The competent authority will then make its decision, taking into consideration the information submitted.

Any information gathered as part of the process under Article 5 must be made available to the public within a 'reasonable timeframe'. Under Article 6(4) the public must also be given 'early and effective opportunities to participate in the environmental decision-making procedures' though the details on how to do this are decided by the member states. Once a decision to grant or refuse development consent has been made, the competent national authority must make certain information available to the public: the content of the decision and any conditions attached thereto; the main reasons and considerations on which the decision is based, including information about the public participation process; and a description, where necessary, of the main mitigation and compensatory measures. The public must also have access to a review procedure (Article 11), in line with the EU's obligations under the Aarhus Convention.

Experience to date with the application of EIA to wave and tidal energy indicates that the requirements have been modelled on those applied to offshore wind, despite the fact that wave and tidal devices are different from wind turbines and indeed each other [78]. Removal of energy directly from the water column differentiates wave and tidal energy generation from offshore wind. Many wave devices also consist of moving parts in the underwater environment, and this in turn introduces new interactions between devices and marine species which need to be considered before a decision to grant consent is made [78]. Devices deployed at smaller scales, or for time-limited periods, mean there is limited data available to assess their potential environmental effects even in real-sea conditions. This type of uncertainty presents difficulties to both regulators and developers and ranges from uncertainty relating to the effects of a single device to effects associated with deployment of multiple devices and their cumulative impacts when combined with other marine activities.

A final aspect of the EIA process relevant to the consenting of wave and tidal projects is that usually it is the outcome of the EIA process that determines the ongoing and required environmental monitoring programme to be implemented at the project site during construction and operation and perhaps even through to the decommissioning stage. In the EU, for example, an analysis of 14 wave and one tidal test sites categorised the ongoing monitoring activities at those sites into nine types: benthos, seabirds, fish and fish habitats, marine mammals, other marine mega-vertebrates (sharks and turtles), physical oceanographic environment, acoustics, terrestrial habitats and socio-economic considerations, plus an additional category for other activities that did not correspond to any of the aforementioned classes [78]. This study revealed that physical oceanography parameters were the most common ones investigated, which is unsurprising given the use proposed for the site. One important finding, however, was that there was widespread variation in the methods utilised, revealing that even when the same ecological question was being studied, very different methods were being employed. This has implications for learning about the impacts of ocean energy devices as results cannot be easily compared or contrasted, even if working from a common EU legislative base.

10.4.4.2 Strategic Environmental Assessment Directive

The Strategic Environmental Assessment (SEA) Directive applies at the strategic level (not the site level like the EIA Directive), to public plans and programmes (not policies). The objective of this Directive is to provide for a high level of protection of the environment and to contribute to the integration of environmental considerations into the preparation and adoption of plans and programmes so as to promote sustainable development. Essentially the Directive applies an impact assessment process to certain plans and programmes which are likely to have significant effects on the environment. An SEA is mandatory for plans/programmes that are prepared for the energy sector (among others) and to those plans and programmes that set the framework for future development consent of projects listed in the EIA Directive or plans and programmes which have been the subject of AA under the Habitats Directive. Steps to be completed include scoping, the consideration of alternatives, the preparation of an 'environmental report' in accordance with Article 5, public consultation (Article 6) and the proposal of mitigation and monitoring measures. Under Article 8, the findings of the environmental report must be taken into account during preparation of the plan or programme.

Usually the findings of an SEA can guide development to locations where environmental effects are minimal or can be avoided. The environmental report produced for the SEA process will also cover any technical and environmental constraints within the area concerned under that plan or programme. Such constraints could include, for example, other uses of the area like aquaculture installations, shipping lanes, pipelines and cables. An added benefit to the environmental report is that the information collated during the process can act as a reference source for use by developers, other regulatory authorities and stakeholders, to support decision-making at project level on future marine or ocean energy developments. Experience of the application of SEA to this sector in particular appears to be mixed. The Scottish Government commissioned a SEA on the environmental effects of developing wave and tidal power in 2007 [79]. This was then used as a basis from which to deliver the Scottish Government's strategy for marine energy development as well as the preparation of regional locational guidance for the sector [80–82]. The Department of Energy and Climate Change in the UK has conducted a number of SEAs relating initially to oil and gas, then offshore wind and most recently a combined offshore energy SEA of a plan/programme to enable future leasing for offshore wind, wave and tidal devices and licensing/leasing for hydrocarbons and carbon dioxide storage [83]. That SEA covers parts of the UK Renewable Energy Zone and the territorial waters of England and Wales; for hydrocarbon gas and carbon dioxide storage it applies to UK waters (territorial waters and the UK Gas Importation and Storage Zone); and for hydrocarbon exploration and production it applies to all UK waters [83]. Both Northern Ireland and the Republic of Ireland have published SEAs for marine renewables (offshore wind, wave and tidal) in their respective waters [84, 85].

10.4.5 Public Participation and Access to Environmental Information

EU advancement on public participation is innately linked to signature and ratification of the UN Economic Commission for Europe Aarhus Convention on Access to Information, Public Participation in Decision-making and Access to Justice in Environmental Matters. This necessitated the enactment of legislation at EU level dealing with the three elements of the Convention. Firstly, a Directive on Public Access to Environmental Information (2003/4/EC) was adopted in January 2003. This was subsequently accompanied by the Directive on Public Participation in respect of the drawing up of certain plans and programmes relating to the environment and amending the public participation and access to justice elements of the EIA and Integrated Pollution Prevention and Control (IPPC) Directives. The third pillar of the Convention, namely access to justice in environmental matters, was the subject of a proposal for a directive in October 2003 (COM(2003) 624) [86] but was not adopted due to strong opposition from some member states. Regulation (EC) No. 1367/2006 on the application of the provisions of the Aarhus Convention in environmental matters to Community institutions and bodies entered into force in September 2006 and covers the institutions, bodies, offices or agencies established by, or on the basis of, the EC Treaty.

The Public Access to Environmental Information Directive (2003/4/EC) has two key objectives. It seeks to guarantee a right of access to environmental information held by or for public authorities and to ensure that environmental information is increasingly made available and disseminated to the public. The term 'environmental information' is expanded in Article 2 and encompasses information in any form on the state of the environment, on factors, measures or activities affecting or likely to affect the environment or designed to protect it, on cost–benefit and economic analyses used within the framework of such measures or activities and also information on the state of human health and safety. The Directive applies to public authorities, including persons who perform public administrative functions. Environmental information should be made available to an applicant on request under Article 3: they do not have to state an interest. There are certain exceptions to that provision in Article 4: a public authority does not have to make the information available if the request is 'manifestly unreasonable', 'too general', 'unfinished' or 'concerns internal communications'.

A request can also be refused if disclosure would adversely affect the 'confidentiality of the proceedings of public authorities', international relations, public security or national defence; the course of justice; the confidentiality of commercial or industrial information, IP rights; confidentiality of personal data/files where consent has not been granted by that person; information given voluntarily; and the protection of the environment to which such information relates (Article 4(2)(b)–(h)). Examination of environmental information *in situ* should be free of charge under Article 5, but for supplying environmental information public authorities may charge which 'shall not exceed a reasonable amount'. The role of technology is recognised in Article 7 and advocates that environmental information becomes progressively available in electronic databases that are easily accessible to the public through public telecommunication networks. Information to be included could be data or summaries of data from the monitoring of activities affecting the environment, EIAs, and/or risk assessments, according to Article 7(2).

The Public Participation Directive (2003/35/EC) transposes the Aarhus Convention requirements relating to public participation in environmental decision-making. Under Article 2, the public must be given 'early and effective opportunities' to participate. Member states must ensure that the public is informed of any plans or programmes, including how they can get involved and comment on the plans or programmes before decisions are made. It also provides that the results of public participation must be taken into 'due account' and that the public are informed of the decision made including how the public consultation elements were considered in the decision-making process. Like all directives, it is up to the member states to transpose these requirements into domestic legislation and detail the operational procedures applicable.

10.4.6 Other Relevant EU Legislation

The EU has a number of legislative instruments applicable to the management of water generally. The key legislative instruments in this area that may impact upon ocean energy development are the Water Framework Directive (WFD, 2000/60/EC) and the Marine Strategy Framework Directive (MSFD, 2008/56/EC), but it should be noted that their objectives and goals relate to a wider geographic area than project level. Both directives have the aim of achieving better water quality. The Water Framework Directive seeks to prevent the deterioration of ecological quality and the restoration of polluted surface and groundwater, with comprehensive environmental objectives detailed in Article 4. The WFD applies to inland surface waters, transitional (estuarine) waters, coastal waters and groundwater. Coastal waters are defined in Article 2(7) as 'surface water on the landward side of a line, every point of which is at a distance of one nautical mile on the seaward side from the nearest point of the baseline from which the breadth of territorial waters is measured, extending where appropriate up to the outer limit of transitional waters'. Member states are required to adopt river basin management plans to achieve the overarching objectives of the Directive. Water quality tends to be included in EIAs of individual wave and tidal energy projects, and while they may have temporary impacts on water quality during the construction phase for example, these effects are expected to be localised and unlikely to impact upon regional water quality which is the focus of the WFD [87].

The Marine Strategy Framework Directive follows a similar layout to the WFD, in that it has the objective of achieving 'good environmental status' (GES) of European marine waters by 2020. The term 'GES' is interpreted according to 11 descriptors contained in Annex I of the Directive and reproduced in Table 10.3. These descriptors

Table 10.3 Qualitative descriptors used in determining GES under Annex I of the MSFD.

Descriptor no.	Narrative
1	Biodiversity is maintained
2	Non-indigenous species do not adversely alter the ecosystem
3	The population of commercial fish species is healthy
4	Elements of food webs ensure long-term abundance and reproduction
5	Eutrophication is minimised
6	The sea floor integrity ensures functioning of the ecosystem
7	Permanent alteration of hydrographical conditions does not adversely affect the ecosystem
8	Concentrations of contaminants give no effects
9	Contaminants in seafood are below safe levels
10	Marine litter does not cause harm
11	Introduction of energy (including underwater noise) does not adversely affect the ecosystem

epitomise what marine waters will look like when GES has been achieved. To help attain GES, the Directive obliges each member state to develop a marine strategy, focused on the protection, preservation and restoration of the marine environment. According to Article 5, these must be developed and implemented on a regional level and, where a marine region is shared by a number of member states, the states concerned must cooperate so as to ensure the strategies are complimentary and coherent.

One of the first deadlines for member states was October 2012, for an initial assessment of the current environmental status of their marine waters, a determination of what GES means for their region and identification of environmental targets and associated indicators to guide progress towards achieving GES by 2020. Under Article 13, a programme of measures designed to achieve or maintain GES must also be developed by 2015, with a view to being operational by 2016 at the latest. According to Annex VI, the programme of measures can include a variety of controls, including both input and output controls as well as temporal and spatial controls, which could include management measures that influence where and when an activity is allowed to occur. Management coordination measures may also be introduced along with measures to improve the traceability, where feasible, of marine pollution. As GES must be achieved at a regional level, in a similar way to the WFD, the actual implications for ocean energy development are minimal. It should, however, be emphasised that as member states are required to report on the different descriptors in their marine waters, it is likely that some of these elements will begin to be reflected in other legislative requirements such as EIA. The MSFD is the first legal instrument in the EU to include underwater noise, for example, and as it has the potential to impact upon GES, member states are now beginning to monitor their baseline marine noise levels but also ensure that new developments do not contribute to excessive levels of noise being introduced to the marine environment, primarily through the EIA process.

10.4.7 Maritime Spatial Planning Directive

The importance of maritime spatial planning (MSP) has been asserted by the EU and its institutions since the publication of the IMP in 2007 [65]. Since then there have been a number of Communications [88, 89] on the subject as well as discussions on creating a legislative basis for MSP. This culminated in July 2014, when the European Parliament and the Council adopted a Directive establishing a common framework for MSP in the EU (2014/89/EU). The purpose of the Directive is to promote growth of maritime economies and sustainable development of marine areas and their resources. The Directive defines MSP, in Article 3, as 'a process by which the relevant Member State's authorities analyse and organise human activities in marine areas to achieve ecological, economic and social objectives'. The geographic scope of application is stated in Article 2 as applying to the marine waters of member states, but not coastal waters or parts thereof falling under a member state's town and country planning system. The effect of this is that, for many coastal member states of the EU, MSP will begin at the low water mark and extend seaward to the limits of national jurisdiction. The term 'marine waters' in the Directive has the same meaning as in other EU legal instruments, specifically the MSFD, while 'coastal waters' has the same definition as that contained in the WFD.

Article 4 obliges member states to establish and implement MSP, taking into account the specificities of their marine regions, existing and future activities and their impacts on the environment, natural resources, and land–sea interactions. Member states are permitted to determine the format and content of their MSP and build upon existing national policies, regulations or mechanisms that already exist, provided those conform with the requirements of this Directive. Member states are requested in Article 5 to consider applying an ecosystem approach and to promote coexistence of activities in their marine space along with the sustainable development of all the marine sectors active in their waters. Article 6 identifies a number of minimum requirements in relation to MSP, namely:

a) to take into account land–sea interactions;
b) to take into account environmental, economic and social aspects, as well as safety aspects;
c) to promote coherence between MSP and the resulting plan(s) and other processes, such as integrated coastal management or equivalent formal or informal practices;
d) to ensure the involvement of stakeholders;
e) to organise the use of the best available data;
f) to ensure transboundary cooperation between member states; and
g) to promote cooperation with third countries.

The maritime spatial plans developed for the purposes of this Directive must be reviewed at least every 10 years.

According to Article 8, when establishing and implementing MSP, member states will set up plans that identify the spatial and temporal distribution of existing and future activities in their marine waters, to include uses such as aquaculture and fishing areas, installations and infrastructures for oil, gas and other energy exploration, exploitation and extraction, renewable energy areas, shipping routes and traffic flows, military training areas, nature and species conservation sites and protected areas, mineral extraction areas, scientific research, cable and pipeline routes, tourism and underwater cultural heritage.

In line with requirements deriving from the public participation legislation, member states are asked in Article 9 to provide a means of public participation so as to inform 'all interested parties' and consult with 'relevant stakeholders and authorities, and the public'. Member states were required to transpose the legislation and administrative provisions necessary to comply with the Directive into their national legal systems by 18 September 2016. The actual maritime spatial plans should be created as soon as possible according to Article 15, but at the latest by 31 March 2021. A number of member states already have some form of MSP in their marine waters. The impact of this on consenting for ocean energy deployments will be referred to by country in the next section.

10.5   National Consenting Systems

10.5.1   Common Consenting Considerations

It is evident from the preceding sections that law and policy systems governing the development of ocean energy projects reflect a variety of legal requirements at all scales of governance (international, regional and EU). The traditional fragmented management of the marine environment, in terms of specific uses or sectors, results in the need to comply with numerous pieces of legislation covering the various elements of an ocean energy project. The following sections outline the consenting process applicable in France, Ireland, Portugal, Spain, all jurisdictions of the UK, and the USA. The term 'consenting process' is used in this section to describe all the necessary permissions or consents to deploy one or more devices in the water. As ocean energy developments are relatively new to regulatory authorities, they tend to be administered under existing legislation which was not created with ocean energy in mind. It is also common for each element of the development or project to be considered separately under different legal instruments. As more MRE projects come to fruition, some regulatory authorities have attempted to streamline their consenting system, recognising that clear and definitive procedures can encourage investment in development.

Typically consenting involves a number of discrete elements. This will include a licence or lease to occupy sea space administered by a central government department or one of its agencies. This arises predominantly because a state proclaims ownership of marine space but is subject to the exercise of certain public rights, such as navigation and fishing. A distinction is usually made between exclusive and non-exclusive uses of sea space, and also according to whether a development is temporary or permanent. For developments or activities that are temporary or for a defined period of time, a licence may be issued. If an activity or development is more permanent or for a longer time, a lease may be granted. Different countries will use different terms depending on the nomenclature used in their governing legislation. The state or its entities has the authority to grant usage rights under different statutes, usually with a provision that such rights must be in the public interest (e.g. in Ireland this appears in sections 2 and 3 of the Foreshore Act, 1933).

10.5.1.1   Occupation of Sea Space
In some countries it is the maritime jurisdictional zone that will dictate if and when an activity can occur as well as who is responsible for its planning and management. In Germany, for example, the Federal Maritime and Hydrographic Agency (BSH) is

responsible for offshore wind energy developments in the EEZ, but from the shore to the limit of the territorial sea (12 miles) it is the coastal states that have the remit for consenting offshore wind energy projects [90]. Where an electrical cable from a wind farm in the EEZ crosses state waters to an onshore grid access point, the adjoining state must grant permission for this [91]. In other countries with a federal system of government, such as the USA, State and federal laws apply to offshore development. This type of governance system has the potential to cause problems in situations where a development straddles a number of planning competencies within different levels of government, in practice resulting in the need for multiple consents from multiple competent authorities.

The issue of straddling projects can also arise when renewable energy produced by an offshore generating station must come ashore, which can trigger the application of requirements deriving from terrestrial planning legislation. Usually ocean energy will come onshore via a cable, a processing plant or an electrical sub-station or perhaps all three. These ancillary structures often require planning permission under the land-based planning system, or equivalent, in operation in that country. In some instances, developments connected to land solely by a cable are exempt from this requirement. Construction on land normally requires a consenting process that operates independently from that applicable to sea-based projects. In Ireland local planning authority jurisdiction ends at the mean high water mark. In contrast, local authorities in the UK have jurisdiction to the low water mark. In the USA, jurisdiction can vary from State to State. Under planning legislation, there may also be a requirement to consult with other regulatory authorities such as nature conservation bodies, health and safety authorities, industry representatives and the public in relation to the proposed works.

10.5.1.2 Connection to the Electricity Grid

Ocean energy is exploited at source, and consequently it needs suitable grid infrastructure that is interconnected to the transmission network. As the resource base for ocean energy is located in remote coastal areas, unless there is a significant population or industrial base at the coast, there is usually only a weak distribution network available. This means that often the grid network needs to be strengthened substantially, raising project costs to a prohibitive level. There are also weak interconnections between EU member states, the power market is generally inflexible, liable to change and fragmented, and there is a lack of offshore electricity grids [92, 93]. Regulatory frameworks and grid codes applicable to ocean energy projects differ from country to country. Grid access and grid connection are also governed separately. In many countries there is no common power market. As will be seen in the country systems described below, developers often require consent to both construct and operate a power generation facility. On occasion these consents might also prompt the need for planning permission, environmental assessment, a connection offer from the applicable operator and a power purchase agreement (PPA). A PPA is a contractual agreement between an electricity generator and a licensed supplier obliging the latter to purchase the output from the renewable energy facility. Grid connection offers can be made by either the transmission system operator, which operates the high-voltage system, or the distribution system operator, which operates the medium- and low-voltage systems, depending on the type of connection needed by the project. These elements are usually governed by organisation-specific administrative provisions rather than a specific legal instrument.

10.5.1.3 Environmental Effects

The environmental impacts of a development must be taken into account by a competent authority before a decision is made to grant consent. This usually takes the form of an EIA, which reflects the individual project characteristics such as the location, scale and type of development. In particular, the EIA must consider the potential interactions between marine species and devices, possible alterations of marine habitats from operational activities and long-term ecosystem effects of deploying and operating arrays of MRE devices in coastal and estuarine waters [94]. Impacts from device deployments can depend on the type of energy being harnessed, the individual device type or the reduction of energy in marine systems. The challenge for consenting authorities with respect to environmental impacts is evidence and certainty. Device deployments to date have tended to be time-limited and only single-unit deployments. This means that there is limited data available to assess potential environmental effects, particularly as proposals are beginning to comprise of more devices and larger spatial scales.

10.5.1.4 Decommissioning

The term 'decommissioning' is not specifically mentioned in any of the main international legal instruments, although the need to address redundant offshore platforms is cited. Legislation prefers the term 'abandonment', though arguably this has a very precise connotation and application to the oil and gas industry, where it refers to permanent closing of a well [95]. The two key legal frameworks relevant to decommissioning are the UN Law of the Sea Convention and the London (Dumping) Convention. The LOSC prescribes, in Article 60, that any installations or structures which are abandoned or disused in the EEZ must be removed to ensure safety of navigation in accordance with any accepted international standards. It is the Maritime Safety Committee of the IMO that has responsibility for such guidelines. In 1989 it published 'Guidelines and Standards for the Removal of Offshore Installations and Structures on the Continental Shelf and in the EEZ' [96]. The Guidelines provide that the coastal state, which has jurisdiction over the installation or structure, must ensure that it is removed in whole or in part 'once it is no longer serving the primary purpose for which it was originally designed and installed, or serving a subsequent new use, or where no other reasonable justification' cited in the guidelines allows it to remain wholly or in part on the seabed [97].

The removal of a structure should take place as soon as reasonably practicable after abandonment or permanent disuse of such installation or structure [97]. Decisions relating to removal are made on a case-by-case evaluation taking into consideration:

- the potential effect on safety of navigation and other uses,
- deterioration of material and future impacts,
- the potential effect on the marine environment including its living resources,
- the risk of movement from its current position,
- the costs, technical feasibility and risk of injury to personnel involved in removal, and
- determination of a new use or other reasonable justification.

A coastal state can decide that a structure does not need to be removed but, if this is the case, then the state must ensure that the remaining parts are stable, indicated on nautical charts and marked by aids to navigation, where necessary. The party responsible for maintaining those aids and for conducting monitoring of the material left behind must also be identified in order to comply with the guidelines. Paragraph 3.4.1 of the Guidelines permits structures to be left *in situ* if the structure would serve a new use, such as an enhancement of the living resource [96]. Artificial reefs are usually human-made and placed on the seabed deliberately to imitate some characteristics of natural reefs. Disused ocean energy devices may introduce new hard substrate that may have artificial reef effects [98], alleviate fishing pressure and potentially allow fish to breed and grow [99] and also act as fish aggregating devices [100]. A degree of caution should, however, be exercised in relation to these aspects as they are not yet fully understood.

The London Convention on the Prevention of Marine Pollution by Dumping of Wastes and Other Matter (1972) was drafted in response to calls for better control of pollution of the marine environment, primarily through greater regulation of deliberate dumping at sea. It defines dumping, in Article III 1(a)(i), as 'any deliberate disposal at sea of waste or other matter from vessels, aircraft, platforms or other man-made structures at sea' and in Article III 1(a)(ii) as 'any deliberate disposal at sea of vessels, aircraft, platforms or other man-made structures'. It provides further that dumping does not include 'placement of matter for a purpose other than the mere disposal thereof, provided that such placement is not contrary to the aims of this Convention' [101]. The 1996 Protocol to the London Convention extends the definition of dumping, but the exemption on placement of matter for a purpose other than disposal remains. In light of concerns that creation of an artificial reef could, in some cases, be used to legitimise dumping at sea, the IMO developed Guidelines for the Placement of Artificial Reefs in 2009 [102]. The definition of artificial reef used in the Guidelines, however, unequivocally excludes 'submerged structures deliberately placed to perform functions not related to those of a natural reef ... even if they incidentally imitate some functions of a natural reef' [102]. National consenting systems for ocean energy developments do not yet tend to explicitly cover decommissioning and abandonment of ocean energy infrastructure, given there have been limited deployments of devices to date, so this topic is given limited consideration in the country-specific examples to follow.

10.5.2 France

France has a national objective to produce 6000 MW of MRE by 2020. To deliver on this objective the French government launched two calls for tenders in 2011 and again in 2013 for fixed offshore wind farms, with an expected capacity of 3000 MW. Also in 2013, the Agency for the Environment and Energy Management (ADEME) launched a call for expressions of interest in tidal turbine test projects [103]. The sites of interest had been pre-selected, to an extent, in a process involving the Ministry of Ecology, the *préfets maritimes* (maritime prefectures) and the *préfets de département* (department prefectures) for each maritime district. Other stakeholders were invited to participate in this process which also involved the collation of large amounts of data including information on the environment, landscapes and historical heritage, defence and security, navigation, physical oceanographic data, electricity grid data and socioeconomic data. The leading tidal stream project in France is led by EDF and involves the installation of the first farm (2 MW) of OpenHydro underwater turbines connected to the power grid in Paimpol Bréhat (northern Brittany). In December 2014, Alstom and GDF Suez were selected to supply the Raz Blanchard tidal pilot farm with four Oceade devices, with a total capacity of 5.6 MW. The project is expected to become operational in 2017.

There is no specific consenting process for ocean energy technologies. France has been working to amend certain legal instruments so as to streamline the licensing process applicable to offshore renewable energy development. The Law Grenelle II (Law no. 2010-788, 12 July 2010) and related Law-Decree no. 2012-41 consolidate the requirements for licensing of generating stations producing power from renewable energy sources. The deployment of offshore renewable energy technologies on the French maritime domain (baseline to territorial sea, including the seabed and subsoil) is subject to two legal authorisations administered by the *préfet de département*:

- A concession to occupy the public maritime domain (PMD) (maritime domain consent) (Article L 2124-3 to L 2124-5, article R 2121-1 and subsequent Public Administration Property Code)
- An authorisation related to water resources protection (Water Resource Protection Licence) (Article L 214-1 and subsequent, article R 214-1 and subsequent of the Environment Code)

The *préfet de département* can grant these authorisations after notice of the *préfet maritime*. The *préfet maritime* is a civil servant who exercises authority over the sea in one particular region known as the *préfecture maritime*. The purpose of both these authorisations is to ensure that occupation of the sea space is compatible with the legal objective of protecting the marine area. The time-scale for administering and granting the maritime domain concession and the Water Resource Protection Licences varies from six to nine months [104].

Granting of both authorisations may be dependent upon the completion of an EIA. Offshore renewable energy developments can be subject to an EIA either on a mandatory or case-by-case-basis (Article R122-2, Environment Code). Even though Article R122-2 does not specify whether an EIA must be systematically conducted, in practice, offshore deployments have been the subject of a systematic EIA. Specifically, applications for a Water Resource Protection Licence are subject to an impact study being conducted (Article R214-6). Conditions attached to the Water Resource Protection Licence once granted, for example, can relate to cetacean, electromagnetic field and noise monitoring [105]. The maritime domain consent application must be accompanied by an EIA in accordance with Article L122-1 and R122-2 and in line with the Environment Code. The EIA required with the application for public domain consent is administered by the Environmental Agency on behalf of the *préfet de département*. The EIA required under the maritime domain consent can be used to fulfil the need for an impact study for a Water Resource Protection Licence if the Environmental Statement contains the same information (Article R214-6, Environment Code). Environmental Statements are submitted to the Environmental Authority for assessment, recommendations and decision. The final environmental approval is given by the *préfet de département* who is under an obligation to take into account the recommendations of the Environmental Authority but is not bound by them.

Under Law Grenelle II and Decree no. 2012-40, electricity generating installations in the public maritime domain and using marine sources are exempt from terrestrial planning permission for all related onshore electrical works – including works related to grid connection (French Town Planning Code: *Code de l'urbanisme*, Articles L421-5 and R421-8-1). In other words, planning permission is deemed to be granted as part of the maritime domain consent. Developers are not required to submit a separate application for associated terrestrial works. Consenting of the electrical aspects of a development varies according to whether the project is in response to a call for tender or is being developed on an individual basis. Projects responding to a call for tender proceed in a more streamlined fashion. Under the Energy Code, authorisation to construct and exploit offshore electricity generating installations is deemed to be granted when a project has been selected after a call for tenders (Article L311-10, Energy Code). This means that the developers do not have to submit an additional application for the construction of offshore stations or national grid connection. Where applications are not a response to a call for tenders, the proponents must obtain a power exploitation permit from the Ministry of Ecology, Sustainable Development and Energy (Articles L311-5 *et seq.*, Energy Code). From experience the power generation and grid connection elements of the project have proved most challenging as this involves a separate administrative process administered by the French utility distribution grid operator *Electricité Réseau Distribution de France* [103].

A central part of the consenting process in France is the creation of a regulatory pathway which essentially facilitates the involvement of strategic stakeholders in the evaluation of a project proposal and subsequent granting of the required consents. Under the Public Maritime Domain Decree, this process operates similarly to a public debate. All development projects over €300 million are subject to this public debate (Article R121-2, Environment Code). The Commission of Public Debate must be consulted by the Public Authority or the project developer. The debate requires a time-frame of four months. As the project progresses, communication becomes more frequent and dedicated meetings with stakeholders are organised on a regular basis. France has no formal MSP process in place at this time.

10.5.3 Ireland

A general policy framework for offshore renewables was published in 2014 in the form of the Offshore Renewable Energy Development Plan (OREDP) [106]. This sets out key principles, policy actions and enablers for delivery of Ireland's significant potential in this area. As part of the development of the OREDP, an SEA for marine renewables was conducted in 2010 [85]. The purpose of the SEA was to evaluate the likely significant environmental effects of implementing plans to develop offshore renewable at low, medium and high scenarios so that areas could be prioritised for development. The implementation of the OREDP is led by the Department of Communications, Climate Action and Environment, but it requires government action across the environmental, energy policy and economic development dimensions and accordingly a dedicated steering group (ORESG) was established to oversee the Plan's implementation. ORESG works on three specific work-streams: environment, infrastructure and job creation. The environment work-stream is tasked with taking forward the findings and recommendations of the SEA and Appropriate Assessment carried out for the OREDP and developing sector-specific guidance to assist developers in this regard.

The consenting process in Ireland involves a number of different consents and authorities. In order to develop, a project proponent requires a foreshore consent, in the form of a licence and lease, to develop in the foreshore area. The foreshore is legally defined as the area between mean high water and 12 nautical mile territorial sea limit. The original Foreshore Act was enacted in 1933 and, while it has been amended on numerous occasions to take account of other legislative changes, it remains the key instrument governing all marine and coastal development. The then Department of Environment, Community and Local Government recognises that although the nature, scale and impact of MRE developments can vary considerably, all require foreshore consent to investigate/survey the site, to construct the development (and cabling), andto occupy the property [107]. In order to undertake site investigation work, including preliminary studies relating to EIA, a foreshore licence is required from the Department. Licences are granted for specific activities and for time-limited periods. The developer will meet with Departmental officials to scope the content of the EIA either prior to applying for a foreshore licence or once they have obtained the licence, as it is usual for the same or similar information to be required. The preliminary stages of the EIA process allow the developer to consult with the adjoining planning authority, for example, as well as other interested parties and relevant regulators with a view to establishing what should be contained in the environmental impact statement (EIS, equivalent to EIA in Irish law).

Following the site investigation works and dependent on their assessment, a developer may subsequently apply for a foreshore lease to undertake further development activities. This gives the developer the right to exclusive use. It is not possible to obtain a lease unless the works referred to above have been completed, though successful completion of site investigation works does not automatically entitle a developer to a foreshore lease. A lease is generally granted for 35 years and is subject to regular review, on a five-year basis, by the responsible authority. In order to obtain a foreshore lease for the construction phase, a developer must be able to show that other relevant licences, permissions and approvals have been obtained from additional regulators. Currently there are no timelines associated with decisions on foreshore licence and lease applications. At all stages, the Minister reserves the right to reject any application for a foreshore licence or lease, to modify the area sought under licence or to allow others to simultaneously investigate the suitability of the licence area [103].

It is important to note that the system for foreshore consenting in Ireland is currently being reformed. This commenced with public consultation on the new process in January 2013. Following consideration of the submissions received, the draft Maritime Area and Foreshore (Amendment) Bill 2013 was published in October that year [108]. The Bill, and the framework proposed therein, seeks to better align the foreshore consent system with the terrestrial planning system and will also introduce a planning system for development activity in the EEZ and on the continental shelf, including strategic infrastructure projects, such as oil and gas, ports and offshore renewable energy. If enacted the proposed legislation will have substantial implications for consenting of marine renewable developments in Ireland. The consenting authority will change from the current Department to An Bord Pleanála, the Irish Planning Board, for developments in the maritime area that are identified as strategic infrastructure projects; that require an EIA or AA; that are located entirely or partially beyond 12 miles; and that are located entirely beyond the nearshore area (high water mark to low water mark). This is intended to streamline the process and it is anticipated that the legislation will be enacted later in 2017. Ireland has no formal MSP system in place yet, but regulations transposing the requirements of the EU Directive on MSP were enacted in 2016.

Along with the foreshore consenting requirements, ocean energy developments are subject to the requirements of the Electricity Regulation Act, 1999, as ocean energy devices qualify as electricity generating stations. There is a complex licensing and authorisation procedure based on installed capacity under that Act [109]. Basically it requires a developer to have a licence to generate and a licence to construct or reconstruct a generating station. The applications can be made separately or jointly and, where applicable, must be accompanied by an EIS, planning permission/exemption for any onshore works, a connection offer from the relevant operator and a PPA. If a wave or tidal energy project involves onshore elements, such as the construction of a substation, planning permission will be required under the Planning and Development Acts, 2000–2010. These are administered by the adjoining local authority (county council). Ireland has a high proportion of coastal designated SACs and SPAs and if the development is likely to impact upon the conservation objectives of such a Natura 2000 site an AA will be required.

10.5.4 Portugal

In Portugal, while there is no single piece of legislation to consent ocean energy developments, all applicable legislation has been modified to better suit the needs of the sector. This has been heavily influenced by the enactment and entry into force of legislation that transposes the EU MSP Directive and national maritime spatial plan for Portugal. There are two types of national MSP instruments which are legally binding on public and private entities. The first is referred to as a Situation Plan which identifies the distribution of uses and resources within the Portuguese maritime space to potentially the (extended) continental shelf limit. The second instrument is the Allocation Plan(s) which identifies and allocates areas for new uses, namely those uses that are not included as potential uses in the Situation Plan. Private use of Portuguese maritime space[1] is assigned through a 'private use title'. Three types of legal permits can be derived from this, depending on the nature and duration of the private use concerned:

1) concession – requires the continuous use (over the entire year) of an area and may have a maximum duration of 50 years;
2) license – for intermittent (or temporary/seasonal) use(s) of the marine area for periods of less than 1 year and up to a maximum of 25 years;
3) authorisation – limited to scientific research projects and/or pilot-projects related to new technologies or non-commercial uses with a maximum duration of 10 years.

If the use to be developed is already identified as a potential use in the Situation Plan, the granting of the title starts with a request by the promoter. If the use is not included in the Situation Plan, however, the assignment of the title depends on the previous development and approval of an Allocation Plan (with the potential exception of a request for using the marine area for scientific research). The Director-General for Natural Resources, Security and Maritime Services (Direcção Geral de Recursos Naturais, Segurança e Serviços Marítimos) is responsible for administering the above forms of

---

1 Defined as 'a utilisation that requires the reservation of an area or volume of the Portuguese maritime space for a use of the marine environment, marine resources or ecosystem services greater than the one obtained by common utilization and which results in a benefit to the public interest'.

permits and acts as a coordinator with other entities that may have a marine remit. In order to compensate for private use of 'common' marine space, a 'utilisation tax' (TUEM) will be applied to all maritime activities that incur a private use of the national maritime space under concessions and licences [110]. Private uses developed under authorisations are 'free' from such a tax as they are deemed to be non-commercial.

Along with the above process, the developer will also have to apply for a licence for a power production installation, administered by the Energy and Geology Directorate-General (Direcção Geral de Engenharia e Geologia, DGEG). This comprises a production permit and an exploration permit. Decree Law 225/2007 amended existing legislation on electricity production so as to accommodate the production of electricity from renewable energy sources. The new legislation also amended the EIA process for projects of this kind. The licence for power production does not include grid connection, but this is necessary if a project is to supply electricity to the grid. Developers requiring a grid connection must first obtain a reservation, from the DGEG, for power supply to the public electrical network from a given point. This requires the submission of an environmental impact study, if the project, or parts of it, is to be located in the vicinity of a national ecological reserve, Natura 2000 site and/or the national network of protected areas. This is the responsibility of the Commission for Coordination and Regional Development (Comissão de Coordenação e Desenvolvimento Regional, CCDR). Outside protected areas and if the project is not covered by national EIA legislation, the CCDR must return a finding of no significant impacts to the DGEG. The EIA procedure for MRE projects (except for wind projects with 20 or more turbines, where a full EIA is required) follows a simplified procedure, led by the CCDR of the area where the project is to be located. This has a shorter time-frame and less formality associated with it. Terrestrial works associated with offshore renewable energy development must be approved by the local planning authority.

10.5.5 Spain

In Spain, the consenting process for ocean energy projects is based on three main legal instruments, each having been amended in recent years to help streamline consenting. The Coast Law of 1988 was the original legal instrument governing occupation of marine space and safety of navigation and was amended by Law 2/2013 (29 May). Management and surveillance competences relating to the Public Maritime Domain on land (MTPD), which incorporates the territorial sea, rests with the Directorate General of Coasts, part of the Ministry of Agriculture, Food and the Environment. This entity is represented at regional level by coastal demarcation departments in each coastal province and autonomous community. Therefore, the development of ocean energy projects in the territorial sea must comply with the legal requirements governing the administrative process for granting titles to territorial occupation (before and during project development) and associated arrangements (e.g. deadlines, transfers and expiry).

Royal Decree 1/2008, of 11 January, provides for EIA of projects to be located in the natural environment. According to this decree, the developer first submits a preliminary document outlining the project and its anticipated environmental impacts. The competent authority analyses this document with input from other administrations and entities with a marine remit so as to enable them to make a decision on whether a full EIA is required. If approved, the environmental authority grants the environmental authorisation and establishes project specific conditions. This legislation covering EIA was amended in 2013 through Law 21/2013 and this now stipulates that all projects producing energy in the marine environment are subject to a simplified EIA process administered by the Ministry of Agriculture, Food and the Environment. The amendment sought to address recognised delays in the consenting process and hence it aims to reduce the time-scale needed for obtaining environmental authorisation, establishing a time period of no more than 4 months, or 6 months if there are exceptional circumstances [103].

The process for consenting electricity generating stations in the territorial sea is covered by Royal Decree 1028/2007. This was originally drafted for competitive off-shore wind proposals but has been expanded to cover other MRE technologies such as wave and tidal in Article 32, as a competitive procedure was not viewed as necessary for ocean energy at this time. Generally, consents for energy production are governed by Royal Decree 1955/2000 which covers energy transport, distribution, commercialisation, supply and operation. This states that all the electricity generating installations listed in Article 111 require the following elements:

- request for administrative authorisation – a technical document relating to the project's installation plan;
- project execution approval – this relates to the commissioning of the project and enables the project proponent to start construction;
- exploitation authorisation – once the project is constructed this allows the development to be 'switched on' and proceed to commercial production.

The Ministry of Industry, through the Directorate General for Energy Policy and Mines, is the decision-making body and responsible for granting the administrative authorisation. Regional governments may be involved in the consenting process if the project is to be located in internal waters. Similarly, if a marine energy development is likely to impact upon maritime safety or navigation, the Directorate General of the Merchant Navy, part of the Ministry of Development, will input into the consenting process.

Spain does not yet have an MSP system although the feasibility of such an approach has been explored for siting wave energy devices on the Basque continental shelf [111]. The Spanish Protection of the Marine Environment Act (2010) also includes principles and procedures for planning in the marine environment and transposes the MSFD into Spanish law. This legislation extends from the internal waters, the territorial sea, the EEZ, the fisheries protection zone in the Mediterranean, to the continental shelf. The Spanish government through its Ministry of Industry, Energy and Tourism conducted an SEA on offshore wind in 2009 [112]. The aim of this work was to determine areas of the public maritime domain that had favourable conditions, including little or no expected environmental effects, for the installation of offshore wind farms. It defined areas according to suitability; unsuitable or 'exclusion zones' are coloured red on a map, suitable areas are coloured green and areas that may be suitable but which are subject to additional requirements or conditions are coloured yellow [112].

10.5.6 United Kingdom

The UK has a devolved system of government, with specific responsibilities retained by the UK government in Westminster and certain other responsibilities devolved to assemblies and governments in Scotland, Wales and Northern Ireland. The Department of Energy and Climate Change (DECC) has a UK-wide remit and coordinates the activity of the devolved administrations. The DECC thus has direct input into policies on marine renewables as an energy source and as a means to mitigate climate change. A Renewable Energy Road Map was introduced in 2011 [113], which followed the 2009 UK Renewable Energy Strategy [114] and the 2010 Marine Energy Action Plan [115]. In this section the jurisdictions of England and Wales are considered together and Scotland and Northern Ireland are dealt with separately. The Department of Environment, Food and Rural Affairs (DEFRA) is the UK government department, legislator and policy-maker for issues relating to the natural environment, sustainable development, and environmental protection.

The Crown Estate acts as the manager of Crown-owned lands and is governed by the Crown Estate Act 1961. It manages approximately half the foreshore and almost the entire seabed in the UK's territorial seas. In the EEZ, the Crown Estate acts as land-owner for all jurisdictions of the UK. The Crown Estate has the power to issue leases, depending on the position of the site and type of renewable energy project in question. The leasing process consists of at least two stages [116]. The first step is an exclusivity agreement, which is a form of 'reservation' over an area of seabed or over seabed rights, that essentially guarantees the developer that the Crown Estate will not permit any other developer of the same technology to use that particular area of seabed for the duration of the exclusivity agreement. During this time a developer may conduct environmental and engineering studies so as to better define the project area. The exclusivity agreement also specifies how the developer can avail of the option to call for an Agreement for Lease (AfL), subject to meeting certain milestones.

The next stage is the AfL, which provides a developer with an enforceable option to call for a lease from the Crown Estate. According to the Crown Estate guidance [116], this gives:

- the developer exclusivity over an area of seabed;
- a restriction on the ability of the Crown Estate to grant rights over the relevant area to other uses;
- an enforceable option right to require the Crown Estate to grant a lease to the developer;
- temporary rights to conduct ancillary activities relating to the potential development and the developer's statutory consent application;
- rights to call for the addition of an export cable route to the offshore AfL at the appropriate stage in development.

The enforceable option to call for a lease from the Crown Estate is conditional on the developer having obtained the necessary statutory consents for the project. If this does not happen then no AfL will be issued. If an AfL is issued it gives investors confidence in the project. The developer will conduct all the necessary technical and environmental work at the project/site level before continuing with the statutory consenting process under other legislation. A marine licence from the Marine Management Organisation (MMO), for example, may be necessary for some of this work. During this period the developer still has no right to begin construction or operation of the project. Once the developer has received the necessary statutory consents from the relevant authorities, the Crown Estate will award a seabed lease to the developer. This is a commercial agreement that enables the developer to construct and operate the project for a set period of time. In return for the lease granting the developer seabed rights, the developer pays the Crown Estate an agreed rent. The lease will also include a number of site-specific controls and restrictions on the developer and oblige him/her to comply with other statutory consents.

The DECC is in charge of the decommissioning of projects under sections 105–114 of the Energy Act 2004. These responsibilities were augmented by sections 69–71 of the Energy Act 2008, which introduced three new provisions into the offshore renewables decommissioning regime in order to give greater clarity to developers and more protection to taxpayers. In 2011, the DECC published guidance notes for industry on the decommissioning of offshore renewable energy installations under the Energy Act 2004 [117]. The guidance notes apply to territorial waters in or adjacent to England, Scotland and Wales (between the mean low water mark and the seaward limits of the territorial sea) and to waters in the UK Renewable Energy Zone (including that part adjacent to Northern Ireland territorial waters). The scheme does not apply to the territorial or internal coastal waters of Northern Ireland. Neither does it apply to intertidal areas (between high water mark and low water mark) of any of the jurisdictions. Removal of structures in that zone must comply with consents granted under the Coast Protection Act 1949 and the Food and Environment Act 1985.

Section 104 of the Energy Act 2004 defines offshore renewable energy installations as those which are (section 4.2):

a) used (or will be used or have been used) for purposes connected with the production of energy from water or winds; and
b) permanently rest on, or are permanently attached to, the bed of the waters; and
c) are not connected with dry land by a permanent structure providing access at all times for all purposes.

It also explicitly states that the provisions apply to all new offshore wind, wave and tidal energy installations regardless of their generating capacity and whether they are commercial or demonstration devices [117]. Developers are requested to discuss their decommissioning plans at an early stage with the DECC, preferably before construction of the installation. Once a developer has secured one of the statutory consents (e.g. marine licence), the Secretary of State will issue the developer with a notice requiring the developer to submit a decommissioning programme. The developer will then draft such a programme outlining what parts and how the project will be decommissioned, an EIA (and AA if necessary) and measures to mitigate impacts on the marine environment, stakeholder consultation, anticipated costs and financial security, seabed clearance and any necessary post-decommissioning monitoring.

DECC makes decisions on decommissioning on a case-by-case and site-by-site basis so that specificities of the project and its location can be thoroughly taken into account [117]. There is a presumption in favour of completely removing the installation and subsequently taking it onshore for reuse, recycling, or disposal. Exceptions to this will only be considered where there are very good reasons, and five such circumstances are specified:

a) the installation or structure will serve a new use (e.g. other energy generation, enhancement of a living resource now (not in future));
b) entire removal would involve excessive cost;

c) entire removal would involve an unacceptable risk to personnel;
d) entire removal would involve an unacceptable risk to the marine environment;
e) the installation or structure weighs more than 4000 tonnes in air or is standing in more than 100 m of water and could be left wholly or partially in place without causing unjustifiable interference with other uses of the sea.

Examples cited in the guidance include a breakwater incorporating a wave energy device (the device could be removed but the breakwater left in place), existing cables that could be used for future projects, or if buried where removal would cause greater disturbance to the environment, and where foundations and other parts of the structure are below the seabed [117]. The DECC's guidance is strongly based on the relevant provisions of the LOSC, the IMO Guidelines and Standards and OSPAR work on this, and related, topics.

The UK Marine Policy Statement (MPS) [118] provides the policy framework for the marine planning system along with the Marine and Coastal Access Act 2010, which provides the legal basis for marine plans. The Act divides UK waters into marine planning regions with an inshore region (0–12 nautical miles) and offshore region (12–200 miles) under each of the four jurisdictions (England, Wales, Scotland and Northern Ireland). The plans developed essentially implement the strategic objectives for the marine environment that are identified in the MPS [118], the National Planning Policy Framework [119] and the Localism Act 2011. Each jurisdiction within the UK is at a different stage in implementing its marine plans (discussed below). If no marine plan is in place for a particular marine area, the MPS is deemed to be the correct policy basis and therefore provides the context for decisions that affect the marine areas. Any public body that diverges from the MPS must justify its decision. From a developer perspective any proposed project must also be in line with both the MPS and local or regional level marine plan.

10.5.6.1 England and Wales

The DECC is the government department and policy-maker for issues related to energy and climate change, and consequently the policies it adopts apply in England and Wales. In England the MMO acts as the decision-maker on behalf of the Secretary of State for Energy and Climate Change for offshore developments generating up to 100 MW and also on behalf of DEFRA for marine licences. For projects larger than 100 MW, the Infrastructure Planning Commission makes the decisions on behalf of the Secretary of State for Energy and Climate Change. In Wales, Natural Resources Wales is responsible for granting marine licenses in inshore waters (up to 12 nautical miles). Beyond 12 nautical miles, in offshore waters, the MMO grants such licences. The MMO process consists of four stages: the applicant registers for an online service account, submits an initial enquiry, a request for pre-application advice (e.g. screening, scoping, environmental statement review) and then the licence application.

Other components of ocean energy projects are licensed under different legal requirements. A marine licence is required under section 66 of the Marine and Coastal Access Act 2009 and available from the MMO. This licence covers the construction of works in the marine environment, the deposit of substances or articles and dredging. To build and operate an energy generation installation, section 36 consent is required under Electricity Act 1989. Some electricity generating installation may be subject to regulation under the

Electricity Works (EIA) (England and Wales) Regulations 2000 (as amended), which would necessitate an EIA to be submitted to the MMO with the main project application. Developers can apply to the MMO for a safety zone declaration around offshore renewable energy generating stations, governed by section 95 of the Energy Act 2004. In Welsh waters the MMO role is limited to sections 36 and 36a of the Electricity Act and Section 95 of the Energy Act.

Depending on the location of an ocean energy project, a developer may require a European Protected Species licence under the Conservation of Habitats and Species Regulations 2010. This gives a developer permission to carry out an activity that would affect a protected animal or plant designated under EU law (Birds and Habitats Directives) that would otherwise be illegal. Licences are only issued for certain purposes, for example in relation to ocean energy; this could involve the disturbance or damage to the habitat of certain protected species. Natural England is responsible for administering these licences in England and in Wales it is within the remit of the Countryside Council for Wales. In a similar fashion to other parts of Europe, local planning authorities or councils are authorised by law to exercise functions relating to town and country planning. This could include electrical sub-stations onshore or overhead cables, covered by section 37 of the Electricity Act 1989.

England has a comprehensive approach to marine planning deriving from the enactment of the Marine and Coastal Access Act 2010. There are currently ten marine plan areas across England, covering an area of about 253,000 km^2. The MMO is responsible for producing these plans and the process will be complete by 2021 [120]. Each marine plan has a strong legal basis and is enforceable. Generally, the marine plan details priorities and directions for future development within the plan area, assists in the sustainable use of marine resources and informs marine users about the best locations for their activities and where new developments may be appropriate. In developing the marine plans for a particular area, the MMO must adopt a Statement of Public Participation outlining the MMO's plan to include stakeholders in the process, to comply with Schedule 6 of the Marine and Coastal Access Act 2010. To date they have worked with a wide range of stakeholders, including local communities, councils, interest and user groups and the general public.

In Wales the Welsh government is currently working on developing a Welsh National Marine Plan which will cover Welsh inshore and offshore waters in a single plan. A public consultation process ran from February until May 2011 on a proposed approach to marine planning in Wales [121]. The final version of the Plan aims to promote suitable marine opportunities and to help to manage marine activities in a sustainable way, while simultaneously informing marine licensing and decision-making. A strategic scoping exercise was carried out to review and analyse the available evidence for Welsh waters [122]. The government has also commissioned a number of research projects to fill specific evidence gaps, such as those relating to aquaculture, seascapes and recreational fishing [123]. A dedicated portal for marine data and information has also been developed as part of this process [124].

10.5.6.2 Scotland

Scotland has invested significant effort and resources into refining its consenting processes to better reflect the needs of the MRE industry in Scottish waters. To date there have been six commercial offshore wind leasing rounds since 2000. Developers wishing to deploy devices in Scottish waters require a lease from the Crown Estate under the process described above. The UK's Marine and Coastal Access Act 2009 sought to modernise and reform marine planning and licensing arrangements, and this has been supplemented in Scotland with the enactment of the Marine (Scotland) Act 2010. This provides for the creation of a national or regional marine plan(s) (Part 3), a rationalised marine licensing process (Part 4), as well as provisions on enforcement (Part 7). The licensing provisions introduce a single marine licence which replaces the licences previously required under the Food and Environment Protection Act 1985 for deposits and under the Coast Protection Act 1949, relating to navigation. A licence to construct an electricity generating station is still required under section 36 of the Electricity Act 1989, but the application for this can now be considered at the same time as the application for the marine licence for an ocean energy project. The Scottish Act also created a pre-application consultation process where prospective applicants must submit a pre-application notice and report before submitting their application.

A new marine management authority for Scottish waters, Marine Scotland, was also created under the Marine (Scotland) Act. Its Marine Licensing Operations Team (MS-LOT) is responsible for all marine licensing functions. Marine Scotland operates as a one-stop shop, which means that in practice there is a single point of contact to administer project development applications. Marine Scotland has a policy target of providing a decision on all applications within nine months, provided the requisite environmental information is available and no local public inquiry is needed [125]. In recognition of the fact that consenting was viewed as complex and confusing by developers, Marine Scotland published a licensing manual to assist them in navigating the process [125]. This manual also includes all the requirements deriving from EU legislation such as EIA and AA under the Habitats Directive, known as Habitats Regulation Appraisal in Scotland. In practice, when a developer has a pre-application consultation with MS-LOT, they will advise the developer on the suitability of their site, whether an AA may be needed, information on the type of environmental data that is required, the stakeholders to be consulted and the expected costs and time-frames associated with the process [125]. The developer can request a formal EIA screening and scoping subsequent to this if it is felt necessary.

After the pre-application consultation, the developer will commence the EIA process and stakeholder consultation. This information coupled with Navigational Risk Assessment, Habitats Regulations Appraisal, and any other information requested will then be submitted to MS-LOT with a completed application form and payment. MS-LOT administers the applications and consultations necessary for the various consents and compiles the feedback and input obtained from statutory consultees and other stakeholders. At this point if information is missing or additional studies are required the developer will be requested to provide that. If the application is unsuccessful, MS-LOT will give reasons and advise the developer on how to proceed. When an application is successful, the various consents will be granted. It is normal for the consents to include standard terms and conditions. Such conditions usually apply to that specific development site and any ongoing or post-consent environmental monitoring requirements. The conditions will not cover decommissioning of the project as it is not a devolved responsibility, and remains the responsibility of DECC.

In Scotland, the National Marine Plan (NMP) was adopted on 25 March 2015 and laid before Parliament on 27 March 2015 [126]. It is a detailed document covering all current marine sectors in Scotland as well as reiterating objectives of EU law such as those of the MSFD. The high-level marine objectives of the plan are to achieve a sustainable marine economy, to ensure a strong, healthy and just society, to live within environmental limits, to promote good governance and use sound science responsibly [127]. The Plan also outlines key objectives for the offshore wind and MRE sector in Scotland covering not only marine planning aspects but also marine licensing and maximising benefits from development of this sector at regional level [127]. The NMP will be supported by Regional Marine Plans to cover 11 marine regions extending to the territorial sea limit (12 miles). These will be developed by local Marine Planning Partnerships that have delegated powers from Scottish Ministers, and will reflect local issues and needs in each region. The first of these regional plans will cover the Shetland Isles and Clyde area [128].

In some respects, the MSP process in Scotland is further forward and more evolved than elsewhere in Britain. Along with the NMP and regional plans to be developed, Scotland already has a series of sectoral marine plans. With respect to marine renewables, Marine Scotland advances this approach through its Sectoral Marine Plan for Offshore Wind Energy in Scottish Territorial Waters [129] and draft plan options for wave [130] and tidal [131]. This was informed by an SEA in 2007 [79] which resulted in separate Regional Locational Guidance for offshore wind [80], wave [81] and tidal [82] respectively in 2012, and deep-water floating wind in 2014 [132]. During the development of the regional guidance, Marine Scotland focused significantly on the environmental effects of offshore energy devices and developed specific management approaches that enabled development while also ensuring environmental protection. This risk-based management approach takes the form of a 'survey, deploy and monitor' policy [133]. It proposes a phased approach to the licensing of wave and tidal developments whereby developers, following pre-deployment monitoring, undertake deployment and monitoring of a test device or demonstration array before seeking consent for a larger array.

The policy is based on three specific aspects: the environmental sensitivity of the site, the scale of the proposed development and the device (or technology) classification. Environmental sensitivity refers to the presence of designated conservation sites and protected species in or near the proposed project. To inform this element of the policy, Marine Scotland has prepared sensitivity maps comprising data from 19 different environmental data sets, enabling areas of relatively higher and lower sensitivity to be identified [133]. Scale of development centres on the proposed installed capacity of the project, sub-divided into low (up to 10 MW), medium (10–50 MW) and high (above 50 MW). Device or technology classification involves a consideration of the risks associated with a particular device or technology. This could include, for example, how it is installed (moorings, piles), how it moves (number of moving parts above or below the water surface), its behaviour and interactions. A score for each of the three elements is determined and totalled, then the geometric mean of that figure is calculated to give a final overall score that can then be classified as a project of high, medium or low risk as shown in Table 10.4 [134].

The final score attributed to the proposed project is used by Marine Scotland to guide the requirements for pre-application site characterisation and assessment of the environmental effects of the devices. Essentially this means that monitoring requirements placed on the developer are appropriate and proportionate to the development proposed.

Table 10.4 Overall device/technology risk.

Geometric mean score	Overall risk
1–1.60	Low
1.61–2.20	Medium
2.21–3.0	High

If, for example, a proposed project were to be sited in an area of low environmental sensitivity and with little device risk it would be assessed as low using the framework described. In such circumstances, if the environmental risk information is viewed as robust, Marine Scotland could fast-track the application and request only 1 year's site characterisation data to inform an EIA, AA (if required) and licence application. If a project were to be located in a site of higher environmental sensitivity and the effects of the actual device to be deployed were unknown or uncertain, then the proposed project would be assessed as high risk and a minimum of 2 years' site characterisation data would probably be necessary to support such an application. The developer would also be expected to monitor the device and its effects elsewhere, perhaps in a test site, and provide the results of that monitoring in support of the application to Marine Scotland [134].

10.5.6.3 Northern Ireland
Northern Ireland has an Offshore Renewable Energy Strategic Action Plan (ORESAP) 2012–2020 [135] following an SEA and associated AA for marine renewables in Northern Ireland waters [84]. The overall aim of the ORESAP is 'to optimise the amount of renewable electricity sustainably generated from offshore wind and marine renewable resources in Northern Ireland's waters in order to enhance diversity and security of supply, reduce carbon emissions, contribute to the 40% renewable electricity target by 2020 and beyond and develop business and employment opportunities for NI companies. The associated development opportunity is for up to 900 MW of offshore wind and 300 MW from tidal resources in Northern Ireland waters by 2020' [135]. Following this, the Crown Estate launched two parallel leasing rounds in December 2011 for offshore wind and tidal developments. Northern Ireland is home to the first grid-connected, commercial scale tidal energy device, the SeaGen device, in Strangford Lough, Co. Down. Following the Crown Estate leasing rounds, two separate commercial scale tidal arrays are being developed off the north-east coast. DP Marine Energy and Bluepower NV have an AfL for a 100 MW tidal array off Fair Head, to be developed on a phased basis, by 2020. The second tidal project off Torr Head is a joint venture between OpenHydro and Brookfield Renewable Energy Group. It will also be a 100 MW project. Both projects are currently at the EIA stage of consenting.

The current operational consenting process for ocean energy in Northern Ireland involves a lease from the Crown Estate, as *de facto* managers of the seabed around Northern Ireland. Under the provisions of the UK Marine and Coastal Access Act, Part 4, a Marine Licence is required from the Department of Agriculture, Environment and

Rural Affairs (DAERA) Marine Division.[2] This covers the marine elements of a project including the placement of anything on, or removing material from, the seabed. Electricity generation consents must be obtained from the Department of Enterprise, Trade and Investment (DETI) and the utility regulator. DETI, now Department for the Economy, is responsible for electricity consents under Article 39 of the Electricity (NI) Order 1992 which enables the construction and operation of generating stations. A licence from the Northern Ireland Authority for Utility Regulation under Article 10(1)(a) and/or Article 10(1)(b) of the Electricity Order is also required. Within their respective legislative frameworks, the Department of Agriculture, Environment and Rural Affairs, Department for the Economy and the relevant local authorities (in respect of any land based development arising from the project) require three separate EIA regulations to be met.

As part of the implementation of ORESAP, the DETI requested the Northern Ireland Assembly to enact new primary legislation, introducing new powers for a regulatory regime for offshore renewable energy projects. A consultation document was published in February 2013 outlining proposed new powers on [136]:

- safety zones around offshore renewable energy installations and prohibition of certain activities in those safety zones;
- mavigation and extinguishing of public rights of navigation in the relevant areas;
- preparation and implementation of decommissioning programmes; and
- consequential amendments to legislation as a result of the above measures.

The main reason for this is that a number of the provisions contained in the UK Energy Act 2004 (as amended in 2008) cannot apply in Northern Ireland waters, due to complex jurisdictional issues between the UK and Ireland. The proposed legislation seeks to align the Northern Ireland consenting system with that applicable in the waters around Great Britain. At the time of writing, the Offshore Renewable Energy (Northern Ireland) Bill has not yet been enacted.

Northern Ireland is part of wider UK initiatives on integrated marine management through the implementation of the Marine and Coastal Access Act 2009. As part of this, the Marine (Northern Ireland) Act entered into force in 2013. The Act provides a framework for developing marine plans covering the Northern Ireland inshore region; marine conservation zones; and streamlining marine licensing for certain electricity works. The Northern Ireland inshore region is defined as the territorial sea and the seabed adjacent to Northern Ireland, out to 12 nautical miles. Work on a marine plan for Northern Ireland commenced in March 2012. As part of this process, the Marine Plan Team published a Statement of Public Participation in June 2012 [137], which was subsequently reviewed and updated in May 2013 [138].

10.5.7 United States

The system of consenting MRE, or marine and hydrokinetic (MHK) technologies as they are known in the USA, follows a federal–state separation according to the location of project. The Federal Energy Regulatory Commission (FERC) exercises regulatory

---

2 Following elections in Northern Ireland on 5 May 2016, certain departmental names, roles and responsibilities changed. The DOENI became the Department of Agriculture, Environment and Rural Affairs.

jurisdiction over MHK projects under an amendment to the Federal Power Act authorising it to regulate and licence hydroelectric projects on navigable waters within 3 nautical miles of the shore and on any projects with an onshore grid connection. The FERC does not have the powers to regulate ocean thermal energy conversion projects, which fall under the remit of the National Oceanic and Atmospheric Administration (NOAA) according to the Ocean Thermal Energy Conversion Act 1980. As the USA is not a signatory to the LOSC it has a different definition of the term 'continental shelf'. In the USA the Outer Continental Shelf (OCS) includes all submerged lands, subsoil and seabed between the seaward extent of the states' jurisdiction, usually 3 nautical miles, to the limit of federal jurisdiction of 200 nautical miles. In the OCS, the Bureau of Ocean Energy Management (BOEM) will administer the granting of leases if a project produces, transmits or transports energy, is located on the OCS and incorporates the temporary or permanent attachment of a structure or device to the seabed. The construction and operation of a MHK project on the OCS also requires a licence from the FERC. Recognising that complex jurisdictional issues impacted upon development of MHK projects, both authorities published comprehensive Guidelines on Regulation of Marine and Hydrokinetic Energy projects on the OCS in 2012 [139].

Three types of leases can be granted by the BOEM for MHK projects. A commercial lease is required for a commercial project. A research lease is issued only to federal agencies or states for renewable energy research activities that support the future production, transportation or transmission of renewable energy. A limited lease applies to projects of limited scope, normally where the activities associated with the project are limited to five years and the power generated by the project is also restricted (e.g. 5 MW) through the terms and conditions attached to the lease [139]. Given the cross-jurisdictional remit fulfilled by the BOEM, it often assists states in the development of their offshore energy resource, through specific development proposals and task-forces [140]. Generally, all lease applications submitted to the BOEM are considered on a case-by-case basis, but in collaboration with other regulatory and state agencies.

Under the Guidelines [139] a developer can develop a project under a BOEM lease but without a FERC licence if certain conditions are adhered to:

1) the technology is experimental;
2) the proposed installation is temporary in preparation for a further licence or for educational purposes; and
3) the power generated would not be transmitted to the grid.

In certain other cases, the FERC is authorised to grant licence waivers or exemptions. To avail of such an exemption, a project must be small, short-term, located outside a sensitive area (in the FERC's opinion), removable and capable of shutdown at short notice, removed before the end of the licence period and initiated by a draft application with relevant supporting environmental information capable of analysis by the FERC [139]. Section 2.13 of the Guidelines provides that if a licence has been granted by FERC for a pilot project, the applicant can usually transition to a standard licence in due course. A limited lease cannot, however, be converted to a commercial lease [139].

The system in the USA is largely similar to that in the UK, in that leasing takes place through competitive rounds instigated by the BOEM and developers can then respond. If there is no suitable licensing round, a developer can submit an unsolicited application to the BOEM stipulating their interest in obtaining a lease for that specific OCS location.

This application should outline the area of interest, describe the project facility and its activities as well as any available resource and environmental data. An applicant must also prove that they are qualified to hold a lease, as governed by the Code of Federal Regulations (CFR). If there are no other applications, the developer will then be requested to provide a Site Assessment Plan (SAP) detailing environmental surveys and resource assessment studies to support the planned project [139]. There is no requirement to submit an SAP if the project proposed does not involve the installation of bottom-founded facilities. For locations where there is a competitive interest in MHK project development in a specific area, BOEM will publish a Proposed Sale Notice followed by a Final Sale Notice, once the requisite environmental review and consultations have been completed. Developers must comply with the Final Sale Notice. Once a developer has completed the necessary documentation and made payment, the BOEM will issue a lease to the successful developer. The successful developer will have to submit a SAP within six months of issuance of the lease, unless there are no bottom-mounted facilities associated with the development.

While the granting of a lease and approval of an SAP enable a developer to conduct certain activities identified therein, the lease does not cover the right to generate power. The latter right is only conveyed once a FERC licence has been obtained. According to the guidelines, where a BOEM lease and FERC licence are required, the two processes can be combined in the interests of time and expediency. This is dependent on the type of licence being applied for and whether it is in response to a competitive leasing round or an unsolicited application. The FERC licensing process can take one of three forms: an Integrated Licensing Process (ILP), a Traditional Licensing Process (TLP) and an Alternative Licensing Process (ALP) [139]. The ILP is the most commonly applied process and involves a pre-application stage, during which any necessary studies are conducted and a licence application is prepared, and a post-application stage, when the application is reviewed, an environmental document is compiled and a decision on licensing is made. During both stages, the FERC will seek the input of stakeholders. In terms of applicable time-frames, the Guidelines specify that for a BOEM lease in an area of competitive interest, the process could take up to 2½ years, including stake-holder consultation and environmental work. Where there is no competitive interest, the time-frame is less, perhaps 1–2 years, depending on the complexity of the project being proposed. Receipt of a FERC licence takes approximately 1 year, but this is largely dependent on the extensiveness of the application submitted. Pilot project licences from FERC take 6 months from the date of application submission.

BOEM customarily issue commercial leases for a period of 25 years, though this can be extended or amended to align with the associated FERC licence. Limited leases have a term of 5 years. The duration of a research lease is negotiated on a case-by-case basis by officials from the BOEM, the applicable federal agency, the state or other authorised representatives. Leases can also be surrendered or cancelled as normal under adminis-trative law. FERC licences have slightly different life spans. The Federal Power Act per-mits the FERC to issue a licence for up to 50 years, with supplementary powers to extend such a licence for between 30 and 50 years. Pilot projects licensed by the FERC have a much shorter existence, usually 5 years, so as to reflect the early stage of the technology, the scale of the project and its overall experimental nature. Occupation of the sea space under a commercial or limited lease requires an initial, one-off payment to the BOEM to obtain the lease as well as ongoing, annual payments once the term of the lease begins.

The amount payable to the BOEM varies according to whether the developer is responding to a competitive leasing auction or a non-competitive leasing process. BOEM does not require payment for research leases. The FERC requires payment for the administrative charges associated with licences for projects over 1.5 MW. It also needs payment for land charges which vary according to whether the land is federal or tribal.

One novel feature of the guidelines published by the BOEM and FERC is its inclusion of hybrid projects, defined as a project that includes technologies that generate electricity from more than one form of renewable energy, one of which is an MHK technology (e.g. wind and wave generation) under the same lease. Calls for projects of this type can occur through either a competitive auction or non-competitive application process. A licence from the FERC is also required for this type of project. 'Straddle projects' are those MHK projects that straddle the boundary dividing state waters and the OCS, and are also covered by the guidelines [139]. In this location, a developer must obtain a lease from the BOEM for the OCS part of the project and a licence from the FERC for both the OCS and state waters part. FERC has a preference to licence the project as one complete project, and this is feasible provided the developer consults with the FERC, BOEM, adjoining state authorities and stakeholders at a sufficiently early stage of the project planning process.

In addition to the BOEM lease and FERC licence, authorisation from the US Army Corps of Engineers is necessary where a structure is deployed in navigable waters, under the Rivers and Harbors Act, §10. If a development requires dredging to facilitate the laying of seabed cables and anchors, these may require a permit under the Clean Water Act, §404. If a device or any of its constituent parts might obstruct navigation, it must be marked by the appropriate navigational aid. This requires a Private Aid to Navigation (PATON) Permit administered by the US Coast Guard under Title 33 of the CFR. The National Environmental Policy Act, which provides a framework for identifying and assessing the environmental effects, also applies to MHK projects. The federal agency will firstly determine if the project can be excluded from a comprehensive environmental review, term a 'categorical exclusion'. If this is not the case, then an environmental assessment (EA) will be conducted by the federal agency. This document will contain sufficient information to conclude whether an EIS is necessary. Where no significant effects have been identified during the EA, federal agency officials will prepare an assessment stating that the project is unlikely to have significant effects, coupled with a finding of no significant impact. Where significant environmental impacts are anticipated an EIS will be produced with the assistance of other regulatory agencies and stakeholders. Other regulatory agencies must comply with their governing legislation and its requirements, which can result in additional consents, as presented in Table 10.5.

The BOEM has published a number of guidance documents relating to different types of environmental information, including spatial data for site characterisation; avian survey information; geological, geophysical and hazard information; fisheries survey information; benthic habitat information; and marine mammal and sea turtle information. All of these guidelines are available to download from the BOEM website [141]. Decommissioning of renewable energy projects is governed by 30 CFR §285, which provides that all facilities, including pipelines, cables and other structures and obstructions must be removed once they are no longer operational. Removal must occur no later than two years after the termination of the related lease (30 CFR §285.902). This could be problematic in future for large projects as the two-year

Table 10.5 Other applicable Federal Statutes and their possible implications for MHK projects.

Legal instrument	Requirement
National Historic Preservation Act	Assess potential effects on historic resources
Endangered Species Act	Evaluate impacts on endangered species and critical habitats
Marine Mammal Protection Act	Assess potential impacts on marine mammals
Migratory Bird Treaty Act	Assess potential impacts on migratory birds
Magnuson–Stevens Fishery Conservation Act	Evaluate actions that may adversely affect essential fish habitat
Clean Water Act	Assess proposed project so as to comply with water quality standards
Clean Air Act	Assess proposed project so as to comply with air quality standards
Coastal Zone Management Act	Review proposal to ensure consistency with state coastal management policies

time-frame applies regardless of the size of the project [142]. According to section 6.2 of the BOEM/FERC guidelines [139], developers are required to provide a decommissioning bond or other acceptable form of financial assurance as part of their BOEM lease and/or FERC licence.

10.6 Gaps and Opportunities

10.6.1 Legal Basis

It is clear from the preceding sections that a significant number of legal instruments, at all governance scales, have the potential to impact upon ocean energy development both now and in future. The LOSC provides a solid legal basis for states to exploit resources within their national jurisdiction, subject to certain conditions and the right of innocent passage. While not economically or logistically possible in 2017, the potential for ocean energy devices to be deployed and operate in areas beyond national jurisdiction is a possibility in future, but currently there is no coherent governance framework to assist development in those areas. It is likely that such developments are more likely to include ocean thermal energy conversion and submarine geothermal energy than wave or tidal energy devices, as presently conceived. Already the deployment of offshore wind farms in the EEZ is common in many countries of the North Sea, but in many other countries there is no operational consenting system in place for marine or ocean energy in that zone or beyond. The priority for development is nearshore areas incorporating internal waters and the territorial sea. In practice this means that national competent authorities have concentrated their effort on streamlining existing consenting processes applicable in these zones and the domestic legal provisions pertaining to them. There has been little or no consideration of how development could, over time, progress to sea areas further offshore.

As ocean energy incorporates a range of regulated activities covered by differing pieces of legislation and administrative authorities, it is somewhat inevitable that the consenting system is convoluted and *ad hoc* in many places. The assortment of consents, licences, leases, permits, authorisations and permissions involved in consenting are individually complex and time-consuming and consequently often difficult to integrate into a single project licence. This results in considerable regulatory uncertainty in many jurisdictions. For developers trying to navigate the process there is often a lack of clear and accurate information on the processes to be followed [1, 143–145]. The number of competent authorities involved and level of communication between them, the time taken to process consenting applications and the applicability of an integrated process which incorporates land, sea and electrical parts of an offshore energy project are the key factors of concern within consenting of ocean energy projects internationally. The result of this is uncertainty for the developer and their investors, but at a broader scale this has significant implications for society's progress towards greater renewable energy production, more efficiency and cost reduction [146]. Evidence from the EU and individual countries suggests that progress is being made in streamlining applicable projects, but this in itself can be difficult in practice, requiring substantial legal amendment in some cases (e.g. the UK) as well as high levels of investment and political commitment, which can sometimes be scarce for new and developing sectors.

In an EU context, the European Commission in its progress report on implementation of the Renewable Energy Directive found that only Denmark, Italy and the Netherlands have a single- permit system for all renewable energy projects [56]. In Denmark the one-stop shop approach was realised by formalising communication channels between existing relevant authorities, thus making it easier and more cost-effective to implement [144]. Article 13 of the Renewable Energy Directive explicitly provides that 'administrative procedures are streamlined and expedited at the appropriate administrative level' and that such procedures are 'clearly coordinated and defined, with transparent timetables for determining planning and building applications' (Article 13(1)(c) and (a)). One adaptive mechanism under this article is the flexibility to apply 'simplified and less burdensome authorisation procedures' to smaller projects and for decentralised devices producing renewable energy (13(1)(f)). In theory this should provide a legal basis for competent authorities to distinguish between the regulatory requirements for demonstration-type projects and larger projects of commercial scale. The 2015 progress report on implementation of the Renewable Energy Directive points out that a majority of member states recognise the need for more improvements in their respective administrative systems as they apply to renewable energy [55]. The Commission has reported on progress made in relation to practical implementation of administrative procedures for renewable energy projects in its 2017 report [60].

The legal basis for consenting of ocean energy is well established, but the procedures involved in its administration remain multifaceted and difficult. This could change as the number of actual operational ocean energy deployments increases and problematic issues become more and more apparent. Institutional fragmentation is a challenge that is very pertinent to ocean energy given the range of elements covered by its operation. While progress has been made in some domestic legal systems an interconnected and consistent framework is an aspiration for many countries and developers alike. The apparent favoured approach towards addressing this issue is the implementation of a 'one-stop shop', as demonstrated by the authorities in both

Scotland and Denmark. These are viewed favourably by developers but, depending on the model adopted, they can also have significant resource implications and may require formal legal amendments.

10.6.2 Environmental Impacts and Assessment

The integrity and protection of the marine environment is of paramount concern to competent authorities who sanction development in that area and can also trigger very strong legal obligations. White [147] states that there is concern that regulatory and governance processes will be relaxed so as to pursue and achieve a reduction in greenhouse gases, thereby causing 'paradoxical harm' to local ecosystems. Conversely, newer forms of renewable energy, such as ocean energy, can result in positive effects that get little or no attention in the consenting process, tending to be overlooked completely. Current impact assessment processes, such as SEA and EIA, concentrate primarily on the negative ecological effects or impacts of a development. Across the EU, there appears to be limited application of SEA to MRE development and ocean energy in particular. In some respects, this could be regarded as a missed opportunity as the documentation compiled for the SEA usually becomes the first source of environmental information that developers consult during site selection and project planning stages. EIA has almost become a prerequisite for consent in most countries. The parameters to be measured and included with the EIA itself are heavily influenced by the proposed location of the project, as they should be, given EIA is a site-specific assessment. What is problematic is that there is little or no consistency in the methodologies applied to the study of specific parameters or questions posed in relation to them. This, in turn, limits the ability to draw inferences, identify trends and increase knowledge on the environmental effects of device deployments as different methodologies may produce different results. The ability of scientists to compare data and results across deployment sites is one way in which knowledge and expertise can be increased, thus advancing learning about these new technologies.

An amended EU EIA Directive (2014/52/EU) entered into force in May 2014, with member states given until 16 May 2017 to transpose its provisions into their domestic legislation. The aim of the revisions was to address the identified shortcomings in the EIA process, to reflect changing environmental and socioeconomic priorities and challenges and to align the EIA Directive and process with the principles of smart regulation as well as reflect recent European Court of Justice findings. One of the changes that could impact upon consenting of ocean energy projects is the provision enabling coordinated and/or joint procedures where a number of assessments have to be completed (i.e. EIA and AA under Birds and Habitats Directives). The effect of this is that a single assessment will be possible once the amended Directive applies. Under Article 4, a revised screening procedure is introduced where member states can set thresholds or criteria to decide when a project does not need to be screened or subject to an EIA. If a competent authority decides that an EIA is not needed, it must make this decision available to the public, along with the reasons why it is not required as well as mitigation measures proposed by the developer to avoid or prevent significant adverse effects on the environment. The competent authority in the member state is required to make its determination (on screening) within 90 days from the date on which the developer submitted the required information, under Article 4(6). In an effort to ensure the quality and completeness of submitted EIAs, Article 5 is amended to provide that the developer must ensure that the EIA report is prepared by 'competent experts' (Article 5(3)(a)), though no definition of this in included in the text of the Directive, and the competent authority must ensure it has, or has access as necessary to, 'sufficient expertise' to examine the EIA report (Article 5(3)(c)).

10.6.3 Public Consultation and Acceptance

Public consultation requirements have also been revised in the amended EIA Directive, recognising the role of technology, with an option for the public to be informed electronically as well as by the more traditional form of public notices. The time-frame for public consultation on the EIA report 'shall not be shorter than 30 days' according to (new) Article 6(7). A new Article 8 provides that the results from public consultation and information gathered during the consultation process 'shall be duly taken into account in the development consent procedure'. In granting consent, the competent authority must provide a reasoned conclusion as well as information on any environmental conditions attached to the decision, including monitoring measures. In a change to the existing EIA process, the revised Directive contains specific provisions relating to post-consent monitoring. The amended Directive provides that measures envisaged to avoid, prevent or reduce significant adverse environmental effects are to be implemented by the developer (new Article 8a(4)). This goes further, stating that the type of parameters to be monitored and the duration of the monitoring 'shall be proportionate to the nature, location and size of the project and the significance of its effects on the environment'. This requirement for proportionality is new and could assist in alleviating the perceived burdensome monitoring requirements placed on small developers when deploying single devices for a time-limited period. There is also an option to use existing monitoring arrangements, from either EU or national legislation, to be used if this avoids duplication of monitoring effort.

Public acceptance is becoming more and more influential in the outcomes of decisions relating to the various consents needed to operate a marine or ocean energy project. This has been colloquially termed a 'social licence to operate'. Responsibility for engaging with the public and communities potentially impacted by a proposed development primarily falls to the developer or project proponent. Acceptance is neither automatic nor unconditional, and significant effort on how best to engage the public and allay any concerns they might have is required. Habitually, in past projects, the involvement of the public was almost entirely limited to the consultation phase of the EIA process or at the behest of the individual site developer if a specific issue arose. This routinely resulted in stakeholders expressing frustration with how they were involved in project planning, either by being consulted too late in the process or having limited influence on the decision made. A social licence to operate is based on a multitude of principles, including legitimacy, credibility and trust, which take time to foster and cannot be neatly slotted into a time-bound consenting process. A concerted effort to promote and understand ocean energy technologies and their associated infrastructural requirements could help mitigate objection, thereby increasing the likelihood of social acceptance. Generally MRE is viewed positively but often there is a lack of familiarity and understanding of wave and tidal energy technologies at local level, given there have been relatively few deployments to date [148, 149].

There are also deeper questions in this context that go to the core of rights and ownership issues. Governments or their agencies are responsible for decision-making in the marine space [150]. Some domestic legal instruments provide that such decisions must be made in the public interest or for the common good, but it is not always acceptable that governments decide on the use or activity that can take place in what is considered by many to be a common resource. Through the implementation of MSP, questions regarding the allocation of marine space and its resources are likely to become more prominent and, depending on the approach taken to MSP at country level, plans developed at regional or local level could be more reflective of the needs and desires of the public in that area for their adjoining marine space.

10.6.4 Maritime Spatial Planning and New Management Approaches

The full implementation of MSP could provide an opportunity to improve consenting for many marine developments, including ocean energy, through increasing transparency and providing greater certainty for both developers and their investors. Apart from Scotland, and to a limited extent maritime spatial plans applicable to certain waters around specific US states, many existing maritime spatial plans include only existing uses, with little or no consideration or inclusion of new or innovative marine activities that could occur in future [151]. As a relatively new approach to planning marine activities, it is vital that MSP, and plans developed as part of that process, promote the coexistence of relevant activities and uses and engage everyone in the process. As a new and somewhat unproven approach to integrated maritime governance, MSP is currently perceived as a panacea and can be 'all things to all people'. Some emphasise the participatory and process aspects with indirect legal ramifications, while others perceive it to prescribe marine zones where only certain activities can occur at specific times [152]. Until such time as there is more implementation of MSP and its progress can be evaluated, it is difficult to ascertain its ability to address challenging management problems. The drivers for MSP to date have been numerous and relate to both economic and environmental considerations. The role of ocean energy in MSP processes to date has been limited either to reiterating that MSP is crucial for ocean energy [153, 154] or alternatively to how ocean energy needs have been reflected in the plans developed [151, 155, 156].

MSP seeks to reflect environmental, social and economic interests in an inclusive way. As such it has the potential to balance precaution and risk so as to provide flexibility, but within a framework that is predictable, consistent and transparent to those involved. The adaptive nature of the MSP process can react to changing circumstances, which is important for developing industrial sectors such as ocean energy. It could promote coherence between terrestrial planning systems and those which operate in the marine space, but this very much depends on how it will be implemented and enforced in each country. It is not yet clear how different competing activities will be accommodated in MSP or how decisions on 'trade-offs' will be made. Coexistence may be advocated as a preferred option, but this is not always technically or legally possible. Realities of health and safety concerns, insurance and liability could stymie coexistence before it even happens. In the majority of jurisdictions, consents granted are limited to a single use in a single location so the possibility of multi-use licences for numerous activities encompassing larger areas also requires further consideration.

Acknowledgement

This material is based upon work supported by Science Foundation Ireland (SFI) under the Charles Parsons Award (grant no. 06/CP/E003) in collaboration with Marine and Renewable Energy Ireland (MaREI), funded by Science Foundation Ireland (12/RC/2302).

References

1 O'Hagan, A.M. 2012. A review of international consenting regimes for marine renewables: are we moving towards better practice? 4th International Conference on Ocean Energy, 17 October, Dublin.
2 O'Hagan, A.M. 2014. Consenting and environmental challenges for ocean energy. Presentation made to the European Commission's Ocean Energy Forum, Dublin, 11 June.
3 Article 8(1), LOSC.
4 Article 18, LOSC.
5 Article 19(1), LOSC.
6 Article 19(2), LOSC.
7 Article 21(1), LOSC.
8 Article 22, LOSC.
9 Article 22(3), LOSC.
10 International Maritime Organization. 2008. Routeing of Ships, Ship Reporting and Related Matters. Amendments to the Traffic Separation Scheme 'Off Land's End, between Longships and Seven Stones', Submitted by the United Kingdom. Sub-Committee on Safety of Navigation, 54th session. IMO Doc. NAV 54/3/5, 28 March 2008. IMO, London.
11 Article 57, LOSC.
12 Article 56(1)(a), LOSC.
13 Article 60(3), LOSC.
14 Article 60(4), LOSC.
15 Article 60(5), LOSC.
16 International Maritime Organization. 2008. Development of Guidelines for Consideration of Requests for Safety Zones Larger than 500 Metres around Artificial Islands, Installations and Structures in the Exclusive Economic Zone: submitted by the United States of America and Brazil. IMO Doc. MSC 84/22/4, 4 February 2008. IMO, London.
17 International Maritime Organization. 2010. Sub-Committee on Safety of Navigation, 56th session, Agenda item 20. IMO Doc. NAV 56/20, 31 August. Report to the Maritime Safety Committee. IMO, London.
18 Ibid. at para. 4.15.
19 Article 77, LOSC.
20 Article 77(4), LOSC.
21 See http://www.un.org/press/en/2015/ga11656.doc.htm (accessed 3 July 2015) and UN General Assembly A/69/L.65, (para. 1) from 11 May 2015.
22 UN General Assembly A/69/L.65, para.2.
23 Article 192, LOSC.

24 Article 194, LOSC.

25 Article 196, LOSC.

26 Article 204, LOSC.

27 Article 206, LOSC.

28 United Nations General Assembly A/67/79 on Oceans and the Law of the Sea: Report of the Secretary-General, 4 April 2012.

29 UN General Assembly A/67/120 on Oceans and the Law of the Sea: Report on the work of the United Nations Open-ended Informal Consultative Process on Oceans and the Law of the Sea at its thirteenth meeting, 2 July 2012.

30 Ibid. at para. 58.

31 Ibid. at paras. 56–57.

32 Article 4(1)(c), UNFCCC.

33 Article 4(5), UNFCCC.

34 UNFCCC Decision 1/CP.16. The Cancún Agreements: Outcome of the work of the Ad Hoc Working Group on Long-term Cooperative Action under the Convention, paras. 113-127.

35 Ibid. at para. 115.

36 Article 12, Kyoto Protocol and http://unfccc.int/kyoto_protocol/mechanisms/clean_development_mechanism/items/2718.php (accessed 3 July 2015).

37 CBD Decision VII/5 Marine and Coastal Biological Diversity, 13 April 2004.

38 CBD Decision VII/5, Elaborated Programme of Work on Marine and Coastal Biological Diversity, Part II: Basic Principles (see https://www.cbd.int/doc/decisions/cop-07/cop-07-dec-05-en.pdf)

39 UNEP/CBD/SBSTTA/14/L.14 on New and Emerging Issues: Draft Recommendations by the Chair, 19 May 2010.

40 UNEP/CBD/SBSTTA/16/INF/12 on Scientific Synthesis on the Impacts of Underwater Noise on Marine and Coastal Biodiversity and Habitats, 12 March 2012.

41 See http://ospar.org/content/content.asp?menu=00010100000000_000000_000000

42 Council Decision 98/249/EC of 7 October 1997 on the conclusion of the Convention for the Protection of the Marine Environment of the North-East Atlantic (OSPAR Convention), OJ L 104 of 3.4.1998.

43 OSPAR Annex V on the Protection and Conservation of the Ecosystems and Biological Diversity of the Maritime Area. See http://ospar.org/html_documents/ospar/html/ospar_convention_e_updated_text_2007_annex_v.pdf

44 OSPAR. The North-East Atlantic Environment Strategy: Strategy of the OSPAR Commission for the Protection of the Marine Environment of the North-East Atlantic 2010–2020. OSPAR Agreement 2010-3. http://ospar.org/html_documents/ospar/html/10-03e_nea_environment_strategy.pdf

45 OSPAR. 2004. Background Document on Problems and Benefits Associated with the Development of Offshore Wind Farms. Publication no. 214-2004. OSPAR Commission, London.

46 OSPAR. 2006. Review of the Current State of Knowledge on the Environmental Impacts of the Location, Operation and Removal/Disposal of Offshore Wind-Farms. Publication no. 278-2006. OSPAR Commission, London.

47 OSPAR. 2008. OSPAR Guidance on Environmental Considerations for Offshore Wind Farm Development. Publication no. 2008-3. OSPAR Commission, London.

48 OSPAR. 2010. Environmental interactions of wave and tidal energy generation devices – special request for advice from ICES. http://www.ices.dk/sites/pub/ Publication%20Reports/Advice/2010/Special%20Requests/OSPAR% 20Environmental%20interactions%20of%20wave%20and%20tidal%20energy.pdf

49 Op. cit. [44], Part II – Thematic Strategies: Biological Diversity and Ecosystems, para. 1.2.e.

50 Op. cit. [47], para. 93.

51 Ibid., para. 98.

52 European Commission. Communication from the Commission to the European Parliament, the Council, the European Economic and Social Committee and the Committee of the Regions. Energy 2020: A strategy for competitive, sustainable and secure energy. COM(2010) 639 final. European Commission, Brussels, 10 November 2010.

53 European Commission. Commission Decision of 30 June 2009 establishing a template for National Renewable Energy Action Plans under Directive 2009/28/EC of the European Parliament and of the Council. Official Journal of the European Union L 182/33. European Commission, Brussels.

54 Osta Mora-Figueroa, V., Huertas Olivares, C., Holmes, B., O'Hagan, A. M., Torre Enciso, Y. and Leeney, R. 2011. SOWFIA Deliverable D2.1: Catalogue of Wave Energy Test Centres.

55 European Commission. Report from the Commission to the European Parliament, the Council, the European Economic and Social Committee and the Committee of the Regions. Renewable energy progress report. COM(2015) 293 final. European Commission, Brussels, 15 June 2015.

56 European Commission. Report from the Commission to the European Parliament, the Council, the European Economic and Social Committee and the Committee of the Regions. Renewable energy progress report. COM(2013) 175. European Commission, Brussels, 27 Match 2013.

57 Ibid. at p. 7.

58 Op cit. [55] at p. 4.

59 Ibid. at p. 11.

60 European Commission. Report from the Commission to the European Parliament, the Council, the European Economic and Social Committee and the Committee of the Regions. Renewable Energy Progress Report. COM(2017)57 final. European Commission, Brussels, 1 February 2017.

61 European Commission. Report from the Commission to the European Parliament, the Council, the European Economic and Social Committee and the Committee of the Regions. A policy framework for climate and energy in the period from 2020 to 2030. COM(2014)15 final. European Commission, Brussels, 22 January 2014.

62 European Council Press Release 139/15, 19 March 2015.

63 Oxfam Press Release http://oxf.am/yN5, accessed 30 March 2015.

64 Friends of the Earth (FOE) Press Release, http://www.foeeurope.org/EU-climate-deal-puts-polluters-before-people-241014, accessed 30 March 2015.

65 European Commission. Communication from the Commission to the European Parliament, the Council, the European Economic and Social Committee and the Committee of the Regions. An Integrated Maritime Policy for the European Union. COM(2007) 575 final. European Commission, Brussels, 10 October 2007.

66 European Commission. Communication from the Commission to the European Parliament, the Council, the European Economic and Social Committee and the Committee of the Regions. Action Plan for a Marine Strategy in the Atlantic Area – Delivering smart, sustainable and inclusive growth (COM(2013) 279). European Commission, Brussels, 13 May 2013.

67 European Commission. Communication from the Commission to the European Parliament, the Council, the European Economic and Social Committee and the Committee of the Regions. Blue Growth: opportunities for marine and maritime sustainable growth. COM(2012) 494 final. European Commission, Brussels, 13 September 2012.

68 European Commission. Communication from the Commission to the European Parliament, the Council, the European Economic and Social Committee and the Committee of the Regions. Blue Energy – Action needed to deliver on the potential of ocean energy in European seas and oceans by 2020 and beyond. COM(2014) 8 final. European Commission, Brussels, 20 January 2014.

69 European Commission. 2000. Managing Natura 2000 sites: The provisions of Article 6 of the 'Habitats' Directive 92/43/EEC. Office for Official Publications of the European Communities, Luxembourg.

70 European Commission. 2002. Assessment of plans and projects significantly affecting Natura 2000 sites: Methodological guidance on the provisions of Article 6(3) and (4) of the Habitats Directive 92/43/EEC. Office for Official Publications of the European Communities, Luxembourg.

71 European Commission. 2006. Nature and biodiversity cases: Ruling of the European Court of Justice. Office for Official Publications of the European Communities, Luxembourg.

72 European Commission. 2007. Guidance document on Article 6(4) of the 'Habitats Directive' 92/43/EEC. Clarification of the concepts of: alternative solutions, imperative reasons of overriding public interest, compensatory measures, overall coherence, opinion of the Commission. European Commission, Brussels.

73 European Commission. 2007. Interpretation manual of European Union habitats. EUR27. European Commission, DG Environment, Brussels.

74 European Commission. 2010. EU Guidance on wind energy development in accordance with the EU nature legislation. European Commission, Brussels.

75 European Commission. 2011. Guidelines on the Implementation of the Birds and Habitats Directives in Estuaries and Coastal Zones with particular attention to port development and dredging. European Commission, Brussels.

76 European Commission. 2012. Guidance on Aquaculture and Natura 2000: Sustainable aquaculture activities in the context of the Natura 2000 Network. European Commission, Brussels.

77 Scottish Government. 2012. A Guide to Marine Licensing. Marine Licensing in Scotland's Seas under the Marine (Scotland) Act 2010 and the Marine and Coastal Access Act 2009. Revised: May 2012. Scottish Government, Edinburgh.

78 Leeney, R.H., Greaves, D. Conley D., and O'Hagan, A.M. 2014. Environmental Impact Assessments for wave energy developments – Learning from existing activities and informing future research priorities. *Ocean & Coastal Management* 99, 14–22.

79 Scottish Government. 2007. Strategic Environmental Assessment for Wave and Tidal Energy. Report prepared for the Scottish Executive by Faber Maunsell and Metoc PLC. http://www.gov.scot/Publications/2007/03/seawave

80 Scottish Government. 2012. Offshore Wind Energy in Scottish Waters – Draft Regional Locational Guidance. Parts 1–11. http://www.scotland.gov.uk/Topics/marine/marineenergy/Planning/windrlg

81 Scottish Government. 2012. Offshore Wave Energy in Scottish Waters – Draft Regional Locational Guidance. Parts 1–6. http://www.scotland.gov.uk/Topics/marine/marineenergy/Planning/waverlg

82 Scottish Government. 2012. Offshore Tidal Energy in Scottish Waters – Draft Regional Locational Guidance. Parts 1–7. http://www.scotland.gov.uk/Topics/marine/marineenergy/Planning/tidalrlg

83 Department of Energy and Climate Change (DECC). 2011. UK Offshore Energy Strategic Environmental Assessment. OESEA2 Non-Technical Summary Future Leasing/Licensing for Offshore Renewable Energy, Offshore Oil & Gas, Hydrocarbon Gas and Carbon Dioxide Storage and Associated Infrastructure. DECC, London. Available from: https://www.gov.uk/government/publications/uk-offshore-energy-strategic-environmental-assessment-2-environmental-report

84 AECOM and Metoc. 2009. Offshore Wind and Marine Renewable Energy in Northern Ireland Strategic Environmental Assessment (SEA) Non-Technical Summary (NTS). AECOM Limited, Cheshire, England.

85 Sustainable Energy Authority of Ireland (SEAI). (2010). Strategic Environmental Assessment (SEA) of the Offshore Renewable Energy Development Plan (OREDP) in the Republic of Ireland. Environmental Report. Produced for SEAI by AECOM, Metoc and CMRC. Available from: http://www.seai.ie/Renewables/Ocean_Energy/Offshore_Renewable_SEA

86 European Commission. Proposal for a Directive of the European Parliament and of the Council on access to justice in environmental matters. COM(2003) 624 final. European Commission, Brussels, 24 October 2003.

87 Copping, A., Hanna, L., Whiting, J., Geerlofs, S., Grear, M., Blake, K., Coffey, A., Massaua, M., Brown-Saracino, J. and Battey, H. 2013. Environmental Effects of Marine Energy Development around the World for the OES Annex IV. Available: www.ocean-energy-systems.org.

88 European Commission. 2008. Communication from the Commission: Roadmap for Maritime Spatial Planning: Achieving Common Principles in the EU. COM(2008) 791 final. European Commission, Brussels, 25 November.

89 European Commission. 2010. Communication from the Commission to the European Parliament, the Council, the European Economic and Social Committee and the Committee of the Regions. Maritime Spatial Planning in the EU – Achievements and Future Developments. COM(2010) 771. European Commission, Brussels, 17 December.

90 O'Hagan, A.M. 2015. Regulation of marine renewable energy, in R. Warner and S. Kaye (eds), *Routledge Handbook of Maritime Regulation and Enforcement*. Routledge, London.

91 Thomsen, K.E. 2012. *Offshore Wind: A Comprehensive Guide to Successful Offshore Wind Farm Installation*. Academic Press, Waltham, MA.

92 Van Hulle, F. 2009. Integrating Wind: Developing Europe's Power Market for the Large Scale Integration of Wind Power. Tradewind Project Deliverable. http://www.uwig.org/TradeWind.pdf

93 Chozas, J.F., Soerensen, H.C. and Korpås, M. 2010. Integration of wave and offshore wind energy in a European offshore grid. *Proceedings of the Twentieth International Offshore and Polar Engineering Conference*, Beijing, 20–25 June 2010.

94 Copping, A., Battey, H., Brown-Saracino, J., Massaua, M., and Smith, C. 2014. An international assessment of the environmental effects of marine energy development, *Ocean & Coastal Management* 99, 3–13.

95 Schlumberger Oilfield Glossary [online edition], http://www.glossary.oilfield.slb.com/ accessed 26 June 2015.

96 International Maritime Organization. Guidelines and Standards for the Removal of Offshore Installations and Structures on the Continental Shelf and in the Exclusive Economic Zone. IMO Resolution A.672(16)). Adopted by the International Maritime Organization on 19October1989.

97 Ibid. at para. 1.2.

98 Linley, E.A.S., Wilding, T.A., Black, K., Hawkins, A.J.S., and Mangi, S. 2007. Review of the reef effects of offshore wind farm structures and their potential for enhancement and mitigation. Report to Department for Business, Enterprise and Regulatory Reform (BERR), No. RFCA/005/0029P. BERR, London, UK.

99 Witt, M.J., Sheehan, E.V., Bearhop, S., Broderick, A.C., Conley, D.C., Cotterell, S.P., Crow, E., Grecian, W.J., Halsband, C., Hodgson, D.J., Hosegood, P., Inger, R., Miller, P.I., Sims, D.W., Thompson, R.C., Vanstaen, K., Votier, S.C., Attrill, M.J. and Godley, B.J. 2012. Assessing wave energy effects on biodiversity: the Wave Hub experience. *Philosophical Transactactions of the Royal Society A* 370(1959), 502–529.

100 Wilhelmsson, D., Malm, T., Öhman, M.C. 2006. The influence of offshore wind power on demersal fish. *ICES Journal of Marine Science* 63, 775–784.

101 London Convention on the Prevention of Marine Pollution by Dumping of Wastes and Other Matter 1972, Article III 1(b)(ii).

102 London Convention and Protocol/UNEP. 2009. London Convention and Protocol/ UNEP Guidelines for the Placement of Artificial Reefs. IMO, London.

103 Simas, T., O'Hagan, A.M., O'Callaghan, J., Hamawi, S., Magagna, D., Bailey, I., Greaves, D., Saulnier, J.B., Marina, D., Bald, J., Huertas, C., and Sundberg, J. 2015. Review of consenting processes for ocean energy in selected European Union Member States. *International Journal of Marine Energy* 9, 41–59.

104 Le Lièvre, C. and O'Hagan, A.M. 2015. Legal and institutional review of national consenting systems, Deliverable 2.2, RICORE Project. Available from http://ricore-project. eu/wp-content/uploads/2016/02/RiCORE-D2.2-Legal-Institutional-Review-Final-1.pdf

105 Kablan, Y. and Michalak, S. 2013. Politiques publiques de soutien aux projets d'énergies marines renouvelables (EMR) et cadres réglementaires régissant le secteur en France. Livrable 4.1.1 du WP4 du projet MERiFIC: Un rapport préparé dans le cadre du projet MERiFIC 'Marine Energy in Far Peripheral and Island Communities'. Available from http://www.merific.eu/files/2012/06/4.1.1-Politiques-publiques-de-soutien-aux-projects-denergies-marines-renouvalables.pdf

106 Department of Communications, Energy & Natural Resources, 2014. Offshore Renewable Energy Development Plan. Department of Communications, Energy & Natural Resources, Dublin. Available from: http://www.dcenr.gov.ie/NR/rdonlyres/ 836DD5D9-7152-4D76-9DA0-81090633F0E0/0/20140204DCENROffshoreRenewable EnergyDevelopmentPlan.pdf

107 See Department of Environment, Community and Local Government website http://www.environ.ie/en/Foreshore/ApplyingforanOffshoreRenewableEnergyProject/

108 Department of Environment, Community and Local Government. 2013. Maritime Area and Foreshore (Amendment) Bill 2013. DECLG, Dublin. Available from http://www.environ.ie/en/Foreshore/PublicationsDocuments/FileDownLoad,34315,en.pdf

109 O'Hagan, A.M. and Lewis, A.W. 2011. The existing law and policy framework for ocean energy development in Ireland. *Marine Policy* 35(6), 772–783.

110 Frazão Santos, C., Orbach, M., Calado, H., and Andrade, F. 2015. Challenges in implementing sustainable marine spatial planning: The new Portuguese legal framework case. *Marine Policy* 61, 196–206.

111 Galparsoro, I., Liria, P., Legorburu, I., Bald, J., Chust, G., Ruiz-Minguela, P., Pérez, G., Marqués, J., Torre-Enciso, Y., González, M. and Borja, A. 2012. A Marine Spatial Planning approach to select suitable areas for installing wave energy converters on the Basque continental shelf (Bay of Biscay). *Coastal Management* 40, 1–19.

112 Ministerio de Industria, Energía y Turismo. 2009. Estudio Estratégico Ambiental del Litoral Español para la Instalación de Parques Eólicos Marinos. Ministerio de Industria, Energía y Turismo, Madrid. Available from: http://www.aeeolica.org/uploads/documents/562-estudio-estrategico-ambiental-del-litoral-espanol-para-la-instalacion-de-parques-eolicos-marinos_mityc.pdf

113 Department of Energy and Climate Change. 2011. UK Renewable Energy Roadmap. DECC, London.

114 UK Government. 2009. The UK Renewable Energy Strategy. Presented to Parliament by the Secretary of State for Energy and Climate Change by command of Her Majesty. UK Government, London.

115 HM Government. 2010. Marine Energy Action Plan 2010 (Executive Summary and Recommendations).

116 The Crown Estate. 2014. The Crown Estate Role in Offshore Renewable Energy Developments: Briefing (January 2014). The Crown Estate, London. Available from: http://www.thecrownestate.co.uk/media/5411/ei-the-crown-estate-role-in-offshore-renewable-energy.pdf

117 Department of Energy and Climate Change. 2011. Decommissioning of Offshore Renewable Energy Installations under the Energy Act 2004: Guidance Notes for Industry. DECC, London.

118 HM Government. 2011. UK Marine Policy Statement. HM Government, London.

119 Department of Communities and Local Government. 2012. National Planning Policy Framework. DCLG, London.

120 See https://www.gov.uk/government/collections/marine-planning-in-england

121 Welsh Assembly Government. 2011. Consultation Document Sustainable Development for Welsh Seas: Our Approach to Marine Planning in Wales. Available at: http://gov.wales/docs/desh/consultation/110216marineconsultationen.pdf

122 Welsh Government. 2015. Marine Planning for Wales – Strategic Scoping Exercise. Welsh Government, Cardiff.

123 Welsh Government. 2015. Marine planning section on Welsh government website: http://gov.wales/topics/environmentcountryside/marineandfisheries/marine-planning/?lang=en Last updated on 25 April 2015, accessed on 19 June 2015.

124 See http://lle.wales.gov.uk/apps/marineportal/#lat=52.5145&lon=-3.9111&z=8

125 Scottish Government. 2012. Marine Scotland Licensing and Offshore Wind Energy Development (Report R.1957). October 2012 version. Scottish Government, Edinburgh.

126 Scottish Government. 2015. Scotland's National Marine Plan (NMP) – A Single Framework for Managing Our Seas. Scottish Government, Edinburgh.

127 Scottish Government. 2015. Scotland's National Marine Plan – A Single Framework for Managing Our Seas. A Summary of Objectives and Policies. Scottish Government, Edinburgh.

128 See http://www.gov.scot/Topics/marine/seamanagement/regional

129 Scottish Government. 2011. Blue Seas – Green Energy: A Sectoral Marine Plan for Offshore Wind Energy in Scottish Territorial Waters. Part A: The Plan. Available from http://www.scotland.gov.uk/Publications/2011/03/18141232/0

130 Scottish Government/Marine Scotland. 2013. Wave Energy in Scottish Waters – Initial Plan Framework (Draft Plan Options). Scottish Government, Edinburgh.

131 Scottish Government/Marine Scotland. 2013. Tidal Energy in Scottish Waters – Initial Plan Framework (Draft Plan Options). Scottish Government, Edinburgh.

132 Scottish Government/Marine Scotland. 2014. Potential Scottish Test Sites for Deep Water Floating Wind Technologies – Draft Regional Locational Guidance. Scottish Government, Edinburgh.

133 Scottish Government. 2012. Survey, Deploy and Monitor Licensing Policy Guidance (August 2012 version). Scottish Government, Edinburgh. http://www.scotland.gov.uk/Topics/marine/Licensing/marine/Applications/SDM

134 Bald, J., Menchaca, P., Bennett, F., Davies, I., Smith, P., O´Hagan, A.M., Culloch, R., Simas, T., and Mascarenas, P., 2015. Review of the state of the art and future direction of the Survey, Deploy and Monitor policy. Deliverable 3.1., RICORE Project.

135 Department of Enterprise, Trade and Investment. 2012. Offshore Renewable Energy Strategic Action Plan 2012-2020. DETI, Belfast. Available from: http://www.detini.gov.uk/ni_offshore_renewable_energy_strategic_action_plan_2012-2020__march_2012_.pdf

136 Department of Enterprise, Trade and Investment. 2013. Energy Consultation: Offshore Renewable Energy Bill. DETI, Belfast.

137 Department of the Environment. 2012. Northern Ireland Marine Plan Statement of Public Participation. DOENI, Belfast. Available from: http://www.planningni.gov.uk/index/policy/final_version-ni_marine_plan_statement_of_public_participation-07_june_2012.pdf

138 Department of the Environment. 2013. Northern Ireland Marine Plan Statement of Public Participation. DOENI, Belfast. Available from: http://www.planningni.gov.uk/index/policy/nimp_spp_reviewed_may_2013.pdf

139 Bureau of Ocean Energy Management and Federal Energy Regulatory Commission (BOEM/FERC). 2012. BOEM/FERC Guidelines on Regulation of Marine and Hydrokinetic Energy Projects on the OCS. Version 2, 19 July. Available from: http://www.ferc.gov/industries//hydropower/gen-info/licensing/hydrokinetics/pdf/mms080309.pdf

140 Atlantic Wind Connection. 2013. Right-of-Way Application Supplement for the Bureau of Ocean Energy Management on the Atlantic Wind Connection Project. Docket No. BOEM–2011–0023. Available from: http://www.boem.gov/uploadedFiles/BOEM/Renewable_Energy_Program/State_Activities/Supplement%20to%20the%20ROW%20application.pdf

141 See http://www.boem.gov/National-and-Regional-Guidelines-for-Renewable-Energy-Activities/

142 Kaiser, M.J. and Snyder, B. 2010. Offshore Wind Energy Installation and Decommissioning Cost Estimation in the U.S. Outer Continental Shelf. US Department of the Interior, Bureau of Ocean Energy Management, Regulation and Enforcement, Herndon, VA.

TA&R study 648. 340 pp. http://www.offshorewindhub.org/sites/default/files/resources/boemre_11-29-2010_installationdecommissioningcostestimation_0.pdf

143 Appiott, J., Dhanju, A., Cicin-Sain, B., 2014. Encouraging renewable energy in the offshore environment. *Ocean & Coastal Management* 90, 58–64.

144 Muñoz Arjona, E., Huertas Olivares, C., O'Hagan, A.M., Holmes, B., O'Callaghan, J., Magagna, D., Greaves, D., Bailey, I., and Conley, D., 2012. Navigating the Wave Energy Consenting Procedure: Sharing Knowledge and Implementation of Regulatory Measures. SOWFIA project report (D2.5 Workshop C report). Available from: https://www.plymouth.ac.uk/uploads/production/document/path/8/8936/WP2_-_Interim_Report.pdf

145 Leary, D., and Esteban, M. 2009. Climate change and renewable energy from the ocean and tides: Calming the sea of regulatory uncertainty. *International Journal of Marine and Coastal Law* 24, 617–651.

146 Dubbs, L., Keeler, A.G., and O'Meara, T., 2013. Permitting, risk and marine hydrokinetic energy development. *Electricity Journal* 26, 64–74.

147 White, R. 2012. Climate change and paradoxical harm, in S. Farrell, T. Ahmed, and D. French (eds), Legal and Criminological Consequences of Climate Change. Hart Publishing, London.

148 Bailey, I., West, J., and Whitehead, I. 2011. Out of sight but not out of mind? Public perceptions of wave energy and the Cornish Wave Hub. *Journal of Environmental Policy and Planning* 13(2), 139–158.

149 Chozas, J.F., Stefanovich, M.A., and Sørensen, H.C. 2010. Toward best practices for public acceptability in wave energy: Whom, when and how to address. Third International Conference on Ocean Energy, Bilbao, Spain.

150 Wright, G., O'Hagan, A.M., de Groot, J., Leroy, Y., Soininen, N., Salcido, R., Abad Castelos, M., Jude, S., Rochette, J. and Kerr, S. 2015. Ocean energy: key legal issues and challenges. IDDRI Issue Brief, No. 4/15 June 2015. IDDRI, Paris, France.

151 O'Hagan, A.M. 2012. Maritime Spatial Planning (MSP) in the European Union and its Application to Marine Renewable Energy. Annual Report of the IEA-OES; 101–109.

152 Wright, G., O'Hagan, A.M., de Groot, J., Leroy, Y., Soininen, N., Salcido, R., Abad Castelos, M., Jude, S., Rochette, J. and Kerr, S. 2016. Establishing a legal research agenda for ocean energy. *Energy Policy* 63, 126–134.

153 Ehler, C., 2011. Marine Spatial Planning – An Idea Whose Time Has Come. Annual Report of the IEA-OES; 96–100.

154 Thoroughgood, C.A. 2010. Marine spatial planning: A call for action. *Oceanography* 23, 9–10.

155 Wagner, A. 2010. Report defining the criteria for assessing national MSP practices affecting the deployment of marine renewable energy sources (D2.1). Seanergy 2020 project. Available from: http://www.seanergy2020.eu/wp-content/uploads/2010/11/101102.Deliverable-2-1_MSP-Criteria_final-version.pdf

156 Copping, A., Sather, N., Hanna, L., Whiting, J., Zydlewsk, G., Staines, G., Gill, A., Hutchison, I., O'Hagan, A.M., Simas, T., Bald, J., Sparling C., Wood, J., and Masden, E. 2016. Annex IV 2016 State of the Science Report: Environmental Effects of Marine Renewable Energy Development Around the World. Available from: http://tethys.pnnl.gov/publications/state-of-the-science-2016

11

The Economics of Wave and Tidal Energy

Gregorio Iglesias[a], Sharay Astariz[b] and Angela Vazquez[b]

[a] *Professor of Coastal Engineering, School of Engineering, University of Plymouth, UK*
[b] *Associate Researcher, University of Santiago de Compostela, Spain*

11.1   Individual Costs

The costs incurred in the lifetime of a wave or tidal energy farm may be classified into four categories: pre-operating cost; capital cost; operation and maintenance (O&M) cost; and decommissioning cost (Figure 11.1) [1]. The pre-operating cost includes the engineering tasks, such as the study of the area, design of the farm, environmental impact studies, and the consenting process. The capital cost includes (Figure 11.1): the structure or rotor, in wave or tidal energy, respectively; the mooring systems (wave energy) or foundations (tidal energy); the power take-off (PTO) and control systems; the connection to the land electricity network; and the installation. In the case of wave energy just over half of the cost is associated with the device structures, the PTO and control systems, whereas in the case of tidal energy the installation of the energy converters is of greater relevance. Other elements such as the grid connection have similar weights in the cost breakdown of both renewables. Finally, O&M costs make up a significant proportion of total lifetime costs, although capital costs represent around 70% of the total cost [2, 3].

Using average cost values normalised by installed capacity (quotations on a per kilowatt or megawatt basis) is an effective approach in view of the scarcity of project- or device-specific cost data in the public domain [4, 7]. This is the common approach in estimating the cost of these renewable energies [4, 8, 9] and performing cross-technological comparisons [2, 10].

The pre-operating cost depends on the location of the farm, since each country has its own regulatory framework and fees, and its size. Overall, the cost of preliminary studies, design and other engineering tasks to start up a wave farm ranges between €550,000 and €2.5 million [11], and the costs related to consenting may be estimated as 2.83% of the nominal power expressed in watts [12].

The capital cost encompasses the purchase of all the equipment and elements of the farm and their installation [4]. Table 11.1 shows capital cost figures used in previous works, which depend on the degree of development of the installation. Importantly, the

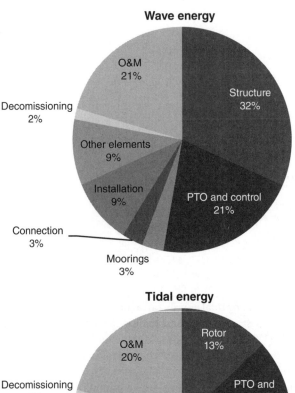

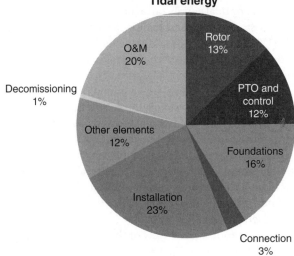

Figure 11.1 Cost breakdown [4–6].

capital investment is incurred before the energy farms start operating, which gives rise to a debt that must be paid back during the lifetime of the installation, estimated at 20 years [9]. Thus, the time-scale for achieving a return on these capital investments is relatively long, and surrounded by uncertainties over market regulation and energy prices, which are quite volatile.

Due to the nascent state of wave technology, with a wide variety of wave energy converters (WECs) under development, it is not straightforward to give a general figure for the WEC cost. Indeed, the literature suggests that the cost of WECs including the PTO system and their installation lies in the broad range of €2.5–6 million per installed megawatt [3, 17–19]. The cost of the WEC structure depends on the material and the total weight. Steel is often the material of choice, with a reference cost of €3400 per ton [20]. As for tidal

Table 11.1 Capital cost quotations used in previous studies.

Category		Installed capacity (MW)	Capital cost	Units	Year	Source
Tidal energy						
	Pre-demonstration	<10	11,200			
Project status	Demonstration	10	4,300	£ kW^{-1}	2010	[13]
	Commercial	>50	3,200			
	High cost	>1	12,300			
	Central cost		10,700	€ kW^{-1}	2013	[14]
	Low cost		9,300			
	First array	n/a	5,600–12,000	$ kW^{-1}	2013	[6]
	First array	n/a	8,700	$ kW^{-1}	2013	[8]
	Second array		6,350			
	First commercial		4,300			
Wave energy						
	Pre-demonstration	5	6,912			
Project status	Demonstration	20	5,035	$ kW^{-1}	2011	[15]
	Commercial	50	4,347			
	High cost		16,050			
Geography	Central cost	>1	8,780	$ kW^{-1}	2013	[16]
	Low cost		5,480			
	High cost		10,700			
	Central cost	>1	9,080	€ kW^{-1}	2013	[14]
	Low cost		7,590			

energy converters (TECs), the vast majority are turbines – although there are some based on other principles of operation. The cost of the turbine rotor is mainly influenced by its diameter, but the relationship is not linear. With increasing diameter the rotational velocity must decrease to avoid cavitation at the tip of the blades. Capturing a relevant proportion of the available tidal stream power in a greater swept (rotor) area at lower rotational velocities requires increased torque. Indeed, the torque increases approximately with the rotor diameter cubed. The rotor, PTO and foundation are therefore subjected to greater loadings, which increases the mass of steel and overall cost of the tidal stream turbine. Indeed, the turbine mass scales with the diameter cubed; by contrast, the power scales with the diameter squared (square-cube law). Thus, increasing the diameter implies reducing the power per unit mass (hence the interest of multi-rotor options). Estimates for a range of diameters have been reported (e.g., [21]); see Figure 11.2.

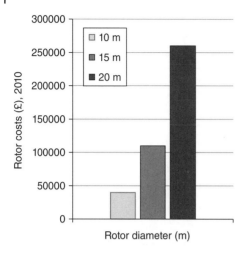

Figure 11.2 Rotor costs as a function of diameter.

The installation cost depends on several factors, not least the proximity to ports with the necessary equipment available during the time required to install all the devices in the farm [22], which is itself a function of the distance to the selected port, its storage capacity, and the transportation means. Moreover, weather conditions can strongly impact installation and maintenance operations. The operating conditions depend on the sea state and wind level, and accessibility to the system. Offshore operations are typically not carried out when wind speeds exceed 12 m s^{-1}, or wave heights 2 m [23]. The O&M cost is estimated at 15–20% of the investment cost (device, mooring, transmission components, etc.) for both wave and tidal energy farms.

Furthermore, it is necessary to consider the cost of the assembly system. Mooring systems are commonly used for WECs and foundations for TECs. Catenary anchor leg mooring is the most usual mooring system; its cost may be calculated as a function of weight, €300 per ton [20], which varies with the diameter and total length of the mooring line. The latter is usually estimated as 3–5 times the water depth [24–26]. For its part, the cost of the foundations is also highly influenced by the water depth and the seabed geology, which determine the choice of foundation [27]. This cost represents 15–25% of the total cost of a tidal energy project [6, 28], and may be estimated as a function of water depth (d) as follows [28]:

$$\text{Water depth} \begin{cases} 0-30\,\text{m} \rightarrow Cost_{\text{foundation}} / \text{MW} = 0.15 + 10^{-5} d^3, \\ 30-60\,\text{m} \rightarrow Cost_{\text{foundation}} / \text{MW} = 0.35 + 4 \cdot 10^{-5} d^3, \\ > 60\,\text{m} \rightarrow Cost_{\text{foundation}} / \text{MW} = 0.15 + 0.016 \cdot d. \end{cases} \quad (11.1)$$

These costs are given in billions of euros per installed megawatt.

The cost of the PTO system – including the generator, power electronics, control and safety systems – may be estimated at €5,000 per kilowatt for both wave and tidal technology [20]. The electricity generated has to be exported onshore. First, the electrical power generated by all the devices must be concentrated at a hub or collection point, where it can be properly conditioned and delivered to the land grid. In early developments individual devices will likely be connected directly to the collection point.

However, as arrays grow, this method of connection becomes too expensive, and tidal array designers will likely follow the wind industry in connecting turbines in more efficient ways – for example, in series [29]. The cost of the submarine intra-array cable consists of the costs of the cables and their installation. Both cost components are a function of the length of the cables ($L_{\text{intra-array}}$), which depends on the layout of the farm and the distance to the transformer. It can be calculated [1] as

$$L_{\text{intra-array}} = \sum_{i=1}^{n} \left(L_{\text{string}} + L_{\text{substation}} \right)_i,$$ (11.2)

where the index i designates a generic string of WECs in which the array is divided to export the power generated, L_{string} is the distance between the WECs of the ith string and $L_{substation}$ is the distance between the end of the ith string and the offshore substation. The cost of the cable, C_{cable} (€/m), is a is a function of the voltage and the maximal current [30]:

$$C_{\text{cable}} = \alpha + \beta e^{\left(\gamma I_n / 10^5 \right)},$$ (11.3)

where α, β and γ are cost coefficients that depend on the operating voltage of the cable. For a 30 kV cable, these coefficients are 54.37 (€/m), 78.83 (€/m) and 234.34 (A^{-1}), respectively. $I_{n,}$ (A) is the maximal current in the medium voltage cable. The specific installation cost for medium voltage (MV) is assumed to be 380 (€/m) [30]. Usually the intra-array cable transports electricity in MVAC (30 kV), which is converted to high voltage (HVAC) to be exported to land. For MV/HV transformers from 50 to 800 MVA the cost can be obtained as a function of the rated power, A_{TR} (MVA) [31]. The cost of the offshore substation platform and its installation can be calculated as 0.0485 million euros per megawatt [32]. This gives

$$C_{\text{MV/HV}} = 44.568 \, A_{\text{TR}}^{0.7513},$$ (11.4)

measured in thousands of euros. Finally, the cost of the export HVDC cable must be accounted for. This can be calculated with the same equation as the intra-array cable (11.3), but with different coefficients: α= 204.97 (€/m), β =47.42 (€/m) and γ =333.587 (A^{-1}). The installation cost for HV cable is €750 per metre [30].

Apart from the initial costs, the O&M cost throughout the lifetime of the installation must be assessed. Calculating this cost is a complex process, since there is not enough experience in wave and tidal energy installations. Nevertheless, it is possible to obtain an approximation based on the experience in oil and gas and in offshore wind energy. Thus, the O&M for offshore installations is often considered to be €30 per megawatt-hour [33–37]. This cost involves scheduled maintenance, unscheduled maintenance, insurance costs and other costs (land rent, administration, etc.) Scheduled maintenance is usually conducted in the summer months to minimise weather downtime and lost production [38]. However, unexpected failures are unavoidable and unscheduled maintenance is also required. The recent experience in offshore wind energy can be taken as a reference [39] to determine the insurance cost and site rental. In Table 11.2, some of these costs are expressed in euros per megawatt-hour and/or as a percentage of the initial investment. Besides, it is necessary to take into account that 10 years after

Table 11.2 Operation and maintenance cost [37, 42–48].

Cost	€/MWh	% of investment cost
O&M tasks	30	3
Revision and time off		10
Spares		90
Insurance	37	2
Public services	3.5	
Site rental	3.3	2.5

installation, WECs have to be removed from the sea for complete revision, which is estimated to cost about 4.2% of the initial expenditures [40]. More specific values may be obtained [12] as a function of the farm attributes (e.g., turbine distribution and turbine configuration), local conditions and maintenance strategy.

The decommissioning cost must also be taken into consideration. The dismantling operation may be considered as the inverse of the installation, and subject to similar constraints [41]. It has been proposed to estimate the decommissioning cost at half the installation cost [22, 41], or at 0.5–1% of the initial investment [11].

11.2 Levelised Cost

The cost of energy production is considered the single most important factor in assessing the economic feasibility of electricity generation systems. Indeed, it is widely used to aid private investments, technological development decisions and policy-making. Due to the particularities of each technology (i.e. different conversion systems, working environments, etc.), a standard that enables various technologies to be compared is of crucial importance, especially if cross-technological studies are to be performed [49].

One such standard is the levelised cost of energy (LCOE), which is widely reported in the energy policy literature [50–52], including economic studies focused on tidal and wave energy [2, 7, 53, 54]. The LCOE is defined as 'the ratio of total lifetime expenses versus total expected outputs, expressed in terms of the present value equivalent' [55], i.e. by discounting future flows of both costs and energy back to the present. Two methods are usually employed to estimate the LCOE, namely the discount method and the annuitising method [2].

In the discount method, the stream of (real) future costs and (electrical) outputs, identified as C_t and O_t in period t, are discounted back to present value (PV) by applying a given discount rate (r):

$$LCOE = \frac{\sum_{t=0}^{n} C_t / (1+r)^t}{\sum_{t=0}^{n} O_t / (1+r)^t}. \tag{11.5}$$

In the annuitizing method, an equivalent annual cost is obtained. To this end, a standard annuity formula converts the present value of the stream of costs over the lifetime of the device. This equivalent annual cost is then divided by the average annual electrical output over the lifetime of the plant [56]:

$$LCOE = \frac{\left(\sum_{t=0}^{n} C_t \Big/ (1+r)^t\right) \times \left(r \Big/ \left(1 - (1+r)^{-n}\right)\right)}{\left(\sum_{t=0}^{n} O_t\right) \Big/ n}. \tag{11.6}$$

These two methods are related in that both yield the same LCOE values given the following conditions: (1) the same rate (r) is used for discounting costs in the discounting method and for calculating the annuity factor in the annuitising method; (2) the annual electrical output is constant over the lifetime of the installation [2].

As mentioned before, one of the advantages of the LCOE approach is that it allows for comparison among technologies e.g. (Figure 11.3). These comparative studies form the basis on which proposals for frameworks concerning the appropriate level of financial support for specific technologies are typically founded [52, 57].

Wave and tidal energy are thus categorised as the most expensive technologies among the renewable sources, whereas hydropower and onshore wind are the most economical (Figure 11.3). In the case of wave energy, a range of values from $213 (€182) to $1037 (€885) per megawatt-hour is found in the literature; in the case of tidal stream energy, the range is from $189 (€161) to $844 (€720) per megawatt-hour[1] [52, 53].

Indeed, comparisons between tidal stream energy and wave energy reveal that the latter is more expensive. There may be various reasons for this difference. First, tidal stream energy benefits from important synergies with offshore wind energy – a relatively mature sector. Both of them employ the same principle of operation and technologies similar to some extent. Wave energy technology is certainly less mature [54]. Second, there is no design consensus among the various wave energy devices and, as a result, the cost scatter is greater than in the tidal stream energy sector (Figure 11.4). Whereas horizontal-axis turbines represent the most common type of tidal energy designs (accounting for more than 70% of R&D efforts in the development of tidal devices worldwide), three completely different types of wave devices, namely point absorbers, attenuators and oscillating wave columns, account for 82% of research efforts.

In order to lower the LCOE of tidal and wave energy it is necessary to pay attention to the different elements that affect its calculation, namely costs (described in the previous subsection), energy output and financial variables (discount rate) [59].

The discount rate takes into account both the time value of the money and the risk of the investment, aspects that vary with the circumstances, location and time period considered [50]. For this reason the discount rate is one of the most popular parameters in

---

1 Values in EUR were obtained by applying an exchange rate of €1 = $1.17, the exchange rate on 21/11/2017. These values do not represent one particular technology or tidal stream site, rather they were obtained by averaging costs across the tidal stream industry as a whole.

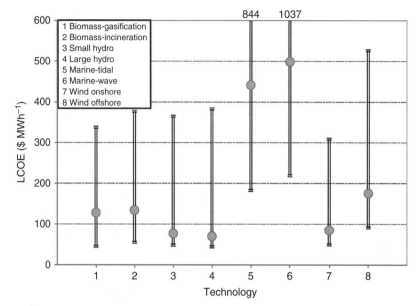

Figure 11.3 LCOE estimates for renewable energy [58].

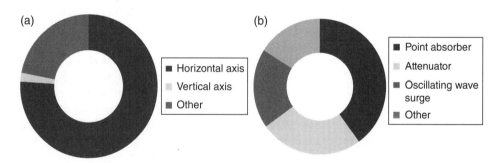

Figure 11.4 R&D efforts in: (a) tidal technology and (b) wave energy types [63].

sensitivity analyses [2]. At present tidal and wave projects are associated with greater technological risks relative to conventional energy plants, which explains the conservative range of discount rate values, between 10% and 12%, used for tidal stream energy and wave energy [60–62]. Reducing the discount rate from 10% to 6% could lower the LCOE of tidal stream and wave energy by ~20% [2].

Related to the discount rate is the project lifetime. For tidal stream energy and wave energy, project lifetimes are estimated at 20 years. After this time, gradual degradation as a consequence of chemical and material processes associated with corrosion and weathering may occur. Maintaining the installation can thus be costly, since older equipment implies a higher O&M cost. However, ongoing research will improve project lifetimes as a result of knowledge gained about failure mechanisms, and this will reduce the LCOE significantly [12].

The LCOE of tidal stream and wave energy is far from the grid costs, which are driven primarily by carbon-based technologies [59]. These technologies have an LCOE below €85 ($100) per megawatt-hour, well below the values of marine renewable energy, and the resulting grid parity gap needs to be bridged.

Concerning costs, the potential for reducing the LCOE of tidal and wave energy lies mainly in changes in capital costs, since they represent over 70% of the costs involved (Figure 11.1). The main drivers in reducing the capital costs of tidal stream energy include the following technological advances [5, 6]:

- Standardisation: efforts need to be concentrated on fewer types of converters.
- Upscaling: larger TECs and WECs will have a lower cost per rated power, since installation and manufacture costs do not increase linearly with energy output. (In the case of TECs this may be realised through multi-rotor systems, to escape the square-cube law trap).
- Economies of scale: if a number of similar devices are produced at scale, individual costs will reduce, i.e. components produced through serial production will result in a lower overall individual component cost when compared to the cost of a one-off design through a better utilisation of equipment, discounts for buying larger batches of components, etc.
- Innovation: alternative materials for devices will offer significant cost reductions, allowing changes in shape and optimisation of weight.
- Learning by doing: experience will bring about optimum routes for production, deployment, and O&M techniques [64, 65]. The reductions in capital costs thanks to learning by doing, economies of scale, etc. are described by means of experience curves, which represent cost reductions as a function of cumulative experience (expressed in terms of cumulative production or capacity). The simple experience curve may be written in the form [64]:

$$C_n = C_0 \left(\frac{P_n}{P_0} \right)^{-b}, \tag{11.7}$$

where C_n is the cost of the unit produced at time n, C_0 is the cost of the unit produced at time 0, P_n is the cumulative capacity or production at time n, P_0 is the cumulative capacity or production at time 0, and b is the learning coefficient. This learning coefficient depends on the learning ratio (LR), which represents the cost savings associated with doubling the cumulative production or capacity [6]:

$$b = \frac{\ln(1 - LR)}{\ln(2)}. \tag{11.8}$$

So far little research has focused on LRs for tidal stream and wave technology, but the reported values are in the range of 6–15% [65], which corresponds to learning factors of 85–90% within the next 10 years [3, 66–70]. According to the experience curve, capital expenditure reductions of 10% and 20% would correspond to cumulative capacity or production increases by a factor of 2 and 4, respectively, with a learning ratio of 0.1 [65].

In the other category of costs, operating expenses, reliability is a key factor, since offshore maintenance and replacement are costly by nature [1, 54]. The higher the reliability, the lower the need for maintenance and repairs. A significant proportion of the total cost of maintenance is the cost of accessing the devices, so any reduction in planned or unplanned maintenance can achieve relevant cost reductions, as well as increase the energy production through increased availability [49]. Additionally, as the distance to a port (or a land-based infrastructure) with suitable facilities for maintenance increases, maintenance costs and the downtime required increase as well. Therefore, suitable local port infrastructure leads to a reduction in maintenance costs [71, 72].

A new line of research is focused on combined marine renewable energies, with promising results [73]. Taking advantage of different marine resources in the same area has emerged as a viable option to enhance the cost-competitiveness of marine energy [74–80]. The deployment of wave and tidal energy technology could be accelerated by incorporating these devices into offshore wind farms, already at the commercial stage, thereby accelerating cost reductions through the learning rate and economies of scale. At the same time, mutual benefits can be realised by means of *diversified* or *combined farms* [78–82]. First, the use of marine space would be optimised [83, 84] by increasing the energy yield per unit area. Second, resource diversity may be used to manage the variability of single-source renewable power and thus lower the system integration costs of renewables [85]. Tidal flow is available for a greater percentage of time than wind, and is also more predictable [86, 87]. As for wave power, the phase lag between wave and wind peaks would be instrumental in smoothing power fluctuations [73, 80–82, 88]. Moreover, a reduced capital cost per megawatt installed might be achieved by sharing the electrical installation or foundations [74, 76, 77, 88, 89]. The submarine cable for connection to the grid would be more reliable and far cheaper if it exported the maximum power that it can carry. A recent study [90] found costs reductions of 12–14% in capital costs by means of combined energy systems. Similarly, cost savings in maintenance tasks may be expected through sharing strategies [74] and other factors such as the shielding effect of WECs [91–95] over the offshore wind farm [79, 96–98], which would increase the weather windows for O&M [1, 48]. In the near term, however, discussion of hybrid offshore wind, wave and tidal projects is limited to demonstration or pilot projects [99, 100]. Certain prototypes [77, 101] exhibit better performance (around 15% more) and lower capital and life cycle costs (30% and 20%, respectively) than standalone wind turbines.

Finally, as regards the energy parameters of LCOE, maximising output is a prerequisite for reducing costs [7]. In this respect the layout of the converters within a tidal or a wave farm can have a significant impact on the power that the array can extract and convert into electricity. For tidal stream energy, negative effects from device interactions (velocity deficit and increased turbulence intensity in the wakes) may be reduced with adequate configurations [9] such as a staggered layout, which reduces the blockage effect between turbines [9, 102–104] and increases the distance available for flow recovery between rows [105]. Power performance can be optimised by modifying lateral and longitudinal spacings, but spatial constraints, bathymetric effects and bottom roughness must be considered [105]. Another tendency in laying out tidal farms is called 'micro-sitting': the array does not follow a predetermined pattern, but the turbines are individually located in positions that deliver the lowest cost [29]. Concerning wave energy farms, the configuration of the farm is highly correlated to the principle of operation, which varies between devices [78].

11.3 Externalities

Externalities are defined as 'the costs and benefits that arise when the social or economic activities of one group of people have an impact on another, and when the first group fail to fully account for their impact' [106]. In the context of electricity pricing,

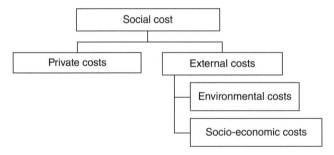

Figure 11.5 Components of social costs.

the term 'externalities' refers to the monetised value of the impacts (either positive or negative) associated with electricity generation which are not incorporated in the cost structure of the electric utility. On this basis, two types of costs may be distinguished: private and social (Figure 11.5). Private costs are a measure of the best use of the resources among the alternatives available to an electricity producer. By contrast, social costs of production, which include both the private and external costs, measure the best alternative use of the resources available to society as a whole.

Since the price of the good or service producing the externality will tend towards equality with the marginal cost to the producer (and the marginal personal utility to the purchaser), electricity prices are biased. Indeed, they are lower or higher than they would be if equality with the marginal social cost of production (and the marginal social utility of consumption) were pursued instead.

This difference between the private and social cost has an effect on electricity market rules, and is particularly relevant for the renewable energy sector. Market prices are key signals in respect of the type of product (electricity from different sources in this case) and the quantity of output (amount of electricity) to be produced. In this regard, the price of electricity generated by conventional sources associated with negative externalities (external costs) is artificially low (Figure 11.6). Therefore, consumers will buy more of this kind of electricity at the artificially low price than at a price that reflects the full production cost. As a result, more electricity from conventional energies will be produced than it would if all costs were included. On the other hand, electricity from tidal and wave energy, which involves positive externalities [107], will be 'underproduced', since the marginal cost is nowadays higher (Figure 11.6).

As an example, oil and coal technologies for electricity production have associated higher external costs (€60 and €58.33 per megawatt-hour, respectively) in comparison with other non-renewable energies like natural gas (€15 per MWh). The difference is even greater compared with wind, with an external cost of a mere €1.75 per MWh [107–110]. The non-inclusion of the externalities in the analysis may indeed prevent an optimal level of production and limit the development of renewable energy, and goes a long way to explaining why all electricity grids worldwide are driven by fuel-based technologies. Indeed, were the externalities included in the studies comparing the economic viability of the various technologies, the cost-competitiveness of renewable energies would change substantially. This point is illustrated in Figure 11.6: coal and fuel technologies experience an important rise in their levelised cost value when their negative externalities are internalised – exceeding the cost of certain renewable energies such as onshore wind.

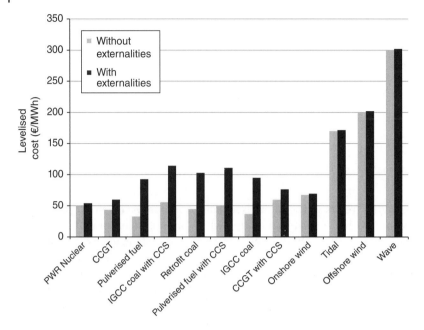

Figure 11.6 Levelised cost of different technologies, including external costs (costs referred to 2006). PWR, pressurised-water reactor; CCGT, combined cycle gas turbine; IGCC, integrated gasification combined cycle; CCS, carbon capture storage.

If externalities can lead to optimal electricity production and prices, why are they often disregarded and not generally internalised into the price of the electricity? The answer is related to the absence of a market for the majority of externalities; as a result, electricity producers have no incentive to integrate costs of this type into their cost structure and decision-making processes [110]. It must be born in mind that externalities are the monetised value of the impacts, and these can be of different nature. The typical impacts associated with wave and tidal energy projects have been investigated in previous works [75, 91, 102, 111, 112]. Although there exists no market for goods such as nature, this is not a reason for not monetising them. Indeed, economists have developed several methods for monetary valuation of externalities, most of which rely on the economic welfare theory – based on the values that individuals ascribe to their preferences regarding an environmental (or social) good with no observable price in the market. The methods can be classified into direct and indirect techniques (Figure 11.7). In direct (stated preference) methods, people are asked directly to state their willingness to pay for a benefit or to avoid a cost, or to rate/rank/choose alternatives regarding a specific hypothetical situation or policy. Travel cost and hedonic prices belong to this group. In indirect (revealed preference) methods, preferences, and thus implicit values of the externalities, are revealed indirectly when individuals purchase market goods and services that are related to the environmental good as complements or substitutes [113]. Both direct and indirect valuations of externalities assume rational individuals with preferences shaped according to neoclassical

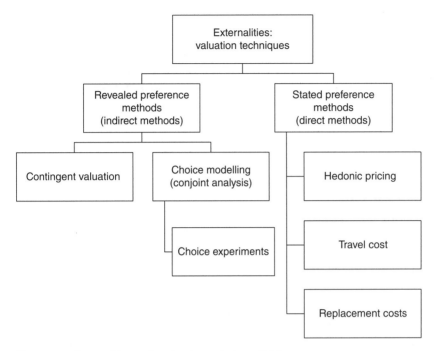

Figure 11.7 Externalities: main valuation techniques [113].

economics [114, 115]. Among the methods, the choice experiment is arguably the most popular technique in the energy sector [116], and has been applied to tidal stream energy [117] and tidal barrage energy [118].

The internalisation of externalities constitutes a strategy to rebalance the social and environmental dimension with the purely economic. Governments, through policies, should ensure that prices reflect the total costs of an activity, including the costs of damages and the benefits of positive effects arising from electricity generation. To this end, different mechanisms can be employed, namely taxes, subsidies, regulations to prevent or curtail the activity that might cause negative externalities, and tradable emission permits.

Tax effectively increases the cost of producing goods with negative externalities. An example is charging polluters for the pollution generated by them, in accordance with the 'polluter pays' principle [119]. By contrast, a subsidy is applied to goods associated with positive externalities, such as renewable energies. Such subsidies provide an incentive for the emerging tidal and wave energy sectors to increase their production. Among the subsidies, feed-in tariffs and renewable obligation certificates are the most important instruments [120, 121], and have underpinned the development of the renewable energy sector in a number of countries [121]. Feed-in tariffs guarantee a fixed price for every unit of electricity from a (certain) renewable source, and thus constitute a subsidy in a cost–benefit analysis. For the tidal and wave energy sector these incentives, which form part of the energy policy in a number of countries worldwide, are necessary to bridge the grid parity gap with conventional sources of energy.

References

1 Astariz S, Abanades J, Perez-Collazo C, Iglesias G, 2015. Improving wind farm accessibility for operation & maintenance through a co-located wave farm: Influence of layout and wave climate. *Energy Conversion and Management* 95, 229–41.

2 Allan G, Gilmartin M, McGregor P, Swales K, 2011. Levelised costs of wave and tidal energy in the UK: Cost competitiveness and the importance of 'banded' renewables obligation certificates. *Energy Policy* 39, 23–39.

3 Carbon Trust, 2006. Future Marine Energy Results of the Marine Energy Challenge: Cost Competitiveness and Growth of Wave and Tidal Stream Energy.

4 Dalton G, Allan G, Beaumont N, Georgakaki A, Hacking N, Hooper T, Kerr S, O'Hagan AM, Reilly K, Picci P, Sheng W, Stallard T, 2015. Economic and socioeconomic assessment methods for ocean renewable energy: Public and private perspectives. *Renewable and Sustainable Energy Reviews* 45, 850–878.

5 Carbon Trust, 2011. Accelerating Marine Energy. The Potential for Cost Reduction – Insights from the Carbon Trust Marine Energy Accelerator.

6 SI OCEAN, May 2013. Ocean Energy: Cost of Energy and Cost Reduction Opportunities. Strategic Initiative for Ocean Energy. http://si-ocean.eu/en/upload/docs/WP3/CoE%20report%203_2%20final.pdf.

7 Vazquez A, Iglesias G, 2015. LCOE (levelised cost of energy) mapping: A new geospatial tool for tidal stream energy. *Energy* 91, 192–201.

8 Ocean Energy Systems (OES), 2015. International LCOE for Ocean Energy Technologies. Report. http://www.juliafchozas.com/projects/iea-oes-cost-energy-assessment/

9 Vazquez A, Iglesias G, 2015. Device interactions in reducing the cost of tidal stream energy. *Energy Conversion and Management* 97, 428–438.

10 Johnstone CM, Pratt D, Clarke JA, Grant AD, 2013. A techno-economic analysis of tidal energy technology. *Renewable Energy* 49, 101–106.

11 Castaño M, 2011. Sistema de monitorización y supervisión de una boya para generación de energía undimotriz. Universidad Politécnica de Cataluña.

12 Li Y, Lence BJ, Calisal SM, 2011. An integrated model for estimating energy cost of a tidal current turbine farm. *Energy Conversion and Management* 52(3), 1677–1687.

13 Ernst and Young, 2010. Cost of and financial support for wave, tidal stream and tidal range generation in the UK. A report for the Department of Energy and Climate Change and the Scottish Government.

14 Magagna D, Uihlein D, 2015. JRC ocean energy status report. European Commission Joint Research Centre. Available from: https://setis.ec.europa.eu/sites/default/files/reports/2014-JRC-Ocean-Energy-Status-Report.pdf

15 Previsisc M, 2012. The Future Potential of Wave Power in the United States. Prepared by RE Vision Consulting on behalf of the US Department of Energy. http://wwwrevisionnet/documents/The%20Future%20of%20Wave%20Power%20MP%209-20-12%20V2pdf (accessed on 25 July 2015).

16 World Energy Council, 2013. World Energy perspective. Cost of Energy Technologies. http://www.worldenergy.org/wp-content/uploads/2013/09/WEC_J1143_CostofTECHNOLOGIES_021013_WEB_Final.pdf (Accessed on: 25 July 2015).

17 BWEA, 2009. Marine renewable energy-state of the industry report.

18 Caballero F, Sauma E, Yanine F, 2013. Business optimal design of a grid-connected hybrid PV (photovoltaic)-wind energy system without energy storage for an Easter Island's block. *Energy* 61, 248–261.

19 Farrell N, Donoghue CO, Morrissey K, 2015. Quantifying the uncertainty of wave energy conversion device cost for policy appraisal: An Irish case study. *Energy Policy* 78, 62–77.

20 Fernández-Chozas J, Jensen NEH, 2014. User guide – COE Calculation Tool for Wave Energy Converters. Technical Report No 161.

21 Bryden G, Naik S, Fraenkel P, Bullen CR, 1998. Matching tidal current plants to local flow conditions. *Energy* 23, 699–709

22 Maslov N, Charpentier J-F, Claramunt C, 2015. A modelling approach for a cost-based evaluation of the energy produced by a marine energy farm. *International Journal of Marine Energy* 9, 1–19.

23 Rademakers L, Braam H, Zaaijer M, Bussel GV, 2003. Assessment and optimisation of operation and maintenance of offshore wind turbines. European Wind Energy Conference, Madrid.

24 Atkins Oil & Gas Engineering Ltd., 1992. A parametric model of costs for the wave energy technology (ETSU-VW–1685).

25 Harris, RE, Wolfram J, 2004. Mooring systems for wave energy converters: a review of the design problems and options. Congress VII World of Renewable Energies, Denver.

26 Veatch N, 2001. The commercial perspectives for the wave energy. ETSU T/06/00209/ REP. DTI Programmes of Sustainable Energy DTI / Pub URN 01/1011.

27 Pérez-Collazo C, Iglesias G, 2012. Integration of wave energy converters and offshore windmills. International Conference on Ocean Energy, Dublin.

28 Serrano Gonzalez J, Burgos Payan M, Riquelme Santos J, 2011. An improved evolutive algorithm for large offshore wind farm optimum turbines layout. PowerTech, 2011 IEEE Trondheim.

29 Culley DM, Funke SW, Kramer SC, Piggott MD, 2016. Integration of cost modelling within the micro-siting design optimisation of tidal turbine arrays. *Renewable Energy* 85, 215–227.

30 Dicorato M, Forte G, Pisani M, Trovato M, 2011. Guidelines for assessment of investment cost for offshore wind generation. *Renewable Energy* 36(8), 2043–2051.

31 Lazaridis, PL, 2005. Economic comparison of HVAC and HVDC solutions for large offshore wind farms under special consideration of reliability. Masters' thesis, Royal Institute of Technology, Stockholm.

32 Douglas Westwood Ltd., 2010. Offshore Wind Assessment For Norway. Douglas Westwood.

33 Bedard R, Previsic M, Hagerman G, Polagye B, Musial W, Klure J, Amsden S, 2007. North American ocean energy status. *Electric Power Research Institute (EPRI) Tidal Power (TP)*, 8, 17.

34 Dunnett D, Wallace JS, 2009. Electricity generation from wave power in Canada. *Renewable Energy* 34(1), 179–95.

35 Oregan Wave, 2010. Value of Wave Power. Oregan Wave Energy Utility Trust.

36 Siddiqui O, Bedard R, 2005. Feasibility assessment of offshore wave and tidal current power production: a collaborative public/private partnership. IEEE Power Engineering, Piscataway, NJ, USA.

37 Waveplam, 2010. Wave energy: a guide for investors and politicians.

38 Hassan GG, 2013. A Guide to UK Offshore Wind. Operations and Maintenance. Scottish Enterprise and the Crown Estate.

39 O'Connor M, Lewis T, Dalton G, 2013. Operational expenditure costs for wave energy projects and impacts on financial returns. *Renewable Energy* 50, 1119–1131.

40 Couñago B, Barturen R, 2010. Estudio técnico-ficanciero sobre la construcción de un parque eólico marino flotante en el litoral español. *Ingeniería Naval* 886, 85–105.

41 Kaiser MJ, Snyder B, 2012. Modeling the decommissioning cost of offshore wind development on the U.S. Outer Continental Shelf. *Marine Policy* 36(1), 153–64.

42 Blanco MI, 2009. The economics of wind energy. *Renewable and Sustainable Energy Reviews* 13(6–7), 1372–1382.

43 Carbon Trust, 2005. Oscillating water column wave energy converter evaluation report.

44 Irish Wind Energy Association, 2016. Planning, Regulations and Administration. Available at: http://www.iwea.com/planning_regulationsandadminis

45 Kabalci E, 2013. Design and analysis of a hybrid renewable energy plant with solar and wind power. *Energy Conversion and Management* 72, 51–59.

46 Lawrence A, Meissner D, 2011. Offshore Wind Farm Operations & Maintenance. Benchmarks, Costs and Best Practices for Current and Future Wind Farms. DVV Media Group, Schiff & Hafen.

47 Milborrow D, 2010. Breaking down the cost of wind turbine maintenance. *Wind Power Monthly* 15.

48 Previsic M, 2004. System level design, performance, and costs of California Pelamis wave power plant. EPRI. Available at: http://www.re-vision.net/documents/System%20Level%20Design,%20Performance%20and%20Costs%20-%20San%20Francisco%20California%20Pelamis%20Offshore%20Wave%20Power%20Plant.pdf

49 Astariz S, Perez-Collazo C, Abanades J, Iglesias G, 2015. Co-located wind-wave farm synergies (Operation & Maintenance): A case study. *Energy Conversion and Management* 91, 63–75.

50 Branker K, Pathak MJM, Pearce JM, 2011. A review of solar photovoltaic levelized cost of electricity. *Renewable and Sustainable Energy Reviews*, 15, 4470–4482.

51 Myhr A, Bjerkseter C, Ågotnes A, Nygaard TA, 2014. Levelised cost of energy for offshore floating wind turbines in a life cycle perspective. *Renewable Energy* 66, 714–728.

52 Ouyang X, Lin B, 2014. Levelized cost of electricity (LCOE) of renewable energies and required subsidies in China. *Energy Policy* 70, 64–73.

53 Astariz S, Vazquez A, Iglesias G, 2014. Evaluation of the levelised costs of tidal, wave and offshore wind energy. 3rd IAHR Europe Congress. Porto.

54 Astariz S, Iglesias G, 2015. The economics of wave energy: A review. *Renewable and Sustainable Energy Reviews* 45, 397–408.

55 IEA, 2005. Projected Costs of Generating Electricity. OECD/IEA, Paris.

56 Gross R, Blyth W, Heptonstall P, 2010. Risks, revenues and investment in electricity generation: Why policy needs to look beyond costs. *Energy Economics* 32, 796–804.

57 Vazquez A, Astariz S, Iglesias G, 2015. A strategic policy framework for promoting the marine energy sector in Spain. *Journal of Renewable and Sustainable Energy* 7, 061702.

58 Randall T, 2016. World energy hits a turning point: Solar that's cheaper than wind. Available at: https://www.bloomberg.com/news/articles/2016-12-15/world-energy-hits-a-turning-point-solar-that-s-cheaper-than-wind

59 Vazquez A, Iglesias G, 2016. Grid parity in tidal stream energy projects: An assessment of financial, technological and economic LCOE input parameters. *Technological Forecasting and Social Change* 104, 89–101.

60 Heptonsall P, Gross R, Greenarcre P, Cockerill T, 2012. The cost of offshore wind: Understanding the past and projecting the future. *Energy Policy* 41, 7.

61 Neuhoff K, Skillings S, Rix O, Sinclair D, Screen N, Tipping J, 2008. Implementation of EU 2020 Renewable Target in the UK Electricity Sector: Renewable Support Schemes. RedPoint Energy Ltd.

62 Oxera Ltd., 2005. What is the potential for commercially viable renewable generation technologies? Available at: http://www.oxera.com/Oxera/media/Oxera/The-potential-for-commercially-viable-renewable-generation-technologies.pdf?ext=.pdf

63 Magagna D, Uihlein A, 2015. Ocean energy development in Europe: Current status and future perspectives. *International Journal of Marine Energy* 11, 84–104.

64 MacGillivray A, Jeffrey H, Winskel M, Bryden I, 2014. Innovation and cost reduction for marine renewable energy: A learning investment sensitivity analysis. *Technological Forecasting and Social Change*, 87, 108–124.

65 Winskel M, Markusson N, Jeffrey H, Candelise C, Dutton G, Howarth P, *et al.*, 2014. Learning pathways for energy supply technologies: Bridging between innovation studies and learning rates. *Technological Forecasting and Social Change*, 81, 96–114.

66 Bahaj WMJ, Bahaj AS, 2006. An assessment of growth scenarios and implications for ocean energy industries in Europe. Report for Sustainable Energy Research Group, School of Civil Engineering and the Environment, University of Southampton.

67 Dalton GJ, Alcorn R, Lewis T, 2012. A 10 year installation program for wave energy in Ireland: A case study sensitivity analysis on financial returns. *Renewable Energy* 40(1):80–9.

68 Hoffmann W, 2006. PV solar electricity industry: market growth and perspective. *Solar Energy Materials and Solar Cells* 90(18–19), 26.

69 Junginger M, Faaij A, 2004. Cost reduction prospects for offshore wind farms. *Wind Engineering* 28,22.

70 Junginger M, Faaij A, Turkenburg WC, 2005. Global experience curves for wind farms. *Energy Policy* 33(2), 133–150.

71 Astariz S, Vazquez A, Iglesias G, 2015. Evaluation and comparison of the levelized cost of tidal, wave and offshore wind energy. *Journal of Renewable and Sustainable Energy*, 7, 053112.

72 Astariz S, Perez-Collazo C, Abanades J, Iglesias G, 2015. Towards the optimal design of a co-located wind-wave farm. *Energy* 84, 15–24.

73 Astariz S, Iglesias G, 2016. Output power smoothing and reduced downtime period by combined wind and wave energy farms. *Energy* 97, 69–81.

74 Astariz S, Iglesias G, 2016. Wave energy vs. other energy sources: a reassessment of the economics. *International Journal of Green Energy* 13, 747–755.

75 Azzellino A, Ferrante V, Kofoed JP, Lanfredi C, Vicinanza D, 2013. Optimal siting of offshore wind-power combined with wave energy through a marine spatial planning approach. *International Journal of Marine Energy* 3–4, e11–e25.

76 Caballero C, 2011. Estudio de plantas de producción de energías renovables con aprovechamiento de la energía del mar. Departamento de Electricidad, Universidad Carlos III de Madrid.

77 Muliawan MJ, Karimirad M, Moan T, 2013. Dynamic response and power performance of a combined Spar-type floating wind turbine and coaxial floating wave energy converter. *Renewable Energy* 50, 47–57.

78 Pérez-Collazo C, Greaves D, Iglesias G, 2015. A review of combined wave and offshore wind energy. *Renewable and Sustainable Energy Reviews* 42, 141–53.

79 Power-technology Ltd., 2010. Green Ocean Energy Wave Trader, United Kingdom. http://www.power-technology.com/projects/greenoceanenergywav/

80 Stoutenburg ED, Jenkins N, Jacobson MZ, 2010. Power output variations of co-located offshore wind turbines and wave energy converters in California. *Renewable Energy* 35(12), 2781–2791.

81 Environmental Change Institute, 2005. Variability of UK Marine Resources. Commissioned by the Carbon Trust. http://www.marinerenewables.ca/wp-content/uploads/2012/11/Variability-of-UK-marine-resources.pdf.

82 Environmental Change Institute, 2006. Diversified renewable energy sources. Commissioned by the Carbon Trust. https://www.carbontrust.com/media/174021/diversifiedrenewableenergyresources.pdf.

83 Graeme RG, Hoste MJD, Mark Z, 2009. Jacobson matching hourly and peak demand by combining different renewable energy sources. Department of Civil and Environmental Engineering, Stanford University.

84 Taniguchi T, Ishida S, Minami Y, 2013. A feasibility study on hybrid use of ocean renewable energy resources around Japan. In ASME 2013 32nd International Conference on Ocean, Offshore and Arctic Engineering. American Society of Mechanical Engineers.

85 Pérez-Collazo C, Jakobsen MM, Buckland H, Fernández-Chozas J, 2013. Synergies for a wave-wind energy concept. In Proceedings of the European Offshore Wind Energy Conference, Frankfurt.

86 Da Y, Khaligh A, 2009. Hybrid offshore wind and tidal turbine energy harvesting system with independently controlled rectifiers. IECON'09 35th Annual Conference of IEEE, pp. 4577–82.

87 Mousavi G SM, 2012. An autonomous hybrid energy system of wind/tidal/microturbine/battery storage. *International Journal of Electrical Power & Energy Systems* 43(1), 1144–54.

88 Fusco F, Nolan G, Ringwood JV, 2010. Variability reduction through optimal combination of wind/wave resources – An Irish case study. *Energy* 35(1), 314–25.

89 Lund H, 2006. Large-scale integration of optimal combinations of PV, wind and wave power into the electricity supply. *Renewable Energy* 31(4), 503–515.

90 Astariz S, Iglesias G, 2015. Co-located wave-wind farms: Economic assessment as a function of layout. *Renewable Energy* 83, 837–849.

91 Abanades J, Greaves D, Iglesias G, 2015. Coastal defence using wave farms: The role of farm-to-coast distance. *Renewable Energy* 75, 572–582.

92 Bento AR, Rusu E, Martinho P, Soares CG, 2014. Assessment of the changes induced by a wave farm in the nearshore wave conditions. *Computers & Geosciences* 71, 12.

93 Millar DL, Smith HCM, Reeve DE, 2007. Modelling analysis of the sensitivity of shoreline change to a wave farm. *Ocean Engineering* 34(5–6), 884–901.

94 Smith HCM, Pearce C, Millar DL, 2012. Further analysis of change in nearshore wave climate due to an offshore wave farm: An enhanced case study for the Wave Hub site. *Renewable Energy* 40(1), 51–64.

95 Tănase Zanopol A, Onea F, Rusu E, 2014. Coastal impact assessment of a generic wave farm operating in the Romanian nearshore. *Energy* 72, 652–670.

96 Matutano C, Negro V, López-Gutiérrez J-S, Esteban MD, 2013. Scour prediction and scour protections in offshore wind farms. *Renewable Energy* 57, 358–365.

97 Moccia J, Arapogianni A, Wilkes J, Kjaer C, Gruet R. Pure Power, 2011. Wind energy targets for 2020 and 2030. European Wind Energy Association, Brussels.

98 Ponce de León S, Bettencourt J.H, Kjerstad N, 2011. Simulation of irregular waves in an offshore wind farm with a spectral wave model. *Continental Shelf Research* 31, 1541–1557.

99 Elefant C, O'Neill S. Hybrid Offshore Ocean Power and Wind Facilities, 2007. http://www.renewableenergyworld.com/articles/2007/03/hybrid-offshore-ocean-power-and-wind-facilities-47873.html (accessed on 27 July 2015).

100 Rahman ML, Shirai SOY, 2009. Hybrid power system using offshore-wind turbine and tidal turbine with flywheel (OTTF) EWEA. Marseille, France.

101 Hanssen J, 2011. Wind wave power. http://www.proexca.es/Portals/0/

102 Ramos V, Carballo R, Sanchez M, Veigas M, Iglesias G, 2014. Tidal stream energy impacts on estuarine circulation. *Energy Conversion and Management* 80, 137–149.

103 Ramos V, Carballo R, Álvarez M, Sánchez M, Iglesias G, 2013. Assessment of the impacts of tidal stream energy through high-resolution numerical modeling. *Energy* 61, 541–554.

104 Turnock SR, Phillips AB, Banks J, Nicholls-Lee R, 2011. Modelling tidal current turbine wakes using a coupled RANS-BEMT approach as a tool for analysing power capture of arrays of turbines. *Ocean Engineering* 38, 1300–1307.

105 Malki R, Masters I, Williams AJ, Croft NT, 2014. Planning tidal stream turbine array layouts using a coupled blade element momentum – computational fluid dynamics model. *Renewable Energy* 63, 46–54.

106 ExternE – External Costs of Energy. http://www.externe.info/externe_d7/

107 Galetovic A, Muñoz CM, 2013. Wind, coal, and the cost of environmental externalities. *Energy Policy* 62, 1385–1391.

108 Mahapatra D, Shukla P, Dhar S, 2012. External cost of coal based electricity generation: A tale of Ahmedabad city. *Energy Policy* 49, 253–265.

109 Gaterell M, McEvoy, 2005. The impact of energy externalities on the cost effectiveness of energy efficiency measures applied to dwellings. *Energy and Buildings* 37(11), 1017.

110 Vazquez A, Iglesias, G. 2015. Should tidal stream energy be publicly funded? Evidence from a choice experiment study. 11th European Wave and Tidal Energy Conference (EWTEC), Nantes, France.

111 Sanchez M, Carballo R, Ramos V, Iglesias G, 2014. Floating vs. bottom-fixed turbines for tidal stream energy: A comparative impact assessment. *Energy* 72, 691–701.

112 Veigas M, Ramos V, Iglesias G, 2014. A wave farm for an island: Detailed effects on the nearshore wave climate. *Energy* 69, 801–812.

113 Eshet T, Ayalon O, Shechter M, 2006. Valuation of externalities of selected waste management alternatives: A comparative review and analysis. *Resources, Conservation and Recycling* 46, 335–364.

114 Birol E, Koundouri P, Kountouris Y, 2010. Assessing the economic viability of alternative water resources in water-scarce regions: Combining economic valuation, cost-benefit analysis and discounting. *Ecological Economics* 69, 839–847.

115 Louviere JJ, Hensher D.A, Swait J.D, 2000. Stated Choice Methods: Analysis and Applications. Cambridge University Press,.

116 Vecchiato D, Tempesta T, 2015. Public preferences for electricity contracts including renewable energy: A marketing analysis with choice experiments. *Energy* 88, 168–179.

117 Vazquez A, Iglesias, G, 2015. Public perceptions and externalities in tidal stream energy: A valuation for policy making. *Ocean & Coastal Management* 105, 15–24.

118 Lee J, Yoo S, 2009. Measuring the environmental costs of tidal power plant construction: A choice experiment study. *Energy Policy* 37, 5069–5074.

119 Glazyrina I, Glazyrin V, Vinnichenko S, 2006. The polluter pays principle and potential conflicts in society. *Ecological Economics* 59, 324–330.

120 García-Alvarez MT, Mariz-Pérez RM, 2012. Analysis of the success of feed-in tariff for renewable energy promotion mechanism in the EU: Lessons from Germany and Spain. *Procedia – Social and Behavioral Sciences* 65, 52–57.

121 Schmidt J, Lehecka G, Gass V, Schmid, E, 2013. Where the wind blows: Assessing the effect of fixed and premium based feed-in tariffs on the spatial diversification of wind turbines. *Energy Economics* 40, 269–276.

12

Project Development

Paul Vigars[a], Kwangsoo Lee[b], Sungwon Shin[c], Boel Ekergard[d], Mats Leijon[e], Yago Torre-Enciso[f], Dorleta Marina[f] and Deborah Greaves[g]

[a] *Director of Teobi Engineering Associates Ltd, UK (formerly Research & Technology Manager at Alstom Ocean Energy, UK)*
[b] *Principal Research Scientist, Institute of Ocean Science & Technology, Korea*
[c] *Research Professor, Kangwon National University, Korea*
[d] *Seabased Industry AB, Sweden*
[e] *Professor, Uppsala University, Sweden*
[f] *BIMEP, Lemoiz, Spain*
[g] *Professor of Ocean Engineering, School of Engineering, University of Plymouth, UK*

12.1   Introduction

This chapter aims to illustrate the different stages of project development through four representative case studies: two based on wave energy conversion technologies and two on tidal energy conversion. With a structure articulated around these case studies, this chapter brings together device design, array planning, policy, regulation and economics, together with detailed discussions of technical aspects (e.g. access and deployment strategies, installation, and reliability and maintenance) and non-technical aspects (e.g. planning, consenting, economic assessment and externalities, financing). The development of the Oceade^{TM} tidal stream turbine developed in the UK is first described in Section 12.2, followed by the Seabased wave energy converter from Uppsala University in Sweden in Section 12.3, the Lake Sihwa tidal power plant of Korea in Section 12.4, and finally the Spanish Mutriku wave energy plant in Section 12.5.

12.2   Alstom Ocean Energy OCEADE™ Tidal Stream Turbine: The Route to Commercial Readiness

12.2.1   Introduction

Exploitation of the power in flowing water has a long history. The attraction of fast tidal currents is difficult to ignore and creating net power from such flows is not difficult *per se*. The challenge comes when the enterprise must compete against alternative sources of energy and ultimately develop to become a bankable asset. The economic pressures of the marketplace, coupled with the relative energy density of the resource and the aggressiveness of the environment, amounts to a significant techno-economic challenge. To embark on the journey of technology development without regard for such realities would be little more than an expensive distraction. What follows is a case study of the development of a tidal stream turbine technology from initial concept through the prototype phases to a first-of-a-kind commercial turbine. Although they are an important component, this involves much more than an efficient set of turbine blades.

The story of the development of a new technology for a new market sector is the story of growing maturity in all aspects of the undertaking: market forecasting, political support, funding mechanisms, financial modelling, supply chain development, resource prediction, load and performance modelling, certification guidelines and standards, marine operations, instrumentation, environmental impact assessment, environmental, health and safety guidelines, device design, not to mention the growth in the required financing and resources.

These many aspects form a complex system, requiring an iterative approach where progress is made on all fronts. From starting assumptions and choices, through learning, to knowledge, more choices, more learning, more knowledge and so on. At each stage more and more aspects become involved, the iterations increase in complexity, duration and use of resources, and the choices of past iterations are increasingly baked into the solution. Figure 12.1 illustrates the different stages of development in terms of Technology Readiness Level and Commercial Readiness Index [1].

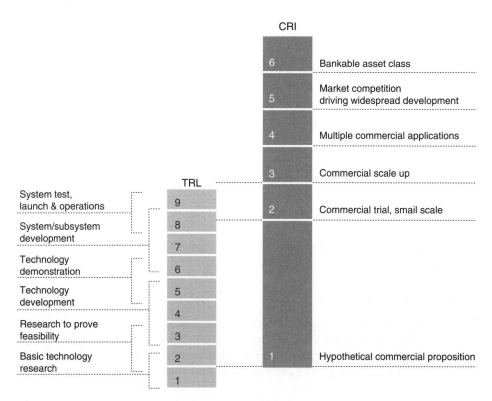

Figure 12.1 Technology Readiness Level and Commercial Readiness Index [1].

Unlike introducing a disruptive novel technology into a mature market, the development of tidal technology has occurred in parallel with the development of the market. This means, of course, that the initial requirements for the technology could not be based on firm market knowledge or data, but on the best predictions and forecasts of how such a market and the access to the resource it seeks to exploit would develop. Continuous reassessment of technology requirements is necessary as the technology and the market develop, but the ability to make significant conceptual changes diminishes as the technology matures, at least in the initial stages as the sector becomes established. The 'selection' of the most viable concepts is left to the vagaries of the developing market and speculative investments. Early government support for a broad variety of technical solutions ensures an initial pool of technologies that then run the test of the difficult investment and technology proving stages. Only the naïve would believe that the 'best' technologies will automatically win through this mechanism. Having the right technology is, of course, necessary, but it is not sufficient to survive the early stages of development.

12.2.2 Alstom Concept

The DNA of the Alstom Oceade™ turbine has its roots in the DEEPGen™ technology conceived and developed by Tidal Generation Limited (TGL), and later as a subsidiary of Rolls-Royce plc. The original philosophy of the TGL DEEPGen technology was driven by the understanding at the time of the nature of the tidal stream resource, particularly the typical speed and depth distribution (from the Black & Veatch study commissioned by the DTI [2]). This highlighted that more than 75% of the economically viable tidal stream resource (for the UK) was in sites deeper than 40 m.

The original guiding philosophy is summarised as follows:

- Simple, cheap and rapid deployment and retrieval of turbine using modestly priced vessels.
- No divers.
- All maintenance onshore.
- All complex systems to be recoverable for maintenance.
- System capable of deployment into deep tidal sites (40–80 m).
- Yaw capability to allow for rotor operation upstream of support structure to minimise fatigue loading and to maximise yield.
- Full power frequency converter on board for maximum load control and grid compliant power at turbine output.
- No surface piercing structures; clearance for marine traffic.

This short list does not fully specify the chosen concept, but these were the guiding principles.

A pragmatic acknowledgement that device reliability will not be markedly different from other similar technologies (such as wind), coupled with the fact that access for *in situ* maintenance is not a practical proposition and the cost of marine operations with large vessels (spot hire rates of the order of $100,000 per day), leads to a need for a cost-effective means of deployment and recovery of turbines for maintenance.

The decision to design a net buoyant turbine that could be deployed using a single winch line [3] achieved multiple aims: deployment and recovery using smaller and cheaper vessels; no heavy lifting at sea (less susceptible to weather downtime for operations); observation class remotely operated vehicle (ROV), no divers; towed transport to and from site; possibility of temporary wet storage for multiple turbines close to site.

Concerns over dynamic loading on blades and power cables leads to favouring a seabed fixed support structure (as opposed to a moored or hinged fixation) and upstream fixed yaw generation with yawing movements only at slack water. This avoids tower shadow fatigue loads on the blades and maximises the available power (yield), particularly where flood and ebb flows at a site are not from directly opposing headings (deviation from directly opposing headings for flood and ebb can be up to 25°) [4]. Fatigue loads are a primary driver of cost. Yield is the single most significant element in the calculation of the levelised cost of electricity (LCOE).

To reduce complexity at the mechanical and electrical interface of the turbine with the support structure, a hydrodynamic thruster was specified to provide the torque to yaw the turbine between flood and ebb positions at slack tide [5].

An initial assessment was made of the relative merits of gravity, monopile or piled jacket foundation structure. Each has merits and difficulties, but the original choice was for a piled jacket structure. The main reasons for this included overall weight per megawatt, complexity/cost of marine operations and ease of cable routeing and replacement. For the piled jacket to be feasible a novel approach to drilling was required. At the depths envisaged for commercial tidal stream farms (greater than 40 m) the use of a jack-up barge as a drilling platform was unproven and believed to be infeasible, particularly for the deeper sites (subsequent trials at EMEC have supported this assumption). The solution was to develop nearshore pneumatic percussion drilling for use at depth, powered by a flexible umbilical from a surface vessel. The use of the flexible umbilical requires lower station-keeping capability of the vessel [6]. At the time of this innovation, the station-keeping capability of moored or dynamically positioned vessels was unknown.

12.2.3 Device Demonstration

The original concept included significant novelty in the installation, deployment, recovery, connection and alignment functions. To lower the risk for prospective investors a full-scale mock-up was built to demonstrate these critical features. This was successfully tested at Plymouth during the summer of 2007. The drilling rig was built and completed a submerged trial in a flooded quarry. As a result of these successes the enterprise became significantly more investable, with a clutch of innovative technologies demonstrated and patented.

The next development vehicle was a 500 kW fully functional prototype installed and tested at the EMEC in the Orkney Islands. The piled tripod jacket support structure was installed using moored barges over two separate campaigns in 2008 and 2009. Difficulties with the performance of the drilling system at depth were eventually overcome. Electrical faults with the cable terminations in the tripod meant that three campaigns of remedial works were required before the installation was ready to receive the turbine. Despite the high day-rate, dynamically positioned (DP) vessels were selected for this work, following a successful trial of the station-keeping capability in the tidal flow. Much was learned during this period, not least the relative strengths and weaknesses of using DP vessels or moored platforms, and the capabilities and limitations of using divers and ROVs for tidal work.

The 500 kW DEEPGen III turbine (part funded by the Technology Strategy Board) was deployed at the Fall of Warness in the Orkney Islands and generated its first power in September 2010. The power take-off was reasonably conventional, with a high-speed induction generator and full-power frequency converter housed on-board. The prototype logged over 3200 running hours and generated nearly 250 MWh. Valuable performance data was gathered during the campaign, to validate numerous design assumptions. Significant learning was garnered in various aspects: hydrodynamic design and load prediction; power take-off and control; grid code compliance; marine operations. The 500 kW turbine has now been recycled (97% by weight). Only the blades have been retained for possible future testing.

Following the success of the 500 kW prototype, a full-scale 1 MWe prototype was designed, built and tested using the same tripod support structure. This was the original intent and the tripod was designed for the duty of the larger turbine. Developments in the analysis tools used and increased knowledge about the environment at the test site meant that the original design loads for the tripod were underestimated. This resulted in the operating envelope of the larger turbine being slightly restricted, specifically when generating in large waves. This did not detract from the research value of the project, indeed it led to further innovative measurement and control strategies [7].

The 1 MW turbine was built for the Energy Technologies Institute (ETI) funded ReDAPT project which aimed to improve the understanding of the influence of a tidal turbine on its environment. This £12.4 million collaborative R&D project included Garrad Hassan (now DNV GL), Plymouth Marine Laboratory, EDF Energy, Tidal Generation/Rolls-Royce, University of Edinburgh, E.On UK and EMEC. The turbine was equipped with a state-of-the-art instrumentation suite for the measurement of flow, performance, loads and equipment health monitoring. In addition, the turbine was used as a platform for materials and coatings tests, blade anti-fouling coatings trials, field measurements of inflow data, wake deficit data and acoustic emissions.

The conceptual design of the ReDAPT turbine (see Figure 12.2) was the same as the previous prototype, with lessons learned being incorporated into the design. The most visible development was to make the winch recoverable [5] after deployment of the turbine so that only one winch system is required for multiple turbine deployments in a farm. For turbine recovery, controlled buoyant ascent [8] was trialled and found to be preferable to use of the winch.

The enhanced measurement capabilities included on and around the turbine were used to validate Garrad Hassan's Tidal Bladed™ tool for the prediction of operating loads [9]. Critical to this was the concurrent measurement of the unsteady inflow, turbine performance and turbine loads. The ReDAPT project presented an ideal opportunity to undertake power curve assessment to the recently published IEC standard [10]. This specifies, among other things, the instrumentation requirements and the test protocol. Without such a standard, comparison of turbine performance curves would be problematic.

During the ReDAPT project, the turbine was deployed for 18 months and experienced a full range of different flow conditions, generating over 1GWh. A partial strip and inspection of key components was undertaken to maximise the learning generated from the project. The tripod support structure is due to be decommissioned and recycled at the end of the berth hire contract at EMEC.

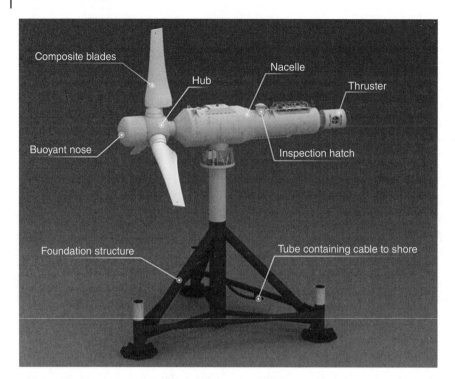

Figure 12.2 The Alstom concept embodied in the 1MW ReDAPT prototype. Courtesy: Alstom Ocean Energy Ltd.

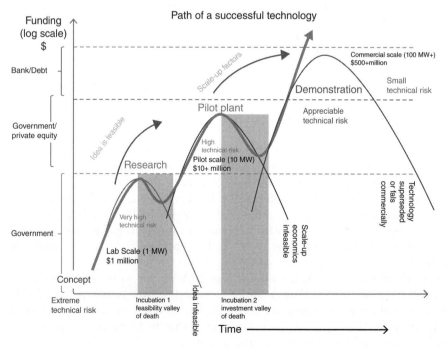

Figure 12.3 Path of a successful technology [11].

12.2.4   First-of-a-Kind Commercial Turbine

Alstom is now taking the learning accumulated over 10 years of research and development, illustrated in Figure 12.4, building on the successful prototype campaigns to design and supply first-of-a-kind commercial turbines. Significant knowledge has been gained not only in device design but also in understanding of sites, understanding of resource, understanding of project economics and knowledge of the market.

The Alstom Oceade™ turbine is conceptually the same as the previous prototypes, redesigned to better match the resource characteristics and economic requirements of the first commercial scale farm installations. The nameplate rating of the turbine has been raised to 1.4 MWe and a monopile has been selected as the most effective support solution. The turbine design has been reconfigured to be modular to minimise assembly and maintenance costs and to maximise productivity through reduced downtime.

12.2.5   Design Iterations

The design of any piece of novel technology is an iterative process, and this has certainly been the case for the development of each of the Alstom tidal turbines. The iterative loops vary in scope and duration depending upon the stage of the development and many are repeated, pulling in ever increasing knowledge from a growing number of fields and influences.

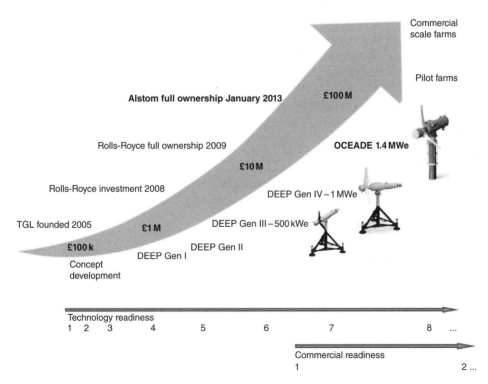

Figure 12.4 Development pedigree of the Alstom Oceade™ tidal turbine. Courtesy: Alstom Ocean Energy Ltd.

Initially, there were a large number of iterations of a techno-economic design to select initial concepts worthy of more detailed work. These were used to assess the technical and economic feasibility and worked with rough order-of-magnitude estimates for a variety of aspects such as cost of manufacture, weight, O&M costs, efficiency and yield. The starting point has typically been to understand rotor loads and yield and then build from there. This gives the impression that blade hydrodynamics is the most important aspect of turbine design. This could be debated at length, but what is clear is that there are very many aspects to 'get right' and there are influences and constraints that need to be considered and incorporated into the hydrodynamic design.

Iterations around blade design leading to rotor and control system design, structural feasibility and initial performance are initially numerous and rapid (1–2 person-days). As the concept crystallises the depth and breadth of analyses increases and the iterations take longer and demand more resources, ultimately leading to fully certified design load assessment and hardware design.

There is a step change in the nature of the design iterations at the point where detailed quotes are required from external suppliers for bespoke components. This adds a significant administrative burden and requires a broader skill set in the team to manage properly. The iteration cycle time is also significantly lengthened, meaning that the number of full iterations is limited by resources and project timescales.

Due to the necessarily holistic nature of a full solution for a tidal turbine, selecting the right concept early on is very important. Conceptual decisions become fundamental to the solution and impose significant constraints on other parts of the complete solution. As the burden of the design iteration becomes ever larger, the concept becomes ever more baked into the design and the changes that are possible become more restricted. As the development progresses, numerous smaller design iterations of subsystems are needed within the framework of the overall design solution.

Breaking out of the existing paradigm becomes ever more difficult, further highlighting the value in getting the right concept early on. Alstom operates a strict gated design process which ensures adherence to business requirements while minimising risk and exposure at each stage. This design process formally defines the iterations at the largest scale.

12.2.6 Managing Uncertainty

Throughout the design process assumptions must be progressively replaced by knowledge. This needs to occur in all fields to ensure that the solution will ultimately be commercially viable. Failing to mature knowledge in any field represents a significant risk, particularly in the stages of the prototype development cycle where hardware is being procured and full-scale trials are being undertaken. A similar risk profile develops for the development of the commercial product, only the risk is now significantly increased (see Figure 12.3). The main point here is that the development of a successful technology is not limited to the development of the device alone – far from it. Knowledge of site characteristics, flow and wave regimes, vessel capability, cable works, grid compliance, ecological effects, device loading, uncertainty in analysis techniques, instrumentation, financial incentives and support mechanisms, legislation – to name but a few – all have to be developed and improved in order to reach a commercially viable solution. The importance of knowledge in each field varies through the timeline of the development process. In the very earliest stages of device concept selection more detail is needed for certain fields and less detail for others (note that *some* knowledge in *all* aspects is needed from the outset). As the development cycle approaches commercialisation, all fields need to be mature as far as possible, to minimise the residual risks.

Management of uncertainty is an important aspect of successful product development. Uncertainty exists in all fields of knowledge required to develop a successful solution. Understanding for each field the existing levels of uncertainty, the impact that uncertainty has and the cost–benefit balance in reducing the uncertainty is key to identifying where to allocate resources to improve the solution.

For the early stages of a new industry sector such as tidal stream, there is a tension between the need to be commercially viable and the need for realistic conservatism to mitigate the risks posed by the unknown and unknowable. One such example is the risk of cable failure in tidal sites due to the high flow. Techniques to guarantee to minimise cable failures due to fatigue loading do exist (e.g. directional drilling, stapling, rock blanketing) but would likely be far too expensive at the outset of a first commercial farm to be viable. While the upfront cost of such techniques can be assessed, the eventual benefits can only be assessed in time, and then only if the technique is not used.

12.2.7 Levelised Cost of Electricity

Cost of electricity or cost of energy is used as the headline metric for assessing the commercial viability of a technology and a project. Other assessments are important such as annual energy production and internal rate of return, but in simplistic terms project goals and engineering challenges during this technology development phase are defined in terms of LCOE. While the concept is very simple (costs divided by energy) the practice involves the modelling of a large number of highly interconnected parameters.

The total cost is broken down into basic elements of up-front capital costs and though-life operational costs. The breakdown varies depending upon the specifics of the technological solution. A breakdown for a 10 MW farm using Alstom Oceade™ turbines is shown in Figure 12.5.

By far the biggest single lever in the LCOE calculation is the yield: how much product (energy) is produced for sale. This is because yield is the only term in the denominator of the calculation. When seeking LCOE improvements, any change that increases yield might also be expected to increase cost, but is unlikely to cause an increase in all of the different categories of cost.

When determining the cost or value of a design concept option it is too simplistic to state that X is better than Y for a subsystem – you need to assess the whole alphabet of the complete solution. For example, a technology that is designed for a zero-intervention maintenance strategy might be expected to be very robust and heavy compared to other concepts, requiring the use of large and expensive vessels for deployment. While concepts with other maintenance strategies could not support the additional expense of marine operations, the low-maintenance concept and a seldom used expensive deployment could work out to be competitive.

As in the example above, concept choices, once fixed, will constrain other parts of the full solution and limit the options available. The optimum LCOE solution is unlikely to arise by optimisation of subsystems without consideration of the whole. Quite possibly, in principle at least, the optimum solution could consist of the second-best options for each subsystem.

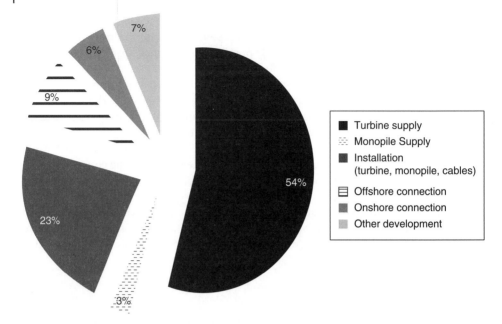

Figure 12.5 Example LCOE contributions for a 10 MW tidal farm.

Optimising for LCOE alone might not be the optimum commercial proposition and it is also important to consider annual energy production. Further, an enterprise that is seeking to sell turbines must seek to make a profit at the same time as remaining competitive enough to win orders.

12.2.8 The Role of Intellectual Property

Intellectual property (particularly patents) has played an important role through the development of the Alstom technology. The patenting profile is shown in Figure 12.6 and has been typical for a start-up business seeking to carve out a unique position, attract investment and progress to commercialisation.

In the early stages, a small number of key innovations were patented. These were the 'big ideas' that made the concept unique: flexible umbilical pile drilling; buoyant winched turbine; combined clamp and bearing; hydrodynamic yaw thruster. Without protection of the concept, the enterprise would have been too risky for investors.

Following the successful proof-of-concept testing of these key innovations at full scale, the investment potential of the enterprise was established. Realisation of the key concepts into working prototypes was an intense period of engineering activity but a quiet period in terms of patenting activity. Budget restrictions limited additional patenting during this period, but many innovations were identified, as might be expected when prototype-testing a non-incremental technology. Further investment allowed for further patenting of the second wave of innovations. This was another important stage in terms of the development of the patent portfolio, as it protected further key intellectual property required for the commercial success of the technology.

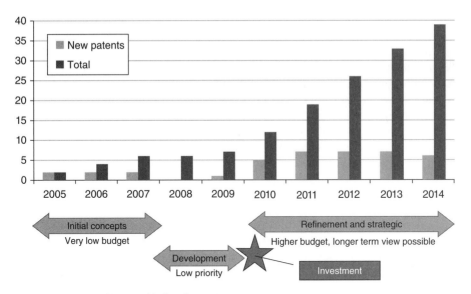

Figure 12.6 Patent filing profile for Alstom Ocean Energy.

Beyond this stage, the ambition was widened, looking to carve an ever larger part of the technology landscape and restrict prospective competitors. Concepts and innovations that would widen a future product range were explored and patented. The 20-year lifetime of a patent allows for early protection of concepts that still require significant development. Some degree of speculation can occur if funding permits, but the prospective market advantages can make this worthwhile. Patents can, of course, be withdrawn or allowed to lapse through non-payment of maintenance fees.

Protection of intellectual property is a balance between benefit, risk, cost, budget and also detectability. This last point is important, particularly in the case of technologies that are difficult to reverse-engineer, for example control system algorithms or manufacturing processes for composite components. The detectability question works both ways: can a technology be protected by keeping the knowledge in house (trade secret)? Can infringement of a patent by a competitor be detected?

Clearly, a patent is not always the right choice of protection. A protection strategy should make best use of alternatives to patents where appropriate, such as trade secrets, publication and design registration. For marketing and messaging purposes, registered trademarks can be valuable.

A patent is specific to a particular country or territory and it grants the owner the right to prevent others from manufacturing or selling an infringing piece of technology in that country. With close to 200 countries worldwide with functioning patent legal systems, a choice must be made in which countries to protect your intellectual property. With upfront costs and ongoing fees, the cost of each country case for each invention can soon add up to a large amount – so having the right strategy is important.

Tidal stream technology is reasonably simple when compared to other technologies such as gas turbines, aerospace and so on. The manufacturing capability requirements do not represent a significant hurdle and, save for some specialist components, are suited to low-cost economies. This means that, quality issues aside, many countries have the skill set needed to manufacture tidal stream devices (indeed, marine renewables more generally). Seeking patent protection in all of these countries would become very expensive.

Unlike other sources of renewable energy such as wind and solar, the number of countries with commercially exploitable tidal stream resources is quite small. This means that a cost-effective patenting strategy can be built by protecting inventions only in the countries with the largest tidal resources (i.e. the biggest markets). Depending upon the available budget and the perceived strength and value of a particular invention, the depth of coverage can be varied, with the core inventions obtaining the largest proportion of the market.

12.2.9 Conclusion

Having a good concept gets you into the game but only for so long. Execution of the 'good idea' is the more substantial contribution to the success or otherwise of the enterprise, notwithstanding the fact that the better the concept the easier the task.

The route from initial concept to commercialisation for tidal stream technology is a journey which has had numerous hurdles, each higher than the next, requiring deep pockets, a good team and a long-term view. The next stages of commercialisation will see the first multi-turbine installations, bringing new challenges and new learning.

Continuous improvement in all technical and commercial aspects of tidal stream exploitation will ensure that tidal energy will be competitive with other renewable energy sources and contribute to a low-carbon future as part of a diversified energy mix.

12.3 Seabased Wave Energy Converter

12.3.1 Strategy

The Seabased wave energy device is developed by the Swedish company Seabased AB. The company's strategy is to develop into a system supplier of turnkey wave energy parks, including pre-studies, permits and maintenance services. This will be enabled through geographically well-situated mass production systems in key markets to ensure high quality and low cost. Seabased has assessed that it is critical to have series production in Sweden and initially produce as many parts as possible in its own factories. When Seabased has gained experience from the production process in Sweden several other places worldwide will be used for series production.

12.3.2 Research and Development

The company is engaged in ongoing R&D through its subsidiary Seabased Industry AB. Research, product development and initial tests are performed internally. R&D is carried out in close cooperation with researchers at the Centre for Renewable Electric Energy Conversion, Ångström Laboratory, Uppsala University. Since its formation, Seabased has been working to build up full-coverage patent protection. A key aspect of patent protection refers to the company's solution for turning the waves constantly varying energy output into electricity that can be fed into the electricity grid. The company holds a large patent portfolio that is constantly increasing.

12.3.3   Park Layout

Deployment of the units containing linear generators on the seabed protects the units from extreme conditions that may occur at the surface of the sea and helps minimise required maintenance. Combined with a mechanically simple (very few moving parts) and robust generator technology, this yields an efficient energy conversion system. In the large-scale project in Sotenäs with Fortum and the Swedish Energy Agency the units have a size of 30 kW. Depending on the wave conditions at the geographical site in question, a normal wave energy park may consist of up to 10,000 wave energy converter (WEC) units. Seabased's generators have today a nominal capacity of 20–200 kW. To have several smaller units enables the park to be less vulnerable to malfunctions in individual WECs.

WECs are deployed effectively with a marine vessel carrying several units that can be sunk into the seabed, and each unit is anchored using a gravitational anchor (concrete base). The WEC can be placed on various types of seabed (preferably not rock if a gravitation anchor is used) and can work satisfactorily with an inclination of up to 15% deviation from the vertical. Luckily, in most areas where the wave climate is good, level seabeds can be found, due to the effect from the moving water, levelling off irregularities. Individual WECs can be disconnected and replaced without affecting the function of other parts of the wave energy park, which simplifies ongoing repairs and maintenance. Due to the technology's scalable and modular design, electricity can be generated from only a few WECs, as construction of the rest of the park is undertaken, beginning cost recovery almost instantly. Due to the scalable system wave energy parks can be expanded in phases, as customers gain confidence in the technology [12].

12.3.4   Development and Collaboration with Uppsala University

For over a decade, a research group at the Division of Electricity at Uppsala University, has been developing a wave energy conversion system. The WEC consists of a direct driven linear generator installed on the seabed, connected by a line to a point absorbing buoy, as illustrated in Figure 12.7. The direct driven magnetic part of the generator, the translator, follows the motion of the heaving ocean waves. Intermediate energy storages and gearboxes are removed, believed to increase the lifetime of the system [13]. The output voltage from the linear direct driven generator varies both in amplitude and frequency. To connect the WECs to the grid, an electrical conversion system is required. A marine substation, located on the ocean floor, is thus integrated in the system [14].

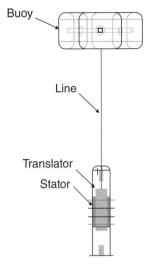

Figure 12.7 The wave energy converter.

The first full-scale WEC was installed in March 2006, and since then additional devices have been added to the site. By the autumn of 2013, 11 different WECs and two marine substations had been deployed, with the latest WEC, L12, installed in July 2013. With the deployed units, experiments on a WEC array of three generators connected to the marine substation have been performed. The power output from each WEC was separately rectified by a passive diode rectifier and connected in parallel to a common DC bus in the substation. After DC/AC conversion and transformation, the electrical power was transmitted to a nearby island, Härmanö, and converted into heat in resistive dump loads [14].

In connection to the development of the WECs and the electrical system, research on environmental impacts [15], hydrodynamics [16, 17], wave resource description [18, 19] and measurement and control systems [18–20] has been carried out.

A further detailed description of the experimental work and main progress within the Lysekil Wave Power Project at the Lysekil research site is presented in [21].

12.3.5 Seabased Technology Concept

The Seabased technology comprises the WEC, consisting of a point absorbing buoy at the ocean surface directly connected to a linear generator placed on the ocean floor, and an electrical system, which enables electrical conversion and grid connection of the system.

12.3.5.1 The Buoy
The buoy (Figure 12.8) is directly connected to the moving part of the generator, the translator, and follows the motion of the heaving buoy in ocean waves. The buoy acts as the first step in the energy conversion chain and the visual impact from the Seabased technology solution is limited to the view of the buoy. The buoy is about 6 m in diameter, painted yellow according to International Association of Lighthouse Authorities (IALA) recommendations for navigational safety. The buoy occupies an area of about 30 m² ocean surface and the part above sea level is about ½ m. Hence, the impact on seascape values will be very limited.

12.3.5.2 The Wave Energy Converter
The WEC, illustrated in Figure 12.9, consists of a direct driven linear generator installed on the seabed, connected by a line to a point absorbing buoy at the ocean surface. The direct driven magnetic part of the generator, the translator, follows the motion of the heaving ocean waves.

Figure 12.8 The Seabased Buoy.

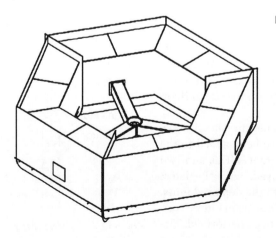

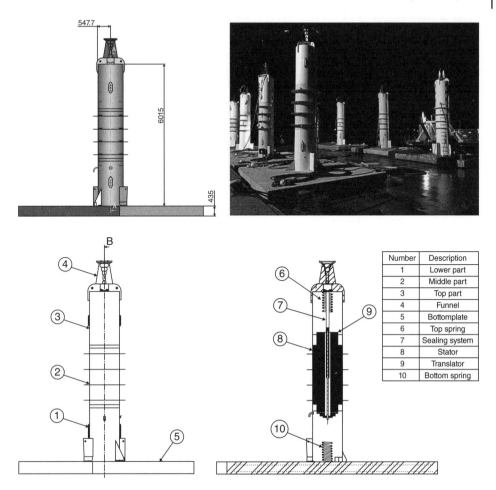

Number	Description
1	Lower part
2	Middle part
3	Top part
4	Funnel
5	Bottomplate
6	Top spring
7	Sealing system
8	Stator
9	Translator
10	Bottom spring

Figure 12.9 The linear generator.

12.3.5.3 Electrical System

The output voltage from the linear direct driven generator varies both in amplitude and frequency. To connect the WECs to the grid, an electrical conversion system is required. A low-voltage marine substation (LVMS), located on the ocean floor, is thus integrated in the system (Figure 12.10).

The control, monitoring and supervisory control and data acquisition (SCADA) philosophy is deliberately minimal. Control inputs include: occasional switching commands as required by operations; inverter DC link voltage set points; and transformer tap changers output voltage set point. Monitoring and SCADA is done by a National Instruments Compact-RIO and measurements include voltage and current in each generator phase at the input to the LVMS, and power metering at each input to the MVMS. An overview of the system is presented in Figure 12.11.

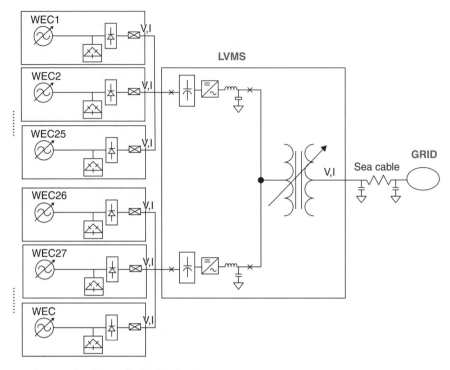

Figure 12.10 The electrical schedule for the wave power park.

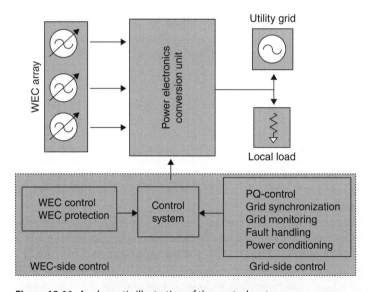

Figure 12.11 A schematic illustration of the control system.

12.3.6 Deployment

The offshore deployment of the WEC is a standard process following the same steps: preparation, transfer of the converter to the location of deployment, deployment and final processing. The *preparation step* comprises assembly of the generator, and performance of the necessary tests (drag tests, calibrating the translator, measuring voltage, leakage testing), as shown in Figure 12.12. To *transfer* the WECs to the location of the deployment, the converter is lifted and placed on a boat/barge (Figure 12.13). During the *deployment* the generators are pressurised with nitrogen. At the final processing step, a diver/ROV connects the cable and unties the slings and chucks (Figure 12.14).

12.4 Lake Sihwa Tidal Power Plant, Korea

12.4.1 Introduction

Lake Sihwa is human-made and was originally planned as a freshwater basin located in the Gyeonggi Bay of the mid-western coast of the Korean peninsula [22]. As a part of the special regional development plan of the early 1970s, in order to supply water to the agricultural and industrial area surrounding the lake and to provide agricultural, industrial and recreational lands near the metropolitan area, the construction of the dyke blocking the entrance of Sihwa Bay began in June 1987 and finished in January 1994. The 12.7-km long seawall holds 332 million tons of water in the lake with an area of 56.5 km^2 (Figure 12.15).

Inside the basin area, 17,300 ha of reclamation to provide space for industrial complexes, agriculture, and recreational facilities is planned [23]. The Sihwa dyke contains two sluices that can allow a maximum 4000 m^3/s to flow out, one fish way, one

Figure 12.12 Measurement of the voltage and leakage testing.

Figure 12.13 Transfer sequence.

Figure 12.14 Deployment of the Seabased device.

Figure 12.15 Location map of Lake Sihwa Tidal Power Plant.

navigation lock, and five wharves, and an 8 m wide 20 km long approach road is constructed on top of the dyke. This project had been expected to have an economic benefit of up to 35.5 billion won (₩) and create 7.8 million jobs along with the building of a suburban leisure and recreation complex.

After the completion of the tidal embankment, however, severe water contamination started due to contaminant accumulation resulting from excessive inflow of polluted waste waters as well as the blocking seawater exchange. Inside the lake, chemical oxygen demand (COD) reached 17.4 mg/l in 1997. The growing urban population in adjacent cities and increasing number of companies in industrial complexes in the vicinity, together with an expanding livestock industry in rural areas, had increased contaminants flowing into the basin where the provision of treatment facilities was limited. In discussion of the potential environmental impacts held in September 1988, the project was approved to discharge treated water into open sea before filling freshwater into the basin; however, the embankment had been completed in January 1994 without sufficient environmental infrastructure to achieve this fully and the water quality had been getting worse ever since.

To counteract this problem, in July 1996, the government confirmed measures to improve the water quality of Sihwa Lake with a budget of ₩449.3 billion to expand water treatment facilities, to maintain sewage lines, to install artificial wetlands, and to exchange seawater inside the basin by discarding the original freshwater reservoir scheme by 2000. The Korea Institute of Ocean Science and Technology (KIOST), formerly the Korea Ocean Research and Development Institute, had suggested the implementation of a tidal power plant as a solution for the water quality problem in 1997, on condition that seawater remains inside the newly made Lake Sihwa. KIOST [24] carried out a study on the possibility of tidal power development in Lake Sihwa with the support of the Ministry of Oceans and Fisheries (MOF) in 2000 [25, 26]. Based on the study results, the Korean government decided to maintain Lake Sihwa as a seawater basin and adjust water quality improvement measures accordingly in 2001, and the development of a tidal power plant was selected as one of the measures for the improvement of water quality inside the Lake.

12.4.2 Planning and Design

According to the conclusion of Korean government and the technical possibility that had been proved by KIOST research work, the feasibility study with basic design works was carried out by the Korea Water Resources Corporation (KOWACO, later Kwater) and KIOST [24], including integrated field surveys, numerical modelling works and plant optimisation processes.

Spring tidal range at the barrage site is 780.4 cm (see Figure 12.16) and the surface area of the basin is 4,300 ha at mean sea level with a water volume of 330 million tons. The water level inside the lake should be controlled at a maximum level of 1.0 m below mean sea level (see Figure 12.17). This water-level limitation allows lower cost as well as safety for land reclamation on the tidal flat inside the lake, but it is undesirable in terms of power generation as the flood generation method was adopted instead of the ebb generation. The turbine generators are operated and produce electricity in the direction from sea to the basin during the flood tide and the turbines work as sluice gates in ebb tide (see Figure 12.18). Based on this one-way generation method using the head difference between inside and outside the barrage, an installed capacity of 252 MW was proposed. The cofferdam method using circular cells was suggested and all the civil works for the construction of powerhouses and sluice structures as well as electro-mechanical works were performed in dry condition and the cofferdam was removed after the completion of power plant (see Figure 12.19).

A numerical model with variable grid system covering the East China Sea region was applied to avoid any possible effect from the barrage to the model boundaries during construction and operation of the plant. As an LNG terminal is located in front of the site, changes to current speed and direction around the terminal by the plant's operation were carefully examined. Water quality was also carefully studied by the particle tracking method. It was expected that the operation of the tidal power plant would contribute to both water level and quality control inside the lake and produce 553 GWh of clean energy annually, which could lead to a 315,000 ton reduction in carbon dioxide emissions. Additionally, the tidal power plant itself could become a popular tourist attraction and could increase employment in the local area.

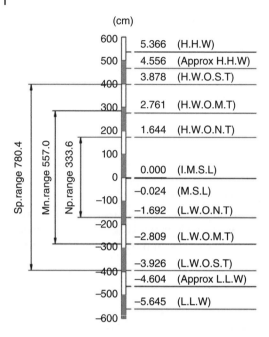

Figure 12.16 Various tide levels referred to the Incheon mean sea level at Sihwa Dyke.

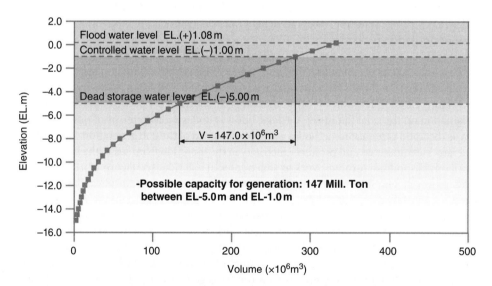

Figure 12.17 Controlled water levels in Lake Sihwa.

The construction method, which links the plant with the existing Lake Sihwa Embankment, yielded significant savings in construction costs, while making significant contributions to the water quality through seawater circulation. The Korean government approved the KOWACO plan for construction of the Lake Sihwa Tidal Power Plant in December 2002.

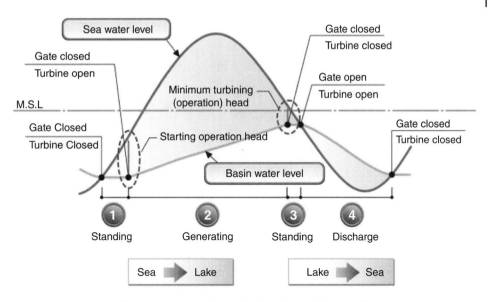

Figure 12.18 One-way flood generation scheme of Lake Sihwa Tidal Power Plant.

Based on the findings from the national research project and the feasibility study [24], the final designed plant would have an installed capacity of 254 MW and comprise 10 one-way bulb turbine generatord of 25.4 MW unit capacity and eight sluices each with openings of 15.3 m × 12.0 m.

12.4.3 Construction

The Lake Sihwa Tidal Power Plant, with a total project cost of approximately $355 million, was the first of its kind in South Korea and the largest in the world since the La Rance tidal power plant in 1965. The construction of the plant was started in December 2004, and completed (including test run) in December 2011, except for the eco-park area. The total construction period was approximately 7 years (2004–2011; Figure 12.20).

The tidal power plant is mainly comprised of powerhouses and sluice structures. In order to build both structures by the *in situ* method under dry condition, a temporary circular cell cofferdam was firstly installed. The outer sea part of the cofferdam was composed of circular cells of 20.4 m diameter and an armoured rock-fill dam in the connecting part. The inner basin part of the cofferdam was erected using steel sheet pile (see Figure 12.21). Cellular cofferdams in the project were built without supporting structures, their stability relying solely by the cell filling. The circular cell cofferdam at Lake Sihwa Tidal Power Plant consisted of 29 primary cells and 28 arc cells of 4.1 m diameter. Welded distribution piles connected the primary cells and spandrel walls.

After the installation of the silt protector, construction of cofferdam and diversion roads was carried out. Ten powerhouses for bulb turbine generators with dimension 19.3 m × 29.0 m × 16.1 m (see Figure 12.22) and eight culvert type sluice structures with dimension 19.3m × 24.0m × 44.3m (see Figure 12.23) were placed along the designated line. The maximum sluicing capacity of the plant is 11,500 m³/s.

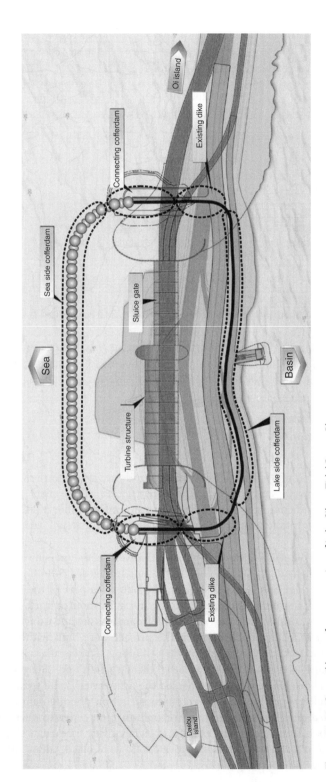

Figure 12.19 General layout for construction of Lake Sihwa Tidal Power Plant.

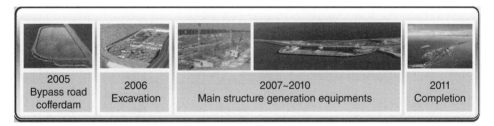

Figure 12.20 Construction schedule of Lake Sihwa Tidal Power Plant.

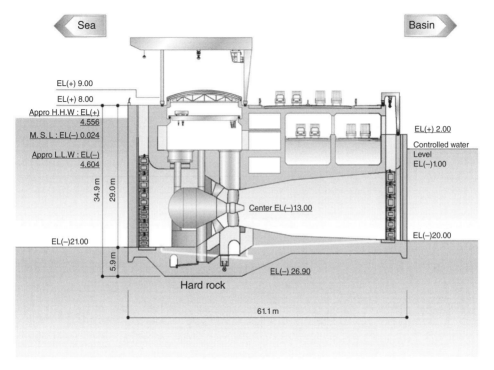

Figure 12.21 Powerhouse of Lake Sihwa Tidal Power Plant.

After dewatering and excavation in October 2007, concrete structures for the power-houses and sluices were built and electrical and mechanical equipment for the turbine generators and gates were installed step by step inside the cofferdam (see Figures 12.24–12.26). The double regulation horizontal-axis bulb-type Kaplan turbine of three blades with runner diameter 7.5 m was chosen, of which the rated head is 5.82 m (maximum head 7.5 m, minimum head 1.0 m), rated discharge 482 m^3/s, and rated speed 64 rpm. The AC synchronous generator was connected directly without step-up gear (direct drive), of which the charging capacity is 25.6 MVAr.

After the new road was constructed over the top of the new powerhouses and sluices, the cofferdam was removed (see Figure 12.27). The turbine generators were tested for normal operation for approximately 6 months and the plant was completed and fully operational from November 2011.

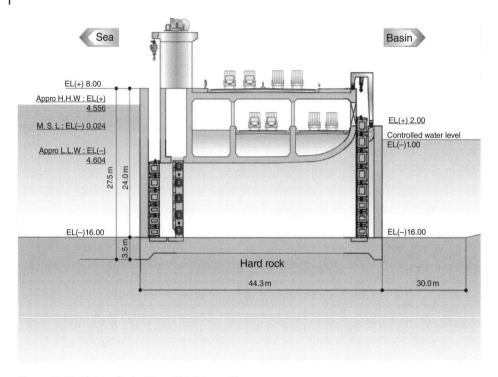

Figure 12.22 Sluice of Lake Sihwa Tidal Power Plant.

Figure 12.23 Installation of draft tubes of turbine generators in the powerhouses.

Figure 12.24 Installation of turbine generators in the powerhouses.

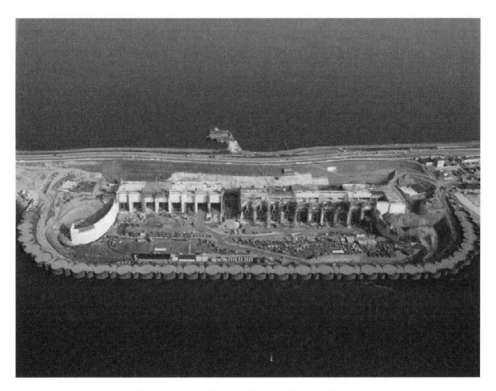

Figure 12.25 Aerial view of construction of Lake Sihwa Tidal Power Plant.

Figure 12.26 Installed turbine generators just before filling with seawater.

Figure 12.27 Aerial view of the Lake Sihwa Tidal Power Plant just after completion.

On the island adjacent to the plant, a new environmentally friendly marine park, T-Light Park, was developed and opened to the public in 2014 [25], including a tidal power culture centre (observation tower, exhibition hall, auditorium, etc.) and waterfront (garden, ocean steps, parking area, etc.).

12.4.4 Economic and Environmental Assessment

The primary purpose of the Lake Sihwa Tidal Power Plant was to improve the water quality of Lake Sihwa by regular exchange and circulation of seawater, and the secondary purpose was to harvest the renewable energy.

Since the completion of the plant in November 2011, it has operated about 16 hours per day on average (8 hours for generation and 8 hours for sluicing). Table 12.1 shows the annual energy output from November 2011 to December 2014. The power generated annually in the three years since 2012 is approximately 480 GWh on average, which is less than expected. However, as operational experience and expertise in environmental effects have been accumulated, annual output has gradually increased year by year.

Since the opening of the plant, the water quality and the species diversity of Lake Sihwa have improved rapidly mainly due to regular exchange and circulation of seawater. The average daily exchange volume of seawater between sea and basin is about 147 million cubic metres, which is equal to approximately 50% of total capacity of Lake Sihwa. The observed mean COD was 2.8 mg/l in the lake and 1.7 mg/l outside the lake in 2014, although the COD in the lake was 17.4 mg/l in 1997. Another reason is that larger tidal action in the lake enhanced the recovery of the ecosystem in restored tidal flats.

The area of tidal flats in the lake was 115 km^2 in 1993 before the dyke construction, and 5 km^2 in 2010 when the exchange of seawater was facilitated by the operation of existing sluices. Since the opening of the plant, water levels in the lake have been maintained between −1 m and −4.5 m, which means the area of tidal flats has increased to 20.3 km^2 (see Figures 12.28 and 12.29). In addition to the increase in tidal flat area, the increased seawater circulation in the basin would also be expected to accelerate the recovery of environmental conditions and function of the restored tidal flat. Although the organic content in the sediment of the tidal flat was over 8% in 2009, it gradually decreased to 1–2% in 2013 after the opening of the plant (see Figure 12.30). This change has accelerated the rehabilitation of the ecosystem in the restored tidal flats. The number of benthic species in the subtidal zone has also increased substantially from 83 in 2005 to 232 in 2014.

Table 12.1 Annual energy output of Lake Sihwa Tidal Power Plant.

Year	Annual energy output (MWh)
2011	52,304
2012	465,924
2013	483,777
2014	492,172

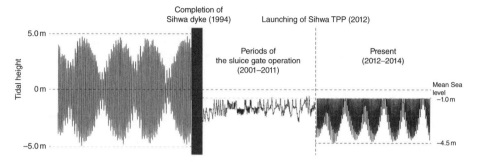

Figure 12.28 Historical water level changes at Lake Sihwa.

Figure 12.29 Restored tidal flat after opening of Lake Sihwa Tidal Power Plant.

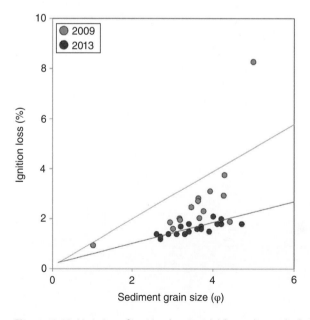

Figure 12.30 Variation of ignition loss in tidal flat sediment before and after the plant opening.

In order to deal with the water quality problem in Lake Sihwa, a comprehensive Lake Sihwa management plan and management committee were established by instruction of the prime minister in 2002, which contributed to the formation of an ecosystem-based, preventative and systematic framework. In addition, the national research programme 'Lake Sihwa marine environment improvement project', supported by the MOF, began in 2003 with a view to securing the systems, policies, and key technologies needed for the smooth execution of the management plan to improve environmental pollution in Lake Sihwa, and to support the implementation of relevant programmes [26, 27]. Major research activities included: supporting the Lake Sihwa Management Committee; monitoring the environment (water quality, sediments, and tidal flat ecosystems, etc.) in Lake Sihwa and its adjacent sea; setting up watershed and seawater quality models; preparing the enforcement plan for the total pollution load management system; and strengthening regional environmental management capabilities for Lake Sihwa and engaging in public relations activities.

12.5 Mutriku Wave Power Plant

12.5.1 Background

Mutriku port is located in a small natural bay (Figure 12.31), protected by the Burumendi headland from north-westerly gales and bounded to the east by the Alkolea headland. Over the years, violent storms blowing in from the Cantabrian Sea have repeatedly damaged the breakwaters of the port, creating rough sea conditions in and around the entrance to the harbour and in the inner harbour itself and making it hazardous and occasionally impossible for vessels to enter the port.

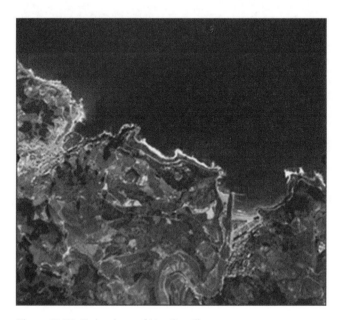

Figure 12.31 Orthophoto of Mutriku village.

To address this problem, the Basque government's Directorate of Ports and Maritime Affairs took the decision to build an outer breakwater. The project was given official approval and put out to tender, and all the necessary administrative authorisations were obtained, including a favourable environmental impact statement (OGBC, 26 October 2004) and a favourable maritime-terrestrial public domain allocation report. The main design features are illustrated in Figures 12.32 and 12.33 and comprised:

- A detached breakwater approximately 440 metres in length with a head at either end: Head A, located closest to Burumendi on the seabed, with a depth of around –2 m; and Head B, at the opposite end, at depths of –17m (all heights relative to the maximum equinoctial spring tide low water (MESTLW).
- The breakwater is accessed by way of a path around 370 m long, protected by rock armour.
- The breakwater is sloped with a concrete haunch and masonry facing running along the entire length.
- The structure of the breakwater comprised a core of quarry run, filter layers of natural stone and an armour layer of concrete blocks of varying sizes (15 t, 25 t and 45 t).

With the initial project defined, and as part of its overall strategy of developing renewable energy sources, the Basque government Department of Transport and Public Works signed a collaboration agreement with Ente Vasco de la Energía (EVE, the Basque Energy Agency) to take advantage of the construction of this breakwater to incorporate an ocean energy power plant. This involved modifying a section equivalent to a quarter of the length of the breakwater to incorporate the infrastructure described below. A photograph taken during construction is shown in Figure 12.34.

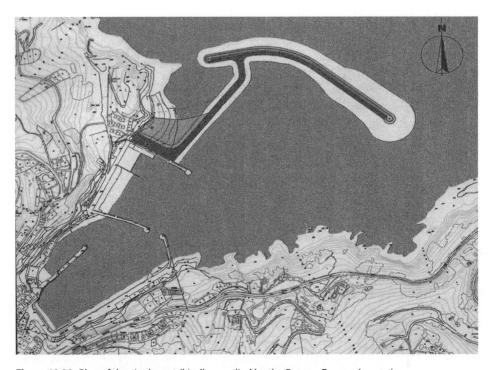

Figure 12.32 Plan of the site layout (kindly supplied by the Basque Energy Agency).

Figure 12.33 Final design (kindly supplied by the Basque Energy Agency).

Figure 12.34 Wave energy plant under construction (kindly supplied by the Basque Energy Agency).

12.5.2 Inclusion of a Wave Energy Plant in the Breakwater

12.5.2.1 Selecting the Technology

Though the breakwater had been designed to solve the problem of vessel access to the harbour at Mutriku, a number of different possibilities were studied on how best to incorporate a wave energy power plant into the breakwater [28].

An analysis was made of the current state of the art of wave energy technologies with a view to identifying one that could be incorporated into a breakwater with a minimum of alterations. It subsequently emerged that an oscillating water column (OWC) type technology could be incorporated into the existing design of the breakwater and its functionality.

There are basically three families of onshore wave energy technology devices: overtopping devices, oscillating devices and OWC devices. Overtopping devices are fitted with a ramp to capture the waves. The seawater is pushed up the ramp and stored in a reservoir, from where it is turbined as in a hydropower plant. These devices are bulky and require very large reservoirs if overflow is to be avoided.

With oscillating devices, the active side of the breakwater is fitted with a vertically tilting structure which, when hit by a wave, drives a piston which compresses a fluid. In turn, the compressed fluid drives a turbine which produces electricity. As most of the machinery in this case must necessarily be exposed to the lashing of the waves, this option was discarded as it was not considered apt for the design of the breakwater and did not offer sufficient guarantee in terms of durability.

Finally, OWC devices are simple and non-disruptive. They use the oscillating movement of the waves, though it is important to note that it is not actually the seawater itself that moves the turbines; indeed the turbines never come into contact with the water. The arrangement consists of a hollow structure, open to the sea below water level with a hole at the top of the chamber. When the wave comes in, the water enters the column, compressing the air inside which is pushed out at high pressure through an opening at the top. This pressurised air turns the turbine, which in turn drives the alternator, generating electricity. When the wave falls, it sucks air through the same opening, again driving the turbine, which continues to generate electricity. The fact that it is the air and not the water that moves the turbine considerably extends the service life of the equipment.

Once the technology had been selected, contact was made with Wavegen, the only company in Europe with experience in building full-scale prototypes using OWC technology. Wavegen had previously worked on both the LIMPET prototype (a facility owned by Wavegen itself on the island of Islay, SW Scotland) and the prototype on the island of Pico in the Azores, Portugal. In 2012, Wavegen became part of the Voith Siemens Group.

12.5.2.2 Consenting Process

In Spain, no dedicated consenting process exists for ocean energy technologies. The process to obtain consent is based on three main legal instruments briefly outlined below [29].

Royal Decree (RD) 1/2008, of 11 January, regulates the need for an environmental impact assessment (EIA) of projects to be developed in the natural environment. In line with this decree, an environmental impact statement has to be compiled and assessed by the relevant agency before granting project approval (or otherwise).

The Coastal Law, 28 July 1988, is the legal framework governing occupation of the territorial sea, together with issues affecting the fishing sector and safety conditions for maritime navigation. Management and surveillance of the marine terrestrial public domain, of which the territorial sea forms part, is the competence of the General Council on Coast and Ocean Sustainability, part of the Ministry for Rural, Marine and Natural Environment. Coastal demarcation departments are their representatives in each coastal province and autonomous region. Therefore, development of electric power projects in the territorial sea must comply with the legal requirements regulating the conditions to process administrative titles granting the occupation of a certain territory (both prior to and during development of the project) and dispositions in terms of deadlines, transference and termination.

Royal Decree 1028/2007 establishes the administrative procedure for processing applications for power generating facilities in territorial waters. Although it is focused on offshore wind, it also encompasses (article 32) power generation from other marine renewable technologies. This decree foresees a simplified procedure regulated by Royal Decree 1955/2000, 1 December 2000, regulating energy transport, distribution, commercialisation, supply and the authorisation procedure for electrical power plants. It also establishes that construction, extension, modification and exploitation of all electric installations listed (in article 111) require the following administrative procedures:

- Request for administrative authorisation: refers to the project's draft installation plan as a technical document.
- Approval of the execution project: refers to the commissioning of the specific project and allows the applicant to start construction.
- Exploitation authorisation: allows the installations, once the project is installed, to be powered up and proceed to commercial exploitation.

A simplified scheme of the Spanish consenting process for ocean energy is presented in Figure 12.35. The total time needed to obtain approval is approximately 2 years, but this time-frame varies between projects. The consenting of the Mutriku wave power plant took less than 2 years as it is located onshore and consequently was subject to the permitting process applicable to an ordinary renewable energy plant.

The reason for such variability in the time taken to obtain the final consent is attributed to whether an EIA is required or not. In Spain, the requirement for an EIA of wave and current technologies is based on a case-by-case analysis. The lack of experience in dealing with offshore renewable energy projects is another contributory factor. In cases where an EIA is required, the process is expected to last at least 2 years plus the additional time needed for other consents to be issued.

The decision on whether an EIA is required or not is taken by the environmental authority after analysing a preliminary document submitted by the developer that summarises the project description together with its main environmental impacts. This decision usually requires advice from administrations and associations that are considered relevant for project approval. If approved, the environmental authority grants the environmental authorisation and establishes project-specific conditions.

However, the Mutriku case was very different from the procedure outlined above. The EIA procedure was initially undertaken only for the construction of the breakwater itself, and an amendment to this procedure was later processed when the decision was taken to incorporate the wave power plant into the breakwater. In the area where the

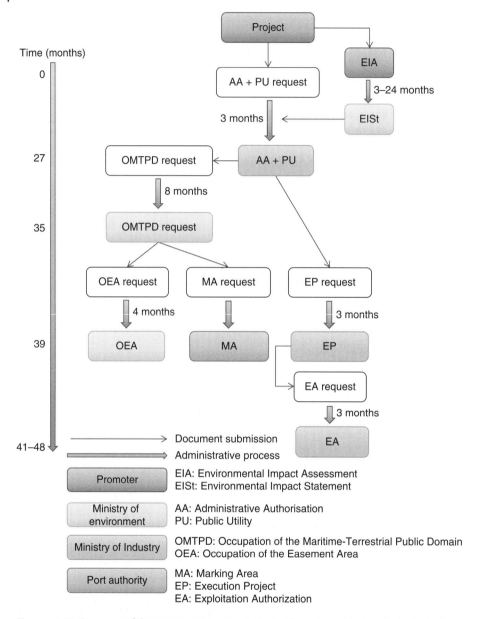

Figure 12.35 Summary of the consenting process in Spain. The coloured rectangles indicate the competent authority for approval or licensing [29].

wave power plant was to be incorporated, the nature of the breakwater itself needed to change from a rockfill breakwater to a vertical seawall: however, the means of construction remained unchanged and the structure itself was to occupy the same amount of space and involved no change of height. Consequently, it was deemed that the amendment would have no impact whatsoever on the EIA regarding breakwater construction. However, the environmental authorities requested a response from the promoters regarding the purpose of the facilities to be installed and the potential environmental impact this could have. A document was subsequently submitted concluding there would be no further impact and that there would only be two emissions from the plant: one was fresh water, used to eliminate the salt deposits from the turbine generators. Taking into account the characteristics of fresh water and that the low associated volumes would be negligible, this was not considered to have any significant environmental impact. Secondly, the power plant generated a certain level of noise which, though not a major factor, required mitigation measures to be taken. An agreement was reached with the local authorities that the noise level would be closely monitored.

12.5.2.3 Pre-design and Design of the Plant

For preliminary design of the plant, a detailed study of the sea climate at the site of the breakwater was required. In terms of the energy resource, three distinct periods need to be distinguished: the winter months (November, December, January, February and March), the summer months (May, June, July, August and September) and the transitional months (April and October).

At undefined depths, the average value of the resource is 26 kW/m and the average direction of the energy flow is N59W (301°). The annual breakdown is as follows:

- average energy flow in winter, 44 kW/m;
- average energy flow in summer, 9 kW/m;
- average energy flow in transitional months, 19 kW/m.

As it approaches the coast, the energy flow falls in a variable manner depending on the orientation of the coast. For example, in the area opposite Cape Machichaco, the energy flow is practically constant until it reaches the coast. However, the protection afforded by Cape Machichaco itself creates an area of shade all along the coast from Bermeo to Deba. Offshore at Mutriku (which is protected by Cape Machichaco), the values obtained at a point 30 m deep were as follows:

- average energy flow in winter, 18 kW/m;
- average energy flow in summer, 4.8 kW/m;
- average energy flow in transitional months, 8.8 kW/m.

In this case, there was no choice regarding plant location as it was to be installed in the breakwater. However, for geometric definition of the plant in the breakwater itself, a wave propagation study was conducted up to the exact location of the Mutriku breakwater at four points along its length (four possible plant sites, three on the main section and one on the breakwater head) and an assessment was made not only to determine best wave capture but also to assess the impact on wave reflections and the effects of the reflected waves on vessel navigation for entering and leaving the harbour. From the study it was deduced that although the positioning of the plant had no impact on the choppiness of the inner waters, it did have an effect on navigability in outer waters.

The four proposed options for positioning of the wave power plant were reduced to two, both in the main section of the breakwater: one in the area of deeper water, a straight section with a less favourable orientation in terms of the prevailing directions of the waves, and another in a shallower area (from 4 to 7 m), in the curved section of the breakwater, better oriented towards the waves.

Once the resource had been measured at each point, the geometry of the columns was determined: the dimensions of the opening between the column and the sea, the straight section of the column, the separation between openings, the ratio between the openings at sea level and the top opening, the interior height of the columns, the thickness of the front wall, etc.

With all these data, the two shortlisted options were tested on a three-dimensional model to corroborate the findings of the studies and determine the best option (shown in Figure 12.36). The final outcome was to position the plant in the shallower area on the curved section of the breakwater, an area of more widespread agitation but of less interference with wave conditions. This location is further from the mouth of the harbour than would have been the case if the plant were located near the breakwater head (in the deeper, straight section), with swell concentrated more in the area of vessel traffic.

From an energy perspective, there is no great difference between the two locations. The deciding factor was therefore to minimise wave reflection into the harbour entrance, a logical criterion taking into account that the whole purpose of the project was to improve vessel access to the harbour.

The Mutriku wave energy plant consists of 16 columns; in each one, the top opening is connected to a turbo-generator set with a rated capacity of 18.5 kW, giving a total capacity of 296 kW. Wells fixed-pitch turbines have been used, which are more robust and simpler with a symmetrical blade design ensuring they always rotate in the same direction, regardless of the direction of the air flow through the turbine, so no device is needed to rectify the air flow.

There are two five-blade rotors that rotate jointly and severally, separated by the generator, which is air-cooled and equipped with an inertia drive to ensure the output capacity curve is as flat as possible. The turbo-generator set, which in Mutriku is

Figure 12.36 Tank testing (kindly supplied by the Basque Energy Agency).

positioned vertically (unlike the two aforementioned prototypes) has a butterfly valve at the bottom so that the column can be isolated if necessary. The valve is electrically activated and has a gravity closure system, thereby ensuring that if connection to the power grid fails, the valve closes automatically. Each turbo-generator is equipped with freshwater injectors which regularly clean the blades of any small accumulations of encrusted salt, and has a noise attenuator fitted at the top.

The turbine is 2.83 m high and a maximum of 1.25 m wide, with a weight of approximately 1200 kg. Assembly and dismantling operations are therefore not complicated, meaning the turbine can be handled in one piece.

The electrical layout of the plant is such that the 16 turbines are separated for control purposes into two groups of eight. The purpose is to help rectify the output capacity curve. Control of each turbine takes into account the pressure reading inside the column at any given moment to fix the rotation speed of each turbine and optimise the power generated. The generator has a voltage of 450 V. Because the turning speed of the turbine, and therefore of the generator, is not fixed but can vary across a wide range, the output signal from the generator is rectified and subsequently converted back to alternating current at 50 Hz in phase with the power grid. It only has to be raised to 13.2 kV to be fed into the local power grid.

The cost of the 16 turbo-generators, the power control and conditioning equipment, the transformer centre and the transmission pine, together with the studies and tests for design of the plant, and alterations to the original breakwater design and subsequent testing, totalled €2 million. The Mutriku wave energy plant was expected to generate 600,000 kWh per year, avoiding 600 tonnes of carbon dioxide emissions per year, the equivalent to the cleansing effect of 80 hectares of woodland. As will be explained later, for various reasons these figures have not been delivered. The power plant was part-funded by the Seventh Framework Programme of the European Commission, under the name of the NEREIDA MOWC project.

12.5.3 Project and Construction of the Plant Infrastructure

The contract for the Mutriku breakwater works was awarded in the summer of 2005 and all design studies to integrate the wave energy plant into the breakwater were concluded by the spring of 2006. This meant that after work on the breakwater had begun, the project had to be amended to integrate the 16 turbines of the OWC plant into the corresponding section of the breakwater.

As explained above, the change had no impact on the alignment of the breakwater itself nor on its protective function. What changed, in broad terms, was a 100 m section of the breakwater (somewhat less than a quarter of the total length) which, to accommodate the OWC plant, changed from a sloped rockfill type to a vertical seawall, as shown in Figures 12.37 and 12.38.

12.5.3.1 Description of the OWC Plant Construction Project

The section of vertical breakwater housing the OWC plant was slotted into 100 metres of the sloped breakwater (Figure 12.39), in a curved section with an outer radius of 220 m and at a depth of 5 m below MESTLW at Mutriku port. This meant replacing the primary armour layer of 25-tonne limestone blocks included in the original breakwater design with a vertical parapet to house the air columns. The vertical section of

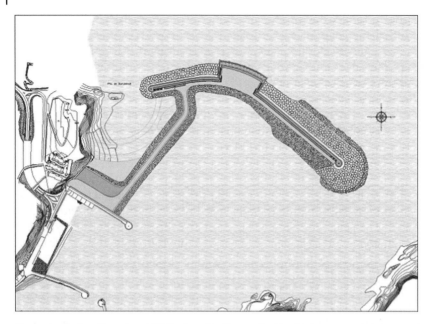

Figure 12.37 Breakwater and OWC plant (kindly supplied by the Directorate of Ports and Maritime Affairs of the Basque government).

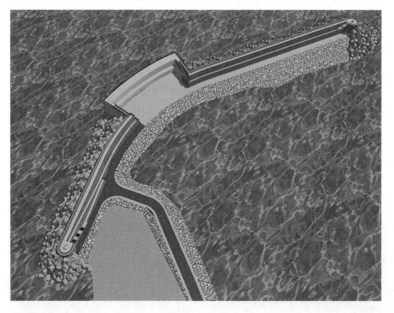

Figure 12.38 Thre-dimensional simulation (kindly supplied by the Directorate of Ports and Maritime Affairs of the Basque government).

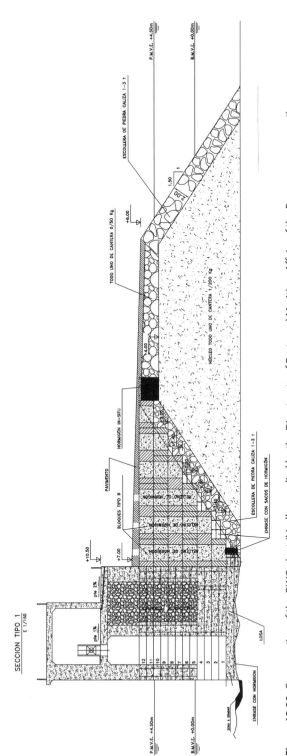

Figure 12.39 Cross-section of the OWC plant (kindly supplied by the Directorate of Ports and Maritime Affairs of the Basque government).

the plant coincides approximately with the foot of the embankment of the rockfill breakwater, so the energy plant itself does not take up any more of the maritime domain surface area. However, the alteration created a large platform in the extrados of around 1600 square metres.

The vertical breakwater housing the 16 air columns was designed with precast parts to facilitate the building process. The power plant was constructed by positioning these precast parts one on top of another to create the vertical front of the breakwater and the oscillating air columns. The opening that transmits the wave oscillations to the air column is 3.20 m high and 4 m wide. The lowest point is at −3.40 m, so that the opening is always below sea level.

These parts are trapezoidal externally, 12.25 m long with a maximum and minimum width of 6.10 m and 5.80 m respectively, and measure 0.80 m along the edge. The difference between the larger and smaller bases provides the whole with the bend radius required to fit perfectly into the line of the breakwater. All parts are made of a framework of HA-35 reinforced concrete with 0.40 or 0.50 m thick walls fitted with two or three lightening cells, plus the cell to house the air column.

Part characteristics vary depending on their vertical position in the breakwater. The first four parts in each column, from −3.40 to −0.20 m, are open on the sea side, to allow oscillation of the waves to be transmitted to the inner air column. The rest of the parts further up are closed. To ensure monolithic performance of the entire structure, the design involved filling some cells with concrete and others with rockfill, except for the 16 chambers containing the air columns where the water oscillates. The plan view is shown in Figure 12.40. The join between the different columns formed by the precast parts was sealed by concreting the gaps, with columns of PVC 400 pipes at the ends to prevent the concrete from coming away.

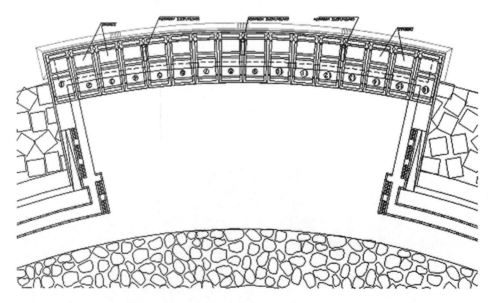

Figure 12.40 Plan of the OWC plant (kindly supplied by the Directorate of Ports and Maritime Affairs of the Basque government).

Construction of the air column and of the breakwater wall continued on top of the precast elements using a reinforced concrete superstructure topped by an 80 cm thick floor, where the turbine room is located. This room has a clearance height of 5.40 m and is 6.10 m wide, with buttresses every 25 m of length up to 16.5 m, the same length as the crest of the rest of the haunch of the breakwater.

The contracted operation budget for the modified project rose to €24.5 million, of which €4.4 million were accounted for by the civil works of the power plant and the remaining €20.1 million by the breakwater. Incorporation of the power plant into the body of the breakwater added seven months to the construction time, extending the total work period to 32 months.

12.5.3.2 Construction of the OWC Plant

Construction work on the wave energy plant began in summer 2006. In order to reach the area of operations from the land, the initial breakwater work (road and embankment) had first to be completed in the extrados of the power plant. Consequently, work on the OWC plant did not actually begin until some 200 of the 440 metres of the breakwater had already been built.

As has been previously mentioned, incorporation of the power plant involved a change in the breakwater design, from the initial sloped design to a vertical section.

To lay the foundations of the power plant, a trench had to be dredged to flatten out the support base of the infrastructure at a level of −4.50 m. Once the dredge trench had been dug, a 20 cm layer of concrete grading was laid, on top of which a 90 cm reinforced slab was put in place as the support base for the precast parts.

The precast parts were built on site on a platform located at the start of the access road to the breakwater (Figures 12.41 and 12.42). The prefabrication and storage area

Figure 12.41 Storage of precast parts (kindly supplied by the Directorate of Ports and Maritime Affairs of the Basque government).

Figure 12.42 Positioning of precast parts (kindly supplied by the Directorate of Ports and Maritime Affairs of the Basque government).

was arranged in such a way that eight units could be built at a time. The time required to transfer each piece from the prefabrication area to the storage area was around 3 days.

Each column consists of four open precast parts and 12 closed parts. A total of 256 parts were built, of which 64 were open and 192 closed. Each of the parts was placed in its final position from the line of the breakwater, using a motor crane on caterpillar tracks with a lifting capacity of 50 t and a reach of 40 m. Divers were also employed to correct piece by piece any deviation in the positional layout. Maximum deviation was of the order of 3–4 cm, and these differences were corrected with concrete at a later phase of the work, when the power plant part had been completed (Figures 12.43 and 12.44).

Once the parts had been placed in their final position, underwater concrete was pumped into the inner cells to ensure that the entire structure acted monolithically. The outer 4.30 × 4.50 metre cells were filled with rockfill, while the rear cells with a gap width of 0.75 m and the front cells with a gap width of 0.85 m were filled with underwater concrete and strengthened with reinforcement formwork. The vertical joins between the different columns were then sealed with concrete, using pipes of PVC 400 filled with reinforced concrete acting as piles as permanent formwork.

Construction of the turbine room began by covering the columns with precast slabs as permanent formwork (Figures 12.45 and 12.46). A slab of reinforced concrete 80 cm thick was laid on top of these slabs as the housing for the 16 0.75 m diameter openings, where the turbines were designed to go. Slab reinforcement had been calculated on the basis of the pressures of the air column and the weight of the turbines. Design features also included cables and pipelines linking all the rooms to optimise communications in the plant.

The walls of the turbine room are 1.65 m wide on the sea side and 0.80 m wide on the inner side. The turbine room, housing a total of 16 turbines, has three buttress walls of reinforced concrete, forming four rooms (Figures 12.47–12.49).

Figure 12.43 Vertical section and breakwater separated (kindly supplied by the Directorate of Ports and Maritime Affairs of the Basque government).

Figure 12.44 Power plant and breakwater joined together (kindly supplied by the Directorate of Ports and Maritime Affairs of the Basque government).

Figure 12.45 Columns before closure (kindly supplied by the Directorate of Ports and Maritime Affairs of the Basque government).

Figure 12.46 Layout of the permanent formwork (kindly supplied by the Directorate of Ports and Maritime Affairs of the Basque government).

Finally, water, electricity and a telecommunications system between the power plant and the municipal connection points were completed after all the required cables and pipelines (two 16 0mm diameter corrugated power lines, four 110 mm diameter tele-communication lines and one 90 mm diameter polyethylene pipe for supply of drinking

Figure 12.47 Construction of the turbine room (kindly supplied by the Directorate of Ports and Maritime Affairs of the Basque government).

Figure 12.48 Wave impact during construction of the turbine room (kindly supplied by the Directorate of Ports and Maritime Affairs of the Basque government).

Figure 12.49 External view of the turbine area, with masonry and impost (kindly supplied by the Directorate of Ports and Maritime Affairs of the Basque government).

water) had been laid along the access road. An aerial view of the breakwater and OWC plant is shown in Figure 12.50.

12.5.4 Start-Up and Operation

12.5.4.1 Operation of the Plant

The Mutriku plant was opened in July 2011 and handed over for operational use in October of the same year, thus bringing the project phase to an end and marking the start of plant operation by EVE [30].

In its first year of operation, the plant proved itself capable of adapting to a wide range of sea conditions. The control programme analyses each operational parameter of the turbines. When any given value exceeds its established limits, an alarm is triggered and the turbine shuts down. As a precautionary measure that first year, whenever an alarm triggered the closure of a turbine a plant operator was required to check the state of the turbine prior to restarting it. This procedure was followed to ensure the alarms work properly. Since that first year, in view of these sporadic shutdowns the plant has automatically attempted to restart the turbine, to try to avoid human intervention in the event of repeated shutdowns.

Mention should also be made here that this need for human intervention whenever an alarm triggers the closure of a turbine has been one of the reasons for a lower than expected level of energy production, as the plant has only been supervised during normal office hours.

Energy production during the first year reached almost 200,000 kWh, as summarised in Table 12.2 [31]. This is a considerably lower figure than that expected for a normal year: 600,000 kWh. However, mention must be made of the following: during the months of June, July, August and September 2011, each turbine was started up and subjected to various operational tests. This same procedure was later repeated for each group of turbines. This accounts for the low amount of energy exported and the need to

Figure 12.50 Breakwater and OWC plant, June 2013 (kindly supplied by the Basque Energy Agency).

Table 12.2 Power production in the first year of operation.

Year	Month	Imported energy (kWh)	Exported energy (kWh)
2011	Jun	2,447	3,888
2011	Jul	2,864	5,919
2011	Aug	2,810	3,994
2011	Sept	2,994	20,808
2011	Oct	1,366	32,603
2011	Nov	599	35,477
2011	Dec	652	34,205
2012	Jan	133	0
2012	Feb	1,708	0
2012	Mar	1,579	20,487
2012	Apr	595	33,403
2012	May	3,954	9,013.60
Total		21,701	190784

Table 12.3 Power production of the Mutriku wave energy plant exported to the grid (kWh, breakdown by month and year of operation.

Year	Jan	Feb	Mar	Apr	May	Jun
2011					2,450	3,888
2012	0	0	20,487	33,403	9,013.6	1,075.62
2013	39,860.13	25,918.68	18,320.1	31,412.54	40,501.34	14,805.3
2014	50,705.35	39,848.26	35,345.91	13,058.79	5,912.42	2,521.32
2015	37,605	28,212,41	32,275,41	10,966,93	19,699,51	13,972,12

Year	Jul	Aug	Sep	Oct	Nov	Dec
2011	5,919	3,994	20,808	32,603	35,477	34,205
2012	200.87	0	8,205.16	1,593.14	17,145.54	35,097.54
2013	6,427.97	8,776.12	15,641.45	10,185.2	29,385.99	37,851.39
2014	6,667.43	8,047.82	9,698.36	29,237.56	29,754.09	48,055.39
2015						

import energy for successive start-ups. It must also be taken into consideration that generated power is one thing, while exported power is another. The difference between the two is, for example, the power consumed in starting up a turbine while others are generating power. It is important to note that, for several reasons, during this period the plant worked mostly with 12 turbines in operation and not the full complement of 16, which has a considerable impact in the total power generation.

Table 12.3 shows monthly power production figures for the four years that the plant has been operational. As can be observed, production figures in recent years are practically double those of the initial years, mainly attributable to the increase in the number of operational hours. As explained above, though the resource is constant, lessons have been progressively learned on how best to control the plant, and fewer and fewer faults have been recorded.

Though production figures have risen significantly, they are still far from the initial estimated target figures of 600,000 kWh per year. This figure was set with no knowledge of the behaviour pattern of a wave power plant and was subsequently reduced to 400,000 kWh per year when the plant became operational. This figure is expected to be delivered in the near future as the learning process is slowly bearing fruit and the number of operational hours of the plant is progressively increasing.

12.5.4.2 Incidents

The plant became fully operational in October 2011, whereupon production figures for October and November began to be significant. The automation of plant start-up, as explained above, considerably improved these figures.

Figures for the first half of December proved that the plant could reach high electricity export levels. The plant was shut down on the afternoon of 15 December due to a severe storm, and that same night the control room doors were damaged and the building flooded, rendering it unproductive.

The efforts involved in setting up provisional protective measures for the control centre, reporting the incident to the insurance company and making a claim for damages and repairs, constructing and fitting outer doors to avoid a repetition of these events, replacing the existing doors and damaged equipment, checking all other power conditioning equipment and, finally, restarting the plant, meant that the facilities were inoperative until 9 March 2012, a period of almost 3 months. In other words, the winter months, the best period of the year in terms of energy resources, were lost.

Production then began to increase until May, when operational problems led to very low production and very high energy consumption levels. A review of the condition of the plant, together with an analysis of why the plant was repeatedly shutting down, revealed an anomaly in the damper. Investigation revealed breakage of the rubber components which prevent displacement of the damper drive motor, leading to an excessive amount of play in the drive motor and thus provoking repeated shutdown of the turbines.

Once again, this illustrates the potential problem a breakdown in a small component part can cause to marine energy facilities. However, in this case, component fault can be easily explained. The supply contract of the plant stipulated that all equipment had to be ready for installation in September 2008, and installed and running by April 2009. The equipment was ready by the agreed date but the breakwater was not, which meant that the equipment was in storage for over 2 years before being installed and becoming operational. The fact that this was clearly detrimental to these rubber components can be readily understood.

12.5.4.3 Social Acceptance

EVE has taken many different actions to raise awareness of the plant in the local community, ranging from the local campaign run during the construction phase to the current initiative of guided tours [32].

In 2007, when the plant was still at the construction stage, a storm arose. The force of the air passing through the hole at the top of each column produced a deafening noise, arousing a level of concern among members of the local community and generally making them unhappy with the idea of a wave energy plant being built in Mutriku. Consequently, EVE took the decision to run an information campaign to reassure the local population that the future plant, once in operation, would not generate that level of noise, and to generally inform on the situation and ongoing progress of the works. Information leaflets were mailed locally and a series of meetings were held with the people of Mutriku.

It would be fair to say that EVE's reaction was triggered by the adverse reaction of the local population to the consequences of the storm. In other words, this was the impetus for all subsequent communication and awareness-raising activities.

An open day was held at the plant on 27 July 2013, to provide the people of Mutriku and any other interested parties with the opportunity to see how the plant works and to better understand why the plant had been built. EVE technicians were on hand to explain everything and access was to the turbine room was provided. Throughout the day, the turbines were intermittently started up for the public to see how they work. At the end of the day, all visitors were asked to fill in a satisfaction survey and, as a gesture of appreciation, EVE gave a specially designed T-shirt to all those who had attended.

Survey findings revealed that 95% of those who participated in the survey believe the construction of the plant will be positive for the town of Mutriku. Additionally, with regard to any possible inconvenience the plant may cause, 78% of survey participants believe the plant will not cause any inconvenience, as opposed to 20% who believe it will. Practically all of the latter believe noise to be the main inconvenience. The survey also revealed most complaints or negative comments to be related to the conclusion of the works on the breakwater, and not to the wave energy plant itself (breakwater construction had yet to be fully completed when the open day was held). The comments section included suggestions to increase the installed capacity of the plant and its production, and to continue offering site visits and providing information on plant operations.

On the basis of growing public interest in the Mutriku plant and the comments made in the open day satisfaction survey, the Information and Tourist Office of the Mutriku Town Council and EVE jointly signed a collaboration agreement to set up a service providing guided tours of the plant. These tours are specifically designed to raise public awareness of the plant which is the first of its kind worldwide, and of the commitment of the Basque government to marine energy.

12.5.4.4 Improvements and Innovation

As well as automating the plant to improve power production, improvements have also been made to the SCADA system. The large number of parameters being recorded was causing the system to collapse and led to a migration from the previously installed CTNet network to an Ethernet LAN data transmission system, thereby solving the communications problem and enabling a stable and continuous flow of data.

Though the facilities at Mutriku have been set up as a commercial plant to generate electricity, EVE intends to take maximum advantage of its investment and welcomes proposals for R&D projects that may be conducted in the Mutriku plant. Involvement in this type of project has been constant since the plant was commissioned: indeed, plant construction itself was part of an R&D project. However, this innovative approach is now being taken one step further and the plant is being prepared as a test laboratory.

It would appear that, as a technology for harnessing wave energy, OWC technology is currently very much on the rise. What Mutriku offers as a test laboratory is the possibility of testing turbine performance onshore but in real sea conditions. Control systems and components may also be tested by connecting them to one of the plant's existing turbines.

Onshore testing under real sea conditions provides developers with invaluable and incomparable information prior to deploying their devices at sea, and could help to reduce the final cost of the energy considerably. Ultimately, this remains one of the primary and basic objectives for the marine energy sector.

References

1 Commercial Readiness Index for Renewable Energy Sectors – ARENA.
2 DTI and ABP Marine Environmental Research (2004) *Atlas of UK Marine Renewable Energy Resources*. London: DTI. Published under funding from the Department of Trade and Industry's (DTI) Strategic Environmental Assessment programme for Offshore Energy. Latest version now available at http://www.renewables-atlas.info/
3 Gibberd *et al.*, UK Patent GB2431628.

4 Hardwick, J., Ashton, I. and Johanning, L. (2015) Field characterisation of currents and near surface eddies in the Pentland Firth. OTE Workshop, Oxford.

5 Vigars, UK Patent GB2441769.

6 Gibberd *et al.*, UK Patent GB2431189.

7 Harper, S., Pittam, G. and Harrison, J. (2015) Tidal turbine foundation load limiting controller. EWTEC 2015, Nantes, France.

8 Hawthorne, UK Patent WO2014199121 A1.

9 Gunn, K. and Stock-Williams, C. (2014) Falls of Warness 3D model validation report ETI ReDAPT – MA1001 PM14 MD5.2. http://www.eti.co.uk/redapt-falls-of-warness-3d-model-validation-report/ (last accessed 10/1/2016).

10 IEC/TS62600-200:2013. Marine energy – Wave, tidal and other water current converters – Part 200: Electricity producing tidal energy converters – Power performance assessment.

11 Novel CO_2 capture taskforce report, 01 Dec 2011, Global CCS Institute. http://decarboni.se/publications/novel-co2-capture-taskforce-report/51-concept-bankability

12 www.seabased.com (accessed 25/3/2016).

13 Leijon, M., Waters, R., Rahm, M., Svensson, O., Boström, C., Strömstedt, E., Engström, J., Tyrberg, S., Savin, A., Gravråkmo, H., Bernhoff, H., Sundberg, J., Isberg, J., Ågren, O., Danielsson, O., Eriksson, M., Lejerskog, E., Bolund, B., Gustafsson, S. and Thorburn, K. (2009) Catch the wave to electricity: the conversion of wave motions to electricity using a grid-oriented approach. *IEEE Power and Energy Magazine* 7(1), 50–54.

14 Rahm, M., Boström, C., Svensson, O., Grabbe, M., Bülow, F., Leijon, M. (2010) Offshore underwater substation for wave energy converter arrays, *IET Renewable Power Generation* 4(6), 602–612.

15 Langhamer, O. (2009) Wave energy conversion and the marine environment: Colonization patterns and habitat dynamics. PhD thesis, Uppsala University.

16 Engström, J. (2011) Hydrodynamic modelling for a point absorbingwave energy converter. PhD thesis, Uppsala University.

17 Gravråkmo, H. (2011) Buoy for linear wave energy converter. Licentiate thesis, Uppsala University.

18 Tyrberg, S., Svensson, O., Kurupath, V., Engström, J., Strömstedt, E., Leijon, M. (2011) Wave bouy and translator motions – on-site measurements and simulations. *IEEE Journal of Oceanic Engineering* 36(3), 377–385.

19 Tyrberg, S., Gravråkmo, H., Leijon, M. (2009) Tracking a Wave Power Buoy Using a Network Camera – System Analysis and FirstResults, Proceedings of the ASME 28th International Conference on Ocean, Offshore and Arctic Engineering (OMAE2009), Honolulu.

20 Svensson, O., Boström, C., Rahm, M., Leijon, M. (2009) Description of the control and measurement system used in the Low Voltage Marine Substation at the Lysekil research site. Proceedings of the 8th European wave and tidal energy conference (EWTEC09), Uppsala.

21 Hong, Y., Hultman, E., Castellucci, V., Ekergård, B., Sjökvist, L.,Soman, D.E., Krishna, R., Haikonen, K., Baudoin, A., Lindblad, L., Lejerskog, E., Käller, D., Rahm, M., Strömstedt, E., Boström, C., Waters, R. and Leijon, M. (2013) Status update of the wave energy research at Uppsala University. Proceedings of the 9th European Wave and Tidal Energy Conference, Aalborg, Denmark, September.

22 Kim, C.W. (2015) Sihwa Tidal Power Plant (Daewoo E&C Co., Ltd.). Personal communication.

23 Koo, B.J. and Kim, K.T. (2014) The story of Sihwa tidal flat. Ministry of Oceans and Fisheries (in Korean).

24 Kwater (2002) Feasibility study and basic plan on Lake Sihwa Tidal Power Plant. Prepared by Hyundai Engineering Co. Ltd. and KIOST (in Korean).

25 http://tlight.kwater.or.kr

26 MOF (2001) Development of utilization technique for ocean energy (I): Tide & Tidal Current Energy. Report No. BSPM 00078-00-1347-2. Prepared by KIOST (in Korean).

27 MOF (2014) Lake Sihwa marine environment improvement project. Report No. BSPG 48391-10564-4. Prepared by KIOST (in Korean).

28 Torre-Enciso, Y., Ortubia, I., López de Aguileta, L.I. and Marqués, J. (2009): Mutriku Wave Power Plant: from the thinking out to the reality, Proc. of the 8th European Wave and Tidal Energy Conference, 7-10 September, Uppsala, Sweden, pp 1-11.

29 Simas, T., O'Hagan, A.M., O'Callaghan, J., Hamawi, S., Magagna, D., Bailey, I., Greaves, D., Saulnier, J-B., Marina, D., Bald, J., Huertas, C., Sundberg, J. (2015) Review of consenting processes for ocean energy in selected European Union Member States. *International Journal for Marine Energy* 9, 41–59.

30 Torre-Enciso, Y., Marqués, J. and López de Aguileta, L.I. (2010) Mutriku: Lessons learnt. Proceedings of the 3rd International Conference on Ocean Energy, 6–8 October, Bilbao.

31 Torre-Enciso, Y., Marqués, J. and Marina, D. (2012) Mutriku: First year review, Proceedings of the 4th International Conference on Ocean Energy, 17–19 October, Dublin.

32 Ajuria, O., Boveda, I., Torre-Enciso, Y. and Marina, D. (2014) Mutriku: A flagship project for driving public engagement, Proceedings of the 5th International Conference on Ocean Energy, 4–6 November, Halifax, NS.

13

Regional Activities

Deborah Greaves[a], Carlos Perez-Collazo[b], Curran Crawford[c], Bradley Buckham[c], Vanesa Magar[d], Francisco Acuña[e], Sungwon Shin[f], Hongda Shi[g] and Chenyu[h]

[a] *Professor of Ocean Engineering, School of Engineering, University of Plymouth, UK*
[b] *PRIMaRE Research Fellow, School of Engineering, University of Plymouth, UK*
[c] *Department of Mechanical Engineering, University of Victoria, BC, Canada*
[d] *Centro de Investigación Científica y Educación Superior de Ensenada (CICESE), México*
[e] *Chief Executive Officer, InTrust Global Investments LLC, Washington, D.C. USA*
[f] *Research Professor, Kangwon National University, Korea*
[g] *Professor, Ocean University of China, Qingdao, China*
[h] *Researcher, Ocean University of China, Qingdao, China*

Since investigation of marine renewable energy (MRE) was initiated in the 1970s with research programmes and testing of some preliminary prototypes in Japan and Europe, it now has an established position in world energy policy. MRE has a fundamental role within the European Strategic Energy Technology Plan (SET-Plan) [1] and the International Renewable Energy Agency (IRENA) [2]. Wave and tidal energy technologies have seen significant development worldwide in recent years, and the first tidal and wave energy arrays have been announced and will be constructed over the next few years [3–6]. In this chapter, an overview of MRE development activities carried out in different geographic regions around the world is presented. Wave and tidal energy developments are described in Europe, North America, Latin America, the Asia-Pacific region and China.

13.1　Europe

Europe is currently at the forefront of the exploitation and development of MRE, hosting more than 50% of tidal energy and about 45% of wave energy developers [7]. At present, much of the infrastructure required for the development of MRE is located in Europe, including MRE test centres, demonstration and pilot zones, research laboratories, and developers. Figure 13.1 shows the location of the wave and tidal developers and the dedicated infrastructure located within Europe.

This section has been structured into four main subsections: European initiatives and policy framework for wave and tidal energy; wave and tidal energy test and demonstration centres; wave energy technology developments; and tidal energy technology developments. Each subsection is divided by country.

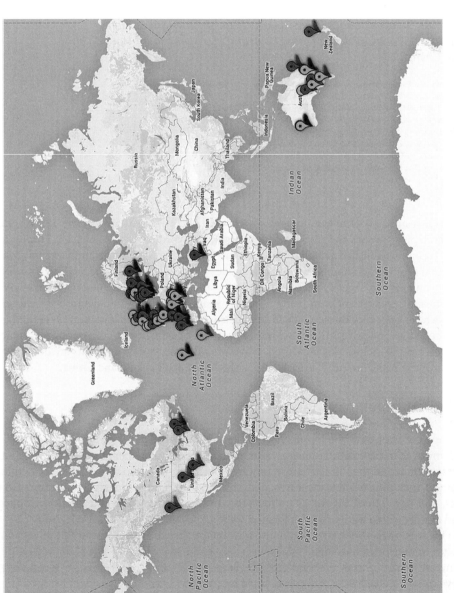

Figure 13.1 (a) Wave energy maps, identifying technology developers (light grey) and dedicated infrastructure (dark grey) existing in Europe [8].

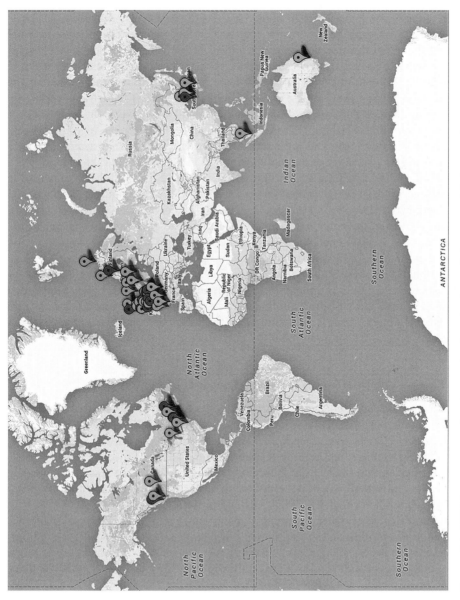

Figure 13.1 (b)Tidal energy maps; identifying technology developers (light grey) and dedicated infrastructure (dark grey) existing in Europe [8].

13.1.1 European Initiatives and Policy Framework for Wave and Tidal Energy

Public support is crucial for the growth and development of the MRE sector, and in European countries research and development (R&D) activities have been mainly funded with governmental support. This support has increased as the technologies at the forefront of the MRE sector are going through the critical 'valley of death' [9, 10]. A range of support mechanisms have been put in place at both national and European level in order to speed up and consolidate the development of this nascent sector.

The European Commission released the Blue Energy Communication [11] in January 2014 in order to highlight the expected contribution that MRE technologies may make in Europe and at the same time to set the framework for the development and consolidation of the technologies by 2020 and beyond. As the first step arising from this communication, the Ocean Energy Forum [12] was created. This is an initiative to bring together the different actors in the sector to discuss challenges and define viable solutions by delivering a strategic roadmap for the successful development of the MRE sector by 2017. The second step aims to develop a European Industrial Initiative for the MRE sector, based on the strategic roadmap and within the SET-Plan framework [13].

The main financial support mechanism at European Union level is the NER 300 programme, a joint initiative of the European Commission, the European Investment Bank and member states [14]. The programme aims to support demonstration projects within emerging carbon capture and storage and innovative renewable energy technologies. Between its two calls for proposals (December 2012 and July 2014) two wave and another two tidal energy projects were selected for NER 300 funding. The two tidal energy projects are the Sound of Islay [15] and Kyle Rhea in the North West coast of Scotland; however, the latter has recently been cancelled by Atlantis, its major developer [16]. The two wave energy projects are the West Wave on the west coast of Ireland [17] and the Ocean SWELL on the coast of Portugal [18].

At national level, there are a wide range of initiatives that the different European governments have adopted to support the development of MRE technologies, from subsidies such as feed-in tariffs and quota systems to investment programmes or setting deployment targets. Table 13.1 summarises the different support mechanisms and investment programs that exist in Europe by country.

13.1.2 Wave and Tidal Energy Test and Demonstration Centres

At sea test centres and dedicated research infrastructures provide essential support for the development of MREs. In Europe the FP7-funded MARINET (Marine Renewables Infrastructure Network for Energy Technologies) project established a network that links all these facilities across Europe, coordinating R&D at all scales (small models through to prototype scales from laboratory through to open sea tests) and allowing researchers and developers easy access to the facilities. Figure 13.2 shows the R&D facilities accessible through the FP7 MARINET project.

Wave and tidal energy test centres may be grouped into three different categories: grid-connected demonstration test centres, where large- and/or full-scale pre-commercial devices are tested; small-scale or nursery test centres, where devices are tested at sea at smaller scale; and demonstration sites, which are first-generation MRE facilities built to prove certain types of technology.

Table 13.1 National support schemes for MRE technologies [19].

Country	Subsidies type	Subsidies amount (€ct/kWh)	Investment programs
Denmark	feed-in tariff	5–8	• Energinet.dk [20] • Energy Technology and Demonstration Project (EUDP) [21]
France	feed-in tariff	73 (plus €200 million capital support)	• ADEME [22] • Feed-in tariff of 17.3 ct/KWh for small arrays
Germany	feed-in tariff	3.4–12.7	
Ireland	feed-in tariff	26	• SEAI prototype development fund [23] • SEAI sustainable RD&D programme [24]
Italy	feed-in tariff	34	
Portugal	feed-in tariff	26	• Fundo de Apoio à Inovação (FAI) for renewable energies [25]
Spain	feed-in tariff	7.65–7.22	• EVE
Sweden	quota system	179	
Norway	quota system	49	
United Kingdom	quota system	50–104	• Renewable Energy Investment Fund (REIF) Scotland [26] • Marine Energy Array Demonstrator (MEAD) [27] • Energy Technologies Institute (ETI) [28] • The Crown Estate [29] • Marine Renewables Commercialisation Fund (MRCF) Scotland [30] • Marine Renewables Proving Fund (MRPF) [31] • Saltire Prize Scotland [32]

13.1.2.1 Denmark

Denmark has a medium to low wave energy resource, but has one of the most active wave energy programmes to foster the harnessing of wave power. Two developments are the wave energy test centres available to wave energy developers in Denmark: a full-scale test centre, the Danish Marine Test Site (DanWEC), located in Hanstholm, and a nursery test centre, the Nissum Bredning Test Station for Wave Energy (NBPB) [34].

13.1.2.2 France

France has vast experience in the testing of tidal energy devices, but does not have a specific tidal energy test centre, although the tidal range demonstrator barrage of La Rance in Brittany is dedicated to tidal energy research and demonstration. On the other hand, the French government has invested in the development of a full-scale wave energy test centre, named SEM-REV (Site d'Expérimentation en Mer pour la Récupération de l'Énergie des Vagues). This site occupies approximately a $1\,km^2$ test

Figure 13.2 Wave and tidal energy research and development facilities accessible through the FP7 MARINET project [33].

zone area and is fully instrumented and monitored. The test site comprises a 2.5 MVA power cable connected to the national grid through an onshore substation [35].

13.1.2.3 Italy

Although Italy has been developing wave and tidal energy devices during the last couple of decades, these have been at small scale and of low profile. A novel full-scale test centre for tidal stream turbines is under development at the Messina Strait, in the South of Italy, with currents just over 2 m/s [36].

13.1.2.4 Ireland

The west coast of Ireland has one of the highest levels of wave power available in the world, as seen previously in Chapter 2. Ireland has two wave energy test sites: the first one, located in Galway Bay on the west coast, is a nursery test site with 37 hectares and water depth ranging from 21 m to 24 m, suitable for testing devices from 1/5 to 1/3 scale; the second is a full-scale and grid-connected test centre, the Atlantic Marine Energy Test Site at Annagh Head, west of Belmullet in County Mayo, which will be used to test the performance of pre-commercial wave energy devices in extreme open ocean conditions [37].

13.1.2.5 Norway

Even though MRE industry is not one of the major priority areas of development for the Norwegian government, support for renewable energy is still included in the general spectrum of innovative projects, making some innovative policy schemes available for wave and tidal companies [38, pp. 9]. Norway has a full-scale wave energy test centre as part of the Runde island environmental centre. The Runde Environmental Centre is an international research station located at the most southerly bird cliff in Norway [39].

13.1.2.6 Portugal

Portugal is one of the pioneer countries in the development of wave energy, as evidenced by the oscillating water colum (OWC) wave energy demonstration site which has been in continuous operation since 1999 at Pico island in the Açores archipelago [40]. Furthermore, Portugal has one full-scale wave energy test centre, Wave Plug, with grid connection and water depth ranging from 30 m to 90 m, located at Figueira da Foz. The country also has two other demonstration sites: one in shallow waters in Peniche, and a deep water one with grid connection at Aguçadoura [34].

13.1.2.7 Spain

The high wave energy resource available on the Atlantic coast of Spain has motivated many wave energy developments. Currently Spain has two full-scale test centres and a demonstration site. BIMEP, a full-scale grid-connected test centre located in Armintza-Lemoiz in the Basque country, has total area of $5.3 \, km^2$ and its water depth ranges from 45 m to 95 m [41]. The second full-scale grid-connected test centre is part of PLOCAM (Plataforma Oceánica de Canarias), which is a multipurpose research platform located on the north-east side of Grand Canary Island; its water depth ranges from 30 m to 200 m and it has a total planned power capacity of 50 MW [42]. Finally, the Mutriku demonstration site is the first commercial OWC system installed on a port infrastructure, which is a 300 kW OWC plant located at the port of the fishermen's village of Mutriku in the Basque country [43].

13.1.2.8 Sweden

Sweden's MRE resources are very limited; however, the Swedish Energy Agency has been very supportive of wave energy development in recent years. Uppsala University has been running a grid-connected demonstration site at Lysekil since 2003 for wave energy arrays, and the Lysekil demonstration zone has been used to demonstrate the Seabased wave energy converter (WEC) [44].

13.1.2.9 United Kingdom

The UK has been at the forefront of MRE developments worldwide, and proof of that is the noteworthy number of development and test centres that there are in the country. At the time of writing, there are 14 test or demonstration wave and tidal energy centres in UK waters in operation or under development [45]. EMEC (the European Marine

Energy Centre), at the Orkney Islands in Scotland, was established in 2003, and has both operational full-scale wave and tidal sites and another two wave and tidal nursery sites are operational. With its 14 grid-connected test berths, EMEC is the largest of the European test centres [46]. At the other end of the country is the fully operational Wave Hub full-scale grid-connected test centre in Cornwall, and also managed by WaveHub are three new demonstration zones under development (the North Devon tidal demonstration site near Lynmouth with a capacity of up to 30 MW, the North Cornwall demonstration zone and the South Pembrokeshire wave demonstration zone) [47]. The tidal stream current test site in Strangford Lough is a full-scale grid-connected site which has been in operation since 2012 [48]. FaBTest (Falmouth Bay Test Site) is a nursery test site located in Cornwall [49]. Furthermore, there are two more tidal energy full-scale test centres under development at the moment, the Perpetuus tidal energy centre in the Isle of Wight in England and the West Anglesey tidal demonstration zone in Wales [45].

13.1.3 Wave Energy Technology Developments

Wave energy attracts much interest and investment and is widely distributed at the global level – currently there are wave developers from 33 countries (Figure 13.3) compared with 18 countries with tidal energy developers (Figure 13.6). However, wave energy technologies have a lower level of technology development, reliability and technology readiness than their tidal counterparts. For example, since 2008 in the UK wave energy devices have supplied a total of 0.8 GWh of electricity to the grid, 8% of the total that tidal devices have generated in the same period [7]. The two recent wave energy competitions – the Wave Energy Prize, supported by the US Department of Energy [50]; and the Wave Energy Scotland (WES) initiative, supported by the Scottish Government [51] – have increased significantly the number of wave energy developers, especially in the USA – the number of wave energy developers globally has risen from 170 in 2014 [7] to 254 in 2015 [52]. These numbers confirm the global efforts and interest in developing a sustainable wave energy sector.

Around 43% of the wave energy developers are based in Europe (Figure 13.3), but only a few of them have reached the full-scale testing phase: recent publications have

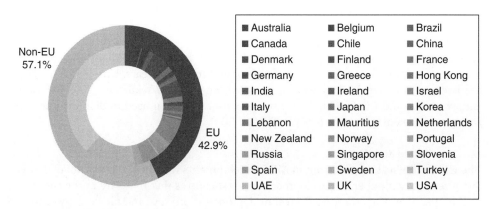

Figure 13.3 Distribution of wave energy developers in the world [52].

identified 46 wave energy developers which have reached or are about to reach the open-sea deployment of their technologies. Since 2009 more than 100 have been announced in Europe alone, amounting to a total installed capacity of 1200 MW; however, over 770 MW of these have already been aborted. Nevertheless, the sector is expected to reach 26 MW of installed capacity in Europe by 2018 by deploying single units and small array demonstration projects [7].

In the following, the different wave energy technology developments are classified by country.

13.1.3.1 Denmark

The strong support from the Danish government for the development of wave energy, despite the country's limited wave energy resource, is the main reason why Demark has over 5% of the global wave energy developers. Of the 12 different concepts under development at the time of writing, four stand out from the others due to their level of technology readiness – all four have been tested at sea as a full- or part-scale prototypes, reaching technology readiness levels of 4–5.

These four devices are: DEXAWAVE, an attenuator type of WEC which has been tested at the DanWEC test centre [53]; Poseidon, a wave–wind hybrid energy converter, which uses an attenuator type of WEC and has been tested at sea in the lee of the Vindeby Danish offshore wind park for a total of 20 months [54]; Wavesta, a point absorber type of WEC which was tested at the DanWEC test centre from 2009 until 2014 [55]; and WaveDragon (Figure 13.4), an overtopping type of WEC which was tested at the Danish nursery centre of Nissum Bredning from 2003 until 2005 [56].

The other Danish devices which are at a lower technology development level are: KNSwing [57], Multi Absorbing Wave Energy Converter (MAWEC) from LEACON Wave Energy [58], Resen Waves LOPF buoys [59], Crestwing [60], WavePiston [61], WavePlane and WEPTOS WEC [62, 63]. These devices cover almost all different types

Figure 13.4 Wave Dragon in operation. Picture by Wave Dragon A.S. [56].

of WEC technologies and are at different stages of development and technology readiness levels 3–4.

13.1.3.2 Finland

Although Finland only has two wave energy developers, both have reached a technology readiness level of 5, and each is being tested at full-scale grid-connected test centres or demonstration zones. The first of them is WaveRoller, an oscillating wave surge type of WEC which is being developed and tested at the Portuguese grid-connected demonstration site in Peniche [64]. The other is the Penguin rotating mass type of WEC developed by Wello (Figure 13.5), which has been tested at the full-scale grid-connected site at EMEC since 2012. Furthermore, Wello has secured an EU grant to deploy an array of three Penguin devices in the UK [65] in 2016.

13.1.3.3 Ireland

Ireland is one of the European countries which has been more active in WEC development during the last decade, and proof of that are the ten Irish WEC concepts which are under development. Two of these concepts stand out from the others due to their development status: the Ocean Energy buoy, a floating OWC type of WEC which was tested as a part-scale prototype at the nursery test site of Galway Bay in Ireland [66]; and the SeaPower device, an attenuator type of WEC which has been tested at the Queen's University Marine Laboratories in Portaferry, Northern Ireland, through the FP7 MARINET project [67]. The remaining wave energy developers are at a lower technology readiness and have only carried out small-scale tests: Blue Power Energy Ltd. [68], Jospa Ltd. [69], Joules WaveTrain from Joules EES Ltd. [70], Limerick Wave Ltd. and WaveBob Ltd [71, 72].

13.1.3.4 Italy

Italy has little experience in the wave energy development sector. However, in the last few years some developments have emerged: Ensea [73], Wave for Energy [74], 40South Energy and Sea Power scrl [67, 75]. Most of these concepts are at medium to low

Figure 13.5 Wello's Penguin WEC in operation at EMEC in Orkney. Picture by Wello [65].

technology readiness levels, with the exception of Wave for Energy with its Inertial Sea Wave Energy Converter (ISWEC), a rotating mass type WEC, which has been recently deployed as a full-scale prototype in the Pantelleria Island [74].

13.1.3.5 Norway

Norway, with its 16 wave energy developers, accounts for around 7% of the sector worldwide. However, only one developer stands out: Fobox AS (a subsidiary company of Fred. Olsen & Co.) has been testing the Fobox device at the Cornish (UK) nursery site of FaBTest for more than two years [76]. The remaining developers either have a smaller technology readiness or they are old projects: Aker Solutions ASA [77], Havkraft [78], Intentium AS [79], Langlee Wave Power [80], Ocean Wave and Wind Energy (OWWE) [81], OWC Power AS [82], Pelagic Power AS [83], Pontoon Power AS [84] and WaveEnergy AS [85].

13.1.3.6 Portugal

Portuguese institutions have been very active in wave energy development, due to the quality and amount of wave energy resource at the Portuguese coast, and proof of that are the numerous test and demonstration centres situated along the coast. However, in terms of technology development, Portugal is below the European average, and only the Instituto Superior Técnico (IST) of Lisbon has been active as a wave energy developer for the last couple of decades. The main developments supported by IST have been the OWC Pico plant, which has been in operation since 1999 and is now run by WavEC Offshore Renewables [40]. More recently, IST has been developing a floating OWC mounted on a Spar buoy but it has not been tested at sea yet.

13.1.3.7 Spain

Even though Spain has 12 different wave energy developers, in recent years, none has been able to secure the R&D funds needed to advance their WECs to a higher technology readiness level. Only the OWC demonstration facility at the port of Mutriku in the Basque country has reached commercial technology readiness. The various wave energy developers are: SDK Marine [86], Abengoa Seapower [87], Innova Foundation [88], Oceantec Energías Marinas SL [89], Rotary Wave SL [90], Tecnalia [91] and PIPO Systems [92].

13.1.3.8 Sweden

Sweden has six wave energy developers, despite the limited resource available at its coasts. There are two that stand out. Seabased AB (a spin-off company of Uppsala University) has been testing a point absorber type of WEC for several years at Lysekil demonstration site, is in the process of deploying a 14 MW wave energy array in Ghana (this will be the first wave energy park in Africa) and is developing, together with Simply Blue, a 10 MW wave energy park to be deployed at the Wave Hub (Cornwall, UK) [93]. Waves4Power AB has tested a part-scale point absorber type of WEC at the Runde Island demonstration site and launched a full-scale deployment in 2016 [94]. The other wave energy developers, all at lower technology readiness levels, are CorPower Ocean AB [95], Ocean Harvesting Technologies [96] and Wavetube AB [97].

13.1.3.9 United Kingdom

The UK is the leading European country in wave energy development with 29 developers, six of which have reached a more advanced technology readiness level. AlbaTERN is developing a point absorber type of WEC, which has been tested in an array of three part-scale units on the Isle of Muck off the west coast of Scotland [98]. Aquamarine Power's Oyster oscillating wave surge type of WEC was tested at full scale at EMEC from 2012, but the company went into administration in 2015 [99]. AWS Ocean Energy Ltd. tested a part-scale prototype at Loch Ness in 2010 [100]. PolyGen Ltd tested its floating oscillating wave surge type of WEC at the FaBTest Cornish nursery test site in 2015 [101]. Sea Wave Energy Ltd is developing a part-scale attenuator type of WEC which was tested in seawater off Larnaca Bay (Cyprus) during 2013. Seatricity Ltd. deployed the Oceanus 2, a full-scale point absorber type of WEC, at the Wave Hub in 2015 [102]. Pelamis Wave Power Ltd, one of the world's leading wave energy developers from when it was founded in 1998 until 2014 when it went into administration, contributed significantly to the development of the sector [103]. Other UK developers, at lower technology readiness, are Checkmate Seaenergy UK Ltd [104], Ecotricity [105], Embley Energy Ltd [106], Greenheat Systems Ltd [107], Lancaster University [108], University of Manchester [109], Marine Power Systems [110], Offshore Wave Energy Ltd (OWEL) [111], Portsmouth Innovation Ltd [112], SEEWEC Consortium [113], Snapper Consortium [114], Cyan Technologies and Trident Energy Ltd [115, 116].

13.1.3.10 Other Developments

With regard to developments in wave energy in other European countries, it is worth mentioning that the German company NEMOS GmbH is testing their device at the Danish nursery site of Nissum Bredning [117]. Other European wave energy developers are FlanSea [118] and Laminaria [119] from Belgium; École Centrale de Nantes [120], Hydrocap Energy [121] and Waves Ruiz [122] from France; Brandl Motor [123] from Germany; and Sigma Energy [124] from Slovenia.

13.1.4 Tidal Energy Technology Developments

Horizontal-axis turbines (HATs) have seen significant development over the last decade and many of the devices under development are at high technology readiness levels. HAT tidal stream energy devices have undergone sustained intensive open-water testing. For example, in the UK, since 2008 MRE technologies have supplied more than 11,000 MWh of electricity to the national grid, of which 10,250 MWh were generated by bottom-mounted HATs [7]. The impressive progress in tidal stream energy technology during the past decade has brought into the sector some key industrial actors, such as Voith Hydro, Kawasaki Heavy Industries, Hyundai Heavy Industries, Torcado, Schottel, and Alstom. Through their involvement either in the direct development of existing devices or in the adaptation of other existing concepts to second- (floating) and third-generation (multi-platform) tidal stream energy devices, these actors are strongly increasing the industrialisation of the sector [7].

Nearly 60% of the world's tidal stream energy developers are based in Europe, and the UK is the world leader in tidal stream energy development (Figure 13.6). The high levels of technology readiness among these developers have meant that, during the past few

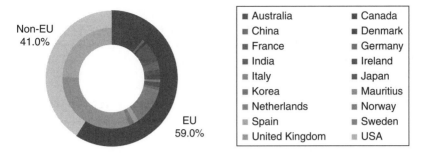

Figure 13.6 Distribution of tidal energy developers in the world [125].

years, multiple tidal energy projects have been proposed, totalling over 1500 MW of projected installed capacity in Europe [126]. However, most of these projects are still a long way from being realised and only the Meygen project has secured the required funding and is under construction [3]. In the UK, the Crown Estate has leased 26 zones for tidal energy development – these leases represent just over 1200 MW of planned installed capacity, and a further announcement was made in July 2014 of three tidal energy demonstration zones, linked to the local communities [127]. The UK government recently approved the Swansea Bay tidal energy lagoon [128], although this has since been put under review [129]. Other projects are under way in France and the Netherlands and, in summary, by 2018 Europe could have a tidal energy installed capacity of about 57 MW, a considerable step forward for the development of the tidal energy sector.

In the following, an overview of the different tidal energy technology developments is presented by country.

13.1.4.1 France

France has pioneered the development of tidal energy, as evident by its La Rance tidal barrage power plant of 240 MW that has been in operation since 1967 and was the first tidal power station in the world [130]. France is now also one of the top countries in the world supporting the development of the tidal stream technology, with multiple tidal current installation projects planned. France has at the time of writing seven tidal stream turbine developers, two of which are at high technology readiness level: Alstom Hydro with its Oceade turbine which has been tested and developed at EMEC for several years [131]; and the Sabella D10 tidal turbine which has been deployed for testing at sea at Ushant Island in Brittany [132]. Other French tidal energy developers are BluStream [133], EEL-Energy [134], Guinard Energies [135], Hydro-Gen and HydroQuest [136, 137].

13.1.4.2 Germany

A number of German companies are supporting the tidal sector as direct investors in developers from other European countries or as major technology providers, as in the case of Siemens or Schottel. Furthermore, Germany has three tidal turbine developers, of which Voith Hydro is the one that has the more advanced technology, after testing its 1 MW turbine at EMEC for several years [138]. Another German tidal developer is Atlantisstrom [139].

13.1.4.3 Ireland

Ireland has two tidal energy developers. OpenHydro is one of the developers leading global development, has been undergoing tests at EMEC for several years, and has a large number of projects under development [140].

13.1.4.4 Netherlands

There are three main tidal stream developers in the Netherlands, among which Torcado Tidal Turbines stands out for its level of technology development. Torcado has two different technologies under development, with its T1 demonstrator being tested at the Afsluitdijk tidal test centre [141]. Other Dutch tidal developers are Bluewater [142], Deepwater Energy BV and Tidalys [143, 144].

13.1.4.5 Norway

Two of the five tidal turbine developers in Norway have made significant progress: Andritz Hydro Hammerfest with its HS1000 1 MW pre-commercial demonstrator being tested at EMEC [145]; and Flumill AS with its dual contra-rotating helix proto-type being tested in open sea at Tromsø [146]. The other Norwegian tidal turbine developers are: Straum AS [147], Norwegian Ocean Power [148] and Tidal Sails AS [149].

13.1.4.6 United Kingdom

The UK has 32% of the global technology developers of tidal energy devices. Among these, five have reached an advanced technology readiness level. Atlantis Resources Ltd has three different tidal turbine technologies under development. Its AR series is the most advanced; it is a 1 MW HAT that has been tested at EMEC for serval years. The company has a number of tidal farm projects under way, including MeyGen [150]. Marine Current Turbines (MCT) was the first company to install a marine current tur-bine, the Seaflow 300 kW near Lynmouth (Devon) in 2003, and since 2008 it has been testing the 1.2 MW SeaGen S turbine in Strangford Narrows in Northern Ireland; in July 2015 MCT was purchased by Atlantis Resources [151]. Nautricity is a spin-off company from Strathclyde University which has been developing its CoRMaT tidal turbine since 2000; CoRMaT has been tested as a full-scale prototype at EMEC [152]. Scotrenewables Tidal Power Limited is developing a floating tidal turbine, the 2 MW SR2000 turbine is soon to be deployed, and the part-scale SR250 250 kW prototype was tested at EMEC for 2½ years [153]. Tidal Energy Ltd has been developing its DeltaStream tidal turbine since 2001, and in 2016 installed its 400 kW prototype at the Ramsey Sound site in Wales [154]. Other UK developers, with a lower technology readiness, are QED Naval [155], Renewable Devices Marine Ltd [156], AquaSientific [157], C-Power [158], Current2Current [159], Flex Marine Power Ltd [160], Free Flow 69 [161], Green Tide Turbines [162], Greenheat Systems Ltd [107], Kepler Energy [163], Lunar Energy [164], Nova Innovation [165], Ocean Flow Energy [166], Robert Gordon University [167], SeaPower Gen [168], SMD Hydrovision [169] and TidalStream Ltd [170].

13.1.4.7 Other Developments

Two developments in tidal energy in other European countries are worth mentioning: the GEM floating contra-rotating tidal turbine prototype developed by Seapower scrl of Italy [67]; and the device being tested by Spanish company Magallanes Renovables at EMEC [171]. Other European tidal energy developers are Centro Technológico SOERMAR [172] of Spain; and Current Power AB and Minesto [173, 174] of Sweden.

13.2   North America

13.2.1   Wave

In contrast to efforts in the UK that were largely motivated by the energy crisis of the 1970s, North American renewable energy development has been driven by recent concerns of energy security and the looming requirements of evolving international greenhouse gas emission standards. Of the continent's renewable options, the density and extent of the wave energy resource along the Pacific coast is prominent. A 2004 survey of selected National Data Buoy Centre wave buoy records by the US Electric Power Research Institute (EPRI) revealed that the coastlines of Washington, Oregon and California received 440 TWh of wave-supplied energy in a typical year [175, 176]. A similar survey of the Canadian Marine Environmental Data Service buoy records, completed in 2006 by the National Research Council (NRC), estimated the annual mean wave power transport onto Canada's Pacific continental shelf to total 37,000 MW – roughly 55% of Canadian peak electricity consumption at that time [177].

Between 2000 and 2007, the enormity of this potential motivated a few feverish industrial efforts in British Columbia, Washington and Oregon to deliver wave-supplied power to North American electrical grids. In British Columbia in 2001, BC Hydro (the government-owned utility) entered negotiations with Ocean Power Delivery (developer of the Pelamis attenuator) and Energetech Australia (now known as Oceanlinx) on the formation of two separate joint ventures aiming to produce 4 MW of wave power for Vancouver Island, a large populated island with average annual demand of 9069 GWh [178]. Also in 2001, Finnavera Renewables, based in Vancouver, declared its intention to develop a 1 MW grid-connected wave power plant in Makah Bay, Washington, delivering 1.5 GWh annually. The Finnavera WEC technology, a point absorber called Aquabuoy, was to be deployed in an initial plant of four 250 kW capacity units, with intent to grow the installation to 75 units [179].

The WEC demonstration projects led by Pelamis, Energetech and Finnavera never delivered electricity to market. The Pelamis and Energetech projects on Vancouver Island were killed by a dramatic shift in British Columbia's energy policy coincident with a change in provincial government in 2002. While Finnavera was able to fund construction and deployment of a test device, named Aquabuoy 2.0, near Newport, Oregon, in September 2007, that device sank in late October 2007, marring the company's attempts to attract power purchase agreements (PPAs). The assortment of political, regulatory and technical struggles experienced in these projects exemplified the North American wave energy sector in the 2000–2007 period. However, lessons learned through these exercises have catalysed broader collaborative policy development, resource assessment, and technological innovation since 2008. In the next sections, we examine these developments and the rebirth of WEC demonstrations in North America that has resulted in recent years.

13.2.1.1   Regulatory Environment

In the period 2000–2007, wave energy development in North America faced stiff resistance in the form of electricity rate considerations. The BC Hydro wave energy demonstration on Vancouver Island ended abruptly after the release of the inaugural BC Energy Plan in 2002. That plan sought to create economic opportunities for independent power producers (IPPs) through call-for-tender processes, but BC Hydro's acquisitions from IPPs were made subject to oversight by the BC Utilities Commission (BCUC), with projects evaluated on a least-cost basis. While the unit cost of the Pelamis and Energetech supplied energy was never publicly released, BC Hydro did state that it was expected to be 'higher than subsequent larger wave energy projects', and that 'the projects would inform study of possible cost recovery mechanisms' [180]. These project characteristics were not accommodated by the BCUC. Public oversight on electricity rates was also a deciding factor in Finnavera ending its wave energy programme in 2008. Finnavera did successfully negotiate a long-term power purchase agreement with Pacific Gas and Electric (PG&E) in 2007 [181]. However, the California Public Utilities Commission (CPUC) rejected PG&E's proposal to use the state's New Renewable Resources Account to subsidise the Finnavera project. The CPUC cited unreasonable costs and the unproven nature of the technology for its decision [182].

Compounding these downward pressures on electricity rates, early wave energy projects faced significant cost uncertainty in the planning stages. Prior to Finnavera's Makah Bay project, there was no precedent for licensing wave energy projects. Between 2001 and March 2003 Finnavera, the Federal Energy Regulatory Commission (FERC) and the Department of Commerce debated the licensing requirements for Makah Bay. Disagreement centred on whether the location of the Aquabuoys (1.9 miles from the shoreline in an established marine sanctuary) and the location of the powerhouse (on tribal lands within the Makah Indian Reservation) made the project exempt from FERC oversight [179, 183]. Ultimately, FERC asserted jurisdiction over the project in 2003. Finnavera finally applied for its FERC licence in November 2006, but the planned 75-unit Aquabuoy farm challenged the existing FERC licence terms. Standard FERC applications were for a 30–50-year term and there was no established track record on which to evaluate wave energy projects over such a duration. Presumably motivated by the Finnavera case, FERC developed a new licensing procedure specific to hydrokinetic pilot projects between 2006 and 2007 [184]. Provided that the pilot technology was small and could be removed quickly and avoided sensitive locations, that stringent environmental monitoring was conducted during the pilot project and that adequate site restoration could be completed, FERC would issue a 5-year license. The final step in the evolution of the FERC licensing process occurred in 2008. Finnavera was awarded a pilot project licence in December 2007, but it was immediately contested by the Washington State Departments of Ecology and Natural Resources on the grounds that the US Clean Water Act, the Coastal Zone Management Act and the Endangered Species Act required state-level authorities to review Finnavera's planned activities inside the state territorial limit prior to FERC approval [185]. To navigate timing conflicts at the state and FERC levels, the FERC hydrokinetic pilot project licences became conditioned licenses that could not independently permit on-site construction or installation. Also, to better coordinate FERC and the various state-level approvals and eliminate any perceived overstepping of bounds, FERC signed memorandums of understanding (MoUs) with Oregon, Washington and California in 2008, 2009 and 2010, respectively.

Oregon has been particularly proactive and in 2013 amended its Territorial Sea Plan (TSP) to align state requirements for permitting activity inside the territorial limit (3 nautical miles) with those of FERC [186]. In the revision of the TSP, the Oregon Wave Energy Trust, a not-for-profit corporation funded by the Oregon Innovation Council, invested approximately $1.2 million in stakeholder engagement, cumulative impacts analyses and marine spatial planning. Two key outcomes from the TSP revisions were the identification of four Renewable Energy Facility Suitability Study Areas (REFSSA) and a simplified permitting process for demonstration projects located in these four spaces that would be administered by a single state agency, the Joint Agency Review Team (JART).

Since 2008, FERC also requires wave energy project developers to first acquire a preliminary permit for proposed development sites. The preliminary permit gives the developer priority to study the wave resource at the site for up to 3 years. The developer is required to submit periodic reports that demonstrate progress in the resource/site assessment. The preliminary permit is analogous to a preliminary patent application; it sets the priority date for the developer in the event of competing interests in the same region. To date, FERC has issued four preliminary permits for wave energy projects to Ocean Power Technologies (Reedsport, OR), Resolute Marine Energy (Yakutat, AK) and two to Dynegy Point Estero Wave Park (Morro Bay, CA).

The licensing procedure for wave energy projects in British Columbia, last revised in 2015, remains untested by commercial applications. The process is two-phased. An initial investigative use permit can extend for up to 5 years and its purpose parallels that of the FERC preliminary permit: it gives the proponent priority to complete data collection, site planning, facility design, surveys and to seek an electricity purchase agreement (EPA) from BC Hydro. In the subsequent operational phase the proponent applies for a Multi-Tenure Instrument (MTI). The MTI evolves with the project, and specific land uses (e.g. powerhouse construction, transmission line rights-of-way, road access) are declared as they develop. The term of the licence is set to match the term of an EPA if in place, or it defaults to a 10-year period. In addition to annual rent for the land use within the project boundaries, an additional ocean energy participation rent is due annually after an initial 10-year grace period. This additional fee is set as a percentage of gross sales and escalates as the capacity factor of the project increases [187].

13.2.1.2   Regional and Community Initiatives

Without acute need for new electricity generation capacity, wave energy projects in North America are primarily motivated by:

1) Transmission constraints – electrically isolated jurisdictions reliant on relatively expensive electricity supply options;
2) Environmental constraints – legislated requirements for a proportion of energy generation to be clean;
3) Financial incentives – targeted funding support for wave energy innovation designed to spur industrial activity and domestic economic growth.

BC Hydro's Vancouver Island wave energy demonstration project was largely driven by system transmission constraints. As early as 2000, BC Hydro was planning for the retirement of one of the two high-voltage DC transmission lines that connected southern Vancouver Island to mainland British Columbia [188]. That retirement, combined with a projected demand growth, led to a projected 30 MW shortfall for Vancouver Island by 2007. The wave energy demonstration project was one entry of a portfolio of wave, wind and micro-hydro projects totalling 20 MW aiming to fill a portion of this shortfall. Eventually, BCUC refusal to accept renewable and natural gas fired options led to a new 230 kV transmission line being installed in 2008.

There remain many small communities on the British Columbia and Alaskan Pacific coast that remain off-grid and entirely reliant on diesel-fuelled electricity generation. In its 2015 Annual Report, the Oregon Wave Energy Trust (OWET) suggested that the typical cost of electricity (COE) in these communities was in the range $0.297–$0.332 per kilowatt-hour [186]. However, the OWET estimates are at the lower end of what is realised in the Northern Pacific region. Remote off-grid villages on the west coast of Vancouver Island have reported COE between $0.50 and $0.60 per kWh [189], and communities on the Alaskan Gulf Coast also report being in this range: False Pass in the Aleutians reported $0.416/kWh in 2010, and Yakutat on the Gulf of Alaska reported $0.507/kWh in the same year [190]. At these COE levels, WECs present potential economic benefits in addition to obvious emissions reductions and protection from diesel price volatility.

In 2009, the borough of Yakutat commissioned an EPRI-led study on the expected COE for a multi-unit wave energy plant installed in 13 m of water off the communities [191]. Yakutat is a community of 650 inhabitants located 200 miles north-west of Juneau on the Gulf of Alaska. In 2007, average electrical demand was 794 kW, with peak demand of 1.5 MW. Diesel for the generators is shipped via barge from Washington State. The EPRI study established a nominal annual wave climate using a SWAN wave propagation model and considered installations of one, two, four and eight 650 kW Aquamarine Oyster surging flap WECs. The study suggested that the initial deployment of a single Oyster WEC would yield a COE of $0.48/kWh and that this cost could be expected to decrease to $0.27/kWh as the plant grew to eight units. At present, the community of Yakutat continues to push for a wave energy plant. In 2013, the community engaged Resolute Marine Energy (RME) to develop a 250–300 kW wave energy plant. RME held a FERC preliminary permit for the Yakutat site between 2013 and 2015, and in 2013 partnered with the Ocean Renewable Power Company to develop the Yakutat wave energy plant. To facilitate such community-driven projects the Alaskan government created an Emerging Energy Technology Fund (EETF) which supplied $3.7 million to hydrokinetic technology, primarily tidal, development in 2012. Despite this funding mechanism and strong community support, no WECs have yet been installed at Yakutat despite plans to be generating by 2015.

Outside of niche market opportunities such as Yakutat, there is no possibility of wave-supplied power competing in the North American marketplace without policy that allows for increased energy costs in support of environmental stewardship objectives. Finnavera's PPA with PG&E was motivated by a 2006 change in the California Renewable Portfolio Standard (RPS). The change advanced the target date of California's goal of growing its renewable portfolio to 20% of annual sales from 2017 to 2010, and the requirement sparked PG&E's rapid search for renewable options.

In British Columbia, the 2007 BC Energy Plan and the 2010 Clean Energy Act have provided some impetus for renewable energy development, including two wave projects on Vancouver Island. The 2007 Energy Plan, 'A Vision for Clean Energy Leadership', called for a review of the BCUC's role in considering social, environmental and economic costs and benefits of IPP-driven generation projects [192]. To support pre-commercial technology demonstrations, the 2007 Energy Plan also established a $25 million Innovative Clean Energy (ICE) Fund. Awards of $2 million from the ICE Fund were subsequently made to two Canadian wave energy projects, one led by SyncWave Systems Inc. focused on providing 100 kW of power to the small community of Ahousat located approximately 50 km north of Ucluelet, BC, on the west coast of Vancouver Island, and the other led by Pacific Coastal Wave Energy (PCWE) Corp. which would supply 5 MW of power directly to the BC Hydro grid at Ucluelet. The SyncWave WEC was a proprietary point absorber design, and the PCWE project licensed technology from Carnegie Wave Energy of Australia. Despite the ICE awards, neither project secured the private sector investment needed to commence construction and the last remnant of either project is an investigative use permit that Carnegie still holds for the coastline of Ucluelet.

British Columbia's Clean Energy Act of 2010 set heightened clear energy goals for the province, including a desire to achieve electricity self-sufficiency, to generate at least 93% of the electricity in British Columbia from clean or renewable resources and, most importantly, to become a net exporter of clean energy [193]. To support uptake of new renewable technologies, the Clean Energy Act provided a feed-in tariff (FIT) regulation for energy projects facilitating achievement of British Columbia's energy objectives under the Act. Projects subscribing to the FIT programme were made exempt from BCUC review. Unfortunately for the wave energy sector, the environmental stewardship measures in the Clean Energy Act were not ambitious enough to trigger significant change in the IPP tender process. In 2010, the provinces heritage hydroelectric generation already accounted for nearly 85% of long term planned capacity and clean energy objectives were achievable through micro-hydro, wind and biomass projects selected through the 2006 and 2008 clean energy calls [194]. Subsequently, in June 2012 the province announced it would not be proceeding with implementation of the FIT due to rate concerns raised by successive years of rate increases. Despite these increases, BC Hydro's electricity rates in 2013 were the third lowest in North America: $0.0797/kWh for first 1350 kWh in an average two-month billing period, and $0.1195/kWh over the 1350 kWh threshold.

13.2.1.3 Government Incentives

To develop wave energy technology for future generations, and to spur economic development through technological innovation, both the USA and Canada have committed financial support to WEC commercialisation. However, in Canada WEC public investment has been very limited: only the SyncWave Systems and PCWE proposed projects selected by the British Columbia ICE Fund have received government investment. In the case of SyncWave Systems, the $2 million of ICE investment was complemented by separate awards of $2.7 million from Sustainable Development Technology Canada (SDTC) in 2009, and $4.6 million from the Natural Resources Canada (NRCan) Clean Energy Fund in 2010. As with the ICE award to SyncWave Systems, the SDTC and Clean Energy Fund conditions on private–public equity sharing in the projects could not be met and the awards were returned.

Since 2007, the US Department of Energy (DOE) has been the primary federal supporter of marine hydrokinetic (MHK) technology development in the USA. This support is provided through the Water Power Program operated by the Office of Energy Efficiency and Renewable Energy. The objective of the Water Power Program is to support commercialisation of emerging MHK technologies, and the programme has invested significantly in a broad range of wave energy related R&D [195]. Between 2008 and 2014, the Water Power Program invested $21 million in US-based WEC technologies. Recipients include Columbia Power Technologies ($6.5 million), M3 Wave Energy

Systems ($0.24 million), Northwest Energy Innovations ($2.5 million), Ocean Power Technologies ($9.3 million), Ocean Energy USA ($1 million), Resolute Marine Energy ($1.2 million), and WaveBob ($2.4 million). More recent funding opportunities presented by the Water Power Program have targeted cross-cutting technologies such as advanced control algorithms and novel mechanical power take-off designs that can be applied across different WECs [196].

In early 2015, the DOE also launched the Wave Energy Prize, a multi-stage competition with cash prizes totalling $2.25 million. The objectives of the Wave Energy Prize are to increase the breadth and conceptual diversity of the WEC technologies being developed in the USA, and to foster innovations that break convention and lead to power performance improvements of 200% over established levels of WEC performance [197]. There were 66 applicants for the Wave Energy Prize, and a contingent of 20 were chosen by a panel of international judges to advance to the competition. The three phases of the competition included numerical simulation and 1 : 50 scale model testing (by January 2016), 1 : 20 scale model testing (by July 2016) and a second round of 1 : 20 scale model testing (by October 2016). The criteria to select who advanced at each stage gate were built on the concept of the technology performance level (TPL) metric developed at the US National Renewable Energy Laboratory (NREL) [198]. In contrast to the conventional technology readiness level (TRL) metric used to measure progress to commercialisation, a device's TPL defines the best COE the device could deliver once commercialised.

To assist early-stage WEC TPL evaluation the DOE has also invested in the creation of WEC-specific computer design tools that will be freely distributed to technology developers. The DOE's WEC-Sim project is led by researchers at NREL and Sandia National Labs and is producing a general-purpose WEC simulator that can capture the multi-body hydrodynamics of a variety of WEC designs [199]. To drive simulation-based studies of WEC farm outputs, NREL and Sandia National Labs also partner on the development of a revised SWAN nearshore wave modelling tool. The new code, referred to as SNL-SWAN [200], uses a performance matrix type representation of a WECs power conversion characteristics within the SWAN code. This additional information is used to adjust the spectral energy balance equation evaluated at each node of a coastal SWAN model to account for the presence of the converter [201].

13.2.1.4 Test Sites, Research Centres and Resource Assessment

The first North American deployment of a grid-connected WEC, a 45 ton 20 kW point absorber developed by Northwest Energy Innovations (NWEI), was in June 2015 at the Wave Energy Test Site (WETS) located at the Kaneohe Marine Corps Base on the northeast coast of the island of Oahu, Hawaii. The WETS facility is operated by the Hawaii National Marine Renewable Energy Center (HNMREC) formed in 2008 through an $8 million grant from the US DOE's Water Power Program to the Hawaii Natural Energy Institute at the University of Hawaii. R&D at the WETS test centre is collaborative: in addition to the companies which test at the centre, NREL and HNEI researchers provide resource assessment for the WETS site, third-party computer WEC modelling and environmental impact analyses. A strategic advantage of the WETS site is that it is entirely controlled by the US Navy – this drastically simplifies the permitting and licensing process that would otherwise accompany the installation of the grid-connected devices. The US Naval Facilities Engineering Command also provides financial support for selected WEC deployments at WETS. Two deeper water grid-connected test locations are planned for 60 m and 80 m deep locations at WETS. The deeper sites are intended for use with higher capacity prototypes (250 kW to 1 MW).

The organisational structure of HNMREC, a collaborative effort encompassing stakeholders from the military, academia, government labs and the private sector, is mirrored by that of the Northwest National Marine Renewable Energy Centre (NNMREC) established through a series of DOE Water Power Program awards to Oregon State University (OSU) in 2008–2010 totalling $13 million. While NNMREC encompasses research at three universities (the University of Washington, OSU and the University of Alaska-Fairbanks (UAF)) wave energy activities are concentrated at OSU; the school has a history of WEC technology research in the Wallace Energy Systems & Renewables Lab extending back to 1998 with a focus on the design of direct drive electricity generation [202, 203].

Researchers at OSU partner directly with WEC developers in the engineering design, analysis and optimisation of innovative technologies through scale model testing in the wave fume (104 m × 4 m × 4.5 m) and directional wave basin (49 m × 27 m × 2 m) at the O.H. Hinsdale Wave Research Laboratory. The OSU wave tank facilities are part of the Pacific Marine Energy Centre (PMEC [204]), which also includes two field test sites, the North Energy Test Site (NETS) and South Energy Test Site (SETS). Both NETS and SETS are located near the community of Newport, OR, within the state's territorial (3 nautical mile) limit – also the region of the Finnavera Aquabuoy 2.0 deployment in 2007. The NETS site has hosted short-term technology demonstrations by Columbia Power Technologies in 2012 and Northwest Energy Innovations in 2014. The NETS site is a deepwater test location between 45 m and 55 m deep, and is not grid-connected.

To emulate a grid connection, a unique moored platform, the Ocean Sentinel buoy produced by Axys Technologies of Victoria, British Columbia, can be deployed at NETS that has a load bank that can dissipate up to 100 kW of WEC power delivery. The Ocean Sentinel is also instrumented to record wind, wave and current conditions and to actively monitor WEC dynamics and performance. The PMEC SETS facilities are still under development and are funded through separate awards from the DOE in 2013 and 2014 totalling $4.75 million. The SETS site is 6–7 nautical miles off the coast of Newport and is planned to be operational in 2018. As a grid-connected centre, SETS is subject to FERC oversight and the permitting process was reported to be under way in 2015 [205].

A third WEC test location was established in Oregon in 2014 at Camp Rilea, an army training base operated by the Oregon Military, as part of the net-zero energy initiative of the Department of Defense (DOD). The Camp Rilea site complements the NNMREC NETS and SETS sites as it is more suitable for shallow-water WEC designs. In 2014, M3 Wave Systems completed a test of its shallow-water OWC device and Resolute Marine Energy plans to test its surging flap technology at that site. While the OMD does not plan on Camp Rilea becoming a dedicated WEC test site in the long term, the site does lie within one of the three REFSSA zones identified in the Oregon TSP amendment and an MoU was signed with Oregon State University in April 2015 to establish how the site will be accessed by the NNMREC in the near term.

While the majority of the funding support for NNMREC has come from the DOE Water Power Program, the OWET has played an important role in seeding academic research and commercialisation activities since 2008 [206]. In addition to the $1.2 million investment in the development of the JART process, OWET funded several environmental assessments of the NETS site that characterised marine mammal activity, sediment transport, ambient noise and benthic habitats establishing the baseline conditions for the test centre. OWET has also directly funded WEC technology development; it provided $194,000 to the M3 wave demonstration at Camp Rilea and invested $210,000 and $215,000 in three phases of design iterations of the Columbia Power and NWEI WEC technologies, respectively.

In total OWET dispersed $6 million between 2008 and 2013. In Canada, there are no direct parallels to the DOE Water Power Program, the DOD net-zero energy initiative or provincial level funding or stakeholder representation akin to that provided by OWET. Without such a broad collection of motivated stakeholders, and with a provincial utility (BC Hydro) that is awash in clean hydroelectricity, public funding for WEC field deployments has not materialised in the period since 2008 as it has in the USA. However, the country has invested smaller pools of funds in building an extensive knowledge foundation on which future WEC deployment programs can be built. Canadian academic research activity has been focused on extensive resource assessment, computer simulations of WEC operations and integration studies examining how WEC-supplied power could be used on British Columbia's electrical grid.

The majority of these studies were executed through the West Coast Wave Initiative (WCWI [207]) between 2008 and 2015. The WCWI was hosted by the Integrated Energy Systems Victoria (IESVic [208]) Institute at the University of Victoria (UVic) located on the southern tip of Vancouver Island. The WCWI was funded by $3.5 million in grants from NRCan, the Natural Sciences and Engineering Research Council and the Pacific Institute for Climate Solutions (PICS). The WCWI programme was primarily computational resource assessment and was driven by a SWAN nearshore wave propagation model that extended from the Haida Gwaii in the north to the mouth of the Columbia River to the south [209]. Resource data was fed into WEC performance estimate studies completed using commercial software in collaboration with Dynamic Systems Analysis (DSA) Ltd. [210]. The WEC performance data was subsequently fed into grid integration studies ranging in scope from remote communities to the provincial electric grid.

The WCWI resource and WEC performance assessment exercises were the Canadian analogue to the 2010 wave resource assessment study conducted by EPRI, NREL and the Virginia Tech Advanced Research Institute that completed a 51-month WaveWatch III hindcast of wave conditions along the entire US coastline [211]. The EPRI hindcast became the basis of the wave component of the Marine Hydrokinetic Database currently maintained by NREL [212]. In contrast to the EPRI study, the WCWI focused on a smaller region, one surrounding only Vancouver Island, with increased detail in the spectral and directional characterisation of the nearshore waves and in the modelling of WEC performance in this region. The goal of the WCWI SWAN modelling programme was to fill a knowledge gap in nearshore (200 m deep water) wave characteristics that was identified in the 2006 NRC-led survey of wave buoy records [177]. That study suggested that third-generation shallow-water wave transformation models, such as SWAN, should be applied to the western shore of Vancouver Island.

The WCWI SWAN model uses an irregular grid with the finest spatial resolution reaching 50 m in the very nearshore region. To validate SWAN model outputs, the WCWI also operates four directional wave monitoring buoys off the coast of Vancouver Island. The buoys were produced by Axys Technologies, the same company that provided the Ocean Sentinel buoy at the NETS test site. The WCWI SWAN model continues to be operated in both hindcast and forecast modes, and a directionally and spectrally resolved hindcast of wave conditions has been generated extending back to 2001 [213].

WCWI researchers use the SWAN model and wave buoy data to drive simulation studies of the expected performance of a number of WEC technologies, including those developed by Resolute Marine Energy, Ocean Energy Ltd and SurfPower [214]. The WCWI software is provided by DSA and is capable of full six-degrees-of-freedom WEC motions including moorings and articulate WEC geometries [215]. For each converter considered, WCWI researchers integrate fully coupled dynamic models of the converter's particular power take-off, creating a complete WEC simulation [216]. The WCWI WEC simulation programme had similar motivations to those of the DOE funded WEC-Sim project led by Sandia National Labs to build affordable computational tools that can be used to accelerate WEC commercialisation. WEC performance data produced in collaboration with DSA has been fed into grid integration studies at kilo-, mega- and gigawatt scales. At the kilowatt scale, the WCWI examined how wave energy could supply remote off-grid villages on Vancouver Island. The megawatt-scale studies used the PLEXOS Integrated Energy Model [217] and considered how 500 MW of WEC capacity would best be used on Vancouver Island [213]. At the gigawatt scale scale, the WCWI partners with the PICS 2060 project to examine how wave energy could be applied to the future (*circa* 2060) Western Canadian transmission system to offset greenhouse gas emissions in the central Canadian provinces [218]. The long-term energy system planning work at the WCWI and PICS follows on independent large-scale integration analysis completed at OSU that examined advantages wave energy provided to the existing US Pacific North-West electrical grid [219].

13.2.2 Tidal

Tidal energy in North America is being pursued on both the west and east coastlines of Canada and the USA. Some of the strongest tides in the world exist in the Bay of Fundy off Nova Scotia, Canada and Maine, USA, which also experiences very large tidal ranges. The Minas Passage/Basin area at the north-east end of the Bay of Fundy, one of the most energetic sites, experiences tidal ranges up to 12 m and tidal currents over 5 m/s [220], with viable power extraction estimates of 2000 MW [221]. The Bay of Fundy is a resonant tidal system such that predictions of even larger power extractions from the system would significantly impact tidal ranges along the east coast. On the west coast, a number of particularly energetic passages exist along the coastline, in particular in the Pacific North-west region around numerous islands between Vancouver Island and the mainland and through various passages in the Puget Sound. A 2006 study of tidal potential along all the coastlines of Canada using oceanographic modelling tools found a potential of 42,000 MW average power [222], including 3580 MW around Vancouver Island, 1008 MW in the Arctic, 29,595 MW in the Hudson Strait, 4112 MW in Ungava, and 2725 MW in the Bay of Fundy. These figures were an upper bound on the resource itself, rather than an amount that could be practically developed, and did not include back-forcing of installed turbines on the predicted flows, so the overly optimistic estimates were based on mean tidal stream kinetic energy. Hydrokinetic turbines, very similar in concept to tidal turbines but utilised in non-tidal forced flows, are also in active development both for harnessing the Gulf Stream off the coast of Florida and in various waterways and rivers throughout North America. The DOE estimated [223] in 2015 the technical resource potentials for tidal, ocean current and river currents (free flowing without dams) as 222–334 TWh/yr [224], 45–163 TWh/yr [225] and 120 TWh/yr [226], respectively. Resource data can also be visualised through the NREL MHK Atlas [212].

The first commercial deployment of tidal energy was the Annapolis Royal tidal power plant in Nova Scotia. This 20 MW plant came online in 1984 and is a traditional barrage-style installation on the Annapolis Basin off the Bay of Fundy. More recent tidal power development activities, resurgent since the early 2000s, have focused primarily on hydrokinetic designs that avoid the environmental impacts and installation costs of barrage designs. To date however, only pre-commercial and development deployments of devices have occurred in North America.

13.2.2.1 Regulatory Environment, Incentives and Initiatives

A number of jurisdictions have introduced policies and support mechanisms to encourage the development and deployment of tidal technology. In Canada, electricity falls under provincial jurisdiction, and many provinces including British Columbia and Nova Scotia have Crown-owned utilities. The USA has local (facility siting), state (retail rates and distribution) and federal (interstate transmission and wholesale power sales) jurisdictional powers, and a large number of private investor-owned utilities [227]. The most prominent federal regulator is the FERC. NRCan in Canada and the DOE in the USA are the main federal bodies tasked with supporting development of energy technologies such as tidal and hydrokinetic power. Provincial and state support for tidal energy development has varied with political direction.

In Canada, Nova Scotia introduced the Community Feed-In Tariff (COMFIT) programme in 2010 which provided a 65.2 cents/kWh guaranteed rate to community-based organisations developing small-scale tidal power projects [228]. The programme was cancelled in August 2015 due to upward pressure on residential power rates; of the five tidal projects submitted, two were withdrawn and three were still in development. Nova Scotia also approved four PPAs in 2015 under a separate Developmental Tidal Feed-in Tariff programme for projects larger than 500 kW. The programme guarantees rates of $455–$530 per MWh for 3–15 years depending on developmental or testing status and annual energy production.

In British Columbia, BC Hydro includes tidal energy in its resource options reports as a technology potentially able to supply the provincial system. Tidal energy has been included since at least 2000 [229] in its Integrated Resource Plan (IRP) and Integrated Electricity Plan (IEP) documents. In 2004, the IEP was updated to include specific site power estimates, namely Discovery Passage and Race Passage on the east coast of Vancouver Island. As of 2015, the IRP includes tidal energy only at the level of 'monitoring technology development' based on uncertain costs and assessment of technology readiness. Tidal power may be compensated if grid-connected using BC Hydro's standing offer programme, if the technical readiness of the technology can be demonstrated to satisfy BC Hydro's requirement for technologies to have demonstrated performance to qualify. A FIT programme was studied in 2012, however the British Columbia government did not proceed to enact the legislation that would have enabled a FIT due to pressure to minimise electricity rate increases.

Marine Renewables Canada (MRC), the successor of the Offshore Renewable Energy Group (OREG) established in 2004, is a non-profit dedicated to advancing the ocean energy industry in Canada. MRC has helped organise numerous conferences, lobby for support and bring together actors in various projects. The Canadian sub-committee for the International Electrotechnical Commission (IEC) TC 114 Marine Energy has been quite active in funding research projects to contributing to the development of the emerging international standards document. The committee secured substantial NRCan funding for the 2011–2016 period to tackle specific pressuring research questions for the standards. NRCan compiled the Canadian Marine Renewable Energy Technology Roadmap in 2011 to help guide the industry [230].

In the USA, there has been no direct FIT programme support for tidal energy, but some state-level RPSs provide indirect incentives for the development of tidal power. Verdant Power received the first FERC pilot commercial licence, FERC No. P-12611, for its Roosevelt Island Tidal Energy (RITE) project in the East Channel of the East River (channel connecting Long Island Sound to the Atlantic Ocean) in New York in 2012. As of October 2015, FERC reported [231] two active preliminary tidal and four inland hydrokinetic permits, two tidal projects in pre-filing (P-13015 Muskeget Channel (MA), 4.9 MW; P-13511 Igiugig Hydrokinetic Project (AK) Inland Pilot, 40 kW), and four licenses issued for pilot projects (P-12611 Roosevelt Island (NY), 1.05 MW (Verdant Power); P-12690 Admiralty Inlet (WA), 1.0 MW; P-12711 Cobscook Bay (ME), 0.3 MW; P-13305 Whitestone Poncelet (AK), in-river, 0.1 MW).

As a primary funding agency for development of hydrokinetic energy in the USA, the DOE has provided significant support, totalling $116 million for FY 2008–2014 for all marine and hydrokinetic activities, with $31 million directly for technology development of ocean/river/tidal current energy systems [195]. The DOE maintains the Marine and Hydrokinetic Data Repository [232], a repository for data collected using funds from the Wind and Water Power Programs of the DOE. The DOE also maintains a worldwide Marine and Hydrokinetic Technology Database [233], an active listing of projects in various states of development all around the world, including North America. Pacific Northwest National Labs maintains Tethys [234], a knowledge management system actively gathering and providing access to information on environmental impacts of marine energy. The NREL, a US government lab which has developed open-source design and analysis software for wind energy for decades, has modified its tools to simulate marine current turbines with appropriate inflow conditions.

As an emerging technology, various efforts have been made to ascertain the potentials and interconnection requirements for widespread deployment of tidal energy on the North American coastlines. In 2007, the EPRI brought together US and Canadian utilities, state and federal agencies, universities and technology developers to work towards demonstrating ocean energy potentials [176]. They provided preliminary site yield and COE estimates for sites in Alaska, Washington, San Francisco, Massachusetts, Maine, New Brunswick and Nova Scotia. Based on the MCT SeaGen dual rotor, they found very favourable COEs in the range of 4.6–10.8 cents/kWh, compared to conventional utility generator costs of 5–12 cents/kWh (2005 values). A PowerTech report in 2008 provided a survey of technologies in development at that point, plus an examination of the interconnection requirements to grid-connect devices [235].

13.2.2.2 Device Development Efforts

A number of companies in North America continue to pursue barrage-style tidal energy devices. Examples of these include Blue Energy and Halcyon Tidal Power. Blue Energy is pursuing a modular vertical-axis turbine enclosed in a rectangular duct, to potentially be built as a series of modules into bridge structures. Starting out as Nova Energy, it carried out flume testing starting in 1982 at the Canadian National Research Council Hydraulics Laboratory in Ottawa. Initial prototypes were installed in 1983 and 1984 (20 kW and 100 kW, respectively), with a test in the Gulf Stream of a 4 kW machine in 1985. Blue Energy carried out a series of tow tank tests at the University of British Columbia (UBC) in 2006–2007. Halcyon is advocating a pile-supported, pre-cast concrete embankment approach to reduce barrage costs, with a bidirectional bulb turbine to generate on the ebb and flood tides in an attempt to avoid disturbing the intertidal zone.

Clean Current Power Systems was formed in 2001 in British Columbia to develop a ducted horizontal-axis machine with a flooded, direct drive generator. The designs were bottom-mounted with bidirectional rotors using symmetric blade profiles. The first operational prototype was 65 kW and 5 m in diameter, deployed in 2006. It is now displayed in the Canadian Science and Technology Museum on Ottawa. The core intellectual property relating to the duct and direct drive technology was licensed to Alstom Power in 2009 with the aim of scaling up the technology to the megawatt scale, but this relationship was terminated in 2012. Clean Current subsequently refocused its efforts on hydrokinetic and shallow-depth (<20 m water) tidal turbines, resulting in a range of prototypes in the 1.5–3.5 m diameter range. Clean Current was forced to cease operations in 2015 due to funding challenges from a main investor, EnCana.

A number of Canadian companies are pursing vertical-axis/cross-flow technologies, many based on floating platforms. New Energy Corporation started operations in 2003 working on freshwater applications with designs locating the gearbox and generator above the water. Initial testing in 2006 at a wastewater treatment plant in Calgary, refined by tests in 2007–2008 in Manitoba, yielded their EnCurrent 5–25 kW devices, installed as initial commercial systems in rivers, and a hydro supply canal. A second generation of EnviroGen systems was designed based on this operational experience, currently offered as a 5 kW unit typically mounted on a floating pontoon boat. Instream Energy Systems has developed a 25 kW vertical-axis system (1.5 m high blades, 3 m diameter rotor), with initial deployment in irrigation canals and a generator/gearbox above the waterline. An initial test of four turbines mounted on floating platforms in an array configuration was carried at Duncan Dam, BC, in 2010. Canal testing in Yakima, WA, was started in 2013. Mavi Innovations has developed a 22 kW Mi1 floating turbine system for river or tidal applications, but utilising a cross-flow turbine with straight blades and a 3 m wide by 2 m tall projected area. This design is being tested in Manitoba, and results from work by the company founders at the UBC test tank on vertical-axis designs in the mid-2000s. Hydrorun developed a unique prototype 40 kW unit for river applications based on a glider concept (3 m span). Based on experience of the proponents with kite-based airborne wind, the river 'glider' with vertical wing and control surfaces would 'fly' in a figure-of-eight profile across a river. A tether attached to a motor/generator would control the reeling in an out of a 40 m tether, and thereby generate electricity as the tether was in the reel-out phase. A functional prototype was developed and tested in British Columbia by 2014; however, the company has since ceased operations due to lack of funding.

A number of European companies have established offices in Canada, primarily with a view to tapping into the subsidies in Nova Scotia and test sites on the Bay of Fundy detailed in Section 13.2.2.3. These companies include Marine Current Turbines, OpenHydro, Tocardo, Atlantis, and Siemens. Descriptions of these technologies, primarily designed in Europe are given in Section 13.1. Their North American technical development work is primarily related to gaining operational experience and establishing markets and supply chains in North America. Some partnerships have also involved setting up local supply chains and local civil structures and deployment vessel work. As part of these transnational activities, in 2014 the Offshore Energy Research Association of Nova Scotia (OERA), the Province of Nova Scotia and Innovate UK signed an MoU to sponsor a research call involving UK and Canadian academic and industrial partners. The initial call resulted in two projects in the areas of environmental and turbulence monitoring at test sites [236]. An MoU was also signed in 2015 between the British Columbia and Nova Scotia governments to collaborate on ocean renewable energy best practices, regulatory frameworks and technology development [237].

In the USA, the Verdant Kinetic Hydropower System (KHPS) is a bottom-mounted tower system employing a three-bladed open rotor. The tower uniquely passively yaws to face ebb and flood tidal directions. The KHPS has 5 m rotors with a nominal power of approximately 30 kW. Improvements from the Gen 4 to Gen 5 included composite blades and improved, redundant sealing for the gearbox, and a shortened nacelle. The Ocean Renewable Power Company is developing a modular cross-flow turbine system with helical blades for river, tidal and Gulf Stream applications. All are based on a turbine generator unit which employs a submerged, gearless permanent magnet motor. The support structures can be configured with tension moorings and buoyancy pods, or ballasted onto the bottom with provision to de-ballast for retrieval. The company was founded in 2004, working through a series of field tests to deliver energy to the grid in Cobscook Bay, Maine, in 2012. In 2014 a 25 kW river variant was successfully deployed to power the remote village of Igiugig, Alaska. Vortex Hydro Energy is working to develop a concept from the University of Michigan, the Vortex-Induced Vibration Aquatic Clean Energy converter, which is based on a series of cylinders placed in crossflow. The cylinders are supported at each end with linear bearings allowing the cylinders to oscillate vertically, and by connecting them to linear generators housed in the support structure, generate electricity. A few prototypes have been tested in a river and canal in 2012, in addition to tow tank testing, with intended markets to be low-flow sites requiring minimal environmental impact from conventional moving blades. Ecomerit Technologies is developing the Aquantis Current Plane (C-Plane) device to harness Gulf Stream currents. It is envisioned as a twin downstream rotor turbine design with a central buoyancy pod, positioned at an intermediate water depth by mooring lines (rather than a bottom-mounted tower). In Alaska, the Oceana Energy Company is developing a hydrokinetic device that uses a duct, but uniquely the front portion of the duct rotates with eight blades that extend both inside and outside the duct. The stated intention is to balance the thrust forces on both halves of each blade. They have been tested at the Alaska Hydrokinetic Energy Research Center (AHERC) river site in 2014.

13.2.2.3 Deployment Activities and Research Centres

In Canada, tidal energy developments have occurred on both the Pacific and Atlantic coasts, in addition to inland rivers. On the west coast, Clean Current carried out a series of successful tests of their prototype machines at the Race Rocks Ecological

Reserve [238]. The first prototype was installed in 2006, removed in 2007 for a bearing change, then redeployed in 2008. The machine was removed in 2011 at the conclusion of the testing programme. This deployment was a great success, as it involved as a partner the Lester B. Pearson College of the Pacific which helps run Race Rocks, and because the deployment was successfully carried out in an ecological reserve. IESVic has worked with various device developers, including Clean Current and Mavi Innovations, on a range of projects, including tank testing of device arrays and coupled CFD-oceanographic numerical modelling. The Bamfield Marine Sciences Centre has also hosted flume testing campaigns in their facilities. The passages in the waterways along the east coast of Vancouver Island are host to numerous potential tidal energy sites. Various companies have investigated commercial opportunities in these waters, including Western Tidal Holdings which has worked with First Nations and various levels of government to secure 14 licences and applications to investigate potential projects. Canoe Pass is a site of particular interest. At present, the pass is blocked by an existing causeway between Maud and Quadra Islands that would be removed to restore flow through the passage. Grid interconnection is also relatively close to the site. Led by New Energy Corporation, with partner Mavi Innovations, the Canoe Pass Tidal Energy Consortium secured Sustainable Technology Development Canada for installation of two 250 kW devices in the pass. UVic has worked on a number of tidal projects, including: development of theory for fundamental analytic limits to tidal energy extraction from passages and basin flows; composite blade investigations for Clean Current; blockage effect investigations utilising porous disk water tunnel investigations with Mavi and Clean Current to support TC 114; and CFD-oceanographic coupling development for array predictions with Triton/Cascadia and Mavi.

Moving east, hydrokinetic turbine development in Canada has been aided by a large grant from NRCan to establish the Canadian Hydrokinetic Turbine Testing Centre (CHTTC) located downstream of the Seven Sisters Generating Station in Manitoba. The CHTTC is operated by researchers at the University of Manitoba, who also have a flume tank facility. A number of testing berths are located at the site, which experiences flows in the range of 1.5–2.5 m/s with seasonal variation, including ice-up, as well as daily variations from the operation of the upstream power dam. The site has been used by a number of developers, including Mavi and Clean Current, to test their devices and utilise acoustic Doppler current profiler (ADCP), acoustic Doppler velocimeter (ADV) and other flow characterisation instruments.

On the east coast, it is interesting to note that tidal energy exploitation has been considered in the Bay of Fundy from at least the start of the last century [239]. This has included a resurgent assessment of large changes in the tidal range resulting from substantial energy extraction from this resonant system in the 1950s and 1960s [240–242]. By the end of the 1960s the locations of interest for modern-day developers, Digby Neck and Minas Passage, had already been studied [243]. More recent estimates have been made of the resource, in particular around Minas Passage [244].

Over the past decade, Nova Scotia has become very supportive of tidal energy development in the Bay of Fundy, resulting in a number of initiatives. Acadia University has been active in oceanographic modelling of the resource, supported by field measurements from researchers at Dalhousie University and detailed CFD investigations by University of New Brunswick researchers. An NRCan funded project by Acadia University from 2013 to 2016 has carried out more detailed site assessment at three

Digby Neck passages: Grand Passage, Petit Passage and Digby Gut. This has included field measurements including turbulence assessment, as well as modelling of various array configurations in concert with a number of device developers to determine expected power output. Minas Passage has continued to attract the most attention, as it is the location of extremely energetic tidal flows in the Bay of Fundy. The Fundy Ocean Research Centre for Energy (FORCE) test site has been established there, with three megawatt-scale cabled testing berths installed since the initiation of the testing site development in 2008.

OpenHydro installed a machine for testing at a first berth from 2009 to 2010. Currently, Atlantis, Siemens/Marine Current Turbines and Schottel, in concert with various local partner companies, are working towards deployment of their pre-commercial devices at the test site. FORCE has developed the Fundy Advanced Sensor Technology platform to deploy at the berths to gather high-resolution flow data during operational testing. FORCE has also developed the Vectron [245], a hybrid ADCP/ADV device capable of resolving turbulence at the hub height of turbines. OERA was established in 2012 by amalgamation of two previous research bodies as a not-for-profit organisation to fund collaborative research in Nova Scotia, linking five provincial universities with the Nova Scotia Department of Energy. Various projects have been funded through OERA to address identified research priorities in geoscience, MRE and marine acoustics. The Fundy Energy Research Network (FERN) was created in 2010 as an independent non-profit to coordinate and foster research on tidal energy development in the Bay of Fundy. It receives funding from FORCE and the Nova Scotia Department of Energy, and logistical housing at Acadia Centre for Estuarine Research. Acadia University hosts the Acadia Tidal Energy Institute which has a focus on community, socioeconomic and marine ecology issues around tidal energy, in addition to work on tidal energy mapping.

In the USA, Verdant Power deployed six grid-connected KHPS (Gen4) at the RITE site in the East River of New York in 2006–2009. It reported 9000 hours of operation and 70 MWh of generation during this period. A particular achievement during this deployment was an extensive underwater marine life monitoring programme. There were initial problems with blade failures on the early deployments of these units that were subsequently remedied. They received a FERC permit in 2012 to install up to 30 Gen 5 units at the site. The NNMREC is a major DOE funded research centre working on both wave and tidal energy, established in 2008 via a $13 million grant. It is a collaboration between OSU, the University of Washington (UW) and the UAF. The UW has a current flume tank, in addition to extensive ocean current measurements and towed tidal turbine testing in Puget Sound. The field testing is carried out by the Applied Physics Laboratory at UW, including development of an instrumentation pod for tidal site assessment. UW has also developed an Adaptable Monitoring Package, a system that can be deployed via remotely operated vehicle to a fixed mount on a turbine installation, and includes various ADCP, ADV, acoustic and marine life monitoring equipment. The UAF established the AHERC in 2010, which in turn operates the Tanana River Hydrokinetic Test Site for hydrokinetic turbine testing.

Puget Sound in Washington State is an estuarine system near Seattle with a network of waterways and basins. A prime tidal energy site is Admiralty Inlet at the inlet to Puget Sound from the Strait of Juan de Fuca. Development of this site was pursued by the Snohomish County Public Utility District #1 through the Admiralty Inlet Pilot Project.

The DOE provided \$12.2 million from 2008 to 2010 to develop the project, which was to involve OpenHydro turbines and be operated as a pilot tidal energy research project. After progressing through extensive licensing and regulatory processes at local, state and federal levels starting with a FERC filing in 2006 [246], the project was cancelled in the face of escalating costs in 2014 [247]. Tacoma Narrows, another energetic site in Washington State, has also been studied for potential tidal energy. In 2006, EPRI carried out a preliminary investigation of the site [248]. Based on the positive findings, Tacoma Power, the local utility for the City of Tacoma, obtained a FERC permit to investigate feasibility in Puget Sound and commissioned a feasibility study [249], with a related study delivered to the Bonneville Power Administration [250]. However, this analysis indicated that tidal generation was not economically viable, in addition to environmental sensitivity concerns, and is presently on hold.

On the US east coast, the Maine Tidal Power Initiative was established at the University of Maine through a \$1.9 million DOE grant in 2009. This initiative includes a number of activities such as resource and environmental impact assessment, community involvement and technical device development. The Ocean Renewable Power Company has deployed its TidGen technology in Maine as part of these activities. The University of Massachusetts, Dartmouth, established in 2007 the New England Marine Renewable Energy Center, with a \$1.7 million grant from the DOE in 2009. The consortium has held a number of conferences and brought together various universities to coordinate research activities and foster ocean renewable energy (including wave and offshore wind) R&D in New England. The University of New Hampshire founded the Center for Ocean Renewable Energy in 2008 with the goal of providing multi-scale testing facilities for tidal and wave energy testing. A 2010 DOE grant of \$750,000 enabled the establishment of its General Sullivan Bridge tidal energy test site, to complement the tow tank already at the University. Florida Atlantic University operates the Southeast National Marine Renewable Energy Center with a mandate to study the Gulf Stream resource and ocean thermal energy conversion (OTEC) technologies. It was established in 2010 from a number of funding sources, including a \$5 million grant from the DOE. It carries out a range of activities including resource/environmental monitoring in addition to education and outreach activities. Through affiliated university labs they also have various technical development projects on land, and they are working to develop small-scale test berths offshore for ocean current prototype deployment.

13.3 Latin America

13.3.1 Introduction

The exploitation of MRE resources in Latin America has not made as much progress as in the USA, and certainly compared to Europe it is in its infancy. However, some efforts towards the development of MRE have been made. This chapter presents some initiatives that have been successful, or at least have been gathering momentum. It is divided into countries, and the emphasis is placed, whenever possible, on R&D for installation of technologies at sea, at technology readiness level 5 and higher [251]. A map of the region is shown in Figure 13.7.

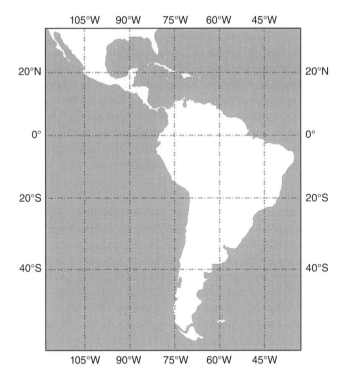

Figure 13.7 Map of the Latin America region (created with Cartopy [252]).

13.3.2 Brazil

Brazil's ocean space is known as the 'blue Amazon', which has been a registered trademark in the National Industrial Property Institute since 2006 [252]. Brazil's economic exclusive zone covers an area of $3{,}539{,}919 \, km^2$. Its MRE resources have been quantified using *in situ* measurements as well as some numerical modelling approaches. It is estimated that at least 114 GW of installed MRE capacity (27 GW of tidal energy in the North East, and 87 GW of wave energy: 22 GW in the North East, 30 GW in the South East, and 35 GW in the South) could be achieved, based on preliminary potential estimates based on the *in situ* measurements alone [253].

13.3.2.1 Tidal and Hydrokinetic Energy

Tidal power can be generated in two ways: using tidal barrages, essentially operating as hydroelectric plants, or using modular technologies, operating as tidal turbine farms. Tidal barrages are the most mature ocean renewable energy technology to date [254], and according to [255], 27 GW of installed capacity could be achieved with tidal barrages in the North East of Brazil. However, because large intertidal areas are inundated after construction, during these projects the feeding and nesting grounds of migrating and local bird populations are, for the most part, destroyed. Hence, a number of mitigation strategies need to be adopted, which usually involve trying to reduce the impact with double dam systems, for example, or other approaches, following the mitigation hierarchy 'avoid, reduce, repair, compensate, enhance', as defined by [256].

Tidal modular turbine systems, on the other hand, have a minor impact on the environment. It is worth noting that since water is 1000 times denser than air, tidal turbines can extract 1000 times more power, for the same flow speeds, than wind turbines with similar cross-section. Apart from the North East of Brazil, the Canales de Guimarães have also been recognised as having enormous potential [257]; however, it is not known whether any initiatives have been undertaken to deploy tidal turbines in this region. A number of numerical studies on the Southern Brazilian Shelf [256] have analysed the potential impacts of marine current turbine installations near the coast. However, it is unclear whether this research is industry-driven, that is, whether the industry has an interest in deploying a tidal farm in this location.

13.3.2.2 Wave Energy

Brazil is at the forefront in Latin America in wave energy technology. Since 2001, the Federal University of Rio de Janeiro (FURJ) has been operating an ocean basin at its Technology Park. The initiative was led by Prof. Segen Farid Estefen and his team, at the Subsea Technology Laboratory, in existence at COPPE-FURJ since 1989. The ocean basin is 40 m long, 30 m wide, and 15–25 m deep. Waves, tides and winds can be simulated. The Subsea Technology Laboratory has been working on wave energy point absorber technologies; a prototype is shown in Figure 13.8. This point absorber has now been tested at full scale, and a number of commercial scale devices have been deployed at Ceará-Pecém harbour. The cost of the pilot plans was US$7 million. Construction began in 2009, and the goal was to have an operational plant by 2012. The pilot plant is a 50 kW WEC consisting of two point absorber modules. Each module consists of a 10 m diameter floater, a 22 m long mechanical arm, and a pump connected to a freshwater circuit that injects this water at high pressure onto a turbine. The turbine activates an electricity-producing generator. This system generated some power for a few minutes in July 2012, but it is still at the pilot stage. The full plant will be generating 1 MW

Figure 13.8 Point absorber developed by the FURJ at scale 1:10 (from [252]).

of energy, and it was originally conceived to power the harbour area, but as the project expands the plant will be connected to the grid.

13.3.2.3 Marine Bioenergy

A bioenergy project based on cultivation of macroalgae between Ubatuba (São Paulo) and Sepetiba Bay (Rio de Janeiro) has been approved. *Kappaphycus alvaresi* has been approved for cultivation as part of a SAGES-COPPE-UFRJ collaboration led by Dejair de Pontes Souza (UFRJ). In [252], the authors show the dry biomass production of different plant species. Three of those shown are terrestrial (eucalyptus, sugar cane and elephant grass), and three of them aquatic (water hyacinth, microalgae and macroalgae). The authors highlight the clear potential for bioenergy production from aquatic species, with dry biomass production for macroalgae being around 300 ton/ha/year, for microalgae 180 ton/ha/year, for water hyacinth 90 ton/ha/year, for elephant grass 80 ton/ha/year, for sugar cane about 40 ton/ha/year, and eucalyptus about 20 ton/ha/year.

13.3.3 Chile

Chile has 4200 km of coastline and around a dozen MRE projects, funded by the Chilean Production Development Corporation (CORFO). Both CORFO and the Chilean Ministry of Energy are encouraging the development of WECs and human capital training through applied research, development and innovation. Recently, the government approved the creation of an International Centre of Excellence, involving the French naval defence and energy Group DCNS. According to [258], the Chilean government should focus on site-specific technology development, including studies, for example, on the impact of long wavelengths on WEC geometry; effects of consistently high waves on wave energy farms' installation, operation and maintenance; novel design mooring development for steep seabed slopes; and designs integrating potential impacts of earthquake and tsunami events.

It is recognised that mapping the seabed at key areas with high renewable energy resource potential would be beneficial, together with a characterisation of wave energy resources in the nearshore region (within 10 km of the coastline). Although a quantification of energy resources has been performed, economically viable sites have not yet been identified. Moreover, regional differences will also influence the WEC development drivers and energy usage. Davies and Wills [258] divide Chile into six main regions: North, Centre/Centre-South, Lagos, Aysen, Magallanes (Magellan) region, and offshore islands. For example, in the Northern part of Chile, a rise in energy and water demand associated with an increase in mining activities, together with tax benefits and a more benign operating environment [258], makes it likely that wave energy projects will become viable sooner in this region than further south. The Centre region is likely to be relevant for the MRE supply chain, rather than for resource exploitation. The Lagos region possesses abundant wave and tidal resources. Here there are many scattered communities and energy costs are high, but projects need to be environmentally and socially acceptable because of the presence of large mammal communities. Aysen, Magellan and the offshore islands have excellent wave energy resource potential, but their remoteness makes it unlikely that these resources will be exploited in the near future. The Magellan Strait is an excellent tidal energy location, and so are the smaller straits in the Aysen region. The resource needs to be quantified in detail for

Table 13.2 Chile's marine energy development potential by region.

Technology	Potential per region
Small-scale and off-grid tidal projects	Good: Lagos, Aysen, Magallanes Some: Offshore islands
Multi-MW tidal projects	Good: Lagos and Magallanes Some: Aysen
Small-scale and off-grid wave projects	Good: North, Centre/Centre-South, Offshore Islands. Some: Aysen and Magallanes
Multi-MW wave projects	Good: North, Centre/Centre-South, Lagos
Device manufacture	Good: Centre/Centre-South, Magallanes Some: North, Lagos, Aysen

all potential tidal energy development sites, in order to rank areas for development based on a number of criteria, including: resource characterisation, environmental impacts, or accessibility to grid connections, among others. The potential for marine energy development per region is shown in Table 13.2 (adapted from [258]).

13.3.4 Argentina

In 2013 the Argentinian government, together with the Consejo Nacional de Investigaciones Científicas y Técnicas (CONICET) and the Argentinian Ministry of Science, Technology, and Productive Innovation approved the creation,in December 2014 of a company called Y-TEC, a spinoff of Yacimientos Petrolíferos Fiscales (YPF) and CONICET. YPF had traditionallybeen an oil company. It is the largest company in Argentina and the third largest oil company in South America, directly or indirectly employing more than 46,000 people across the country. From April 2013, YPF transferred all its human resources, equipment and infrastructure, including its Applied Technology Centre, to Y-TEC, with the aim of relying on the new company for all its research, development and technology needs and services [259]. Y-TEC incorporates researchers from different institutes, with the goal of attaining around 250 researchers working on different projects, articulated through CONICET and the institutes the researchers belong to. Y-TEC's mission is to generate and apply knowledge acquired through R&D in different energy areas, including non-conventional energy resources and the diversification of the energy matrix. It is this mission that led Y-TEC into ocean energy and the exploration of Argentina's MRE potential.

In 2014, Y-TEC deployed monitoring equipment to quantify the tidal stream energy potential at the Gallegos Estuary and at the mouth of the Magellan Straits, in Santa Cruz. The Magellan Straits have the second largest tidal range in the world (after the Bay of Fundy), with a difference of 14.6 m between low and high tide. Scientists also have been monitoring the wave energy potential in Austral Patagonia. The project is being led by the Facultad Eegional Santa Cruz of the Universidad Tecnológica Nacional, and the Centro Nacional Patagónico [260]. The studies involve environmental monitoring, in order to assess the potential impacts on the local fauna and their migratory patterns; and oceanographic and morphodynamic analyses, to assess not only the energy potential, but also the bathymetry and its evolution at short and long time-scales.

In relation to technological innovation, an Argentinian company called INVAP has been developing, through links with Y-TEC, a hydrokinetic turbine that is initially to be installed in shallow rivers, with an installed capacity of 9 kW. INVAP's goal, however, is to develop turbines with up to 1 MW of installed capacity [261], for tidal energy extraction. Argentina's objective is to produce 15% of its energy from renewables, such as wind, solar, and tidal energy resources. However, the ocean energy projects have objectives that go beyond achieving a certain goal in relation to renewable energy generation. As Gustavo Bianchi (Y-TEC's General Director) stated for Energía Estratégica [255], it is also about opening new doors, generating new knowledge, creating the infrastructure, training highly skilled staff, and including ocean technologies in the political agenda.

13.3.5   Mexico

According to [262], Mexico could satisfy up to 46% of its annual electricity requirement from renewable energy sources. However, if current renewable energy development trends remain the same in the country, then it would only cover 18% of its energy needs from renewables. As Mexico's renewable energy production target to combat climate change is 35% by 2024, as established in the Law for the Exploitation of Renewable Energies and Energy Transition Financing (known by its Spanish acronym, LAERFTE), current trends necessarily need to change. The government expects that the energy reform of 2014 will play a strategic role in the achievement of the renewable energy production target. Furthermore, the Ministry of Energy (SENER) is aggressively pursuing various strategic programs to achieve such a goal, for example, through the creation of Mexican Innovation Centres for Renewable Energy (CeMIEs). Three CeMIEs for geothermal, wind and solar energy, have been approved to date [263]. Proposals for two more CeMIEs, for bioenergy and MRE, are currently under evaluation. SENER has also been supporting innovation and human resource capacity building, through initiatives led by the private sector. For example, InTrust Global Investments and the Center of Public Health and Global Environment, of Harvard University, have been providing support for the development of more than a hundred commercially driven projects on renewable energy (see www.intrustglobal.com for more details).

Mexico's territory is 38.4% land and 61.6% water [264], and the country has an 11,592 km coastline [265], with important wave and tidal energy potential in the Northern Mexican Pacific and the Gulf of California, as well as some marine current potential in some regions in the Caribbean. A map of the country is shown in Figure 13.9. The development of MRE would lift pressures on a territory that is much needed for other uses (such as agriculture, terrestrial species conservation and restoration), as well as satisfy the need for diversification of primary energy sources and to contribute to a healthy and sustainable future for the country.

13.3.5.1   Tidal and Hydrokinetic Energy

The Gulf of California, between the Baja California Peninsula and the states of Sonora and Sinaloa, and the Yucatán current near the shore of the states of Quintana Roo and Yucatán, are two regions in Mexico that are well known for their high tidal and hydrokinetic energy potential. These two regions have been monitored and analysed extensively for oceanographic purposes, but it is only recently that their potential for tidal and hydrokinetic energy extraction has started to be considered.

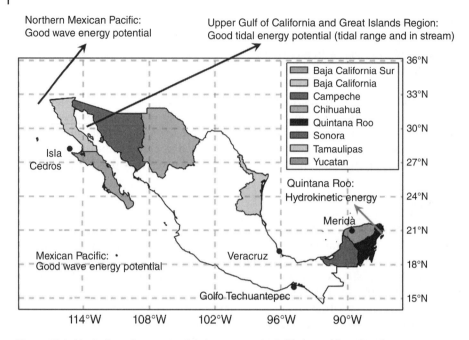

Figure 13.9 Mexico's marine energy development potential (adapted from [266]).

The Gulf of California, in Mexico's North West, is a key area for energy extraction from tidal range and tidal current infrastructure [254]. The Gulf of California is Mexico's only marginal sea; it is around 1500 km long and its width varies between 92 km and 22 km [267], with tides that co-oscillate with the M_2 tide component [268, 269]. A number of studies have been done in the deep trenches within the Gulf of California, called *umbrales*, where flow speeds within commercially viable limits have been found [270]. However, such locations are quite far from the coast, which limits the potential for exploitation. There also has been significant interest in developing a tidal barrage in the Colorado Delta region, and in some of the coastal lagoons in the Upper Gulf of California. The Canal del Infiernillo (Infiernillo Canal), between Isla Tiburón (Shark Island) and mainland Mexico, in the state of Sonora, has also attracted some attention [267]; however, the average current speeds measured in this location were found to be under 1 m/s, below the limit for viable commercial exploitation.

The Yucatán current, flowing along the states of Quintana Roo and Yucatán [271], at the southern end of the country, is also recognised as another important source of hydrokinetic energy. In the case of the Yucatán current, some of the first direct current measurements were taken by [257], who reported currents on the western side with average speeds of 1.7 m/s at 6.3 m depth on the western side of the Channel, and with other researchers reporting southerly flows on the eastern side [272]. However, it is unclear whether the sites with technical potential satisfy current technological and economic constraints.

In relation to technological developments, a number of turbine prototypes have been developed, with national funding from CONACYT. The Programa de Investigación Multidisciplinaria de Proyectos Universitarios de Liderazgo y Superación Académica, or projects IMPULSA from UNAM, in particular IMPULSA IV (2005-2010), allowed for some of the first technology developments in the country. Some recent calls for funding from the Comisión Federal de Electricidad (CFE), through the CFE-CONACYT funding managed by CONACYT, included a hydrokinetic turbine project [273]; however, no marine energy applications were successful in that round. It is worth noting that CFE used to have the monopoly in energy generation and distribution in Mexico, but with the energy reforms the market has opened up to other local, national and international power producers.

13.3.5.2 Wave Energy

Mexico's wave energy potential is localised principally on the Pacific coast, with some hot spots in the northern part (within Mexico's territory) of the Gulf of Mexico. Although many projects have focused on technological proof-of-concept developments, some have been targeting technology deployment at sea and wave energy commercialisation projects. For example, Energy Forever, in Ensenada, Baja California, has been developing a point absorber for wave energy exploitation in the Port of El Sauzal, Ensenada. Figure 13.10 shows one of the point absorber prototypes at 1 : 4 scale being constructed at Energy Forever's workshop in Ensenada [274]. Another company, Marersa, was in the process of developing a wave energy farm in Rosarito, a few miles from Tijuana, in 2012. However, the farm did not materialise, leaving Energy Forever as the leading company in Mexico in this renewable energy sector in the Baja California peninsula. The suitability of other regions of the Pacific coast is being analysed for exploitation of wave energy by point absorber technologies.

Indeed, although it is well known that the Baja Californian Pacific has excellent wave energy resources, several other locations on the Pacific coast would be exploitable, for example in Oaxaca. Some other coastal locations in Tamaulipas, on the Gulf of Mexico side, also have potential.

Figure 13.10 Point absorber for deployment at El Sauzal, Ensenada. Prototype II, 1 : 4 scale (from [274]).

13.3.5.3 Offshore Wind Energy

Although at present there are no offshore wind farms planned in Mexico, it has been of interest in recent years to explore the potential of an offshore wind farm installation, in particular in the Gulf of Tehuantepec or in the North of the state of Yucatán. Figure 13.9 highlights states for which the National Renewable Energy Laboratory (NREL) and other researchers have focused efforts on onshore wind energy analyses for Mexico, including the coastal areas of those states when applicable (see http://www. nrel.gov/ wind/pdfs/ and [275] for further information).

The most comprehensive analysis of Mexico's offshore wind energy potential to date was performed only recently by [266], who identified several offshore and coastal regions suitable for wind energy resource exploitation, and published a clickable map for interested end-users (see www.usuario.cicese.mx/~mgross). They have also quantified the impacts of climate change under scenario RCP8.5 on wind energy resources in the region [252]. Research on this topic is ongoing, strongly driven by commercial wind energy developers' needs.

13.3.6 Colombia

Colombia, at the north western end of South America, is the only South American country with coasts on the Caribbean and on the Pacific Ocean [276]. It has 2900 km of coastline, with major coastal developments on the Caribbean side of the country. In contrast, the Pacific coast is quite isolated, not grid-connected, very green and humid because of the considerable rainfall in the region, and apart from two large cities (Tumaco and Buenaventura), small rural communities are the norm. Some MRE initiatives have focused on small renewable energy devices that potentially could make rural communities more independent and less vulnerable [277]. Most prominent is Project Aquavatio (or aquacharger), led by Miguel Borbón of the Universidad de los Llanos Villavicencio de Bogotá, and representing a company called APROTEC. In Aquavatio, a river current turbine was designed for installation in poor rural regions of Colombia. The three-bladed, 1.8 m diameter, fibreglass turbine has an installed capacity of 2.4 kW. The turbine can supply energy for 10–15 families living in riverine and coastal rural areas [278]. The turbine concept technology was one of the winners of an IDEAS Energy Innovation Contest in 2011.

In 2009 a Centre for Research and Innovation (CIIEN) was created to promote collaborations in renewable energy R&D. The research areas promoted by the CIIEN included the evaluation of Colombia's MRE resources, the development of hybrid renewable energy systems, the development of state-of-the art hydroelectric systems, and the development of bioenergy technologies. The evaluation of the MRE resources performed within the CIIEN project was led by the National University of Medellín, specifically by the OCEANICOS group. The results of the evaluation, presented at the Seminario Nacional de Economía [279], included, for example, a wave energy resource characterisation in the Colombian Caribbean. This evaluation showed there is some potential for wave energy extraction in the country. Most importantly, however, a roadmap has been proposed for the implementation of MRE that includes waves, tides, marine currents, and saline and thermal gradients [276]. The roadmap proposed is as follows:

- marine energy resource characterisation for all five forms of energy identified in the roadmap;
- feasibility studies including technological, social, economic, and environmental aspects;

- start-up and construction of one or more pilot plants;
- monitoring, and operation and maintenance optimisation of the pilot plant.

It is noteworthy that this roadmap is applicable to any renewable energy development in any country, not only to marine energy resources or to Colombia.

13.3.7   Other Initiatives

Although some OTEC assessments have been carried out in some countries, for example for Isla San Andrés in Colombia [276], the only OTEC project being developed at present in the tegion is led by DCNS France and is on the island of Martinique/Bellefontaine [280]. It is noteworthy that no other Caribbean country is considering ocean energy as a potential renewable energy resource, and they are focusing instead on wind and solar energy exploitation projects [281]. This may be due to a number of reasons, but possibly is mostly related to the availability of good wind and solar energy resources, and availability of mature wind and solar technologies.

Venezuela, despite being one of the main oil producers and exporters in the world, started a programme of development and adoption of renewable energy in order to reduce its carbond dioxide emissions per capita, the highest in Latin America between 1970 and 2001 [282]. This programme focuses primarily on solar and wind energy. A number of coastal wind energy projects have been proposed since 2006, within the national wind energy development plans, particularly in the Venezuelan North West. In this region, bordering the Gulf of Venezuela, wind speed measurements have shown high wind energy potential. Insular regions, such as Nueva Esparta, also appear to be commercially viable for wind energy exploitation. These are also arid and semi-arid regions, with rural communities in need of socioeconomic development. Hence, commercially driven coastal wind energy projects would provide further income and development opportunities in these regions [283].

13.3.8   Synthesis and Recommendations

The development drivers for wave and tidal energy farms in most Latin American countries are:

- secure clean and economic energy supplies;
- the creation of new employment opportunities and wealth through a locally developed industry;
- diversification of the energy mix with efficient low-carbon technologies; and
- integration of renewable energy policies within wider regional development policies.

Energy markets include not only electricity production, but also fluid power (pumping capacity), water desalination and heat generation. Wave and tidal energy technologies are developing rapidly, but generally remain at a pre-commercial stage of development. Some technologies may become competitive with other forms of renewables by the mid-2020s. Technological advances are essential for the emergence of a viable marine energy industry throughout Latin America. The following recommendations, drawn up by [258] for the case of Chile, apply to all Latin American countries:

- Technology development, and cost reduction and adaptation for regional conditions. Regional differences will influence development drivers and energy usage.

- Site-specific resource assessment and environmental impact assessment.
- Development of systems and operations and maintenance strategies for isolated communities.
- Synergies with other industries, for example water pumping and desalination strategies, or aquaculture and coastal management and protection.

Infrastructure, including electricity grid and supply chain, could be an important barrier to the development of commercial projects. In [251] a 'global grid' for Latin America was proposed as a possible solution. In Chile, the Ministry of Energy plans to identify sites with existing infrastructure, or with the potential to develop capacity that could support the marine energy industry. Such a strategy is advisable for all of Latin America, otherwise development will be hampered. Also, technology transfer and the development of local industry will foster and promote the development of marine energy. Many site-specific factors will require adaptations to existing technology, new technology developments, and novel operation and maintenance procedures to be adopted [258]. Devices or foundations suitable for steeply sloping or deeper seabeds are another feature in Latin America, because the continental shelf is in most places narrower than in the North Sea or the UK's western approach. In both Chile and Mexico, seismic and tsunami risks are high. Another important barrier has in some cases been government approval for developments and permits for installation of technologies in demonstration sites, causing important delays to *in situ* testing. Lack of funding and government incentives has also been a major barrier to development.

It is believed that the technology is still at the pilot stage, in order to understand better the effects of waves on energy yield, to develop the interconnections and technology standards, as well as the supply chains and the type of certifications. The pilot projects will need capital expenditure and production support, such as feed-in tariffs or other government subsidies, before commercialisation can be achieved.

There are several initiatives in Mexico, Brazil, Argentina, Chile and Colombia, as well as some in other countries in Latin America, related to device and prototype development. There are a couple of initiatives for MRE centres to be developed in Mexico and Chile, with some other centres in Colombia and Brazil.

Finally, a number of the conclusions and recommendations made by [284] for renewable energy technology in Latin America apply, in particular, to MRE. For example, innovation needs in the region should be met not only by considering novel technological developments, but also by improved goods and services, or alternatives to produce these goods and services. Crucial for rapid development will be cooperative relationships, partnerships between the public and the private sectors, and technology transfer agreements.

13.4 Asia-Pacific

In this section, we discuss the activities in wave and tidal energy in the Asia-Pacific region. China has a significant programme in marine energy, which has supported a number of developments in wave and tidal energy, and this is dealt with separately in Section 13.5.

13.4.1 Wave

Wave energy development in the Asia-Pacific region is led by Australia and Japan, which have been very active in supporting MRE research, although Australia has been most active in wave energy development during the last decade. Governmental support for MRE research, channelled through the Australian Renewable Energy Agency, has been crucial in positioning some of the Australian developers at the forefront of the world's wave energy sector.

An overview of the different wave energy developments taking in place in the Asia-Pacific region (excluding China) is presented by country in the following.

13.4.1.1 Australia

With its eight wave energy developers, Australia has around 4% of the world's sector and leads wave energy development in the Asia-Pacific region. The present front runner of the Australian developers is Carnegie Wave Energy Ltd [285]. Carnegie has successfully deployed an array of three full-scale units of the CETO 5 device at Garden Island in Western Australia and is currently developing the CETO 6 device, which it plans to install at the Wave Hub test site in Cornwall (UK). BioPower Systems Pty Ltd has tested a demonstrator 250 kW part-scale unit of its bioWave, an oscillating surge type of device, at Port Fairy [286]. Oceanlinx was initially founded in 1997 as Energetech Australia Pty Ltd and since then it has developed and tested various iterations of its part-scale floating OWC type device in Port Kembla [287]. Bombora Wave Power is currently developing a full-scale unit of its mWave device which will be installed in Western Australia and off the coast of Portugal [288]. The other Australian wave energy developers are Perpetuwave Power Ltd [289], Protean Wave Energy Ltd [290], AquaGen Technologies Pty Ltd [291] and Global Renewable Solutions Ltd [292].

13.4.1.2 Japan

Even though Japan was among the pioneers of wave energy development, successfully testing full-scale technologies such as the Mighty Whale floating OWC in the late 1990s [293], it has now lost its leading position. However, two wave energy developers should be mentioned: the Murotan Institute of Technology, with its Pendulor device [294]; and Gyrodynamics Co Ltd which has recently developed a gyroscopic WEC [295].

13.4.1.3 South Korea

Two Korean government agencies – the Ministry of Oceans and Fisheries (MOF) and Ministry of Trade, Industry and Energy (MOTIE) – have established a strategic plan clean ocean energy development for the medium and long term. The MOF and MOTIE are providing R&D and installation funds, including provision for ocean renewable energy. There are six active projects for developing WEC operated by research institutes and industries as shown in Table 13.3. Although there is no active test site in Korea, a wave energy test bed of 5 MW capacity with five berths is under construction. Recently the Korea Research Institute of Ships & Ocean Engineering (KRISO) has been trying to develop an ocean wind–wave energy combined platform to maximise power out (Figure 13.11).

Table 13.3 Research projects for wave energy converters, Korea [296].

Project (charged by, funded by)	Type of converter	Power capacity	Project period
Yongsoo OWC with dual system (KRISO, MOF)	OWC with impulse turbines (gravity caisson)	500 kW	2003–2016 (started test in 2015)
Pendulum WEC utilising standing waves (KRISO, MOF)	Oscillating surge (floating twin hull)	300 kW	2010–2018 (sea test of pilot plant in 2017)
Swinging semi-sphere with hinged arm (Hwa Jin Co., MOTIE)	Floating point absorber (jack-up platform)	Expandable 15 kW units	2013–2016 (sea test in 2016)
Controllable resonant WEC with yoyo oscillator (iKR, MOTIE)	Point absorber with variable spring stiffness (moored array of cylinders)	10 kW	2013–2016 (sea test in 2016)
WEC for navigational lighting buoy (KPM, MOTIE)	Point absorber with solenoid (single point moored buoy)	50 W	2013–2016 (sea test in 2016)
INWave WEC with multi-degree of motion converting pulleys (Ingine Inc., MOTIE)	Point absorber with pulleys (floating disk with touted mooring lines)	135 kW	2014–2017 (sea test in 2016)

KPM: Korea Plant Management Company.

13.4.1.4 Russia

There are three wave energy developers in Russia active at the time of writing: OceanRus Energy, with its 3 kW spar buoy under development [298]; Applied Technologies Company [299]; and Vortex Oscillation Technology Ltd [300].

13.4.1.5 Other Developments

Elsewhere in the Asia-Pacific region, the New Zealand company, Power Projects Ltd [301], has been developing and testing, under the Wave Energy Technology – New Zealand (WET-NZ) R&D programme, different versions of a part-scale WEC in open waters. Other wave energy developers in the region are Nualgi Nanobiotech [302] from India and Hann-Ocean Technology Pte Ltd [303] from Singapore.

(a)

(b)   (c)

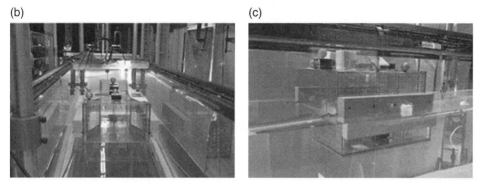

Figure 13.11 Floating pendulum wave energy converter (KRISO, MOF) [297]. (a) artist impression of the device (b) front view of a scale physical model of the device (c) side view of a scale physical model of the device.

13.4.2 Tidal

R&D activity in tidal energy in the Asia-Pacific region has been concentrated in three countries: Australia, South Korea and Japan; the region contains around 30% of the estimated global tidal energy exploitable resource [304]. Governmental support via research funding or allocation of renewable energy production, such as the Korean RPS or the Australia Renewable Energy Agency, provides the main support for the sector in this region.

The different tidal energy developments are presented by country in the following.

13.4.2.1 Australia

Two tidal turbine developers have made significant progress in Australia: Mako Turbines Ltd, with its Mako and SeaUrchin (owned by the company subsidiary Elemental Energy Technologies) pre-commercial demonstrators [305]; and Tidal Energy Pty Ltd with its Davidson-Hill Ventury enclosed vertical-axis turbine [306].BioPower System Pty Ltd [307] and Cetus Energy [308] are also developing tidal energy turbines in Australia.

13.4.2.2 Japan

Since the 2011 earthquake and the nuclear accident at Fukushima, Japan has increased its support for MRE, especially for tidal and ocean current developments. Kawasaki Heavy Industries Ltd has proposed a prototype horizontal-axis tidal turbine [309], and Modec Inc has also proposed the SKWID, a hybrid wind and current energy converter [310].

Table 13.4 Research projects for tidal energy, Korea [296].

Project (charged by, funded by)	Type of converter	Power capacity	Project period
MW Class Tidal Current Device (HHI, MOTIE)	Pitch control (pile type structure)	500 kW	2010–2015 (sea test in 2014)
Active Control Tidal Current System (KIOST, MOF)	HAT with pitch control (Caisson)	200 kW	2011–2018 (sea test in 2016)
Semi-active Flow Control Turbine (Inha Univ., MOTIE)	HAT with flow control (moored submersible)	10 kW	2013–2016
Active Impeller Tidal Turbine System (Daum Eng., MOTIE)	Vertical impeller with flow control (pre-cast concrete)	50 kW	2013–2016 (sea test in 2016)

13.4.2.3 South Korea

South Korea has recently become interested in tidal energy, as a way to exploit its own renewable energy resources. The Shiwa Lake tidal barrage power station, commissioned in 2011 as the world's largest tidal barrage plant, with a total power output capacity of 254 MW [311], is proof of this and its development is described in Chapter 12. Furthermore, Hyundai Heavy Industries (HHI) is developing a 500 kW pre-commercial tidal turbine which has been tested at the Uldolmok Passage in Jeollanam-do [312]. Table 13.4 shows the active R&D projects for tidal energy in Korea.

13.4.2.4 Other Developments

There are two developments in tidal energy in other Asis-Pacific countries that are worth mentioning: the Indian company Kirloskar Integrated Technologies Ltd, with its small-scale enclosed turbine, is proposing turbines ranging from 2 kW to 1 MW [313]; and Balkee Tide Ltd of Mauritius is developing the TWPEG hybrid wave and tidal energy converter [314].

13.5 China

13.5.1 Marine Energy Research and Development Programmes

13.5.1.1 Special Fund for Marine Energy

In May 2010, the Ministry of Finance of China allocated part of the special fund for renewable energy to MRE (hereinafter referred to as the Special Fund for Marine Energy), aiming to facilitate the development of MRE in China. The major projects supported in this programme are:

- The demonstration project of an independent power system aiming to increase the power supply on remote islands and provide access to electricity, including wave energy, tidal current energy, and various renewable and complementary energy resources.

- The demonstration project of a large interconnected power system of marine energy in the regions rich in marine energy, including tidal, tidal current and wave energy power stations.
- The demonstration project of the industrialisation of key technologies through the use of marine energy, including local production of key technologies such as key equipment of power generation systems, whole machines and supporting parts, maritime installation and construction, and (maritime) storage and transportation of electricity.
- The research and test projects of the technology for comprehensive development of marine energy, including research and test projects of new technologies, devices and methods such as ocean thermal (temperature-difference) energy and marine bioenergy, as well as of the technology or the comprehensive utilisation of marine energy.

Part of this work has been the establishment of standards for marine energy exploitation and supporting service systems, including the exploration and evaluation of marine energy, power products, technical standards, and criterion formulation of a connected grid as well as the establishment of comprehensive laboratory and marine tests and the inspection and certification system. Since the establishment of the Special Fund for Marine Energy, four batches of projects have been set up and implemented. There have been 93 projects with 800 million yuan of funding, as shown in Figures 13.12 and 13.13.

Thanks to the support of the Special Fund for Marine Energy, there is a general understanding of the marine energy in China's offshore waters and initial breakthroughs have been made in some key technologies, such as wave and tidal current energy. Additionally, increasing attention has been paid to the exploitation of marine energy, and China's marine energy is now ushering in a new age of exciting breakthroughs.

13.5.1.2 National High-tech Research and Development Program of China (863 Program)

Marine energy technology has gained much support as one topic within the technology of renewable energy, and during the 11th Five-Year Plan, nearly 10 million yuan was allocated for the study of new technologies and devices for electricity generation from

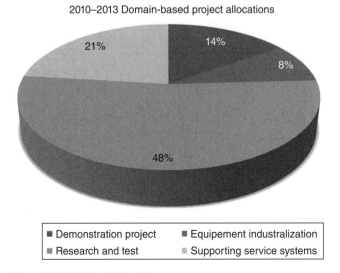

2010–2013 Domain-based project allocations

Figure 13.12 Project allocation.

2010–2013 Domain-based fund allocations

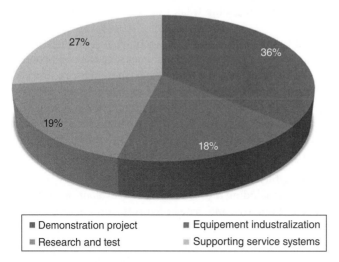

Figure 13.13 Funding allocation.

tides, tidal currents, waves and temperature difference. From the 12th Five-year Plan, the theme of renewable resource technology has emphasised support for the study of biomass-based liquid fuel preparation and energy generation, island-based energy generation with multi-energy combined systems, gigawatt super-experimental energy stations, and wave and tidal current energy generation technologies. In addition to this, the plan emphasises research such as integrated energy supply system research and demonstration of marine energy, as well as new energy algae cultivation and energy product transformation technology.

13.5.1.3 National Sci-tech Support Plan

In May 2008, the 11th Five-Year National Sci-tech Support Plan established a key project titled 'Research and Demonstration of Key Exploitation Technology of Marine Energy' (with 32 million yuan of national funds). To address the energy shortage on the remote islands of China, the plan was devoted to the development of a 100 kW floating wave energy station, a 100 kW pendulum wave energy station, a 20 kW ocean current energy device, a 150 kW tidal current energy station, research and experimentation on a 15 kW ocean thermal energy converter, and study of the comprehensive test technology of a marine energy generation system.

13.5.1.4 National Natural Science Foundation of China

In the early 1980s, in order to facilitate sci-tech system reform and change, 89 members of the Chinese Academy of Sciences (committee members of the Department of Science) wrote to the Party Central Committee and the State Council to propose the establishment of a country-oriented natural science fund, which was approved. Under the guidance of Deng Xiaoping, the State Council approved the establishment of the National Science Foundation of China on 14 February 1986. In recent years, the National Science Foundation of China has subsidised the study of scientific problems in the field of emerging marine energy. Since 2010, the Foundation has successively supported different projects and various topics of key projects such as applied fundamental research on the utilisation of machinery equipment by the kinetic energy of marine fluid, control and grid-connected operation of a direct-driven wave energy generation system, and study of water turbine hydrodynamic characteristics of tidal current energy of a flexible blade. This has contributed largely to the development of the fundamental study of marine energy and improved the ability to research and innovate in this field.

13.5.1.5 China Renewable Energy Scale-up Programme

As a policy development and investment project concerning renewable energy jointly carried out by Chinese government, the World Bank and the Global Environmental Facility, the China Renewable Energy Scale-up Programme is aimed at making a survey of the renewable energy in China, and drawing inspiration from developed countries to formulate renewable energy development policies for China. Based on pilot work, the development of renewable energy is gradually being achieved in China, which will provide efficient commercial renewable energy for the energy market.

13.5.1.6 The 12th Five-year Plan of Renewable Energy Development

On 1 January 2013, the State Council issued the 12th Five-Year Plan of Renewable Energy Development. According to the overall judgement on the trend of economic and social development and the general requirements during the 12th Five-Year Plan period, full consideration is given to factors such as safety, resource availability, environmental impact, and technological and economical concerns to make clear the major goals of energy development in 2015.

To accelerate the technical improvements in marine energy, exploitation has been listed as one of the nine major tasks in the 12th Five-Year Plan which aims to improve the technical level of marine energy exploitation, to construct marine energy demonstration projects, and facilitate technical improvements in marine energy exploitation as well as improvement of the marine energy equipment industry. As required by the 12th Five-Year Plan, it is necessary to establish independent demonstration stations of multi-marine-energy implementation in remote islands, such as a gigawatt-scale tidal energy station, and several grid-connected tidal current energy stations. Additionally, different types of 5 GW ocean energy stations are planned in order to lay the foundations for development on a larger scale.

13.5.1.7 Outline for the Development of Renewable Marine Energy

Marine renewable energy refers to tidal energy, tidal current energy, wave energy, ocean thermal energy (temperature-difference energy), and salinity gradient energy in the ocean, all of which are characterised by sustainable utilisation and clean green energy generation. China is abundant in marine energy, with promising exploitation prospects. To deploy the strategy proposed by the central government to guide and facilitate marine energy development and the implementation of the Special Fund for Marine Energy in China, the Outline for the Development of Renewable Marine energy (2013–2016) has been formulated in compliance with the 12th Five-Year Plan of National Marine Cause Development as well as the 12th Five-Year Plan of Renewable Energy Development.

13.5.2 Development of Wave Energy Technology

China's wave energy exploiting technology, which can be traced back to the 1970s, has experienced rapid development since the 1980s. For instance, the microwave energy converters developed by the Guangzhou Institute of Energy Conversion for buoy lights have become commercial products. In recent years, with the support of relevant national plans and special funding, wave energy technology has undergone rapid development. In particular, the Special Fund for Marine Energy established in 2010 has been supporting the study of wave energy technologies such as floating duck-shape WECs, buoy-array WECs, and raft-type WECs. Laboratory model tests of these technologies have been completed and some engineering prototypes have been constructed and used in sea trials. The technical process of independent innovation has been implemented to remove the technical difficulties that negatively impact the reliability, practicality, and efficiency of wave energy conversion.

13.5.2.1 Oscillating Buoy Array Wave Energy Converter

The Ocean University of China team, with Professor Shi Hongda as chief scientist, has tackled issues concerning the fundamental science and key technologies of 10 kW combined oscillating buoy wave energy conversion with the support of the Special Fund for Marine Energy. In addition, industry–university research collaboration has been conducted to finish the production as well as the marine trial operation of device engineering prototypes. In the future, with the support of the National 863 Program, the converter will be upgraded to a 2×100 kW device engineering prototype for marine trial operation.

The key technical aspects of the WEC shown during deployment in Figure 13.14 are: gyrostat buoy and dual track hydraulic system for energy transferring and converting; submerged floating body cooperating with tension rope for on-site installation and

Figure 13.14 Marine trial operation of engineering prototype.

anchoring; two heaving body system overcoming tidal difference effects for continuous autonomous operating; and automatic online monitoring and operating system for remote controlling and data acquisition.

13.5.2.2 Duck-shape Wave Energy Converter

The Guangzhou Institute of Energy Conversion is developing duck-shape WECs with the support of the 863 Program, National Sci-Tech Support Plan, and Special Fund for Marine Energy. Since the first duck-shape WEC was put into operation in 2009, several devices, namely Duck-Shape Wave Energy Converters I, II, and III have gone through sea trials.

The duck-shape WEC is made up of the body, an external floating body, a mooring system, a hydraulic conversion system, and an energy generation and distribution system. The relative motion between the duck-shape body and floating body can capture wave energy. The floating body, with its horizontal plate and vertical plate, is capable of mobilising the seawater around to form added mass to inhibit the motion of the floating body while moving.

In 2009, the first 10 kW duck-shape converter underwent sea trials in the Wanshan Islands, Guangdong. As revealed by the experiment in real-sea conditions, the device has high energy capturing efficiency but still faces such problems as unstable flotation, a fragile submerged cable, and waterlogged energy generation devices. In 2010, the second 10 kW duck-shape converter was launched. As shown by experiments in real-sea conditions, the converter has made improvements in floating stability but still encountered such problems as damage to the mooring system, a fragile submerged cable, and the self-locking of the energy conversion system. In 2012, the third 10 kW duck-shape converter was launched. Due to its improvements in shape, floating body, and energy conversion system, the third duck-shape converter is superior to Duck-Shape Wave Energy Converter I in terms of wave energy capture and conversion. In 2013, a 100 kW floating duck-shape converter underwent sea trials, during which the device generated energy with interruptions in the presence of small waves but continuously in the presence of large waves, with the maximum output energy reaching 25 kW. Figure 13.15 shows a 100 kW duck-shape converter.

Despite its simple structure, as well as high efficiency of energy capture and conversion, the duck-shape WEC faces constraints that hinder its development, including wave absorption at the end of the rounded duck-shape body, huge rotational inertia and weight of the duck-shape body, frequent breakdowns, and maintenance difficulties. The duck-shape wave energy technology has accumulated a large amount of experience in conversion efficiency, system operation monitoring technology, device typhoon-resistant technology, system stability, and attitude, which has laid a favourable foundation for the development of the subsequent eagle WEC.

13.5.2.3 Eagle Wave Energy Converter

Drawing inspiration from the R&D experience of the duck-shape converter, the Guangzhou Institute of Energy Conversion optimised the eagle WEC with the characteristics of a semi-submerged ship through a large number of physical model experiments and theoretical analysis with the support of the Special Fund for Marine Energy.

The eagle device is composed of a light floating body that absorbs wave energy, a hydraulic cylinder, an energy accumulator, a hydraulic motor, and a generator. The light,

Figure 13.15 The duck-shape wave energy converter.

specially designed wave-absorbing body looks much like a beak, able to make energy from fewer waves by absorbing the incident waves since its motion track coincides with the motion of a wave. The device uses a semi-submerged ship as a marine towing carrier and stable floating body during operation, which has to a large extent decreased the operation cost of the device.

In December 2012, the 10 kW eagle WEC underwent sea trials in the Wanshan Islands. After several recycling, maintenance, and redistribution operations, the failure-free operation lasted for 5990 hours. The results show that the eagle WEC runs stably with great reliability and high energy conversion.

With the support of the Special Fund for Marine Energy in 2013, Guangzhou Institute of Energy Conversion, based on the 10 kW eagle device, built the 100 kW Wanshan eagle WEC, which increased the number of wave-absorbing buoys from one to four, as shown in Figure 13.16.

Two-stage open-sea tests took place in the waters near the Wanshan Islands from November 2015 to June 2016. During the open sea tests, the energy conversion efficiency was measured, and results show that the capture width ratio of the Wanshan remains higher than 20% in the wave period between 4 and 6.5 seconds. It was found that the Wanshan achieves intermittent power generation under mild waves only 0.5 metres high and survives in place when the wave height is up to 4 metres. The device operates under a wide range of wave conditions and demonstrates good adaptability.

13.5.2.4 Buoyant Pendulum Wave Energy Converter

With the support of the National Sci-tech Support Plan, the National Ocean Technology Centre developed a 100 kW buoyant pendulum WEC and carried out a device test in actual sea conditions, earning 11 authorised patents. The buoyant pendulum WEC relies on modular design, including two independent energy generation systems and electronic control systems. In each independent energy generation system, there is a

Figure 13.16 The 100 kW eagle wave energy converter under working conditions.

pendulum plate, a hydraulic gear and an electric converter. Under the action of waves, the pendular plate will transform wave energy into pendular mechanical energy. The output end of the pendular plate is connected to a hydraulic pump which can transform mechanical energy into hydraulic energy. The converter generates energy when being driven by a hydraulic motor and then outputs electric energy to the electric control system.

Against the backdrop of low wave energy flux density in China, with the support of the Special Fund for Marine Energy in 2013, Qingdao Haina Heavy Industry Co., Ltd and National Ocean Technology Centre began a design programme for a 50 kW pendulum WEC and optimised the key technological parameters of the pendulum and the hydraulic gear system to improve the wave energy transformation efficiency of the device and improve its service life and reliability.

13.5.2.5 Floating Hydraulic Wave Energy Generation Station
Developed by Shandong University
With the support of the Special Fund for Marine Energy in 2010, Shandong University developed a 120 kW oscillating buoy WEC and trialled it at sea, securing one invention patent and two utility model patents. This project was accepted by the State Oceanic Administration in 2014. This device, shown in Figure 13.17, is made up of six modules, including a hard tank and a heave plate module, a main body pillar module, a submersible buoyancy tank, a converter room support module, a converter room module, and a mooring arrangement module. The equilibrium position is 30.77 metres in height, the gross weight 93 tons.

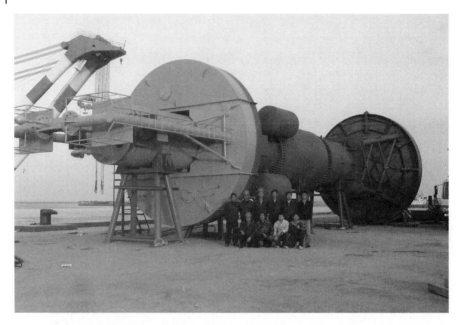

Figure 13.17 120 kW oscillating buoy wave energy converter.

In November 2012, a further sea trial was conducted in Shantou waters of Guangdong, and now the wind and wave resistance and erosion prevention of the device are being improved.

13.5.2.6 Nezha II Marine Instrument Wave Energy Base Station

With the support of the Special Fund for Marine Energy in 2010, the Guangzhou Institute of Energy Conversion has developed a 20 kW marine instrument wave energy base station (Figure 13.18), and obtained an invention patent. This project passed the inspection of the State Oceanic Administration in 2014.

Nezha II is jointly developed by the Guangzhou Institute of Energy Conversion and No. 712 Research Institute of China Shipbuilding Industry in 2009, supported by the 863 Program in the 11th Five Year Plan. The installed capacity of the developed sample is 10 kW (No. 1 Nezha). It has been installed in the Big Wanshan sea of Zhuhai city, and maximum recorded power generation is 7.17 kW. In February 2013, with the support of the Special Fund for Marine Energy, another Nezha II with an installed capacity of 20 kW was successfully installed to generate power in the same sea area and the maximum recorded power generation is 11.5 kW.

Nezha II is a point absorber WEC and mainly composed of three parts, including an underwater damping plate with central tube, a water surface dish float and a linear generator. The top part of the central tube is above water, the linear generator is installed within the tube and the dish float is set on the central tube. Under the action of waves, the dish float produces vertical reciprocating movements on the central axis and causes the linear generator to output electric energy, which supplies power to a battery after rectification or uses a DC bus to supply power to the shore.

Figure 13.18 Nezha II marine instrument wave energy base station.

13.5.2.7 Efficient and Stable Wave Energy Converter Developed by Zhejiang Ocean University

With the support of the Special Fund for Marine Energy in 2011, Zhejiang Ocean University accomplished the project of a self-protecting WEC in Poor sea conditions, earning two invention patents and 12 pragmatic model patents. This device is made up of a buoy, an energy storage base, a hydraulic cylinder, and an accumulator (Figure 13.19). At the end of 2012, a 3-month sea trial was carried out. Under the unfavourable winter wave conditions in the test waters, only 503 kWh of energy were generated. In the course of generating energy, the wave drives the heaving oscillation of a wave energy collection system along the pile of the self-elevating lift platform. The hydraulic system begins to work and produces electric energy after transmitting energy to an energy converter. Meanwhile, it outputs electric energy through the energy output system.

13.5.3 Development of Tidal Current Energy Technology

The tidal current energy exploitation technology in China, which can be traced back to the 1980s, has experienced rapid development since the 1990s. The 70 kW Wanxiang I experimental tidal current energy station, developed by Harbin Engineering University and the Kobold Energy Station of Italy, was built at the same period: the 863 Program in the 10th Five-Year Plan supported not only the Harbin Engineering University in researching and developing a 40 kW vertical tidal current energy converter, but also Northeast Normal University in researching and developing a 1 kW horizontal-axis tidal current converter. In recent years, with the support of the relevant national programs and special funds, tidal current energy technology has experienced rapid

Figure 13.19 Zhejiang Ocean University wave energy converter.

development. In particular, the Special Fund for Marine Energy has facilitated the booming development of tidal current energy technology. The participation of domestically renowned enterprises and institutes of higher learning, such as the Harbin Electric Corporation, the China Energy Conservation and Environmental Protection Co., Ltd, the Ocean University of China, the Harbin Institute of Technology, and the Dalian University of Technology, has injected vigour into tidal current energy converter R&D.

13.5.3.1 Haineng I Vertical-axis Tidal Current Converter

With the support of the National 863 Program and the National Sci-tech Support Program, Harbin Engineering University, together with such institutes as Shandong Electric Energy Engineering Consulting Institute Corp., Ltd, has developed the Haineng I 2×150 kW tidal current energy station, which passed inspection in April 2013.

The Haineng I floating tidal current energy converter with dual vertical-axis impellers and a direct drive motor is made up of five parts, including a floating carrier, an anchor system, a water turbine energy generation system, an electric energy converter and controller, and an energy transmission and load system. The diameter and blade numbers of the impellers are 4 m and 2×4 respectively, and the initial and maximum flow velocities are 0.8 m/s and 2.5 m/s respectively.

In July 2012, Haineng I underwent sea trial under the influence of the strong typhoon Haikui in Guishan Waterway, Daishan County, and Zhejiang. In November 2013, at the same location, Haineng I underwent a 3-month sea trial (see Figure 13.20), but due to mechanical failures it stopped energy generation.

Figure 13.20 Haineng I vertical-axis tidal current energy converter sea trial.

13.5.3.2 Haiming I Horizontal-axis Tidal Current Energy Converter

Haiming I, the small experimental tidal current energy station, is the first long-term running, submersible and independent horizontal-axis tidal current energy generation system developed independently by China. Developed by Harbin Engineering University, the system is either equipped with fairing or free from fairing. At the current velocity of 2 or 2.3 m/s, the fairing or fairing-free device achieves energy generation of 10 kW.

In September 2011, Haiming I underwent sea trials in Xiaomentou Waterway, Daishan County, Zhejiang (see Figure 13.21). At present, it is still being tested in this area. The energy generated is transmitted to Xianzhou Bridge Lighthouse via submarine cable for the purpose of providing heat and light for the lighthouse.

Haiming I, which utilises a submersible enhanced diversion energy station as its main body, consists of a horizontal paddle impeller and a permanent magnet (PM) direct-driven converter. The installed capacity is 10 kW and the rated flow velocity 2 m/s. Due to the three-phase energy supply and the voltage of 400 V/50 Hz, the system can deal with over-current, over-voltage, and under-voltage, is hurricane-resistant, and is adapted to the tidal range of 4 m. In addition, the system can run at variable speeds automatically and realise two-way auto-convection. It is estimated that the demonstration will last 7 years. According to observations of actual sea conditions, energy is devoted to further testing the technical performance and operational reliability of the key devices such as the submersible structure, water turbine energy converter set, and electric energy conversion and control system, which has accumulated experience for system operation, technical development and engineering application in the future.

Figure 13.21 Haineng I horizontal-axis tidal current energy converter sea trial.

13.5.3.3 Horizontal-axis Pitch-varying Tidal Current Converter Developed by Northeast Normal University

With the support of National Sci-tech Support Plan, Northeast Normal University developed a 20 kW horizontal pitch-varying tidal current converter, earning an invention patent and a utility model patent. This project passed inspection in April 2013.

The turbine of this device consists of a large solid blade of symmetrical airfoil (NACA0012), the blade root is equipped with a pitch-varying hydrodynamic compensation winglet, and the joint between the blade and the hub with the location control system enables the blade to adjust its angle within ±14° for the purpose of catering to the incoming flow. When the incoming flow changes direction, the hydrodynamic complementation winglet will accommodate the change, and the location controller of the blade root will control the angle within 14°. At this point, the blade will stay at the optimal angle of attack, achieving the highest conversion efficiency. In the presence of the opposite incoming flow, the rotational direction of the blade will remain unchanged, which can avoid direction adjustment of the whole converter set. As a result, there is no need for an energy transmission slip ring and a direction adjustment control device, and there is no intertwining of the energy transmission cables.

The blade diameter of this device is 5 m, with an initial flow velocity of 0.7 m/s and rated flow velocity 1.5 m/s. Recently, this device has undergone sea trials in Zhaitang Island, Qingdao (see Figure 13.22). In 2013, with the support of the Special Fund for Marine Energy and based on the technologies of this device, a finalised prototype of a 300 kW horizontal-axis tidal current energy converter set was built.

13.5.3.4 Qingdao Histro Steel Tower Tidal Current Energy Converter

With the support of the Special Fund for Marine Energy in 2010, the research and development of a 10 kW new PM direct-driven tidal current energy converter was jointly carried out by Qingdao Histro Steel Tower Co., Ltd and the Harbin Institute of Technology at Weihai. The converter set has undergone a 3-month sea trial.

This project draws inspirations from previous European experience and assimilates the latest technology to realise local production of the components and a complete

Figure 13.22 Horizontal-axis pitch-varying tidal current converter.

machine for tidal current energy conversion. The efficiency of energy conversion of this tidal current energy converter is greater than 30%. Due to the use of the PM direct-driven technology, the device is characterised by low energy consumption, with the energy generation efficiency 5–10% higher than that of the common speed-varying gearbox energy converter.

The water turbine of the converter set is more than 7 metres in diameter and the wing is made of nanometre carbon fibre material. In the R&D process, the technical difficulties such as sealing-up and seawater corrosion resistance have been removed. As a result, the device can run at sea for more than a year without breaking down. This device is placed about 16–40 metres away from the offshore sea by ship. As long as the flow velocity reaches 0.6–1.3 m/s, the device can generate energy.

13.5.3.5 Vertical-axis Direct-driven Tidal Current Energy Converter
Developed by Dalian University of Technology

With the support of the Special Fund for Marine Energy, a study of a new vertical-axis direct-driven tidal current energy converter was carried out by the Dalian University of Technology. In mid-September 2013, the sea trial of the 15 kW energy converter was implemented in the waterway between Da Changshan Island and Xiao Changshan Island. This experimental energy station is the fithe technical foundations for the commercial development of tidal current energy generation in seawater.

The project team studied the water turbine (the key part of the tidal current energy converter), came up with the coaxial bi-rotor scheme and gave priority to the parameters of the blade and supported arm, such as wing profile and chord length. This has improved the self-starting properties of straight-blade and vertical-axis water turbines. With the plug baffle used to inhibit the trailing vortex of the structural supporting pile,

Figure 13.23 Dalian University of Technology 15 kW tidal turbine.

the flow field around the water turbines was improved. The project team has designed and produced a 15 kW low-speed PM converter and inhibited the harmonic impact through a fractional slot and concentrated winding to reduce winding resistance and heighten efficiency. The sealed magnetic pole structure has enhanced the anti-corrosion ability of the converter at sea. Furthermore, a study of the method to regulate converter energy was conducted to make the converter adaptable to tidal current variations, and this has achieved the highest energy generation efficiency and longest energy generation duration. Additionally, research on the anti-corrosion and sealing technologies of the converter and its parts such as the supporting frame was done to acclimatise the converter to the complicated and hostile marine environment (Figure 13.23). This project passed inspection on 20 November 2014.

13.5.3.6 Semi-direct-Driven Horizontal-axis Tidal Current Energy Converter Developed by Zhejiang University

With the support of the Special Fund for Marine Energy in 2010, Zhejiang University developed an engineering prototype of a 60 kW semi-direct-driven horizontal-axis tidal current energy converter. Tasks such as impeller production, processing and assembly of energy transmission system, and no-load and loaded test have been accomplished, and the sea trial is set to be carried out.

With the support of the Special Fund for Marine Energy in 2013, a project to design and manufacture a $2 \times 300\,kW$ tidal current energy converter was jointly initiated by Guodian United Energy Technology Co., Ltd and Zhejiang University based on the technology of the horizontal-axis tidal current energy converter.

13.5.3.7 Coupling Energy Generation System of Developed by Zhejiang University

With the support of the Special Fund for Marine Energy in 2010, the Ningbo Institute of Technology of Zhejiang University started the development and establishment of an independent island energy generation and desalination system based on the coupling between tidal current energy and wave energy. The generation system adopts a floating catamaran platform, with the horizontal-axis tidal current energy-capturing device in the middle and a wave buoy on both sides. The device was manufactured in early May 2013 and was used for tidal current energy generation and desalination in the Port of Shipu in May and September (see Figure 13.24).

13.5.3.8 The Multi-energy Complementary Energy Station at Zhaitang Island

With the support of the Special Fund for Marine Energy in 2010, China National Offshore Oil Corporation led the development and establishment of a 500 kW independent ocean energy generation demonstration project. The energy station consists of a 200 kW floating horizontal-axis turbine, a 150 kW fan, and a 50 kW solar energy generation system. The debugging and installation of a solar and wind energy generation system has been completed, and tidal current energy is undergoing sea trials in the Zhaitang Island, Qingdao, Shandong.

Figure 13.24 Coupled tidal current energy and wave energy system.

Figure 13.25 Haiyuan horizontal-axis tidal current energy generation device.

Figure 13.26 Sea trial of Haineng II.

Haineng II, a 2×100 kW horizontal-axis tidal current energy generation device designed by Harbin Engineering University has a turbine of 12 m in diameter and achieves an initial flow velocity of 0.6 m/s, a rated flow velocity of 1.7 m/s, and a maximum flow velocity of 2 m/s. In November 2013, the device underwent a sea trial at

Zhaitang Island, and was repaired afterwards because of a broken paddle (see Figure 13.25). The Haiyuan horizontal-axis tidal current energy generation device designed by the Ocean University of China has a turbine 10.5 m in diameter and achieves an initial flow velocity of 0.9 m/s, a rated flow velocity of 1.5 m/s, and a maximum flow velocity of 2.5 m/s. In August 2013, the device was installed for sea trials at Zhaitang Island (see Figure 13.26).

References

1 European Commission: A European strategic energy technology plan (SET-Plan): 'Towards a low carbon future' (European Commission, 2007)

2 Mofor, L., Goldsmith, G., and Jones, F.: Ocean energy: Technology readiness, patents, deployment status and outlook (International Renewable Energy Agency, 2014)

3 http://www.offshorewind.biz/2014/11/25/meygen-enters-construction-phase/?utm_source=emark&utm_medium=email&utm_campaign=Daily%20update%20Offshore%20Wind,%202014-11-26&uid = 3513, accessed 01/10/2015

4 http://renews.biz/79763/ceto-5-comes-alive-for-carnegie/, accessed 01/10/2015

5 http://media.fortum.se/2014/07/01/fortum-och-seabased-installationen-av-varldens-storsta-vagkraftspark-paborjades-idag-utanfor-sotenas/, accessed 01/10/2015

6 Hogan, O., Sheehy, C., Ismail, O., Raza, M., and McWilliams, D.: The Economic Case for a Tidal Lagoon Industry in the UK (Centre for Economics and Business Research (CEBR), 2014)

7 Magagna, D., and Uihlein, A.: Ocean energy development in Europe: Current status and future perspectives, *International Journal of Marine Energy*, 2015, 11, 84–104

8 Silva, M., Raventos, A., Uihlein, A, and Magagna, D: JRC Ocean Energy Database: Technology characterization at sub-assembly level, Proceedings of the European Wave and Tidal Energy Conference, 2015.

9 Leete, S., Xu, J., and Wheeler, D.: Investment barriers and incentives for marine renewable energy in the UK: An analysis of investor preferences, *Energy Policy*, 2013, 60, 866–875

10 Weller, S., Thies, P.R., Godelier, T., Harnois, V., Parish, D., and Johanning, L.: Navigating the valley of death: Reducing reliability uncertainties for marine renewable energy (ASRANet, 2014), pp. 10

11 European Commission: Blue Energy Action needed to deliver on the potential of ocean energy in European seas and oceans by 2020 and beyond (European Commission, 2014)

12 https://webgate.ec.europa.eu/maritimeforum/en/frontpage/1036, accessed 31/05/2016

13 MacGillivray, A., Jeffrey, H., Hanmer, C., Magagna, D., Raventos, A., and Badcock-Broe, A.: Ocean energy technology: Gaps and barriers (SI Ocean, 2013)

14 http://ec.europa.eu/clima/policies/lowcarbon/ner300/index_en.htm, accessed 31/05/2016

15 https://islayenergytrust.org.uk/tidal-energy-project/, accessed 31/05/2016

16 http://www.bbc.co.uk/news/uk-scotland-highlands-islands-35874248, accessed 31/05/2016

17 €23 million EU funding for ESB wave power project. https://www.ouroceanwealth.ie/news/%E2%82%AC23-million-eu-funding-esb-wave-power-project, accessed 31/05/2016

18 http://www.marine-renewables-directory.com/en/ocean-energy-europe/en/europe-s-ner300-programme-renewal, accessed 31/05/2016

19 Magagna, D., and Uihlein, A.: 2014 JRC Ocean Energy Status Report (European Commision: Joint Research Centre – Institute for Energy and Transport, 2015)

20 http://www.energinet.dk/EN/Sider/default.aspx, accessed 31/05/2016

21 http://www.ens.dk/en/policy/energy-technology/danish-funding-programmes, accessed 31/05/2016

22 http://www.ademe.fr/en

23 http://www.seai.ie/Renewables/Ocean-Energy/Prototype-Development-Fund/, accessed 31/05/2016

24 http://www.seai.ie/Renewables/Energy_Research_Portal/National-Energy-Research/, accessed 31/05/2016

25 http://fai.pt/, accessed 31/05/2016

26 http://www.scottish-enterprise.com/services/attract-investment/renewable-energy-investment-fund/overview, accessed 31/05/2016

27 http://www.wavehub.co.uk/latest-news/marine-energy-array-demonstrator-mead-capital-grant-scheme, accessed 31/05/2016

28 http://www.eti.co.uk/, accessed 31/05/2016

29 http://www.thecrownestate.co.uk/energy-minerals-and-infrastructure/wave-and-tidal/, accessed 31/05/2016

30 https://www.carbontrust.com/client-services/technology/innovation/marine-renewables-commercialisation-fund/, accessed 31/05/2016

31 https://www.carbontrust.com/client-services/technology/innovation/marine-renewables-proving-fund, accessed 31/05/2016

32 http://www.saltireprize.com, accessed 08/10/2015

33 http://www.fp7-marinet.eu/, accessed 02/10/2015

34 Osta Mora-Figueroa, V., Huertas Olivares, C., Holmes, B., O'Hagan, A.M., Torre Encirso, Y., and Leeney, R.: SOWFIA Deliverable D2.1: Catalogue of Wave Energy Test Centres (2011), pp. 87

35 Mouslim, H., Babarit, A., Clément, A., and Borgarino, B.: Develoment of the French Energy Test Site SEM-REV (2009), Proceedings of the 8th European Wave and Tidal Energy Conference, pp. 31–35

36 Coiro, D.P., De Marco, A., Nicolosi, F., Melone, S., and Montella, F.: Dynamic behaviour of the patented kobold tidal current turbine: numerical and experimental aspects, *Acta Polytechnica*, 2005, 45(3), 77–84

37 http://www.seai.ie/Renewables/Ocean_Energy/, accessed 08/10/2015

38 Sølvskudt, I.Ø., and Sønning, B.: Norwegian Marine Energy Industry: To what extent can Norway develop a marine energy industry with a limited home market? Norwegian University of Science and Technology, 2012

39 http://www.rundecentre.no/en/, accessed 08/10/2015

40 http://www.pico-owc.net/, accessed 03/01/2014

41 http://bimep.com/, accessed 08/10/2015

42 http://www.plocan.eu/index.php/en/, accessed 08/10/2015

43 http://www.fp7-marinet.eu/EVE-mutriku-owc-plant.html, accessed 08/10/2015

44 Leijon, M., Boström, C., Danielsson, O., Gustafsson, S., Haikonen, K., Langhamer, O., Strömstedt, E., Stålberg, M., Sundberg, J., Svensson, O., Tyrberg, S., and Waters, R.: Wave energy from the North Sea: Experiences from the Lysekil Research Site, *Surveys in Geophysics*, 2008, 29, (3), pp. 221–240

45 http://www.thecrownestate.co.uk/energy-and-infrastructure/wave-and-tidal/wave-and-tidal-current/our-portfolio/test-and-demonstration/, accessed 08/10/2015

46 http://www.emec.org.uk/, accessed 08/10/2015

47 http://www.wavehub.co.uk/, accessed 05/05/2014

48 http://www.seageneration.co.uk/, accessed 15/10/2015

49 http://www.fabtest.com/, accessed 30/03/2016

50 http://waveenergyprize.org/, accessed 08/10/2015

51 http://www.hie.co.uk/growth-sectors/energy/wave-energy-scotland/, accessed 30/03/2016

52 http://www.emec.org.uk/marine-energy/wave-developers/, accessed 08/10/2015

53 http://www.dexawave.com/, accessed 10/10/2015

54 http://www.floatingpowerplant.com/, accessed 10/10/2015

55 http://wavestarenergy.com/, accessed 09/11/2013 2013

56 http://www.wavedragon.net/index.php, accessed 09/11/2013

57 http://www.fp7-marinet.eu/access-menu-post-access-reports_KNSWING.html, accessed 30/03/2016

58 http://www.leancon.com/technology.htm, accessed 30/03/2016

59 http://www.resenwaves.com/, accessed 30/03/2016

60 http://crestwing.dk.html, accessed 30/03/2016

61 http://www.wavepiston.dk/, accessed 30/03/2016

62 http://www.waveplane.com/, accessed 30/03/2016

63 http://www.weptos.com/, accessed 10/10/2015

64 http://aw-energy.com/, accessed 10/10/2015

65 http://www.wello.eu/en, accessed 10/10/2015

66 http://www.oceanenergy.ie/, accessed 10/10/2015

67 http://www.seapower.ie, accessed 10/10/2015

68 http://bluepowerenergy.ie/, accessed 30/03/2016

69 http://www.jospa.ie/, accessed 30/03/2016

70 http://www.fp7-marinet.eu/access-menu-post-access-reports-wavetrain.html, accessed 30/03/2016

71 http://limerickwave.emergemediaireland.com/, accessed 30/03/2016

72 http://www.wavebob.com/09/11/2013

73 http://waveenergyprize.org/teams/ensea-inc#, accessed 30/03/2016

74 http://www.waveforenergy.com/, accessed 10/10/2015

75 http://www.40southenergy.com/, accessed 30/03/2016

76 http://www.fobox.no/, accessed 10/10/2015

77 http://www.fp7-marinet.eu/access-menu-post-access-reports-wec-pto.html, accessed 31/03/2016

78 http://www.havkraft.no/, accessed 31/03/2016

79 http://www.intentium.com/, accessed 31/03/2016

80 http://www.langleewavepower.com/, accessed 31/03/2016

81 http://www.owwe.net, accessed 31/03/2016

82 OWC Power AS, accessed 31/03/2016

83 http://www.pelagicpower.no/, accessed 31/03/2016

84 http://www.pontoon.no/, accessed 31/03/2016

85 Vicinanza, D., Margheritini, L., Kofoed, J.P., and Buccino, M.: The SSG wave energy converter: Performance, status and recent developments, *Energies*, 2012, 5(2), 193–226

86 http://www.fp7-marinet.eu/access_completed-projects_sdk_waveturbine.html, accessed 31/03/2016

87 http://www.abengoa.com/web/en/innovacion/areas_de_innovacion/energias_marinas/, accessed 31/03/2016

88 MARINET: Aker WEC website, accessed 31/03/2016

89 http://www.oceantecenergy.com/, accessed 31/03/2016

90 http://www.rotarywave.com/, accessed 31/03/2016

91 http://www.tecnalia.com/en/energy-and-environment/news/wind-and-sea-our-future-for-energy.htm, accessed 31/03/2016

92 http://www.piposystems.com/EN/, accessed 31/03/2016

93 http://www.seabased.com/en/, accessed 10/10/2015

94 http://www.waves4power.com/, accessed 10/10/2015

95 http://www.corpowerocean.com/, accessed 31/03/2016

96 http://www.oceanharvesting.com/, accessed 31/03/2016

97 http://www.thinkwavetube.com/, accessed 31/03/2016

98 http://albatern.co.uk, accessed 10/10/2015

99 http://www.aquamarinepower.com, accessed 10/10/2015

100 http://awsocean.com, accessed 10/10/2015

101 http://www.polygenlimited.com/, accessed 10/10/2015

102 http://seatricity.com/, accessed 10/10/2015

103 http://www.emec.org.uk/about-us/wave-clients/pelamis-wave-power/, accessed 31/03/2016

104 http://www.checkmateukseaenergy.com/, accessed 31/03/2016

105 http://www.ecotricity.co.uk/our-green-energy/our-green-electricity/and-the-sea, accessed 31/03/2016

106 http://www.sperboy.com/index.html?_ret_=return, accessed 31/03/2016

107 http://www.greenheating.com/, accessed 31/03/2016

108 Aggidis, G.A., Bradshaw, A., French, M.J., McCabe, A.P., Meadowcroft, J.A.C., and Widden, M.B.: PS frog MK5 WEC developments and design progress, Proceedings of the World Renewable Energy Congress, 2005

109 http://www.mace.manchester.ac.uk/our-research/research-themes/offshore-coastal/specialisms/wave-energy/, accessed 31/03/2016

110 http://marinepowersystems.co.uk/, accessed 31/03/2016

111 http://www.owel.co.uk/, accessed 31/03/2016

112 http://homepage.ntlworld.com/b.spilman/wavestorejune27_003.htm, accessed 31/03/2016

113 http://www.seewec.org/, accessed 31/03/2016

114 http://www.snapperfp7.eu/, accessed 31/03/2016

115 http://www.fp7-marinet.eu/ccess-menu-post-access-reports_cyanwave4.html, accessed 31/03/2016

116 http://www.tridentenergy.co.uk/, accessed 31/03/2016

117 http://www.nemos.org/english/news/, accessed 10/10/2015

118 http://www.flansea.eu/, accessed 31/03/2016

119 http://www.laminaria.be/, accessed 31/03/2016

120 http://www.ec-nantes.fr/version-francaise/recherche/projets-d-application/searev-50100.kjsp?RH=ECN-FR/, accessed 31/03/2016

121 http://www.hydrocap.com/, accessed 31/03/2016

122 http://www.fp7-marinet.eu/access_user_project_revagues.html, accessed 31/03/2016

123 http://brandlmotor.de/index_eng.htm, accessed 31/03/2016

124 http://www.sigma-energy.si/technology/, accessed 31/03/2016

125 http://www.emec.org.uk/marine-energy/tidal-developers/, accessed 08/10/2015

126 Sheng, S.: Report on Wind Turbine Subsystem Reliability – A Survey of Various Databases (2013)

127 http://www.thecrownestate.co.uk/news-and-media/news/2014/further-uk-wave-and-tidal-opportunities-unlocked, accessed 15/10/2015

128 http://www.bbc.com/news/uk-wales-33053003, accessed 15/10/2015

129 http://www.bbc.co.uk/news/uk-wales-politics-35541862, accessed 01/04/2016

130 https://tethys.pnnl.gov/annex-iv-sites/la-rance-tidal-barrage, accessed 15/10/2015

131 http://www.alstom.com/products-services/product-catalogue/power-generation/renewable-energy/ocean-energy/tidal-energy/tidal-power/, accessed 10/10/2015

132 http://www.sabella.fr/index.php?lg=gb, accessed 15/10/2015

133 http://www.pole-mer-bretagne-atlantique.com/en/ressources-energetiques-et-minieres-marines/project/2251, accessed 01/04/2016

134 http://www.eel-energy.fr/, accessed 01/04/2016

135 http://www.guinard-energies.bzh/home/, accessed 01/04/2016

136 http://www.h2oceanpower.com/index.htm, accessed 10/10/2015

137 http://www.hydroquest.net/en, accessed 10/10/2015

138 http://voith.com/en/products-services/hydro-power/ocean-energies-587.html, accessed 15/10/2015

139 http://www.atlantisstrom.de/index_english.html, accessed 01/04/2016

140 http://www.openhydro.com/home.html, accessed 15/10/2015

141 http://www.tocardo.com/, accessed 15/10/2015

142 http://www.bluewater.com/new-energy/bluetec-phil/, accessed 01/04/2016

143 http://www.tidalys.com/, accessed 01/04/2016

144 http://www.english.oryonwatermill.com/, accessed 01/04/2016

145 http://www.hammerfeststrom.com/, accessed 15/10/2015

146 http://www.flumill.com/, accessed 15/10/2015

147 http://www.straumgroup.com/hydratidal, accessed 01/04/2016

148 https://www.norwegianoceanpower.com/, accessed 01/04/2016

149 http://tidalsails.com/, accessed 01/04/2016

150 http://atlantisresourcesltd.com, accessed 16/10/2015

151 http://www.marineturbines.com/, accessed 16/10/2015

152 http://www.nautricity.com/, accessed 16/10/2015

153 http://www.scotrenewables.com/, accessed 16/10/2015

154 http://www.tidalenergyltd.com/, accessed 16/10/2015

155 http://www.qednaval.co.uk, accessed 01/04/2016

156 http://renewabledevices.com.html, accessed 01/04/2016

157 http://aquascientific2.moonfruit.com, accessed 01/04/2016

158 http://www.c-power.co.uk/, accessed 01/04/2016

159 http://www.current2current.com, accessed 01/04/2016

160 http://www.flexmarinepower.com, accessed 01/04/2016

161 http://www.freeflow69.com/, accessed 01/04/2016

162 http://www.green-tide.com/, accessed 01/04/2016 2012

163 http://www.keplerenergy.co.uk/, accessed 01/04/2016

164 http://www.lunarenergy.co.uk/, accessed 01/04/2016

165 http://www.novainnovation.co.uk/, accessed 01/04/2016

166 http://www.oceanflowenergy.com/, accessed 01/04/2016

167 https://www4.rgu.ac.uk/cree/general/page.cfm?pge=10769201601/04/2016

168 http://www.seapowergen.com/, accessed 01/04/2016

169 http://www.reuk.co.uk/TidEl-Tidal-Turbines.htm, accessed 01/04/2016

170 http://www.tidalstream.co.uk/, accessed 01/04/2016

171 http://www.magallanesrenovables.com/en/proyecto, accessed 15/10/2015

172 http://www.upm.es/internacional/UPM/UPM_Channel/News/fcf37479424af310VgnV CM10000009c7648aRCRD, accessed 01/04/2016

173 http://www.currentpower.eu/, accessed 01/04/2016

174 http://minesto.com/, accessed 15/10/2015

175 Electric Power Research Institute: Epri 2004 wave power feasibility study – final project briefing. http://oceanenergy.epri.com/waveenergy.html\#reports, March 2005.

176 Bedard, R., Previsic, M., Hagerman, G., Polagye, B., Musial, W., Klure, J., von Jouanne, A., Mathur, U., Collar, C., Hopper, C., and Amsden, S.: North American ocean energy status, March 2007, 7th European Wave and Tidal Energy Conference, Porto, Portugal, 11–13 September 2007.

177 Cornett, A.: Inventory of Canada's marine renewable energy resources, Technical Report CHC-TR-041 (Canadian Hydraulics Centre National Research Council Canada, April 2006).

178 Moazzen, I., Robertson, B., Wild, P., Rowe, A., and Buckham, B.: Impacts of large-scale wave integration into a transmission-constrained grid, *Renewable Energy*, 88, 2016, 408–417.

179 Federal Energy Regulatory Commission: Order ruling on declaration of intention and finding licensing required. FERC Docket No. DIO2-3-000, October 2002.

180 BC Hydro Green & Alternative Energy: Vancouver island ocean wave development project – call for proposals, July 2001.

181 http://www.theglobeandmail.com/report-on-business/jostled-by-wave-power-a-bc-firm-hopes-for-calm/article700082/

182 http://www.greentechmedia.com/articles/read/ california-sinks-its-first-wave-energy-project-5059

183 Federal Energy Regulatory Commission: Order denying rehearing. FERC Docket No. DIO2-3-001, February 2003.

184 Federal Energy Regulatory Commission: Licensing hydrokinetic pilot projects. http:// www.ferc.gov/industries/hydropower/gen-info/ licensing/hydrokinetics/pdf/white_ paper.pdf, April 2008.

185 Federal Energy Regulatory Commission: Order on rehearing and clarification and amending license. Project No. 12751-001, March 2008.

186 Dragoon, K., Eckstein, J., and Patton, L.: Wave energy industry update – a northwest US perspective. http://www.oregonwave.org, August 2015.

187 BC Ministry of Forests Lands and Natural Resource Operations: Land-use operational policy – ocean energy projects. http://www2.gov.bc.ca/gov/ content/industry/ natural-resource-use/land-use/crown-land/ crown-land-policies, August 2011.

188 2004 Integrated Electricity Plan. Technical report, BC Hydro, 2004.

189 Crompton, J. and Van Dijk, G.: Community electrification plan for refuge cove. Technical Report Revision R00, BC Hydro, October 2013.

190 Alaska Energy Authority: Power cost equalization program – statistical data by community. http://www.akenergyauthority.org/Content/Programs/PCE/ Documents/ FY13StatisticalRptComt.pdf, February 2014.

191 Previsic, M., and Bedard, R.: Yakutat conceptual design, performance, cost and economic wave power feasibility study. Technical Report EPRI-WP-006-Alaska, Electric Power Research Institute, November 2009.

192 BC Ministry, Mines Energy, and Petroleum Resources: The BC energy plan: A vision for clean energy leadership. http://www2.gov.bc. ca/assets/gov/farming-natural-resources-and-industry/ electricity-alternative-energy/bc_energy_ plan_2007.pdf, 2007.

193 Clean Energy Act. British Columbia, Canada, 2010. http://www.bclaws.ca/civix/ document/id/consol24/consol24/00_10022_01, accessed 06/09/2017.

194 BC Hydro: 2008 long-term acquisition plan. https://www.bchydro.com/ energy-in-bc/ meeting_demand_growth/irp/past_plans.html, February 2009.

195 US Department of Energy: Marine and hydrokinetic energy projects, fiscal years 2008–2014. Technical report, US Department of Energy, Energy Efficiency & Renewable Energy, Wind and Water Power Technologies Office, 2014.

196 http://energy.gov/eere/articles/ energy-department-awards-74-million-develop-advanced-components-wave-and-tidal

197 http://waveenergyprize.org/about

198 Weber, J., Costello, R., and Ringwood, J.V.: Wec technology performance levels (TPLs) – metric for successful development of economic WEC technologies. In Proceedings of the European Wave and Tidal Energy Conference, 2013.

199 http://energy.gov/eere/articles/ revamped-simulation-tool-power-wave-energy-development

200 http://energy.sandia.gov/energy/renewable-energy/water-power/market-acceleration-deployment/snl-swan-sandia-national-laboratories-simulating-waves-nearshore/

201 Ruehl, K., Porter, A., Chartrand, C., Smith, H., Chang, G., and Roberts, J.: Development, verification and application of the SNL-SWAN open source wave farm code. In Proceedings of the 11th European Wave and Tidal Energy Conference, 2015.

202 Von Jouanne, A. and Brekken, T.: Wave energy research, development and demonstration at Oregon state university. IEEE Power and Energy Society General Meeting: the Electrification of Transportation and the Grid of the Future, 2011.

203 A. Von Jouanne, T. Lettenmaier, E. Amon, T. Brekken, and R. Phillips. A novel ocean sentinel instrumentation buoy for wave energy testing. *Marine Technology Society Journal*, 2013, 47(1), 47–54.

204 http://nnmrec.oregonstate.edu/pmec-facilities

205 http://nnmrec.oregonstate.edu/announcement/ pmec-sets-licensing-process-update2016

206 http://oregonwave.org/oceanic/wp-content/uploads/2013/05/Comprehensive-OWET-Funded-projects_August-2013.pdf

207 http://www.uvic.ca/research/projects/wcwi/

208 http://www.uvic.ca/research/centres/iesvic/

209 Robertson, B., Hiles, C. and Buckham, B.: Characterizing the near shore wave energy resource on the west coast of Vancouver Island, *Renewable Energy*, 2014, 71, 665–678.

210 www.dsa-ltd.ca

211 Hagerman, G. and Scott, G.: Mapping and assessment of the united states ocean wave energy resource, Technical Report 1024637, Electric Power Research Institute, 2011.

212 http://maps.nrel.gov/mhk_atlas

213 Robertson, B., Hiles, C., Luczko, E., and Buckham, B.: Quantifying wave power and wave energy converter array production potential. *International Journal of Marine Energy*, 2016, 14, 143–160.

214 www.surfpower.ca

215 http://dsa-ltd.ca/software/proteusds/description/

216 Bailey, H., Ortiz, J., Robertson, B., Buckham, B., and Nicoll, R.: A methodology for wave-to-wire WEC simulations, Proceedings of the 2nd Marine Energy Technology Symposium, 2014.

217 http://energyexemplar.com/software/plexos-desktop-edition/

218 http://pics.uvic.ca/2060-project-integrated-energy-pathways-british-columbia-and-canada

219 Parkinson, S., Dragoon, K., Reikard, G., Garcia-Medina, G., Ozkan-Haller, T., and Brekken, T.: Integrating ocean wave energy at large-scales: A study of the US Pacific Northwest, *Renewable Energy*, 2015, 76, 551–559.

220 Oceans Ltd.: Current in Minas Basin May 1, 2008 – March 29, 2009. Technical report, 2009.

221 Karsten, R., Culina, J., Swan, A., O'Flaherty-Sproul, M., Corkum, A., Greenberg, D., and Tarbotton, M.: Assessment of the potential of tidal power from minas passage and minas basin, Technical report.

222 Tarbotton, M. and Larson, M.: Canada ocean energy atlas (phase 1) potential tidal current energy resources analysis background, Technical report, Triton Consultants Ltd., Vancouver, BC, May 2006.

223 US Department of Energy: Quadrennial technology review an assessment of energy technologies and research opportunities., Technical report, 2015.

224 Georgia Tech Research Corporation: Assessment of energy production potential from tidal streams in the United States, Technical Report DE-FG36-08GO18174, 2011.

225 Georgia Tech Research Corporation: Assessment of energy production potential from ocean currents along the United States coastline, Technical Report DE-EE0002661, 2013.

226 Assessment and mapping of the riverine hydrokinetic resource in the continental United States, Technical Report 1026880, Palo Alto, CA, 2012.

227 Regulatory Assistance Project: Electricity regulation in the US: A guide,Technical report, 2011.

228 Nova Scotia Department of Energy: COMFIT Nova Scotia Community Feed-In Tariff, Technical report, 2011.

229 BC Hydro: 2000 integrated electricity plan, Technical report, 2000.

230 Natural Resources Canada: Charting the course: Canada's marine renewable energy technology roadmap, Technical report, 2011.

231 Federal Energy Regulatory Commission: Marine & hydrokinetic projects (as of October 9, 2015). http://www.ferc.gov/industries/hydropower/gen-info/licensing/hydrokinetics/hydrokinetics-projects.pdf, accessed 28/10/2015.

232 https://mhkdr.openei.org/

233 http://en.openei.org/wiki/Marine_and_Hydrokinetic_Technology_Database

234 http://tethys.pnnl.gov/

235 Khan, J., Bhuyan, G., and Moshref, A.: An assessment of variable characteristics of the Pacific Northwest region's wave and tidal current power resources, and their interaction with electricity demand & implications for large scale development scenarios for the region – phase 1, Technical Report 17458-21-00, Powertech Labs, January 2008.

236 Canadian and UK researchers to collaborate in developing innovative sensor technologies for tidal, valued at $1.4 million. http://www.oera.ca/marine-renewable-energy/tidal-research-projects/canadian-and-uk-researchers-to-collaborate-in-developing-innovative-sensor-technologies-for-tid, accessed 28/10/2015.

237 Nova Scotia and British Columbia collaborate on tidal energy. http://novascotia.ca/news/release/?id=20150721003, accessed 28/10/2015.

238 The integrated energy project at race rocks. http://www.racerocks.ca/ energy/, accessed 28/10/2015.

239 Hachey, H.B.: Utilization of tidal energy, *Journal of the Franklin Institute*, 1934, 217(6), 747–756.

240 McLellan, H.J.: Energy considerations in the Bay of Fundy system, *Journal of the Fisheries Research Board of Canada*, 1958, 15(2), 115–134.

241 Trites, R.W.: Probable effects of proposed Passamaquoddy power project on oceanographic conditions,*Journal of the Fisheries Research Board of Canada*, 1961,18(2), 163–201.

242 Duff, G.F.D.: Tidal resonance and tidal barriers in the Bay of Fundy system, *Journal of the Fisheries Research Board of Canada*, 1970, 27(10), 1701–1728.

243 Godin, G.: Theory of the exploitation of tidal energy and its application to the Bay of Fundy. *Journal of the Fisheries Research Board of Canada*, 1969, 26(11), 2887–2957.

244 Karsten, R.H., McMillan, J.M., Lickley,M.J., and Haynes, R.D.: Assessment of tidal current energy in the minas passage, Bay of Fundy. *Proceedings of the Institution of Mechanical Engineers, Part A*, 2008, 222(5), 493507.

245 Fundy Ocean Research Centre for Energy: The Vectron project: Advancing the science of tidal power, Technical report, 2013.

246 Admiralty inlet pilot tidal project. https://tethys.pnnl.gov/ annex-iv-sites/admiralty-inlet-pilot-tidal-project, accessed 28/10/2015.

247 Pud tidal project not to advance. http://www.snopud.com/PowerSupply/ tidal/tidalpress.ashx?756_na = 277, accessed 28/10/2015.

248 Polagye, B., and Previsic, M.: System level design, performance, cost and economic assessment – Tacoma Narrows Washington tidal in-stream power plant, Technical report, Electric Power Research Institute, 2006.

249 Hamner, B.: Tacoma Narrows tidal power feasibility study: Final report, Technical report, Puget Sound Tidal Power LLC for Tacoma Power, Tacoma Public Utilities, City of Tacoma, Washington, 2007.

250 Amsden, S.: Tacoma narrows tidal power feasibility study, Technical report, Tacoma Power, December 2007.

251 Mitchell, J.: Mitigation in Environmental Assessment – Furthering Best Practice, Environmental Assessment, 1997, 5(4), 28–29.

252 Met Office: Cartopy: a cartographic Python library with a matplotlib interface, 2010–2015. Available at: http://scitools.org.uk/cartopy, accessed 27/02/16

253 Gross, M.S., and Magar, V.: Offshore wind energy climate projection using UPSCALE climate data under the RCP8.5 emission scenario. *PLoS One*, 2016, 11(10).

254 Le Bert, G.H.: Potencial energético del alto Golfo de California. *Boletín de la Sociedad Geológica Mexicana*, 2009, 61(1), 143–146.

255 Fenés, G.: Gustavo Bianchi: 'la energía del mar podría generar 10.000 MW de potencia', Energía Estratégica, 2014. http://www.energiaestrategica.com/gustavo-bianchi-la-energia-mareomotriz-podria-generar-10-000-mw-de-potencia/, accessed 18/06/15

256 Marques, W.C., Fernandes, E.H., Rocha, L.A.O., and Malcherek, A.: Energy converting structures in the Southern Brazilian Shelf: energy conversion and its influence on the hydrodynamic and morphodynamic processes, *Journal of Earth Sciences and Geotechnical Engineering*, 2012, 1(1), 61–85.

257 Pillsbury, J.E.: The Gulf Stream – A description of the methods employed in the investigation, and the results of the research, USCC-Geodetic Survey, Silver Spring, MD, 1890.

258 Davies, G. and Wills, T.: Recommendations for Chile's Marine Energy Strategy – a roadmap for development, Aquatera Environmental Services and Products Project no. P478, March 2014.

259 CONICET: Y-TEC es un ejemplo de cómo el CONICET puede organizarse y contribuir al impulso de sectores estratégicos para el país, 2013. http://www.conicet.gov.ar/y-tec-es-un-ejemplo-de-como-el-conicet-puede-organizarse-y-contribuir-al-impulso-de-sectores-estrategicos-para-el-pais/, accessed 18/06/15

260 Clarín: Estudian con boyas el Mar Argentino en busca de nuevas fuentes de energía, 2014. http://www.clarin.com/sociedad/mar_Argentino-ciencia-Y-Tec-Galuccio-YPF_0_1267073684.html, accessed 18/06/15

261 INVAP: Hydrokinetic turbine, 2014. http://www.invap.com.ar/en/2014-05-12-14-56-12/projects-industrial/hydrokinetic-turbine.html, accessed 18/06/15

262 IRENA: Renewable energy prospects: Mexico. REmap 2030 Analysis, IRENA, Abu Dhabi, 2015. http://www.irena.org/DocumentDownloads/Publications/IRENA_REmap_Mexico_report_2015.pdf, accessed 13/07/15

263 CONACYT: Formalización de los convenios de asignación de recursos para tres centros mexicanos de innovación en energías renovables (CeMIE's), 2014. http://www.conacyt.mx/index.php/comunicacion/comunicados-prensa/312-formalizacion-de-los-convenios-de-asignacion-de-recursos-para-tres-centros-mexicanos-de-innovacion-en-energias-renovables-cemie-s, accessed 17/07/15

264 INEGI: Extensión territorial de México, 2015. http://cuentame.inegi.org.mx/territorio/extension/, accessed 13/07/15

265 INEGI: Datos básicos de la geografía de México, 1991. http://www.inegi.org.mx/, accessed 06/09/2017.

266 Gross, M.S., and Magar, V.: Offshore wind energy potential estimation using UPSCALE climate data. *Energy Science and Engineering*, 2015, 3(4), 342–359.

267 López-González, J.: Desarrollo de un dispositivo de conversión de energía de las corrientes marinas, PhD thesis, UNAM, Mexico, 2011 (in Spanish).

268 Ketchum, B.H. (ed.): *Estuaries and enclosed seas*. Amsterdam: Elsevier Scientific, 1983.

269 Marinone, S.G.L., and Lavín, M.F.: Mareas y corrientes residuales en el Golfo de California. In M.F. Lavín (ed.), *Contribuciones a la Oceanografía Física en México*. Unión Geofísica Mexicana, 1997.

270 López Mariscal, J.M., and García Cordova, J.A.: Moored observations in the northern Gulf of California: A strong bottom current. *Journal of Geophysical Research*, 2003, 108(C2), 3048.

271 Gyory, J., Mariano, A.J., and Ryan, E.H.: The Yucatan Current, 2015. http://oceancurrents.rsmas.miami.edu/caribbean/yucatan.html, accessed 16/07/15

272 Ochoa, J., Sheinbaum, J., Badan, A., Candela, J., and Wilson, D.: Geostrophy via potential vorticity inversion in the Yucatan Channel. *Journal of Marine Research*, 2001, 59, 725–747.

273 CFE-CONACYT: Fondo sectorial para investigación y desarrollo en energía, anexo de demandas específicas, 2014. http://www.conacyt.gob.mx/index.php/sni/convocatorias-conacyt/convocatorias-fondos-sectoriales-constituidos/convocatoria-cfe-conacyt-1/convocatorias-cerradas-cfe/2014-01-cfe/4483-anexo-demandas-especificas/file, accessed 17/07/15

274 Arroyo, J.: Energy Forever SA de CV: Empresa dedicada a la planeación, construcción, y comercialización de centrales eléctricas que transforman la energía de las olas en energía eléctrica. Invited presentation, TMRES14 workshop: Towards Marine Renewable Energy: Achievements, Challenges and Networks, CICESE, Ensenada, Mexico, 14 August 2014.

275 Carrasco-Díaz, M., Rivas, D., Orozco-Contreras, M., and Sánchez-Montante, O.: An assessment of wind power potential along the coast of Tamaulipas, Northeastern Mexico. *Renewable Energy*, 2015, 78, 295–305.

276 Osorio-Arias, A.F., Ortega, S., Agudelo, P., Velez, J.I., Montoya, R., Mesa, J., Otero, L., Davis, A., Lonin, S., Ruiz, M., and Bernal, G.: Potencial de producción de energías marinas en Colombia. Seminario Nacional de Economía, Villavicencio, Colombia, May 2011. www.conalpe.gov.co/oficial/images/stories/archivos/andres_osorio.pdf, accessed 25/07/15

277 Street, K. (2014). Alstom International Tidal Projects. 6th Tidal Energy Forum – Bristol, UK, June 2014.

278 Zileri, D.: Aquavatio: a hope for river communities. *GVEP International Magazine*, 15 March 2011. http://www.gvepinternational.org/en/business/news/aquavatio-hope-river-communities, accessed 25/07/15

279 Osorio-Arias, A.F., Agudelo, P., Correa, J., Otero, L., Hernandez, J.A., and Restrepo, J.P.: Building a roadmap for the implementation of marine renewable energy in Colombia. OCEANS, Santander, June 2011.

280 IRENA: Ocean Thermal Energy Conversion. IRENA Ocean Energy Technology Brief 1, 2014.

281 IRENA: Renewable energy country profiles for the Caribbean, 2012. http://www.irena.org/DocumentDownloads/Publications/_CaribbeanComplete.pdf, accessed 25/07/15.

282 Kuntsi-Reunanen, E.: A comparison of Latin American energy-related CO_2 emissions from 1970 to 2001. *Energy Policy*, 2007, 35, 586–596.

283 FIDA: Proyecto de desarrollo integral y sustentable para las zonas semiáridas, áridas y en transición en los estados de Nueva Esparta y Sucre. Informe del diseño del proyecto, 2012.

284 IRENA: RD&D for renewable energy technologies: cooperation in Latin America and the Caribbean. http://www.irena.org, accessed 25/07/15

285 http://carnegiewave.com/, accessed 06/06/2016

286 http://www.biopowersystems.com/, accessed 06/06/2016

287 http://www.oceanlinx.com/, accessed 06/06/2016

288 http://bomborawavepower.com.au/projects/, accessed 06/06/2016

289 http://www.perpetuwavepower.com/, accessed 06/06/2016

290 http://www.proteanwaveenergy.com/, accessed 06/06/2016

291 http://www.aquagen.com.au/technologies/surgedrive, accessed 06/06/2016

292 http://www.globalrenewablesolutions.org/, accessed 06/06/2016

293 Washio, Y., Osawa, Y., Nagata, F., Fujii, F., Furuyama, H., and Fujita, T.: The offshore floating type wave power device 'Mighty Whale': Open sea tests, Proceedings of the Tenth International Offshore and Polar Engineering Conference (ISOPE), Seattle, USA, June 2000.

294 Watabe, T., Yokouchi, H., Kondo, H. Inoya, M., and Kudo, M.: Installation of the new Pendulor for the 2nd stage sea test, Proceedings of the Ninth International Offshore and Polar Engineering Conference (ISOPE), Brest, France, 199.

295 Hanki, H., Arii, S., Furusawa, T., and Otoyo, T.: Development of advanced wave power generation system by applying gyroscopic moment, Proceedings of the Eighth European Wave and Tidal Energy Conference (EWTEC), Uppsala, Sweden, 2009.

296 OES (Ocean Energy Systems) annual report 2015

297 Hong, K.: Update on marine renewable energy activities in Korea, Country report of OES 26th Exco. Meeting, Paris, France, 2014.

298 http://oceanrusenergy.com/production/ocean-3-kw/, accessed 06/06/2016

299 http://atecom.ru/wave-energy/, accessed 06/06/2016

300 http://www.vortexosc.com/index.php?newlang=english, accessed 06/06/2016

301 http://www.powerprojects.co.nz/marine-energy, accessed 06/06/2016

302 http://rocknroll.nualgi.com/, accessed 06/06/2016

303 http://www.hann-ocean.com/products/drakoo-wave-energy-converter/introduction/, accessed 06/06/2016

304 http://atlantisresourcesltd.com/marine-power/global-resources.html, accessed 02/06/2016

305 http://makoturbines.com/, accessed 02/06/2016

306 http://tidalenergy.com.au/, accessed 02/06/2016

307 http://www.biopowersystems.com/biostream.html, accessed 02/06/2016

308 https://www.facebook.com/CetusEnergy, accesses 02/06/2016

309 http://www.power-eng.com/articles/2014/02/design-study-for-kawasaki-heavy-industries-tidal-turbine-completed.html, accessed 02/06/2016

310 http://www.modec.com/fps/skwid/index.html, accessed 02/06/2016

311 http://english.kwater.or.kr/eng/busi/project03Page.do?s_mid=1192, accessed 03/06/2016

312 http://hyundaiengineering.blogspot.co.uk/2011/06/hyundai-heavy-industries-500-kw-tidal.html, accessed 02/06/2016

313 http://kitl.kirloskar.com/TIDAL-MARINE-Products_and_Solutions, accessed 02/06/2016

314 http://peswiki.com/index.php/Directory:Balkee_Tide_and_Wave_Electricity_Generator, accessed 02/06/2016

Epilogue

The Future of Wave and Tidal Energy

Deborah Greaves

Professor of Ocean Engineering, School of Engineering, University of Plymouth, UK

In this book, we have attempted to give a broad overview of marine energy, and specifically the generation of renewable energy from the waves and tides. We discuss all aspects of wave and tidal energy, starting with an analysis of the resource available for exploitation, then describing the theory of wave and tidal energy conversion, together with examples of the technology. We consider the tools available to inventors and engineers for the design of this new technology, encompassing numerical modelling and laboratory modelling, the development pathway and instigation of design standards, together with aspects of the design, including mooring and foundations. Alongside the technological solutions for wave and tidal energy conversion, the options for electricity generation and the theory of power take-off and integration with the electrical grid are discussed. Interaction of the new wave and tidal technologies with the marine environment is a subject of particular concern as the first devices are deployed in the sea. Potential environmental impacts of wave and tidal energy are discussed in a review of studies into the biological and physical aspects. The economics of wave and tidal energy together with consenting and legal aspects of the new industry are then presented. To demonstrate how the different components are integrated in the development of a wave or tidal energy project, four example case studies of technology development projects are given in Chapter 12. Finally, in Chapter 13, a review of wave and tidal energy developments and current status world-wide is presented.

It is evident that there are clear differences between wave and tidal energy industries and their rate of development. The technological challenges in mechanical design of devices to convert energy through the use of tides and waves are different. Tidal energy technologies are generally much further advanced than wave energy technologies and arguably have converged to a single design category, namely the horizontal-axis turbine. The wave energy sector, on the other hand, still includes many highly varied design types and different approaches to wave energy conversion and there is no convergence in design foreseeable in the short term.

However, there are also some common challenges between the wave and tidal sectors, and these concern operating in the marine environment and all that entails. Interaction with the marine environment means surviving environmental action on the device, such as wave loading, current loading and sediment dynamics and abrasion; it also means the effect of the device on the environment, such as impacts on marine ecosystems and physical processes. The public are rightly concerned with protecting the marine environment, and so marine consenting, environmental impact assessment and social impact assessment are developing rapidly for the new wave and tidal industries. Early deployments of wave and tidal technologies are closely monitored to build up knowledge and experience of environmental impact, and marine demonstration sites have an important role to play in building the knowledge to aid future consenting processes. The tidal industry, being more advanced than wave energy, is likely to pave the way through some of these hurdles and difficulties, smoothing the path for wave energy when it reaches the same stage.

Tidal energy has seen the greatest advance in development over recent years and the first large arrays of tidal stream converters are under development in the UK and Canada. The MeyGen project in Scotland is under construction and the first array of three devices is installed and operational in 2017, and the FORCE project in the Bay of Fundy in Canada will soon see four different devices being deployed. Alongside tidal stream developments, the concept of tidal lagoons is growing in interest, with a number of tidal lagoons in planning for the Severn estuary in the UK. Tidal range projects have been in use since the 1960s in France, the 1980s in Canada and more recently the Lake Sihwa tidal barrage power station in Korea that was opened in 2011 and is described in Chapter 12. Some of the main challenges remaining in tidal stream energy are the reliability of devices operating in the marine environment, their installation and maintenance, and better understanding of their interaction with the environment – in particular, the environmental effects of the presence of the devices and also the array interaction of multiple devices and their impact on the resource and environment.

In Spain the first commercial wave power plant is built into a breakwater at Mutriku, and this example paves the way for future wave plants that may be integrated into new or existing structures. After many years of experience of deployments at sea, the companies developing the Pelamis and Oyster devices both went into administration recently (2014 and 2015). The knowledge they have accrued through their experience in developing the technologies is not lost and new and similar concepts are under development with lessons learnt from the experiences of the industry taken into account in their design. Failure of these commercially leading technologies has been met with a resurgence of activity, fundamental research and new innovation in the wave energy sector with Wave Energy Scotland in the UK stepping in to provide funding streams to develop innovation as well as focus on cross-cutting technology, such as power take-off and electrical systems. Similarly, the NREL in the USA has recently run a competition for low TRL wave energy solutions. Main challenges in wave energy include resolving reliability and survivability at reasonable cost; research is focused on reducing costs while maintaining performance through innovative use of materials and design solutions and by investigating routes to reduce cost through shared infrastructure and exploring dual benefits of wave energy conversion. For example, by extracting energy from the marine environment, wave farms can be seen as part of a coastal defence scheme and built into coastal management plans and strategies, as discussed in Chapter 9.

There is also potential dual benefit to be found through co-location of wave and wind energy farms in the same sea space and this may lead to some cost savings through sharing of infrastructures, such as electrical cables, foundations and maintenance infrastructure. Additional benefit is afforded from the wave farm protecting wind farm structures from extreme wave loading by extracting some of the wave energy. Infrastructure sharing between devices within a wave or tidal farm is recognised as a key area for potential cost savings in marine energy. Furthermore, new and radical designs for hybrid systems of wave energy converters integrated within offshore wind structures and foundations, as well as innovative floating wind and wave hybrids, are being explored in the research domain.

The review of regional activities in wave and tidal energy given in Chapter 13 demonstrates the wealth of innovation and development under way in the sector. Wave and tidal energy as part of the wider ocean energy area is securely embedded in the funding priorities of the EU, the Americas, China and the pacific region, with active research and development and new initiatives under way in all regions.

The recent climate change conference agreed global targets for climate change, and through these drivers of policy, world-wide commitment to reducing carbon emissions and sourcing renewable energies is strengthened year on year. Marine energy and specifically wave and tidal energy are a significant part of the renewable energy of the future, as a significant part of the overall energy infrastructure, and as a key clean and secure component of the energy mix. Marine energies can also be a bespoke solution for remote island communities and form part of rapid deployment emergency relief in some situations.

Wave and tidal energy is a fast moving and exciting sector, with complex unresolved research challenges and many unknown and open questions remaining. It is a fascinating sector to be involved in and what we will see in twenty years is likely to be very different from the technologies we see under development today. Despite these complex challenges, we can expect to see the wave and tidal industry move through and beyond research and into development in the next decades and be part of the future for us all.

Index

Wave and Tidal Energy, First Edition. Edited by Deborah Greaves and Gregorio Iglesias.
© 2018 John Wiley & Sons Ltd. Published 2018 by John Wiley & Sons Ltd.

Table 9.1 Model configurations, wave conditions and values of the transmission coefficient (α = angle between hulls; F = freeboard; H_s = significant wave height; T_e = energy period (in model values); K_t = transmission coefficient) [7].

Test no.	α (°)	F (m)	H_s (m)	T_e (s)	K_t
1	30	0.04	0.083	1.807	0.806
2	30	0.04	0.1	1.972	0.777
3	45	0.04	0.083	1.807	0.776
4	45	0.04	0.1	1.972	0.756
5	60	0.04	0.083	1.807	0.507
6	60	0.04	0.1	1.972	0.760
7	90	0.04	0.083	1.807	0.749
8	90	0.04	0.1	1.972	0.772

To compute the effects of wave farms on the nearshore wave climate, SWAN, a coastal wave model successfully used for wave energy resource assessments, was applied. It computes the evolution of the wave spectrum based on the conservation of wave action. The wave action balance equation reads

$$\frac{\partial N}{\partial t} + \nabla \cdot \left(\vec{C} N \right) + \frac{\partial \left(C_\theta N \right)}{\partial \theta} + \frac{\partial \left(C_\sigma N \right)}{\partial \sigma} = \frac{S}{\sigma}, \tag{9.2}$$

where N is the wave action density, t the time, $\vec{C}$ the propagation velocity in the geographical space, θ the wave direction, σ the relative frequency, and C_θ and C_σ the propagation velocity in the θ- and σ-space, respectively. Therefore, on the left-hand side of equation (9.2), the first term represents the rate of change of wave action with time, the second term describes the spatial propagation of wave action, and the third and fourth terms stand for the refraction and changes in the relative frequencies respectively induced by depth and currents. Finally, on the right-hand side, S is the source term representing the generation and dissipation of energy density by the different processes involved.

For the accurate representation of the wave farm effects, it is essential to use a fine computational grid in the farm area and its lee. For this purpose, a high-resolution nested model was implemented with two computational grids (Figure 9.15): a coarser grid, extending from deep water to the area of interest, and a finer (nested) grid covering the area of interest, with a resolution of 18 m. The resolution of the nested grid increased progressively from its lateral boundaries (NE and SW) towards the wave farm area. The bathymetry was digitised from nautical charts of the Spanish Hydrographic Institute and interpolated onto the grids.

Regarding the wave farm layout, two configurations with 12 WECs were considered: configuration I with the WECs arranged on one row, and configuration II with the WECs on a two-row staggered layout (Figure 9.16). The comparison between configurations allowed the assessment of the effects of the energy absorption in the first row of devices on the overall wave farm performance. A spacing of 2.2D – with D = 90 m, the